Advances in

Bistatic Radar

Advances in
Bistatic Radar

Edited by

Nicholas J. Willis

and

Hugh D. Griffiths

SciTech PUBLISHING, INC.

SciTech Publishing, Inc
Raleigh, NC
www.scitechpub.com

Printed in the U.S.A.
10 9 8 7 6 5 4 3 2 1

ISBN: 1891121480
ISBN13: 9781891121487

SciTech President: Dudley R. Kay
Production Director: Susan Manning
Production Coordinator: Robert Lawless
Cover Design: Kathy Gagne

Cover images, used with permission, clockwise from upper left: Courtesy of P. Howland and D. Maksimiuk, NATO C3 Agency; Copyright QinetiQ Ltd., 2003, all rights reserved; Courtesy of JPL/NASA/Goldstone; Courtesy of Dr. John Sahr, University of Washington; Courtesy of Dr Hugh Griffiths, DCMT Shrivenham.

Library of Congress Cataloging-in-Publication Data.
Advances in bistatic radar / edited by Nicholas J. Willis, Hugh D. Griffiths.
p. cm.
Includes bibliographical references.
ISBN-13: 978-1-891121-48-7 (hbk. : alk. paper)
ISBN-10: 1-891121-48-0 (hbk. : alk. paper)
1. Bistatic radar. I. Willis, Nicholas J., 1934- II. Griffiths, H. (Hugh), 1956-

TK6592.B57A38 2007
621.3848--dc22

2007013159

This book is dedicated to our wives, who have soldiered through yet another cloistering.

Ling Chan (Wurman)
Morag Griffiths
Millie Lyon
Tonette Melvin
Ann Reisenauer (Simpson)
Eliza Sutton (Sahr)
Maria Rigling
Sandra Weiner
Carlaine Willis

Contents

Foreword

Bistatic radars have fascinated surveillance and tracking researchers for decades. Despite evolution from the early Chain Home radars in Britain to today's coherent multimode monostatic radars, there remains a rich research in bistatic and multistatic applications. The promise of quiet receivers, aspect angle diversity, and improved target tracking accuracy are what fuel this interest.

Nicholas Willis led the modern day interest in bistatic radars during his long tenure at the Defense Advanced Research Projects Agency in the late 1970's and early 1980's. His 1991 book *Bistatic Radar* has remained the principal reference book for researchers. He has often emphasized that bistatic radar systems should not be the system of choice when a monostatic radar can accomplish the detection and ranging of targets more efficiently. However, there are a number of applications where a monostatic radar cannot accomplish the user's requirements. One such application is passive bistatic radar, which Hugh Griffiths pioneered with his seminal analysis and field measurements starting in 1985. It is encouraging to see a new collection of scientific and engineering details to motivate this rich radar waveform, its signal processing, and the challenge of system architecture.

Nicholas Willis and Hugh Griffiths have collected a comprehensive set of monographs that provide insight into the more recent history of bistatic radar design and implementation. It combines improved details of the environment along with novel system applications. Each chapter is authored by radar experts who had the vision and motivation to tackle a challenging development problem and share their view of the successes and remaining implementation challenges. This new book is more than a sequel to the original *Bistatic Radar*; it is a picture of where the advent of better time and frequency control and digital signal processing will take us in the future.

Mark E. Davis
Defense Advanced Projects Research Agency
Arlington, Virginia USA

Preface

When embarking on a project to edit or write a book, one has to be persuaded of the justification for such an enterprise. Of course, it is ultimately satisfying and enjoyable to see the finished product; however, it inevitably takes more time and effort than originally foreseen. In the course of the work, one's knowledge of the subject advances, too.

But it is also necessary to be convinced that there is a need for such a book. In this case, interest in bistatic radar over the past decade has been significant. We have observed that interest has varied cyclically, with a period of 15–20 years. The very first radars were bistatic, until pulsed waveforms and T/R switches were invented. Interest was revived in the 1950s with semi-active homing missiles and the SPASUR system, and then died away. The second resurgence occurred in the 1970s with planetary exploration and continued with air surveillance systems and the first experiments in passive bistatic radar using broadcast transmitters of opportunity. The third resurgence started in the mid-1990s and continues, with persuasive reasons.

The consequences of Moore's law mean that digital processing power, which is essential to realize practical systems, especially multistatic systems, increases inexorably. The advent of GPS solves many stressing problems of bistatic timing and synchronization. Consequently, bistatic systems may be able to address new military surveillance requirements such as the detection and tracking of low-signature targets by passive bistatic radars and of ground targets from unmanned air vehicles. And, of equal significance, bistatic radar continues to be used for civil applications such as planetary surface exploration, ionospheric turbulence measurements, and three-dimensional vector wind field measurements.

The principal reason for this book, then, is to report recent work in this third resurgence. A second reason is that a significant number of publications on systems developed in past decades have recently been declassified, and can now be reported in the open literature.

The book is divided into two sections: the first part documents both new and resurrected information about the development, testing, and fielding of bistatic and multistatic radar systems for military, scientific, and commercial use; the second part updates and publishes the previously limited bistatic clutter database and its analysis, and reports the development of (i) image focusing and motion compensation methods for bistatic SAR and (ii) adaptive cancellation methods for bistatic MTI.

This book evolved from the bistatic SAR material contained herein that was first proposed to SciTech Publishing by Brian Rigling as a stand-alone monograph. Subsequent SciTech-requested reviews by Nicholas Willis (author of *Bistatic Radar*) and Hugh Griffiths (editor of the IET's *Special Issue on Bistatic Radar*) led to the conclusion that bistatic SAR was one of many significant bistatic and multistatic radar developments occurring in the third resurgence, which should logically be assembled in a full report to the radar community. Dr. Rigling graciously deferred in having his monograph subsumed by the larger effort. Thus began the search, solicitation, and confirmation of a remarkable group of experts on the subject. With both surprise and satisfaction, we can say that the sustained efforts of authors, contributors, editors, and the publisher brought forth this book in less than two years' time.

The writing and editing of this book has been constrained by the fact that the two coeditors are separated by eight time zones, so the work has necessarily been done via the Internet and e-mail, which we can report worked splendidly. We are indebted to our families—particularly our wives, Carlaine and Morag—for their forbearance, when we may have appeared to be too engrossed in our work to give them the time and attention that they deserve. We thank all authors and contributors for producing their material, often to tight deadlines, and for accepting our edits with good grace. We thank the numerous people who have helped us by providing material, assisting with experimental work, and constructively reviewing early drafts of the manuscript. It is always invidious to mention individuals since we are bound to offend by omission, but we are grateful particularly to Anthony Andrews, Chris Baker, Alan Bernard, Paul Howland, Daniel O'Hagan, Marc Thomas, and M. C. Jackson, who truly codified bistatic radar analysis. Finally we gratefully acknowledge the encouragement and support of SciTech's CEO, Dudley Kay, and Chief Editor, Susan Manning, in enabling and executing this project.

Hugh Griffiths
London, England

Nicholas Willis
Carmel, California

February 2007

PUBLISHER'S NOTE:

Additional files, errata, and commentary—including full color versions of many of the figures in the text—can be found at the companion website www.scitechpub.com/willisgriffiths.htm.

List of Contibutors

David K. Barton
Hugh Griffiths
Edwin Lyon III
William Melvin
Brian D. Rigling
Richard Simpson
Merrill I. Skolnik
John D. Sahr
Melvin M. Weiner
Nicholas J. Willis
Joshua Wurman

Advances in Bistatic Radar

1

Introduction

Nicholas J. Willis

1.1 DEFINITIONS

A *bistatic radar* is a radar that uses two antennas at separate locations, one for transmission and one for reception. Usually the transmitter and receiver accompany the antennas at these locations. A variation of the bistatic radar is the *multistatic radar*, which uses multiple antennas at separate locations, one antenna for transmission and multiple antennas at a different location, for reception, or vice versa. Again, transmitters or receivers can accompany the antennas. Multistatic radar can use *multilateration* for target state estimates (i.e., target position, velocity, and acceleration). Multilateration combines simultaneous range and/or doppler data from multiple transmitter/receiver pairs having overlapping spatial coverage to estimate the target state without using range-dependent angle data.

Bistatic (and multistatic) radars can operate with their own *dedicated transmitters*, which are specially designed for bistatic operation, or with *transmitters of opportunity*, which are designed for other purposes but can be suitable for bistatic operation. When the transmitter of opportunity is from a monostatic radar, the bistatic radar is often called a *hitchhiker*. When the transmitter of opportunity is from—sources other than a radar, such as a broadcast station or communications link, the bistatic radar can also be referred to as *passive radar*, *passive coherent location*, *parasitic radar*, and *piggyback radar*. In this book it is called *passive bistatic radar* (*PBR*), with *piggyback radar* used to refer to planetary exploration in recognition of their independent development. Finally, transmitters of opportunity in military scenarios can be designated as either *cooperative* or *noncooperative*, where cooperative denotes an allied or friendly transmitter and noncooperative denotes a hostile or neutral transmitter. PBR operations are more restricted when using the latter.

1.2 APPLICATIONS

In nearly all cases of bistatic and multistatic operation, antenna separation is selected to achieve an operational, technical, or cost benefit and is usually a significant fraction of the target range. One operational benefit is to covertly exploit a noncooperative surveillance radar transmitter for surveillance. A classic example is the German *Klein-Heidelberg-Parasit* hitchhiking off of the British *Chain Home* radars in World War II to warn of Allied bombing raids from England. A second operational benefit is to have an air-to-ground attack aircraft hitchhike off a cooperative, standoff, monostatic radar when penetrating the radar-illuminated target area in radio frequency silence.

Other transmitters of opportunity illuminating the target area might also be exploited. Another operational benefit is to counter retrodirective jammers and radar cross-section cancellation techniques that retransmit the incident radar signal back to the transmitter, which is now separate from the receiver.

A technical benefit is to improve target location accuracy with a multistatic radar operating in a multilateration configuration when *geometric dilution of precision* conditions are favorable. Examples of two such applications are improving the accuracy of range instrumentation systems and tracking ballistic missile launches. Another technical benefit is to combine vector wind field measurements from a monostatic doppler weather radar and multiple bistatic receivers hitchhiking off of the monostatic radar to provide an accurate estimate of the full vector wind field. Semiactive homing missile accuracy during endgame can also be improved by increasing the *bistatic angle*, which reduces *target glint*.

Cost benefits include piggybacking on existing satellite data link transmitters for planetary exploration and exploiting terrestrial broadcast transmitters for ionospheric measurements, as does the *Manastash Ridge Radar*. Also, very high transmitter power can be provided at the lowest cost by a continuous wave (CW) transmitter, which in turn requires spatial isolation from the receiver, conveniently provided by bistatic operation. The space surveillance systems *SPASUR* and *Graves* use this approach.

Finally, a combined operational, technical, and cost benefit is to use a multistatic PBR to exploit for example, a commercial FM broadcast station for military air surveillance. This configuration can (a) operate both covertly and in the resonance region of most air vehicles, including stealthy air vehicles, where target radar cross section is enhanced; (b) generate accurate location estimates (in two dimensions) via multilateration; and (c) eliminate the cost of the transmitter and associated equipment. However, these benefits come with both technical and operational limitations, which are detailed subsequently.

While these types of benefits are both credible and useful, they remain niche applications when compared to the ubiquitous capabilities of monostatic radars, which remain the paramount method for *radio detection and ranging*.

1.3 PURPOSE, SCOPE, AND HISTORY

The history of bistatic and multistatic radars is well documented [1–15]. However, a brief summary will be provided here to establish the purpose and scope of this book. With a nod to Christian Hülsmeyer's early 1904 *telemobilskop* demonstrations, all radar observations and experiments in the 1920s and early 1930s were of the bistatic type. Many countries deployed CW bistatic radars in a fence configuration primarily for air-defense alerting and cueing prior to and during World War II. With the advent of pulsed operation and the duplexer in the late 1930s, the monostatic radar, with its single-site operational advantage, became the configuration of choice and all bistatic radar work ended after World War II.

Since then, bistatic and multistatic radars have had periodic resurgences when a specific bistatic application was found attractive or when the concept was simply rediscovered. That observation was made in 1991 [12], which documented two such resurgences. The first started in the 1950s, when (a) tactical semiactive homing missiles, (b) bistatic fences for air defense and ballistic missile launch warning, and (c) multistatic radars for test range instrumentation and satellite tracking were developed and deployed. Both the tactical missiles and satellite trackers remain operational.

The second resurgence began in the late 1960s when data link transmitters on satellites and ground-based receivers were used for moon and planetary surface measurements in ten missions over a span of 40 years (That work is ongoing, and now includes ground-based transmitters and receivers on satellites.) A multistatic radar hitchhiker, called the *Multistatic Measurement System*, was deployed on a ballistic missile test range to improve the measurement accuracy of reentry vehicles. It was finally retired after significant improvements were made to the monostatic radars. In contrast, other bistatic radar concepts were developed and tested, but not deployed, to counter the new, antiradiation missile and emitter locator threats. The bistatic fence was redeveloped and tested to protect high-value ground targets, but not deployed. Willis [12] then concluded the introduction to his history chapter with:

> This chapter details bistatic radar history in an attempt to illuminate… special, potentially worthwhile bistatic [and multistatic] applications as well as bistatic "dead ends," and to ease the process of rediscovery in the next resurgence cycle.

We are well into that next cycle, which establishes the purpose of this book: to report events in the third resurgence. These events are divided into two parts:

Part 1: Bistatic/Multistatic Radar Systems, which documents both resurrected and new information about the development, testing, and fielding of bistatic and multistatic radar systems for military, scientific, and/or commercial use in approximate chronological order.

Part 2: Bistatic Clutter and Signal Processing, which updates and publishes the previously restricted bistatic clutter database and its analysis, and reports (a) the development of autofocus and image formation methods to improve bistatic synthetic aperture radar (SAR) performance and (b) the development of adaptive cancellation methods to improve bistatic moving target indication (MTI) performance.

1.4 SUMMARY

The following summarizes events occurring in the third resurgence, outlined in the table of contents.

Part 1: Bistatic/Multistatic Radar Systems

Chapter 2: History Update As usually happens during continuing and even intermittent radar developments over many years, old information is released and new information from earlier events surfaces.[1] Such is the case for bistatic radars and that information is reported by Nicholas J. Willis in this chapter as an update to Chapter 2. The update includes new information on some very old bistatic radars, including Christian Hülsmeyer's 1904 *telemobilskop*, the World War II Japanese *Type A* fence, and the German *Klein-Heidelberg-Parasit*, along with some obscure British *Chain Home* bistatic and multistatic radar modes.

Information on bistatic radars developed in the first resurgence, including the *AN/FPS-23 Fluttar* and *440-L*, has now been released and is the subject of Chapters 3 and 4. An example of bistatic glint reduction for semiactive homing missiles surfaced along with new information on bistatic radar developments in the United Kingdom and the United States during the second resurgence. Finally, all available information is summarized on bistatic radar developments during the third resurgence.

Chapter 3: Fluttar DEW-Line Gap-Filler *Fluttar* was an experimental bistatic radar developed by MIT Lincoln Laboratory in the 1950s which led to the *AN/FPS-23*, the first operational bistatic radar deployed by the United States. It was used in the *Distant Early Warning* (*DEW*) *Line* as a

1 Often the new information surfaces after publishing a book or giving a lecture, which triggers a memory from some reader or listener. That person then approaches the author with the new data, which in turn restarts the documentation cycle. This happened many times to the editors, with many thanks for the offerings.

gap-filler radar to detect low-flying aircraft in the late 1950s. It has been briefly described in early literature, but until the release of the 1956 report "Prediction of Bistatic CW Radar Performance" [16], little information was known about its design and performance. That problem has now been resolved by Merrill I. Skolnik in this chapter.

Chapter 4: Missile Attack Warning In a similar vein, little information had been made available on HF over-the-horizon (OTH) and surface wave (SW) bistatic radars, notably just a brief description of the *440-L* forward-scatter OTH system for ballistic missile launch detection deployed in 1967. That status was considerably improved with new information [17–22] further describing the *440-L* and also describing the *Sugar Tree*, an HF SW/OTH PBR for detecting ballistic missile launches. Edwin Lyon redacts this newly available information, augmenting it with photos from his private collection.

Chapter 5: Planetary Exploration Measurements of moon and planetary surfaces by bistatic radars started in 1967 and have been ongoing for nearly 40 years. The first successful *piggyback* operation exploited data link signals transmitted from the Luna-11 probe, scattered off the moon's surface and then received by an earth-based station, in a *downlink* mode. These simple and inexpensive surveys have included measuring surface roughness and density, searching for clean-water ice, and probing planetary atmospheres and plasmas. A reciprocal but more complex and costly *uplink* mode uses a bistatic receiver carried by the probe, collecting high power, earth-transmitted command signals after scattering from the planet. It has a greater link margin, which will be needed to measure Pluto's tenuous atmosphere by the *New Horizons* probe in 2015.

A comprehensive review of these continuing operations and measurements was published by the IEEE in 1993 [23] and hence was not available in [12], first published in 1991. That omission has now been rectified by Richard A. Simpson, who extended and updated his seminal 1993 paper for this chapter.

Chapter 6: Air Surveillance PBR exploitation of commercial TV and FM broadcast transmitters for bistatic and multistatic air surveillance began in the 1980s. Initial U.K. field trials in 1985 concluded that doppler processing of TV waveforms was more promising than range processing. Later U.K. PBR field trials used TV transmitters for doppler and coarse bearing measurements and FM transmitters for range, doppler, and coarse bearing measurements, yielding aircraft detection ranges of 160–175 km from the receiver. Real-time target location was elusive, primarily due to a lack of sufficiently accurate bearing estimates.

In the U.S. PBR, experiments were being conducted at the same time, but none were published in technical journals. One PBR operating with an FM transmitter and multiple receivers in a multistatic multilateration mode, called *Silent Sentry*™, showed promise for air surveillance. It was evaluated by radar engineers from government and industry assembled in various government-sponsored committees over a period of about 5 years starting in the late 1990s.[2]

The charter of these committees was then expanded to evaluate PBRs that exploited other transmitters of opportunity, including satellite broadcast and navigation transmitters, high-definition TV transmitters, cell-phone towers, and HF (high frequency) broadcast and radar emitters. A U.K. technical journal and a U.S. conference published much of this *passive radar* (i.e., PBR) research and testing [24, 25], which forms the basis for this chapter. These data are extended by papers in later journals and conferences and by further analysis, including an updated method for quantifying bistatic radar performance. Nicholas J. Willis, Hugh D. Griffiths, and David K. Barton author this chapter.

Chapter 7: Ionospheric Measurements Quite independent of the air surveillance PBR work reported in Chapter 6, John D. Sahr developed a second PBR variant, the *Manastash Ridge Radar*

2 Six authors contributing to this book served on one or more of these committees.

(*MRR*), at the University of Washington in the mid-1990s. This PBR exploits FM broadcast transmitters to study turbulence in the ionosphere, specifically auroral E-region irregularities, using range, doppler, and direction-of-arrival (via interferometry) measurements. Motivation for *MRR* development included lower cost, increased safety, spectrum availability, and pedagogical opportunity. Although it is not subject to the stringent air surveillance requirements for accurate location of multiple targets in real time, it routinely detects meteors and aircraft in the course of normal operations. Dr. Sahr later became a member of the government-sponsored PBR committees and authors this chapter based on publications and recent observations [26–30].

Chapter 8: Wind Measurements At about the same time and again independent of other bistatic radar developments, Joshua Wurman and researchers at the U.S. National Center for Atmospheric Research developed and tested a small, nonscanning, inexpensive bistatic receiver hitchhiking off of a monostatic weather research radar to measure *dual-doppler* vector wind fields. Subsequent tests using a multistatic network measured full, three-dimensional vector wind fields. Such measurements are useful for research, aviation, weather prediction, and severe weather detection. In the ensuing ten years Dr. Wurman commercialized this hitchhiking concept and delivered nine *bistatic network receivers* to weather research centers around the world for testing and evaluation. With Dr. Wurman's permission and review Nicholas J. Willis has redacted information from his publications [31–34].

Part 2: Bistatic Clutter and Signal Processing

Chapter 9: Clutter Land and sea clutter degrades bistatic radar performance, just as it does monostatic radar performance. And like monostatic radar, both the bistatic radar's resolution cell and the clutter scattering coefficient must be known to calculate the degradation. Bistatic land and sea clutter measurements, analysis, and modeling began in the 1960s and continued intermittently through the turn of the century. Data published between 1965 and 1979 was summarized and evaluated by Melvin M. Weiner in a comprehensive MITRE report, *Multistatic Radar Phenomenology, Vol. 1: Terrain and Sea Scatter*, in September 1980 [35], but with distribution limited to government agencies only. That restriction was lifted in 2003. Subsequently, Dr. Weiner updated his 1981 report with data taken through 2002 and extended his analysis, all of which are included. This chapter documents data on both *in-plane* and *out-of-plane*[3] bistatic clutter scattering coefficients and extends development of the bistatic radar resolution cell from the basic calculations made in [12].

Chapter 10: Spotlight SAR The last chapter of [12] evaluated the phase noise characteristics of dual oscillators needed to support bistatic SAR and MTI operation. The final paragraph of that chapter concluded with the following:

> Finally, the monostatic single oscillator configuration allows reasonably long ($T < 10$ sec) SAR coherent integration times without the use of autofocus [automatic focusing] techniques, whereas the bistatic dual oscillator configuration is currently constrained to relatively short ($T < 3$ sec) coherent integration times [and thus of limited utility]. Perhaps this constraint can be relaxed through the use of the bistatic autofocus, and the results will be tested during the next bistatic resurgence.

Bistatic autofocus and related processes including two-dimensional and three-dimensional image formation were indeed developed and tested in this third resurgence. They are reported by

3 In-plane constrains the transmitter, receiver, and clutter patch geometry to one plane, sometimes called the *bistatic plane*. Out-of-plane constraints allows all other geometries.

Brian D. Rigling in this chapter. Specifically, local oscillator phase instabilities have been reduced and bistatic autofocus algorithms were developed to improve range measurement accuracy from the antenna phase center to the imaged scene. This problem is exacerbated in bistatic SAR where the position of both the transmit and receive platforms must be tracked with subwavelength relative accuracy to correct slow time-varying phase errors as part of the image formation process. The result allowed an increase in coherent integration times comparable to monostatic SAR, typically >10 sec. Consequently, bistatic SAR image quality has been greatly improved, extending its tactical applications.

Chapter 11: Adaptive MTI As discovered during the second bistatic resurgence, when both transmitter and receiver are airborne and moving, the ground clutter becomes nonstationary, in that it skews and spreads depending on the platform geometry, kinematics, and look angles [12, 36]. It was also found that the clutter could be cancelled by counterrotating the transmitter and receiver about the target area. However, that technique often placed unacceptable constraints on airborne platform operation. Consequently, adaptive clutter cancellation using bistatic space–time adaptive processing (STAP) was developed to improve clutter-limited detection performance independent of airborne platform operation. That development is reported in this chapter by William L. Melvin as a summary and extension of his many publications [37–39].

Bistatic STAP methods are not simple applications of monostatic approaches, but a new class of algorithms. Specifically, they apply a data-dependent weighting to echo signals collected over multiple pulses and receive antenna channels configured for digital beamforming. This weighting dynamically tailors a filter response in angle and bistatic doppler to suppress ground clutter reflections. Other requirements include estimating the spatio-temporal clutter covariance matrix and hypothesis of the target steering vector. With compensation for nonstationary behavior, bistatic clutter suppression has been greatly improved, extending its tactical applications.

References

1. R. I. Wilkinson, "Short Survey of Japanese Radar-II," *Electrical Engineering*, Vol. 65, 1946, pp. 455–463.
2. Pierre David, *Le Radar*, Presses Universitaires de France, Paris, 1949 (5th ed., 1969).
3. A. Price, *Instruments of Darkness: The History of Electronic Warfare*, Charles Scribner's Sons, New York, 1978.
4. M. I. Skolnik, *Introduction to Radar Systems*, McGraw-Hill Book Company, New York, 1980.
5. D. K. Allison, "New Eye for the Navy: The Origin of Radar at the Naval Research Laboratory," NRL Rep. 8466, Naval Research Laboratory, Washington, DC, 1981.
6. J. E. Summers and D. J. Browning, "An Introduction to Airborne Bistatic Radar," in *IEE Colloquium on Ground Airborne Multistatic Radar*, London, 1981, pp. 2/1–2/5.
7. S. S. Swords, *Technical History of the Beginnings of RADAR*, IEE History of Technology Series, Vol. 6, Peter Peregrinus, London, 1986.
8. H. E. Guerlac, *Radar in World War II*, Vols. I and II, Tomask/American Institute of Physics, New York, 1987.
9. H. W. Cooper, "Radar: History of a Need-Fostered System," Invited paper, *Microwave Journal*, May 1989, p. 213.
10. M. Prichard, *The Radar War: The German Achievement, 1904–1945*, Patrick Stephens Ltd., Somerset, England, 1989.
11. M. R. B. Dunsmore, "Bistatic Radars," *Alta Frequenza*, Vol. LVII, No. 2, March–April 1989, p. 54.
12. N. J. Willis, *Bistatic Radar*, 2nd ed., Technology Service Corp., Silver Spring, MD, 1995, corrected and republished by SciTech Publishing, Inc., Raleigh, NC, 2005.
13. Y. Nakagawa, *Weapons of World War II*, Aegean Park Press, Laguna Hills, CA, 1997.
14. B. T. Neal, "CH—The First Operational Radar," in Russell Burns (ed.) *Radar Development to 1945*, Peter Peregrinus Ltd: London, 1997 Chapt. 8.
15. V. S. Chernyak, et al., "Radar in the Soviet Union and Russia: A Brief Historical Outline," *IEEE AES Magazine*, December 2003, pp. 8–12.

16. M. I. Skolnik, J. Salerno, and E. S. Candidas, "Prediction of Bistatic CW Radar Performance," in *Symposium on Radar Detection Theory*, ONR Symposium Report ACR-10, Washington, DC, March 1–2, 1956, pp. 267–278.
17. State Department Airgram, Subject: "Project CLEAR SKY," to Amembassy ROME from Sec'y Rusk, U.S. State Department, February 26, 1968 (Declassified), available at http://www.gwu.edu/~nsarchiv/NSAEBB/.
18. State Department Memorandum, To: Leonard Weiss, From: Frank Perez, Subj.: "Launch on Warning," dated January 29, 1971, p. 2 (Declassified), available at http://www.gwu.edu/~nsarchiv/NSAEBB/.
19. J. Richelson, *Wizards of Langley*, Westview Press, Boulder, CO, 2001.
20. Memorandum, Chief of Naval Research to Chief of Naval Operations, Subj.: CW Transmit Siteat Spruce Creek, Florida, dated April 29, 1966.
21. P. J. Klass, "HF Radar Detects Soviet ICBMs," *Aviation Week and Space Technology*, Vol. 95, No. 23, 1971, pp. 38–40.
22. *Backscatter Literature Survey*, Project 3311, Communication and Propagation Laboratory, Stanford Research Institute, Menlo Park, CA, January 1961.
23. R. A. Simpson, "Spacecraft Studies of Planetary Surfaces Using Bistatic Radar," *IEEE Transactions on Geoscience and Remote Sensing*, Vol. 31, No. 2, March 1993, pp. 465–482.
24. *IEE Proceedings—Radar, Sonar Navigation*, Special Issue on Passive Radar, Vol. 152, No. 3, June 2005.
25. *AOC 4th Multinational PCR Conference*, Syracuse, NY, October 6, 2005.
26. J. D. Sahr and B. G. Fejer, "Auroral Electrojet Plasma Irregularity Theory and Experiment: A Critical Review of Present Understanding and Future Directions," *Journal of Geophysical Research*, Vol. 101, 1996, pp. 26893–26909.
27. J. D. Sahr and F. D. Lind, "The Manastash Ridge Radar: A Passive Bistatic Radar for Upper Atmospheric Radio Science," *Radio Science*, Vol. 32, 1997, pp. 2345–2358.
28. J. D. Sahr and F. D. Lind, "Passive Radio Remote Sensing of the Atmosphere Using Transmitters of Opportunity," *Radio Science Bulletin*, Vol. 284, 1998, pp. 4–7.
29. J. D. Sahr and M. G. Meyer, "Opportunities for Passive VHF Radar Studies of Plasma Irregularities in the Equatorial E and F Regions," *Journal of Atmospheric and Solar-Terrestrial Physics*, Vol. 66, No. 17, 2004, pp. 1675–1681.
30. J. D. Sahr, "Remote Sensing with Passive Radar at the University of Washington," *IEEE Geoscience and Remote Sensing Society Newsletter*, December 2005, pp. 16–21.
31. J. Wurman, S. Heckman, and D. Boccippio, "A Bistatic Multiple-Doppler Radar Network," *Journal of Applied Meterology*, Vol. 32, December 1993, pp. 1802–1814.
32. J. Wurman, "Vector Winds from a Single-Transmitter Bistatic Dual-Doppler Radar Network," *Bulletin of the American Meterological Society*, Vol. 75, No. 6, June 1994, pp. 983–994.
33. J. Wurman, M. Randall, C. L. Frush, E. Loew, and C. L. Holloway, "Design of a Bistatic Dual-Doppler Radar for Retrieving Vector Winds Using One Transmitter and a Remote Low-Gain Passive Receiver" invited paper, *Proceedings of the IEEE*, Vol. 82, No. 12, December 1994, pp. 1861–1872.
34. S. Satoh and J. Wurman, "Accuracy of Wind Fields Observed by a Bistatic Doppler Radar Network," *Journal of Atmospheric and Oceanic Technology*, Vol. 20, 2003, pp. 1077–1091.
35. M. M. Weiner, *Multistatic Radar Phenomenology, Vol. 1: Terrain and Sea Scatter*, MTR 8140, MITRE Corp., Bedford, MA. September 1980 (Cleared for public release on June 30, 2003).
36. D. C. Lorti and M. Balser, "Simulated Performance of a Tactical Bistatic Radar System," in *IEEE EASCON 77 Record*, Publication No. 77, Arlington, VA, 1977, Chapts. 4-4A–4-40, pp. 1255–1259.
37. W. L. Melvin, M. J. Callahan, and M. C. Wicks, "Bistatic STAP: Application to Airborne Radar," in *Proceedings of the 2002 IEEE Radar Conference*, Long Beach, CA, April 22–25, 2002.
38. W. L. Melvin, "A STAP Overview," *IEEE AES Magazine—Special Tutorials Issue*, Vol. 19, No. 1, January 2004.
39. W. L. Melvin, M. J. Callahan, and M. E. Davis, "Comparison of Bistatic Clutter Mitigation Algorithms for Varying Geometries," in *Proceedings of the 2005 IEEE International Radar Conference*, Alexandria, VA, May 9–12, 2005, pp. 98–103.

PART I

Bistatic/Multistatic Radar Systems

2

History Update[1]

Nicholas J. Willis

2.1 BEGINNINGS

Radar was first patented and demonstrated in 1904 by the German engineer Christian Hülsmeyer. As is well documented [2–5], his *telemobilskop* used a special spark-gap transmitter operating on a 40–50-cm wavelength and a separate receiver that rang a bell when detecting ships up to 5 km from the receiver. Figure 2–1 shows photographs of Hülsmeyer and his invention.

The first fully documented demonstration of the *telemobilskop* was in Cologne, in May 1904, with both the transmitter and receiver located on a platform under the Rhine Bridge (a chain suspension bridge destroyed during World War II) in a monostatic configuration to detect ships on the Rhine River [5].[2]

Purists might argue that Hülsmeyer's *telemobilskop* was not a radar because it did not directly measure range, as in "radio detection and ranging," and technically they would be correct. However, his invention included the essential elements of a radar, all reduced to practice: antennas, transmitter, receiver (with adequate shielding), and indicator, in this case an audio alarm to signal when a target was sufficiently close in range to require attention. More sophisticated indicators would have to await the development of timing circuits and displays.

Now for an excursion to follow the monostatic *telemobilskop* story a little further. Hülsmeyer's *telemobilskop* was first rejected by none other than Admiral von Tirpitz of the German Navy—who was namesake of the battleship *Tirpitz*, sister ship to the *Bismarck*—with the comment "Not interested. My people have better ideas!" [5].[3]

Although the admiral's "better ideas" were not revealed at the time, one was likely the optical rangefinder, a device that could measure both range and bearing to another ship at horizon distances (15–20 nmi) under good visibility conditions. This development was a major advance in naval fire control at the turn of the nineteenth century because target range had always been a critical but poorly estimated parameter in gun fire control calculations. Figure 2–2 shows one of

1 This chapter updates Chapter 2 in *Bistatic Radar* [1]. It adds new information and enough previous data to present a coherent story. However, not all photographs and details have been replicated here.

2 Prichard devoted a full chapter in his book [5] to the telemobilskop, documenting events with Hülsmeyer's German patent, letters and an oral history from his daughter Annelise. All these events showed a monostatic configuration with substantial shielding between an adjacent transmitter and receiver. In contrast, Cooper [4] reported the May 1904 Cologne tests with the transmitter on a high tower in a bistatic configuration and Swords [2] reported the first public demonstration was again in May 1904, but on the Hohenzollern bridge (a steel arch bridge that is still standing). Thus it is possible that Hülsmeyer conducted a preliminary bistatic experiment on a different bridge, which may not have used shielding and thus necessitated transmitter-receiver separation for adequate isolation.

3 Apparently the British Royal Navy also rejected it.

Figure 2–1 Christian . The spark-gap transmitter is on the left; receiver, dipole antennas, and bell on the right. Left photo was taken in 1904; right photo, considerably later. The *telemobilskop* is now on display at the Deutsches Museum, Munich. Courtesy of University College, London.

the first optical rangefinders used by the U.S. Navy. Figure 2–3 shows its evolution 30 years later in the German Navy.

Unfortunately, while Hülsmeyer's device was advertised to give a coarse indication of bearing and a very coarse indication of range (by sliding the telemobiloscope up and down the mast for triangulating the target [5]—not unlike the optical rangefinder), and most importantly, could operate at night and under zero visibility conditions, its range was simply inadequate for naval gunfire control, the paramount objective of a warship.

Prichard [5] continues Hülsmeyer's story:

> But reports had reached Director Wierdsma of the Holland-Amerika-Line who invited the inventor to demonstrate his discovery at a Technical Nautical Meeting at Scheveningen, and on Thursday, 9 June 1904, assisted by students of Delft University, Hülsmeyer assembled his equipment in the tender *Columbus* and cruised up and down Rotterdam Harbour detecting vessels at up to 5 kilometers' range with unerring accuracy...
>
> The representatives were... enthusiastic about the demonstration and showered Hülsmeyer with praise, and after a lecture in which he explained that even longer ranges were quite possible he waited for orders.
>
> But none came. Not even a trial order.

Prichard [5] then analyzed the reasons for rejection, first dismissing technical objections and then concentrating on the maritime issue:

> In the first place, world shipping in 1904 was going through a bad patch and expenditure on even the bare essentials of safety procedures was viewed with reluctance...
>
> Second, and more important, the arrival of wireless telegraphy was a hindrance to the acceptance of Hülsmeyer's discovery. By 1904 it had been installed in a number of vessels which were now able to communicate with others at distances far beyond the range of visual signalling. In emergencies the wireless operator could quickly call for help, and this had been possible as early as 1901. Moreover, direction-finding by the use of cross-bearings from shore stations enabled a master to fix his position with a fair degree

Figure 2–2 "In the maintop of the [U.S. battleship] North Dakota," circa 1910 [6]. The rangefinder is the long, horizontal tube in front of the seamen standing on the platform at the head of mainmast, in this case built as a cage-mast. It consists of lenses, reflectors, and prisms at each end and center of the tube, which display two target images, one above the other, to an observer at the center. The observer aligns the two images with a vernier calibrated in range, solving the triangle with the target at the apex. The range is then phoned to the fire-control station below. The observer also reports subsequent shell splashes as *short* or *over* the target. The system operated only in daylight and fair weather. It (and the seamen) was also handicapped by stack gas (background) and gunfire smoke (foreground). Rangefinders remained on most U.S. warships until well after World War II. It was not one of the coveted battle stations. The addition of radar—when it was working—rectified all these problems. (Photo courtesy of the U.S. Navy.)

of accuracy. Ship-owners were therefore reluctant to part with money for a new-fangled system when they were already paying enough for the present one.

Following unsuccessful attempts to sell the device to major electrical companies, including the Wireless Telegraphy Company, now known as Telefunken, Hülsmeyer abandoned his quest for other interests. Prichard observes in hindsight that the *telemobilskop* might have prevented the *Titanic* disaster had the maritime industry embraced it. And in further hindsight, even the U.S. Navy may have been able to stave off its worst disaster in peacetime history, which was caused by a flawed implementation of that same type of wireless telegraphy direction-finding [7]. Figure 2–4 shows a photograph of the disaster.[4]

Clearly even warships need autonomous methods to safely navigate near land, for example devices like Hülsmeyer's *telemobilskop*. Of course such a device must be (1) installed, (2) working,

4 One line of bearing to a ship's transmitter was measured by the U.S. Navy's radio-compass station at Pt. Arguello, CA, and then transmitted to the ship, but with a 180° measurement ambiguity. A squadron of 19 destroyers was en route from San Francisco to San Diego on a speed run in limited visibility conditions due to fog. The squadron navigator, the only one in the squadron with authority to navigate, mistakenly plotted the ambiguity, which put the squadron south of Pt. Arguello, rather than north of it. The squadron in single-file, follow-the-leader, formation then turned east, landing the first seven ships on the rocks at Honda. Compounding the disaster was the fact that the bearing ambiguity could have been resolved by installing a switchable reflector behind the radio-compass antenna at Pt. Arguello, as done on the British *Chain Home* radars (Figure 2–9).

Figure 2–3 Optical rangefinder, still in a tube configuration but now much larger and with the operators encased in a turret, on the German pocket battleship *Admiral Graf Spee*, 1939 [5]. The 3 × 6 m antenna covered by canvas just above the rangefinder is the *FuMO 22 Seetakt*, a 361–429-MHz monostatic radar used for both air defense and gunfire control. The latter was useful up to 20 km with a range accuracy of about 70 m and a direction-finding accuracy of 3°. Data from the radar and the optical rangefinder were readily compared. Early versions of these radars often failed after a few salvos from the main batteries, which violently rattled the ship. A modified version of *Seetakt* was installed on the *Bismarck* but not on the *Tirpitz*, which apparently remained satisfied with optical rangefinders. (Photo courtesy of Fritz Trenkle [5].) © Patrick Stephens, Ltd., Cambridge, England.

Figure 2–4 Seven 1,200-ton U.S. destroyers on the rocks at Honda, 10 miles north of Pt. Arguello, CA, September 8, 1923. Five destroyers are aground but upright; two have capsized. Twenty-three seamen died. In a period of five minutes the U.S. Navy lost more combatant ships than she had lost by enemy action during World War I. Material loss was over 13 million U.S. dollars or roughly 200 million U.S. dollars in today's currency. (Photo courtesy of the U.S. Navy. Information courtesy of *Ships and the Sea* [7].)

(3) turned on, (4) manned, and most importantly (5) heeded.[5] The probability of all those events occurring was infinitesimally small in the 1920s. For example, the *better idea* optical rangefinder had not yet been installed on any of those destroyers. But the potential was there—just not appreciated for either maritime or military use.

As a result, those rejections ended all radar work for the next 18 years. That, in final retrospect, was the major tragedy of Hülsmeyer's work: it was forgotten. There was no documentation of his successful experiments in technical journals—just a few newspaper articles and obscure patents buried in the files. So his legacy was lost to the engineering community, which was then forced to start over with new and, as it turned out, quite different approaches.

This *radar effect* was rediscovered by A. H. Taylor and L. C. Young in 1922 when they were carrying out 60 MHz propagation experiments using a superheterodyne receiver and a 50-W transmitter amplitude-modulated at 500 Hz at the U.S. Navy Aircraft Laboratory, Anacostia, DC. They first detected bistatic reflections from buildings, trees, and other stationary objects and then a wooden steamer traveling down the Potomac River, which cut the beam in what is now known as the forward-scatter mode [2, 8]. Again and likely for similar reasons such as no range measurements, the concept was rejected by the U.S. Navy.

Then in the next 15 years both this forward-scatter effect and the doppler beat frequency generated between the direct path signal and echoes from moving targets were observed by many engineers and scientists conducting RF (radio frequency) propagation experiments. Both effects could be observed together, as shown in Figure 2–5.

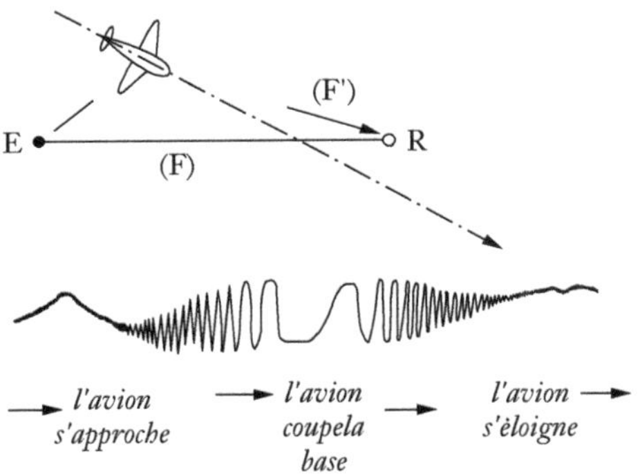

Figure 2–5 1937 drawing by Pierre David, showing an aircraft approaching, crossing, and then departing his *electromagnetic barrier* [9]. (F) is the *barrier*, also called the *fence* or *baseline*. The doppler beat frequency decreases during the approach, is zero during the crossing, and increases during the departure. The increase in amplitude is the forward-scatter radar cross-section enhancement. (Pierre David, *Le Radar* © PUF, coll "Que sais-je?" No. 381, 1969.)

These experiments used new, commercial radio transmitters or similar experimental equipment operating between 30 and 80 MHz.[6] Table 2–1 summarizes these events.[7]

---

5 On December 7, 1941, an *SCR-270* air surveillance radar at Pearl Harbor was installed, working, turned on, and manned. But its data was not heeded.

6 An exception is the 1935 U.K. experiment, which exploited the 6-MHz BBC transmitter station at Daventry.

7 There is a sketchy report [10] about A. S. Popov detecting the Russian warship *Lieutenant Il'in* when it crossed a radio communication link between two other Russian ships *Europe* and *Africa* on the Baltic Sea in 1897. The report then credited him with discovering radiolocation—yet another father of radar. (Such claims lead Lamont Blake, nominated by this author as father of the definitive radar range equation, to conclude that radar was one of those bastard inventions: one mother, many fathers.) Unfortunately the Popov et al.'s 1945 report (in Russian) is not accessible. Since nothing further appears to have been accomplished or published, we shall let it be.

Table 2–1 Summary of Early Bistatic Radar Discoveries and Developments All Prior to the Invention of the Duplexer.

Country	Researchers	Date	Transmitter	Target	Results
United States (Naval Research Laboratory)	Taylor, Young, and Hyland	1922–1936	Communications Landing beacons Radio stations	Ships Aircraft Dirigible	Forward-scatter detection Beat frequency detection Rejected by the U.S. Navy Patent in 1934
United Kingdom	Appleton and Barnett	1924–1925	BBC station at Bournemouth	Ionosphere (Heaviside–Kennelly layer)	Doppler beat frequency using stepped frequency waveform for ranging
United States	Breit and Tuve	1925	Station NKF, Washington, DC	Ionosphere (Heaviside–Kennelly layer)	Pulsed (1-ms) transmissions for ranging
United Kingdom (Post Office)	Nancarrow, Mumford, Carter, and Mitchell	1932	Communications	Aircraft	Beat frequency detection Reported but overlooked
United States (Bell Telephone Laboratories)	Englund, Crawford, and Mumford	1933	Communications	Aircraft	First widely published (Institute of Radio Engineers) beat frequency report
France	David (predicted in 1928; then read BTL-IRE paper)	1933	Own design	Aircraft	Forward-scatter fences Deployed ~20 sets in 1938

These early bistatic radars were typically configured as fixed, ground-based fences to detect the presence of aircraft as they crossed the fence, which could then alert and cue air defenses. The problem of extracting target position along the fence could not be solved since range estimates become indeterminate on the fence. However, Pierre David configured multiple fences in a multistatic configuration to estimate speed, direction, and altitude of nonmaneuvering aircraft based on the time they crossed each fence. The most complex was called the *maille en Z*, roughly translated as a *mesh in (the shape of) a Z*, which in tests estimated speed within 10%, direction within 10–20°, and altitude within 1 km [2]. An example calculation is given in [1]. Figure 2–6 shows the deployment of David's electromagnetic fences around Brest, France. Figure 2–7 shows a photograph of *RAPID*, the first experimental U.S.S.R. forward-scatter fence that evolved into *RUS-1*.

Under favorable propagation conditions, some VHF fences could operate over extraordinarily long baseline ranges. For example, the Japanese *Type A* fence, also called *Wan Wan Shiki* (*Bow Wow Type*), owing to the sound generated by the doppler beats, operated between 40 and 60 MHz at 10, 20, 100, and 400 W transmitter power, which established fence lengths of 80, 120, 200, and

Country	Researchers	Date	Transmitter	Target	Results
Italy	Marconi	1933	Microwave telephone link	Steam roller, automobiles	Forward-scatter detection Awaited monostatic radar
Germany	Unknown	1934	Decimeter radio link	Ship	Beat frequency detection Reverted to monostatic
Union of Soviet Socialist Republics	Oshchepkov (who was then imprisoned in the 1937–1938 Great Purge)	1934	Own design	Aircraft	Forward-scatter detection Deployed ~45 *RUS*-1 fences starting in 1939
United Kingdom	Wilkins and Watson-Watt (+ Appleton's 10-year-old receiver)	1935	BBC station at Daventry (GSA)	Aircraft	Forward-scatter detection plus angle of arrival Evolved into *Chain Home*
Japan	Okabe, Yagi, Satake, and Kobayashi	1936–1938	NEC (Sumitomo) design, not specified	Aircraft	Beat frequency detection Built 400–500 and deployed >100 *Type A* fences starting in 1941

350 km, respectively [11]. Another report [12] placed a *Type A* fence between Formosa (Taiwan) and Shanghai, a range of ~800 km.

Invention of the radar duplexer in July 1936 by Young and Page of the Naval Research Laboratory provided a means of using pulsed waveforms with a common transmit and receiver antenna. This single-site configuration is the familiar monostatic radar, which greatly expanded the utility of radar, particularly for use by aircraft, ships, and mobile ground units. Consequently, by the end of World War II all bistatic radar fences had been abandoned.

During World War II, the Germans developed a totally new bistatic radar concept, called *Klein-Heidelberg-Parasit* (Small Heidelberg Parasite), which *hitchhiked* off of the British *Chain Home* (*CH*) radar transmitters. Both Prichard [5] and Price [13] provide details about *Klein Heidelberg* (*KH*); the following summarizes and evaluates their information.

This bistatic hitchhiker, also called a *half-active* radar system [5], was developed in 1942 by Dr. Wächter of Telefunken in cooperation with the Central Research Establishment of the Post Office. A number of research models were tried near Cherbourg, with the first system sited on the island of Roms off the west coast of Denmark. It was later moved to Oostvorne, the Netherlands, when the Royal Air Force bombing raids moved south.

Two antennas were used, one called a synchronized antenna measured the transmitter's time of arrival and bearing and a larger antenna measured the target's echo time of arrival and bearing. No further details or photographs of these antennas apparently survived. Figure 2–8 shows the geometry.

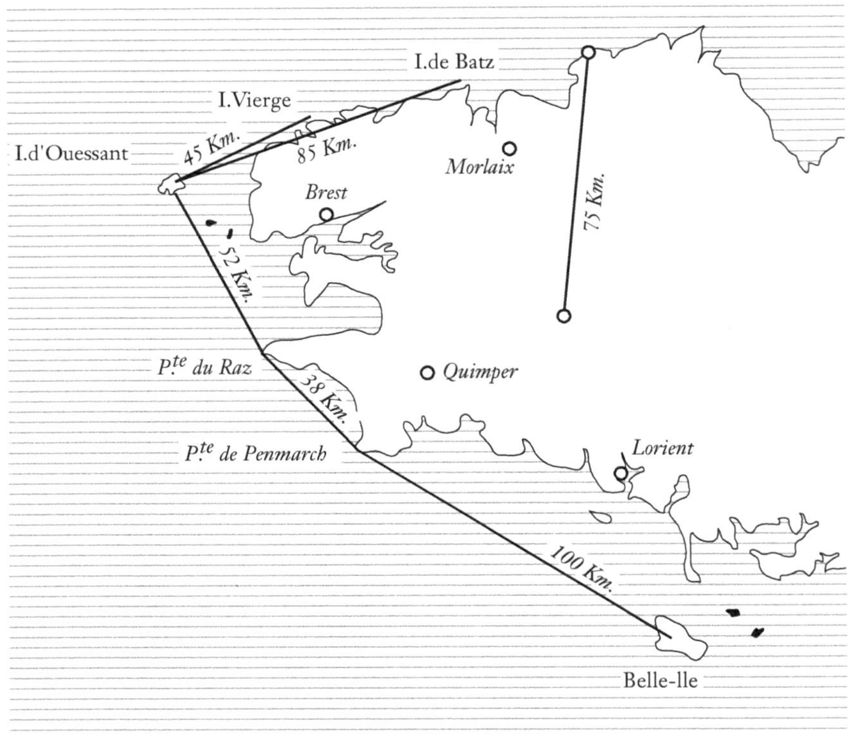

Figure 2–6 Six of Pierre David's 30-MHz forward-scatter fences deployed around Brest, France, in 1939 [9]. It is not clear why two gaps remained. Note the near-parallel fences in the northwest coverage, which suggest partial elements of his more complex configuration called the *maille en Z*, or a *mesh in the form of a Z*. Under some conditions it could give an estimate of aircraft course and speed. (Pierre David, *Le Radar* © PUF, coll "Que sais-je?" No. 381, 1969.)

The receiver used a cathode ray tube indicator that displayed two blips, one from the direct path transmitter and the second from the range-sum target echo. Prichard [5] reports that:

> A circular scale on the screen had graduated marks from 0 to 40, which correspond to the number of ellipses for every Chain Home station, and a special map with accurately registered ellipses was also provided. The expected [target-to-receiver] range of 200 km turned out to be 400 km with a range of accuracy of between 1 and 2 km and a bearing accuracy of 1°. This system gave impeccable results even when other radars were severely affected by jamming transmissions or by *Window* [chaff].

Alternatively, Price [13] reports that under ideal circumstances an aircraft's position could be determined to within 6 miles up to 280 miles (460 km) from *KH*.

These surprisingly large ranges can be evaluated by comparing *KH* equivalent monostatic range performance, R_M (Section 6.5), to *CH* range performance. Typical *CH* performance achieved by experienced operators on a medium bomber flying at an altitude of 20–25 kft was ~180 nmi (330 km) [14]. For *KH*, $R_M = (R_T R_R)^{1/2}$, where R_T is the transmitter-to-target range and R_R is the target-to-receiver range. From Figure 6–28, R_R is a maximum when the target lies on the baseline extended beyond the transmitter, which from Figure 2–8 is the baseline from Oostvorne to Dover, extended past Dover.[8] Under these conditions, $R_T = R_R - L$. Since the

8 This geometry can be achieved by rotating the line of shoot assumed for the Dover transmitter in Figure 2–8 about 30° clockwise.

из приемных устройств опытной аппаратуры радиообнару жения самолетов (лето 1934 г.)

Figure 2–7 Photograph of the *Rapid* receiver from P. K. Oshchepkov's *Life and Dreams*, Moscow Worker, Moscow, 1967, but now unavailable. Caption is translated as: *Experimental equipment from an airplane radio reception and detection device (Summer 1934).* Aircraft were detected by a doppler beat note in the headphones. With a baseline range of 11 km, aircraft up to heights of 1 km were detected 3 km from the receiver. Later tests established that 75-km ranges were possible. Subsequently five factory-produced experimental sets, called *REVEN* (*RHUBARB*), were produced. Then all work stopped after Oshchepkov was imprisoned. In 1938, work was restarted and 45 *RUS-1* sets were produced using a 35-km baseline. Many were deployed to the Soviet Far East and Transcaucasus starting in 1939 [2]. (A second photograph showing the transmitter and receiver has also been circulated, but in marginal condition.) Courtesy of Moscow Worker, Moscow.

baseline range L from Oostvorne to Dover is approximately 210 km and R_R is given as 400–460 km, R_T = 190–250 km, yielding R_M = 275–340 km, which brackets the reported performance of *CH*.[9]

Such *KH* performance requires a receiving system with capabilities (antenna aperture, bandwidth, losses and gains, etc.) matching those used by *CH*. Certainly, antenna matching would have been possible since *CH* used pairs of orthogonal, horizontal, center-fed dipoles mounted on a wooden tower [2, 14], which were observable and thus reproducible, as shown in Figure 2–9. In fact, Prichard [5] reports that the Oostvorne *KH* antenna was camouflaged at the rear of a *Wassermann S* monostatic radar antenna. That antenna was also 200 ft high, so the argument is plausible. In addition, Prichard [5] reports that German receiver technology in some areas was superior to that of the Allies. Thus such a match appears possible.

A second *KH* calibration can be made by assuming that the ellipse in Figure 2–8 represents the maximum range-sum that can be achieved using the Dover transmitter and then scaling range values from the ellipse. In this case L = 210 km, R_R = 305 km, and R_T = 95 km, yielding R_M = 170 km, certainly a more achievable and less controversial result. While these results are roughly half the performance of the earlier estimates, they remain a significant capability as shown in the figure.

However, the bearing accuracy estimates are suspect, since the *CH* receiving antennas were estimated [2] to measure target bearing with errors commonly as large as 12°, an order

9 Using equation (6.5.6) of Section 6.5, a *KH* receive antenna at 600-m MSL will have a 460-km line of sight to an aircraft flying at 25 kft, assuming a smooth, 4/3 earth.

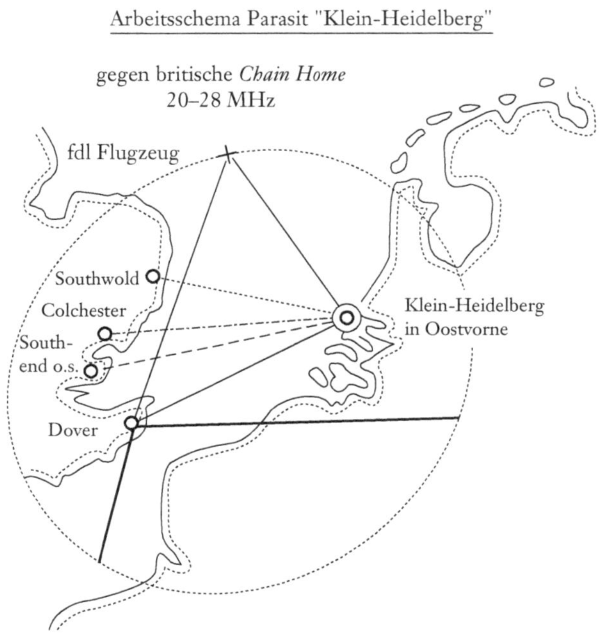

Figure 2–8 Sketch of the *Klein Heidelberg* (*KH*) geometry, as reproduced by M. R. B. Dunsmore [15] from K.O. Hoffman's, *The History of Air Intelligence Unit, Book II, the World War*, 1968, in German, but not available. The constant range-sum ellipse shown in the figure invites discussion: (1) A bistatic target detection area, or coverage region, is defined by an oval of Cassini; but since the shapes of an oval and ellipse are not too dissimilar in this co-site region (e.g., Figure 4–8 in [1]), this ellipse approximates the potential bistatic coverage. (2) An East Coast *CH* transmitter had a floodlight beamwidth of approximately 100° in azimuth [14]; assuming the transmit beam at Dover is oriented in a *line-of-shoot* direction toward the southeast, it will illuminate an area bounded by the ellipse and dark lines shown in the figure. That area is suitable for bistatic operation. The remaining area in the ellipse yields no detections using the Dover site. (3) However, *KH* could also work off the other sites shown in the figure. Since their transmitters will have lines of shoot more to the east, bistatic coverage is considerably broadened. © Alta Frequenza.

of magnitude greater than reported by Prichard [5] and Price [13]. Typically that reduction in measurement error requires an order-of-magnitude increase in the antenna's horizontal aperture for equal signal-to-noise ratios, an unlikely prospect owing to the reported shape of the antenna. Consequently, caution should be exercised when quoting these estimates.

One final caveat must be observed: *CH* was a massive U.K. program of the highest priority, with virtually unlimited finance and with a full development and testing cycle starting in the mid-1930s [14]. In contrast, while *KH* was a one- (or at best, a few-) of-a-kind, receive-only system, it was started in 1942 *after CH* was discovered, with little time for development, testing, or training. Thus, such a *KH* deployment task would have been formidable indeed, requiring its own high priority and funding. But given the *CH* effectiveness, which was painfully apparent to the *Luftwaffe* at the time, such priority could have well been forthcoming.

Unfortunately, while no further information about this first bistatic exploitation of a noncooperative radar transmitter has surfaced, unjammed early warning and possibly ground-controlled intercept of Allied bombing raids, conveniently cloaked in a covert operation, appears to be the principal motivation for its use.

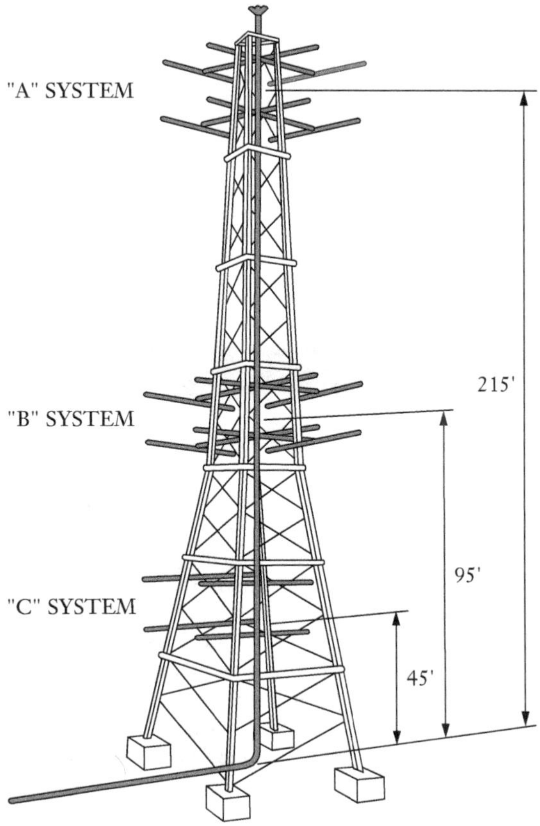

Figure 2–9 Orthogonal, half-wave, dipole arrays on a *CH* receive tower [14]. The "A" system was the main array; an identical "B" system was used to fill the "A" system's multipath null and, with the A system, to estimate target height. The "C" system was used with the "B" system to measure height when the target was in the main array's null. A *goniometer* compared signals from the two arrays to measure height and signals from the orthogonal dipoles in either array to measure azimuth. The measurements had a 180° ambiguity, which was resolved by placing remotely switchable reflectors $\lambda/4$ behind the dipoles, as shown in the figure, and noting whether the signal strength increased or decreased when they were switched in. Reprinted with permission of the IET.

The *CH* radars themselves operated with transmitter and receiver sites separated for isolation (Figure 2–10). However, they had a standard, reversionary mode which, in the presence of electronic countermeasures or a transmitter failure, a receiver could operate with a transmitter at an adjacent site about 40 km away in a bistatic configuration [16].

CH also developed two range-only multilateration location techniques, the first in a multistatic radar configuration for improving direction-finding (D/F) accuracy and the second in a multiple-monostatic radar configuration for tracking German *V2* rockets. Neal elaborates:

> It is worth mentioning at this point that the accuracy of the goniometer method of D/F [Figure 2–9 caption] was earlier thought likely to be inadequate especially in a heavy jamming environment. An alternative scheme, known as the "Chapman" method (after the inventor, Corporal Chapman) using a system of range cuts to fix the target position, was proposed. One station would display echoes from its own

Figure 2–10 Typical East Coast *CH* station. The station employed three in-line, 109.7-m steel transmitter towers spaced approximately 60 m apart. The transmitter "curtains" were slung between towers and fed by transmission lines leading from the heavily protected transmitter building nearby. Typical operating conditions were 20–30 MHz frequency, 350 kW (later 750 kW) peak power, 25 and 12.5 pulses/s PRF and 20 μs pulse length. Four 73.2-m wooden receiver towers, usually placed in rhombic formation, are shown on the right. (See text and Figure 2–9 for their operation.) These towers and the associated receiver building were some hundreds of meters from the transmitter buildings and in some cases were in a separate compound. The dipole antennas on each receive tower were cut for a specific frequency, different from the others, as an antijamming measure. This plan was later abandoned. Important considerations for siting *CH* stations, circa 1936, included being well back from the coast and inconspicuous from the air, having a smooth slope between the site and the sea, soil suitable for supporting the steel masts, and finally, not gravely interfering with grouse shooting [14]. Reprinted with permission of the IET.

transmitter plus echoes from the transmitter of an adjacent station on a second display. The operator had to swing his goniometer and note the two echoes, one on each c.r.t. [cathode ray tube], which attained minima at the same goniometer setting to resolve ambiguity. The position of the target giving rise to the two echoes could then be found by range cuts, the goniometer [i.e., angle measurements] playing no part in the position fixing. To prevent jitter between the two sources of signals, both locked to the grid system... [In any] event, the goniometer method of D/F, after careful calibration, was found to give sufficient accuracy for all practical purposes and the Chapman was abandoned.

CH played a vital part towards the end of World War II in tracking V2 (A4) rockets launched from occupied Europe towards London and the southern counties. The problem was to detect rockets soon after launch so that the likely point of impact could be estimated and early warning given. Additionally, the point of launch could be pin-pointed to enable Mosquitoes of Bomber Command to make a precision attack on the launching site. A system of simultaneous range cuts from five stations: Bawdsey, Great Bromley, High Street, Dunkirk and Swingate (Dover) enabled the trajectory to be plotted as the rockets passed through the vertical lobes of each station, with sufficient accuracy to fulfill both these requirements.

The system was quite complex and relied on the integrity of the communication links and the high performance of the CH stations involved. The CH wavelength was most favourable for detection of the rocket which behaved roughly as a quarter-wave dipole with a very good response and provided detection ranges in excess of 160 km. The system was code-named "Big-Ben." [14]

While these multilateration modes were retired with *CH* after World War II, their legacy reemerged in early German and U.S. range instrumentation systems (using doppler multilateration), the U.S. *Multistatic Measurement System*, and nearly all of the passive bistatic radars (Chapter 6) developed in bistatic radar's third resurgence.[10]

2.2 FIRST RESURGENCE

It was not until the early 1950s that interest in bistatic radars was revived, principally for aircraft, missile, and satellite locations [17–20]. The name *bistatic radar* was coined by K. M. Siegel and R. E. Machol in 1952 [21]. Bistatic radar system [22] and bistatic radar cross-section theories were established, and measurements were taken [23–29]. Bistatic clutter measurements were also started [30, 31], culminating in the public release of all unclassified data and their analysis in 2005, which are reported in Chapter 9.

The first textbook on bistatic radar, *Dispersed Radar Stations and Systems*, was authored by V. Ya. Aver'yanov and published by Nauka I Tekhnika, Minsk, in 1978. It was machine-translated into English in 1980 but never distributed outside U.S. government agencies. The first English language textbook *Bistatic Radar* [1] was published in 1991, expanding earlier chapters in M. I. Skolnik's *Radar Handbook*, 1st and 2nd editions.

The United States considered and Canada tested forward-scatter bistatic radars as gap-fillers for their northern air defense: the Distant Early Warning (DEW) Line and the McGill fence, respectively [17] (see also Chapter 3). However, the radar became overloaded by radar echoes from birds, especially during the migrating season. This false-alarm problem was compounded by the use of continuous wave (CW) transmissions, which could not resolve multiple targets within its coverage.

As a result, the forward-scatter concepts were abandoned and replaced in the United States by the AN/FPS-23 *Fluttar* bistatic radar fence, which used facing transmit and receive beams offset in azimuth to allow CW doppler measurements for false-alarm discrimination. CW operation was continued because of its low costs. It became operational in 1957 but did not fully solve the false-alarm problem due to two unanticipated factors. First, many large birds were drawn to the warm diesel power generators at the sites, producing very strong echoes at short range from either the transmit or the receive site, as described in Section 6.5. Second, small supply aircraft that regularly flew along the DEW Line were detected and while they could be resolved in doppler there was no other means to obtain target resolution, location, and track. Thus they also caused unacceptable false alarms and the radar was removed. Chapter 3 provides a detailed history of *Fluttar.*

During the 1960s several HF over-the-horizon (OTH) radar research programs to detect ionospheric disturbances from ballistic missile launches and nuclear weapon blasts were initiated. One OTH *forward-scatter* system, called *440-L*, was deployed in 1967 using transmitters in the Western Pacific and receivers in Europe, using sky-wave propagation to cover most of the Sino-Soviet landmass [32]. However, the system used an unmodulated CW waveform that was easily spoofed, signaling the end of its utility. It was dismantled beginning in 1975 [33–35].

A second OTH system, called *Sugar Tree*, was configured as a *passive bistatic radar* (PBR; Chapter 6). *Sugar Tree* exploited Soviet-operated, HF radio broadcast signals, which illuminated Soviet missile launching areas via surface-wave propagation. A remote receiver then detected the ionospheric disturbances via a sky wave to warn of a missile launch. It became operational in the 1960s and then was dismantled a few years later [37, 38]. Further details of these HF bistatic OTH programs appear in Chapter 4.

In the United States, range instrumentation systems used both mutilateration and interferometric multistatic radars. Most systems included a beacon transponder on the test

10 In fact, one might nominate Corporal Chapman (tragically of no first name) as the father of multilateration.

vehicle, but two doppler multilateration systems were designed for skin track: the VHF *Passive Ranging Doppler* (*PARADOP*) and the *Microwave Doppler* (*MIDOP*) [39]. As with all doppler multilateration systems, initial target location must be established *a priori*. With these instrumentation systems, initial conditions were satisfied by inputting launch coordinates of the test vehicle. However, if a track were interrupted during flight, the systems had no way of reinitializing the target state estimate and subsequent data was lost. These multistatic radars were replaced by precision monostatic radars and optical trackers, now the instruments of choice on all U.S. test ranges.

In a related development, the *Space Surveillance* (*SPASUR*) 217-MHz multistatic radar fence was deployed starting in 1958 and 1959 at seven sites spanning the United States to detect and track noncooperative satellites [40, 41]. Transmitters are located at three sites, the largest of which transmits 1-MW CW from a linear array about 3 km long generating a fixed fan beam. The six receive sites consist of seven or eight phase-measuring linear arrays with dimensions on the order of 1 km, also generating fixed fan beams coplaner with the transmit beam.

Target location is established by triangulation, that is, the intersection of direction of arrival (DOA) measurements from two or more receive sites. While target doppler is measured by the large alert (or acquisition) array, it is only used to tune the smaller baseline receivers for more accurate DOA measurements. While *SPASUR* uses multiple interferometer antennas, they are not phased together to form one large aperture. Subsequently, a three-site fence in south Texas was deployed to evaluate the use of bistatic range measurements to improve location accuracy [42], but never entered continuous operation.

According to Easton [40] the *SPASUR* design was driven by cost: a 15,000 nmi detection range required very high average power, which was satisfied by the lowest cost CW operation. But this solution in turn required separate sites for isolation—hence multistatic operation. The stationary beam linear arrays also minimized cost when compared to scanning arrays or reflector antennas. In short, fix the beams and let the satellites orbit through them. *SPASUR* is approaching 50 years of continuous use, and remains the most successful multistatic radar application to date.

In 1962 the VHF CW forward-scatter radar fence was resurrected and tested using the doppler beat frequency to detect personnel intrusion into defended areas, which in turn triggered an alarm circuit [43]. In field tests the simple, inexpensive, battery-powered, and transistorized laboratory model provided reliable detections at baseline ranges of 10 m, with projections to 100 m depending on the permitted complexity. It was not deployed.

A major development during the 1950s and 60s was the semiactive homing missile seeker, in which the large, heavy, and costly transmitter could be off-loaded from the small, expendable missile onto the launch platform. While these seekers are clearly a bistatic radar configuration, missile engineers have developed a different lexicon to describe their technology and operation, for example, semiactive versus bistatic, illuminator versus transmitter, rear reference signal versus direct-path signal, and so on. The missile and radar communities continue to go their separate ways with only occasional technical interchanges. One such interchange, *target glint reduction*, summarized in [1], can significantly reduce the missile's endgame miss distance by approaching the target at bistatic angles of 20–30°, which in turn reduces the target's glint, a major contribution to miss distance. Figure 2–11 illustrates such a geometry and Figure 2–12 illustrates a successful implementation.

2.3 SECOND RESURGENCE

Measurements of moon and planetary surfaces by bistatic radars signalled the start of the second resurgence in 1967 and have been ongoing for nearly 40 years. The first successful *piggyback*

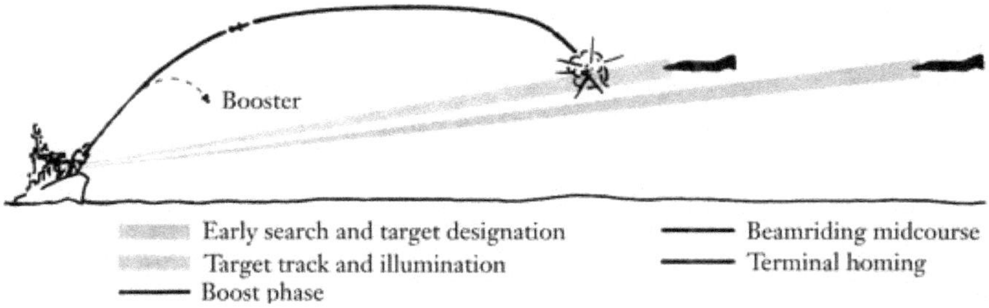

Figure 2–11 Operational concept of the U.S. Navy's Unified Talos air defense system [44]: "[Talos is launched] ...at elevation angles from 25 to 50°. After the boost phase, the booster rocket was automatically separated from the missile and the ramjet engine was ignited. The missile was then controlled by midcourse beamriding [cruising at an altitude of 60–70 kft for long-range engagements.] Near the end of the midcourse phase, the missile was sent a signal that activated the semiactive homing system and armed the warhead." Since aircraft flew at significantly lower altitudes, Talos dove on its target, which generated a large bistatic angle, an exaggerated ~60° in the figure. This geometry in turn reduced target glint, which reduced the miss distance, often to the point of causing direct hits. Ironically, the concept of glint reduction at moderate bistatic angles was unknown at the time; engineers and operators, including a coeditor, just attributed it to good design, maintenance, and operation. (Reprinted from *Johns Hopkins APL Technical Digest*, Vol. 3, No. 2, by permission. © The Johns Hopkins University Applied Physics Laborartory.)

Figure 2–12 A Talos missile launched from the U.S. Navy's *Desert Ship* at White Sands Missile Range, NM [44]. The ramjet missile was 21 ft long and weighed 3,360 lb. The booster was 11 ft long and weighed 4,360 lb. It cruised at >M2 between 60 and 70 kft altitude and had an engagement range of >100 nmi. The four probes on the front of the missile just outside the inlet diffuser are the semiactive homing interferometer antennas. Over 2,400 Talos missiles were built, with 887 fired from the fleet between 1958 and 1978. Missile success rate was >80%, while system success rate was ~40%. (Reprinted from *Johns Hopkins APL Technical Digest*, Vol. 3, No. 2, by permission. © The Johns Hopkins University Applied Physics Laborartory.)

operation exploited communication signals transmitted from the Luna-11 probe, scattered off the moon's surface, and then received by an earth-based station, in a *downlink* mode. Such stations include the Arecibo Observatory and the NASA Deep Space Network. Subsequent measurements were made using Lunar Orbiter-1, Explorer-35, and Apollo 14–16. Mars bistatic radar measurements were made using Mariner-6, Mariner-7, Viking-1, Viking-2, and Mars Express. Venus measurements were made by Veneras-9, Veneras-10, and Magellan [45].

Even though two legs of the bistatic triangle are extraordinarily long ($>10^5$ miles), the third, transmitter-to-target leg is sufficiently short (<10 miles) to produce strong echoes at the receiver. This unique bistatic phenomenon, which is characterized by ovals of Cassini (Section 6.5), also caused the bird false-alarm problem in the AN/FPS 23, *Fluttar*, only now it is exploited as the bistatic radar's operating mode. A reciprocal but more complex and costly *uplink* mode uses a bistatic receiver carried by the probe, collecting high-power, earth-transmitted signals first scattered off the planet's surface. It has a ~30-dB greater link margin, which was first used for *Mars Odyssey* and will be needed for the *New Horizons* Pluto probe in 2015.

These piggyback, bistatic radars have provided useful data in simple, inexpensive surveys of planetary surface properties as a prelude to robotic or human exploration, specifically centimeter-to-meter-scale roughness and material densities in the top few centimeters of regolith. Special geometries, such as near-backscatter to identify deposits of clean-water ice, are also a unique advantage of bistatic radar. The ability to observe forward scattering at latitudes away from planetary equators is also advantageous for probing neutral planetary atmospheres and the plasma environments. These ongoing experiments are detailed in Chapter 5 as an update of [45].

Bistatic radars received renewed interest for military applications starting in the 1970s as counters to retrodirective jammers and electronic support measures- (ESM-) directed attacks, for example from antiradiation missiles (ARMs) and emitter location-directed artillery.[11] Retrodirective jamming levels can be reduced by selecting a geometry such that the receive site lies outside the jammer's main beam, which is directed at the transmit site. ESM-directed attacks can be mitigated by either removing the transmitter from the battle area into a less vulnerable *sanctuary* or deploying multiple transmitters separated from the receiver in the battle area, thus increasing survivability and availability.

Several air defense field test program explored these capabilities and the problems inherent in bistatic operation, such as time synchronization, coverage, clutter suppression, and range performance [46–54]. In the United States, the *Sanctuary* program was field tested with generally successful results [1, 47, 48]. However, it was not developed further. Apparently decoys were selected to protect air defense radars such as Hawk and the AN/TPS-75(V) from ARM attack.[12,13]

In the United Kingdom, bistatic radars for air defense were analyzed, developed, and selectively tested in a systematic set of programs called *Bistatic Experimental Array Receiving System* (*BEARS*) and *Range Ambiguous Bistatic Experimental System* (*RABIES*), culminating in the *Argus* technology demonstrator program [54]. The *Byson* monostatic research radar at the Royal Signals and Radar Establishment (RSRE) was used for target acquisition and tracking and as an illuminator for the *Argus* receiver. *Pulse-chasing* [1] was a principal mode for *Argus* and was successfully demonstrated at hundreds of kilometers using commercial airline targets flying over the south of England. These programs ended in 1993, not long after the collapse of the Soviet Union.

---

11 A real attention-grabber occurred during the 1973 Yom Kippur War when a Soviet *AS-5 Kelt* air-to-surface missile, configured as the *KSR-11 ARM*, was launched by an Egyptian *Tu-16* bomber against the U.S.-made, Israeli-manned *AN/TPS-43* ground-based air surveillance radar, scoring a near miss but shutting down the radar (www.vectorsite.net/twcruz_6.html, www.angelfire.com/nb/eaf/11.html).

12 hawk.legends.dk/hawk.asp -42k

13 Commerce Business Daily, Nov. 13, 2001, PSA# 2976: TPS-43 Radar System Upgrades to TPS-75(V) Provide ARM Decoys.

Figure 2–13 Photograph of the U.S. Army's Advanced Technology Demonstration *Bistatic Radar for Weapons Location*, circa 1996. The unmanned, low-cost, 4-kW average power, S-band transmitter is on the right; the bistatic receiver with multibeam antenna is on the left. Both antennas are electronically scanned. The ATD goals were "as a minimum, to provide the performance of the AN/TPQ-37 Firefinder radar with the inherent survivability of a bistatic radar." Photos courtesy of the U.S. Army

The concept of using a small bistatic receiver that hitchhikes off cooperative but nondedicated airborne radars, called *Bistatic Alerting and Cueing*, was also developed and successfully tested in the United States [55, 56]. It alerts and cues autonomous, short-range air defense and ground surveillance systems to enhance survivability via silent operation and acquisition performance via real-time target location with minimum registration errors. It was not developed further. In particular, the *Global Positioning System* (*GPS*) became ubiquitous, thus eliminating registration errors.

In a later U.S. program a bistatic radar was designed and tested for weapons (artillery, mortar, and rocket) location using a low-cost, expendable transmitter and antenna displaced up to several kilometers from the receiver [57]. Pulse chasing [1] was incorporated to fill in coverage at short (4–10 km) target ranges. The radar (Figure 2–13) was successfully field tested, with performance emulating current *fire-finder* monostatic radars. In subsequent competition with other sensors the bistatic concept, although rated "best technical," lost to a "best value" monostatic radar.[14]

The *Multistatic Measurement System* (*MMS*) was installed at the U.S. Kwajalein Missile Range in 1980 as a hitchhiking adjunct to the *TRADEX* L-band monostatic radar [58]. (Apparently the planned adjunct to the UHF *ALTAIR* was never implemented.) *TRADEX* operated in its normal monostatic mode, acquiring, tracking, and illuminating ballistic missile reentry vehicles (RVs). Two unmanned, slaved, receive stations located about 40 km from *TRADEX* received echoes bistatically scattered off the RV and recorded bistatic range, doppler, and signature data from RVs. These data were used to calculate RV dynamics near the atmospheric pierce point, or start of reentry. The system was projected to measure three-dimensional (3-D) position and velocity with accuracies better than 4 m and 0.1 m/s, respectively, throughout reentry [59].

14 L. Bovino, private communication, July 6, 2006

Field tests showed that *MMS* bistatic range data combined with *TRADEX*'s monostatic range data in a trilateration net provided the most accurate estimate of exoatmospheric RV positions obtained by any of the range radars. *MMS* operations were concluded in 1993 [60]. Although no reasons were given, a likely cause is that continuing, aggressive improvements in the metric accuracy of monostatic radars was sufficient to satisfy user requirements, which in turn eliminated the need for *MMS*.[15,16]

A sophisticated version of the forward-scatter fence was developed and tested to protect military aircraft on the ground from intruders [61]. It was configured for pulse doppler operation at 5.8 GHz to minimize nuisance alarms from nearby traffic and small targets such as birds and rain. Five small, battery-powered, portable transmitter–receiver units, typically separated by 65 m, were located around the aircraft, with one transmitter servicing an adjacent receiver. In field tests the radar detected moving targets, including high-speed vehicles and intruders creeping at 2 cm/s. It was not deployed, losing out to a pulsed infrared intrusion-detection system, which now protects classified aircraft such as the *B-2* bomber [62].

Other bistatic radar concepts were identified and tested during the second resurgence, including *clutter tuning* from an airborne transmitter and receiver [63–65]. One implementation allows the receiver to generate a synthetic aperture radar (SAR) map of nearly equal down-range and half the crossrange monostatic resolution directly on its velocity vector—an impossible task for the monostatic SAR. Clutter tuning combined with the sanctuary concept protects the transmitter while allowing the receiver platform to fly toward the target with no radar emissions—the RF term in the *stealth equation*, another emerging concept at the time. Real-time field-tests were successful, and are detailed in Chapter 10. Figure 2–14 shows the concept. Space-borne illuminators have also been studied and tested [64, 65].

2.4 THIRD RESURGENCE

These early bistatic SAR concepts spawned considerable research starting in the 1990s to improve the quality of bistatic SAR images, which were limited by coherent integration times of a few seconds in the 1980s and consequently were of limited tactical interest [1]. Specifically, local oscillator phase instabilities have been reduced and the tricky bistatic automatic focusing (*autofocus*) algorithms were developed to improve range measurement accuracy from the antenna phase center to the imaged scene. This problem is exacerbated in bistatic SAR where the position of both transmit and receive platforms must be tracked with subwavelength relative accuracy to correct slow time-varying phase errors as part of the image formation process. The result allowed an increase in coherent integration times comparable to monostatic SAR, typically >10 sec. Consequently, bistatic SAR image quality was greatly improved, extending its tactical applications, as reported in Chapter 10.

Bistatic space–time adaptive processing (STAP) was also developed to improve clutter-limited detection of moving targets from moving transmit and receive platforms. Unlike monostatic clutter, which generally exhibits stationary space–time characteristics, bistatic clutter is often nonstationary. Thus bistatic STAP methods are not simple applications of monostatic approaches, but a new class of algorithms. Specifically, they apply a data-dependent weighting to voltages collected over multiple-receiving antenna channels and pulses. This weighting dynamically tailors a filter response in angle and bistatic doppler to suppress ground clutter reflections. Digital beamforming—

15 A. Bernard, private communication, July 24, 2006.

16 In a case of deja vu all over again, the demise of *MMS* is virtually identical to the fate of the *Chapman* range-only multistatic technique to improve D/F measurement accuracy of the World War II *CH* radars: improved calibration of the existing goinometers for beam splitting was deemed sufficient for acceptable D/F measurements.

Figure 2–14 Artist's rendition of the *Tactical Bistatic Radar Demonstration* (*TBIRD*) concept. A stand-off SAR-equipped aircraft (upper right) detects and illuminates a target (lower right) while designating an attack aircraft (left) equipped with a bistatic receiver to the target area. The attack aircraft acquires the target using SAR imagery directly on its velocity vector in RF silence. Further details of the program are in Chapter 10. (Operational concept from Ref. 64; painting replicating the concept from Goodyear Aerospace Corp.)

also useful for pulse chasing [1]—is required for this operation. Other necessary elements include a means to estimate the spatio-temporal clutter covariance matrix (the data-dependent element of the filter weighting) and hypothesis of the target steering vector. In general, auxiliary data taken from range bins other than the cell under test is used to estimate the unknown, but critical, clutter covariance matrix. With compensation for nonstationary behavior, bistatic clutter suppression has been greatly improved, extending its tactical applications, as reported in Chapter 11.

Exploiting commercial broadcast transmitters for bistatic and multistatic operation, a concept called *passive bistatic radar* (Chapter 6) was started in the 1980s. The first published account in 1985 documents air surveillance field trials using the UHF Crystal Palace TV transmitter in London and a modified commercial TV receiver about 12 km away. Initial attempts to detect aircraft were reported to be "...negative for the most part...," the first such report in bistatic radar annals [66]. The principal reason was the "...unsuitability of using a TV waveform for pulsed ranging operation." Spectral lines in such a waveform were estimated to be better suited for doppler operation, and indeed found to be the case.

Subsequent U.K. bistatic radar trials at the turn of the century used both UHF-TV and VHF-FM broadcast transmitters, measuring doppler and coarse bearing with the TV transmitter and range, doppler, and coarse bearing with the FM transmitter. Results demonstrated aircraft detection ranges of 160–175 km from the receiver. Real-time target location was not achieved, primarily due to a lack of sufficiently accurate bearing estimates [67–69].

In the United States, multistatic experiments were being carried out at the same time, unfortunately with nothing published in technical journals. One multistatic PBR, called *Silent Sentry*™, used range multilateration with multiple VHF-FM broadcast transmitters and one receiver to track (real time in 2-D) air targets out to about 100 km from the receive site. Integrating doppler data from VHF-TV transmitters to improve the location estimates was attempted but not achieved. However, missile launches were doppler-tracked in real time and in 3-D (due to a favorable *geometric dilution of precision*, Section 6.6), out to 150 km, again using launch coordinates as the initial conditions [70–72]. It remains under demonstration and testing as of this writing.

A later multistatic PBR, called *HDTV-Based Passive Radar*, also used range multilateration techniques and reported test results in the open literature [73]. It used a wide-band, high-definition UHF-TV transmitter and four receivers located within 10 km of the transmitter to track (real time in 2-D and coarse elevation) low-flying aircraft and helicopters as a gap-filler for air surveillance radars. Predicted detection and tracking ranges of 30 km on a 1 m^2 target were demonstrated in real time with 2-D tracking errors generally less than 50 m. Doppler data were used to resolve *ghosts*, that is, false detections that inevitably arise when using multilateration on unaugmented targets (Section 6.6). It also remains under evaluation as of this writing.

Broadcast-band PBRs can be made covert and those that operate at VHF or low UHF are inherently counter-stealth due to resonance scattering from air targets [74]. Hence, they have been called a *stealth trap*. They also have the potential of countering active and passive cancellation techniques used against VHF/UHF monostatic radars [74] (Section 6.3). No reports of deployment have been published.

In the mid-1990s, a variation of the FM broadcast PBR was developed by the University of Washington called the *Manastash Ridge Radar* (*MRR*). It is designed to study turbulence in the ionosphere, specifically auroral E-region irregularities, using range, doppler, and DOA (via interferometry) measurements [75]. Motivation for *MRR* development included lower cost, increased safety, spectrum availability, and pedagogical opportunity. *MRR* provides range-time intensity and range-doppler plots to the World Wide Web every half-hour. Although it is not subject to the stringent air defense requirements for accurate location of multiple targets in real time, it has detected meteors and aircraft in the course of normal operations. It is reported in Chapter 7.

As in the previous resurgences, the forward-scatter fence was resurrected in the 1990s, now known as the Russian *Struna-1* or *Barrier* [76, 77]. Figure 2–15 shows the antenna. The system is advertised for guarding regions including residential areas, narrow straits, canyons, and frontiers against low-flying air targets, ranging from 30 m minimum to 7 km maximum altitude. A one-man service crew is specified, with unattended operation possible.

Struna-1 has also demonstrated a limited capability to estimate the state (i.e., position and velocity vector) of a nonmaneuvering target crossing the barrier. Both doppler and azimuth measurements are used for the estimate. Although not described in [78], the azimuth angle is apparently measured with multiple, ~12° beams in some type of beam-splitting configuration to yield ~1° measurement errors. Then by using a maximal likelihood estimation algorithm, initialized by approximating a set of nonlinear equations relating the measurement set to the state vector, the target state is estimated. Initialization uses 24 doppler and azimuth measurements taken over a 1-sec interval. When compared to target *GPS* data, results showed position errors ranging from 0 to 3 km for target locations within about 7 km of the baseline. Estimates for either maneuvering or multiple targets were not reported.

In 1993 a U.S. National Center for Atmospheric Research *CP-2* monostatic weather research radar coupled with a small, nonscanning, and inexpensive (~$50,000) bistatic hitchhiker successfully measured *dual-doppler* vector wind fields [79, 80]. When extended to a *multiple-doppler*, multistatic network, full vector wind fields can be measured and are useful for research, aviation, weather prediction, and severe weather detection. In the ensuing ten years, the hitchhiking concept has been commercialized by Binet Inc., which, according to a recent website, has delivered nine *bistatic network receivers* to weather research centers around the world for test and evaluation. It is reported in Chapter 8.

Figure 2–15 Russian *Struna-1* receiving antenna, which shows the receiver unit attached to the back of the antenna [76, 77]. Both transmit and receive antennas are mounted on 25–30-m towers separated by a baseline range of 40–50 km. The system operates at VHF, typically near 230 MHz. Transmit and receive antenna gains are 11 and 23 dBi, respectively; transmit power is 2–3 W using harmonically modulated CW [78]. Image courtesy of NNIIRT, Russia.

Bistatic and multistatic radars have been analyzed, proposed, and in some cases developed for other scientific and commercial applications. They include high-resolution imaging at short ranges (in the near field of the antenna) for use by robotics in an industrial environment [81]; geological probing of horizontally stratified, underground layers from a transmitter and receiver on the surface, usually operating at frequencies from 100 to 1,000 MHz [82]; ocean wave spectral measurements (wavelength, frequency, and direction of travel) using a Loran-A system [83]; airport ground vehicle and aircraft collision warning and avoidance using a base-band bistatic radar [84]; automotive collision warning and avoidance using bumper-mounted radars that measure range and velocity monostatically and azimuth via multilateration [85]; and vehicle speed measurement over large angular areas on freeways, which can often be made more accurately bistatically than monostatically [86].

Other RF transmitters have been evaluated for bistatic/multistatic operation over the past 50 years. They include radio stars [87], very low frequency lightning [88, 89], navigation satellites (*GPS*, *GLONAS*) [90, 91], communication satellites (low-earth to geosynchronous orbits) [92–95], and land mobile communications (cell phones) [96, 97]. Although most of the evaluations concluded that their concept was feasible with "more research or development," they generally suffered from an overly optimistic technical or operational assessment of its utility for military operations. Only the high-powered (~1-MW), Ku-band *Direct Broadcast Satellite-Television* (*DBS-TV*) transmitter launched into geosynchronous orbit in the 1990s offers the potential for short-range detection and tracking of ground targets from airborne platforms. Its usefulness for aircraft or ship surveillance is not promising, as reported in Chapter 6.

References

1. N. J. Willis, *Bistatic Radar*, 2nd ed., Technology Service Corp., Silver Spring, MD, 1995, corrected and republished by SciTech Publishing, Inc., Raleigh, NC, 2005.

2. S. S. Swords, *Technical History of the Beginnings of Radar*, IEE History of Technology Series, Vol. 6, Peter Peregrinus, London, 1986.
3. H. E. Guerlac, *Radar in World War II*, Vols. I and II, Tomask/American Institute of Physics, New York, 1987.
4. H. W. Cooper, "Radar: History of a Need-Fostered System," Invited paper, *Microwave Journal*, May 1989, p. 213.
5. D. Prichard, *The Radar War: The German Achievement, 1904–1945*, Patrick Stephens Ltd., Cambridge, England, 1989.
6. F. J. Reynolds, *The United States Navy, From the Revolution to Date*, P. F. Collier & Son, New York, 1916.
7. E. F. Snow, "Follow the Leader," *Ships and the Sea*, Fall 1958, p. 26.
8. D. K. Allison, "New Eye for the Navy: The Origin of Radar at the Naval Research Laboratory," NRL Rep. 8466, Naval Research Laboratory, Washington, DC, 1981.
9. P. David, *Le Radar*, Presses Universitaires de France, Paris, 1949 (5th ed., 1969).
10. V. S. Chernyak, et al., "Radar in the Soviet Union and Russia: A Brief Historical Outline," *IEEE AES Systems Magazine*, December 2003, pp. 8–12.
11. Y. Nakagawa, *Weapons of World War II*, Aegean Park Press, Laguna Hills, CA, 1997.
12. R. I. Wilkinson, "Short Survey of Japanese Radar-II," *Electrical Engineering*, Vol. 65, 1946, pp. 455–463.
13. A. Price, *Instruments of Darkness: The History of Electronic Warfare*, Charles Scribner's Sons, New York, 1978.
14. B. T. Neal, "CH—The First Operational Radar," in R. Burns (ed.) *Radar Development to 1945*, The Institution of Engineering and Technology, London, 1998, Chapt. 8.
15. M. R. B. Dunsmore, "Bistatic Radars," *Alta Frequenza*, Vol. LVII, No. 2, March–April 1989, p. 54.
16. J. E. Summers and D. J. Browning, "An Introduction to Airborne Bistatic Radar," in *IEE Colloquium on Ground Airborne Multistatic Radar*, London, 1981, pp. 2/1–2/5.
17. M. I. Skolnik, *Introduction to Radar Systems*, McGraw-Hill Book Company, New York, 1980.
18. L. G. Eon, "An Investigation of the Technique Designed to Provide Early Warning Radar Fence for the Air Defense of Canada," Defense Research Board (Canada), Rept. TELS 100, December 1, 1952.
19. E. A. Sloane, J. Salerno, E. S. Candidas, and M. I. Skolnik: "A Bistatic CW Radar," MIT Lincoln Laboratory Technical Report 82, AD 76454, Lexington, MA, June 6, 1955.
20. M. I. Skolnik, J. Salerno, and E. S. Candidas: "Prediction of Bistatic CW Radar Performance," in *Symposium on Radar Detection Theory*, ONR Symposium Report ACR-10, Washington, DC, March 1–2, 1956, pp. 267–278. (Approved for public release: distribution unlimited.)
21. K. M. Siegel, "Bistatic Radars and Forward Scattering," in *Proceedings of the National Conference of Aeronautical Electronics*, May 12–14, 1958, pp. 286–290.
22. M. I. Skolnik, "An Analysis of Bistatic Radar," *IRE Transactions*, Vol. ANE-8, March 1961, pp. 19–27.
23. K. M. Siegel, et al., "Bistatic Radar Cross Sections of Surfaces of Revolution," *Journal of Applied Physics*, Vol. 26, March 1955, pp. 297–305.
24. K. M. Siegel, "Bistatic Radars and Forward Scattering," in *Proceedings of the National Conference of Aeronautical Electronics*, May 12–14, 1958, pp. 286–290.
25. J. W. Crispin, Jr., et al. "A Theoretical Method for the Calculation of Radar Cross Section of Aircraft and Missiles," University of Michigan, Radiation Laboratory Report 2591-1-H, July 1959.
26. R. E. Hiatt, et al. "Forward Scattering by Coated Objects Illuminated by Short Wavelength Radar," *Proceedings of the IRE*, Vol. 48, September 1960, pp. 1630–1635.
27. R. J. Garbacz and D. L. Moffett, "An Experimental Study of Bistatic Scattering From Some Small, Absorber-Coated, Metal Shapes," *Proceedings of the IRE*, Vol. 49, July 1961, pp. 1184–1192.
28. C. R. Mullin, et al., "A Numerical Technique for the Determination of the Scattering Cross Sections of Infinite Cylinders of Arbitrary Geometric Cross Section," *IEEE Transactions*, Vol. AP-13, January 1965, pp. 141–149.
29. R. E. Kell, "On the Derivation of Bistatic RCS from Monostatic Measurements," *Proceedings of the IEEE*, Vol. 53, August 1965, pp. 983–988.
30. S. T. Cost, "Measurements of the Bistatic Echo Area of Terrain of X-Band," Ohio State University, Antenna Laboratory Report 1822-2, May 1965.
31. V. W. Pidgeon, "Bistatic Cross Section of the Sea," *IEEE Transactions*, Vol. AP-14, May 1966, pp. 405–406.

32. T. Greenwald, "Reconnaissance and Arms Control," *Scientific American*, Vol. 228, No. 2, February 22, 1973, p. 22.
33. State Department Airgram, Subj.: Project CLEAR SKY, To: Amembassy Rome from Sec'y Rusk, U.S. State Department, dated February 26, 1968 (Declassified) available at http://www.gwu.edu/~nsarchiv/NSAEBB.
34. State Department Memorandum, To: Leonard Weiss, From: Frank Perez, Subj.: Launch on Warning, dated January 29, 1971, p. 2 (Declassified). available at http://www.gwu.edu/~nsarchiv/NSAEBB.
35. Private communication from O.G. Villard to E. Lyon, August 1973.
36. *Backscatter Literature Survey*, Project 3311 Communication and Propagation Laboratory, Sanford Research Institute, Menlo Park, CA, January 1961.
37. J. Richelson, *Wizards of Langley*, Westview Press, Boulder, CO, 2001.
38. Memorandum, To: Chief of Naval Operations, From: Chief of Naval Research, Subj.: CW Transmit Site at Spruce Creek, Florida, April 29, 1966.
39. J. J. Scavullo and F. J. Paul, *Aerospace Ranges: Instrumentation*, D. Van Nostrand Co., Princeton, NJ, 1965, p. 7.
40. R. L. Easton and J. J. Fleming, "The Navy Space Surveillance System," *Proceedings of the IRE*, Vol. 48, 1960, pp. 663–669.
41. J. T. Mengel, "Tracking the Earth Satellite and Data Transmission by Radio," *Proceedings of the IRE*, Vol. 44, June 1956, pp. 755–760.
42. P. J. Klass, "Navy Improves Accuracy, Detection Range," *Aviation Week and Space Technology*, August 16, 1965, pp. 56–61.
43. W. A. Visher, J. O. Wedel, and M. I. Skolnik, "Bistatic Radar Personnel-Intrusion Detector," in *National Convention on Military Electronics*, MIL-E-CON 7, 1963.
44. F. A. Dean, "The Unified Talos," *Johns Hopkins APL Technical Digest*, Vol. 3, No. 2, April–June 1982, p. 124 and front cover.
45. R. A. Simpson, "Spacecraft Studies of Planetary Surfaces Using Bistatic Radar," *IEEE Transactions on Geoscience and Remote Sensing*, Vol. 31, No. 2, March 1993, pp. 465–482.
46. E. F. Ewing and L. W. Dicken, "Some Applications of Bistatic and Multi-Bistatic Radars," in *International Radar Conference*, Paris, 1978, pp. 222–231.
47. R. J. Lefevre, "Bistatic Radar: New Application for an Old Technique," in *WESCON Conference Recordings*, San Francisco, 1979, pp. 1–20.
48. F. L. Fleming and N. J. Willis: "Sanctuary Radar," in *Proceedings of the Military Microwaves Conference*, London, October 22–24, 1980, pp. 103–108.
49. J. R. Forrest and J. G. Schoenenberger, "Totally Independent Bistatic Radar Receiver with Real-Time Microprocessor Scan Correction," in *IEEE International Radar Conference*, 1980, pp. 380–386.
50. C. Pell, et al., "An Experimental Bistatic Radar Trials System," in *IEE Colloquium on Ground Airborne Multistatic Radar*, London, 1981, pp. 6/1–6/12.
51. J. G. Schoenenberger and J. R. Forrest, "Principles of Independent Receivers for Use with Co-operative Radar Transmitters," *Radio and Electronic Engineer*, Vol. 52, February 1982, pp. 93–101.
52. T. A. Soame and D. M. Gould, "Description of an Experimental Bistatic Radar System," in *IEE RADAR-87*, IEE International Radar Conference Publication No. 281, 1987, pp. 12–16.
53. M. R. B. Dunsmore, "Bistatic Radars for Air Defense," in *IEE RADAR-87*, IEE International Radar Conference Publication No. 281, 1987, pp. 7–11.
54. M. R. B. Dunsmore, "Bistatic radars," in G. Galati (ed.) *Advanced Radar Techniques and Systems*, Peter Peregrinus, London, 1993, Chapt. 11.
55. "Bistatic Radars Hold Promise for Future Systems," *Microwave System News*, October 1984, pp. 119–136.
56. E. C. Thompson, "Bistatic Radar Noncooperative Illuminator Synchronization Techniques," in *Proceedings of the 1989 IEEE National Radar Conference*, Dallas, TX, March 29–30, 1989.
57. L. Bovino, Brochure on *Bistatic Radar for Weapons Location*, U.S. Army Communications Electronics, Command, Fort Monmouth NJ, circa 1994.
58. J. E. Salah and J. E. Morriello, "Development of a Multistatic Measurement System," in *IEEE International Radar Conference*, 1980, pp. 88–93.
59. "Multistatic Mode Raises Radar Accuracy," *Aviation Week Space Technology*, July 14, 1980, pp. 62–69.

60. *Technology in the National Interest*, MIT Lincoln Laboratory, Lexington, MA, 1995.
61. B. C. Walker and M. W. Callahan, "A Bistatic Pulse-Doppler Intruder-Detection Radar," in *IEEE International Radar Conference*, 1985, pp. 130–134.
62. Pulsed Infrared Perimeter Intrusion Detection System, *National Defense*, October 1994, p. 31.
63. D. C. Lorti and M. Balser, "Simulated Performance of a Tactical Bistatic Radar System," in *IEEE EASCON 77 Record*, Publication No. 77, Arlington, VA, 1977, Chapts. 1255–1259, pp. 4-4A–4-40.
64. F. Johnson, "Synthetic Aperture Radar (SAR) Heritage: An Air Force Perspective," Air Force Avionics Laboratory, Wright Patterson AFB, Ohio, June 2003.
65. J. L. Auterman, "Phase Stability Requirements for a Bistatic SAR," in *Proceedings of the IEEE National Radar Conference*, Atlanta, GA, March 1984, pp. 48–52.
66. H. D. Griffiths and N. R. W. Long, "Television-Based Bistatic Radar," *Proceedings of the IEE*, Vol. 133, Pt. F, December 1986, pp. 649–657.
67. P. E. Howland, "Target Tracking Using Television-Based Bistatic Radar," *IEE Proceedings – Radar, Sonar and Navigation*, Vol. 146, No. 3, June 1999, pp. 166–174.
68. P. E. Howland, D. Maksimiuk, and G. Reitsma, "FM Radio Based Bistatic Radar," *IEE Proceedings – Radar, Sonar and Navigation*, Vol. 152, No. 3, June 2005, pp. 107–115.
69. P. E. Howland, "FM-Radio Based Bistatic Radar," in *AOC 4th Multinational PCR Conference*, Syracuse, NY, October 6, 2005.
70. "Passive System Hints at Stealth Detection Silent Sentry—A New Type of Radar," *Aviation Week and Space Technology*, November 30, 1998, pp. 70–71.
71. J. Baniak, G. Baker, A. M. Cunningham, and L. Martin, "Silent Sentry™ Passive Surveillance," *Aviation Week and Space Technology*, reprint. June 7, 1999.
72. "Land-Based Air Defense Radars, United States," *Jane's Radar and Electronic Warfare Systems*, October 22, 2004, Jane's Information Group, Martin Streetly.
73. A. Andrews, "HDTV-Based Passive Radar," in *AOC 4th Multinational PCR Conference*, Syracuse, NY, October 6, 2005.
74. D. C. Jenn, *Radar and Laser Cross Section Engineering*, 2nd ed., AIAA Education Series, Reston VA, 2005.
75. J. D. Sahr, "Remote Sensing With Passive Radar at the University of Washington," *IEEE Geoscience and Remote Sensing Society Newsletter*, December 2005, pp. 16–21.
76. *Russia's Arms Catalog, Vol. 5: Air Defense*, Military Parade Ltd., Moscow, 1997.
77. *"Barrier" Bistatical Low Flying Target Detection System*, Nizhny Novgorod Scientific-Research Radiotechnical Institute, Moscow, circa 2000.
78. A. G. Blyakhman, et al., "Forward Scattering Radar Moving Object Coordinate Measurement," in *IEEE International Radar Conference*, 2000, pp. 678–682.
79. J. Wurman, M. Randall, C. L. Frush, E. Loew, and C. L. Holloway, "Design of a Bistatic Dual-Doppler Radar for Retrieving Vector Winds Using One Transmitter and a Remote Low-Gain Passive Receiver," *Proceedings of the IEEE*, Vol. 82, No. 12, December 1994, pp. 1861–1871.
80. S. Satoh and J. Wurman, "Accuracy of Wind Fields Observed by a Bistatic Doppler Radar Network," *Journal of Oceanic and Atmospheric Technology*, Vol. 20, 2003, pp. 1077–1091.
81. J. Detlefsen, "Application of Multistatic Radar Principles to Short Range Imaging," *Proceedings of the IEE*, Vol. 133, Pt. F, December 1986, pp. 658–663.
82. Zhou Zheng-Ou, et al., "A Bistatic Radar for Geological Probing," *Microwave Journal*, May 1984, pp. 257–263.
83. A. M. Peterson, et al., "Bistatic Radar Observation of Long Period, Directional Ocean-wave Spectra with Loran-A," *Science*, Vol. 170, October 1970, pp. 158–161.
84. A. M. Nicholson and G. F. Ross, "A New Radar Concept for Short Range Application," in *IEEE International Radar Conference*, 1975.
85. H. Rohling, A. Hoss, U. Lubbert, and M. Schiementz, " Multistatic Radar Principles for Automotive RadarNet Applications," Project No. IST-14031, RadarNet Project, 5th Framework Programme, European Commission, circa 2004.
86. K. Tomiyasu, "Conceptual Performance of Bistatic Doppler Radar for Vehicle Speed Determination," *IEEE Transactions on Vehicular Technology*, Vol. VT-30, No. 3, August 1981, pp. 130–134.

87. G. W. Preston, "The Theory of Stellar Radar," Rand Corporation Memorandum RM-3167-Pr, May 1962.
88. E. A. Lewis, et al., "Hyperbolic Direction Finding with Sferics of Transatlantic Origin," *Journal of Geophysical Research*, Vol. 5, July 1960, pp. 1879–1905.
89. R. H. Espeland, "Experimental Evaluation of SCORDES Resolution Capabilities," Naval Ordnance Laboratory Report 652, May 1966.
90. V. Koch and R. Westphal, "A New Approach to a Multistatic Passive Radar Sensor for Air Defense," in *IEEE International Radar Conference*, 1995, pp. 22–28.
91. X. He, M. Cherniakov, and T. Zeng, "Signal detectability in SS-BSAR with GNSS non-cooperative transmitter," *IEE Proceedings – Radar, Sonar and Navigation*, Vol. 152, No. 3, June 2005, pp. 124–132.
92. J. Fawcette, "Bistatic Radar May Find a 'Sanctuary' in Space," *Electronic Warfare/Defense Electronics*, January 1978, pp. 84–88.
93. M. Cherniakov, D. Nezlin, and K. Kubik, "Air Target Detection via Bistatic Radar Based on LEOS Communication Signals," *IEE Proceedings – Radar, Sonar and Navigation*, Vol. 152, No. 3, June 2005, pp. 33–38.
94. H. D. Griffiths, A. J. Garnett, C. J. Baker, and S. Keaveney, "Bistatic Radar Using Satellite-Bourne Illuminators of Opportunity," in *Proceedings of the RADAR-92 Conference*, Brighton, IEE Conference Publication No. 365, October 12–13, 1992, pp. 276–279.
95. R. F. Bernstein, Jr. and N. J. Willis, *Bistatic Radar Project*, Department of Electrical and Computer Engineering, U.S. Naval Postgraduate School, Monterey, CA, 2000.
96. D. K. P. Tan, H. Sun, Y. Lu, M. Lesturgie, and H. L. Chan, "Passive Radar Using Global System for Mobile Communication Signal: Theory, Implementation and Measurements," *IEE Proceedings – Radar, Sonar and Navigation*, Vol. 152, No. 3, June 2005, pp. 116–123.
97. http://www.roke.co.uk/sensors/stealth/cell_phone_radar_concept.asp and many news/magazine articles.

3

Fluttar DEW-Line Gap-Filler

Merrill I. Skolnik

3.1 BACKGROUND

This is a brief history of the bistatic radar known as Fluttar, developed by MIT Lincoln Laboratory under the sponsorship of the U.S. Air Force. The experimental Fluttar system led to the AN/FPS-23, which was the first operational bistatic radar the United States deployed. It was used in the Distant Early Warning (DEW) Line in the late 1950s as a gap-filler radar to detect low-flying aircraft approaching North America from the north.

In the early 1950s, the United States and Canada were concerned about airborne attack from the northern polar regions. Air defense of North America was of major importance at that time since nuclear weapons were no longer a monopoly of the United States and long-range bomber aircraft were available to other countries. A significant part of the North American air defense system planning in the early 1950s was the design, development, and installation of radars located in the northern regions to give early warning of the approach of hostile bombers. The northernmost line of warning radars, known as the DEW Line, extended 2,500 mi[1] along the 70°N latitude (more or less) from western Alaska to eastern Canada.

The idea for the DEW Line arose in 1952 during a summer study conducted by Lincoln Laboratory [1–3]. Lincoln Laboratory and the Bell Telephone Laboratories did much of the radar conceptual development for the DEW Line, and Western Electric was responsible for its construction. Canada also participated significantly. The main radars of the DEW Line were long-range L-band monostatic radars that were installed at 58 stations, spaced approximately 100 mi apart. Their designation was the AN/FPS-19. Each monostatic radar consisted of two back-to-back rotating antennas, with each antenna generating a high beam and a low beam. Such redundancy was important because of the need for highly reliable radar coverage.

It was recognized that even with 100-mi separation between sites, that coverage at a low altitude between sites was not adequate to ensure detection of low-flying aircraft. Halfway between stations, the low-altitude gap, below which an aircraft might not be detected, could extend from about 2,000 to 4,000 ft in altitude, depending on the site [2]. The gap in low-altitude coverage was filled by a bistatic CW (continuous wave) radar generating a “fence” coverage that would indicate

1 The abbreviation *mi* here means statute miles. That was the unit of miles usually used at the time. The nautical mile did not become popular until later.

a detection when aircraft flew through the fence. *Flutter Radar*, or *Fluttar* for short, was the name given to this bistatic radar. It was also known as *Type F*. When operationally deployed, the radar was designated the AN/FPS-23. The term "flutter" was used because the picture on the TV screens of that time period would flutter as an aircraft flew over or near the TV receiver. This flutter was due to the doppler frequency shift of the bistatic echo of the TV transmitter signal scattered from the aircraft, which interfered with the direct signal arriving from the TV transmitter. At the time this was a well-known effect that is no longer experienced because most TV is now obtained via cable; but even before cable, the flutter effect was eliminated in well-designed TV receivers.

3.2 EARLY THOUGHTS ABOUT A DEW LINE BISTATIC GAP-FILLER RADAR

The conception and development of a new radar system (or most other engineering capabilities) for a new application is seldom done in a direct, step-by-step manner. One usually starts out full of enthusiasm in a certain direction, only to discover problems that require one to think again and start over. This can occur more than once until a suitable design is obtained. This is what happened with the Dew Line gap-filler radar. Although the bistatic radar was known in the early 1930s and it was operationally employed to some extent by Japan, France, and the Soviet Union during World War II [4], it generally did not perform well as compared to the monostatic radars at that time, and interest in bistatic radar waned after World War II.

However, there were two aspects of bistatic radar that were of interest. One was the very large radar echo that was produced when a target was directly on the "line" between the bistatic transmitter and receiver. This is the so-called *forward scatter* that occurs when the scattering angle is exactly equal to or very near to 0° or, alternatively, as the bistatic angle approaches 180°.[2] The other potential advantage was that a CW transmission was usually employed in a bistatic system so that a simpler transmitter could be used than was found in a comparable monostatic pulse radar since it did not require a modulator or high voltage. Both of these advantages resulted in a simpler radar system compared to the monostatic pulse radars available at that time. (It might be noted that target detection with a CW signal usually depends on recognition of the doppler frequency shift produced by a moving target.) These advantages were potentially important for a radar that had to reliably operate unattended in the extremes of the arctic region. There are, however, some disadvantages with both forward-scatter and CW systems.

It might be mentioned, as an aside, that one of the original bistatic early warning concepts considered during the early 1950s for use in the more populated regions of Canada was to employ small, short-range CW radars mounted on existing telephone and power-line poles [3]. These lines would supply the relatively low prime power for the radar. The separation between transmitter and receiver might be a few miles so that a very low power system could be obtained. It was envisioned that the information about target detections could be transmitted back to a central station via the telephone line. For reasons that are now obscure, but most likely included the need to build over 1,000 of these radars, this particular concept never materialized, but its inherent simplicity helped to sell the bistatic radar as an unattended gap-filler system.

3.2.1 Forward-Scatter Approach

Because of the very large target cross section produced by a forward-scatter radar, this type of system was initially considered for the role of gap-filler in the DEW Line. The forward-scatter radar cross section of a target of projected area A that lies on the line joining the transmitter and

---

2 Some theorists, such as V. I. Tatarskii and his followers, use the scattering angle which is equal to 180° minus the bistatic angle. Others, such as K. M. Siegel and N. J. Willis, prefer to use the bistatic angle. A monostatic radar has a bistatic angle of 0° but a scattering angle of 180°.

receiver can be very large. With a radar of wavelength λ (where λ is small compared to the target dimensions), the forward-scatter cross section is given by [5]

$$\sigma = 4\pi A^2/\lambda^2 \tag{3.1}$$

From this equation it can be calculated that the forward-scatter cross section can be many orders of magnitude greater than the monostatic (backscatter) radar cross section. This can be a great advantage for a radar operating in the remote north since far less prime power is needed with the extremely large radar cross section offered by forward scatter. Thus, the large radar cross section available with forward scatter was attractive, but there was a serious problem that eventually precluded its use for the DEW Line or similar applications.

To take advantage of the large forward-scatter cross section, the transmitting antenna, target, and receiving antenna all have to be located on a straight line. Therefore the separation between bistatic radar terminals on a curved earth had to be much less than the 100-mi separation between the main monostatic radars of the DEW Line. A suitable separation for achieving forward scatter was about 25 mi; therefore, three additional sites would be required between the 100-mi separation between adjacent main monostatic radar sites.

The Canadians thought it was worth trying to develop a forward-scatter bistatic system with a baseline as short as 25 mi that might be used for its Mid-Canada air-defense line, also known as the McGill Fence, located along the 55th parallel. When an experimental forward-scatter system was built and tested, the real problem with forward-scatter systems was found, which was that it was susceptible to unwanted echoes from birds. Monostatic radars also detect birds, but in some monostatic radars means have been successfully applied to reduce their unwanted effects [6]. However, in a forward-scatter radar, especially one that uses CW transmissions, it is more difficult to deal with birds, as detailed next.

The simple form of the bistatic radar equation can be written as

$$P_r = \frac{P_t G_t G_r \lambda^2 \sigma_b}{(4\pi)^3 D_t^2 D_r^2 L_p(t) L_p(r)} \tag{3.2}$$

where P_r is the received signal power, P_t is the transmitter power, G_t is the transmitting antenna gain (in direction of target), G_r is the receiving antenna gain (in direction of target), λ is the radar wavelength, σ_b is the target bistatic radar cross section, D_t is the distance from transmitter to target, D_r is the distance from target to receiver, $L_p(t)$ is the propagation loss (due to diffraction) over the transmitter to target path, and $L_p(r)$ is the propagation loss over the target to receiver path. It is seen from this equation that when the target is a short distance from either the transmitter site or the receiver site, the received radar signal can be very large even when the bistatic cross section of the target is small. Because of the particular geometry of a forward-scatter radar that produces a large target echo, birds can be more of a problem than with a monostatic radar. (With a monostatic radar the problem with birds is at short range near the radar site, a relatively small volume. With the forward-scatter radar, the bird problem is extended and occurs along the entire baseline between the transmitter and receiver, as well as in the vicinity of the sites.)

In an attempt to reduce the bird problem of the forward-scatter radar, a lower frequency was investigated for the forward-scatter radar in the hope that the radar cross section of a bird might be in the Rayleigh scattering region where the cross section is proportional to the fourth power of the radar frequency. If so, the cross section of birds at a lower radar frequency might be much smaller than that of aircraft and so would not be as detectable. Experimental measurements made by the Air Force Cambridge Research Laboratory [7], however, found that even at VHF frequencies the radar cross section from the large birds, as are found in the arctic, was still not within the Rayleigh region, and so operating at a low radar frequency did not help. Large birds in the arctic might typically have a cross section at radar frequencies in the vicinity of perhaps 0.01 m^2, whereas birds likely to be found around a household bird feeder might have a cross section more or less in the vicinity of 0.001 m^2.

The critical test of forward-scatter radar came when an experimental 25-mi radar was built to explore the concept. Basically the unwanted radar echoes from birds were so numerous, especially during the migrating season, that it was not able to perform its job. The overload problem was compounded by the use of CW transmissions, which could not readily resolve multiple targets within its coverage. Monostatic radar, on the other hand, can sometimes use its range resolution to resolve multiple targets, but the CW forward-scatter radars of the mid-1950s could not. In spite of its advantage of providing a large target cross section, the forward-scatter bistatic radar was abandoned as a viable approach for the DEW Line gap-filler radar.

Before leaving this discussion of forward-scatter radar, other limitations should be mentioned. When the large forward-scatter echo is obtained, there is no way to determine at what point the target crossed a single transmit–receive baseline. In this way, the forward-scatter system acts like a trip wire; one knows that something has been detected, but does not know where. A monostatic pulse radar that measures range, however, with similar fixed fence coverage can determine where the target crossed the fence. Also, the doppler frequency shift is zero when the forward-scatter signal is obtained, and so doppler cannot be used to separate moving targets from stationary targets.

3.2.2 Lincoln Laboratory Doppler Approach

Once the limitations of the forward-scatter radar were fully recognized, Lincoln Laboratory approached the gap-filler radar problem for the DEW Line by reverting to the classical bistatic radar, *but* one designed to employ doppler to separate birds from the faster moving aircraft. Instead of the transmitting and receiving antennas pointing at each other, as was typically done in prior bistatic radars, the beams were *offset* from the baseline, as depicted in Figure 3–1, by 15° so as to

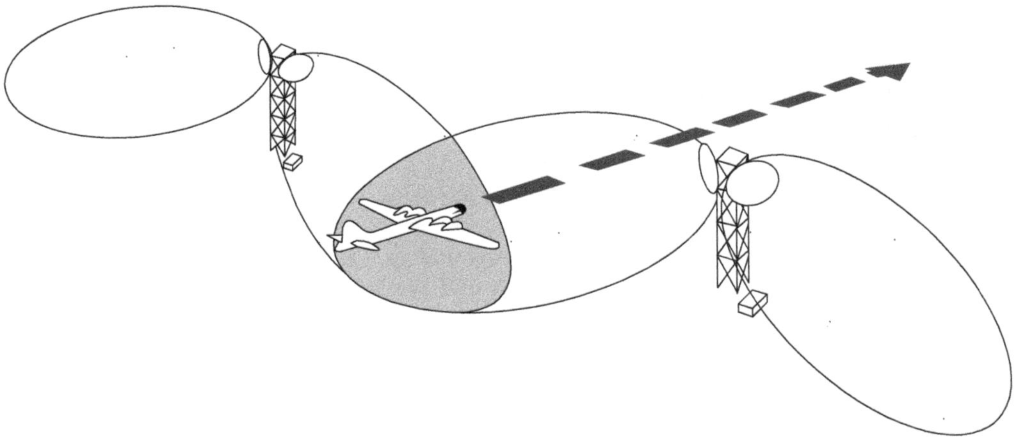

Figure 3–1 Plan view of the horizontal offset orientation of the transmit and receive antenna beams of Fluttar illuminating only one side of the fence [8].

not illuminate the forward-scatter region but to view the region where a doppler frequency shift is obtained from a moving target [8, 9]. Since forward scatter was not being employed, separation between transmitter and receiver could be increased. The separation was taken as 50 mi, so that only one intermediate bistatic site was needed between the 100-mi separation of the monostatic DEW Line radars. (The other two bistatic sites were located at the adjacent monostatic radar sites.) This longer range and the absence of the large forward-scatter cross section significantly increased the size of the transmitter, but it reduced the number of gap-filler radar sites needed. It was considered a worthwhile trade. The offset beams provided doppler information that allowed the system to reject slow-flying targets such as birds.

3.3 FLUTTAR SYSTEM CONSIDERATIONS

Special consideration had to be given to the design and construction of the bistatic radar equipment because of its need for highly reliable operation with minimum number of personnel in isolated northern regions. The transmitter was a 1-kW CW, three-cavity tunable klystron amplifier that could operate within the frequency range from 450 to 500 MHz. The UHF band was chosen for Fluttar because of propagation conditions, tolerable antenna size, low receiver noise (at that time), and availability of suitable CW power sources. The antenna polarization was vertical. The antennas were located on towers that ranged in height from about 100 ft to over 400 ft, depending on location. The separation between AN/FPS-23 radar sites in the DEW Line varied from 40 to 70 mi.

3.3.1 Bistatic Radar Constant Power Contours

In a monostatic radar the contours of constant power are concentric circles centered on the radar, when it is assumed the antenna gain is independent of the viewing aspect. (This, of course, usually never happens, but the assumption is made for illustrative purposes.) These contours of constant power, however, are quite different in a bistatic radar. The simple form of the bistatic radar equation as given by equation (3.1) shows that the received power is a more complicated function of the bistatic geometry than it is for the conventional monostatic geometry, especially when propagation effects due to diffraction occur. In designing Fluttar, the bistatic radar range equation was solved for a given D_t and a given D_r to yield a constant receive power. The results were plotted as constant power contours as shown in Figure 3–2 for a 50-mi baseline [8, 9]. Three examples are shown for two different target altitudes and two different antenna heights. These contours take into account diffraction loss due to the earth's surface, since there will be diffraction loss with the baseline distances, antenna tower height, and target altitudes that were of interest for the DEW Line. Comparison of Figures 3–2*a* and 3–2*b* illustrate the effect of the target height. Comparison of Figures 3–2*b* and 3–2*c* illustrate the effect of antenna tower height.

The contours of constant receive power shown in Figure 3–2 do not resemble the classical ovals of Cassini, which apply to a flat earth and assumes all parameters are constant. It can be said to apply to the *bistatic plane*, which is the plane containing the transmitter, receiver, and target, but this situation rarely occurs on a round earth. Figure 3–2, on the other hand, takes account of the diffraction loss due to a round earth, as well as differences in antenna gain with target location. Note that the Fluttar contours shown in Figure 3–2 do not shrink around the transmitting and receiving sites as would be expected for classical ovals. In fact, the radar engineers involved with the development of Fluttar for the DEW Line were never aware of the ovals of Cassini. Even if they had been, it would have not been of help in the design of the Fluttar bistatic radar. The lesson is that those involved with bistatic radar design should determine whether the ovals of Cassini can be applied to their particular radar before depending on them for design guidance.

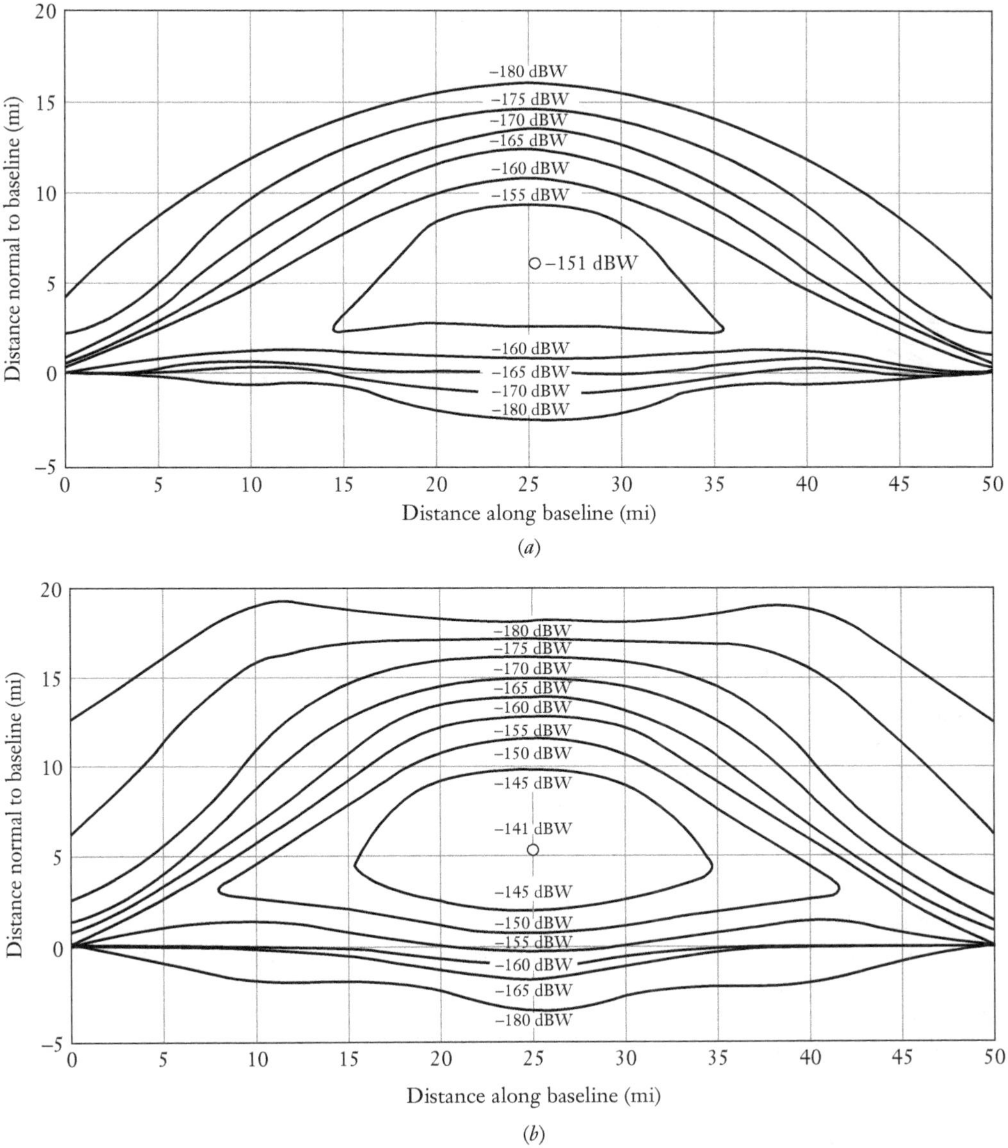

Figure 3–2 Calculated plot of constant power contours (in dBW, or decibels of receive power relative to 1 W) based on equation (3.2) for a bistatic radar with a 50-mi baseline, 20° antenna beamwidths, antenna gain of 23 dB, offset angle of 15°, transmitter power of 1 kW, and a target cross section of 25 m^2 [8]. (Note that a target cross section of 25 m^2 might seem large, but this is what the cross section of aircraft at that time period was thought to be at the UHF frequency of this radar.)
(*a*) Target altitude = 100 ft, transmitter tower height = receiver tower height = 200 ft;
(*b*) target altitude = 200 ft, transmitter tower height = receiver tower height = 200 ft;
(*c*) target altitude = 200 ft, transmitter tower height = receiver tower height = 300 ft.

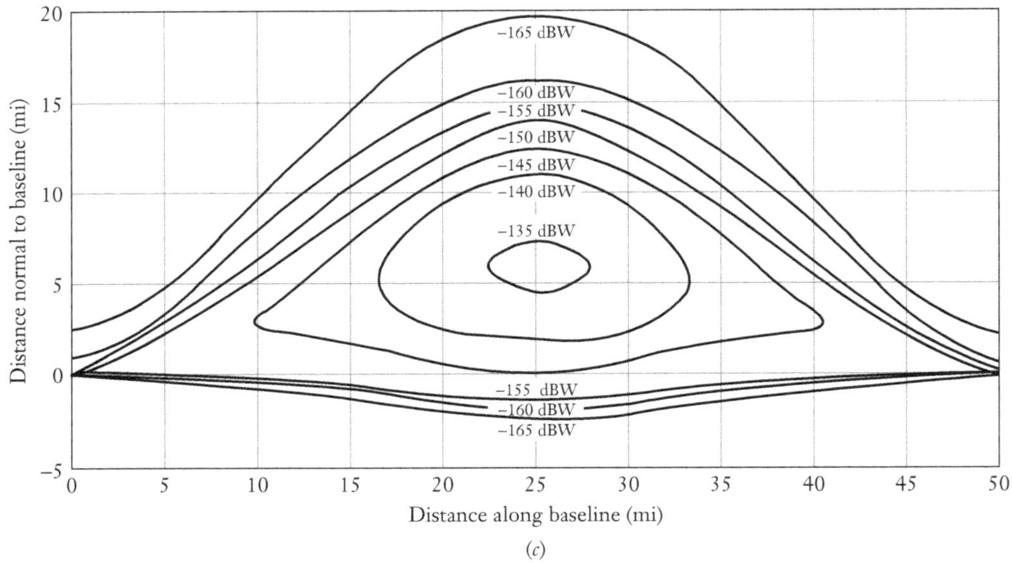

Figure 3–2 *(Continued)*

3.3.2 Bistatic Radar Doppler Shift

There is a doppler frequency shift with a bistatic radar just as there is with a monostatic radar. The monostatic doppler frequency is the familiar $f_d = 2v_r/\lambda$, where v_r is the relative velocity or rate of change of range R with time. The corresponding doppler frequency shift for a bistatic radar is [10]

$$f_d = \frac{1}{\lambda} \times \frac{d}{dt}(D_t + D_r) \tag{3.3}$$

That is, the doppler shift is proportional to the rate of change of the *sum* of D_t and D_r with time, where D_t is the distance from transmitter to target and D_r is the distance from target to receiver. As the bistatic configuration approaches a monostatic configuration, D_t approaches D_r and they both approach the monostatic radar range R. A more complete expression for the doppler shift in terms of the geometrical parameters can be found in Reference [8] or [10], but for brevity it is not reproduced here.

An example of the contours of constant bistatic doppler frequency is shown in Figure 3–3. Based on such curves, a series of narrow band audio filters with different center frequencies were designed to cover one side of the bistatic radar baseline. As the target approaches the coverage of the filters, the doppler frequency associated with the target will decrease in some monotonic fashion. The series of narrowband audio filters with predetection integration circuits detects the presence of the doppler signal. If the filters are properly arranged, a target detection is indicated when two or three signals are excited in sequence. Since the coverage of the offset antennas is asymmetric, the direction of penetration can be determined, that is, whether the target approaches from the north or from the south.

With doppler filters it was expected that unwanted slow moving targets (such as birds) would be separated from fast moving targets (such as aircraft). In one experimental version of Fluttar, targets with velocities less than about 45 kt were rejected, which was thought to be sufficient for eliminating most birds. The actual implementation of the detection was a bit more complicated

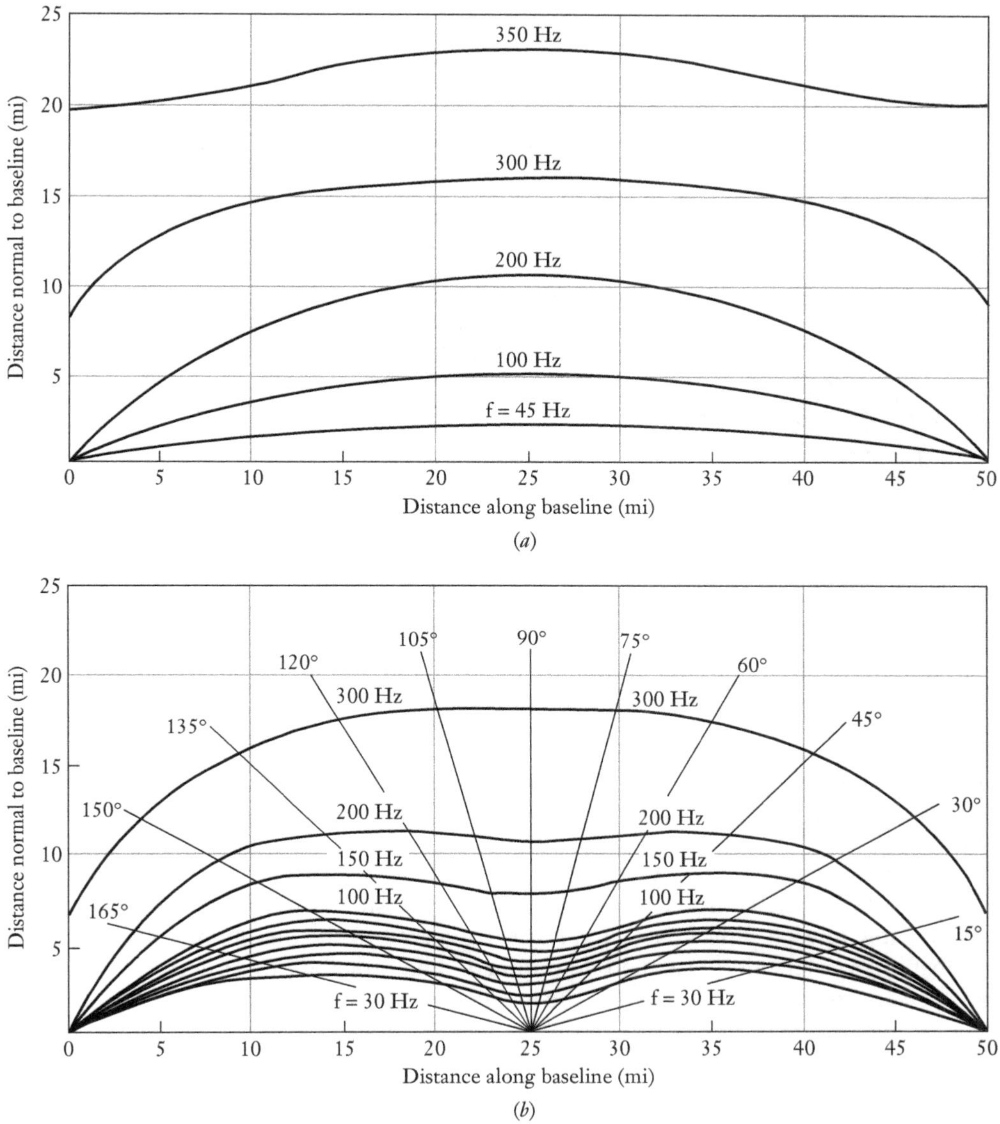

Figure 3–3 Calculated plot of constant frequency contours for a 50-mi baseline for a 500-MHz bistatic radar. Aircraft velocity = 300 kt or 345 mph. Target is flying 100 ft above the antenna height [8]. (*a*) Target crossing normal to the baseline; (*b*) target crossing through the midpoint of the baseline; (*c*) target crossing over one of the terminals.

than described here [8], but the objective was to obtain a high degree of sensitivity with minimum complexity and operator interpretation.

3.3.3 Probability of Detection

Because a bistatic radar such as Fluttar had to detect targets flying at low altitude, propagation losses due to diffraction had to be taken into account. Examples can be found in either [8] or

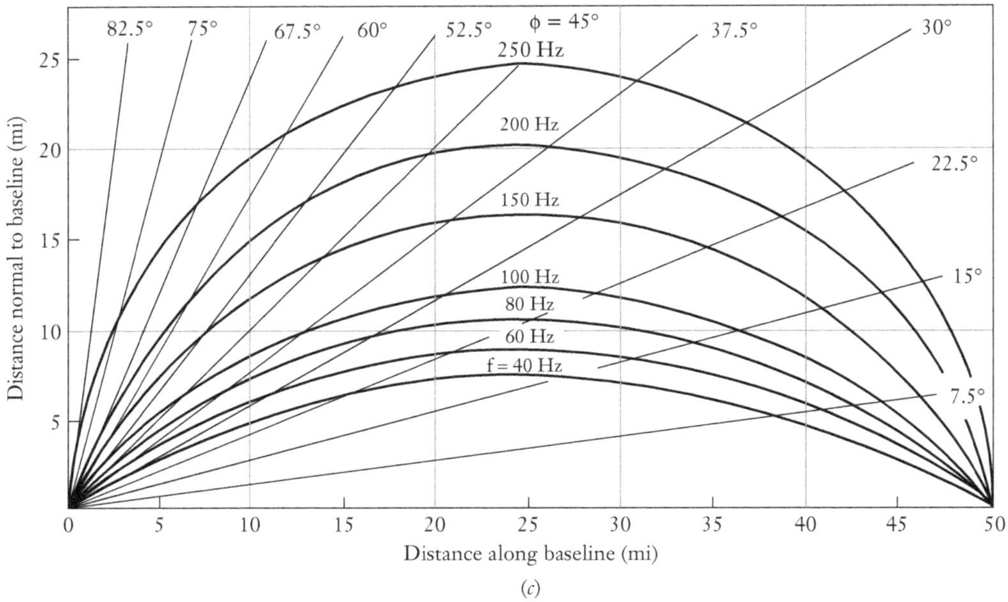

Figure 3–3 *(Continued)*

[9]. An example is shown in Figure 3–4. The numbers within the circles are the probability of detection for a given radar design. The spokes illustrate the direction of the aircraft trajectory. The trajectories are seen to pass through the center of the baseline, over one terminal of the link, and normal to the baseline. These figures were obtained by determining the signal power for each of the several doppler filters by placing a transparent set of constant doppler contours (similar to Figure 3–3) over the constant received power curves (similar to Figure 3–2). Once the signal power available in each filter was found, it was compared to the minimum detectable signal and the probability of detection was then determined. The probabilities of detection in Figure 3–4 included the propagation loss as well as the loss in the transmission lines to the elevated antenna. Tests were performed on a 50-mi experimental Fluttar link, and the results compared favorably with the predictions given in [8].

3.4 UNEXPECTED TROUBLE

The DEW Line, including the AN/FPS-23 gap fillers, became operational in 1957. However, the AN/FPS-23 was later removed because the false-alarm problem was not fully solved. Two unanticipated factors were involved. One was that there were numerous large birds in northern regions and they apparently liked to congregate where it was warm, such as near the diesel power generators at the sites. As mentioned previously, the bistatic range equation given by equation (3.2) shows that a very strong echo signal can be obtained even from small targets when they are at short range from either the transmitter or receiver sites.

The other unanticipated problem was that the sites along the 2,500 mi extent of the DEW Line were supplied by small aircraft that regularly flew along the line from one site to another. These were no problem for the monostatic pulse radar since they could be detected, recognized, and be continuously tracked. The bistatic radar, however, was operated by a CW and although it

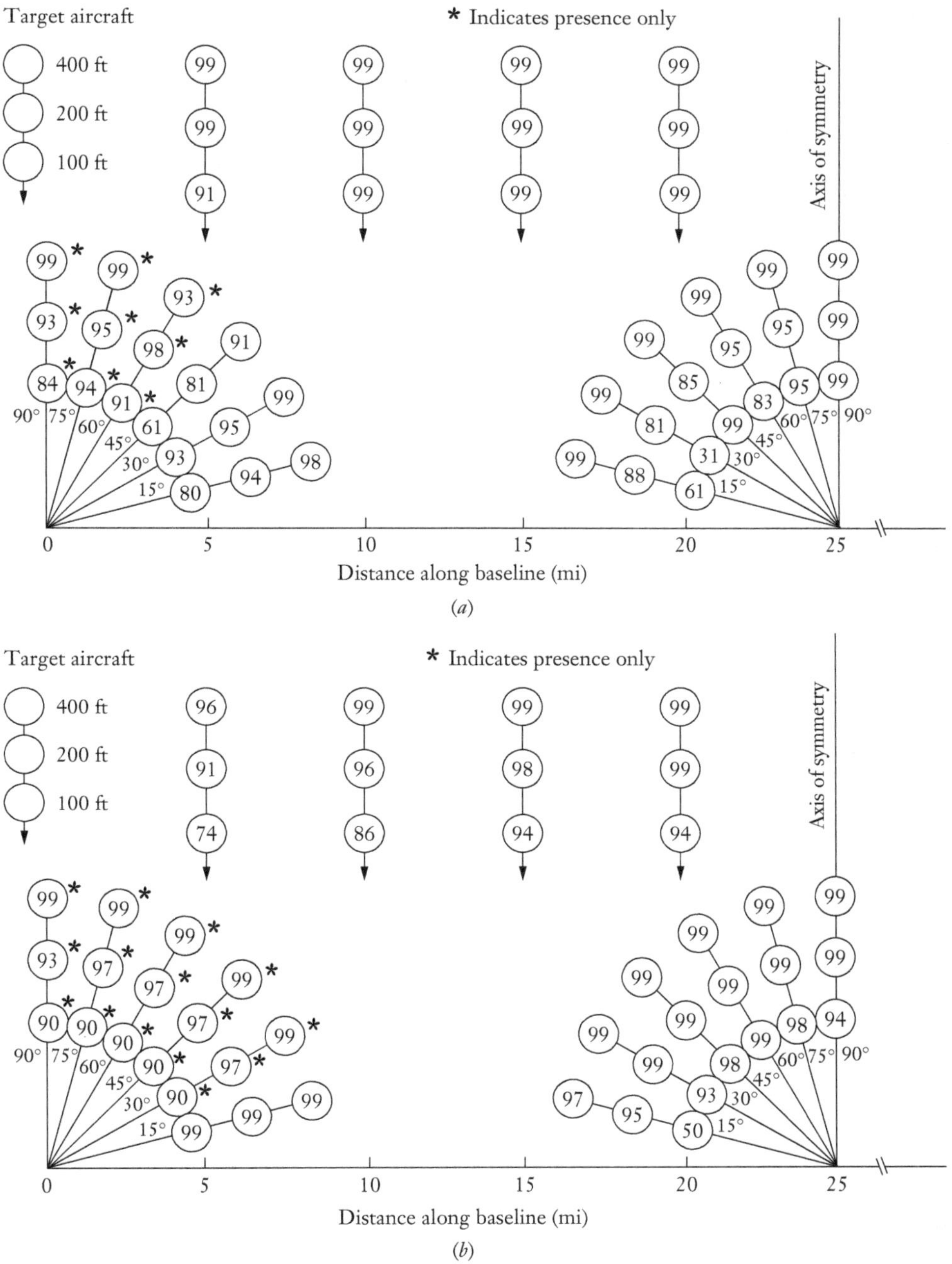

Figure 3–4 Calculated probability of detection for a 50-mi Fluttar link. Target cross section = 25 m^2, transmitter power = 1 kW, and tower height = 200 ft [8]. (*a*) Target velocity = 125 kt; (*b*) target velocity = 450 kt. *The term "Indicates presence only" in the figure is explained by [8] as follows: "In a limited number of cases, the geometry of the fluttar system is such that it is difficult, without a prohibitive number of decision-making devices, to obtain the sense of the crossing. For such situations, a "presence filter" is used to detect a crossing of the line, the direction of the crossing being unknown. In these cases, detection can be obtained by using one broad-band predetection filter."

could resolve targets in doppler, it had no other means to obtain target resolution, location, and track. Thus, multiple, unresolved detections by the bistatic radar were made as a supply aircraft flew from one site to another or because of large numbers of birds within the coverage. The monostatic pulse radars in the DEW Line did not suffer the problem of large numbers of birds in the same way as did the CW bistatic radar designed for a very low density target environment. Years later, an analysis of the numbers and species of birds in the northern regions found [11] that the total number of ducks, swans, and geese (the larger-sized birds present in the north) passing through the DEW Line area in western Canada might reach a total of almost 2 million birds during each of the migrating seasons in the spring and fall.

Since the radars of the DEW Line were designed for a low-target-density environment, a serious problem addressed during development of both bistatic and monostatic radars was how to ensure that the radar operators could remain alert when so few targets entered the coverage each day. To solve this problem both the monostatic and the bistatic radars were probably the first to employ automatic target detection without the assistance of human operations in the detection process [2]. With Fluttar, as mentioned above, the problem was not too few targets, but too many unwanted clutter targets.

3.5 MONOSTATIC PULSE RADAR FOR FENCE COVERAGE

The bistatic radar approach was considered for fence, or "trip wire," coverage because its geometry was well suited for the purpose. Since its antenna beams were fixed and did not scan, a bistatic radar could observe a target over a long interval of time so that its average transmitter power could be less than that of a radar that searches by continuously rotating its antenna over 360° in azimuth (assuming the same data rate). Also, the fixed antennas could be mounted on high towers to ensure better low-altitude coverage. CW transmitter power was used because it was thought to be cheaper and less complex than a pulse system.

A CW radar, on the other hand, cannot obtain the type of target information available with a pulse radar, but as said in [8] in 1955, "for low-traffic-density warning fence application, a large amount of information is not necessary." This quote was written before the recognition of how serious the false-alarm problem was with a bistatic CW radar, especially since birds were not always of low-traffic density. Being bistatic or CW, however, is not especially advantageous to the type of capability required of a gap-filler radar. If a gap-filler fence coverage were required today, it is likely that one would seriously investigate the use of a pulse radar. Two monostatic pulse radars with nonscanning antennas looking in opposite directions could be located halfway between each pair of main monostatic radar stations to act as gap-fillers. Such monostatic pulse radars could probably be designed to deal with the bird problem better than the CW bistatic radar. They might also employ a clutter map to eliminate those resolution cells containing slowly moving objects such as birds [12].

3.6 LOOKING BACK

Fluttar did what it was designed for: provide a highly reliable gap-filler to detect low-flying aircraft in a low-traffic-density environment. It did not work as well as expected because the density of unwanted targets turned out to be larger than originally anticipated.

One of the interesting things learned from the DEW Line radar experience is that it does not take a large number of engineers to perform a conceptual design of a new radar system, even though it might take a large number of engineers to perform detailed design, development, and manufacture.

The conceptual design of Fluttar was under the full-time leadership of Edwin Sloan and carried out by Sloan, Edward Candidus, Joseph Salerno, and Merrill Skolnik, with the overall guiding direction of Herbert G. Weiss, all of MIT Lincoln Laboratory at that time. The author of this chapter wishes to note that all figures in this chapter were from [8], published by the MIT Lincoln Laboratory, Lexington, MA. The author also wishes to express his gratitude for the assistance of Roger Sudbury, MIT Lincoln Laboratory Director's Office, in obtaining approval for the public release of [8].

References

1. M. D. Fagen (ed.), *A History of Engineering and Science in the Bell System*, Vol. 2, Bell Telephone Laboratories, Inc., Murray Hill, NJ, 1978.
2. F. R. Naka and W. W. Ward, "Distant Early Warning Radars: The Quest for Automatic Signal Detection," *MIT Lincoln Laboratory Journal*, Vol. 12, No. 2, 2000, pp. 181–204.
3. J. R. Zacharias and A. D. Hill, *Final Report of Summer Study Group 1952*, Vol. 2, Lincoln Laboratory, Bedford, MA, February 10, 1953.
4. N. J. Willis, *Bistatic Radar*, Technology Service Corporation, Silver Spring, MD, 1995, Sect. 2.1.
5. J. W. Crispin, Jr. and K. M. Siegel, *Methods of Radar Cross-Section Analysis*, Academic Press, New York, 1968, p. 156.
6. G. V. Trunk, W. B. Gordon, and B. H. Cantrell, "False Alarm Control Using Doppler Estimation," *IEEE Transactions*, Vol. AES-26, January 1990, pp. 146–153.
7. P. Blacksmith, Jr. and R. B. Mack, "On Measuring the Radar Cross Sections of Ducks and Chickens," *Proceedings of the IEEE*, Vol. 53, August 1965, p. 1125.
8. E. A. Slone, E. S. Candidus, J. Salerno, and M. Skolnik, "A Bistatic CW Radar," MIT Lincoln Laboratory Technical Report No. 82, June 6, 1955.
9. M. I. Skolnik, J. Salerno, and E. S. Candidus, "Prediction of Bistatic CW Radar Performance," in *Symposium on Radar Detection Theory*, sponsored by the Office of Naval Research and the Naval Research Laboratory, Washington, DC, March 1–2, 1956 (Approved for Public Release, Distribution Unlimited).
10. M. Skolnik, "An Analysis of Bistatic Radar," *IRE Transactions*, Vol. ANE-8, March 1961, pp. 19–27.
11. J. Antonicci, "A Statistical Model of Radar Bird Clutter at the DEW Line," N.Y. RL-TR-91-85, Air Force Systems Command Rome Laboratory, Rome, May 1991.
12. N. I. Franzen, "The Use of a Clutter Map in Artillery Locating Radar ARTHUR," in *IEEE International Radar Conference*, Arlington, VA, May 7–9, 1990, pp. 207–210.

4

Missile Attack Warning

Edwin Lyon

4.1 INTRODUCTION

Radar systems designed to operate at high frequencies between 5- and 30-MHz are often classified as *over-the-horizon* (OTH) radars, because at such frequencies the signals can travel via ionospherically refracted paths, called *sky waves*. Such radar-to-target paths can be from 500 to 3,600 km in overall length (per *hop*), allowing surveillance of targets over large areas of the Earth from relatively few radar sites. Targets include low-altitude missiles and aircraft, ships under way, and ballistic missiles during their boost phase.

A second mode of HF (high frequency) radar operation is *surface-wave* propagation, where HF energy propagates along the Earth's curvature by diffraction. In this mode, propagation signal loss increases exponentially with range; thus, ranges against these targets are reduced to 100–300 km. An HF bistatic radar might exploit a surface-wave mode on one path while using a sky-wave mode for the other path.

With a few early exceptions, HF radars use separated transmitters and receivers to reduce the complexity and cost of the receiving array, as well as to establish receiver isolation from the direct path transmitted signal. Site separation is typically on the order of 100 km. When an HF radar uses sky-wave propagation for both paths, this separation is small compared to the target range; thus, the radar operates with single-site characteristics, for example, with small bistatic angles, and is usually called a *monostatic* (or *near-monostatic*) OTH radar.

Nearly all HF radar literature has focused on these monostatic systems. Recently, however, considerable information on bistatic HF radars has become available, which describes their military applications, specifically for ballistic missile launch warning. The purpose of this chapter is to document this new information, prefaced by a review of relationships between HF and VHF/UHF (very-high frequency/ultra-high frequency) bistatic radar operation.

4.2 HF–VHF/UHF RADAR RELATIONSHIPS

The theory and phenomenology of HF radars is well documented in the literature [1–4]. This section summarizes basic relationships between HF and VHF/UHF bistatic radar operation, the latter being detailed in Chapter 6.

4.2.1 Target Free-Space Radar Cross Sections

Like targets at VHF (Section 6.3), most military targets at HF have large radar cross sections (RCS), with absolute free-space RCS running from a few square meters (m^2) to well over 1,000 m^2. Specifically, at wavelengths of 10–60 m, all manned aircraft fall into the Mie or resonance regime, and many such targets exhibit spans (wingspan or length) of one-half wavelength, resulting in RCS values that can approach λ^2, where λ is the wavelength. However at HF, the RCS of small air targets such as unmanned air vehicles can fall into the Rayleigh regime where their RCS values drop precipitously.

Free-space, anechoic-chamber measurements (at scaled frequencies) and method-of-moments modeling (e.g., using the Numerical Electromagnetic Code (NEC)) have been made of air targets at the frequencies of interest, 5–30 MHz, and for all polarizations. Typical modeling results are shown in Figure 4–1.

The RCS of other targets, such as ships and boats as well as ballistic and cruise missiles, has also been modeled and observed by HF bistatic radars. Typical RCS values are those of resonant-sized objects for most targets. For small ships and boats, while their RCS magnitudes may be quite high, separation of the target echo from the surrounding, large sea clutter returns may be difficult, as outlined in Section 4.2.4. However, large ships exhibit RCS values close to the empirical expression used at microwave frequencies: $52f^{-1/2}D^{3/2}$, where f is the frequency in MHz and D is the ship's full load displacement in kT [5]. Thus, they are nearly always detectable in sea clutter.

4.2.2 Rocket-Ionized Exhaust RCS

The earliest HF sky-wave radar operations against ballistic missiles sought to exploit an expected RCS enhancement due to ionization in the engine exhaust plume. Indeed, such an enhancement was seen, but was significant only after the missile had reached altitudes of about 100 km, with the engine still burning [6, 7].

The exhaust plume phenomenon was studied by Jackson et al. [8] and others, with observations taken at line-of-sight ranges from Cape Canaveral launches generally confirming the theory. Two ballistic missile echoes have been observed, one having the missile doppler characteristics and comprising the missile skin plus the exhaust plume boundary shock wave that travels with the missile. The other echo is the turbulent exhaust ionization itself, evident only in the 100–200 km altitude region where the ambient air pressure allows sufficiently long electron mean free path lengths to provide reflections at the operating frequency. This latter echo exhibits diffuse doppler characteristics, with HF-RCS values measured at 30–60 dBsm. It is usually classified as an ionospheric disturbance, owing to its altitude and ion/electron content. Thus, it was easily detected by early HF sky-wave radars without doppler processing, and is readily pattern-recognized by HF sky-wave doppler radars.

When HF radars are used for early warning of missile launch, the low-altitude echo is of greater interest than the high-altitude echo. At low altitudes the primary effect of the exhaust plume is absorption of the radar signals whenever the skin-echo ray paths must pass through the exhaust plume. Kolosova [3] shows the low-altitude plume-attenuation and the high-altitude plume-enhancement effects in a figure attributed to Fenster [4]. This figure compares the direct, downcoming sky-wave RCS with the after-earth-reflection RCS for a near-vertical ballistic missile launch, both as functions of time after launch (or effectively altitude). The after-Earth-reflection ray path suffers a 6-dB RCS exhaust plume obscuration loss compared to the direct downcoming ray path. The RCS on both paths then show a 15–20-dB enhancement at 150 sec after launch, corresponding to 95-km altitude, for a medium-range ballistic missile.

In the low-VHF regime, Jackson et al. [8] calculated threshold missile velocities (and altitudes) required to obtain overdense electron concentrations at the shock boundary for intercontinental and medium-range ballistic missiles. They conclude that neither missile reaches sufficient velocity to produce an echo from the exhaust shock boundary above 45 MHz. Consequently, absorption of radar signal energy can be expected at frequencies above 45 MHz. This calculation was corroborated in line-of-sight measurements of plasmas created by carbon/metal arcs at

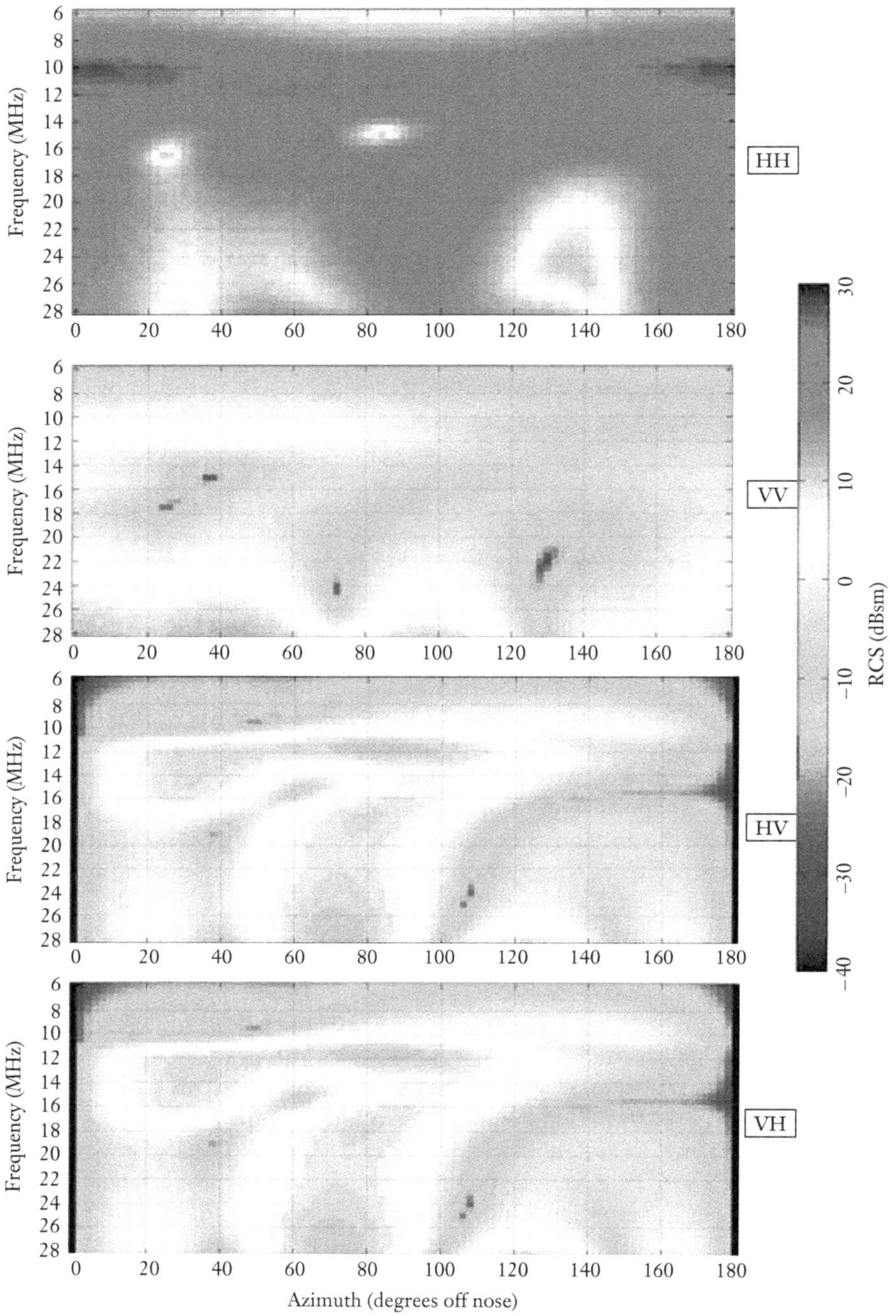

Figure 4–1 Results from method-of-moments modeling of the HF RCS of small aircraft for four-polarization combinations, HH, VV, HV, and VH. H indicates horizontal and V indicates vertical polarization. The first letter signifies the *E*-vector polarization of the illumination, and the second letter signifies the *E*-vector polarization of the received echo. The legend bar is common to all plots. These plots are for the downcoming ray path (from 10° above the horizon) and are virtually identical to the upcoming ray path (from 10° below the horizon), covering the most probable ray-path angles for all HF radars.

ionospheric altitudes. The resulting ionized trails were detected by HF radars, but only when the arc was extinguished did microwave radars detect the payload echo. These results were confirmed in telemetry measurements from ICBMs under testing at Cape Canaveral [9].

4.2.3 Polarization

Both sky-wave propagation via refraction and surface-wave propagation via diffraction are unique characteristics of HF radars. As with monostatic radar sky-wave propagation, the polarization effects of Faraday rotation must be considered for either link in a bistatic radar. Faraday rotation, a result of birefringence of the ionosphere in the Earth's magnetic field, causes the plane of polarization to rotate as the radar rays are refracted in the ionosphere. Typical rotations are many hundreds of degrees, depending on the ionospheric electron density, the radar operating frequency, and the distance traveled in the ionosphere. As a result, an aircraft target flies through a field of rotating polarization, typically with a full rotation taking place in 10–50 km of travel. Consequently, the radar signal arrives at the receiver with arbitrary polarization, which in turn generates the requirement to model targets in all polarizations.

Only components of the target echo that arrive at the receive antenna matched to the antenna's polarization are detected. Thus as an aircraft target flies toward or away from the transmitter or receiver, this time-dependent polarization mismatch will cause the target echo to fluctuate in amplitude. Furthermore, the fluctuation is compounded by ever-present multipath. Since the direct path and the multipath are seldom range-resolved in HF radars, they interfere both constructively and destructively with each other. The interference can become periodic as the target moves along. Hence, very large target echo fluctuations are often observed, caused by both polarization mismatch and multipath. These fluctuations in turn limit the receiver's signal processing gain, coherent integration, and doppler extraction.

For surface wave propagation beyond the local horizon, vertical polarization is used, since only vertically polarized waves exhibit diffraction, or retardation, at the Earth–atmosphere boundary needed for propagation beyond the local horizon. HF surface-wave propagation losses over land at any frequency or polarization are prohibitively high; consequently only sea-surface-wave propagation is considered. Propagation losses over the ocean are the usual function of range, but also of frequency. At 100-km range, for example, the path loss at 20 MHz is 25 dB higher than at 5 MHz for a sea state 3, further increasing at higher sea states.

For a bistatic radar using sky-wave illumination and surface-wave echo reception beyond the local horizon, the VV and HV plots of Figure 4–1 are relevant, because only the illumination signal undergoes polarization rotation, and the receiving polarization must be vertical. However, for sky-wave illumination and free-space, line-of-sight reception of echoes within the local horizon, the most beneficial receiving antenna polarization is horizontal. This configuration exploits the significantly greater HH-RCS whenever it occurs. Finally, for beyond-the-local-horizon surface-wave illumination and sky-wave reception, the transmitting antenna must be vertically polarized but the receiving antenna can be of any convenient polarization; thus the VV and VH plots apply. Usually VH and HV RCS plots are similar, as shown in the figure.

4.2.4 Clutter

Most targets moving on the Earth's surface are seen by HF radars only with vertical polarization, since the scattering response of these objects generates a null at the horizon for horizontal polarization. The same effect applies to most types of land and sea clutter, the exception being tall structures such as aerial wires and buildings. As with monostatic radars at any frequency, bistatic radars can use doppler processing to separate these targets from clutter.

However, for all HF radars operating against a sea surface background, typical sea clutter is an enormous signal—often an order of magnitude greater than land clutter—with a scattering coefficient (σ^0) of −23 to −26 dB. Since HF antennas have modest electrical dimensions, at least relative to those possible at VHF and above, the resulting received clutter echo can be some 50–60 dB higher in power than the target echo.

Consequently, the demands on clutter-rejection signal processing are severe. Typically, range and doppler processing employs window weighting that reduces side lobes 80–90 dB below the central response. Receiving antennas use beam-former weighting to achieve at least −60 dB side lobes, especially the near-in lobes that are likely to be receiving clutter from the broader illumination lobe. Attaining such side lobe control at HF, with antennas that occupy acres of land and in surroundings that offer abundant secondary scatter, is not a simple task.

Furthermore, detection of ships and boats at HF must contend with nonstationary clutter, caused by ocean wave motion. Upon long-dwell coherent integration, the sea clutter decomposes into approaching and receding *Bragg grating* spectral lines and second-order scattering that raises the clutter floor in the vicinity of zero doppler. To further aggravate this nonstationary clutter problem, the ionospheric path can become unstable, especially near the auroral zones and the equatorial belt, where the resultant ionospheric motion and turbulence spectrally broaden both sea clutter and target echoes. These clutter returns can hide all but the fastest and largest seagoing targets, requiring heroic processing measures by the receiver.

Of course, for surface-wave operation on both paths, these ionospheric spectral broadening problems are largely avoided, but not entirely. In some cases, the radar can receive sea clutter from beyond the surface-wave surveillance zone via incidental sky-wave propagation, with the resultant clutter processed via range ambiguities. This problem has been solved by using patterns of interleaved, orthogonally coded pulses, which largely attenuated range-ambiguous clutter echoes.[1]

4.2.5 Losses and Noise Figure

Absorptive (ohmic) losses occur for sky-wave propagation in the D region of the ionosphere (from 20 to 80 km altitude) during sunlight hours. These losses can amount to 1–12 dB per D-region crossing depending on the sun's zenith angle and the spectral distance between operating frequency and maximum ionospherically propagated frequency. The loss is higher for lower zenith angles because the D region is ionized by UV energy from the sun. Also, the higher the operating frequency, the lower the losses. At night, the D region ceases to exist.

The predominant contribution to receiver noise figure at HF is atmospheric noise caused by lightning strokes in the tropics, which are a near continuum. They can occur elsewhere as well. This source generates a noise figure of 10–50 dB, with the lower values occurring at high frequencies (15–30 MHz) during daylight and the higher values occurring at lower frequencies (<15 MHz) at night.

4.2.6 Radio Frequency Interference

At HF it is hypothetically possible to operate with a direct-conversion receiver. Such a receiver uses an analog-to-digital (A/D) converter at the output of each element in a receiving array, and

1 Since 1994, Raytheon (Canada) and Defence Research Establishment Ottawa (DREO) have collaborated in the development of a series of near-monostatic HF surface-wave radars (HFSWR) for harbor defense and Exclusive Economic Zone (EEZ) surveillance. They have fielded systems at 2, 4, and 13 MHz for detecting and tracking icebergs, ships, boats, and aircraft, the last two with the 13-MHz system. A 13-MHz HFSWR was tested in the Bahamas by the DoD Counter-Drug Technology Development Program Office in 2003–2004 against very small boats and light aircraft. Further details are on the Raytheon HFSWR website.

then carries out all signal and data processing functions cheaply in digital computers. However, such a configuration must contend with radio frequency interference from extraneous signals, which in turn drive the dynamic range of the A/D. As expected, other HF broadcast and radar transmitters generate the strongest signals, with ambient power densities in the United States running from −30 to +10 dBm/m^2. They are numerous and always spectrally near the radar's optimum operating frequencies, which in turn generate spurious input responses within the A/D. Typical requirements are for third-order intercept levels, referred to the input, of 35–40 dBm and second-order intercepts of 80–90 dBm, thus calling for well-designed preselector filters.

Without such filters placed ahead of any active (potentially nonlinear) component, it has thus far been impossible to design an HF radar receiver of adequate sensitivity. All successful HF radar receivers to date therefore use such preselector filters, typically in front of an upconversion superheterodyne receiver. Such a filter would also be required in front of the hypothetical direct-conversion receiver, significantly raising its costs and diminishing its flexibility.

4.3 440-L FORWARD-SCATTER OTH BISTATIC RADAR [10-14]

The *440-L* system was briefly described in 1973 [10] as follows:

> ...The currently deployed "forward scatter" OTH radar detects the disturbances in the ionosphere caused by the ionized jet of gas emanating from a rocket's motor. Since each type of missile disturbs the ionosphere somewhat differently, a detected missile can be identified by its characteristic OTH signature. In the currently operational system three transmitters are deployed in Taiwan, Japan and the Philippines. These transmitters are matched with corresponding receivers in Italy, Germany and ... [the United Kingdom]. Although the system was originally intended as an early-warning system for a massive missile attack, it has detected a high percentage of the known single events [missile launches] since 1968. All long range missiles fired from test sites in the U.S.S.R. are detectable.

Since then new material has become available [11-14] to produce the following account of the *440-L* saga.

From 1960–1970 OTH radars were being developed in monostatic pulse, near-monostatic CW (continuous wave), and bistatic configurations. From 1962–1963, Stanford University conducted two forward-scatter propagation research projects that tested the manageability of HF propagation over long paths and the feasibility of detecting missile- and nuclear weapon-induced path anomalies over these paths. Basic research was conducted across the U.S. well into the Caribbean and to Cape Canaveral (later Cape Kennedy) to detect missile launches. Receivers were located in California with transmitters in the Bahamas. Short-range, line-of-sight HF bistatic radar work was also conducted by erecting temporary transmitter facilities at Spruce Creek, FL, while the receiving capability existed in the Bahamas. The objective was to obtain direct measurements of the effective RCS[2] of missiles under power at high (>100 km) altitudes, where the engine burn creates the ionospheric anomalies [14].

By 1964 the long-range portion of the project was moved to a Western Pacific-to-Europe path complex. A second path continued in parallel, using an Okinawa-to-Brindisi (Italy) link. Radar propagation paths passed over the entire breadth of the Sino-Soviet landmass, with the objective of detecting doppler-induced ionospheric anomalies (the expanding electron-density gradients caused by a rocket engine plume).

2 As outlined in Section 4.2.2, effective RCS is the complex combination of missile skin RCS and exhaust plume RCS. It depends on the missile orientation with respect to the ray path and the altitude, which establishes the electron mean free path.

Early tests were successful, and by 1967 a contract was established to field the *440-L* system. HF transmitters were sited in the Western Pacific (Japan, Guam, Philippines, and Okinawa) and corresponding receivers in Europe (Italy, Germany, and England). Other receiving sites were sought, with temporary success in some instances, but by the time these sites were ready for operation as receiving stations, the entire system was reaching the end of its operational life. The system used unmodulated CW transmissions with triangulation providing a measure of target range along many of the paths. Detection of missile launches was by doppler-time signature recognition.

In early 1968 operators' positions, receiving equipment, and doppler-time pattern-recognition schemes were added to some of the *440-L* receive sites so that the system would also be sensitive to Chinese high-altitude nuclear weapon tests [12]. While the *440-L* system was of a type that had relatively small target coverage areas per signal path, it had many such paths, most of which crossed over the suspected Russian missile testing sites in the Ukraine and Chinese missile and weapon test sites in Asia.

At transmitting sites in the Far East, the *AN/FRT-80* transmitters fed imbricated sloping rhombic antennas. The receiving sites used special low-noise, low-spurious receivers and a variety of different antennas, including eight-element-phased monopole arrays, as shown in Figure 4–2. Phase-coherent receivers were used to drive their doppler processors and pattern-recognition equipment. *Barry Chirp*-sounders (forerunners of *AN/TRQ-35* path sounders) were used at both ends of the links to facilitate optimum operating frequency selection. Eventually output data were networked for prompt reporting to alerting centers. By 1970, the *440-L* system was declared operational [13].

One of the original concept developers was Stanford University Professor O. G. Villard who concluded that electronic countermeasures could easily spoof the radar since modulation of the CW waveform by missile and weapon effects was so simple. Specifically, he and OTH radar scientists had argued for complex waveforms and frequency hopping to provide a measure of jam resistance, but those recommendations were rejected. By the time the *440-L* system reached operational status, he had become greatly concerned about this vulnerability to spoofing.

On a site visit to Okinawa on April 1, 1971 (year is approximate), Dr. Villard fed a programmable signal synthesizer into the Stanford transmitting antenna. The synthesizer was programmed

Figure 4–2 Two of the eight-element phased-array receiving antennas used at a *440-L* receiving site. The two arrays allowed beam forming over about 100° of azimuth coverage, from 5 to 25 MHz. (Photo courtesy of the E. Lyon archives.)

to produce offsets in frequency versus time that clearly spelled out "April fool" on the *440-L* receive-site strip-chart records to show how easily an adversary could provide the system with any doppler-time signature.[3] While Stanford was chastised for the trick, that event demonstrated a critical system vulnerability, which in turn signaled the end of the system's utility. By 1975, *440-L* was being dismantled and in 1979, 20 of the *AN/FRT-80* transmitter modules were shipped to Australia where they became the basis for an Australian OTH radar system.

4.4 SUGAR TREE OTH PASSIVE BISTATIC RADAR [15]

Monostatic and multistatic HF radar experiments were often conducted together. One project of the 1962–1970 period was called *Sugar Tree*, a project in which passive receiving stations might detect missile launch signatures from noncooperative transmitters. Early *Sugar Tree* experiments took place in a line-of-sight configuration on the old Atlantic Missile Range.

When expanded for overseas testing, *Sugar Tree* deployed passive receiving sites with suitable sky-wave propagation to suspected Soviet missile launch sites. At these sites, HF receiving equipment sought signals from high-powered transmitting stations that illuminated the missile launching sites. While theoretically possible that the illumination path (transmitter-to-missile launch area), pilot signal path (transmitter-to-receiver), and echo path (missile launch area-to-receiver) could all be sky-wave paths, frequency management issues favored the simpler situation of a surface-wave illumination path.

Consequently, the most useful transmitter signals became Soviet-operated international broadcast (HF) signals. Some of these broadcast stations were not too far from suspected launch sites and were very high powered (several hundred kilowatts average power into 20-dBi antennas) [15]. But because most such stations illuminated areas other than in the direction of the missile launch sites, the actual effective illumination power density was much lower. Circuit analysis estimated isotropic antenna response in the direction of the missile launch, a reasonable assumption.

The *Sugar Tree* receiving system used phased-arrayed monopoles as antennas feeding special synthesizer-controlled receivers and doppler spectrum analyzers. Only the carrier of the broadcast transmitter signal was employed, since most of the double-sideband AM program material was at voice frequencies, with a small modulation index.

The pilot signal path was used as a reference signal against which missile echoes were correlated. This process removed some of the incidental transmitter-signal drift and ionospheric modulation effects. In particular, extended ionospheric gravity waves that occurred near the transmitter/target area often imparted common phase changes in both pilot signal and echo paths and could be compensated this way. But other ionospheric anomalies remained uncompensated, causing some degradation in the echo's signal-to-noise ratio (SNR), just as in any correlator using a noisy reference. *Sugar Tree* transitioned to an operational system and then was dismantled a few years later [15].

References

1. J. M. Headrick and M. I. Skolnik, "Over-the-Horizon Radar in the HF Band," *Proceedings of IEEE*, Vol. 6, 1974, pp. 664–672.
2. J. M. Headrick, "HF Over-the-Horizon Radar," in M. I. Skolnik (ed.) *Radar Handbook*, 2nd ed., McGraw-Hill, New York, 1990, Chapt. 24.

3 Private communication of O. G. Villard to E. Lyon, ITT Electro-Physics Laboratories, August 1973

3. A. A. Kolosova, *Osnovy Zagorizontnoi Radiolokatsii*, Radio i Svyaz, Moscow, 1984. [The Fenster data are described in his Chapter 4, using a graphic, Figure 4–2.]
4. Warren Fenster, "The Applications, Design, and Performance of an Over-the-Horizon Radar," in *Proceedings, Radar-77, International Radar Conference*, London, 1977, pp. 36–44.
5. M. I. Skolnik, *Introduction to Radar Systems*, 2nd ed., McGraw-Hill, New York, 1980, p. 44.
6. Anon., "Project Tepee May Cost Less, Outperform BMEWS," *Electronic Design*, Vol. 5, August 17, 1972, p. 45. (Tepee was the nickname for *Thaler's Project*, named for Dr. William J. Thaler, ONR scientist who first proposed using sky-wave HF radar for missile exhaust plume detection.)
7. A. Shuster, "Down-to-Earth Wizard of Outer Space," *New York Times*, September 13, 1959. (This article was the first unclassified description of Thaler's HF radar concept for detecting ballistic missiles; it included several photographs of Thaler at a blackboard showing schematically how sky-wave radar could detect missile launches by their engine exhaust plume disturbances in the ionosphere, thousands of miles from the radar.)
8. J. E. Jackson, et al., "Local Ionospheric Disturbance Created by Burning Rocket," *Journal of Geophysical Research*, Vol. 67, No. 5, 1962, pp. 2059–2061.
9. E. J. Bagdady, et al., "Effects of Exhaust Upon Signal Transmission to and from Rocket-Powered Vehicles," *Proceedings of IEEE*, Vol. 54, No. 9, 1966, pp. 1134–1146.
10. T. Greenwald, "Reconnaissance and Arms Control," *Scientific American*, Vol. 228, No. 2, February 22, 1973.
11. P. J. Klass, "HF Radar Detects Soviet ICBMs," *Aviation Week and Space Technology*, Vol. 95, No. 23, 1971, pp. 38–40.
12. State Department Airgram, Subj.: Project CLEAR SKY, To: Amembassy Rome from Sec'y Rusk, U.S. State Department, dated February 26, 1968 (Declassified available at http://www.gwu.edu/~nsarchiv/NSAEBB).
13. State Department Memorandum, To: Leonard Weiss, From: Frank Perez, Subj.: Launch on Warning, dated January 29, 1971, p. 2 (Declassified available at http://www.gwu.edu/~nsarchiv/NSAEBB).
14. *Backscatter Literature Survey*, Project 3311, Communications and Propagation Laboratory, Stanford Research Institute, Menlo Park, CA, January 1961.
15. J. Richelson, *Wizards of Langley*, Westview Press, Boulder, CO, 2001.

5

Planetary Exploration

Richard Simpson

5.1 INTRODUCTION

As the space age moved into full swing in the 1960s, bistatic radar became one of several tools for exploring remote planetary surfaces. Wavelengths (λ) commonly used for space telecommunications and research were in the meter and centimeter ranges. These interact with surface structure having similar scales; with proper interpretation the results of radar scattering observations can provide unique information on root-mean-square (rms) slopes (ζ) and near-surface dielectric constant (ε) and density (ρ) that are important in selecting sites for planetary landers and rovers. In addition, the unique microwave properties of clean-water ice make bistatic radar potentially one of the most important tools in locating sites for future habitation.

Monostatic radar has been employed for studying the moon since shortly after World War II [1, 2]; Mars [3] and Venus [4] had also been detected using Earth-based systems. But the huge cost of overcoming the R^4 factor in the denominator of the monostatic radar equation, where R is the radar-to-target range, made the bistatic geometry attractive if a spacecraft could be used on the shorter of the two legs of the path (see Section 5.2).

Originally envisioned as an *uplink* experiment with high-powered transmitters on Earth providing illumination, planetary bistatic radar needed a spacecraft receiver either in orbit or flying nearby to sample and record echo signals reflected from the target. The radio data could then be returned in the spacecraft science telemetry stream along with camera images, infrared spectra, magnetic field measurements, and other data acquired at or around the same time. Unfortunately, the added costs associated with uplink spacecraft radio instrumentation and the scheduling conflicts introduced by adding another investigation with unique attitude requirements (pointing the spacecraft antenna toward the planet) have largely relegated bistatic radar to *add-on* or *piggyback* status—an experiment of opportunity, which more readily could be conducted in a *downlink* configuration using existing hardware.

In downlink (Figure 5–1) the spacecraft transmitter, already on board for telecommunications (Figure 5–2), provides the signal source while ground receivers (Figure 5–3), fed by antennas such as the U.S. National Aeronautics and Space Administration (NASA) Deep Space Network 70-m dishes (Figure 5–4), collect the surface echoes. As summarized in Table 5–1, downlink experiments have been carried out on the moon, Venus, Mars, and Titan using Soviet, U.S., and European Space Agency (ESA) spacecraft and/or ground stations. The first successful uplink experiment is also shown. Echoes of as little as 1 zW (10^{-21} W) have been reported. Additional historical details are reported by Simpson [5].

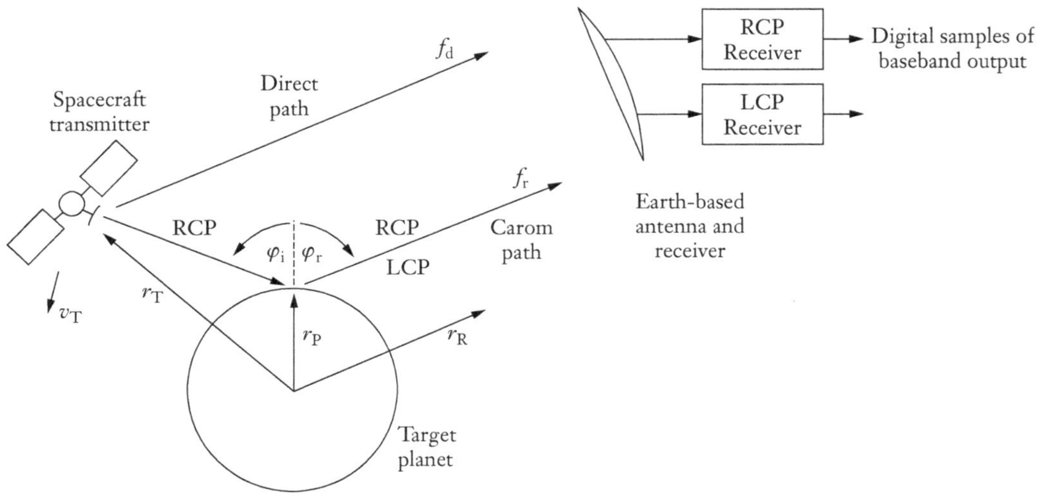

Figure 5–1 Downlink bistatic radar geometry with RCP transmissions from the spacecraft. During the surface reflection process, an LCP echo component is generated, necessitating two (coherent) receivers on Earth for complete characterization of the signal. Vectors from the target center to the transmitter, surface point, and receiver are r_T, r_p, and r_R, respectively. At the specular point all vectors are in the same plane and $\varphi = \varphi_i = \varphi_r$. Rate of change of the carom path length $|r_p - r_T| + |r_R - r_p|$ yields a doppler shift f_r which differs from the doppler shift along the direct path f_d; the direct signal strength depends on the side lobe pattern of the spacecraft antenna. The spacecraft velocity v_T is shown; in general, all elements are in motion. For arbitrary target points on the surface r_P, the bistatic angle β is the included angle between $r_T - r_P$ and $r_R - r_P$.

Table 5–1 Milestones in Bistatic Radar Studies of Planetary Surfaces

Spacecraft	Target	Wavelength λ (cm)	Reference(s)	Notes
Luna 11	Moon	170	[6]	First estimates of ε and ρ from reflected power
Lunar Orbiter 1	Moon	13	[7]	Detection of "scattering centers"
Explorer 35	Moon	220	[8]	Brewster angle detection; correlation of ζ with surface type and surface statistics
Apollos 14-16	Moon	13, 116	[9]	λ variation, echo polarization properties
Veneras 9, 10	Venus	32	[10]	$1° < \zeta < 7°$; surface altitudes
Mariners 6, 7	Mars	13	[11]	Grazing incidence prior to occultation
Vikings 1, 2	Mars	3.6, 13	[12]	Surface variability
Magellan	Venus	13	[13]	Detection of semiconductor surface
Clementine	Moon	13	[14, 15]	Search for water ice at poles
Mars Express	Mars	3.6, 13	[16]	Systematic use of polarization ratio
2001 Mars Odyssey	Mars	68	[17]	First successful uplink experiments
Cassini	Titan	3.6	[18]	Most distant detection to date

Key: ε, near-surface dielectric constant; ρ, surface density; ζ, rms surface slope; λ, wavelength.

Figure 5–2 Artist's view of *Mars Express* (*MEX*) in orbit. The spacecraft high-gain antenna (HGA) is a body-fixed 1.6-m parabolic reflector, shown pointing toward the upper right, and has gains of 29.6 dB at S-band (λ = 13 cm) and 41.4 dB at X-band (λ = 3.6 cm). During bistatic radar experiments the entire spacecraft is reoriented so that the HGA is aimed toward the specular point on the surface. *Venus Express* was assembled from many duplicates of *MEX* (and *Rosetta*, the European mission to comet 67P/Churyumov-Gerasimenko) components except that the HGA reflector is slightly smaller. (Image courtesy of the European Space Agency (ESA).)

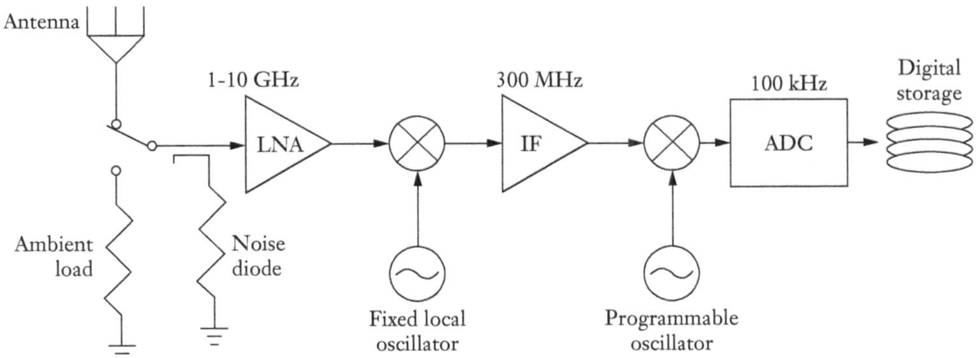

Figure 5–3 Block diagram of a typical Earth-based system used for downlink bistatic radar (an expansion of the "receiver" shown in Figure 5–1). The low-noise amplifier (LNA) is typically a cooled maser and may be switched between the antenna and an ambient load, allowing amplitude calibration of the input. During bistatic operations a signal from the low-level noise diode, previously calibrated against the ambient load, may be injected to monitor real-time performance of the system. Microwave inputs (1–10 GHz) are mixed to a 300-MHz intermediate frequency (IF) for amplification. A programmable oscillator, which can correct for first-order doppler effects, then mixes the IF signal to baseband and digital samples are stored for later processing. Although an analog-to-digital conversion is shown at the output, the ADC may actually take place at any point in the system.

Figure 5–4 Seventy-meter antennas of the NASA Deep Space Network have been used for most planetary bistatic radar campaigns. Receiving gains are about 63 dBi at S-band (λ = 13 cm) and about 74 dBi at X-band (λ = 3.6 cm). This is the antenna at the Goldstone complex; note the truck for scale at the lower left of the pedestal. (© Richard Simpson)

The net difference in signal-to-noise ratio favors uplink (with as much as 1 MW transmitted) over downlink (with rarely more than 100 W available on the spacecraft) by roughly 30 dB, after including a 10-dB penalty for poor receiver performance on the spacecraft. Because downlink technology does not provide enough sensitivity at the edge of the solar system, the first dedicated planetary uplink hardware has now been launched on the *New Horizons* spacecraft, which may also observe transient surface echoes in 2015 during its encounter with Pluto.

The simplest bistatic radar experiment requires only that an unmodulated spacecraft carrier be reflected from the planetary surface. The longer wavelength of the radio signal (compared with light, for example) interacts with the surface as though it were gently undulating. Fjeldbo [19] published a physical optics theory for planetary bistatic radar, explaining the echoes as aggregations of specular (mirror-like) reflections from properly oriented elements of the undulating surface. Except for the bistatic geometry, Fjeldbo's model had much in common with contemporary proposals by Beckmann and Spizzichino [20], Hagfors [21], Muhleman [22], and Kodis [23] who were primarily concerned with explaining monostatic radar echoes from the moon.

Differential doppler between the *direct path* and the *carom path* (defined in Figure 5–1) usually ensures that the surface echo can be separated from the carrier. Further, dispersion of the echo itself (caused by differential doppler within the reflecting region) provides a measure of the rms slopes of the undulations via processing analogous to *doppler beam sharpening* [24, 25]. Where echo dispersion is not large, the doppler difference between the two primary paths can be used to infer large-scale topography on the planet.

In general, an incident signal with known polarization will generate a reflected signal with both the original and orthogonal components; application of the Fresnel reflection theory to absolutely calibrated echoes in one polarization or to relatively calibrated echoes in orthogonal pairs can be used to recover the dielectric constant of the surface material. Extremely rough surfaces on the scales of the radar wavelength (e.g., certain types of lava or dense boulder fields) invalidate the physical optics assumptions and make inference of surface properties more difficult. Determining the horizontal and vertical scales to which the radar-derived estimates of surface properties apply and the conditions under which the models either break down or need to be adjusted remains an active area of research today (see, e.g., [26]).

Most bistatic geometries used for planetary exploration in the past have exploited forward, near-specular (e.g., Fresnel), surface reflection. In Figure 5–1, r_p and r_T define the plane of incidence, while r_p and r_R define the plane of reflection. In *specular reflection*, the two planes are identical and $\varphi_i = \varphi_r$. *Forward specular reflection* occurs when $\varphi_i = \varphi_r > 0°$; *specular backscatter* ($\varphi_i = \varphi_r = 0$) is the remaining case. Real planets do not reflect specularly; roughness on various scales breaks up the echo into a set of *glints* clustered around the specular point of the mean (e.g., perfectly spherical) surface. At wavelengths of centimeters to meters the aggregate of these glints can be comparable to the single return expected from the mean surface. The process is often called *quasi-specular scattering*, and the surfaces are called *quasi-specular surfaces*. The spatial distribution of the quasi-specular glints is directly responsible for the doppler dispersion in the echo noted above.

For geometries other than specular the advantages of bistatic radar are rapidly overcome by increasingly complex signal processing and the fact that there are few suitable models for interpretation. Mapping bistatic radar measurements to surface strips or points is simply more difficult when symmetries do not exist. Also, as is the case for monostatic radar remote sensing, models for interpreting the results are not as well founded when the geometry deviates from specular. Even the scattering function derived by Hagfors [21], which is based on physical optics and fits both backscatter and bistatic data from many planetary surfaces, relies on a Taylor series expansion, which is, at best, unorthodox [27]. The situation only gets worse for surfaces that produce *diffuse* radar echoes (echoes typically attributed to components with edges, cracks, buried rocks, and voids—that is, components that are not gently undulating) and that are viewed at arbitrary combinations of incidence and scattering angles. One exception to the gloomy view for nonspecular bistatic radar is in the near-backscatter search for water ice, a subject that will be addressed in Section 5.4.

The possibility of modulating bistatic transmissions has been proposed in various scenarios (see, e.g., [28]); echoes could then be demodulated using techniques analogous to those developed for imaging radars [25]. The resolution provided by such modulation degrades as the *bistatic angle* β (see Figure 5–1) increases, however. Thus users could be faced with having to choose between low resolution in a forward-scattering geometry, poorly understood scattering behavior at nonspecular angles, or both.

In parallel with its application to surface studies, the bistatic radar method was also developed for probing neutral planetary atmospheres and the plasma environments of the Sun, Earth, and other planets. In these experiments spatial differences in refractive index are inferred from changes in the frequency of the propagating radio wave. For example, as a spacecraft moves toward occultation by a planet, the ray path intercepts the atmospheric envelope—first tenuously, but then more substantially. The wave is retarded by the ever-increasing atmospheric density; at the receiver, this effect appears as a reduction in the signal frequency. The amount of frequency change depends on the velocity of the spacecraft and the gradient of the atmospheric refractivity. With suitable assumptions and modeling, the frequency signature can be inverted to give the refractivity as a function of altitude and then profiles of atmospheric temperature (T) and pressure (p) [29]. Over 10,000 Mars T–p profiles have been derived from *Mars Global Surveyor* occultation data [30].

Plasmas differ from neutral atmospheres in that they have an index of refraction less than unity, rather than greater than unity. Furthermore, hydrostatic equilibrium, which can be assumed for a neutral, spherically symmetric atmosphere, does not apply in many plasma environments, such as the solar wind. However, the underlying strategy of linking spatial refractive index variations to observed frequency changes applies equally well when propagating through plasmas [31]. For example, Hinson [32] interpreted ionospheric radio wave scintillations as being caused by structure aligned with the magnetic fields of Jupiter and Saturn. Occultations by planetary rings have yielded information on the density of ring material and size distribution of the particles. Classical knife-edge diffraction is seen with some rings. Additionally, radio data have been used to unravel intricate wake patterns caused by resonances between moons and bulk ring material [33].

Recent advances in planetary bistatic radar have occurred in three primary areas: collection and analysis of polarization data from bistatic experiments, searches for enhanced near-backscatter echoes (a possible indicator of clean-water ice), and implementation of uplink bistatic radar experiments. In addition, there has been preliminary work on interpretation of grazing incidence echoes (e.g., at angles of incidence $\varphi_i > 85°$), such as those observed during the *Mariner 6* and 7 flybys. In the remainder of this chapter we review the principles behind planetary bistatic experiments, and then summarize recent results and future experiments.

5.2 PRINCIPLES OF OPERATION

Quantities of interest in planetary bistatic experiments are the amplitude, frequency, polarization, and (to some extent) the time of the echo signal.

5.2.1 Echo Power

Incremental echo power dP_R from an unresolved surface element is given by the radar equation as

$$dP_R = (P_T G_T / 4\pi R_T^2)\sigma(A_R / 4\pi R_R^2) \tag{5.1}$$

where P_T is the transmitted power, G_T is the transmitting antenna gain in the direction of the surface element, R_T is the distance from the transmitter to the surface element, A_R is the effective area of the receiving antenna aperture (which may, like G_T, be directional), and R_R is the distance from the surface element to the receiver. Monostatic radar is the special case in which $R_T = R_R$ and $\beta = 0°$. For typical values of the parameters in (5.1), see Table 5–2.

Table 5–2 Bistatic Radar Nominal Performance

	Spacecraft			
	Mars Express	*Venus Express*	*Cassini Orbiter*	*New Horizons*
Target	Mars	Venus	Titan	Pluto
Date	2004-	2006-	2006-	2015
Mode	Downlink	Downlink	Downlink	Uplink
Bands	S, X	S, X	S, X, Ka	X
λ (m)	0.036	0.13	0.036	0.042
P_T (W)	60	5	20	10^5–10^6
G_T (dB)	41	26	47	73
R_T (km)	10,000	7,050	10,000	4.9×10^9
ϕ (deg)	60	60	60	85
σ (m^2)*	3.7×10^{13}	2.3×10^{14}	2.7×10^{13}	1.4×10^{13}
G_R (dB)#	74	63	74	41
R_R (km)	1.5×10^8	1.5×10^8	1.3×10^9	60,000
P_R (W)	2.0×10^{-16}	7.0×10^{-18}	2.6×10^{-18}	1.8×10^{-8}
$P_n = kT_{SYS}$ (W/Hz)	0.35×10^{-21}	0.35×10^{-21}	0.35×10^{-21}	3.11×10^{-21}

* Calculated at ϕ for a perfectly reflecting sphere from Reference [19].
# $G_R = 4\pi A_R/\lambda^2$ is the gain of the receiving antenna.

In equation (5.1), σ is the bistatic radar cross section of the surface element, which can be written more conveniently as

$$\sigma = \sigma_0 \, dS \tag{5.2}$$

where σ_0 is the specific radar cross section (radar cross section per unit area) and dS is the area of the surface element. In general, σ_0 is a function of the directions of both the incoming and outgoing wave vectors (three angles each). For quasi-specular conditions (the mirror-like reflections assumed by Fjeldbo), these reduce to a single angle $\varphi = \varphi_i = \varphi_r$ measured with respect to the local mean surface (Figure 5–1).

For many quasi-specular planetary surfaces, an expression for σ_0 derived by Hagfors [21] provides good agreement with measurements:

$$\sigma_0 = (\rho_i C/2)(\cos^4\varphi + \sin^2\varphi)^{-3/2} \tag{5.3}$$

where ρ_i is the Fresnel power reflection coefficient appropri ate to the transmitted and received polarization combination and C is a parameter interpreted as the inverse squared surface roughness in radians [34].

5.2.2 Specular Reflection Coefficients

The classical derivations for reflectivity ρ_i begin with the Fresnel voltage reflection coefficients for horizontally and vertically polarized electromagnetic waves at a planar interface between free space and the planetary surface:

$$R_{HH} = (\cos\varphi - (\varepsilon - \sin^2\varphi)^{1/2})/(\cos\varphi + (\varepsilon - \sin^2\varphi)^{1/2}) \tag{5.4a}$$

$$R_{VV} = (\varepsilon\cos\varphi - (\varepsilon - \sin^2\varphi)^{1/2})/(\varepsilon\cos\varphi + (\varepsilon - \sin^2\varphi)^{1/2}) \tag{5.4b}$$

where ε is the dielectric constant of the surface material. The reflection coefficients for circularly polarized waves can be derived from R_{HH} and R_{VV}:

$$R_{SC} = (R_{VV} + R_{HH})/2 \tag{5.5a}$$

$$R_{OC} = (R_{VV} - R_{HH})/2 \tag{5.5b}$$

where R_{SC} is the voltage reflection coefficient for the same sense circular polarization—for example, right-hand circular polarization (RCP) transmitted and RCP received—and R_{OC} is the coefficient for the opposite sense—for example, RCP transmitted and left-hand circular polarization (LCP) received.

The power reflection coefficients ρ_i for circular polarizations are derived from (5.5a, b):

$$\rho_{SC} = |R_{SC}|^2 \tag{5.6a}$$

$$\rho_{OC} = |R_{OC}|^2 \tag{5.6b}$$

The ratio ρ_{SC}/ρ_{OC} is sometimes easier to measure than either component by itself. Figure 5–5 shows the ratio $|R_{SC}|^2/|R_{OC}|^2$ plotted against angle of incidence and dielectric constant.

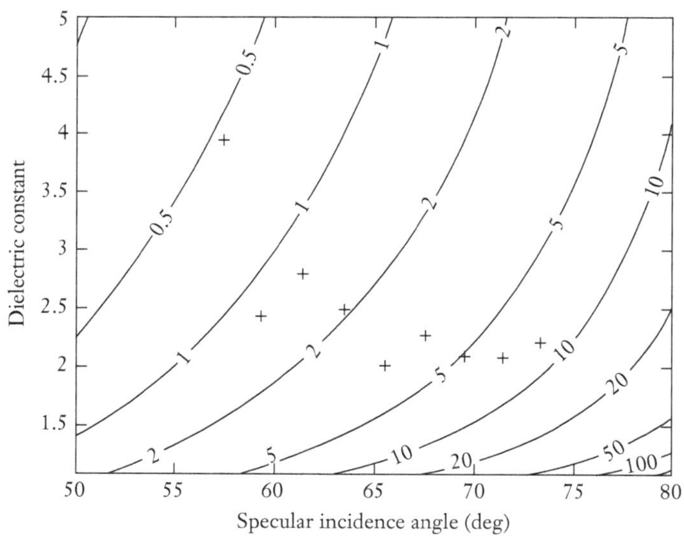

Figure 5–5 Contours of RCP to LCP power ratio plotted against specular incidence angle and dielectric constant for RCP transmitted. The unity contour defines the Brewster angle; for example, $\varphi_B = 60°$ for $\varepsilon = 3$. Individual points ("+") are from the *Mars Global Surveyor* bistatic radar experiment on May 14, 2000.

Table 5–3 Electrical Properties of Surface Materials and Analogs

Rock Type	Wavelength λ (cm)	Dielectric Constant (Real Part, ε')	Loss Tangent ($\tan\delta$)	Reference(s)
Pumice	68	2.5	0.007	[35]
Tuff (three types)	68	2.6–6.1	0.006–0.06	[35]
Volcanic ash	68	3.4	0.07	[35]
Andesite	68	5.1	0.004	[35]
Granite (seven types)	68	5.0–7.0	0.004–0.034	[35]
Basalt (nine types)	68	5.6–9.6	0.0103–0.09	[35]
Carbondale Red clay	60	2.49–3.82	0.108–0.169	[36]
Mars JSC-1 soil simulant	60	3.08–3.28	0.065	[36]
Holbrook meteorite	68	7.8	0.015	[35]
Bonita springs meteorite	68	43-81	0.13–0.19	[35]
Indarch meteorite	68	130–150	0.065–0.117	[35]
Apollo 11 powder (1 g/cm^3)	68	1.8	0.007	[37]
Lunar samples (68)	68	1.69–6.87	0.0015–0.139	[38]
Solid CO_2 (1 g/cm^3)	2–14	1.7	<0.005	[39]
Solid water ice	3.2	3.15	0.0007–0.0026	[40]

Example planetary rocks, soils, terrestrial analogs, simulants, frozen volatiles and their electrical properties in the 2–70-cm wavelength range. Electrical behavior varies depending on wavelength, detailed composition (including fractional metal and water content), and porosity. As a general rule, ε decreases slowly and $\tan\delta$ increases slowly with decreasing λ. Campbell and Ulrichs [35] concluded that density ρ of powdered rocks was more important than their original composition. The tabulation by Heiken et al. [38] includes a range of Apollo samples from dust to rock under many laboratory conditions; the 68 results summarized here were reported by Gold and coworkers at λ = 68 cm in a series of papers and reports throughout the 1970s beginning with Gold et al. [37].

For general interfaces—surfaces composed of lossy or conducting materials, equations (5.4) and (5.5) may be used if it is understood that the dielectric constant is replaced by its complex equivalent; that is, $\varepsilon \rightarrow \varepsilon' - i\sigma_C/2\pi f = \varepsilon'(1 - i\tan\delta)$, where ε' is the real part of the dielectric constant, σ_C is the surface material conductivity, f is the radar frequency, and $\tan\delta$ is known as the loss tangent. Table 5–3 shows (real) dielectric constant and loss tangent for several materials, which have been of interest in interpreting planetary radar data.

5.2.3 Doppler Shift

The doppler shift for a spacecraft transmitter at r_T and receiver at r_R is the rate of change of the path length measured in wavelengths

$$f_d = (d/dt)|r_R - r_T|/\lambda \tag{5.7}$$

where the vectors must be defined in an inertial frame and light time adjustments must be included but are not shown explicitly. The doppler shift for a signal reflected by the specular point on a smooth planet is, similarly

$$f_r = (d/dt)|r_R - r_p|/\lambda + (d/dt)|r_p - r_T|/\lambda \tag{5.8}$$

In the case of rough surfaces, the single reflection shown in Figure 5–1 is dispersed into a set of glints generated by small surface scattering elements distributed around the central point r_p. Since the bistatic geometry is slightly different for each of these glints, the echo spectrum is broadened. For a surface in which the roughness is not too severe, the total echo power in all of the glints is the same as the power from the single specular point on the perfectly smooth sphere. The degree to which the glints are concentrated is inversely proportional to the rms surface roughness; under certain conditions empirical scattering functions, analogous to (5.3), can be derived from measurements using inversion procedures [41]. As roughness increases, the spatial distribution of the reflecting facets becomes broad enough that some of the glints fall outside the illumination pattern of the spacecraft antenna, and the observed echo power is reduced accordingly. In the extreme case, only a small fraction of the possible glints are illuminated, and the distribution of incremental echo power mimics the antenna pattern G_T rather than the surface scattering function σ_0.

Since the glints on a rough surface are distributed around the central point r_p, each has a slightly different doppler shift. The actual doppler frequency dispersion of the surface echo is determined by (i) the spatial distribution of reflecting facets which are actually illuminated by the spacecraft antenna and (ii) the relative velocities expressed in terms of the wavelength (Figure 5–6). Broader

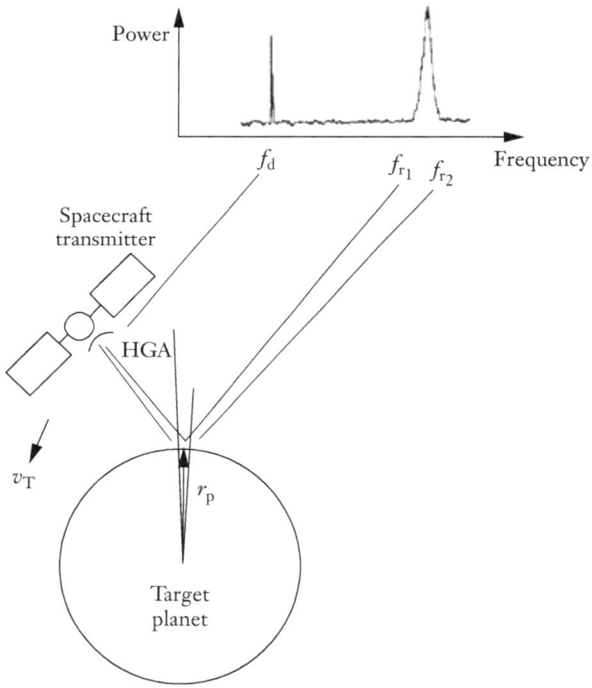

Figure 5–6 For real planetary surfaces there will be a region contributing to the echo rather than the single point shown in Figure 5–1. The region illuminated by the spacecraft antenna is between the outer radials extending upward from the target center. Surface facets need to be tilted with respect to the mean spherical surface in order to satisfy the specular condition locally. Since the tilt requirement increases with separation from r_p, the density of contributing facets also decreases away from r_p. Each facet has its own doppler shift f_{ri} so that the surface echo is dispersed in frequency with the taper depending on the distribution of surface tilts modulated by the high-gain antenna (HGA) illumination pattern. In the figure, f_d is the frequency of the directly propagating signal and f_{r1} and f_{r2} denote the limits on doppler spreading in the echo received via the carom path(s).

echoes come from rougher surfaces and/or more rapidly moving spacecraft. For experiments that are not "beam limited," the half-power width of the echo B and the rms roughness of the surface ζ are related by

$$B = 4(\ln 2)^{1/2}(V\zeta/\lambda)\cos\varphi \tag{5.9}$$

where V is the velocity of the specular point across the surface and ζ is the rms surface roughness [5].

5.3 POLARIZATION MEASUREMENTS

5.3.1 Power Ratios

As noted above, the ratio ρ_{SC}/ρ_{OC} is sometimes easier to measure than either component by itself. This is because a number of terms in the radar equation (5.1) are known at best, imperfectly, but they are common to the signal path when a pair of orthogonal polarizations is received. Although the absolute calibration of quantities such as P_T, G_T, and A_R may be off by tens of a percent, their relative uncertainty is much smaller. In fact, the only term in (5.1) which is not common to both polarizations is σ, the surface radar cross section; and one can argue that the only part of σ that is not common is the Fresnel power reflection coefficient $|R_{SC}|^2$ or $|R_{OC}|^2$, in the case of circular polarization. Thus, careful measurement of echo power in two orthogonal polarizations very quickly yields an estimate of the surface dielectric constant through the ratio of the Fresnel power reflection coefficients (shown graphically in Figure 5–5).

The special case of a Brewster angle detection on the Moon was reported by Tyler [42]; Simpson et al. [43] used power ratios for part of their analysis of early *Viking* (Mars) bistatic radar data. However, the systematic application of power ratios did not begin until 2000. Part of the delay has been due to the difficulty in actually achieving the necessary precision in ground calibration to make the measured ratios meaningful. Despite the early success reported by Simpson et al. [43], only one of the remaining 28 *Viking* experiments yielded data of sufficient quality to support publication of dielectric constant (ε) values.

A limited number of *Mars Global Surveyor* quasi-specular experiments were conducted starting in May 2000. The area of interest was the intended landing site for the *Mars Polar Lander/Deep Space 2* (*MPL/DS2*) mission (76°S, 195°W); all spacecraft were lost during descent to the surface. An independent investigation concluded that a spurious internal signal was the most likely cause of the *MPL* failure; but the possibility of hostile surface conditions could not be ruled out. The investigating panel explanation for the loss of *MPL* does not explain the loss of the *DS2* probes, which were deployed separately.

The results from one bistatic ground track, which passed almost directly over the *MPL/DS2* site, are shown in Figure 5–7. X-band (λ = 3.6 cm) RCP was transmitted from the spacecraft; both RCP and LCP were received. The calibration procedure failed for these observations, but there is good reason to believe that the two X-band channels had very nearly the same sensitivity. A nominal value T_{SYS} = 30 K, where T_{SYS} is the receiver's equivalent noise temperature, was assumed to allow derivation of the noise and echo power. Half-power bandwidth estimates B were derived manually from the X-RCP channel and are consistent with the echo's being beam-limited.

Measured power ratios from this *Mars Global Surveyor* bistatic radar track have been inverted to obtain dielectric constant (Figure 5–5). Most of the values are low compared with values derived previously for Mars, but they are not unreasonable. At the time these measurements were made, relatively few radar observations had been converted to ε at λ = 3.6 cm, where less than a centimeter of loosely compacted veneer could mask much larger values below. More importantly for *MPL/DS2*, there is no "hole" in the data near the landing site; the surface properties appear to

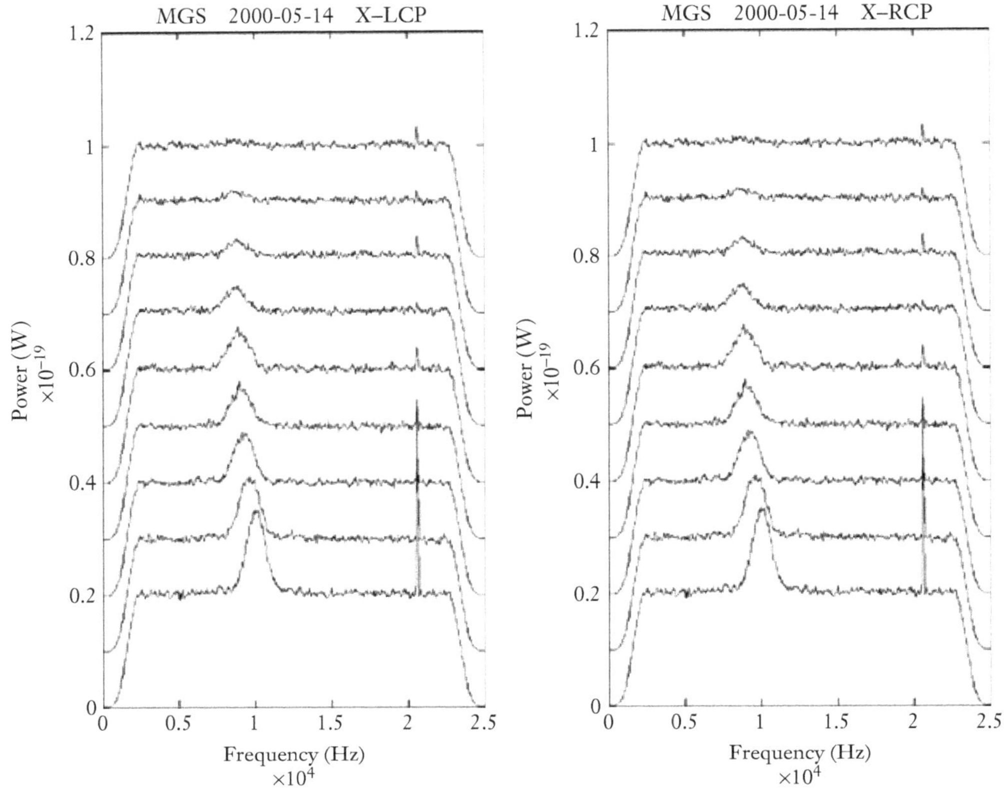

Figure 5–7 X-band ($\lambda = 3.6$ cm) 60-sec average power spectra showing *Mars Global Surveyor* LCP and RCP surface echoes (left and right panels respectively) collected on May 14, 2000. Time increases from bottom to top at 60 seconds per spectrum. The *MPL/DS2* target area (76°S, 195°W) would have been probed in the third pair from the bottom. The narrow signal near the right-hand edge of most spectra is the carrier signal propagating directly from the spacecraft; it provides a convenient frequency reference.

be consistent over at least 10° of latitude (600 km of distance). Where the dielectric constant rises ($\varphi < 60°$ in Figure 5–5), the echo signals are very weak; additional measurements are needed to characterize the surface properly at latitudes equatorward of 65°S.

Since 2004 new dual-frequency measurements of power ratio have been obtained using the *Mars Express* (*MEX*) spacecraft at $\lambda = 3.6$ and 13.1 cm. Two experiments in 2004 were only partially successful because of calibration and receiver tuning problems, but a comprehensive new calibration protocol was developed and the 16 experiments conducted during 2005 were largely successful. A significant limitation with *MEX* has been the restriction to inertial antenna pointing—that is, the spacecraft attitude is fixed so that the specular point target drifts through the beam. There is only one target per experiment and it satisfies the specular condition at only one instant; the remaining measurement time is with less well optimized geometries. Four of ten experiments during 2006 were conducted with the spacecraft's high-gain antenna tracking the moving specular point, allowing dozens of spatially independent measurements over an hour of observation. Specular track experiments are expected to be the default beginning in 2007.

The 2005 *MEX* experiments were conducted primarily over targets at southern latitudes. As expected, dielectric constants fell within the range $2.0 < \varepsilon < 4.0$ with 10–50% higher values at the longer wavelength—consistent with deeper penetration in soil that is more densely packed [16]. Doppler sorting of X-band ($\lambda = 3.6$ cm) echoes was used to achieve one-dimensional surface resolutions of about 20 km; some spatial variation in ε can be seen in the data (e.g., Figure 8 in Reference [16]) despite the limitation to inertial pointing. Over the South Pole, where there is good evidence for permanent deposits of solid CO_2, the S-band values of ε are too large to be consistent with CO_2. But the bistatic data were collected during southern summer when the permanent cap is smallest; the doppler sorting algorithm has not been rigorously applied to sharpen the spatial resolution [44].

5.3.2 Coherency

When orthogonally polarized receiver channels are sampled coherently, the results may be combined to allow synthesis of any receiver polarization. For example, the sum of equal RCP and LCP signals yields a linearly polarized signal while their difference is the orthogonal linear. Born and Wolf [45] defined the coherency matrix

$$\mathcal{J} = \begin{vmatrix} \langle V_R V_R^* \rangle & \langle V_R V_L^* \rangle \\ \langle V_R^* V_L \rangle & \langle V_L V_L^* \rangle \end{vmatrix} = \begin{vmatrix} \mathcal{J}_{11} & \mathcal{J}_{12} \\ \mathcal{J}_{21} & \mathcal{J}_{22} \end{vmatrix} \tag{5.10}$$

from which a number of quantities may be derived:

$$\text{Total power:} \quad P_{RT} = \text{Trace}(\mathcal{J}) = \mathcal{J}_{11} + \mathcal{J}_{22} \tag{5.10a}$$

$$\text{Degree of polarization} \quad \Gamma = [1 - 4\,\text{Det}(\mathcal{J})/\text{Trace}^2(\mathcal{J})]^{1/2} \tag{5.10b}$$

$$\text{Total polarized power:} \quad P_{RP} = \Gamma\,\text{Trace}(\mathcal{J}) \tag{5.10c}$$

$$\text{Total unpolarized power:} \quad P_{RU} = (1 - \Gamma)\,\text{Trace}(\mathcal{J}) \tag{5.10d}$$

where $\text{Det}(\mathcal{J}) = \mathcal{J}_{11}\mathcal{J}_{22} - \mathcal{J}_{12}\mathcal{J}_{21}$.

On June 5, 1994, bistatic radar experiments were conducted using the *Magellan* spacecraft, then orbiting Venus. Experiments were attempted at two wavelengths, but surface echoes were detected only at $\lambda = 13.1$ cm (S-band), presumably because of the much higher atmospheric absorption at $\lambda = 3.6$ cm. Unlike most other spacecraft, *Magellan* used linear polarization for its S-band transmissions; its antenna was canted 45° so that equal illumination in the horizontal and vertical planes would reach the surface. Orthogonal circular polarizations were received on Earth, but calibrations before and after the experiments allowed determination of their relative phases so that linears with known orientations could be synthesized later.

The ground tracks on three orbits transected Maxwell Montes, the highest point on Venus and known from a number of monostatic radar observations to be unusually bright in backscatter. In the forward-scatter geometry the echoes were reduced to less than 10% of their normal intensity because of the extreme roughness within the mountains. More surprising was the discovery of unusual echo polarization when the specular point was above 6,054-km radius—the same altitude at which the backscatter echoes became unusually bright. Using a Stokes vector analysis—equivalent to (5.10) above, Pettengill et al. [13] concluded that only a semiconductor coating on the surface could explain the data—that is, the dielectric constant for surface materials within Maxwell Montes was $\varepsilon = -i100$.

Pettengill et al. [13] suggested deposits of elemental tellurium, a trace constituent of terrestrial volcanic eruptions, for the semiconductor. Interestingly, tellurium undergoes a phase change at temperatures corresponding to the 6,054-km altitude. Based on observations collected prior to the bistatic experiments, Shepard et al. [46] had proposed a ferroelectric model for the highlands and Brackett et al. [47] had proposed cold trap deposition of high-dielectric volatiles. Schaefer and Fegley [48] have shown that elemental tellurium is not stable on Venus and is not expected to condense below 6,100 km; instead, they favor condensation of certain lead and bismuth compounds with $\varepsilon \sim 100$–200. The jury remains out on what really controls the microwave properties of the high mountains on Venus; new radar observations and laboratory work can both play a role.

5.4 COHERENT BACKSCATTER OPPOSITION EFFECT AND THE SEARCH FOR LUNAR WATER ICE

Monostatic radar backscatter from the icy Galilean satellites—Europa, Ganymede, and Callisto—has proven to be unusual in two respects: (i) the integrated echo for each of the targets is larger than would be expected from a smooth, perfectly conducting sphere of the same size, and (ii) the circular polarization sense of the echo is the same as that transmitted [49]. The latter characteristic is puzzling because echoes from nominally flat surfaces have reversed senses of circular polarization; in the case of the icy Galilean satellites, the incoming wave appears to have been turned 180° without undergoing a Fresnel-like reflection.

In fact, even numbers of Fresnel reflections will also preserve polarization. Indeed, one early model [50] explained these echoes as coming predominantly from multiple reflections by crater walls, but slope requirements for the craters were severe. Other models, based on phenomena such as critical internal reflection [51], have been proposed. But models in which the wave is repeatedly deflected by inhomogeneities within a low-loss icy regolith are now favored [52].

Reciprocity is a corollary of the current theory: photons following one path through the ice will return to a monostatic radar with exactly the same phase as photons that follow the same path but in the opposite direction [53]. Because of coherent addition, they will appear with four times (rather than two times) the echo power expected from photons following completely different paths. Of course, if the radar moves (as must be the case when round-trip light time to Jupiter is an hour), the coherent addition is degraded, but only to the extent that the two paths differ by a significant fraction of a wavelength. Thus, the motion of the radar is important, but so also is the length of the path within the ice. The higher the purity of the ice, the longer the potential travel time of each photon before it is either absorbed or expelled from the regolith, and the sharper the angular dependence of the coherent enhancement will be. This enhancement is known as the *coherent backscatter opposition effect* (*CBOE*).

CBOE is thus an ideal application for bistatic radar. With the *Clementine* spacecraft in a polar orbit around the moon, a bistatic experiment in 1994 was designed to explore the near-backscatter geometry emphasizing regions near the poles (e.g., deep craters), which are permanently shaded from the Sun. It has been hypothesized that water vapor from impacting comets and other sources would migrate to these cold traps, where substantial quantities of relatively clean water ice could accumulate over time [54, 55].

The *Clementine* antenna beam was broader than any single-shaded feature, but data were collected during several experiments over each pole and processed using doppler filtering to localize the echo to particularly promising targets. The results were inconclusive. Nozette et al. [14] claimed detection of a 1 dB enhanced echo in the same sense of circular polarization (SC) near the South Pole at bistatic angles $\beta < 1°$. Simpson and Tyler [15], working with the same raw data, were unable to reproduce those results. In both cases, surface resolution was provided only by doppler beam sharpening of the echo and corresponded to about 4,000 km^2 per frequency bin.

Earth-based monostatic radar has since been used to explore the polar regions at view angles which see as much as 5° deeper into permanently shadowed craters than the Sun; at pixel sizes much smaller than were possible in the *Clementine* experiments, these data also show no enhancements that cannot be explained by more conventional means [56].

5.5 TRANSIENT SURFACE ECHOES AT OCCULTATION

In the few seconds before an ingress occultation, a small area around the occultation point falls within the beams of both the transmitting and receiving antennas and can support generation of a near-forward-scattering surface echo. The same is true immediately after an egress occultation. The first detections of surface echoes from such regions, called *transient surface echoes*, were reported by Fjeldbo et al. [11] during the *Mariner 6* and 7 flybys of Mars; the authors reported rms surface roughness $\zeta \sim 1.5°$ at $\lambda = 13$ cm, but they made no attempt at estimating dielectric constant.

The observation of transient surface echoes in conjunction with occultations has become routine in the era of multiyear orbiters; Simpson [57] reported over 4,000 detections after 4 years of *Mars Global Surveyor* operations (Figure 5–8). But theoretical developments that

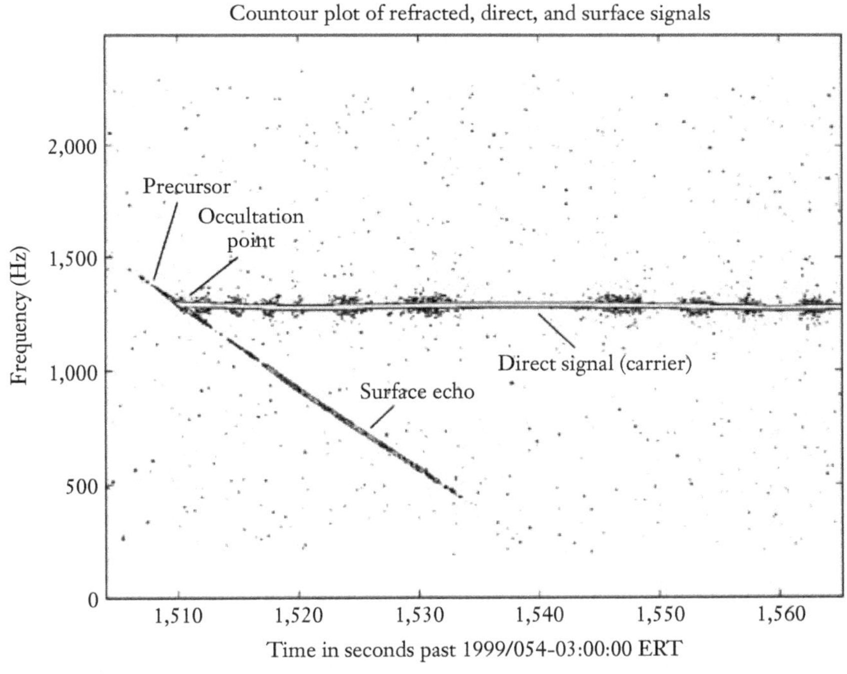

Figure 5–8 Time–frequency contour plot of signal intensity during a *Mars Global Surveyor* egress occultation on February 23, 1999 ($\lambda = 3.6$ cm). The directly propagating carrier is the solid, approximately horizontal line. Receiver tuning was referenced to the *predicted* frequency of the carrier, but small bursts of processing noise show where the observed carrier moved from one digital frequency bin to the next. The surface echo separates from the carrier at egress, with an approximately linear, negative frequency drift over the interval shown here. The echo initially gains strength, peaking at about 1,520 seconds, as more surface area becomes mutually visible to the receiver and transmitter. At about 1,530 seconds, the decreasing incident flux as the spacecraft antenna boresight moves away from the limb overcomes the increasing area illuminated and the echo quickly disappears. Note the "precursor," a diffraction-related signal that appears before both the carrier and the surface echo and on the same frequency trajectory as the latter.

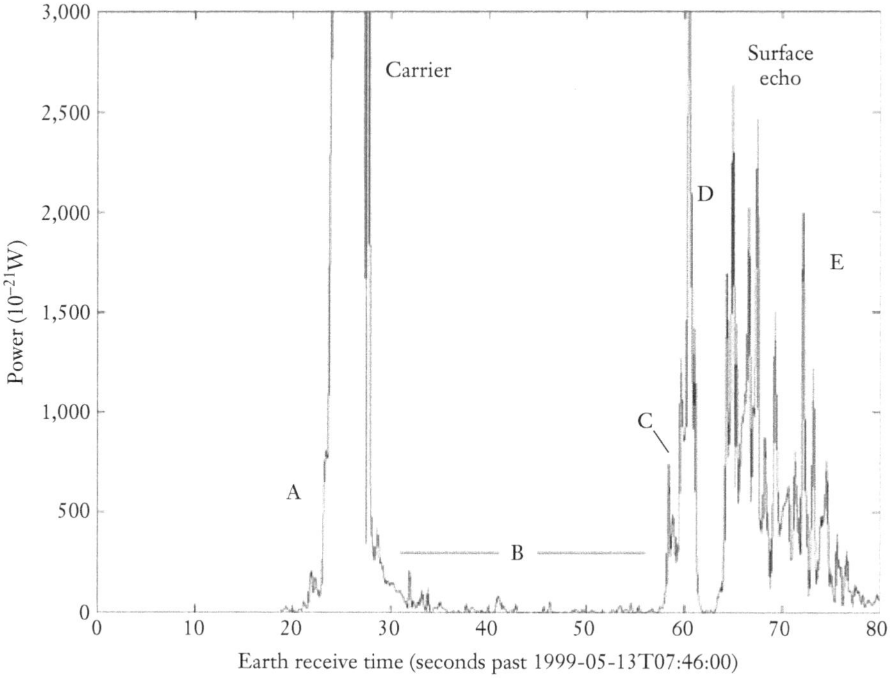

Figure 5–9 Signal power vs. time for a *Mars Global Surveyor* (λ = 3.6 cm) occultation, illustrating the effects of limb topography (Figure 5–10) on echo strength. The directly propagating carrier dominates at egress (*A*). Echo strength would ordinarily peak during interval *B*, and then taper off as the spacecraft antenna beam left the surface. But topographic features on the limb at *B* and *D* kept part of the contributing surface in shadow.

would be useful in interpretation have been painfully slow. The strength of the surface echo depends on a number of factors including the style and amount of roughness at the limb. Several electromagnetic phenomena (shadowing, diffraction, and surface wave propagation) are involved. Vesecky et al. [58] did some preliminary studies, focusing on the shadowing question; but much remains to be done.

Since the objective in occultation experiments is usually a characterization of the atmospheric column above the occultation point, the spacecraft antenna continues to remain pointed toward Earth. The surface echo results from illumination by the side of the main beam (or side lobes) and becomes very sensitive to the angular displacement of the target from the boresight. As the amount of area mutually visible to the transmitter and receiver increases, the surface echo should become stronger; at the same time, however, the incident flux from the spacecraft antenna is tapering off.

Figure 5–9 shows the observed echo power as a function of time from one *Mars Global Surveyor* egress occultation. In this case the signal strength has been modulated additionally by limb topography (Figure 5–10). The sequence is roughly as follows:

A. Occultation emersion. The directly propagating signal from the spacecraft appears over the limb of Mars. There is very little mutually visible surface area. The surface echo is too weak to be detected.

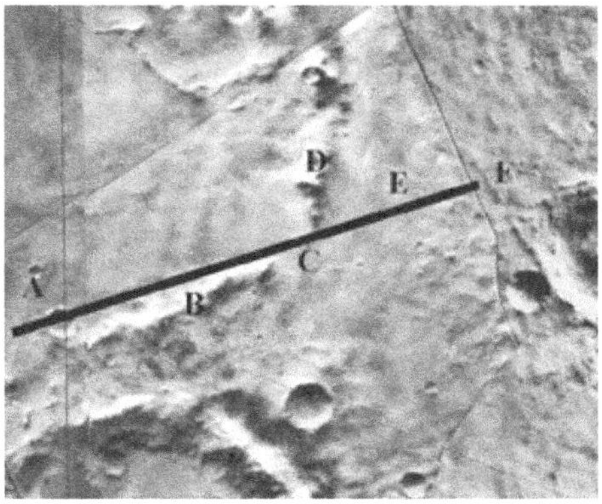

Figure 5–10 Image mosaic showing features responsible for echo blockage in Figure 5–9. North (and the spacecraft) is toward the top. The vertical line near *A* is the 90°W meridian and point *C* is at about 68.3°S, 86.4°W. The solid line running diagonally across the image is approximately the track of the specular point; the length shown is about 160 km. (Figure by Richard Simpson, using U.S. Geological Survey, "Controlled Photomosaic of the Mare Australe D Quadrangle of Mars," map I-1634 (MC-30D), 1985 and U.S. Geological Survey, "Controlled Photomosaic of the Mare Australe E Quadrangle of Mars," map I-1635 (MC-30E), 1985.)

B. During this time, the surface echo strength should be building. But a mountain ridge keeps the surface which would be reflecting in shadow (from either the spacecraft, the Earth, or both) and no echo is observed.

C. A gap in the ridge lets echo through.

D. Another local topographic feature interrupts the reflection process.

E. A final clear area allows echo to be observed. But now the spacecraft antenna beam is rising above the surface. Incident power reaching the surface drops rapidly and the echo vanishes at F.

Although it is difficult to find a "typical" surface on Mars, values of rms surface roughness in the range $3° < \xi < 5°$ are common when studying quasi-specular echoes. The fact that these transient echoes at occultation are viewed at incidence angles $\varphi > 89°$ means that a relatively small fraction of the physical surface is illuminated based on simple geometrical arguments. The quasi-specular theory developed by Fjeldbo [19] is simply inadequate to describe this situation; at best, only the topographically highest surface elements contribute to the scattering. On the observational side, Simpson [57] concluded from a statistical analysis of 5 months of transient echoes that even though some general patterns of "smooth" versus "rough" surfaces were emerging, uncertainties in the spacecraft antenna pointing were probably more important in determining the strength of the surface echoes. A large number of these "spot" measurements now exist in the *Mars Global Surveyor* data collection; more work is needed to extract their geophysical significance.

5.6 UPLINK BISTATIC RADAR

The earliest proposals for planetary bistatic radar assumed that transmission would be from Earth, that the surface echoes would be captured by a spacecraft near the target, and that the raw data would then be returned to Earth for processing. In fact, obtaining the satellite resources for this *uplink* configuration has been nearly impossible; instead, bistatic radar has piggybacked on missions designed with other objectives in mind and using hardware that was not primarily intended for radio science.

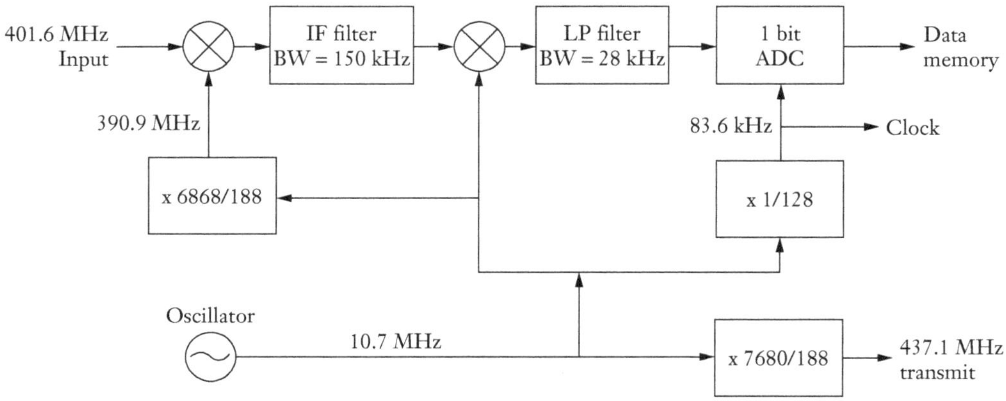

Figure 5–11 Block diagram of *2001 Mars Odyssey* receiver and sampling system for uplink bistatic radar. The input signal is downconverted to 28 kHz, and then digitized using one-bit sampling. The results must be read out from each experiment before the memory is overwritten on the next orbit (by activities which may have nothing to do with radar).

Because of the large distances, the tenuous atmosphere of Pluto can *only* be characterized using uplink bistatic radar, and the *New Horizons* mission to Pluto and the Kuiper Belt is equipped with the first receiver and signal processor designed for this type of experiment. The performance of uplink is ~40 dB better than downlink because Earth-based transmitter powers are available in the 1-MW range, whereas spacecraft are typically limited to 100 W. On the negative side, a spacecraft receiver will never be as sensitive as a cryogenically cooled front end on the ground, but the difference may be as little as 10 dB. *New Horizons* prelaunch tests measured $T_{SYS} < 250$ K (I. R. Linscott, private communication 2006), which compares favorably with $T_{SYS} \sim 25$ K for the state-of-the-art receivers at major ground-based observatories. The net performance advantage for uplink over downlink in planetary bistatic radar is therefore ~30 dB.

The first successful uplink planetary bistatic radar observations were conducted by Callas et al. [59] using an ad hoc arrangement powered by 3-30-kW transmissions from Earth, a receiver at $\lambda = 75$ cm wavelength, and a one-bit sampling system on the *2001 Mars Odyssey* spacecraft (Figure 5–11). Example output is shown in Figure 5–12 where the direct "carrier" signal-to-noise ratio (SNR) is ~40 dB while the surface echo SNR is ~15 dB. Striations in the echo signal result from passage of the specular point over inhomogeneous surface structure. Striations in the carrier signal are signal-processing artifacts. Interpretation is hampered by the weak echo strengths, the imperfectly understood spacecraft antenna characteristics, and the grazing incidence geometry for most of the observations. But potentially useful information is being obtained in the search for landing sites at high northern latitudes on Mars [17].

5.7 RECAPITULATION AND FUTURE EXPERIMENTS

Bistatic radar cannot compete with monostatic synthetic aperture radar mapping or precision altimetric radar systems when science or exploration goals are compelling and funding agencies are willing to invest the necessary resources. In simple, inexpensive surveys, however, bistatic techniques that piggyback on existing communications links can provide useful information on planetary surface properties that are relevant to robotic or human exploration—centimeter- to meter-scale roughness and material densities in the top few centimeters of regolith. Spacecraft

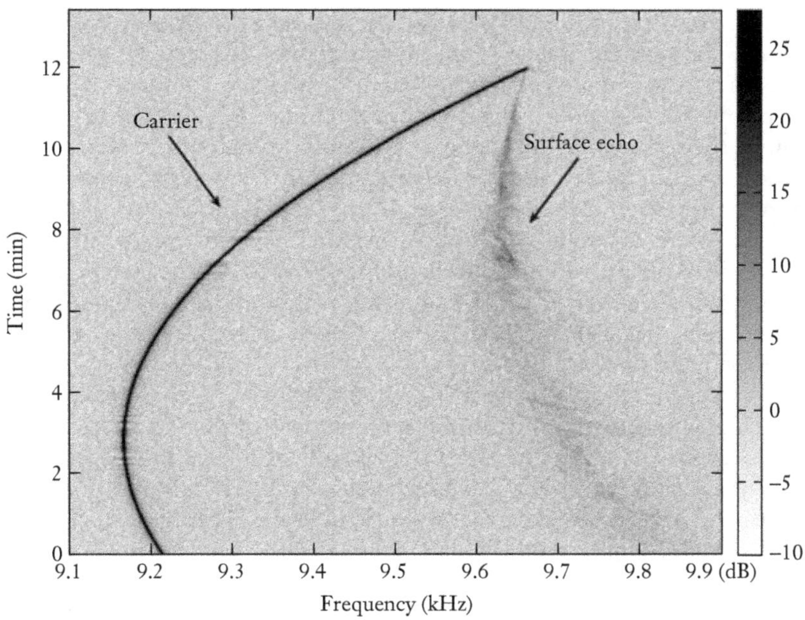

Figure 5–12 Uplink bistatic radar data from *2001 Mars Odyssey* on August 2, 2005. The strong signal on the left is the 30-kW carrier transmitted from Earth; the dispersed signal on the right is the surface echo, which merges with the carrier at occultation ingress. (Image courtesy of H. Gunnarsdottir.)

proximity to the surface is a major advantage in comparison with entirely Earth-based systems; bistatic transmitters need far less power. Special geometries such as near-backscatter, which may help identify deposits of clean-water ice, are also unique advantages of bistatic radar. The ability to observe quasi-specular forward scattering at latitudes away from planetary equators is also advantageous.

Table 5–2 summarizes several current and upcoming bistatic radar experiments in terms of the radar equation (5.1) and receiver noise power density $P_n = kT_{SYS}$. Typical values have been used; comparison of P_R against P_n provides a relative measure of performance. Note that σ has been calculated for a perfectly reflecting sphere; a real surface is likely to have a smaller radar cross section by a factor of 10, reducing P_R accordingly. Also, real echoes will be spread in frequency (Figure 5–6)—often by kilohertz (equation (5.9))—so that the noise power density needs to be multiplied by at least 1,000 before being compared with the echo power.

Mars Express experiments in progress are expected to continue through at least the end of 2007. The values in Table 5–2 are for orbital positions near apoapsis (the high point of its orbit); at other points in the orbit R_T will be smaller and P_R larger, albeit with more doppler spreading. *MEX* experiments are being conducted simultaneously at S-band, albeit at roughly 20 dB lower performance (P_T = 5 W and lower G_T).

Venus Express was inserted into its 24-h orbit in April 2006. The values shown in Table 5–2 are appropriate for low-altitude passes over Maxwell Montes, which was found to be anomalous in 1994. Maxwell observations were conducted in June 2006 at both S- and X-band; doppler effects were severe compared with *MEX*, and atmospheric absorption reduced X-band echo powers to below the power levels at S-band. But echoes were detected at both wavelengths. Observations over other high-altitude targets are planned, but most of the other targets are at equatorial latitudes.

Periapsis (closest approach) is at about 70°N and the spacecraft is severely limited in the attitudes it can adopt because of thermal concerns; so scheduling these other targets will be challenging.

Cassini carries an onboard monostatic radar system, but a limited number of bistatic experiments have also been planned for Titan using the telecommunications/radio science system. These will be used to confirm the reflectivity and dielectric constant estimates by observing three-frequency reflections at incidence angles near the Brewster angle [18]. As can be seen from Table 5–2, the large Earth–Saturn distances make these measurements very difficult.

Finally, the impressive capabilities of uplink bistatic radar can be seen in the *New Horizons* column of Table 5–2. This instrument is designed for refractivity measurements of the tenuous atmosphere of Pluto. But the ratio P_R/P_n is 6 orders of magnitude higher than for any of the other experiments, so it seems likely that at least some transient surface echoes will be captured along with the atmospheric signatures.

For *New Horizons*, the plan is to transmit multiple X-band carriers, carefully spaced in frequency, using high-power transmitters at 70-m antennas of the NASA Deep Space Network (Figure 5–4). The multiple carriers first provide redundancy, but their different paths through the Earth's atmosphere and ionosphere reduce the chance that minor perturbations in the phase path on the Earth end will be mistaken for structure in Pluto's atmosphere.

References

1. J. H. DeWitt and E. K. Stodola, "Detection of Radio Signals Reflected from the Moon," *Proceedings of the IRE*, Vol. 37, 1949, pp. 229–242.
2. Z. Bay, "Reflection of Microwaves from the Moon," *Hungarica Acta Physica*, Vol. 1, No. 1, 1947, pp. 1–22.
3. R. M. Goldstein and W. F. Gillmore, "Radar Observations of Mars," *Science*, Vol. 141, 1963, pp. 1171–1172.
4. R. M. Goldstein and R. L. Carpenter, "Rotation of Venus," *Science*, Vol. 139, 1963, pp. 910–911.
5. R. A. Simpson, "Spacecraft Studies of Planetary Surfaces Using Bistatic Radar," *IEEE Transactions on Geoscience and Remote Sensing*, Vol. 31, 1993, pp. 465–482.
6. O. I. Yakovlev and A. I. Efimov, "Studies of Reflection of Meter-Length Radio Waves by the Lunar Surface," *Doklady Akademii Nauk SSSR*, Vol. 174, 1967, pp. 583–584.
7. G. L. Tyler, V. R. Eshleman, G. Fjeldbo, H. T. Howard, and A. M. Peterson, "Bistatic-Radar Detection of Lunar Scattering Centers with Lunar Orbiter 1," *Science*, Vol. 157, 1967, pp. 193–195.
8. G. L. Tyler and R. A. Simpson, "Bistatic Radar Measurements of Topographic Variations in Lunar Surface Slopes with Explorer 35," *Radio Science*, Vol. 5, 1970, pp. 263–271.
9. G. L. Tyler and H. T. Howard, "Dual-Frequency Bistatic-Radar Investigations of the Moon with Apollos 14 and 15," *Journal of Geophysical Research*, Vol. 78, 1973, pp. 4852–4874.
10. M. A. Kolosov, O. I. Yakovlev, A. I. Efimov, S. S. Matyugov, T. S. Timofeev, E. V. Chub, A. G. Pavelyev, A. I. Kucheryavenkov, I. E. Kalashnikov, and O. E. Milekhin, "Investigation of the Venus Atmosphere and Surface by the Method of Radiosounding Using Venera-9 and Venera-10 Satellites," *Acta Astronomica*, Vol. 7, 1980, pp. 219–234.
11. G. Fjeldbo, A. Kliore, and B. Seidel, "Bistatic Radar Measurements of the Surface of Mars With Mariner 1969," *Icarus*, Vol. 16, 1972, pp. 502–508.
12. R. A. Simpson, G. L. Tyler, and G. G. Schaber, "Viking Bistatic Radar Experiment: Summary of Results in Near-Equatorial Regions," *Journal of Geophysical Research*, Vol. 89, 1984, pp. 10385–10404.
13. G. H. Pettengill, P. G. Ford, and R. A. Simpson, "Electrical Properties of the Venus Surface from Bistatic Radar Observations," *Science*, Vol. 272, 1996, pp. 1628–1631.
14. S. Nozette, C. L. Lichtenberg, P. Spudis, R. Bonner, W. Ort, E. Malaret, M. Robinson, and E. M. Shoemaker, "The Clementine Bistatic Radar Experiment," *Science*, Vol. 274, 1996, pp. 1495–1498.
15. R. A. Simpson and G. L. Tyler, "Reanalysis of Clementine Bistatic Radar Data from the Lunar South Pole," *Journal of Geophysical Research*, Vol. 104, 1999, pp. 3845–3862.
16. R. A. Simpson, G. L. Tyler, M. Pätzold, and B. Häusler, "Determination of Local Surface Properties Using Mars Express Bistatic Radar," *Journal of Geophysical Research*, Vol. 111, 2006, E06S05, doi:10.1029/2005JE002580, pp1–10.

17. H. M. Gunnarsdottir, I. R. Linscott, J. L. Callas, and M. D. Cousins, "Surface Properties of the 2007 Phoenix Landing Site B Using Bistatic UHF Radar," in *Lunar and Planetary Science Conference XXXVII*, Paper No. 2421, 2006.
18. E. A. Marouf, F. M. Flasar, R. G. French, A. J. Kliore, A. F. Nagy, N. J. Rappaport, P. J. Schinder, C. A. McGhee, R. Simpson, A. Anabtawi, S. Asmar, E. Barbinis, G. Goltz, D. Fleischman, D. Kahan, and D. Rochblatt, "Cassini First Radio Science Observations of Titan's Atmosphere and Surface," in *38th Meeting of the AAS Division for Planetary Sciences*, Pasadena, CA, 2006.
19. G. Fjeldbo, *Bistatic-Radar Methods for Studying Planetary Ionospheres and Surfaces*, Ph.D. thesis, Stanford University, 1964.
20. P. Beckmann and A. Spizzichino, *The Scattering of Electromagnetic Waves from Rough Surfaces*, Pergamon, New York, 1963.
21. T. Hagfors, "Backscattering From an Undulating Surface With Applications to Radar Returns from the Moon," *Journal of Geophysical Research*, Vol. 69, 1964, pp. 3779–3784.
22. D. O. Muhleman, "Radar Scattering from Venus and the Moon," *Astronomical Journal*, Vol. 69, 1964, pp. 34–41.
23. R. D. Kodis, "A Note on the Theory of Scattering from an Irregular Rough Surface," *IEEE Transactions*, Vol. AP-14, 1966, pp. 77–82.
24. D. C. Schleher, *MTI and Pulsed Doppler Radar*, Artech House, Norwood, MA, 1991.
25. R. J. Sullivan, *Microwave Radar*, Artech House, Inc., Norwood, MA, 2000.
26. A. K. Sultan-Salem and G. L. Tyler, "Generalized Fractal-Based Laws for Scattering from Planetary Surfaces: A Unifying Scale-Explicit Paradigm," *Journal of Geophysical Research*, Vol. 111, 2006, E06S08.
27. D. E. Barrick, "Unacceptable Height Correlation Coefficients and the Quasi-Specular Component in Rough Surface Scattering," *Radio Science*, 5, 1970, pp. 647-654.
28. C. Elachi, R. Goldstein, and D. Held, "Spaceborne Bistatic Synthetic Aperture Imaging Radar," in *Proceedings of 1981 International Geoscience and Remote Sensing Symposium*, Washington, DC, Publication 81CH1656-8 of Institute of Electrical and Electronics Engineers, 1981, pp. 944–950.
29. G. Fjeldbo, A. J. Kliore, and V. R. Eshleman, "The Neutral Atmosphere of Venus as Studied with the Mariner V Radio Occultation Experiments," *Astronomical Journal*, Vol. 76, 1971, pp. 123–140.
30. G. L. Tyler, G. Balmino, D. P. Hinson, W. L. Sjogren, D. E. Smith, R. Woo, J. W. Armstrong, F. M. Flasar, R. A. Simpson, S. Asmar, A. Anabtawi, and P. Priest, *MGS RST Science Data Products (MGS-M-RSS-5-SDP-V1.0)*, Vol. USA_NASA_JPL_MORS_1xxx, NASA Planetary Data System, 2006.
31. R. Woo, "Spacecraft Radio Scintillation and Solar System Exploration," in *Wave Propagation in Random Media (Scintillation)*, Society of Photo-Optical Instrumentation Engineers Optical Engineering Press, Bellingham, Washington, 1993.
32. D. P. Hinson, *Radio Scintillations Observed During Atmospheric Occultations of Voyager: Internal Gravity Waves at Titan and Magnetic Field Orientations at Jupiter and Saturn*, Ph.D. dissertation, Stanford University, 1983.
33. G. L. Tyler, "Radio Propagation Experiments in the Outer Solar System with Voyager," *Proceedings of the IEEE*, Vol. 75, 1987, pp. 1404–1431.
34. J. V. Evans and T. Hagfors, *Radar Astronomy*, McGraw-Hill, New York, 1968.
35. M. J. Campbell and J. Ulrichs, "Electrical Properties of Rocks and Their Significance for Lunar Radar Observations," *Journal of Geophysical Research*, Vol. 74, 1969, pp. 5876–5881.
36. K. K. Williams and R. Greeley, "Measurement of Dielectric Loss Factors Due to a Martian Dust Analog," *Journal of Geophysical Research*, 109, 2004, E10006, doi:10.1029/2002JE001957, pp. 1–9.
37. T. Gold, M. J. Campbell, and B. T. O'Leary, "Optical and High-Frequency Electrical Properties of the Lunar Sample," *Science*, Vol. 167, 1970, pp. 707–709.
38. G. H. Heiken, D. T. Vaniman, and B. M. French, *The Lunar Sourcebook: A User's Guide to the Moon*, Cambridge University Press, Cambridge, UK, 1991, p. 736.
39. R. A. Simpson, H. T. Howard, and B. C. Fair, "Microwave Properties of Solid CO_2," *Journal of Geophysical Research*, Vol. 85, 1980, pp. 5481–5484.
40. W. A. Cumming, "The Dielectric Properties of Ice and Snow at 3.2 Centimeters," *Journal of Applied Physics*, Vol. 23, 1952, pp. 768–773.
41. R. A. Simpson and G. L. Tyler, "Radar Scattering Laws for the Lunar Surface," *IEEE Transactions*, Vol. AP-30, 1982, pp. 438–449.

42. G. L. Tyler, "Brewster Angle of the Lunar Crust," *Nature*, Vol. 219, 1968, pp. 1243–1244.
43. R. A. Simpson, G. L. Tyler, J. P. Brenkle, and M. Sue, "Viking Bistatic Radar Observations of the Hellas Basin on Mars: Preliminary Results," *Science*, Vol. 203, 1979, pp. 45–46.
44. R. A. Simpson, G. L. Tyler, M. Pätzold, and B. Häusler, "Bistatic Radar Exploration of Mars' South Polar Region," *American Geophysical Union Fall Meeting*, Paper No. P13C-07, San Francisco, 2005.
45. M. Born and E. Wolf, *Principles of Optics*, Cambridge University Press, Cambridge, 1999.
46. M. K. Shepard, R. E. Arvidson, R. A. Brackett, and B. Fegley, "A ferroelectric model for the low emissivity highlands on Venus," *Geophysical Research Letters*, Vol. 21, 1994, pp. 469–472.
47. R. A. Brackett, B. Fegley, and R. E. Arvidson, "Volatile Transport on Venus and Implications for Surface Geochemistry and Geology," *Journal of Geophysical Research*, Vol. 100, 1995, pp. 1553–1563.
48. L. Schaefer and B. Fegley, "Heavy Metal Frost on Venus," *Icarus*, Vol. 168, 2004, pp. 215–219.
49. D. B. Campbell, J. F. Chandler, S. J. Ostro, G. H. Pettengill, and I. I. Shaprio, "Galilean Satellites: 1976 Radar Results," *Icarus*, Vol. 34, 1978, pp. 254–267.
50. S. J. Ostro and G. H. Pettengill, "Icy Craters on the Galilean Satellites?" *Icarus*, Vol. 34, 1978, pp. 268–279.
51. V. R. Eshleman, "Mode Decoupling During Retrorefraction as an Explanation for Bizarre Radar Echoes from Icy Moons," *Nature*, Vol. 319, 1986, pp. 755–757.
52. K. J. Peters, "The Coherent Backscatter Effect: A Vector Formulation Accounting for Polarization and Absorption Effects and Small or Large Scatterers," *Physical Review B*, Vol. 46, 1992, pp. 801–812.
53. B. Hapke, "Coherent Backscattering and the Radar Characteristics of the Outer Planet Satellites," *Icarus*, Vol. 88, 1990, pp. 407–417.
54. K. Watson, B.C. Murray, and H. Brown, "The Behavior of Volatiles on the Lunar Surface," *Journal of Geophysical Research*, Vol. 66, 1961, pp. 3033–3045.
55. J. R. Arnold, "Ice in the Lunar Polar Regions," *Journal of Geophysical Research*, Vol. 84, 1979, pp. 5659–5668.
56. N. J. S. Stacey, D. B. Campbell, and P. G. Ford, "Arecibo Radar Mapping of the Lunar Poles: A Search for Ice Deposits," *Science*, Vol. 276, 1997, pp. 1527–1530.
57. R. A. Simpson, "Highly Oblique Bistatic Radar Observations Using Mars Global Surveyor," in *XXVIIth URSI General Assembly*, Paper No. 1635, Maastricht, 2002.
58. J. F. Vesecky, E. J. Sperley, and H. Zebker, "Electromagnetic Wave Scattering at Near Grazing Incidence from a Gently Undulating, Rough Surface," in *Proceedings of the International Geoscience and Remote Sensing Symposium*, Edinburgh, IEEE Publication 88CH2497-6(3), 1988, pp. 1579–1583.
59. J. L. Callas, I. R. Linscott, H. M. Gunnarsdottir, M. D. Cousins, and G. L. Tyler, "Bistatic UHF Radar Experiments at Mars Using the SRI 150-ft Dish and the Mars Odyssey Spacecraft," in *American Geophysical Union Fall Meeting*, Paper No. P31C-0216, 2005.

6

Air Surveillance

Nicholas J. Wills, Hugh D. Griffiths,
and David K. Barton

6.1 INTRODUCTION

As outlined in Chapter 1, *passive bistatic radars* (*PBRs*) are a subset of bistatic radars that exploit nonradar *transmitters of opportunity* as their sources of radar illumination. They are often designed for military or civil air surveillance. When more than one transmitter is simultaneously exploited, the configuration becomes *multistatic*. In this case, measurements from transmit–receive pairs with overlapping coverage can be combined to locate a target usually via *multilateration*, for example, by determining the intersection of *isorange contours* generated by each pair. One transmitter operating with multiple receivers is also multistatic and can be used for multilateration.

The most common transmitters exploited by PBRs are terrestrial broadcast transmitters used by TV and FM stations. Other potential transmitters include satellite TV broadcasts, HF radio broadcasts, and even cell phone communication networks. Communications and navigation satellite transmitters have also been considered. These transmitters can be either *cooperative* (allied or friendly) or *noncooperative* (hostile or neutral), depending on the scenario of interest to the PBR. In either case, the transmitter has no knowledge that it is being exploited. The PBR is just another receiver in its net. This feature is one of the operational benefits of a PBR.

These exploited transmitters are characterized by broad, fixed, transmit antenna coverage and CW operation, using either analog or digital modulation dedicated to information transmission. These characteristics distinguish the PBR from *hitchhikers*, which exploit monostatic radar transmitters as their illuminator. Such transmitters typically use a high gain, scanning antenna, pulsed modulation, and waveforms tailored to radar operation. Thus, hitchhikers must contend with antenna scan-on-scan problems, and with their cost and performance penalties, whereas PBRs do not. In contrast, PBRs must contend with less-than-optimum waveforms, which cause a host of performance penalties, whereas hitchhikers do not. As a result, hitchhikers are usually treated separately, for example in [1], and are not considered here.

This chapter documents PBRs that have been proposed, designed, tested, and evaluated for military and civil air surveillance, starting with a PBR review in Section 6.2. It then assesses their utility in Section 6.3, specifically when compared to the entrenched, 70-year benchmark: monostatic radars. In subsequent sections, it documents the theory and available data to predict and assess the performance of a PBR, and then does so for a generic set of PBRs.

PBR is but one of many names used for this subset of bistatic radars. Other names include *passive coherent location* (*PCL*), *passive radar*, *passive* and *covert radar* (*PCR*), *covert radar*, *noncooperative radar*, *broadcast radar*, *parasitic radar*, *and opportunistic radar*. PCL has been subdivided into *narrowband*

PCL and *wideband* PCL, where narrowband characterizes the video or audio carrier of a TV signal and wideband characterizes the full modulation spectrum of an FM signal.

The term *passive* is somewhat of a misnomer since all radars, including bistatic, require an *active* transmitter. Thus, only in the sense that the PBR's transmitter is unaware it is being used for another purpose is there anything passive about the operation.[1] The term *covert* is too restrictive. It suggests that since the receiver has no radio frequency (RF) signature it is difficult to detect and locate. However, some PBRs are used for scientific experiments and have no need to be covert. The term *noncooperative* is also too restrictive in that cooperative, that is, friendly, transmitters have often been used for radar illumination. The term *broadcast* suffers the same restriction since it covers only one type of transmitter. The term *parasitic* is rather pejorative, suggesting an organism living in or on another organism, which it usually injures. The term *opportunistic* has the same problem: taking advantages of opportunities with little regard for principles or consequences. Thus, no term fully captures this type of bistatic radar operation.

PCL has been favored in the United States, parasitic radar in the German Democratic Republic, and passive radar in the United Kingdom. One selection criterion could be based on the amount of country interest, as evidenced by its publications. Since a full *IEE Proceedings–Radar, Sonar and Navigation* issue, (Vol. 152, No. 3, June 2005) was devoted to passive radar in the United Kingdom, this type of radar would be favored. A second criterion could be based on the opinion of radar lexical experts. One expert expressed serious reservations about the term *passive radar*, along with the other candidate terms. After much dialogue, some of which is reflected in the preceding paragraphs, he suggested what appears to be a reasonable compromise: *PBR*.[2] So PBR it is.

A complete taxonomy of passive systems includes intercept receivers and emitter locators, sometimes called *electronic support measures* (ESM). A principal objective of the emitter locator is to detect, identify, and geolocate monostatic radars. As such, it often has characteristics similar to a PBR including signal sorting, classification, location algorithms, and multiple receive sites.

In fact, some PBR and hitchhiker concepts were derived from the ESM community. For example, when an ESM system achieves sufficient receiver sensitivity to detect a radar's side lobe emissions, it can usually detect large objects illuminated by the radar's main lobe and then scattered into the ESM's receiver—hence becoming a bistatic hitchhiker. Then when the search for suitable illuminators broadened to include first space-borne and then terrestrial transmitters, PBRs appeared. While both the hitchhiker and the PBR share an ESM legacy, ESM systems are not radars and are not considered here.[3]

However, another multistatic configuration is germane to PBRs: multilateration systems on military test ranges. These systems measured range and/or doppler and/or angle of arrival (AOA) of air targets, principally missiles, using a transmitter, multiple receive sites, and often a transponder on the target. Then the *target state* (typically target location and velocity vector, occasionally target acceleration) was estimated by combining both simultaneous and time-sequenced measurements from each site at a central node. They were the instruments of choice following World War II at U.S. missile and aircraft test ranges. But, because of fundamental limitations, including the inability to reestablish the target state if track was lost during flight, they were replaced by optical trackers and monostatic radars.

1 In contrast, *radiometers* are passive sensors that detect and measure electromagnetic radiation. While they are truly passive, they are not a radar.

2 Merrill I. Skolnik, private communication, October 2005.

3 Rediscovery of the bistatic radar by the ESM community fueled much of its second resurgence, starting in 1970s. It has continued into the third resurgence, starting in the early 1990s and focuses primarily on passive bistatic radars.

While there are differences in the configuration and operation of PBRs and these instrumentation systems—with all differences favoring the latter: smaller target densities, less stringent receive antenna constraints, and transponder-aided tracking (which reduces transmit power, extends range, and eliminates ghosting)—their similarities are sufficient to require documentation. Principal similarities include multistatic architecture, measurement parameters, real-time operation, state estimation multilateration techniques and their limitations, and, probably most importantly, lessons learned. These relationships are documented in Section 6.6, Target Location.

Other controversial topics include the ability to bistatically capture specular flashes from air targets, in particular *stealth* targets, which is evaluated in Section 6.3, Military Utility. Unavoidable time outages due to fluctuations in the broadcast transmitter bandwidth and the effects of radio frequency interference on receiver noise figure are quantified in Section 6.4, Waveforms and Interference. *Geometric dilution of precision* (*GDOP*) advantages and constraints on PBR systems using multilateration and little or no receive spatial directivity—the typical case for systems working at VHF/UHF—are included in Section 6.6, Target Location. Effects of *electronic countermeasures* (ECM) deployed against military PBRs and an assessment of exploiting a jammer as the illuminator for a PBR are developed in Section 6.7, ECM. The parallel subject of HF bistatic radars is covered in Chapter 4.

6.2 PBR REVIEW

The first published description of a bistatic radar exploiting a transmitter of opportunity that satisfies the definition of a PBR appears to be the 1960 *IRE* paper by Rittenbach and Fishbein, [2].[4] An annotated summary is reproduced as follows:

> This paper describes a semi-active radar system [the U.S. Army's preferred nomenclature, although the term 'bistatic' had been coined by Seigel and Machol in 1952] in which the transmitter is carried in a [geosynchronous] satellite. The satellite transmits a randomly modulated signal [proposed at 100 W-CW, illuminating a ground area of 7,000 miles in diameter]. On the ground the radar has two antennas and receivers. One antenna points at the satellite, the other at the target [a ground vehicle]. The signal from the satellite-oriented receiver is delayed and [time-] correlated with the satellite signal reflected from the target. The delay corresponding to the peak of the correlation function is used to determine range [1,000 yds for a 1 m^2 target, 10,000 yds for 100 m^2]. It is planned to test this system with various communications satellites [once they are orbited].

Two significant events clearly triggered this concept: (1) the launch of the first U.S. satellite, *Explorer 1*, on January 31, 1958, 4 months after the launch of the U.S.S.R.'s *Sputnik*, and (2) the advent of digital processors, which in turn enabled the development of their correlator, a correlator almost made-to-order for bistatic radar operation.[5]

Although nothing appears to have been published on their planned tests, many similar tests have been carried out by both government and industry over the next half century. Unfortunately, the early work has seldom been published, because of either security restrictions or discouraging test results.

---

4 While most bistatic radar events of the 1930s described in Table 2–1 also used transmitters of opportunity, including commercial broadcast transmitters, that would be stretching the PBR definition since there were not many opportunities for other choices. (Only Hulsmeyer, Oschepkov, and David crafted their own transmitters, with the latter two most likely using existing communication equipment.) Thus, recognizing this tenuous connection, we press on with more current events.

5 Although not germane to this discussion, it is worth recalling that *Explorer* 1 weighed a total of 18 lbs. It carried a small package of instruments that produced the first and probably most significant Space Age discovery up to the *Hubble* launch in 1990: the Van Allen radiation belts surrounding the earth. However, it would be many years before payloads large enough to accommodate a 100 W-CW transmitter, antenna and power supply could be orbited.

One exception, and probably the first PBR to be built in the 1960s, is the *Sugar Tree* system [3, 4]. *Sugar Tree* was a high frequency (HF), *over the horizon* (*OTH*) bistatic receiver that exploited Soviet-operated, international HF broadcast transmitters to detect missile launches from the Soviet Union. The transmitter, which was located close to the missile launch area, illuminated the missile target via a surface wave; the target echo arrived at the receiver via a sky wave and was then cross-correlated with the direct path sky wave for detection. Further details are given in Chapter 4.

The PBR concept gained interest when the USAF *DSCS II* and *NATO III B* geosynchronous communication satellites were launched in 1977. For example, The Aerospace Corporation in El Segundo, CA, ran a series of tests using these satellites, detecting commercial aircraft approaching or departing the Los Angeles International Airport with a receiver located a few miles away. But again results were limited to military-industrial audiences, and only briefly alluded to in the open press [5].

Then as succeeding satellites designed for joint military and civilian use were orbited, studies and tests were again conducted (and now published) to assess the feasibility of exploiting them for PBR operation. For example, when *global navigation satellite systems* (*GNSS*) became operational in the 1990s, they were evaluated for PBR air defense applications [6]. However, in limited tests they were found to work only in a forward-scatter mode, that is, when the air target flew between a satellite and the ground station—not a particularly useful air surveillance configuration.

A later study [7] evaluated GNSS illumination for synthetic aperture PBR short-range search and rescue, natural disaster damage assessment, and change detection. It concluded that receiver-to-target ranges of a few kilometers were possible with integration times ranging from 30 to 100 sec. If those times were extended to 1,000 seconds using a stationary receiver, ranges could be extended to the horizon. The feasibility of achieving these integration times was not assessed.

Similarly, when the *low earth orbiting* (*LEO*) communication satellites became operational in the1990s, they were evaluated for PBR air defense operation [8]. The study and some very limited tests suggested that (a) with postdetection integration times of 15–20 seconds, which required a multibeam antenna and multichannel receivers, (b) with unusually large estimates for the bistatic radar cross section of air targets, and (c) ignoring the doppler-shifted clutter and direct-path breakthrough, again doppler-shifted due to illuminator motion, receiver-to-target ranges of possibly 30 km could be achieved. (Ranges on the order of a few kilometers might be a more robust expectation.) While this optimistic result was a little more encouraging, it is not particularly persuasive in view of the satellite motion effects and anticipated receiver complexity. More useful bistatic and multistatic configurations could and would be developed.

Probably the satellite system most analyzed for PBR operation is the high-powered (~1 MW), Ku-band, *Direct Broadcast Satellite* Television (*DBS* TV), launched into geosynchronous (~40,000 km) orbit in the 1990s. This system yields approximately 20-dB and 40-dB greater power densities on the ground than do the LEOS and GNSS systems, respectively. Two coauthors were independently involved in DBS analysis and testing for PBR surveillance of ship targets [9, 10]. In their proof of concept demonstrations, they developed virtually identical test configurations and arrived at nearly identical test results and conclusions.

For example, Griffiths et al. [9] ran the bistatic range equation and determined that more than 75 dB of processing gain would be required to detect a 1,000 m^2 target such as a ship at 100 km. They then evaluated methods for achieving such processing gain and concluded that both coherent and noncoherent integration were needed, which in turn would require both doppler- and range-walk compensation even for slowly moving ships.

They then assembled an experimental PBR receiver from both laboratory and commercial equipment available ca 1991. An example of the latter was use of the commercial, 38-dBi gain receiving antenna. Because only 29 dB of processing gain (via video pulse-compression and noncoherent integration) was planned, they used a flat plate reflector located within 60 m of the receiver and oriented so that the bistatic specular was pointed at the receiver. Its bistatic RCS was estimated to be 670 m^2.

Test results were just 5 dB lower than test predictions, but clearly confirming the more than 75-dB processing gain requirement. Additional means of satisfying this requirement were identified but not assessed. They included multiple bistatic receivers to capture and noncoherently integrate multiple DBS channels, along with larger receiver apertures.

The 2000 Bernstein and Willis work [10] independently duplicated the Griffiths et al. experiment (including use of the commercial receiving antenna and the flat plate reflector) and reached identical results. However, further work on both projects ended, essentially for two reasons: (1) achieving the approximately 45-dB additional link margin, via processing gain, larger apertures, and multiple channels, was and continues to be a daunting and expensive task; and (2) more importantly, operation of this PBR configuration is limited to littoral waters, since DBS TV transmissions are directed over land areas, not oceans.

An alternative mode of PBR-DBS operation might be considered: short-range air defense, say at <30 km receiver-to-target ranges. But PBR requirements become only more stringent: while the range is reduced by a factor of 3.3 (100–30 km), that reduction is more than offset by a loss in the RCS by a factor of 100 (1000 m^2 for ships to 10 m^2 for aircraft), thereby reducing the range by a factor of ten.[6] Furthermore, the doppler- and range-walk problems become more severe because of increased target velocities.

However, a PBR-DBS configuration may well find application for very short range (~5 km), ground surveillance from an airborne platform such as an *unattended air vehicle* (*UAV*) employing spotlight bistatic SAR (Chapter 10) or bistatic MTI (Chapter 11). But DBS use for air or even ship surveillance is very problematic in the foreseeable future. Hence, exploitation of current satellites for such PBR operation will not be considered further in this chapter.[7]

Nevertheless, Rittenbach and Fishbein [2] were the first to (a) publicly define the concept, (b) devise a practical implementation, (c) predict performance, and (d) identify many of the advantages, including minimizing mutual interference, developing a simple equation for calculating receiver-to-target range (identical to (5.4) in [11], which was developed 30 years later) and predicting the 6-dB/octave range relationship.

In parallel with these attempts to exploit space-borne microwave transmitters—and quite independent of that work—was a Griffiths and Long's 1985 attempt at exploiting terrestrial TV broadcast transmitters [12]. These U.K. fund-limited, air defense experiments used the very powerful (~62 kW available effective radiated power per channel) and very close (12-km baseline) *Crystal Palace* TV transmitter in London.

The bistatic receiver was located at University College London, and partially shielded by a building from the direct-path interference. It used four 17-element Yagi antennas and modified commercial TV tuners and IF circuits, which generated very high noise figures and thus quite modest receiver-to-target detection ranges. TV test patterns emulating a pulsed waveform were used, which yielded a less-than-optimum autocorrelation function: high-range side lobes (~5 dB), range ambiguities every 9.6 km, and modest range resolution (~4 km). A photograph of the experimental system is shown in Figure 6–1. More details are summarized in the third column of Table 6–1.

6 In this geometry, doubling or halving the receiver-to-target range requires only a 6-dB change in system parameters. This subject is covered in Section 6.5.6.

7 It might be argued that this conclusion is shortsighted—similar to looking for a lost wallet only under the streetlight—and that other less intense illumination sources are being ignored. And, indeed that is precisely the point: laws of physics and operational requirements constrain such operation to streetlights. In short, "What you see is all you get."

Figure 6–1 Unpublished photo of 1985 passive bistatic radar experimental hardware at University College London. From the archives of H. Griffiths.

Table 6–1 Parameters of Significant Passive Bistatic Radar Programs Designed and Tested for Air Surveillance*

System	Silent Sentry™	TV-Based Bistatic Radar (I)	TV-Based Bistatic Radar (II)	FM Radio–Based Bistatic Radar	Multistatic HDTV–Based Radar
Developer	IBM, now Lockheed Martin [13–15]	Univ College London [12]	DERA, United Kingdom [16]	NATO [17,18]	SAIC, U.S. Army [19]
Decade	1980 ≥ 2000	1980	1990	2000	2000
Configuration	Multistatic: Rx: 1 Tx: up to 6	Bistatic	Bistatic Near-forward scatter	Bistatic	Multistatic: Rx: 4 Tx: 1
Transmitter operation	FM: Real time TV: Nonreal time	TV: Nonreal time (543 MHz)	TV: Near-real time (543 MHz)	FM: Real time	HDTV: Real time
Baseline	100 km, typical	12 km	150 km	50 km	10 km, typical
Target	Aircraft Missile launches	Aircraft	Aircraft	Aircraft	Aircraft, below ~ 5 k ft
Target data	range doppler bearing	range bearing	doppler bearing	range doppler bearing	range doppler
Measured performance	R_M = 100–150 km 2-D tracks on A/C 3-D tracks on missile launches	R_M ~ 25 km Occasional A/C detections, but mostly negative	R_M ~ 160 km Detections on high and medium altitude A/C but only 1/3 tracked	R_M ~ 175 km Achieved with innovative direct path excision	R_M ~ 30 km Target location via range multi-lateration Ghost excision via doppler association

continued

Table 6–1 *continued*

Comments	Version 3 system potentially useful for covert air surveillance, intelligence collection, instrumentation	Good start on a small budget Negative report a first in bistatic radar annals	Well documented New algorithms for initiating doppler track, but time latency > 60 s	Well documented Integrating bearing and doppler into Kalman filter not yet successful	Designed as a low-altitude gap-filler for U.S. FAA air traffic control radars Maximum time-latency 10 s for 200 targets
Status	Version 3 for sale to U.S. Government for < 1 million dollars Work continues	Program ended	Program ended	Program continuing possibly in a multistatic mode	Test phase complete Awaiting evaluation/ funding

* R_M is the equivalent maximum monostatic range defined in Section 6.5 as $(R_R R_T)_{max}^{1/2} = R_M$, where R_R is the receiver-to-target range and R_T is the transmitter-to-target range. The remaining entries are detailed subsequently. Although *TV-Based Bistatic Radars I* and *II* were associated with the United Kingdom's DERA Malvern, there is no formal link between them. From [20].

In field tests, ground targets were detected on an A-scope and commercial aircraft were detected with a *moving target indication* (*MTI*) canceller at 10–20 km, although not consistently. And in what is clearly a first in the annals of bistatic radar history, the authors concluded [12] that "The experimental results proved negative for the most part, and it is necessary to account for this." They cited difficulties in capturing adequately long data records, limited dynamic range of the MTI cancellation system, and, probably most importantly, the unsuitability of using a TV waveform for pulsed ranging operation. They then concluded that such a radar is best suited for doppler rather than range measurements. This conclusion pointed the way for all subsequent PBR developments: doppler exploitation of stable, narrowband, carrier lines in TV transmissions and range/doppler exploitation of the wider band, noise-like spectrum of FM transmissions.

These results dampened much of the interest in using terrestrial broadcast transmitters over the next ten years. Occasionally the concept would be proposed in research and development (R&D) communities and submitted to government R&D agencies. But without the prospect of solutions to the U.K.-identified PBR problems, the proposals would usually be rejected.

But not totally. In the United States two remarkable phenomena evolved during the Cold War to shelter truly radical, controversial, and sometimes unpopular ideas: *special access* or *black* programs and congressional *earmarks*. The former is often used within the government to hide a significant breakthrough in weapon technology until it has been developed, tested, and prepared for deployment. Shielding it in such a program had two benefits: (1) denying potential enemies knowledge of the program and its progress, thereby preventing exploitation and countermeasures, and (2) denying other government agencies knowledge of the program and its progress, thereby preventing scrutiny and cancellation. Stealth aircraft are an example of the first benefit. Examples of the second benefit include all the failures, which by definition remain unknown.

Congressional earmarks can be invoked as a measure of last resort after overtures to all government agencies have been rejected. Then the inventor lobbies his congressman and staff with the idea, arguing that hide-bound government engineers are simply incapable of recognizing a new and radical idea. Their prime example is the invention (and reinvention) of radar, which as reported in Chapter 2 was first rejected by the German and British navies in 1904 and then by the U.S. navy for roughly ten years starting in 1922. Once convinced, the congressman then earmarks the annual appropriations bill directing a particular government agency to spend x dollars on y idea and report back in z months. Use of the ultra-wideband radar as a counter-stealth measure received such treatment [21, 22].[8]

The IBM, now Lockheed Martin, *Silent Sentry*™ PBR [13–15] received both treatments. After being rejected by main line, U.S. government R&D organizations, it found a small government R&D laboratory sponsor with a modest exploratory development budget, who immediately sheltered the program. In this way *Silent Sentry*™ survived and evolved into a modest multistatic configuration, principally constrained to non–real-time operation.

At some point in this development, Lockheed Martin committed significant corporate funds to the program, speeding its development progress notably into real-time operation for some modes. However, further development required funding levels beyond the charter of the initial government sponsor. But again, no other sponsor stepped up to the next, more expensive phase: advanced development. Then in the 1990s, a congressional earmark was obtained, mandating both independent analyses and field tests by the government.

Over a span of about five years the government oversaw both analyses and tests of *Silent Sentry*™ and related PBRs. In fact, during that period six of this book's authors and contributors served on various study panels and test review boards. And while nearly all results were unclassified, they are not available for public release due to government-only distribution and Lockheed Martin proprietary caveats. Those caveats persist to this writing, which unfortunately preclude a meaningful description of *Silent Sentry*™ and its performance. A few magazine articles and a series of websites, along with data in Table 6–1, characterize available information on the system. Figure 6–2 shows photographs of *Silent Sentry*™, Version 2.

As of this writing, *Silent Sentry*™ continues in development with the addition of a collapsible Adcock antenna, upgraded digital signal processors and data recorders, improved interference suppression algorithms, and an extension to digital TV processing. Evaluation and testing continues, with one unit sold to the U.S. Navy.[9]

In the United Kingdom, interest was revived through a series of increasingly complex bistatic radar trials by P. E. Howland first at the Defence Research Agency (DRA) Malvern, United Kingdom, and then at NATO C3 Agency in the Netherlands, starting in the mid-1990s [16–18, 23]. In the first trial, reported in 1995 [23], doppler echoes were received from an aircraft about 30 km from Malvern, United Kingdom and illuminated a television transmitter located in Rennes, France, about 500 km away. Just the vision carrier from the transmitter was used, as suggested by Griffiths in his 1986 paper [12]. Even with separate doppler, and coarse direction of arrival (DOA) (azimuth) tracking via a Kalman filter, target state estimates were found to be quite noisy.

The second trial, reported in 1999 [16], used the *Crystal Palace* TV transmitter in London and a receiver located in Pershore, United Kingdom, 150 km away. It is summarized in column 4 of

8 The ultimate shelter is, of course, the *black earmark*, where even the earmark disappears from scrutiny. Such earmarks have received significant press exposure, for example, in *The Wall Street Journal*, November 3, 2006.

9 Mr. Dennis Freeman, Lockheed Martin Co., private communication, September 8, 2006.

Figure 6–2 *Silent Sentry*™, showing the fixed-site receiver subsystem on the left and the mobile (Version 2, ca 1999) system with mounted antennas on the right. A later, ca 2002, Version 3 used optimized algorithms and increased processing power to reduce the size of the receive equipment [14].

Table 6–1. Again both vision carrier doppler and coarse DOA were measured. The conversion of doppler and DOA measurements into target state estimates was improved by initializing the process with a genetic algorithm. Howland found that while nearly all high-altitude targets were detected, only about one third could be tracked, with the remaining two thirds either lost in the CFAR and initial Kalman filtering, or having ambiguous or too inaccurate bearing estimates. Furthermore, the signal and data-processing scheme, including the new genetic algorithm, caused a significant time latency of at least one minute in the target state estimation.

The third trial, reported in 2005 [17, 18], switched to an FM broadcast transmitter located in Lopik, the Netherlands, which now allowed range, doppler and coarse DOA of air targets to be measured by a receiver located at the NATO C3 Agency in The Hague, about 50 km away. It is summarized in column 5 of Table 6–1. Now lower altitude air targets could be detected but at a cost of greatly increasing the direct path breakthrough in the receiver. Howland then developed a sidelobe canceller that suppressed this narrow-band, stationary, direct path interference by approximately 75 dB. That suppression coupled with physical shielding was adequate to allow target detections 150 km from the receiver.

Howland also significantly reduced the processing time latency to less than one second by adding an initial decimation technique and "FMCW-like processing" running on a Pentium 4, 2.4 GHz PC. Figure 6–3 shows a photograph of his third-trial system.

In spite of this innovative work, satisfactory conversion to target state estimates was not accomplished. Howland [17] explains:

> At the time of writing, the integration of bearing into the system is very new. We therefore use the simple trigonometrical transformation from bistatic range and bearing into a local Cartesian coordinate scheme for displaying the target on a geographical map… Whilst this serves to quickly show the detections on a map, it effectively discards all the excellent doppler information provided by the system.

Part of the problem appears to lie in establishing initial conditions necessary for integrating very accurate doppler data into the Kalman filter, as detailed in Section 6.6.

Finally, Howland outlined efforts to extend the current algorithms, exploring nonlinear estimation techniques including a particle filter, and going to a multistatic configuration using

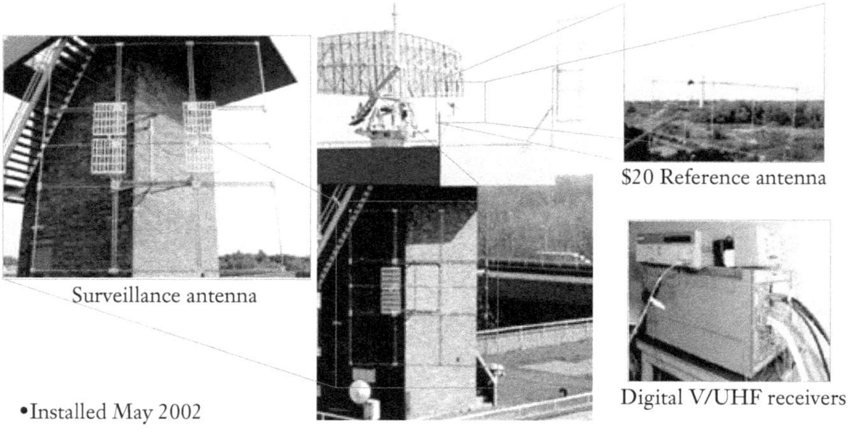

Figure 6–3 *FM Radio-Based Bistatic Radar*, showing surveillance and direct signal antennas and digital receivers [17, 18]. Courtesy of Paul Howland and Darek Maksimiuk, NATO C3 Agency.

multiple FM transmitters, which "...promises to offer far greater target location accuracy and more robust detection [17]." All three of Howland's trials demonstrate the immense difficulty of generating useful air surveillance target state estimates with a single bistatic radar that measures range, doppler, and coarse DOA. As shown in Section 6.6, the problem is virtually intractable.

In the mid-1990s, a PBR, called the *Manastash Ridge Radar* (*MRR*), was independently developed by J.D. Sahr et al. at the University of Washington for studying turbulence in the ionosphere [24]. It has been supported primarily through grants from the National Science Foundation. Up to this time, monostatic radars had been successfully used for these measurements. Motivation for the MRR development included lower cost, increased safety, spectrum availability, and "pedagogical opportunity." Furthermore, the resulting data products were expected to approximate those from the current monostatic radars: fine range, time, and doppler resolution.

The MRR exploits FM broadcast transmitters to measures range, doppler, and azimuth AOA (via interferometry) on auroral E-region irregularities, the primary subject of research. Meteors and aircraft have also been detected. And while the stringent, air defense requirements for accurate location of multiple targets in real time are not necessary for these geophysical applications, MRR does provide range-time-intensity and range-doppler plots to the web every half hour. A photograph of the system is shown in Figure 6–4. The system is described in Chapter 7.

The latest (2003–2005) documented system to appear on the PBR scene is the Science Applications International Corp. (SAIC)–developed and U.S. Army–funded *HDTV-Based Passive Radar* [19]. This multistatic system detects and tracks low-flying aircraft and helicopters as a gap-filler for other air-surveillance systems, principally air traffic control radars. It currently exploits one HDTV transmitter and four receivers, all deployed within short range (5–10 km) of the transmitter for range-only multilateration, and designed with an equivalent monostatic range, R_M = 30 km on a 1 m^2 target. Doppler is measured in a unique process to resolve ghosts.[10]

Because transmitter-receiver separations are small, significant direct path breakthrough occurs. It is reduced with a brute-force, least-squares channel estimator, yielding a 20- to 25-dB noise figure. The cost of this process is to more than double the computational load to 20.5 B floating-point

10 A. Andrews, private communication, January 13, 2006.

Figure 6–4 *Manastash Ridge Radar* receiver, showing the simplicity of the digital receiver and global positioning system (GPS) synchronization hardware [24]. Photo courtesy of Hugh Griffiths.

Figure 6–5 Receiver and signal processor subsystem of the HDTV-Based Passive Radar. The 5-node, dual-Pentium Xeon cluster used to reduce direct path breakthrough is shown at the bottom of each bay [19]. (Photo courtesy SAIC.)

operations per second. Total system latency was found to be a maximum of 10 sec with an update rate of 5 sec. A photograph of the receiver hardware is shown in Figure 6–5. Further details are summarized in column 6 of Table 6–1.

Tests conducted in the Washington, DC, area in 2005 demonstrated that low-altitude commercial air traffic could be detected and tracked beyond the design range. Tracking errors were generally less than 50 m. However, considerable ground traffic was also detected and not always filtered out, especially close-in traffic.

While the system was designed and tested primarily for air traffic control, implications for air defense are clear. Furthermore, when VHF HDTV transmissions become available, their inherent counter-stealth capabilities can also be realized, as detailed in Section 6.3. As of this writing, the system continues in testing, data analysis, and evaluation.[11]

11 J. O. Gobien, private communication, August 29, 2006.

Attending the development of any new technology is the usual supporting cast of enhancement research. The PBR supporting cast includes exploiting the received target echo for image formation and target identification. Cetin and Lanterman [25] have shown that the conventional image formation approach of direct Fourier transformation yields poor images due to sparse and irregular data samples. They assert that a region-enhanced, inverse synthetic aperture radar (ISAR) image reconstruction technique can improve this image quality. They illustrate their assertion with a method of moments simulation of the technique, which uses prior information and constraints about the observation process and the scene.

A second supporting study developed a new approach to ghost excision and target state estimation [26]. Tobias and Lanterman attacked the multistatic, multiple target tracking problem by extending the *probability hypothesis density* (*PHD*) approach developed by Mahler [27] to PBRs. PHD is the factorial moment density found in point process theory, which provides a straightforward method of estimating the number of targets in a region under observation, and thus is a useful tool for tracking multiple targets and identifying multistatic ghosts. Simulation of the PHD using particle filters showed mixed but encouraging results. Details and plans for further development are given in Section 6.6.

As part of the U.S. Air Force Institute of Technology master's thesis program ca 2001, students could select PBR topics. One topic, minimizing the effects of terrain shadowing and multipath nulls on receiver coverage, was analyzed by Hoyuela et al. [28]. Hoyuela and colleagues developed computer-based methods to assess coverage capabilities of real world, receiver sites, using the *Advanced Refractive Effects Prediction System* (*AREPS*) and the *digital terrain elevation database* (*DTED*) models. Since the principal author was with the Chilean Air Force at the time, he chose a model of the Andes and a flight altitude of 100 m as an example of particularly stressing terrain masking. Useful receiver sites were indeed located for this scenario.

The last illuminator of opportunity currently to appear on the PBR scene is the mobile telephone (cell phone) transmitter. These terrestrial towers are found in increasing numbers throughout the world. They operate primarily in L-band, originally with analog waveforms and now with digital waveforms usually in the *Global System for Mobile* (*GSM*) communications standard TDMA format. This format is much preferred for radar operation due to its noise-like spectrum.

Two concepts have been reported, one called *CELLDAR*™ for air defense and other purposes [29] and the second for tracking ground-moving targets [30]. Considerable press is available on the potential of *CELLDAR*™ for air defense and related operations including counter-stealth, but nothing has been published in technical journals so that the concept remains obscure and difficult to evaluate.[12]

However, details of the second concept were published by Tan et al. in the special passive radar edition of the June 2005 *IEE Proceedings*. Reference [30] concisely and objectively assesses, via testing, the feasibility of exploiting cell phone transmitters for PBR operation. It starts with an analysis of a typical GSM 900 (MHz) waveform, and concludes that the 81.3-kHz bandwidth of the GSM signal "...is insufficient to provide an acceptable range resolution."

Specifically, the smallest bistatic range resolution, equivalent to the monostatic range resolution of $c\tau/2$, is 1.845 km, which is greater than the expected target detection range. Thus, all detected targets will fall into the first—and only—range cell. Consequently, Tan et al. revert to doppler measurements, using 0.2-sec coherent integration times, yielding 5 Hz, and

12 As detailed in Section 6.3, counter-stealth operation against most air vehicles of interest is constrained to VHF and low UHF, where body and body–element resonances occur. Furthermore, an early claim for locating a stealth vehicle based on detecting its bistatic specular reflections does not appear useful, also detailed in Section 6.3. Thus these *CELLDAR*™ counter-stealth claims must be considered of a most tenuous nature.

approximately 1 m/sec velocity resolution, depending on the geometry "...very suitable for moving target indication applications."

Tan et al. then report experiments detecting moving vehicles and people on semirural roads in Singapore. These targets were illuminated by a 938-MHz, 25-W, GSM transmitter located 180 m from the receiver. Tracking data were presented on doppler vs time or *waterfall* charts. Results mark only the second discouraging report in recorded bistatic history: "In the above experiments, the maximum operation range of the GSM-based passive radar prototype is not al all impressive, i.e., several hundred meters to 1 km for ground-moving vehicles [trucks]." [30]

They also reported less than 100 m for people. Thus the best case (over-the-shoulder) equivalent monostatic range for moving trucks, $R_M = (1.18 \times 1.0)^{1/2} = 1.1$ km. They cited the low-powered GSM transmitter as a primary cause of the short detection ranges and suggested that in larger countries transmitter powers can be as high as 640 W. Use of such a transmitter will roughly double the detection range, again depending on geometry. While this concept might have application for very short range velocity measurements of moving ground objects, it does not appear to have much application for air surveillance.

Subsequent tests conducted by Sun et al. (unpublished data) used a higher gain, multihorn, monopulse antenna and multichannel receiver, now with commercial airliners as an approximately 40 m^2 target. With a 16-W GSM transmitter approximately 1 km behind the receiver, results showed increased detection ranges: 4 km, extrapolated to 6 km, for an estimated $R_M = 6.5$ km. This result is slightly more than would be anticipated on the basis of increased receiver gain and RCS but lower transmitter power, but still within experimental limits. Target-bearing measurement accuracies were in the 2°–4° range. No target state estimates were attempted.

Separate calculations by Griffiths and Baker [31] were made for an 1800 MHz, 400W GSM transmitter, located 200 m from a receiver with a 25-dB measured noise figure. Results confirm the initial results of Tan et al.: maximum receiver-to-target range of 1.2 km for a 1 m^2 target, yielding a best case equivalent monostatic range, $R_M = (1.4 \times 1.2)^{1/2} = 1.3$ km.[13] At these limited ranges the authors suggested the following applications, given a sufficiently dense network of base stations: "...counting of vehicles for traffic flow management and remote monitoring of movement around buildings as a security device, possibly acting as a cue for a camera system." [31]

In contrast, a second, upper-limit estimate was made for a *CELLDAR*™-like system working with a 100-W transmitter, which yielded $R_M \sim 12$ km, more in line with predictions from Sun et al. (unpublished data). Furthermore, if 28 such transmitters were spaced evenly at 10-km intervals on a grid, the receiver's coverage could be extended to 30–35 km. This estimate used the following, quite optimistic assumptions: 12 dBi electronically scanned antenna pointed directly at the target; adaptive cancellation of all interfering transmitters to achieve 10-dB noise figure plus other losses; perfect reference signal; 10-dB postprocessing SNR. Even so, the estimate concluded with "... claims for the multistatic benefits of the system must be made with great caution. The engineering challenges are immense and the promised performance benefits not great."[14]

Finally, a major processing issue remains to be addressed for these doppler-only systems: integrating doppler measurements into target state (position and velocity) estimates. As detailed in Section 6.6, this integration requires initializing the doppler track with independently derived target coordinates (the pesky constant of integration requirement). One initialization method is to use range-only multilateration measurements from multiple transmitters or receivers. But these measurements are usually orders of magnitude less accurate than that needed for the doppler integration, and are not useful. A second method is to initiate track beginning when the target is

13 And, 3.3 km for a 40 m^2 target.

14 P. Howland, private communication, ca 2003.

Table 6–2 Summary of Passive Bistatic Radars Exploiting Other Illuminators of Opportunity, None With Much Promise for Military Operations, Particularly Air Surveillance

System Concept	Stellar radar	Sferics Correlation Detection System (SCORDS)	Communication Satellite Exploitation	Navigation Satellite Exploitation	Land Mobile Communication Exploitation
Developer	United States [32]	United States (Naval Research Laboratory) [33, 34]	United States (Aerospace Corp.) United Kingdom Russia Australia [5, 8–10]	Germany (DIEHL) United Kingdom China [6, 7]	United Kingdom (Roke Manor) Singapore France (Sun et al., unpublished data) [29, 30]
Decade	1962	1960	1970–2000	1990–2000	2000
Transmitter	Sun and radio stars	VLF lightning	Low earth-orbiting Geosynchronous	GPS GLONASS	Digital (and analog) cell phone towers
Target	Space objects	Ballistic missile flight through ionosphere	Aircraft Ships	Aircraft Tactical missiles	Aircraft (+ stealth) Moving ground vehicles
Results	Forward-scatter detection only	Transatlantic ballistic missile location	Very short range (<10 km) operation	Forward-scatter detection only	Very short range (<10 km) operation
Assessment	Very marginal utility	Thunderstorms required	Insufficient link margin	Marginal utility	Insufficient link margin when using practical estimates

stationary and that position is known a priori, for example, at a stop sign or a missile launch.[15] Other than this second, and quite limited method, the integration process is usually not successful.

Table 6–2 summarizes PBR concepts that have been tested over the last half century and found not well suited for military operations. To the authors' knowledge, the first two concepts have not been revived. The three recent concepts (last three columns) were discussed in the foregoing paragraphs.

6.3 MILITARY UTILITY

When PBRs are considered for air surveillance, specifically in support of air defense with occasional excursions to air traffic control and range instrumentation, they not only offer operational and technical advantages but also come with operational and technical limitations. This section details these advantages and limitations.

15 As reported in Section 6.6, this missile launch technique was successfully used in early, doppler-only instrumentation systems. But, if track was subsequently broken during missile flight, reestablishing target state estimates was nearly impossible.

6.3.1 Military Advantages

The military advantages of PBRs were succinctly summarized by Howland in the conclusion to his 1999 report on the television-based bistatic radar [35] as follows:

> Although the demonstrated accuracy [of single transmitter and receiver] is less than that expected from a dedicated monostatic radar system, the approach is believed to offer sufficient accuracy to cue other sensors, or to provide a low-cost means of surveillance in regions where deployment of a dedicated radar system may not be possible. In a military scenario, the system also offers a covert means of surveillance, together with operation in a frequency band which offers performance benefits against low observable [i.e., *stealth*] targets. Exploiting existing broadcast transmissions circumvents the need to try to obtain military frequency allocations in bands already allocated for civilian purposes.

These fundamental advantages of covert operation, counter-stealth operation, frequency utility, and lower costs (and prime power) have spurred interest in this technology for the last 20 years in the United Kingdom and the United States and more recently in many other countries. These advantages are detailed, expanded, and evaluated below.

6.3.1.1 Covert Operation A PBR is covert in the RF domain, since its exploited transmitter is "up and running" for an entirely different purpose and thus provides no clue to bistatic radar operation. And, as stated at the outset, even the transmitter has no knowledge that it is being exploited. This assertion requires that the PBR receiver is indeed silent. Specifically, the receiver's local oscillator must be properly shielded and the site must have adequately secured its communications links.[16]

In this case, an adversary's *ESM* have no useful signals to work with and are rendered ineffective. *Antiradiation missiles* (*ARM*) also fall into this category, although this observation is of little importance since ARMs are seldom designed to operate at the low frequencies most PBRs exploit. (Receive aperture requirements to achieve the necessary homing accuracy cannot be supported by small airframes.) In short, all VHF/UHF radars are usually immune to ARM attack.

ECM would not be used against a PBR until an adversary (a) identifies the PBRs receive site through visual reconnaissance, infiltration, or compromise, or (b) deduces that a covert air defense net must be operating in the battle area on the basis of unexplained aircraft losses and a priori knowledge that PBRs have been deployed. In either case, broadband, broadcast noise jamming would be the most likely ECM response [36]. The effectiveness of such ECM is discussed in Section 6.7.

Finally, as Howland [35] implies, an RF-covert PBR can "...provide a low-cost means of surveillance in regions where deployment of a dedicated radar system may not be possible." Such surveillance could include (1) monitoring air traffic from military airfields deep in hostile territory, (2) following military exercises in hostile or neutral areas, and (3) tracking missile launches and stealth vehicle flights on foreign test ranges during periods when satellite coverage is absent. However, this type of intelligence gathering lies outside the purview of the book, and will not be considered further.

6.3.1.2 Counter-Stealth Operation Three bistatic radar cross-sectional (RCS) phenomena have been identified to counter-stealth aircraft: *resonance scatter* [37–39], *forward scatter* [40–44], and *specular reflection* [39, 40, 45]. Resonance scatter and specular reflection also apply to monostatic

---

16 Local oscillator (LO) shielding is common in *fuzz-buster* intercept receivers used by U.S. auto-enthusiasts to counter radar speed traps, particularly in states where fuzz-busters are illegal. In a continuing measure–countermeasure game, the police have developed their own intercept receivers to detect the fuzz-buster's LO, which, when detected, allows the police to stop the miscreant, confiscate the fuzz-buster, *and* issue a ticket, along with a second ticket for any speeding infraction. Fuzz-buster manufacturers were delighted to provide the shielding because it made their current equipment obsolete, thereby doubling sales.

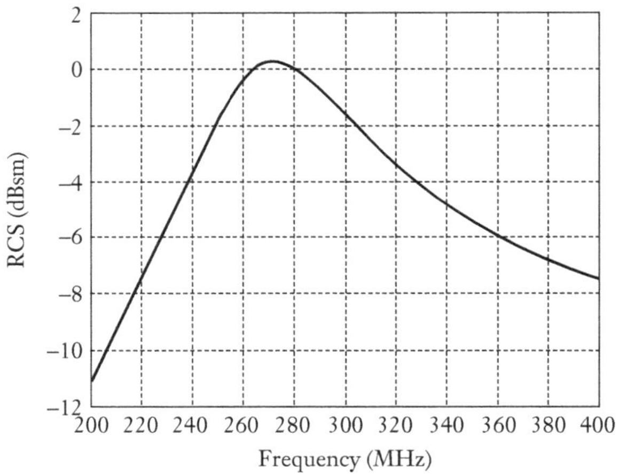

Figure 6–6 Monostatic radar cross section of a simple, wire-grid aircraft model using method-of-moments computed patterns, TE-polarized incident plane wave [38]. Jenn reports that the peak in the curve around 270 MHz is due to a resonance condition for the fuselage (D. Jenn, private communication, November 4, 2005). Resonances occur at several frequencies because of various aircraft parts and are most pronounced for slender metallic shapes. The resonance effect for this model fades starting at about 400 MHz. From Jenn [38]; with permission of the AIAA Inc.

radar; forward scatter applies only to bistatic radar. This section summarizes and reviews these sources within the context of PBR operation.

6.3.1.2.1 Resonance Scatter The resonance scatter effect for monostatic radars has been well documented for conventional (i.e., nonstealth) targets [37–39, 46, 47]. For example, in the simplest case of a conducting sphere of radius a, resonance occurs in the region $0.5 < 2\pi a/\lambda < 10$, where λ is the wavelength [46]. Barton [46] then observes that "Physically, the resonant region can be explained by the interference between the incident wave and the creeping wave, which circles the sphere and either adds to or subtracts from the total field at the leading surface."

Jenn [38] expands the resonance effect for smooth targets of extent or length d, which generate standing waves when $d = \lambda/2, \lambda, 3\lambda/2, 5\lambda/2,\ldots$ Standing waves are a resonant current condition in which the current distribution matches the incident field and at the same time boundary conditions are satisfied (zero current at the ends of the body). Consequently, for this special case the body becomes a resonant structure.[17] The *chaff dipole* is a classic example, which yields a maximum RCS of 0.88 λ^2 when viewed broadside [46]. For example, at 1.5 m (200 MHz), a single chaff dipole (sometimes called rope at these low frequencies) will exhibit a maximum RCS of 2 m^2.

The net result of these additive effects is that when wavelengths are of the order of discrete aircraft dimensions, for example, fuselage, wing, tail, inlet, and exhaust ducts, the resulting resonance significantly enhances RCS when compared to the optical region, which for the sphere starts at $2\pi a/\lambda > 10$. Figure 6–6 shows an example of RCS enhancement in the resonance region.

Appendix 6.3–A summarizes laboratory and field RCS data for monostatic radars at VHF/UHF and bistatic radars at VHF. This data illustrates the monostatic and bistatic resonance effect at VHF for conventional aircraft. Monostatic resonance extends into the UHF region, in some cases up to 600 MHz, although the latter remains controversial. A 450-MHz bound is more

17 D. Jenn, private communication, November 4, 2005.

supportable. While there is no comparable UHF bistatic data, extrapolation of these UHF results to the bistatic case should be relatively uncontroversial at moderate bistatic angles ($\beta < 90°$) because of the broad resonance scattering lobes.

Technical information about stealth aircraft, specifically aircraft with low RCS, is beginning to emerge in the open literature. Significant sources include Kuschel's *VHF/UHF radar, Part 1: characteristics* [39], Lynch's *Introduction to RF Stealth* [37], and again, Jenn's *Radar and Laser Cross Section Engineering*, 2nd ed. [38].

In his comprehensive 2002 paper [39], Kuschel summarizes the state of RCS modeling and measurements of stealth targets in the VHF/UHF region. He first confirms that dielectric *radar absorbing material* (*RAM*) optimized for X-band (approximately 10-dB absorption) shows negligible effect at VHF/UHF (<0.5 dB). Furthermore, this type of RAM optimized for VHF/UHF would be "...generally too thick to apply to air vehicles since it would change their aerodynamic properties." He observes that magnetic absorbers can be made thinner, but are heavier and thus significantly reduce the target's payload. Finally, structural absorbers, such as composite materials, can be made both thin and absorbing at VHF/UHF, but may not be strong or stable enough to replace metal fins, for example. Hence, RAM is currently not considered effective at VHF/UHF.

Kuschel then considers *vehicle shaping* "...so as to deflect the electromagnetic energy impinging on it into directions other than the direction of [monostatic] illumination."[18] He reports Deutsche Aerospace, Bremen anechoic chamber measurements of a faceted, metallized 1:10 scale model of an F-117 aircraft: "The aircraft geometry was obtained from open literature and hence the target model does not take into account fine structure details and surface materials such as RAM, which are of less importance at VHF/UHF..."

Kuschel then shows results in his Figure 4, which is reproduced as Figure 6–7. Kuschel continues:

> The scaled measurement results... show that the attempt to reduce the target's RCS has been successful in the ± 70° section around the nose-on aspect and for the frequency range above 400 MHz. High RCS values covering the whole frequency range occur when the direction of illumination is perpendicular to the front or back edges of the wings or other dominant structures of the fuselage. In the nose-on section, however, an increase in the RCS can be seen at VHF around 100 MHz [6 to10 dBsm] and UHF around 400 MHz [0 to 6 dBsm, > 6 dBsm nose on] due to resonance effects. Hence, such stealth techniques can be efficient at high radar frequencies but are ineffective at VHF/UHF [39].

While Kuschel's absolute RCS values might be questioned because of limited modeling fidelity, relative effects in terms of RCS changes as a function of frequency appear more robust, which would indeed confirm the resonance effect for stealth targets. A similar plot would be expected for the bistatic case extending into the UHF region. Specifically, using Figure 6–A.1 of the Appendix as the model, *bistatic specular flashes* should occur at the same aspect angles when referenced to the bistatic bisector, and relative RCS values should be comparable again for moderate (<90°) bistatic angles.

Such results, in full fidelity, must certainly be known to—and rationalized by—stealth designers. Lynch [37] provides a clue by comparing the detection performance of four monostatic air defense radars operating against a "stealth platform." He defines typical parameters of a 200-MHz long-range early warning, a 900-MHz surveillance and target acquisition, a 7.5-GHz fire control, and a 20-GHz fire control radar, and then plots their performance as detection range versus RCS. His results show that the 200-MHz detection performance is much greater than that of the others at comparable RCS. For example, detection ranges of 130, 50, 30, and 10 km, respectively, are achieved against a target of 0.01 m^2.

---

18 This off-normal specular reflection, while countering a monostatic radar, might be exploited by a bistatic radar and is considered subsequently.

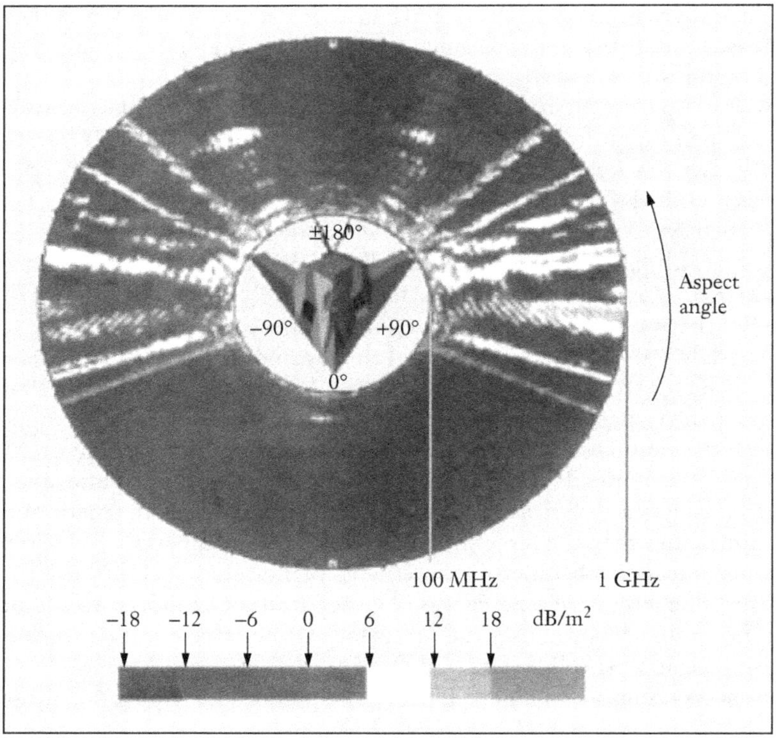

Figure 6–7 K-plane monostatic RCS of F-117-like target. Data is presented in the horizontal plane. RCS values are indicated by scale below. Frequency ranges from 100 MHz on the inner circle to 1 GHz on the outer circle. Reproduced from figure 4 of Kuschel [39]; with permission of the IET. (To see the figure in color, visit http://www.scitechpub.com/willisgriffiths.htm)

These results are due to the much greater link-margin (excepting RCS) assumed for the 200-MHz radar in the radar range equation: 17, 25, and 45 dB, respectively. So Lynch's 200-MHz radar started off with a large advantage in detection range, ignoring both RCS resonance enhancement and stealth treatment. This advantage reflects the well-established principle that long-range surveillance is best accomplished at low frequencies where high power and large antennas are easier to achieve and clutter and weather are less severe [47]. (However, multipath becomes more severe.) Then adding the RCS disparity—RCS reduction at high frequencies and RCS enhancement at low frequencies—significantly increases the VHF/UHF link advantage. Lynch then makes his point: "The good news for stealth platforms is that, even though detectability increases with wavelength, [low frequency, early warning] systems usually have such poor position accuracy that exploitation of the detection in a dense target environment is limited [37]."

In short, stealth platforms are typically designed to counter the lethal, fire control element of the air defense threat. They worry much less about the early warning element for two reasons: (1) these radars, by themselves, are not a lethal threat to stealth aircraft; and (2) even if they were, there is no easy method of reducing the resonance effect of air targets generated at these low frequencies, as outlined next.

Jenn [38] describes two RCS reduction methods for countering these low-frequency radars: passive and active cancellation.

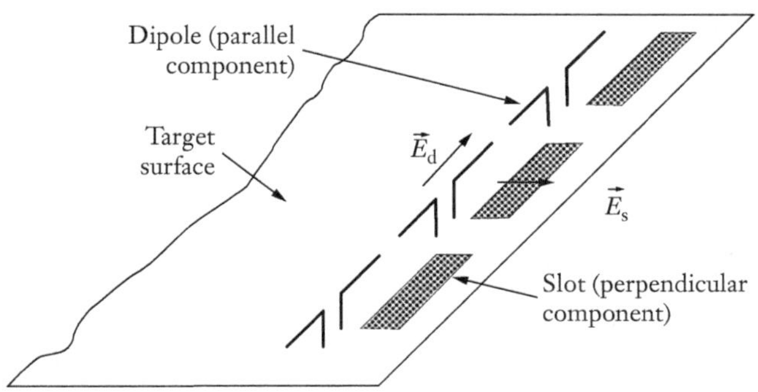

Figure 6–8 Monostatic radar cross-section cancellation using a combination of slots and dipoles. From Jenn [38]; reprinted with permission of AIAA Inc.

> *Passive cancellation* refers to RCS reduction by introducing a secondary scatterer [slot, dipole, parasitic wire, lumped load] to cancel with the reflection of the primary target...
>
> As with any cancellation or tuning method, this technique is effective over only a narrow frequency band and is usually limited to a small spatial sector. If large parasitic elements are to be avoided, the magnitude of RCS that can be canceled is relatively small. Thus, passive cancellation is used to supplement shaping and absorbers. *The exception to this is the treatment of traveling waves, in which case passive cancellation by parasitic structures is often the primary means of RCS reduction*...[19]
>
> A simple application for a solid surface is shown in...[Figure 6–8]. Two orthogonal elements are required to cancel and incident wave of arbitrary polarization. These elements could be slots, dipoles, or a combination of both. The terminals are shorted, and the transmission lines are adjusted to provide the proper phase for the field reflected at the short. The number of elements is chosen to provide sufficient gain to cancel the RCS in the direction of interest. Thus, the field reradiated by the network cancels the RCS of the bare target. It is evident that a fair amount of information about the threat and the target itself is required: the threat radar frequency, direction, and polarization and the target RCS in the direction of the threat.
>
> *Active cancellation* is the extension of passive cancellation to handle dynamic threat scenarios... Two levels of sophistication are considered.
>
> 1. *Fully active:* The cancellation network receives, amplifies, and retransmits the threat signal such that it is out of phase with the static RCS of the target...
> 2. *Semiactive:* No boost in threat signal energy is provided by the cancellation network, but adjustable devices in the network allow the reradiated signal to compensate for limited changes in the threat signal parameters...
>
> The demands for a fully active system are almost always so severe as to make it impractical. It requires a transmitter and antennas that cover the anticipated threat angles, frequencies, incident power densities, and polarizations. A knowledge of the threat direction is required as well as the target's own RCS. A semi-active system is not as complicated in terms of hardware, but the use of adjustable devices still requires bias lines, controller units, and a computer with the appropriate data bases.

In essence, these cancellation schemes *retrodirect* the incident radar signal with a predetermined amplitude and phase back to the radar, in an attempt to cancel the target's echo received by that radar. The cancellation signal is, perforce, retrodirected toward the radar's transmitter, which in the monostatic case is collocated with the receiver. This cancellation operation is similar to that of a retrodirective jammer, which directly or indirectly measures the direction of arrival (DOA)

19 Emphasis added. Specifically, traveling waves can be the primary scattering mechanism in the resonance region, depending on target shape, composition, aspect angle, etc.

of a radar signal and then uses a directional antenna to transmit maximum or minimum power at that DOA. (For example, maximum power for a noise jammer; minimum power for a *cross-eye* repeater.)

However, in the bistatic radar case when the receiver is sufficiently separated from the transmitter, it will fall outside the retrodirected antenna's main beam, thus reducing the effect of either countermeasure. A useful rule of thumb is to anticipate enhanced bistatic radar performance whenever the bistatic angle is greater than the estimated 3-dB width of the retrodirected beam [40].

Implications for a PBR are clear. First, many of its transmitters of opportunity operate at frequencies where enhanced RCS occurs: 100–450 MHz. Second, with proper receiver siting the effectiveness of cancellation schemes is significantly reduced, if not negated. Hence, the PBR becomes an attractive candidate for the early warning element in a counter-stealth air defense (or surveillance) system whenever suitable transmitters of opportunity are operating in the battle area. Third, this capability is enhanced by the covert nature of a PBR: aircraft of any type have no initial knowledge that such an early warning element is operating. In short, a stealth platform could be detected by a PBR, without the stealth platform knowing it. This feature has been called a *stealth trap* or *stealth ambush* [36].

6.3.1.2.2 Forward Scatter A second counter-stealth capability is available only to bistatic radars: forward scatter. This phenomenon occurs when the bistatic angle, β, approaches 180°. It was first identified by Siegel et al. in 1955 [41], subsequently reported by Siegel in 1958 [42], and then expanded upon in many papers and books [40, 43, 44, 46, 48–50]. In essence, when $\beta = 180°$, Siegel showed by using physical optics that the forward-scatter RCS, σ_F, of a target with *shadow* (or *silhouette*) *area A* is $4\pi A^2/\lambda^2$, when the wavelength λ is small compared to the target dimensions. The targets can be either smooth or complex structures, and from the application of Babinet's principle, can be totally absorbing [51]. Hence conventional and stealth targets with the same silhouette will generate the same forward-scatter RCS: σ_F.

While the forward-scatter RCS, σ_F, can completely negate the low RCS treatment of stealth vehicles, for example, at 100 MHz yielding a 21 dBm^2 RCS on a stealth vehicle with a shadow area of 10 m^2, it comes with severe, well-known limitations: (1) since bistatic range resolution is inversely proportional to cos ($\beta/2$), range measurement becomes indeterminate as $\beta \geq 180°$, and (2) since bistatic doppler is directly proportional to cos ($\beta/2$), doppler measurement approaches 0 as $\beta \geq 180°$. Hence, the system is reduced to determining the fact that the target lies somewhere on the baseline, between transmitter and receiver in a simple fence-type operation. As outlined in Section 2, these fences (and grids of fences) were deployed by many countries in World War II, including Japan, USSR, and France for early warning and alerting of air defenses.

When forward scatter occurs at these low frequencies, the forward-scatter lobe width becomes broad, which can allow system operation with enhanced RCS off the baseline. For example, at $\beta = 180°$ the angular width of the scattered signal, θ_F, in the horizontal or vertical plane approximates λ/d, where d is the target linear dimension in the appropriate plane. Figure 6–9 plots the dependence of σ_F and θ_F on frequency, for a simple target with projected area $A = 10$ m^2 and projected linear dimension $d = 10$ m, both perpendicular to the baseline.

At 100 MHz, a typical FM broadcast frequency, θ_F is approximately 17°, as shown in Figure 6–9. If the forward-scatter system is designed with sufficient sensitivity to detect this forward-scatter main lobe and its next three side lobes, it can operate within a region roughly 60° on each side of the main lobe axis. That region is bounded by $\beta = 120°$, which is an ogive (opposing sectors of two circles both intersecting the transmit and receive sites) with centers located on the perpendicular bisector of the baseline and radius $r_\beta = L/(2 \sin \beta)$. Since the distance between the center of a circle and the baseline, d_β, is $L/(2 \tan \beta)$ [40], the maximum separation of the ogive from the baseline is $r_\beta - d_\beta = (L/2)$ ctn ($\beta/2$) for $\beta > 90°$.

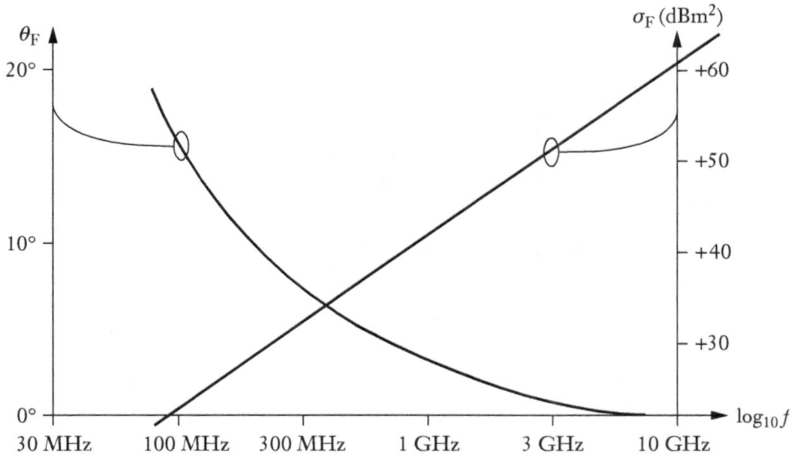

Figure 6–9 Forward-scatter RCS σ_F and angular width θ_F of a target.

For example, if the baseline $L = 50$ km, the separation is 14 km, yielding a maximum distance on both sides of the baseline of 28 km. At this point the typical 3-km range resolution cell for FM broadcasts is increased by a factor of 1/cos (120°/2) or 2, to about 6 km, which may allow coarse target location estimates for limited cueing of air defenses. However, the coverage of such an ogive is neither large enough nor oriented properly to provide useful area defense surveillance. Thus, a PBR configured for forward-scatter, counter-stealth operation is inherently limited to fence-type alerting and coarse cueing.

6.3.1.2.3 Specular Reflection A third counter-stealth capability has sometimes been ascribed to bistatic radars in general: the ability to detect specular reflections that have purposely been directed away from anticipated monostatic radar threat locations, for example, by using tilted surfaces or facets on stealth platforms. Figure 6–10 is an example of this specular reflection geometry that appeared in the press just after the *Advanced Technology Bomber* (*B*-2) was announced [45]. The assertion is that through judicious location of multiple bistatic receivers, these off-normal speculars of large amplitude can be detected and tracked or networked together to support some level of engagement.

Unfortunately the bistatic radar configuration shown in Figure 6–10 is flawed. It shows a narrow-beam antenna for both the transmitter and the receiver, which generates a severe beam *scan-on-scan* problem for bistatic surveillance. To further compound the problem, only when the target's aspect angle is such that its incidence and reflection angles are aligned with the illuminator-to-target and target-to-receiver look angles, respectively, will the specular flash be oriented toward the receiver. These compounded effects are called *scan-on-scan-on-flash*. While beam scan-on-scan remedies are available, they are costly, are complicated, or compromise performance [40]. The scan-on-scan-on-flash problem has not been solved for this bistatic configuration.

When a PBR uses broad, fixed beams for both transmitter and receiver, it avoids the scan-on-scan problem. Hence, it depends only on the specular flash to occur. The occurrence of this flash can be visualized for PBRs using ground-based transmitters and receivers by considering the underside of typical stealth aircraft, which fuze the underside of its wing and fuselage into a virtually flat plan-form with no wing dihedral, generating a large specular reflection area.

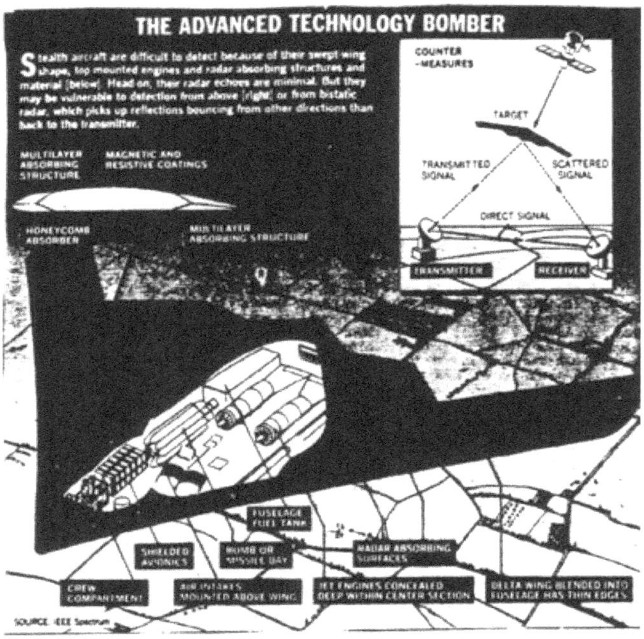

Figure 6–10 Article from the *Washington Post*, May 1, 1988 [45], citing the *IEEE Spectrum* as its source, but with no further attribution. The caption said in part: "Another possible stealth countermeasure is bistatic radar: Unlike conventional radars, the receiver is placed at some distance from the transmitter. Consequently, stealth configurations designed for deflecting energy away from the transmitter source may instead direct it to a bistatic receiver..."

Thus when these aircraft fly a straight and level path, a very high amplitude, narrow lobe, specular flash will be generated for a monostatic radar only when the aircraft flies over the radar, in regions near the radar's zenith. Now most air defense radars and systems are not designed to operate near their zenith. First, there is little operational requirement to do so, since virtually all threats first appear near the radar's horizon. Second are physical constraints. Mechanically driven antennas are not configured to scan through very high elevation angles due to elevation gimbal limits.[20] Faces of electronically scanned array radars are usually oriented vertically to maximize antenna gain for long-range operation near the horizon. As such, scanning toward the zenith approaches end-fire conditions. In addition, the engagement envelope of many air defense missiles does not cover regions near the zenith due to maneuver constraints. Finally, nearly all air defense artillery are proscribed from firing near the vertical to prevent recoil damage to the weapon carriage.

20 Exceptions are the early *AN/SPQ-5* and *AN/SPG-49* fire control radars for the U.S. Navy's *Terrier* and *Talos* shipboard air defense systems. Each used a three-gimbal system—azimuth, elevation, and traverse—which allowed scanning and tracking throughout the hemisphere. The trade-off for this third gimbal was increased complexity, cost and topside weight, and reduced reliability. As depicted in Figure 1.8 of [61], each mount weighed 13 tons, and was driven in azimuth by a 100 HP electric motor. A coauthor cut his radar teeth on the *SPG-49*.

Under these conditions, implications for PBR specular scatter operation now become clear. To generate a bistatic specular flash from its horizontal underside, a level-flying stealth aircraft must pass between a PBR transmitter/receiver pair; that is, it must cross the bistatic baseline. But this geometry is nearly identical to the forward-scatter geometry. The principal difference is that the forward-scattered lobe is directed outward, along the transmitter-to-target LOS, whereas the specular scattered lobe is directed toward the receiver. In fact the specular flash amplitude is similar to that of forward scatter: $(4\pi A_s^2/\lambda^2)$ $(\sin\varphi)$, where A_s is the specular reflection area and φ is the incident grazing angle. Thus the specular flash and the forward-scatter effects will combine in some complex manner to form a signal at the receiver roughly equivalent to the stronger of the two signals, depending on A, A_s, their sidelobe structures, and the geometry, including φ. The resulting engagement scenario closely follows the forward-scatter example (Section 6.3.1.2.2) with identical large bistatic angle measurement constraints and an identical conclusion: a PBR configured for forward-scatter/specular-scatter, counter-stealth operation is inherently limited to fence-type alerting and coarse cueing.

This foregoing discussion assumed a single stealth aircraft present in the scenario. The problem is greatly exacerbated when multiple stealth and/or conventional aircraft are present, since each aircraft will generate its own data set at various times at each receiver. Then when all sets are combined, ambiguities or false targets are generated. This problem is addressed in Section 6.6, and has no apparent solution when the data are intermittent. Thus it would appear optimistic to ascribe more than a fence-type alerting and coarse cueing capability when exploiting specular reflections from stealth aircraft.

6.3.1.3 Frequency Allocation Frequency allocation for a monostatic radar operating in the VHF/UHF region is severely restricted in many countries. For example in the United States, only the 216- to 225-MHz and 420- to 450-MHz bands are designated for *radio location*, that is, radar, by the Federal Communications Commission (FCC). Furthermore, the 216- to 225-MHz band must be shared with fixed, mobile, land, and amateur communications. The 420- to 450-MHz band must be shared with amateur communications. Thus, opportunities exist for *radiofrequency interference (RFI)* between users in these bands, raising inevitable protests and operating restrictions and even litigation. In fact, the first clear U.S. band for radio location is the 3.1- to 3.3-GHz band.

In the United Kingdom, the 420- to 450-MHz and 590- to 598-MHz bands are designated for radar use. The former band is used for airborne early warning and the ballistic missile early warning radar at RAF Fylingdales in North Yorkshire. The latter band is used by airport radars, but must be shared with TV Channel 36. Thus again RFI can become a problem.

Although no reference to VHF radar frequency allocation appears on the U.K. OfCOM website, some VHF radar experiments have been carried out in Europe in recent years. However, it can be very difficult to get approval for such tests, especially for any appreciable bandwidth. Even with approval it may be necessary to operate far from centers of population and only for restricted periods of time. Testing of the Swedish *CARABAS VHF foliage penetration (FOPEN) synthetic aperture radar (SAR)* is typical, with several papers published on techniques used to suppress interference from broadcast and communication signals [39, 53].

PBR completely avoids this allocation problem simply by exploiting transmitters already approved for and operating anywhere in the VHF/UHF bands. Furthermore, as the RF spectrum becomes increasingly congested with more broadcast and communications frequencies, the PBR has a greater selection of transmitters to exploit. The *new high definition TV (HDTV)* transmissions, with their much preferred, noise-like waveform, are a prime example of this trend.

6.3.1.4 Prime Power and Costs Reduced prime power consumption and lower costs are other, often-cited advantages for PBRs. T. A. Weil in the introduction to his transmitter chapter in the second edition of Skolnik's *Radar Handbook* [54] has this to say about radar transmitters:

> The transmitter is usually a large fraction of radar system cost, size, weight, and design effort, and it typically requires a major share of system prime power and maintenance. It generally ends up being a big box that sits in the corner of the radar equipment room, hums to itself, and has a big sign on it that says, "Danger, High Voltage"; so most people prefer to keep away from it. Its insides tend to look peculiar, more like a brewery than a TV set or a computer….

Unfortunately, Weil did not quantify what "large fraction" or "major share" were, probably because they vary depending upon a specific application. At UHF and below where most PBRs currently operate, both radar and broadcast transmitters typically use grid-control tubes (i.e., the old *bottles*) or solid-state devices using silicon bipolar transistors. For high-power applications, tubes are usually more efficient and less costly than transistors; transistors generate more average power and are more reliable. The two transmitter types are occasionally interchangeable. For example, a solid-state transmitter successfully replaced a tube transmitter in the U.S. Navy's *AN/SPS-40* UHF air surveillance radar.[21]

In the coauthors' experience, "a large fraction or radar system cost" ranges from 30% to 50%, with—surprisingly—the higher number representing older radars and the lower number representing modern radars. The reason is that most modern radars are multifunction systems that require large digital signal processors and computers and massive amounts of software to write, maintain, and upgrade. Simply put, transmitters are a mature technology, so their costs have remained relative constant, whereas software—a minor player 30 years ago—is the current cost driver. A similar rule of thumb is to allocate costs for conventional radars (i.e., radars not using active apertures) as 1/3 transmitter, 1/3 antenna, and 1/3 receiver/processing.[22]

In contrast, "major share of system prime power" has stayed relatively constant over the years, since transmitter average power and efficiency have not changed much. A typical example is the *AN/SPS-40's* tube transmitter, which generates 4-kW average RF power. The overall efficiency of a grid-control tube is typically 50% [55], so approximately 8 kW is required to power the transmitter. Prime power input to just the radar (excluding support functions to operate a radar) is 27 kVA, or approximately 22 kW.[23] Thus, "major share of system prime power" for the SPS-40 is 36%, a number typical of many surveillance radars.

Thus, the PBR's power and cost savings are roughly one third of the total. That said, the cost for some radar applications are driven by factors completely outside the radar designer's control. For example, real estate costs can trump all other costs for an air traffic control radar located in an urban area. And, personnel costs can trump all other costs for a surveillance radar located in a remote area such as the Arctic. In any case, these power and cost claims appear to be modest.

6.3.2 Military Limitations

PBRs considered for military operations, most notably for air defense, suffer the following limitations, some of which have only recently been identified and quantified.

6.3.2.1 Autonomous Transmitter Exploiting a transmitter of opportunity by the PBR is a double-edged sword. While covert and counter-stealth operation, expanded frequency usage, and reduced power and costs are principal advantages, loss of control of the transmitter is a serious

---

21 The *AN/SPS-40* is a 2D shipboard air surveillance radar used mainly by U.S. Navy antisubmarine warfare destroyers [47]. It operates at 400–450 MHz with an 11° × 19° beamwidth, and has a design detection range of 200 nmi on a 1 m^2 target (P. Buczynski, private communication, November 22, 2005). As such, it becomes a useful benchmark for comparing performance of a conventional monostatic radars and passive bistatic radars.

22 D. K. Barton, private communication, November 12, 2005.

23 P. Buczynski, private communication, November 22, 2005.

disadvantage. First, the PBR can operate only in regions illuminated by the transmitter. And, multiple transmitters must illuminate a common region for multistatic PBR operation. Second, the transmitter must be up and running, that is, *on*.[24]

Third, the passive receiver is forced to operate with suboptimal transmitted waveforms, waveforms that are not designed for radar operation. For example, the coauthor reported in his 1985 PBR experiment using the *Crystal Palace* TV transmitter in London [56] that

> In most of these respects the television waveform is not ideal. The autocorrelation function for a TV waveform shows broad peaks at 64 μs intervals corresponding to the line sync pulses; if the sync pulses are gated out, the auto-correlation function peaks will be sharper (depending on the picture content), but will still recur at 64 μs intervals. In the case of any quasi-CW illuminator, the range sidelobe level will be high, and the radar best suited for Doppler rather than range measurements...

This observation about only using TV waveforms for narrowband doppler measurements established the precedence since nearly all subsequent PBRs have done so. On the other hand, FM broadcasts with their noise-like modulation—and thumb-tack-like autocorrelation function—have been found more suitable for range measurements of modest accuracy. Ambiguity properties of these types of waveforms are detailed in Section 6.4.

A second, waveform problem is that the modulation on many transmitters—particularly broadcast transmitters—changes as a function of time, thereby complicating range measurements using matched filtering by the PBR's receiver. The receiver typically samples and stores a segment of the direct path waveform and then cross-correlates it with the returned echo for each range measurement, all in real time. Such processes are now feasible to implement, but complicate an otherwise well-developed operation in the stable, more predictable monostatic world.

A related waveform problem is that at times the broadcast transmitter modulation can go to zero, as when broadcasting a talk show or classical music. (Popular music seldom exhibits this effect [57].) When this occurs, range measurements are lost. The frequency of this type of outage is not insignificant and is also detailed in Section 6.4.

6.3.2.2 Radio Frequency Interference (RFI) As detailed in Section 6.3.1.3, the avoidance of RFI in designated radar frequency bands was cast as a PBR advantage. But when working outside those bands, RFI can also plague PBR operation. In this case, RFI comes from two principal sources: the exploited transmitter and adjacent transmitters. This RFI can arrive via a direct path or multipath, and include scattering from terrain or sea surfaces, also called *clutter*. The direct path signal from an exploited transmitter is the dominant RFI source. Multipath signals from that transmitter are less severe but can significantly contribute to RFI. Multipath over the transmitter-to-target and target-to-receiver paths is nearly always present at VHF/UHF, and can cause both signal enhancement and deep fades, which must be accounted for in any air surveillance design, for example, as done by Howland et al. [58, 59].

Monostatic surface clutter is more severe at microwave frequencies than at VHF/UHF, since clutter-scattering coefficients, σ_0, are inherently lower at these long wavelengths, especially at low grazing angles typical of air surveillance operation [55]. For the bistatic case, Chapter 9 shows that out-of-plane, microwave bistatic σ_0 can be tens of decibels lower than in-plane σ_0. (In-plane specifies operation on or near the baseline and extended baseline, and includes the monostatic geometry. Out-of-plane specifies operation off the baseline.) While no VHF/UHF bistatic clutter measurements have been published in the open literature, extrapolation of bistatic σ_0 data from

24 At first reading this observation appears obvious since most broadcast transmitters operate continuously. However, those in a battle area are both observable and subject to direct attack or collateral damage. Both occurred during the initial phases of the war in Iraq, in which case they were *off*.

microwaves to UHF/VHF using the monostatic model should not be too controversial. Thus for typical, out-of-plane PBR operation, bistatic clutter should be minimal, especially at VHF/UHF and indeed has not been reported in PBR field tests.

Direct path *breakthrough* from the exploited transmitter is common to all but the simplest, low-power CW radars: without mitigation the received signal is always masked in range and often masked in doppler by part of the simultaneously arriving transmitted signal. A brute-force mitigation remedy is to physically block the transmit signal from the receiver through shielding or separation. The *AN/MPQ-46* CW tracker-illuminator of the *Hawk* missile system is an example of the former. High-Frequency Radars, Chapter 4, and the Manastash Ridge Radar, Chapter 7, are examples of the latter.

Physically adjacent transmitters can significantly raise the receiver noise level, simply through spectral spill-over from adjacent or nearby bands. In the United States, the FCC mandates a Gaussian spectral roll-off for broadcast transmitters, which is sufficient to prevent interference in one-way home receivers from adjacent transmissions. However, it is not sufficient for two-way radar receivers, which necessarily must work much farther into the noise.

Griffiths and Baker [53] recently demonstrated the severity of these problems. They took in-situ measurements of several VHF and UHF broadcast and communications bands in a dense urban environment. They found ambient noise levels typically 45 dB greater than thermal noise and direct path illuminator signals some 45 dB greater still. Even with robust mitigation, the residue of this unsuppressed RFI can increase the PBR receiver's noise figure by many tens of decibels: a 25-dB noise figure in urban and semirural areas is common. In some geometries, cosmic noise can further increase the receiver's noise figure. Section 6.4 details measurements and candidate mitigation factors for these RFI sources.

6.3.2.3 Limited Receive Aperture Monostatic radars operating at VHF/UHF frequencies have by necessity used large antennas, usually as large as their operational platform would allow. For example, the Russian *NEBO 55G6U VHF* (180–220 MHz) 3D monostatic air surveillance radar is shown in Figure 6–11. As reported by Skolnik [47], the horizontal antenna structure is 27 m wide by 6 m high and obtains the range (100-m accuracy) and azimuth angle (0.2° accuracy) of detected targets. The vertical antenna structure is 30 m high by 6 m wide and is electronically steered to provide the elevation (and height, 400 m accuracy) of targets. Such a radar is quite suitable for a long range, 3-D air surveillance of both conventional and stealth air vehicles. Furthermore, with such a large aperture—hence directivity—it has a significant capability to resolve multiple targets in reasonably dense environments. However, such an antenna constrains the radar to semifixed locations and generates large optical and RF reconnaissance signatures. Covert operation is clearly not a consideration for these radars.

In contrast, when a PBR is designed for covert, mobile, and/or low-cost operation, it is precluded from using such an antenna. Smaller antennas, such as *Yagis* or compact *circular arrays*, are typically employed in these systems. While these types of antennas significantly reduce the optical signature and increase mobility, they also impose a significant radar performance penalty through lower directivity and gain and higher side lobe levels. Consequently, detection range is reduced, RFI and multipath levels are increased, ECM vulnerability is increased, and target state estimates become more difficult to make, when compared to its monostatic counterpart.

The impact on detection range of these small antennas is assessed in Section 6.5. The problems associated with estimating target state when DOA measurements are degraded or not available are detailed in Section 6.6, along with the elegant, but complicated, bistatic and multistatic schemes developed in an attempt to mitigate these problems. The associated impact when operating in an ECM environment is assessed in Section 6.7.

Figure 6–11 Photo of the Russian *55G6U* long range, VHF air surveillance radar antenna. The radar is transportable and can be deployed and redeployed in 22 hours [47]. Courtesy of Dr. A. A. Zachepitsky, Designer General, Nizhny Novgorod, Russia.

While this limited aperture constraint applies to virtually all PBRs, with early versions of *Silent Sentry*™ employing the largest apertures to date (6- to 12-element VHF arrays), the constraint can be eliminated in the special scenario of defending an urban area. In this case, the PBR designer might exploit existing buildings as the platform for a large receive phased array antenna, where the building itself might be a suitable ground plane at VHF. While mobility and cost are obviously sacrificed, a significant level of covertness could be retained with judicious placement of elements, say within window openings.

This concept was first identified by Dr. Aaron Lanterman ca 2002, and was subsequently dubbed the *Lanterman Tower*. Antenna apertures approaching or even exceeding that of the Russian *NEBO 55G6U* VHF radar shown previously are possible, given large buildings in suitable locations. In fact, with such an aperture the PBR can operate much like the Russian monostatic radar over an approximately 120° azimuth sector, with the phased array replacing the mechanically rotated antenna. It will also provide DOA data for target state estimates—a most useful addition to PBR operation. Alternatively, smaller but still capable PBR receive antennas might be camouflaged in artificial trees as cell phone towers sometimes are to minimize urban blight.[25]

While this hypothetical concept is attractive in some static scenarios, it is not appropriate for mobile military operations. However, it should be of concern to strike-planners, since it would represent a formidable stealth trap as outlined earlier. It will be carried along in subsequent sections as special case, called the fixed-site PBR excursion.

25 Specifically, as John Chastain, an old side-wise thinking SRI colleague of the coauthor, would cast the problem: design the antenna and wrap the building (or a tree) around it.

6.4 WAVEFORMS AND INTERFERENCE

6.4.1 Illuminator Waveform Properties

The performance of PBR systems is significantly influenced by the properties of the illuminating waveforms. By definition, signals and their modulation properties are determined by factors totally independent of those desired in a PBR, and so it is important for the designer of such a system to understand these limitations and to utilize the illuminating signals appropriately. The important properties of an illuminator are (i) power density at the target, (ii) coverage, and (iii) the form of the signal and its ambiguity properties.

The power density Φ (i.e., W/m^2) at the target is readily evaluated from

$$\Phi = \frac{P_T G_T}{4\pi R_T^2} F_T^2 \tag{6.1}$$

where P_T is the transmit power, G_T the transmit antenna gain, R_T the transmitter-to-target range, and F_T the transmitter-to-target path pattern propagation factor. It is important that the transmit power is evaluated only for that part of the signal spectrum that is used; for example, it may be desirable to use only part of the spectrum of an analog television transmission in order to reduce the ambiguities associated with the line repetition rate. Although this improves the ambiguity performance of the signal, it also reduces the usable transmit power. This reduction is defined by the power utilization factor ρ in equation (6.8).

Coverage can be subdivided into spatial coverage and temporal coverage. The spatial coverage will be given by the transmit antenna gain and pattern propagation factor, but will also depend on whether the transmit platform is stationary or moving, and whether it radiates continuously or for only a proportion of the day or year. If the platform is moving, what is the revisit time? If there is more than one such transmitter, how many are there and what are their spacing? Will the geometry permit the special case of forward scatter, with the associated enhancement of target signature but loss of target localization?

Key parameters of the signal are the frequency, bandwidth, and, in some cases, the presence or absence of a CW carrier (which may be important if doppler is to be measured). The ambiguity properties of the signal depend both on the signal itself and on the bistatic geometry. The ambiguity properties of radar signals are expressed using Woodward's ambiguity function [63], which expresses the point target response of the signal $u(t)$ as a function of delay τ and doppler shift ν (or equivalently target range and velocity)

$$|\chi(\tau,\nu)|^2 = \left|\int u(x)\,u^*(x+\tau)\exp(j2\pi\nu x)\,dx\right|^2 \tag{6.2}$$

Broadcast, communications, and radio-navigation transmissions that are candidates for PBR illuminators will, in general, have time-varying modulation whose ambiguity properties will vary significantly, both from one illuminator to another and for an individual illuminator as a function of time. Examples are given below.

6.4.1.1 FM Radio FM radio transmissions lie in the 88- to 108-MHz VHF band in most countries. The modulation is broadband FM, with a channel bandwidth B of typically 50 kHz (corresponding to a monostatic range resolution $c/2B$ = 3000 m). The transmitters are usually sited on tall towers or masts in high locations. The radiation patterns are usually omnidirectional in azimuth, although the elevation-plane patterns are usually shaped to avoid wasting power above

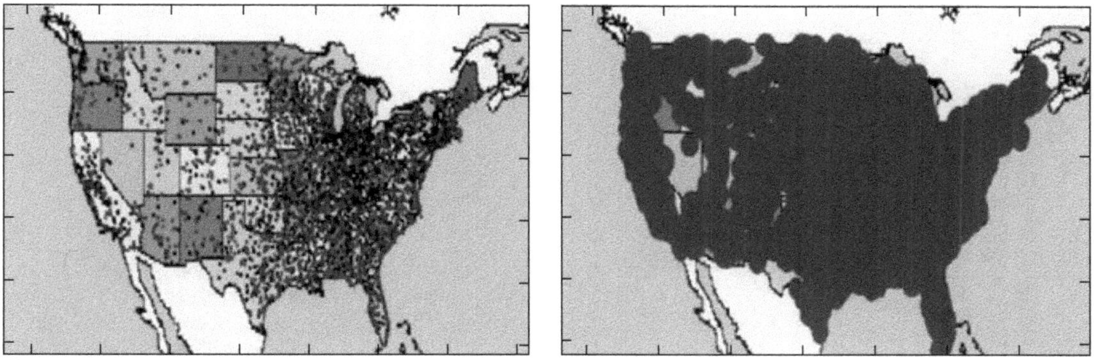

● 4,000 FM transmitters ≥5 kW
● 3,600 FM transmitters ≥1 kW

(*a*) (*b*)

Figure 6–12 Coverage of continental United States by FM broadcast stations: (*a*) locations of stations ≥ 5 kW and ≥ 1 kW; (*b*) proportion of area (shown in dark shading) that is line-of-sight to high-power transmitters for targets at 1000′ altitude (Copyright 2006, Lockheed Martin Corporation. All rights reserved.)

the horizontal. In the United Kingdom [64] and the United States, the highest power transmitters are 250-kW EIRP, which from equation (6.1) yields a power density (under the assumption of free-space propagation) of $\Phi = -57$ dBW/m^2 at a target range of 100 km.

This value is a substantial power density, and may be explained by the fact that broadcast receivers often have poor noise figures and inefficient antennas, and may be sited in poor locations, so many tens of decibels of link margin need to be built into the link budget to assure full coverage. This factor works in favor of the passive radar designer, of course.[26] Since most FM radio transmitters are located near urban and suburban areas, PBR receivers operating in these areas will be within range of at least four or five transmitters at substantial power density, which in turn provides reasonable coverage of aircraft targets in both bistatic and multistatic modes of operation.

To illustrate, Figure 6–12(*a*) shows the location of continental United States by FM transmissions of ≥ 5 kW and ≥ 1 kW transmit power. Figure 6–12(*b*) shows the proportion of area that is line-of-sight to high-power transmitters for targets at 1000′ altitude, demonstrating that existing commercial FM transmitters provide low-to-medium altitude coverage for virtually all areas of interest in the United States. Coverage maps for Europe show essentially similar results.

It is also useful to consider the coverage in littoral regions. Broadcast transmitters will, in general, be sited inland to maximize their coverage of land. If the coastal region is mountainous, there may be blockage so that extended coverage out to sea is not achieved. In such cases topographic maps can be used to evaluate the available coverage.

Over the ocean, atmospheric and precipitation losses can usually be ignored at VHF and UHF frequencies, but interference between the direct path signal and the reflected signal from the sea surface (multipath or the *Lloyd's mirror* effect) can cause deep nulls in the receiver's antenna

---

26 R. Lodwig, private communication, 2001.

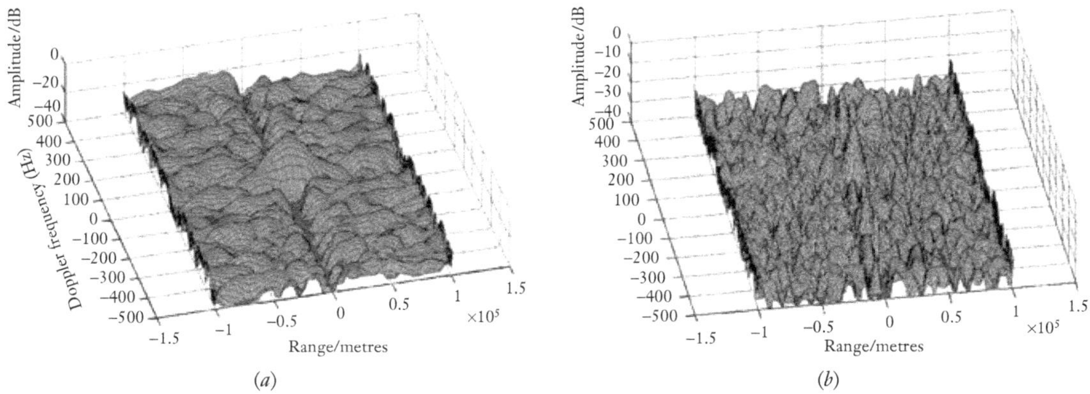

Figure 6–13 Typical off-air ambiguity functions from (*a*) speech (BBC Radio 4) and (*b*) fast-tempo jazz music (Jazz FM).

pattern. For both of these reasons, coverage in the littoral region against low-altitude targets may not be complete.

The ambiguity performance of FM transmissions will depend on the instantaneous modulation, which will depend on the program content, in other words, the spectral content of the modulation and how it varies with time. It is found, not surprisingly, that music with high spectral content, such as rock music, gives the narrowest ambiguity function peak and hence best range resolution. With speech, the width of the peak of the ambiguity function, and hence the range resolution, becomes very poor during pauses between words [65–67]. Of course, the majority of FM radio channels—even music channels—will broadcast speech, in the form of news bulletins on the hour, or speech from the program-host or an advertisement will interrupt the music every few minutes.

These points are illustrated in Figures 6–13 and 6–14. Figure 6–13(*a*) shows the ambiguity function of a station with speech modulation (BBC Radio 4). The peak and the side lobe structure are well defined, although the peak is relatively broad, as a consequence of the low spectral content of the modulation. Figure 6–13(*b*) shows the equivalent result for a station with fast-tempo jazz music modulation (Jazz FM). The peak and the side lobe structure are correspondingly sharper because of the higher spectral content of the modulation. In both cases, the floor of the ambiguity function is down by a factor of $(B\tau)^{1/2}$, rather than by $(B\tau)$, which would be expected for coherent waveforms.

Figure 6–14 compares the range resolution against time (sample) for a number of differing transmission types, over a time interval of approximately 2 sec. The two news channels (BBC Radio 3 and Radio 4) show a high degree of temporal variability in range resolution in comparison with the music channels, since for speech the range resolution will be poor during pauses between words. Overall, the range resolution varies approximately between 1.5 km and 16.5 km. The pop and dance music channels exhibit the least variation, rock music and jazz music have slightly poorer performance, and classical music is degraded a little further, reflecting the spectral content of these different types of music.

6.4.1.2 Analog Television The majority of analog television transmissions lie in the UHF band around 500–600 MHz. Some countries also use VHF bands for television; in the United States,

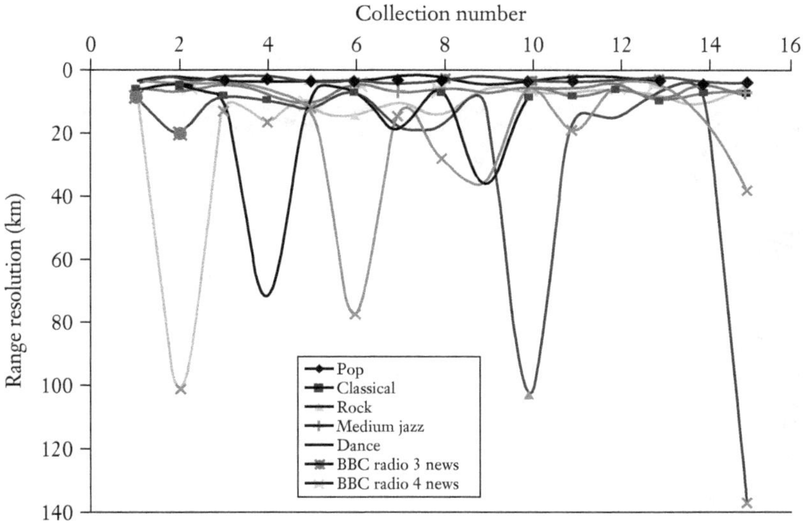

Figure 6–14 Variation in range resolution as a function of time for seven types of VHF FM radio modulation over a 2-s interval.

the band allocations are 54–88, 174–216, and 470–806 MHz. In the United Kingdom, the PAL (Phase Alternating Line) modulation format is used, in which the video information is coded as two interlaced scans of a total of 625 lines at a frame rate of 50 Hz. The start of each line is marked with a sync pulse, and the total duration of each line is 64 µs. The video information is modulated onto a carrier as vestigial-sideband AM, coded as *luminance* (red + green + blue) and two *chrominance* signals (green – blue) and (red – blue). The two chrominance subcarriers are in phase quadrature so that they can be separately recovered. The sound information (including stereo information) is frequency modulated onto a second carrier. Variants of this basic scheme are used in different countries; in the United States, the NTSC (National Television System Committee) format is used; in France and in Eastern Europe, the SECAM (Sequentiel Couleurs avec Memoire) format.

Figure 6–15 shows the measured spectrum of an analog TV signal (PAL modulation format) with the various components of the spectrum identified. For comparison, on the left-hand side of the spectrum is the corresponding digital TV signal, which has a flat spectrum with a bandwidth of 7 MHz.

The bandwidth of the analog video modulation is typically 5.5 MHz (corresponding to a monostatic range resolution $c/2B = 30$ m). As with FM transmissions, the radiation patterns are usually omnidirectional in azimuth, although the elevation-plane patterns are usually shaped to avoid wasting power above the horizontal. In the United Kingdom, and in most other countries, the highest power transmitters are 1-MW EIRP, which corresponds to a power density $\Phi = -51$ dBW/m^2 at a target range of 100 km, under the assumption of line-of-sight propagation.

It can be appreciated that there will be pronounced range ambiguities associated with the analog line and frame scan rates. In particular, since in general one line of a TV picture will be very similar to the previous one, there will be strong range ambiguities corresponding to the line scan period of 64 μs, equating to a bistatic range of 9.6 km. Figure 6–16(a) shows the measured

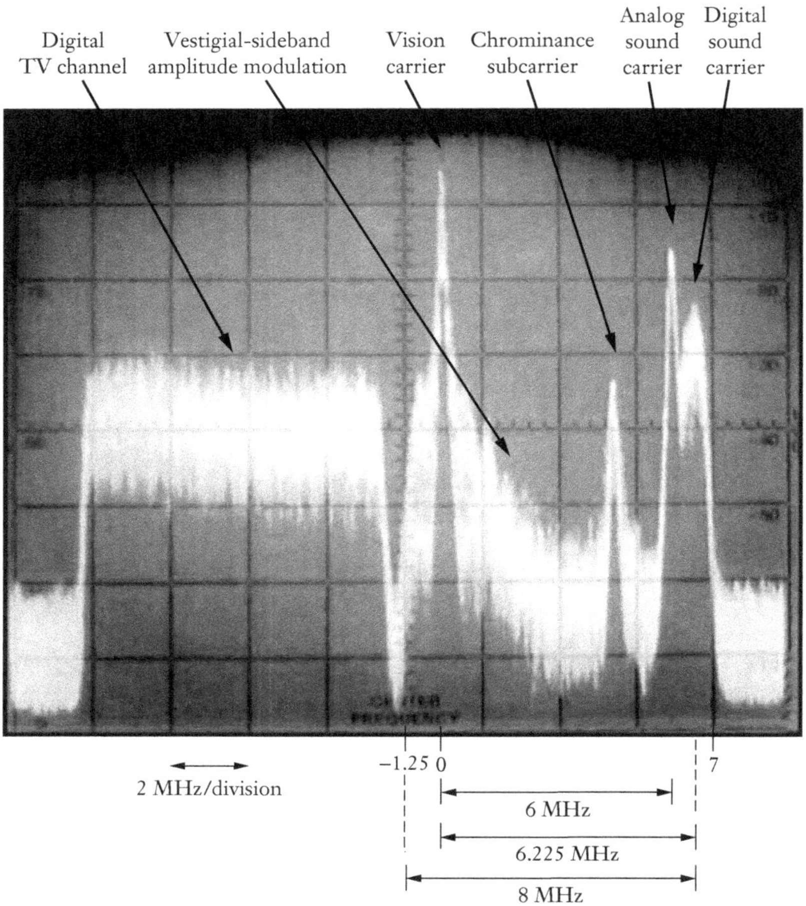

Figure 6–15 Spectrum of typical PAL analog TV signal (right of center) and digital TV signal (left of center). Horizontal scale 501–521 MHz; vertical scale 10 dB/division.

ambiguity function corresponding to the chrominance subcarrier of an analog TV signal. The ambiguities associated with the frame scan rate are easily visible, appearing as rapid modulation on the basic peak of the ambiguity function. Figure 6–16(*b*) shows the measured ambiguity function for the FM sound carrier and its modulation, which is similar in most respects to that of Figure 6–13(*a*).

6.4.1.3 Digital Radio and TV Many countries are now introducing digital radio and television. These transmissions use coded orthogonal frequency division multiplex modulation, in which all transmitters for a given station use the same frequency (so-called "single-frequency networks"). Details of this modulation format may be found in [68], but an essential feature is that the information is transmitted in synchronized frames. Each frame contains a large number of orthogonally coded subcarriers, which carry the modulation information. The receiver samples each frame only after a guard interval delay, whose duration is greater than the maximum delay of the

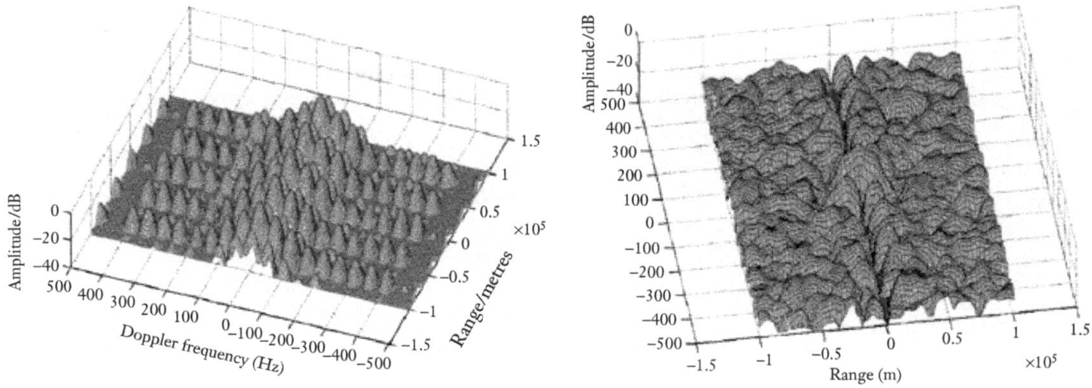

(*a*) (*b*)

Figure 6–16 Ambiguity functions for components of analog TV signal: (*a*) chrominance subcarrier and (*b*) FM sound carrier.

propagation path. This means that any multipath or signal from another cochannel transmitter will be stationary.

According to Poullin [69], typical parameters of a DAB modulation scheme are as follows:

- Symbols are of 1-ms useful duration with a guard interval of 0.246 ms.
- 1,536 subcarriers are transmitted simultaneously per symbol.
- Quadrature phase shift keying (QPSK) modulation is used for each subcarrier.
- Symbols are organized into frames of 77 symbols.
- The first symbol is null (with no-frequency transmitted or only the center frequency).
- The second symbol is a reference, where all the subcarriers are transmitted with reference code elements. This symbol is used for the propagation channel estimation, and hence equalization.

Since this type of modulation is more noise-like and does not show the same dependence on program content or variability with time as FM radio, it has potentially favorable PBR properties. Offsetting this advantage is the lower radiated power for DAB transmitters, which at about 1 kW is significantly less than the equivalent VHF FM transmissions.

6.4.1.4 Cell Phone Networks Cell phone networks are now ubiquitous in most countries [70, 71]. The GSM system uses bands centered on 900 MHz and 1.8 GHz, and 1.9 GHz in the United States [72]. The uplink and downlink bands each have a bandwidth of 25 MHz, split into 125 FDMA (Frequency Division Multiple Access) carriers spaced by 200 kHz. A given base-station will use only a small number of these channels. Each of these carriers is divided into eight TDMA (Time Division Multiple Access) time slots, with each time slot of the duration of 577 μs. Each carrier is modulated with using GMSK (Gaussian Minimum-Shift Keying) modulation. A single bit corresponds to 3.692 μs, giving a modulation rate of 270.833 kbits/sec. Figure 6–17(*a*) and (*b*) show time-domain and frequency-domain representations of these signals.

The third-generation (3G) system uses a band in the region of 2 GHz. The UMTS (Universal Mobile Telecommunication System) is the main implementation of 3G, with the following characteristics [73–75]:

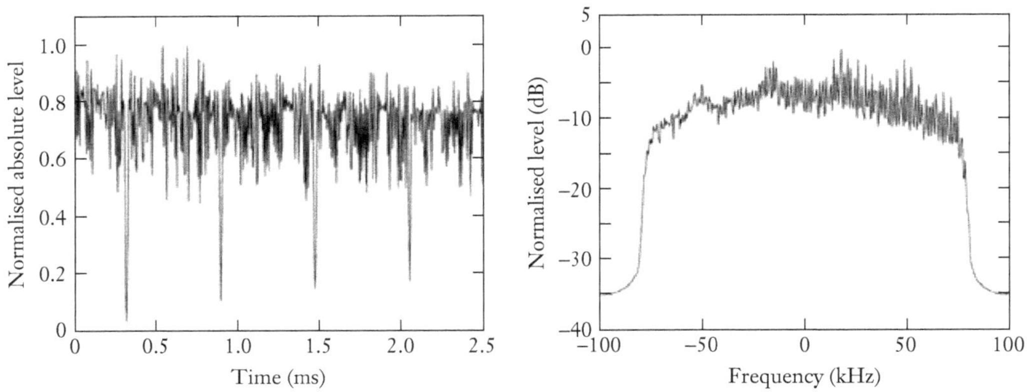

Figure 6–17 (*a*) time-domain representation of part of one TDMA-modulated carrier, showing the 577-μs slots; (*b*) frequency-domain representation, showing the 200-kHz channel (after Tan et al. [71]).

- There are two forms: Frequency Division Duplex (FDD) and Time Division Duplex (TDD). FDD requires two frequency bands (one for the up-link and one for the down-link); TDD requires a single band. A given band (or pair of bands) is allocated to a particular operator.
- FDD and TDD bands are of 5-MHz nominal width/channel spacing. The width can be reduced (in 200 kHz steps) to 4.4 MHz if operators wish.
- The transmission is Wideband CDMA (WCDMA) using Walsh-Hadamard coding. The transmission rate is always 3.84 Mchips/sec. The data rate may be varied, which means that the selected spreading code length is dependent on the data rate. The codes used are referred to as Orthogonal Variable Spreading Factor Codes (OSVF). Code length may vary from 4 (giving data rate of 960 kbit/sec) up to 512 (giving data rate of 7.5 kbit/sec). Data is also scrambled, but this does not affect the rate.
- The modulation used is QPSK. The null-to-null bandwidth is effectively 3.84 MHz, hence the 4.4 MHz minimum channel spacing. The signals are shaped with a 0.2 Root Raised Cosine Filter.

The choice of frequency band for UMTS in Europe and Asia is consistent, but in the United States these bands were not available. At the World Radio Conference (WRC-2000) in Istanbul, Turkey, in May 2000, three bands were suggested for the implementation of UMTS in the United States: 806–890 MHz (used for cellular and other mobile services, 1710–1885 MHz (used by Department of Defense), and 2500–2960 MHz (used commercially for instructional TV and wireless data providers). However, the fact that these bands are already used for other purposes led to further consultation, with the result that 45 MHz of bandwidth in the 1710–1755 MHz band, and 45 MHz of bandwidth in the 2110–2170 MHz, are to be made available for 3G services.

All of these cell phone bands are above the resonance band of most air vehicles, as outlined in Section 6.3. Hence, cell phone networks are not suited for counter-stealth, air defense operation.

The radiation patterns of cell phone base-station antennas are typically arranged in 120° sectors, with the vertical-plane radiation pattern shaped to avoid wasting power above the horizontal. Typical base-station separations are of the order of 10 km, with transmit powers of the order of 100 W, though with closer spacing and lower powers in cities. Future trends will be to more base-stations, with lower transmit powers and the use of "smart antennas."

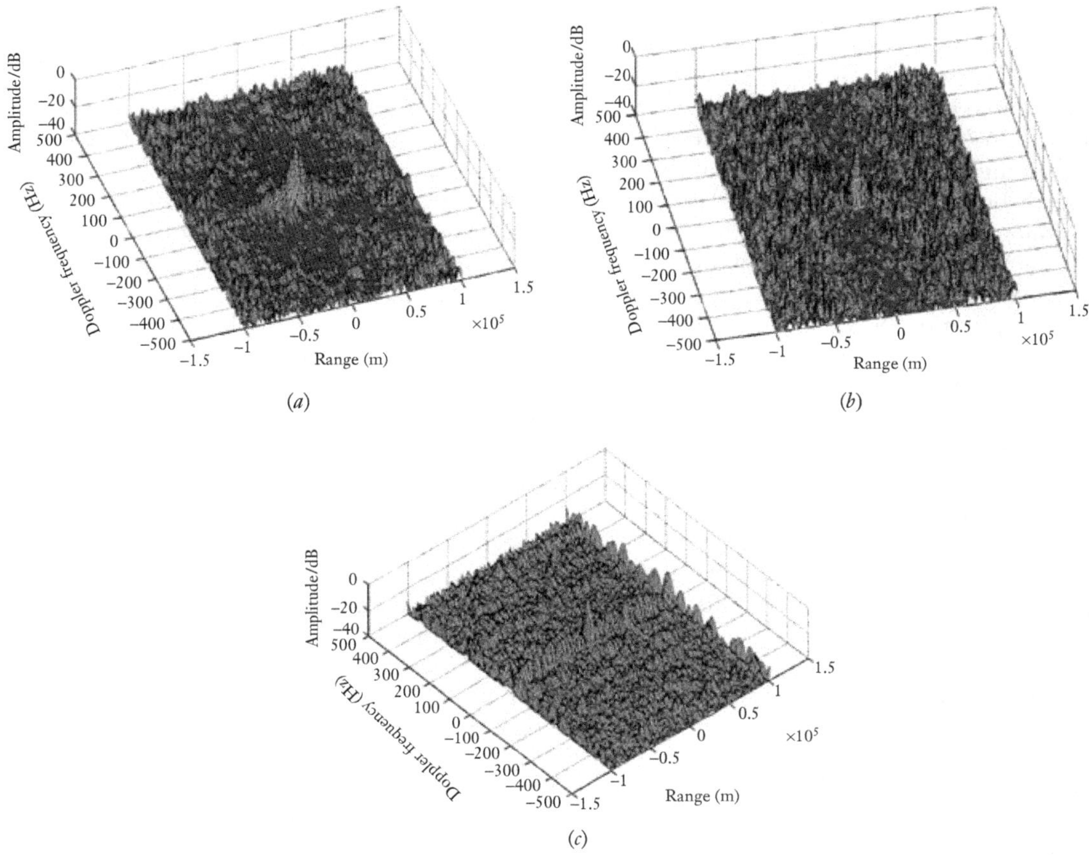

Figure 6–18 Ambiguity functions for three digital transmissions: (*a*) digital audio broadcast (DAB) at 222.4 MHz; (*b*) digital video broadcast (terrestrial DBV-T) at 505 MHz; (*c*) Global System for Mobile (GSM) 900 at 944.6 MHz.

Figure 6–18 shows typical ambiguity functions for digital transmissions (DAB, DBV-TV, and GSM, respectively). These functions are more favorable for PBR purposes than signals with analog modulation (Figures 6–13 and 6–16), since the peak of the ambiguity function is narrower and the side lobes are lower. Also, they are much more constant with time.

6.4.1.5 Other Transmissions Other transmitters have been considered as illuminators for PBRs, principally satellite-borne transmitters. They include broadcast TV (DBS, Echo Star), communications (INMARSAT, IRIDIUM), and navigation (GPS, GLONASS, GALILEO). Satellites in geostationary orbit give continuous coverage, but the power density at the Earth's surface is very low: many tens of decibels below that of terrestrial emitters. In some cases, for example, DBS, the antenna footprint is arranged to give coverage only over land. Satellites in low earth orbit (LEO) give somewhat higher-power density, but only illuminate a given target scene for a very brief period.

As detailed first in Section 6.2, and then in Section 6.5, exploiting any of these low EIRP satellite transmissions is constrained to either very short range operation or forward-scatter fences, neither particularly suited for air surveillance. One potential, short-range application is coupling the more powerful geostationary DBS transmitters with a bistatic MTI or a SAR receiver carried by an UAV flying at low altitudes to conduct ground target surveillance and imaging. Chapter 10 on spotlight SAR and Chapter 11 on adaptive MTI, provide design details.

HF broadcasting is another class of transmission that had been used in the 1960s for PBR illumination; it is reported in Chapter 4. A new, very powerful HF broadcast signal is in Digital Radio Mondiale (DRM) format and of particular interest to PBR. In DRM, the digitized audio stream is source coded using a combination of advanced audio coding (AAC) and spectral band replication (SBR) to reduce the data rate before time division multiplexing with two data streams (which are required for decoding at the receiver). A Coded Orthogonal Frequency Division Multiplexing channel coding scheme is then applied, nominally with 200 subcarriers and a quadrature amplitude modulation (QAM) mapping of these subcarriers is used to transmit the encoded data [76]. The effective bandwidth is 10 kHz. This scheme is designed to combat channel fading, multipath, and doppler spread, enabling reception of data in the most demanding of propagation environments.

Figure 6–19 shows the unweighted ambiguity function and range and doppler cuts of a typical DRM signal after 80 ms of integration time. There are no ambiguities in either domain within practical ranges and doppler shifts. The DRM signal does in fact exhibit range ambiguities at multiples of 60,000 km, a result of the 400-ms frame synchronization of the signal transmission, but these are significantly beyond the detection ranges of interest. The side lobe structure of the ambiguity function is flat, as would be expected for a noise-like signal, and the side lobe level, which is proportional to the bandwidth and the integration time, is approximately 25 dB below the peak. In this example, the range resolution of the signal, determined from the range cut, is 16 km and the doppler resolution is 3.4 Hz (equivalent to a velocity resolution of 39.2 m/sec). Analysis was performed on a variety of speech and music signals. The resulting ambiguity functions had very similar properties, indicating that they are virtually independent of the broadcast content and are essentially a function of the modulation format.

The range resolution of DRM signals (and indeed of all HF signals) is poor compared with that of higher-frequency radars, but for PBR purposes doppler resolution is also important as an input to localization and tracking algorithms. In HF radar it is common to have integration times of many tens of seconds for air and surface targets; thus, in an HF passive radar application, similar integration times are likely to be used. In Figure 6–20, the doppler cut for a more practical integration time of 5 s is presented. The doppler resolution is inversely proportional to the integration time and has improved significantly; the interference floor has also decreased to approximately –40 dB. A doppler resolution of 0.2 Hz and a corresponding radial velocity resolution of 2.3 m/sec is recorded, sufficient for many radar applications. Experimental results have shown that for 30-s integration times (appropriate for sea surface target detection) the interference floor approaches –50 dB and the doppler resolution has improved still further.

6.4.1.6 Summary of Transmitters Table 6–3 summarizes properties of transmitters that have been considered for passive bistatic radar operation. Figure 6–21 arranges some of them in a "league table" in order of power density at representative target ranges. Satellite transmitters have been added for comparison.

6.4.1.7 Geometry-Dependent Ambiguity Performance The preceding ambiguity functions were developed independent of the bistatic geometry, that is, in a monostatic configuration. However, a

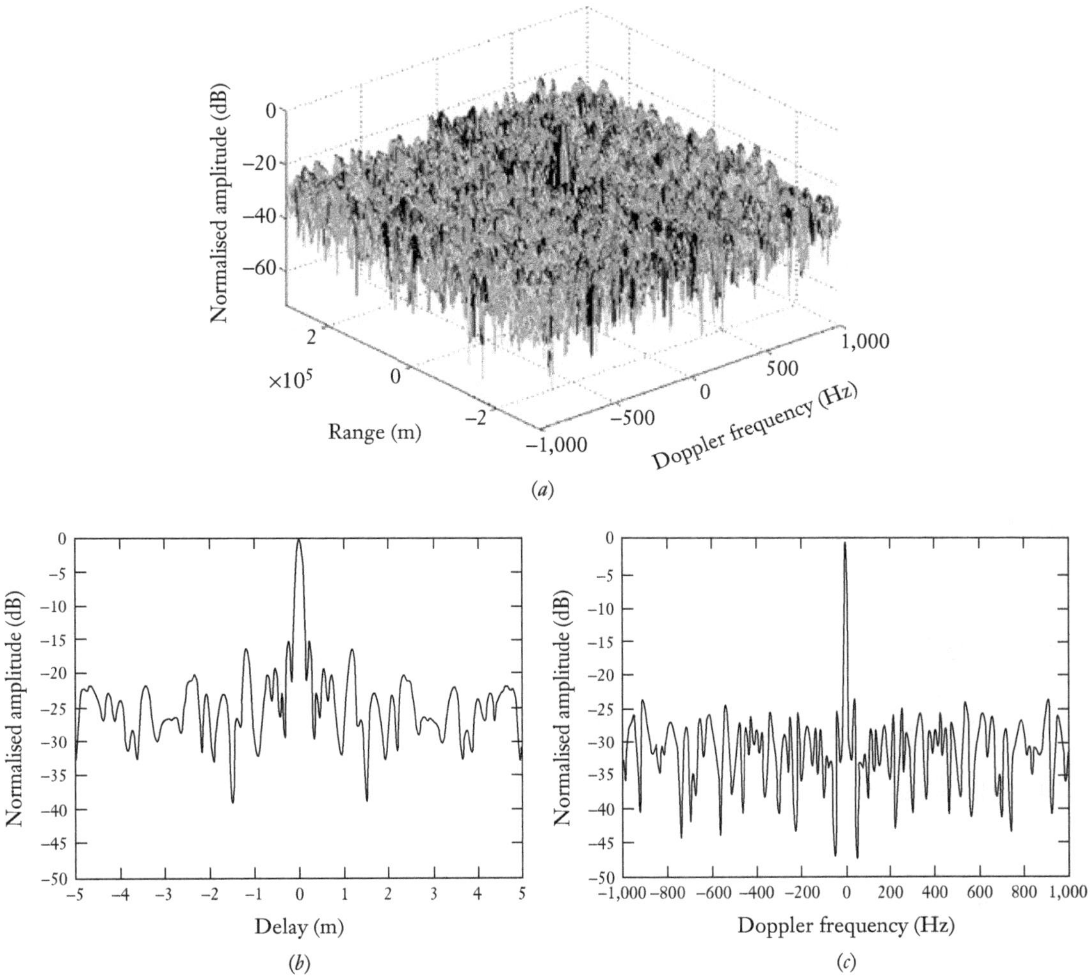

Figure 6–19 (*a*) Normalised ambiguity function for DRM signal with 80-ms integration time; (*b*) range cut; and (*c*) doppler cut.

bistatic radar ambiguity function is geometry-dependent. This dependence can be visualized by considering a moving target crossing the transmit–receive baseline: range information becomes indeterminate and the doppler shift becomes zero [77]. Hence, the function "blows up" on the baseline, giving neither range nor doppler resolution.

If the ambiguity function is evaluated for target locations elsewhere on the bistatic plane, the peak of the ambiguity function in both range and doppler broadens with respect to the monostatic peak, and is equal to the monostatic peak only when the target lies on the extended baseline where the bistatic angle is zero. The effect was investigated by Tsao et al. [78], who noted the nonlinear relationships between target velocity and doppler shift, and between target range and delay. They proposed that the ambiguity function for a bistatic radar should be

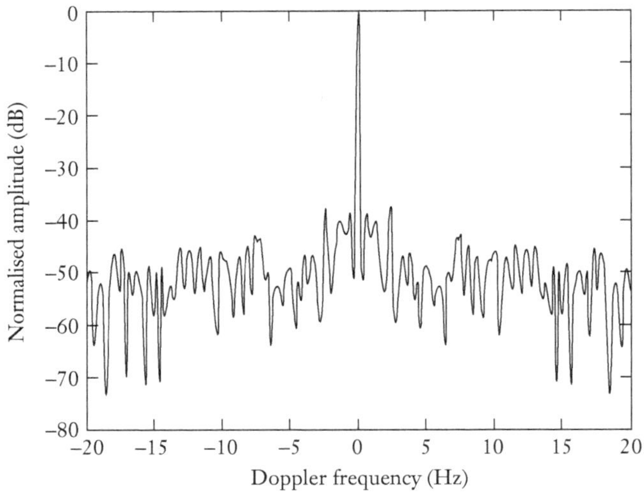

Figure 6–20 Doppler cut of ambiguity function of DRM signal with 5-s integration time.

Table 6–3 Signal Parameters for Typical Passive Radar Illumination Sources

Transmission	Frequency	Modulation	Bandwidth	P_tG_t (typical)
HF broadcast	10–30 MHz*	DSB AM	9 kHz	50 MW
		DRM	10 kHz	
VHF FM (analog)	~100 MHz	FM	50 kHz	250 kW†
UHF TV (analog)	~550 MHz	Vestigial-sideband AM (vision); FM (sound)	5.5 MHz	1 MW
Digital Audio Broadcast	~220 MHz	Digital, COFDM	220 kHz	10 kW
Digital TV (U.K.)	~750 MHz	Digital	6 MHz	8 kW
Cell-phone Networks (GSM)	900 MHz, 1.8 GHz	GMSK, FDM/TDMA/FDD	200 kHz	100 W‡
Cell-phone Networks (3G)	~2 GHz	CDMA	3.84 MHz	100 W

* Appropriate frequency will depend on time of day.
† http://www.bbc.co.uk/reception/.
‡ http://www.sitefinder.radio.gov.uk.

written as follows:

$$\left|\chi(R_{R_{\rm H}}, R_{R_{\rm a}}, V_{\rm H}, V_{\rm a}, \theta_R, L)\right|^2 = \left|\int_{-\infty}^{\infty} \tilde{f}(t-\tau_{\rm a}(R_{R_{\rm a}}, \theta_R, L))\tilde{f}^*(t-\tau_{\rm H}(R_{R_{\rm H}}, \theta_R, L)) \exp[-j(\omega_{D_{\rm H}}(R_{R_{\rm H}}, V_H, \theta_R, L)-\omega_{D_{\rm a}}(R_{R_{\rm a}}, V_{\rm a}, \theta_R, L))t]\,{\rm d}t\right|^2 \tag{6.3}$$

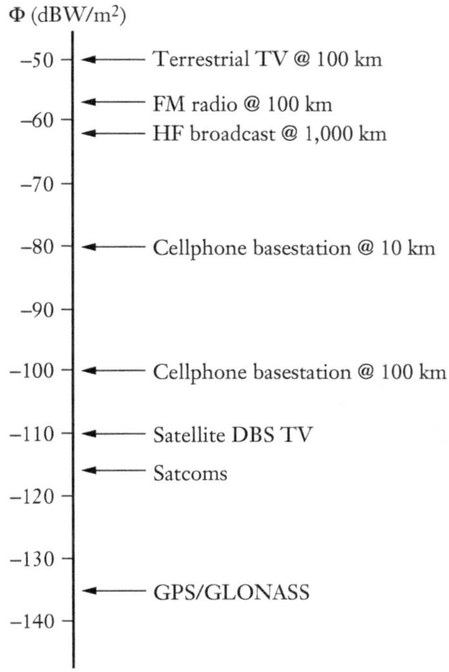

Figure 6–21 A "league table" of some passive bistatic radar illumination sources, arranged in order of power density at the target. These power densities have been calculated on the basis of a single channel, full-signal bandwidth, no processing gain, and free-space propagation (except for the HF broadcast signal, for which the propagation, and hence the power density, will depend on time of day, season, and time of sunspot cycle).

where R_R and R_T are the ranges of the target from the transmitter and receiver, V is the target velocity vector, θ_R is the angle of the target measured from the receiver, L is the bistatic baseline, τ is the transmitter–target–receiver delay time, and the subscripts H and a denote hypothesized and actual values, respectively.

Evidently this expression depends on rather more than two variables, so it is not straightforward to plot and display it in a simple manner, as is readily done for the monostatic ambiguity function. They show further, by means of simulation, that a signal whose monostatic ambiguity function has a simple Gaussian shape has a bistatic ambiguity function whose shape depends significantly on the bistatic geometry, and degrades badly when the target is close to the bistatic baseline.

To illustrate this effect, Figure 6–22 shows the ambiguity function for four different target locations and directions of motion. The waveform in each case consists of a short sequence of three rectangular pulses. In Figure 6–22(*a*), the target is on the baseline approaching the receiver and the ambiguity function is essentially the same as for the monostatic configuration. In Figure 6–22(*b*), the target approaches the baseline obliquely, but the ambiguity function is little altered. In Figure 6–22(*c*), the target approaches the baseline from a perpendicular direction, which broadens the main peak and alters the position of the side lobes in the ambiguity function. Finally in Figure 6–22(*d*), the target crosses the baseline, giving effectively no resolution in either range or doppler.

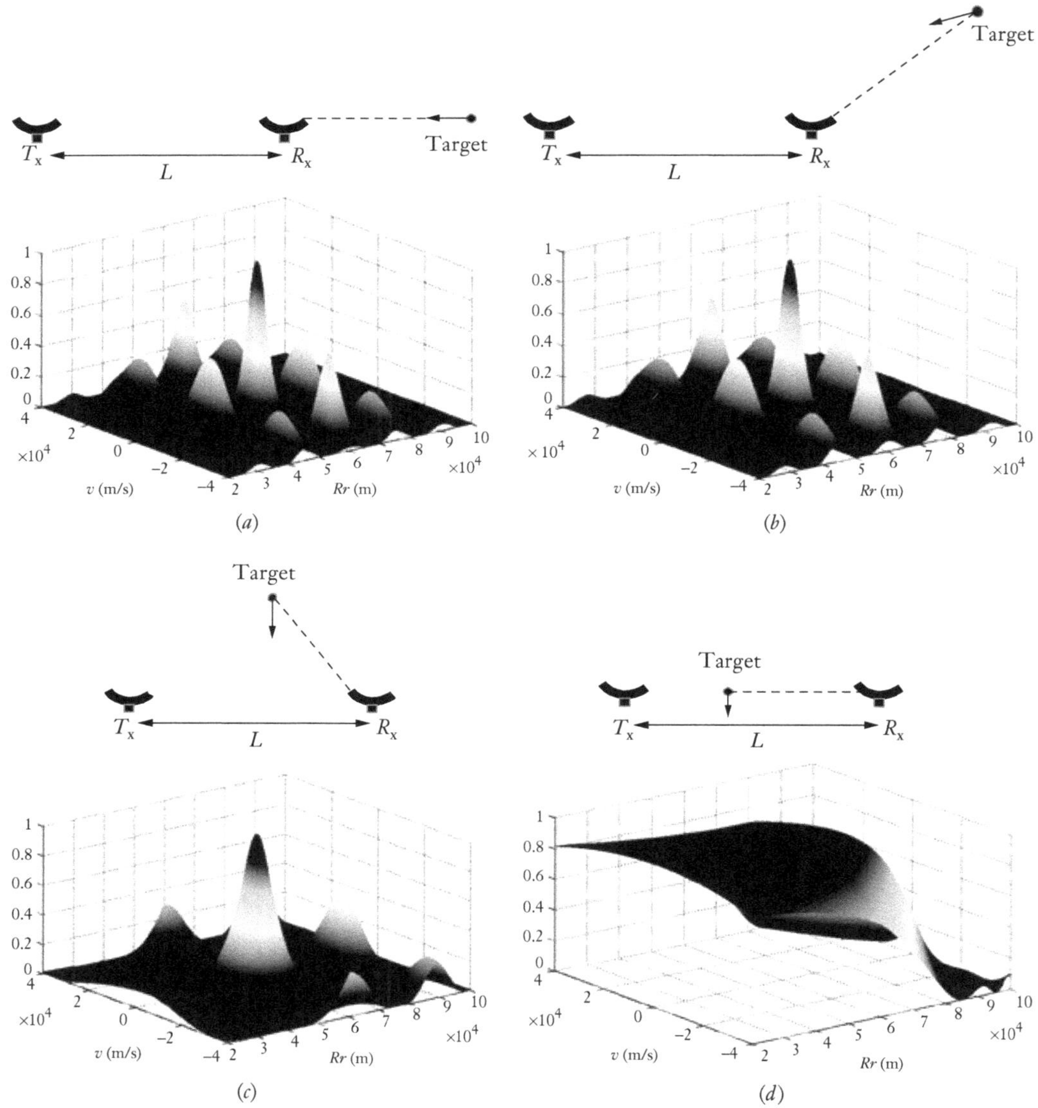

Figure 6–22 Bistatic ambiguity functions for four different target locations and directions of motion.

6.4.2 Radio Frequency Interference

The RF environment in which PBR receivers operate includes two types of RFI: sky noise and man-made noise. The former consists of sun, galactic, and atmospheric noise. The latter consists of signals from broadcast, communications, and navigation transmitters and of impulsive noise from older automobile ignitions, power tools, fluorescent lights, cooling fans, etc. This interference can arrive via a direct path or by multipath from terrain or buildings and can be severe in dense urban environments.

The effect of this interference on a PBR receiver can be quantified using the method of Blake [79] and expanded by Barton [80], as follows. All interference components are combined into a single source at the antenna output terminal, represented by an input termination resistor with a temperature T_i, which makes available to the receiver a noise with the spectral density (in W/Hz):

$$N_0 = kT_i \tag{6.4}$$

where k is Boltzmann's constant and T_i is the system input noise temperature:

$$T_i = T_a + T_r + L_r T_e \tag{6.5}$$

where

$$T_a \text{ is the antenna contribution} = T_s + T_m \tag{6.5a}$$

$$T_r \text{ is the RF component contribution} \cong T_o (L_r - 1) \tag{6.5b}$$

$$T_e \text{ is the receiver contribution} = T_o (F_n - 1) \tag{6.5c}$$

T_s is the temperature of the sky-noise interference (ignoring the normally small dissipative loss within the antenna), T_m is the temperature of the man-made interference, L_r is the RF component loss, F_n is the receiver's noise factor, and $T_o = 290$ K. For example, assume a well-designed receiver in which $L_r = 1$ dB and $F_n = 5$ dB. Thus $T_r + L_r T_e = 290\,[(1.26 - 1) + 1.26\,(3.16 - 1)] = 865$ K. To this temperature must be added the contributions from interference, T_a, which is considered next.

6.4.2.1 Sky Noise The contributions of sun, galactic, and atmospheric noise are plotted as a function of frequency by Barton in Figure 1.2.2 of [80]. The maximum contributions between 1 GHz and 10 GHz (where galactic noise is maximum, sun noise is 100 times the quiet level, and the antenna is at 0° elevation) are small: <200 K. The minimum contributions (minimum galactic noise, zero sun noise, and 90° elevation angle) are smaller still: <100 K over all frequencies of interest, 100 MHz to > 10 GHz. In contrast, the maximum contributions below 1 GHz rise monotonically to 800 K at 400 MHz and 3,000 K at 200 MHz.

Thus, the effects of sky-noise interference are minimal at frequencies above 1 GHz, when not pointed near the center of the galaxy and during periods of low sun spot cycles. At other times and when conducting air surveillance near the local horizon, which happens to be aligned near the galaxy center, and when operating at frequencies below approximately 400 MHz, the noise temperature of the above example can increase by a factor of 2–4. While these occasional increases are vexing, they are not nearly as vexing as the huge increases caused by intentional jamming (Section 6.7) or by the next topic, man-made noise. Thus, until these interferences are suppressed, there is no motivation for attempting to mitigate the lesser effects of sky noise.

6.4.2.2 Man-Made Noise Spectrum measurements of signal and noise levels in an urban environment were made from the 10th floor of the Engineering Faculty Building of University College London in Central London [81]. The measurements included a number of broadcast and communication sources that might be used for PBR.

Figure 6–23(*a*) shows the thermal noise level in a 50-kHz bandwidth in the VHF FM radio band corresponding to a 5-dB receiver noise figure ($kT_0BF = -122$ dBm) and a system input noise temperature $T_i = 626$ K. Of course, the thermal noise level depends on the bandwidth, and the value of 50 kHz has been chosen to correspond to typical FM signal bandwidths.

Figure 6–23(*b*) shows the RF environment consisting of both FM signals and ambient interference received on a vertically polarized dipole antenna. The two spectra have been offset

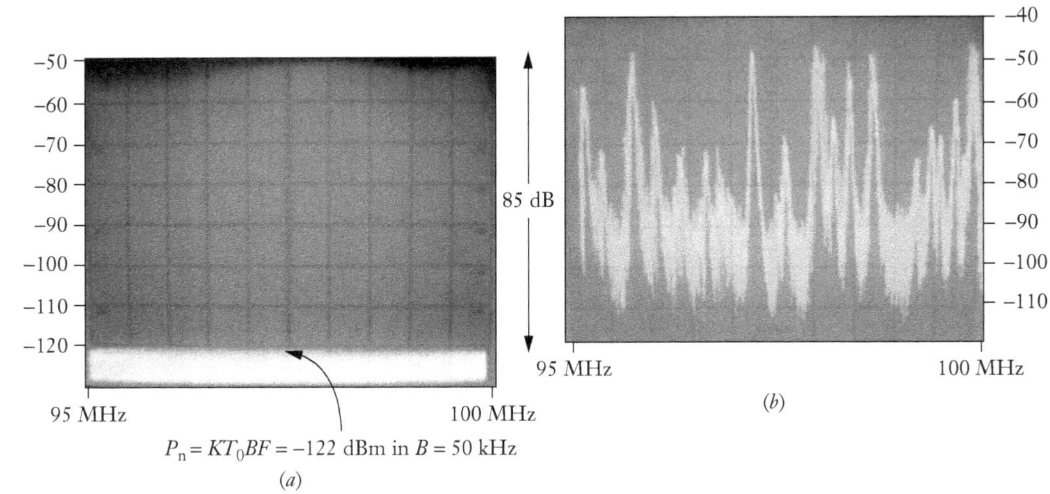

Figure 6–23 Noise levels in the FM radio band: (*a*) thermal noise in 50-kHz bandwidth corresponding to 5-dB receiver noise figure; (*b*) signal and impulsive noise levels measured from 10th floor of UCL in Central London on vertically polarized dipole antenna. Horizontal scale 95–100 MHz; vertical scale 10 dB/division.

vertically so that they may directly be compared. The strong signals at near –45 dBm in the figure are typical of FM broadcast transmitters that have been partially shielded from the dipole antenna by buildings or terrain. Smaller peaks represent transmissions that are lower-powered or have more shielding from the antenna. While these signals represent transmitters to be exploited by a PBR, the reception of their transmissions over the direct transmit-to-receive path raises the system input noise temperature by four to six orders of magnitude and clearly must be suppressed.

How much direct path suppression can be achieved depends on the spatial and spectral cancellation algorithms (which are described later) and the interference from adjacent signals. As reported by many PBR developers, such as R. Lodwig,[27] Howland [82], and Andrews [83], their direct path cancellation schemes could be made effective down to levels 20–30 dB above the receiver thermal noise level. Attempts at further reduction have not been successful, because of cochannel interference, which is not cancelled. This interference is principally caused by the broad spectrum roll-off of an interfering signal, which in the United States is mandated by the FCC as Gaussian. Under the FCC-specified channel separation, this roll-off causes no interference between adjacent channels for one-way broadcast reception—for which they were designed. But for two-way reception, which requires working many tens of decibels further down in amplitude, such roll-off broadens and causes interference in the adjacent channel, and in fact even in separated channels.[28]

An example of this effect is shown in Figure 6–23(*b*): the width of a typical signal near –45 dBm is approximately 50 kHz; the width of the signal at –90 dBm has broadened to 200 and even 300 kHz, which causes interference in adjacent channels, effectively setting the

27 Private communication, 2001.
28 R. Lodwig, private communication, 2001.

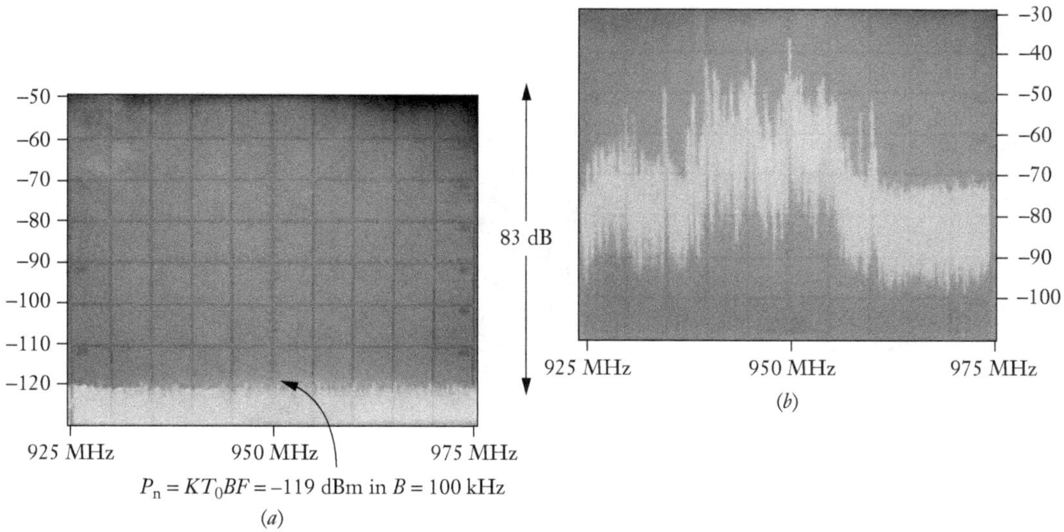

Figure 6–24 Noise levels in a passive bistatic radar receiver in the GSM 925- to 975-MHz radio band: (*a*) thermal noise in 100-kHz bandwidth corresponding to 5-dB receiver noise figure; top of screen = –50 dBm; (*b*) signal and noise levels measured from 10th floor of UCL in Central London on vertically polarized dipole antenna; top of screen = –30 dBm. Horizontal scale 925–975 MHz; vertical scale 10 dB/division.

cancellation floor. In this example, a –100-dBm floor corresponds to an increase in the system input noise temperature T_i of 1,450,000 K, or an equivalent increase of 31 dB in the receiver's noise figure, yielding F_n = 37 dB. This floor, in turn, limits the detection performance of a PBR system operating in such a severe urban environment.

This by no means represents a worst case, since the receiver is not line-of-sight to the transmitter. If the power density Φ from equation (6.1), and the receiver antenna gain G_{Rt} in the direction of the transmitter are known, then the direct signal level can be evaluated directly:

$$P_{\text{direct}} = \Phi \frac{G_{Rt}\lambda^2}{4\pi} \tag{6.6}$$

Figure 6–24 shows a similar measurement of the signal environment in the GSM cell phone radio band, 925–975 MHz, where the two spectra have been offset vertically so that they may directly be compared. Here the spectral roll-off is not as apparent due in part to the increased levels of ambient interference, which by itself will increase the system input noise temperature and noise figure. No reports of achieved cancellation levels using these waveforms have been published.

This man-made noise environment can be visualized as a two-dimensional function $P(\theta, f)$ of direction ($-1 \le \sin\theta \le +1$), and frequency (the VHF FM band), as shown in Figure 6–25. The signal environment consists of (1) thermal noise; (2) FM radio signals, each with its own (time-varying) modulation and bandwidth with a particular spectral roll-off, and each with its own set of multipath signals; plus (3) broadband noise sources such as computers and impulsive noise. In

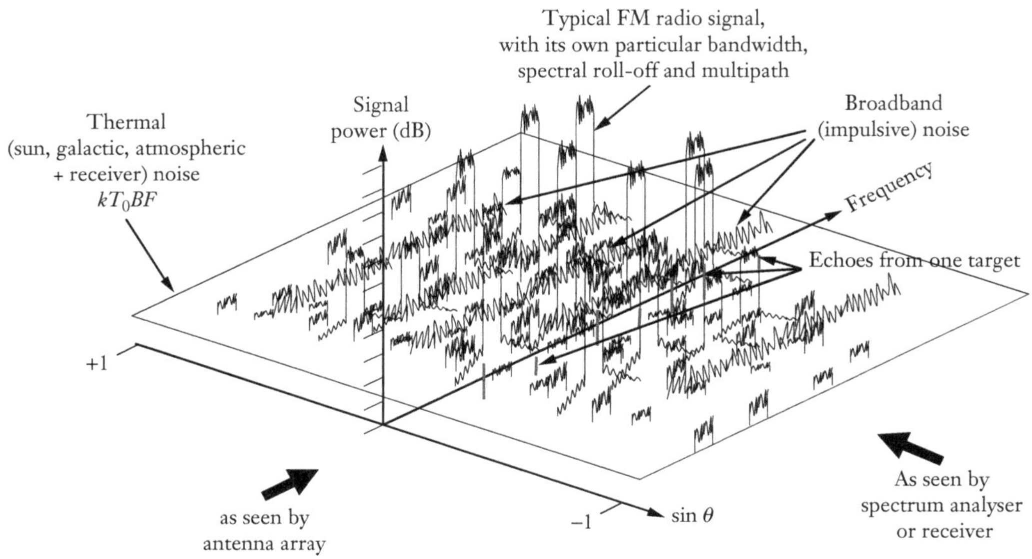

Figure 6–25 Schematic of the signal and noise environment as a two-dimensional function $P(\theta, f)$ of direction ($-1 \leq \sin\theta \leq +1$), and frequency (the VHF FM band).

conventional radar terms, the direct signal and its multipath components might be regarded as clutter, but in view of all of the other sources of signals and noise this is not a particularly helpful distinction.

An individual target will result in a set of echoes from each transmitter, all at the same direction of arrival (DOA) from the receiver. A receiver can filter this signal and noise environment in the frequency domain; an antenna array can filter in the angular domain. As discussed in the next section, a combination of approaches will be necessary to suppress the unwanted components to an acceptably low level, and even then the residue is likely to be some 20–30 dB above receiver thermal noise.

6.4.2.3 Interference Suppression Various techniques have been reported to suppress this RFI. First, and most attractive because of effectiveness, simplicity, and cost, is to site the receive antenna so that it is physically shielded from the direct path signal, using topography, buildings or shrouds. This technique alone can often provide adequate suppression, but in turn limits the receive antenna's field of view (FOV). In some cases, this trade-off is entirely acceptable. Both OTH radars, Chapter 4, and the *Manastash Ridge Radar* at the University of Washington, Chapter 7, are examples. In each case, the receiver's line-of-sight to the transmitter is blocked by terrain—earth's curvature for the former and the Cascade Mountains for the latter. And, in both cases the FOV is not hindered since propagation paths of interest are directed toward the (elevated) ionosphere.

But for air surveillance, terrain blockage of the receiver's FOV is nearly always unacceptable, especially for low-altitude surveillance. However, judicious siting of the receive antenna in some geometries can still provide acceptable physical shielding. The classic example is to site the receiver (a) between the transmitter and the surveillance region with the transmitter looking over the receiver's shoulder and (b) with a large building or other structure located directly behind the receiver, that is, along the baseline, to shield the direct path.

Even so, the shielding may not be sufficient depending on edge diffraction around the structure.[29] For example, this over-the-shoulder geometry and a small intervening building were used in Howland's *FM radio-based bistatic radar* NATO trials in 2000 [84, 85]. While providing some suppression, it still required other measures as reported below. In any case the technique is constrained to areas where large structures are present, typically urban areas. It is further limited when multiple transmitters are exploited in a multistatic configuration, since each transmitter must be so shielded. A multistatic system using one transmitter and multiple receivers might provide more flexible siting opportunities.

Spatial cancellation of RFI is a second principal option. An array antenna at the receiver can be configured to steer nulls at discrete RFI sources, including the direct path signal. Inspection of Figure 6–23 shows that radiation pattern nulls should ideally be broadband so that multiple signals from a given transmitter may be suppressed by a single null. Furthermore, the figure depicts the signal environment over 180° of space, which is appropriate if a linear antenna array is to be used. However, in many cases coverage will be required over the full 360°. This may be achieved using a circular antenna array, excited using phase modes [85], and allows broadband nulls to be formed, potentially over bandwidths of an octave or more [86].

Of course, an adaptive antenna and its associated signal processing introduces significant complexity, and, to some extent, negates one of the claimed advantages of PBR—that the receiver is simple and cheap.

To add additional complexity, if the environment is nonstationary, adaptive control of the antenna array is required. Adaptive arrays and null steering techniques have been studied and developed extensively over several decades, both for mobile and personal communications and for radar applications [85, 87–91]. The first of these (in the 1970s and early 1980s) were analog, but these were soon overtaken by digital approaches as fast analog-to-digital converters and processing power became readily available.

Typical of the earliest analog approaches is the Howells-Applebaum correlation loop shown in Figure 6–26. Here, a version of the signal received by a second, auxiliary antenna is weighted in amplitude and phase by a vector modulator, and added to the signal from the primary antenna in the hybrid combiner Σ. Suppose that there is an interfering signal incident from a particular direction. The interfering signal received by the auxiliary antenna is weighted by the vector modulator so that at the hybrid combiner it is equal in amplitude and opposite in phase to that received by the primary element, and hence is cancelled. The correct amplitude and phase weight are derived by correlating the cancelled signal with the original version received by the auxiliary antenna (both having been down converted to a suitable IF and band pass-filtered to match to the bandwidth of the interfering signal). This produces in-phase (I) and quadrature (Q) signals, which control the vector modulator. The low-pass filters in these paths determine the dynamic behaviour of the cancellation loop. The basic scheme can readily be extended to more than one degree of freedom, so that more than one source of interference can be simultaneously cancelled [92].

Although such analog approaches have largely been superseded, it is suggested that a stage of analog cancellation of this kind may usefully be used prior to digital processing so that the dynamic range requirement on the analog-to-digital converters (which might otherwise approach 100 dB) is not so stringent.

In most cases the direct path signal should receive primary attention. Howland succinctly detailed this problem and then his solution [82]:

> The greatest limitation on system performance is the interference received from the transmitter being used to detect aircraft. This unwanted direct signal correlates perfectly with the reference signal and pro-

29 A similar problem occurs with attempts to adequately shroud a VHF/UHF antenna: the shroud becomes excessively large and cumbersome.

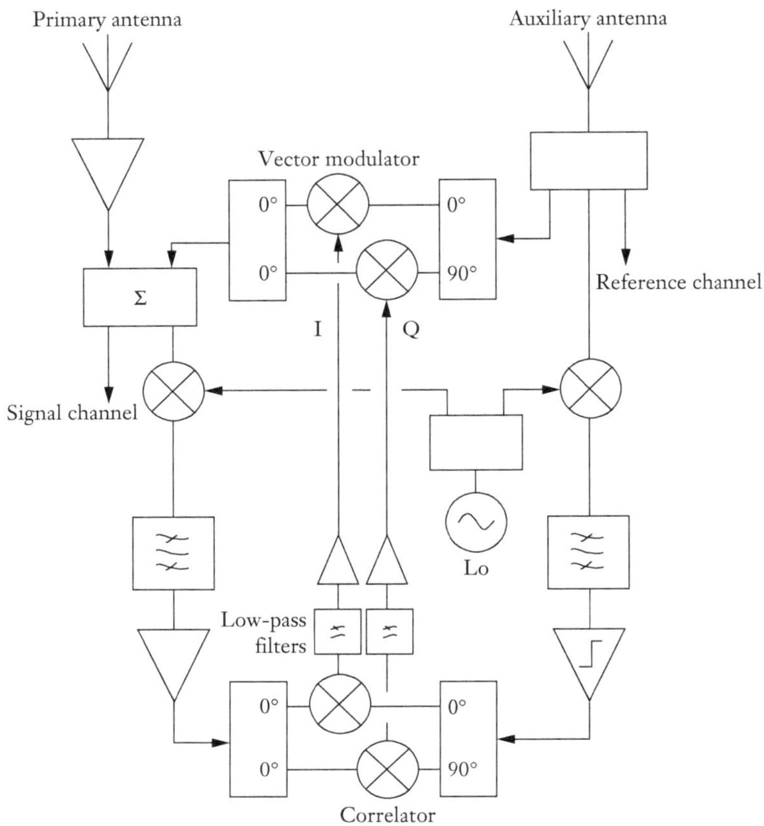

Figure 6–26 Howells-Applebaum analog cancellation loop. The basic scheme may readily be extended to more than one degree of freedom.

duces range and Doppler sidelobes that are several orders of magnitude greater than the echoes that are sought. To detect anything but the closest of targets it is necessary to remove this signal, by both angular nulling with the antenna and adaptive echo cancellation in the receiver. However, eventually the dynamic range of the receiver limits the cancellation and so the principal limitation on system performance lies with the analogue-to-digital converter technology.

Howland details his development of a two-stage adaptive noise canceller. The first stage is an adaptive M-stage lattice predictor, with prediction order $M = 50$, and the second an adaptive tapped delay line (Figure 6–27). The first is equivalent to the Gram–Schmidt algorithm and the second to a multiple regression filter. Howland reports [84] that this noise canceller suppressed the narrow-band, stationary, direct path interference by approximately 75 dB. That suppression coupled with the physical shielding was adequate to allow target detections 150 km from the receiver.

A spectral attack on this direct path cancellation problem was reported by Andrews [83]. The SAIC-developed, *HDTV-Based Passive Radar* system currently uses one HDTV transmitter, which broadcasts a finite-length pseudo-noise (PN) waveform at 600–800 MHz and four receivers, which are deployed within short ranges (5–10 km) of the transmitter for range-only multilateration. Doppler is also measured to resolve ghosts that inevitably arise in any multilateration system. The SAIC system is designed to detect and track low-flying aircraft and helicopters as a gap-filler (typically < 5 kft altitude) for other air-surveillance systems, principally air traffic control radar systems.

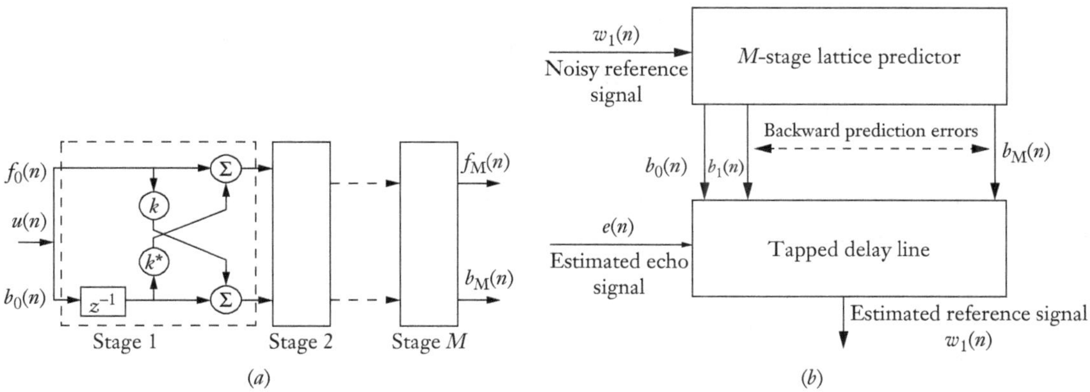

Figure 6–27 Two-stage adaptive noise canceller used by Howland [84]. The first stage *(a)* is an adaptive *M*-stage lattice predictor with prediction order $M = 50$, equivalent to the Gram–Schmidt algorithm, and the second *(b)* is equivalent to a multiple regression filter.

Because transmitter–receiver separations are small, 5–10 km, significant direct path breakthrough occurs. It is reduced with a brute-force, least-squares channel estimator, yielding a 20- to 25-dB noise figure. SAIC selected this approach when they found that a cross-correlator was roughly 40 dB inferior in reducing the noise floor because it did not account for side lobes of the PN waveform in the ambiguity plane. The cost of this process is to more than double the computational load to 20.5 B floating-point operations per second at each receiver.

It is clear that a combination of techniques will often be needed to achieve adequate direct path suppression, and even with such suppression the achieved value of receiver noise figure will be greater than 20 dB.

6.5 RANGE PERFORMANCE

6.5.1 Introduction

The range performance of PBRs operating in the VHF through microwave bands is quantified in this section. The basic tool for this task is the bistatic radar equation in a slightly revised format. Traditionally, this bistatic radar equation has been developed exactly as a monostatic radar equation and then solved for received signal power [93–95] or received signal-to-noise ratio (SNR) [96]. Then, to complete the development, concentric *ovals of Cassini* (defined subsequently) are drawn around a fixed transmitter-to-receiver distance, or baseline L, as a function of either signal power levels or SNRs [97–100]. While these signal-dependent contours provide a sense of bistatic radar performance, they do not show maximum/minimum detection ranges and coverage for different baselines, all parameters of operational interest.

To remedy this situation, the bistatic radar equation is developed in two parts. In the first part (Section 6.5.2), the canonical bistatic radar range or link equation is solved for the *bistatic maximum range product*, $(R_T R_R)_{max}$, where R_T is the range between transmitter and target and R_R is the range between receiver and target. Then an *equivalent monostatic maximum range*, $(R_M)_{max}$, is defined, omitting the subscript max for convenience, as:

$$R_M = (R_T R_R)^{1/2} \tag{6.7}$$

This equivalent monostatic maximum range, also known as the *geometric mean range*, represents performance of the bistatic radar when transmitter and receiver are colocated; that is, when $L = 0$. It is defined as the bistatic radar's *benchmark range* or simply *benchmark*.[30] Since this benchmark is geometry-invariant, it becomes useful when comparing bistatic to monostatic range performance—certainly more so than when using the geometry-sensitive term R_R.

This benchmark range is artificial in the sense that the receiver cannot be physically colocated with the transmitter unless special provisions are implemented. The first provision is providing sufficient isolation between transmitter and receiver through, for example, antenna shielding or shrouding and/or receiver cancellation or blanking. The second provision requires that transmitter and receiver antennas be properly aligned in space, which is also a requirement when $L > 0$, but may impose additional requirements or restrictions on the system, for example, adding more flexible beam pointing logic or selecting unimpeded receive antenna siting. Of course, an analytical colocation simply assumes that these requirements are satisfied, and proceeds to use the benchmark as a convenient tool in the assessment of bistatic radar performance.

The second part of this development (Section 6.5.3) plots ovals of Cassini as a function of baseline range, L, normalized to the benchmark range, R_M. Then, when a baseline L is selected, maximum and minimum ranges and coverage areas can readily be calculated as a function of the previously determined R_M. Line-of-sight (LOS) constraints are also quantified.

This method of developing the bistatic radar equation also allows definition of PBR operating regions (Section 6.5.4). This development immediately leads to a range assessment of PBRs using specific transmitters of opportunity (Section 6.5.5). A related range topic, the *6-dB per octave* assertion, is quantified in Section 6.5.6, which in turn leads to a simple method for optimizing multistatic coverage.

6.5.2 Range Component of the Bistatic Radar Equation

Since all illuminators of interest to PBRs operate with modulated continuous wave (CW) waveforms, the bistatic radar range equation in its CW form [93, 97] will be used and solved for the bistatic maximum range product, $(R_T R_R)_{max}$:

$$(R_T R_R)_{max} = \left[\frac{P_T \rho G_T G_R t_c \lambda^2 \sigma_B F_T^{\,2} F_R^{\,2}}{(4\pi)^3 k T_i (S/N)_{min} L_T L_R L_{sp}} \right]^{1/2} \tag{6.8}$$

where R_T = transmitter-to-target range (m), R_R = receiver-to-target range (m), P_T = transmitter average power output (W), ρ = power utilization factor, t_c = coherent processing time (sec), G_T = transmitting antenna power gain, G_R = receiving antenna power gain, λ = wavelength (m), σ_B = bistatic RCS (m^2), F_T = pattern propagation factor for transmitter-to-target path, F_R = pattern propagation factor for receiver-to-target path, k = Boltzmann's constant [1.38×10^{-23} W/(Hz K)], T_i = receiving system input noise temperature (K), $(S/N)_{min}$ = SNR required for detection, L_T = transmitting system losses (>1) not included in other parameters, L_R = receiving system losses (>1) not included in other parameters, L_{sp} = signal processing loss.

Equation (6.8) is related to the corresponding monostatic maximum range equation by the following: $R_T R_R = R_M^{\,2}$, $\sigma_M = \sigma_B$, and $L_T L_R L_{sp} = L_M$. The power utilization factor ρ is a quantity, ≤ 1, that specifies the fraction of the total transmitter power utilized by the PBR receiver. It is usually accomplished by band-limiting the PBR receiver, and becomes significant when narrowband

30 Note that assuming $L = 0$ also establishes $R_T = R_R$. But the converse is not true: assuming $R_T = R_R$ does not establish $L = 0$. In short, $L = 0$ is the necessary condition.

doppler operation with analog TV transmissions is exploited. In this case, ρ can range from 0.1 to 0.5. In most other cases, ρ can approach 1.0.

The coherent processing time, t_c, also called *coherent processing interval* (CPI), is defined as the inverse of the noise bandwidth of receiver's predetection filter, B_n.[31] The time available for coherent processing is typically set by the amount of doppler spreading or velocity walk, Δf_d generated by the target. Specifically, $\Delta f_d = (t_c)^{-1} = B_n$. In the monostatic case doppler spreading is given as

$$(\Delta f_d)_{mono} = [2a_r/\lambda]^{1/2} \tag{6.9}$$

where a_r is the radial component of target acceleration. Equation (6.9) also applies to the bistatic case at small bistatic angles, β. However, in the general case of a larger β, the radial component, which is aligned with the bisector of the bistatic angle, will be reduced. A rule of thumb for these large β conditions is

$$(\Delta f_d)_{bist} = [a_r/\lambda]^{1/2} \tag{6.10}$$

Equation (6.10) is used to set the coherent processing time t_c and hence the noise bandwidth of receiver's predetection filter B_n.

Noncoherent integration of multiple CPIs can be used when the dwell time on target t_o exceeds the CPI and the range walk is not excessive. Range walk, or the time a return remains within a range resolution cell, depends on the receiver's RF bandwidth, target kinematics, and the geometry, and can vary from a few meters to many kilometers. Since a CPI is usually of the order of t_o for many PBR designs, little noncoherent integration is available.

As in the monostatic equation, the transmitting and receiving pattern propagation factors, F_T and F_R, each consist of two terms: the propagation factors, F_T' and F_R', and the antenna pattern factors, f_T and f_R, respectively [101]. The antenna pattern factors are the relative strength of the free-space field radiated by the transmitting and receiving antennas as a function of their pointing angles. These factors are applied whenever the target is not at the peak of a beam. They seldom are an issue for PBRs that use broad, fixed beam transmitting and receiving antennas, particularly for multilateration.

Propagation factors customarily include the effects of multipath, diffraction and refraction, with atmospheric absorption effects included in the loss terms. As with a monostatic radar, bistatic radar propagation requires a suitable path from the transmitter to the target and the target to the receiver. In contrast to a monostatic radar, however, propagation effects can be significantly different over the two bistatic paths and must be treated separately.

Multipath is the primary example, where the target can be in a multipath lobe on one path and a multipath null on the other, depending on antenna and target altitude and terrain conditions [97]. Multipath is the dominant factor when operating at VHF/UHF. Object multipath, such as specular reflections from artificial surfaces and buildings, is of particular concern in urban environments. The periodic fading, or *picket fence* reception, of FM broadcast signals by an automobile radio is a typical example of such (one-way) multipath.

A signal processing loss term, L_{sp}, is included in equation (6.8) to account for imperfect matched-filter detection performance. Specifically, a cross-correlator is nearly always used as

---

31 Coherent processing time t_c is used in place of the more commonly used noise bandwidth B_n to discourage attempts at achieving extraordinarily long detection ranges simply by reducing B_n (D. K. Barton, private communication, March 2006). An alternative method uses the ratio of processing gain G_p, also called the time-bandwidth product, to RF input bandwidth B. Since $G_p = t_c B$, both yield the same result: t_c.

the PBR matched filter, where the reference signal is the demodulated, direct path RF signal. As such, it is subjected to interference (multipath and RFI), which is different from interference affecting the target echo path. If the correlator operates in its linear region the echo and its interference signals convolved with the reference plus its interference signals produce the desired echo with full matched-filter gain plus interference signals with gain reduced by mismatch. These signals add vectorially to produce a modified pattern-propagation factor. If the correlator operates in its nonlinear region, which frequently happens, cross-products are generated, which reduce the echo's matched-filter gain. The amount of loss depends on the magnitude of the interference, which varies depending on the scenario and is accounted for in the signal processing loss term.[32]

6.5.3 Geometric Component of the Bistatic Radar Equation

The bistatic radar benchmark range, $R_M = (R_T R_R)^{1/2}$, was developed in Section 6.5.2. The free space, maximum detection contour of this benchmark is a circle of radius R_M, just as in the monostatic case. Such a circle assumes constant RCS, pattern propagation factors, and receiving system noise temperature, all of which are scenario- and geometry-dependent. For the general bistatic case, where $L > 0$, the free space, maximum detection contour becomes the familiar oval of Cassini, again with all the monostatic caveats cited above.[33] Thus, this oval (or ovals) provides a convenient but at times oversimplified view of bistatic range coverage and must be used with care.[34]

A further bistatic caveat is necessary. When the target is on or near the baseline, that is, located between receiver and transmitter where the bistatic angle $\beta \to 180°$, a completely different environment is generated: *forward scatter* from both target and clutter. In this case the target RCS and clutter-scattering coefficient (σ_o) are greatly enhanced, while range and doppler measurements are greatly degraded.

However, when the target lies on the *extended baseline*, the baseline extended beyond either the transmitter or the receiver, $\beta \sim 0°$, and the bistatic radar operates with many monostatic characteristics, including target RCS, clutter σ_o, range and doppler measurements, etc. This configuration is often called *pseudomonostatic*, and when the target lies on the receiver's extended baseline, the geometry is called *over-the-shoulder*.[35]

The revised method of presenting these ovals is to plot them as a function of the baseline range, L, normalized to the benchmark range, R_M, shown in Figure 6–28. Four cases are shown: (*a*) benchmark: $L = 0$; (*b*) one-oval: $L < 2R_M$; (*c*) lemniscate: $L = 2R_M$; (*d*) two ovals: $L > 2R_M$. In all cases the transmitter is located at the left oval focus (0). The receiver is located at the right oval focus (0,1,2,3). Table 6–4 lists specific values for oval area and maximum and minimum receiver ranges for these four cases, again referenced to the benchmark range, R_M.

Table 6–5 summarizes expressions for calculating these values. The one-oval area formula, also used for calculating the lemniscate area, is derived from equation (D.7a) [97]. Formulas for the two-oval area and range terms for $L > 3R_M$ are derived from equation (D.11a) [97]. More terms in these series can be used if more accuracy is required. Note that the area of every bistatic oval is always less than the monostatic circle.

32 D. K. Barton, private communication, March 2006.

33 An *oval of Cassini* is defined as the locus of the vertex of a bistatic triangle when the product of the sides adjacent to the vertex (R_T, R_R) is constant and the length of the opposite side (L) is fixed.

34 A striking example of how these free space detection contours can be distorted by terrain-induced multipath, diffraction, and/or masking effects is given in Chapter 3, Fluttar DEW-Line Gap-filler.

35 Another characterization of this over-the-shoulder geometry is to say that the receiver lies between the transmitter and target.

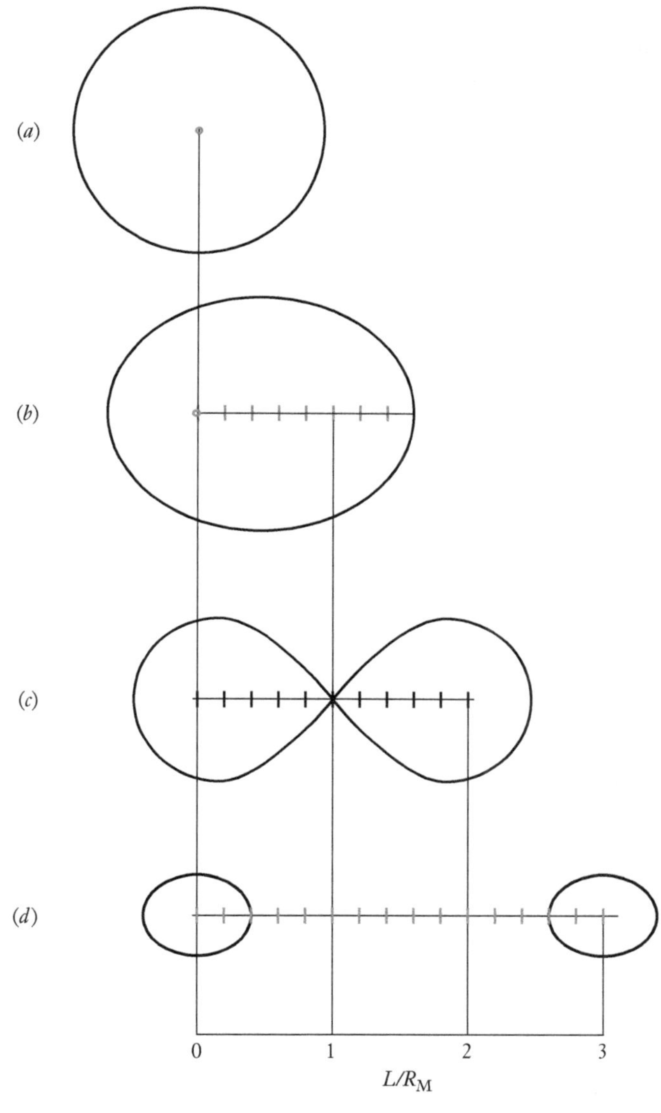

Figure 6–28 Normalized ovals of Cassini, lying in the bistatic plane, the plane containing transmitter, receiver, and target. T_x = Transmitter (left focus); R_x = receiver (right focus); T_{gt} = target (lying on oval); L = baseline (T_x-to-R_x distance); R_M = equivalent monostatic range; (benchmark case where $L = 0$); $R_{R(max)}$ = maximum R_x-to-T_{gt} range permitted on oval, when T_{gt} lies on the baseline extended beyond T_x; $R_{R(min)}$ = minimum R_x-to-T_{gt} range available on oval when T_{gt} lies on the baseline extended beyond R_x (over-the-shoulder geometry); and $R_{R(av)}$ = radius of a circle with area equal to area of one oval in the two-oval case.

Because of mirror-image symmetry in the two-oval case, maximum/minimum receiver ranges to targets lying on the transmitter-centered oval, $R_{R(max/min)}$ (on T_x oval), are calculated as:

$$R_{R(max/min)}\,(\text{on } T_x \text{ oval}) = L \pm R_{R(min/max)}\,(\text{on } R_x \text{ oval}) \tag{6.11}$$

Table 6–4 Area and Range Calculations for the Four Specific Cases, Where Case A Is the Equivalent Monostatic Range, or Benchmark

Case	T_x	R_x	Area (of one oval)	$R_{R(max)}$ (on R_x oval)	$R_{R(min)}$ (on R_x oval)
A	0	0	$1.00\,\pi\,R_M^2$	$1.00\,R_M$	$1.00\,R_M$
B	0	1	$\sim 0.98\,\pi\,R_M^2$	$1.62\,R_M$	$0.62\,R_M$
C	0	2	$\sim 0.38\,\pi\,R_M^2$	$2.41\,R_M$	$0.41\,R_M$
D	0	3	$\sim 0.11\,\pi\,R_M^2$	$0.38\,R_M$*	$0.30\,R_M$*

*$R_{R(av)} = 0.33\,R_M$.

Table 6–5 Area and Detection Ranges for General Ovals of Cassini, Where Terms Are Defined With Figure 6–28

Case	L	Area (of one oval)	$R_{R(max)}$ (on R_x oval)	$R_{R(min)}$ (on R_x oval)
Circle (Benchmark)	0	$\pi\,R_M^2$	R_M	R_M
One oval	$<2R_M$	$\sim\pi\,[R_M^2 - L^4/(64R_M^2)]$	$(R_M^2 + L^2/4)^{1/2} + L/2$	$(R_M^2 + L^2/4)^{1/2} - L/2$
Two ovals	$>2R_M$	$\sim\pi\,R_M^2\,[R_M^2/L^2]$	$L/2 - (L^2/4 - R_M^2)^{1/2}$	$(R_M^2 + L^2/4)^{1/2} - L/2$
	$>3R_M$	$\sim\pi\,R_M^2\,[R_M^2/L^2]$	$\sim R_M^2/L$	$\sim R_M^2/L$

Again, because of oval symmetry, corresponding transmitter ranges are the reverse of receiver ranges. With one exception, all expressions in Table 6–5 reduce to the monostatic case, the benchmark case, when $L \rightarrow 0$. The exception is in the two-oval case when L is constrained to be greater than $2R_M$, and thus cannot approach zero.

The expressions in Table 6–5 can also be used to assess LOS constraints of any bistatic radar. Specifically, for given target, transmitter and receiver altitudes, the target must be simultaneously within LOS to both the transmitter and receiver sites [97]. For a smooth, 4/3 earth model and ignoring multipath lobing, the LOS range r_R between a receive antenna of altitude h_R and target of altitude h_t is

$$r_R = 130\,(\sqrt{h_R} + \sqrt{h_t}) \tag{6.12}$$

where all units are in kilometers.

Similarly, the LOS range r_T between a transmit antenna of altitude h_T and target of altitude h_t is

$$r_T = 130\,(\sqrt{h_T} + \sqrt{h_t}) \tag{6.13}$$

Thus, to prevent LOS truncation of the ovals,

$$r_R \geq R_{R(max)} \text{ and } r_T \geq R_{T(max)}.$$

A more common PBR task is to select a baseline L so that a receiver with antenna altitude h_R will match existing LOS coverage of a transmitter with antenna altitude h_T. For a typical PBR over-the-shoulder geometry (receiver between transmitter and target) the requirement would be to match LOS coverage on the extended baseline, $r_T = r_R + L$, so that

$$L = 130\,(\sqrt{h_T} - \sqrt{h_R}) \tag{6.14}$$

For example, when $h_T = 0.1$ km and $h_R = 0.01$ km, $L = 28$ km, which from equations (6.12) and (6.13) will provide LOS to an 8.5-m altitude target flying over the extended baseline, 25 km from the receiver and 53 km from the transmitter.

Baselines greater than 100 km can pose severe target LOS problems. For example, if $L = 120$ km and $h_T = 300$ m, then from equation (6.13) the transmitter would only illuminate a target flying directly above the receiver at an altitude $h_t > 140$ m. Thus, the target could readily underfly the illumination and low-altitude air surveillance capability is lost. As a consequence, the bistatic radar must either employ greatly elevated (~1 km) transmitters at these long baselines or operate with shorter baselines to achieve acceptable low-altitude coverage.

Similar expressions can be developed when $h_R > h_T$ and for the mirror–image geometry. Note that the two-oval case also requires a very high altitude for the site located in the oval not under surveillance—so high that the site must often become airborne.

Finally, the transmit antenna will be in direct LOS of the receive antenna when $L \leq r_T + r_R$ with $h_t = 0$, so that

$$L \leq 130\,(\sqrt{h_T} + \sqrt{h_R}) \tag{6.15}$$

If equation (6.15) is satisfied, extraordinary measures are usually required to suppress the direct path signal to a level where targets can be detected. Examples are given in Section 6.4.

While this special geometry was developed to simplify the assessment of PBRs, it also applies to other bistatic radars, especially those that hitchhike off monostatic radar transmitters. An exception is the forward-scatter case. Even though ovals of Cassini apply here, they become greatly distorted by the very large changes in target cross section at receiver look angles near the transmitter. (Recall that invariant target cross section was a necessary condition in constructing the ovals.) Consequently, they are seldom used to assess forward-scatter bistatic radar performance.

6.5.4 Operating Regions Ovals of Cassini define three distinct operating regions for a bistatic radar: cosite, receiver-centered, and transmitter-centered [97]. In Figure 6–28, cosite corresponds to Case B, receiver-centered to Case D right oval, and transmitter-centered to Case D left oval. The type of transmitter exploited by a PBR—cooperative (allied or friendly) or noncooperative (hostile or neutral)—completes the taxonomy. Table 6–6 summarizes this taxonomy, along with the types of PBR operation that have been considered in each operating region.

Table 6–6 Passive Bistatic Radar

Receiver Operating Regions	Range Relationships	Cooperative Transmitter	Noncooperative Transmitter
Cosite	$L < 2R_M$	Air surveillance Intrusion detection Range instrumentation Ionospheric measurements	Air surveillance
Receiver-centered	$L > 2R_M$ $R_T >> R_R$	Short-range surveillance Planetary surface measurements	Short-range surveillance
Transmitter-centered	$L > 2R_M$ $R_R >> R_T$	Planetary surface measurements	Air threat monitoring

The term cooperative transmitter is actually a misnomer in the sense that seldom will a cooperative transmitter be inclined to *cooperate* with the PBR. For example, it would likely resist replacing broadcast material with special waveforms, or even superimposing these waveforms on broadcast material. A possible exception might be in time of war when a TV or an FM station could be persuaded to stay on the air after its normal sign-off time and transmit a CW or noise-like waveform to support, for example, intrusion detection. But even this possibility is problematical since most stations now broadcast continuously. As a result, the PBR is free to exploit cooperative or noncooperative transmissions if they are suitable; but to date, commerce totally controls their use, with the bistatic radar remaining a user of opportunity.

The exploitation of noncooperative transmitters requires elaboration. If a noncooperative transmitter and a PBR receiver were located in or near a battle area, a PBR could use that transmitter to conduct air surveillance in the cosite region ($L < 2R_M$), just as it would for a cooperative transmitter. A PBR receiver could also exploit a noncooperative transmitter located significantly behind a battle area to monitor hostile or neutral air activity near that transmitter in the transmitter-centered oval ($L > 2R_M$, $R_R >> R_T$). Finally, a PBR receiver implanted in or flying over hostile areas could use any high-powered satellite transmitter illuminating that area to conduct short-range surveillance in the receiver-centered oval ($L > 2R_M$, $R_T >> R_R$).

An example of exploiting a cooperative transmitter is the Lockheed Martin *Silent Sentry*™ system [102]. In field tests it has demonstrated 2D air surveillance performance ($R_M \sim 100$ km) using multiple FM broadcast transmitters (as reported in Table 6–2). This capability, especially the covert, counter-stealth features of the system, can be used for medium-range air defense.

Other PBRs have demonstrated aircraft detection in field tests, including two versions of the British TV-based Bistatic Radar [103, 104], and the University of Washington's *MRR* [105]. However, the primary function of this latter radar is for ionospheric measurements. Such measurements were successfully made in 2005 and are reported in Chapter 7.

Silent Sentry™ has also demonstrated a range instrumentation function by 3D tracking of space shuttle launches to ranges > 100 km (Table 6–2). Here, geometry—in terms of a favorable *GDOP*—allowed 3D measurements, again using multiple FM and/or TV transmitters surrounding the launch site.

Finally, moon and planetary surface measurements by (passive) bistatic radars have been ongoing for nearly 40 years. The first successful operation was in 1967, using Luna-11 communication signals scattered from the moon. Similar measurements were made using Lunar Orbiter-1 and Explorer-35. Venus and Mars bistatic radar measurements were made using such satellites as Venera-10 and Mariner-6 and -7 [106].

The primary bistatic radar configuration for these measurements used the transmitter-centered oval region. Communication signals first transmitted from the probe vehicle, then scattered off the planet's surface, were received by one of the large, earth-based stations, such as the NASA Deep Space Network or the Arecibo Observatory. A second, more complex configuration used the receiver-centered oval, where the bistatic receiver was carried by the probe, collecting earth-transmitted signals first scattered off the planet's surface. Details of these PBR systems appear in Chapter 5.

6.5.5 Air Surveillance Performance Assessment

The following assessment of PBR air surveillance performance details the range performance that can be expected from a generic class of PBRs, for example, PBRs exploiting analog FM or TV broadcast transmitters, GSM cell phone transmitters, etc. It does not purport to emulate specific systems since some of their design parameters have yet to be published. Nor does it emulate experimental systems since parameters of these systems can be expected to change as a result of new testing.

In all cases, equation (6.8), the bistatic range equation solved for the benchmark range $R_M = (R_T R_R)^{1/2}$, is used for this assessment of PBR capability to conduct air surveillance, with emphasis on air-defense and air-traffic control operations. In addition, since many PBR parameters in the range equation have not been rigorously quantified in laboratory or field measurement programs, they are established through extrapolations of similar, measured parameters or as a last resort, estimates based on other documented radar programs.

While such a process initially appeared to be a rationale surrogate for actual measurements, the extrapolations and estimates showed a surprising divergence among *PBR experts*. Specifically, the value of many parameters was found to differ by as much as 5 dB. And, equally surprising was the final result, which nearly always converged to a unified value of the benchmark range. One could postulate that since many of these benchmark ranges (or similar measures of PBR range performance) were first calculated or measured, and then reviewed, assessed, and generally agreed to by these experts, the final result was known *a priori*. Thus, our experts (including many authors in this book) simply backed into their own, and preferred, set of parameters. Whatever the process, the following assessment must be considered an initial estimate of PBR air surveillance performance, to be revised whenever measured data becomes available.

6.5.5.1 Transmitters Properties of typical transmitters that can be exploited for PBRs are shown in Table 6–7. The transmit power P_t is substantial for broadcast systems, since their commercial receivers typically have inefficient antennas and poor noise figures; furthermore, the transmission paths are often far from the line of sight. Thus, the transmit powers have to be significantly higher to overcome these inefficiencies and losses. Transmit antennas for broadcast systems are omnidirectional in azimuth, and are sited on tall masts on high locations to maximize coverage

Table 6–7 Parameters of Typical Broadcast Transmitters*

Transmitter	Frequency (MHz)	Modulation	Available RF Bandwidth, B (kHz)	Maximum ERP, P_tG_t (kW)
FM (analog)	~100 (U.K.) 88–108 (U.S.)	FM	50	250
TV (analog) (Power utilization factor, $\rho = 0.5$)	~550 (U.K.) 76–88, 174–216 (U.S.)	Vestigial-sideband AM (vision); FM (sound)	5,500	1,000
Digital audio broadcast	~220	Digital, COFDM	220	10
HD-TV (U.S.)	600–800	Digital	5,000	1,000
	200 (future)	Digital	5,000	1,000
Digital TV (U.K.)	~750	Digital	6,000	8
Cell phone (GSM)	900, 1,800	GMSK, FDM/TDMA/FDD	200	0.4
Cell phone (3G)	~2,000	CDMA	3840	0.4

*Details and acronyms are given in Section 6.4.1 and Table 6–3.

for ground-based receivers. The vertical-plane radiation patterns are tailored to avoid wasting power above the horizontal [96]. The result is a modest transmit antenna gain, G_t.

The radiation patterns of GSM cell phone base station antennas are typically arranged in 120° azimuth sectors, and again shaped in the vertical plane to avoid wasting power, so transmit antenna gains are similarly modest. Their pattern of frequency reuse means that there will be cells using the same frequencies within quite short ranges of each other. Licensed ERPs are typically in the region of 400 W, although in most cases implemented transmit powers are lower. While the figures quoted in the paragraph above relate to the United Kingdom, the figures for other countries will be comparable.

As argued in Section 6.2, satellite transmitters are omitted from the list and subsequent calculations since they currently cannot support useful air (or ship) surveillance capability.

6.5.5.2 Bistatic RCS As detailed in Section 6.3.1.2.1 and the associated Appendix A, conventional aircraft were found to exhibit monostatic RCS resonance enhancement at VHF and low UHF, typically 100–450 MHz, occasionally extending up to 600 MHz. As also detailed in that section, stealth vehicle modeling tends to confirm this effect, particularly at 100 and 400 MHz [107]. Available bistatic data showed nearly identical enhancement at 250 MHz, which can reasonably be extrapolated to the higher frequencies for moderate bistatic angles ($\beta < 90°$). The enhancement is typically 7 dB when compared to monostatic microwave RCS data for conventional aircraft, with excursions as high as 12 dB.

At microwave frequencies the bistatic RCS of complex targets, that is, targets with multiple scattering centers, should be reduced. Specifically, the bistatic RCS can roll off as much as 1 dB per 3°, for $0° < \beta < 30°$, giving a 10-dB loss of RCS at $\beta = 30°$ [108]. This effect is caused by the loss or shadowing of discrete scattering centers including resonant cavities and retroreflectors [109]. While this rather severe loss was measured for coastal freighters with many flat surfaces, dihedrals, and trihedrals, similar effects have been observed on other complex targets. However, since most aircraft have blended surfaces and are less complex, a more modest, 5-dB "bistatic penalty" should be added for most bistatic geometries, for example, $\beta > 10°$, in the microwave region.

Since single values of the bistatic RCS, σ_B, are required for benchmark calculations, a small fighter (or 4-passenger jet) with a 2 m^2 monostatic microwave RCS [110] will be designated as the critical target of interest for air surveillance. It will be enhanced by 7 dB to 10 m^2 for bistatic operation at VHF and low UHF. It will be reduced by 5 dB to 0.6 m^2 for bistatic operation at higher frequencies and $\beta > 10°$. For the special case of exploiting commercial airliners (or light bombers) in PBR test scenarios at VHF and low UHF, a 100 m^2 bistatic RCS—based on measured data—will be used. At higher frequencies, Skolnik's 20 m^2 monostatic RCS, again reduced by 5 dB to 3 m^2, will be used for bistatic operation. The VHF and low UHF estimates also apply to stealth aircraft. No reliable data is available on these aircraft at higher frequencies. Table 6–8 summarizes these estimates, where the transition between low and high UHF is taken at 450 MHz.

6.5.5.3 Noise Temperature As detailed in Section 6.4.2 a receiving system noise figure of 20–25 dB appears to be the minimum achievable using suppression (shielding, filtering, and cancellation) in RFI environments encountered at VHF and UHF bands. Assuming a 1-dB RF component loss, these noise figures map into a receiving system noise temperature, T_i, of 36,250 K to 115,260 K. For the benchmark case, assume T_i = 100,000 K in the VHF/UHF bands. At frequencies above UHF, the RF environment has not been characterized, but should be not as severe: possibly 10–15 dB noise figures, which map into noise temperatures of 3,365–11,265 K. Assume T_i = 5,000 K for the benchmark case at frequencies above UHF.

Table 6–8 Summary of Radar Cross Section Estimates With Values in Bold Used in the Benchmark Calculations*

	Conventional		Stealth	
	Monostatic	**Bistatic**	**Monostatic**	**Bistatic**
VHF/Low UHF				
Fighter	10	**10**	10	**10**
Bomber	100	100	100	100
High UHF/Microwaves				
Fighter	2	**0.6**	Unknown	Unknown
Bomber	20	3		

*Low UHF < 450 MHz; High UHF > 450 MHz.

6.5.5.4 Coherent Processing The coherent processing time t_c, and hence the noise bandwidth of receiver's predetection filter B_n, is usually set by the amount of doppler spreading or velocity walk by the target, equation (6.10). For example, at 100 MHz a 1 *g* radial component of acceleration (along the bistatic bisector) will allow 0.4 sec of coherent processing; 3 *g*, 0.23 sec. At 500 MHz, a 1 *g* component will allow 0.18 sec; 3 *g*, 0.1 sec. In recent PCR test programs using commercial aircraft as their targets of opportunity, the doppler spread was found to be smaller than these values—as expected. Hence, the coherent processing time could be longer, ranging from 0.5 s to 1.0 s. These conditions, while suitable for commercial air surveillance, are not so for air defense. Hence, a 0.3-sec coherent processing time and a corresponding receiver bandwidth, $B_n = 3.3$ Hz will be assumed for the benchmark calculations. For reasons cited earlier, noncoherent integration will be neglected.

6.5.5.5 Receive Antenna As outlined in Section 6.3, PBR receivers designed for covert, mobile, and/or low-cost operation are nearly always precluded from using large antennas, such as the one used by the Russian *55G6U* long-range, VHF air surveillance radar (Figure 6–10). Smaller antennas, such as a Yagi or compact circular arrays, are typically employed by these receivers. Their coverage can be tailored in elevation to match that of a broadcast transmitter, and in some scenarios can be constrained to a 90° or 120° azimuth sector. The resulting receiving antenna power gain G_R will be modest, typically 6–10 dBi. A value of 8 dBi is used for these mobile PBRs.

For the special case of urban air defense using suitably located buildings as platforms for large phased array receive antennas, a 15 m × 60 m array will be assumed, which at 100 MHz will support about 400 elements for an approximately 26-dBi gain. This limited excursion is called the fixed-site PBR.

6.5.5.6 Propagation Factors, Losses, and SNR Transmitter and receiver pattern propagation factors F_T and F_R are assumed to be unity, with the caveat that a realistic scenario must include multipath effects, which can be severe at VHF and UHF frequencies. Transmitting and receiving system losses are assumed to be 6 dB. Since most of the transmitting system losses L_T should be included in Table 6–7, receiving system losses $L_R L_{sp}$ will comprise the bulk of 6 dB, which may be a bit conservative.

Finally, the SNR required for detection $(S/N)_{min}$ is assumed to be 13 dB, which closely corresponds to a 0.5 probability of detection (P_d) and 10^{-6} probability of false alarm for all Swerling target fluctuation models [94]. This assumption is somewhat of an artifact, but is selected for two reasons: (1) it finesses the issue of which fluctuation model to use at which frequency, especially in the VHF resonance region, since the SNR for all models is within 1 dB at a 0.5 P_d, and (2) many air surveillance radars are designed with a 0.5 P_d, under the assumption

Table 6–9 Summary of Calculations for Free Space, Equivalent Monostatic Range, or *Benchmark Range* of Generic Passive Bistatic Radars*

Transmitter	Frequency Used in Calculations (MHz)	Benchmark Range, $R_M = (R_T R_R)^{1/2}$ (km)	Typical $T_x - R_x$ Baseline range, L (km)	T_x power spectral density @ min L, Φ' (dBW/Hz)	Required Direct Path Cancellation, Φ'/N_0 (See text) (dB)
FM (analog)	100[a]	120	50–120	−91	88
FM (analog, fixed site excursion)	100	340	50–120	−108	71
TV (analog)	550 (U.K.)	30	10–30	−107	72
	200 (U.S.)[a]	100	30–100	−107	72
Digital audio broadcast	220[a]	36	10–40	−105	74
HD-TV (U.S.)	700	31	10–30	−108	84
	200 (future)[a]	120	50–120	−111	68
Digital TV (U.K.)	750	9	10	−131	61
Cell phone (GSM)	900	4	5	−125	67
	1,800	6	5	−131	61
Cell phone (3G)	2,000	5.5	5	−146	46

* Frequencies with a superscript are candidates for the fixed-site PBR excursion. The FM (analog) PBR is shown as an example. (See the text.) Typical maximum levels of direct path cancellation are also shown.

that sufficient time is available for multiple dwells or scans, which can raise the cumulative P_d to a more conventional 0.9 value.

6.5.5.7 Equivalent Monostatic Range Summary Column 3 of Table 6–9 summarizes results of the free space, PBR benchmark range calculations using the assumptions and parameters listed in Sections 6.5.5.1–6.5.5.6. Frequency is given in column 2 and is roughly the midpoint of the broadcast range.

As anticipated, and ignoring the fixed-site excursion for a moment, PBRs exploiting high-power, VHF terrestrial broadcast transmitters achieve the greatest benchmark range performance: ranges near 100 km for analog FM, TV, and the future U.S. HD-TV. These ranges represent a medium-range, 2D air surveillance capability against both conventional and stealth aircraft. A PBR exploiting the VHF DAB transmitter achieves a short-range capability against both conventional and stealth aircraft. PBRs exploiting high UHF terrestrial broadcast transmitters achieve ranges near 30 km, representing a short-range capability, but only against conventional aircraft.[36] PBRs that exploit the maximum cell phone ERP of 400 W require an additional 30–35 dB of link margin to offer even a short-range air surveillance capability against conventional aircraft.

36 The 550-MHz U.K. TV transmitter might offer a counter-stealth capability but RCS measurement data remains controversial in this frequency range.

The fixed-site excursion is a limited urban defense scenario using a very large, building-mounted (or tree-hidden) receive antenna as a PBR stealth trap. Any of the VHF (and low UHF when available) PBR systems might exploit this configuration, including FM, TV, DAB, and the future HD-TV. When exploited, the PBR benchmark ranges will be extended by a factor of $\log^{-1}[(26 \text{ dBi} - 8 \text{ dBi})/40]$, or 2.8. The analog FM PBR is shown as an example and now achieves a long-range air surveillance capability. Furthermore, it can autonomously provide accurate 3D data for target state estimates. The TV systems achieve similar performance with the DAB's benchmark range extended to 100 km—all formidable capabilities against stealth aircraft. The trade-off for this capability is significantly increased cost, total loss of mobility, and possible compromise of covertness.

Note that these calculations assume RFI levels typical of urban areas, which generate noise figures near 25 dB at VHF/UHF and closer to an estimated 10 dB at frequencies above UHF. These values are expected to decrease in suburban and rural areas. But so will the number of available transmitters, particularly in rural areas. Thus, great care must be taken in siting PBRs in specific scenarios and then selecting a realistic noise figure for each PBR in that scenario.[37]

Column 4 of Table 6–9 shows typical transmitter–receiver baseline ranges. For air surveillance operation, the PBR designer would be inclined to select a maximum baseline that satisfies two conditions. (1) A baseline that generates ovals approximating circular monostatic coverage, that is, operation in the cosite region. From Figure 6–28, such operation requires $L \le R_M$. This condition is often required for multistatic PBRs, which operate with overlapping coverage from multiple sites. (2) A baseline that generates acceptable low altitude coverage around the receive site. From equations (6.12) to (6.14), such coverage requires the transmit site be elevated to ~1 km altitude for >100 km baselines. Otherwise, shorter baselines are required. To reflect these conditions the maximum baseline shown in the table, and particularly for the fixed-site excursion, is constrained to 120 km. In contrast, the minimum baseline is usually selected as a tradeoff between low altitude air coverage and the amount of direct path cancellation that can be achieved, as detailed next.

Column 5 of Table 6–9 calculates the illuminator's free-space, power spectral density at the PBR receiver in dBW/Hz, as follows:

$$\Phi' = P_t\, G_t\, (G_r)_t\, \lambda^2 \,/\, (4\pi)^2 B\, L_{min}^{\ 2} \tag{6.16}$$

where $P_t G_t, \lambda$ (*c*/frequency) and B (input RF bandwidth) are given in Table 6–7, L_{min} is the minimum value of L shown in the table, transmitter pattern propagation factors and losses are assumed to be unity, and $(G_r)_t$ is the receive antenna's gain in the direction of the transmitter. For normal PBR operation $(G_r)_t$ = 8 dBi; for the fixed-site PBR excursion $(G_r)_t = -9$ dBi, assuming the 26-dBi main lobe gain and typical -35-dB side lobes. When normalized by $N_o = kT_i$, the receiver's input noise power spectral density, it indicates the maximum amount of direct-path cancellation needed to reduce the direct path signal to a level of the receiver's input noise temperature, T_i.

Column 6 of Table 6–9 gives the ratio, Φ'/N_o, for the benchmark example, $N_o = -179$ dBW/Hz at VHF/UHF and -192 dBW/Hz above UHF. At this level of cancellation, N_o will be doubled and the benchmark range R_M will be reduced to R_M' in a manner analogous to the noise jamming case [97, eq. 10.5]:

$$R_M' = R_M\, [N_o/(N_o + \Phi')]^{1/4} \tag{6.17}$$

37 A coauthor lives in Carmel Valley, a semirural area of California, which does not receive broadcast TV and only a few low-powered FM stations, even though communities *over the hill* have had full broadcast service for more than 50 years. It was permanently populated only when cable TV arrived in the 1970s.

For example, when $\Phi' = N_o$, $R_M' = 0.84\ R_M$. So there is an advantage to increasing direct path cancellation to levels below ambient input noise. In fact, some PBR test programs have strived for 10–20 dB of additional cancellation at VHF/UHF.

Some PBRs have generated equivalent monostatic ranges during field tests that significantly exceed the calculated values shown in Table 6–9. For example, Table 6–1 shows Howland's *FM Radio-Based Bistatic Radar* [111] with R_M = 175 km and an early version of *Silent Sentry*™ near 150 km. Much of this difference can be ascribed to the targets of opportunity used in the tests: commercial jet airliners, which generate a large bistatic RCS, for example 100 m^2 at VHF. Thus, when working with the smaller, 10 m^2 targets assumed for the benchmark range calculations, the Howland FM system's benchmark range would decrease to approximately 100 km and the *Silent Sentry*™ TV-mode, to approximately 85 km, which more closely match the values in Table 6–9.[38]

In another example, Andrews predicted a benchmark range of 30 km on a 1 m^2 target for the *Multistatic HDTV-Based Radar* [112], which approximates the 31 km on a 0.6 m^2 target given in Table 6–9. However, in instrumented field tests it exceeded those predictions by detecting a small Cessna aircraft out to approximately 40 km, roughly equivalent to a 5-dB increase in link margin. Usually these small differences are considered within the uncertainty level of field tests and ignored. But in this case they can (probably fortuitously) be rationalized by observing that the tests were conducted at small bistatic angles where the 5-dB *bistatic penalty* would not apply. Hence, the additional detection range on the small aircraft.

6.5.6 The 6-dB per Octave Path Loss Issue and a Multistatic Application

From time to time it is asserted that doubling the range of a PBR requires only a 6-dB increase in SNR. More specifically, what is meant is that any 6-dB change in the variables of the bistatic radar range equation will double (or halve) the receiver-to-target range R_R. It is clear that assertion is correct if one of the target legs of the bistatic triangle is much greater than the other, for example, if either the transmitter or the receiver were located on a satellite. But it is not obvious when the legs are more nearly the same length. This section quantifies the issue and then applies the result to optimize multistatic coverage.

Consider a simple variation of bistatic radar range equation (6.8):

$$(R_T R_R)^2 = \kappa \tag{6.18}$$

where R_T = transmitter-to-target range, R_R = receiver-to-target range, κ = bistatic maximum range product-squared, consisting of all terms within brackets on the right side of equation (6.8).[39]

Suppose in Figure 6–29 that the receiver look angle α and the baseline range L are held constant. Then when R_R is changed by some amount ΔR_R, R_T will be changed by some other amount ΔR_T.

These changes in turn will require a corresponding change δ in κ to maintain adequate detection performance:

$$(R_T + \Delta R_T)^2 (R_R + \Delta R_R)^2 = \delta\kappa = \delta (R_T R_R)^2 \tag{6.19}$$

38 Howland's earlier *TV-Based Bistatic Radar* also generated large benchmark ranges at UHF, but some of this increased performance is ascribed to operation at very large bistatic angles, thereby generating significant forward-scatter enhancement.

39 The term κ is called the *bistatic maximum range product* in [97], a confusing designation since $R_T R_R$ is also called that in this section—and a more logical descriptor. To resolve the ambiguity, κ will now be called the *bistatic maximum range product-squared.*

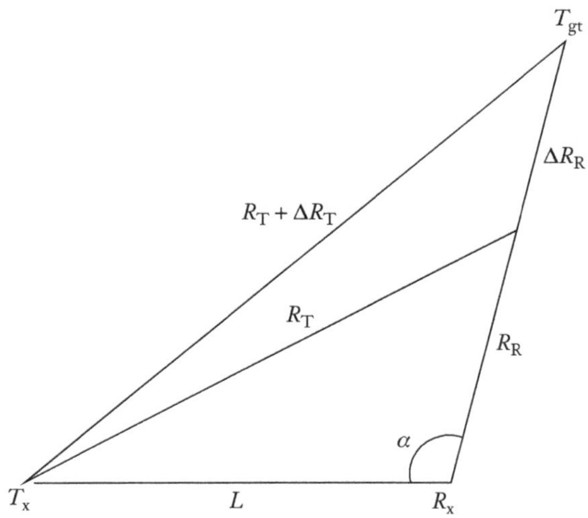

Figure 6–29 Geometry for the 6-dB per octave issue.

Let $\Delta R_R = R_R$, so that R_R is doubled. Then solving for δ yields:

$$\delta = 4[(R_T + \Delta R_T)/R_T]^2 \tag{6.20}$$

As expected, when $R_T >> R_R$ (and consequently $>> \Delta R_R$ and ΔR_T) terms in the brackets approach 1 and $\delta = 4$, or the claimed 6-dB change. (A similar case can be made for receiver on satellite and transmitter on ground.) But when R_T and R_R are more nearly the same, terms within brackets, which establish the geometry-dependent change in transmit path loss, must be accounted for.

Values of R_T and ΔR_T are determined by applying the law of cosines in Figure 6–29. For example, assume $L = 50$ km, $R_R = 50$ and 100 km for $0° \le \alpha \le 180°$. Table 6–10 summarizes results.

At large values of α the transmitter is behind the receiver, in an over-the-shoulder geometry, where $R_T > R_R$. Since R_T started at a large value, the change is relatively small. Thus, the transmitter's contribution to δ is relatively small, < 6 dB.[40] At values of α near 90° coverage is on both sides of the baseline where $R_T \approx R_R$. Consequently $\delta \rightarrow 12$ dB, the monostatic case. At small values of α the receiver is now looking back toward the transmitter and $R_R > R_T$. Since initial values of R_T are small, the transmitter's contribution to δ is now much larger, >6 dB—in fact increasing without limit at $\alpha \rightarrow 0°$.

These examples show that the required change, δ, in the bistatic maximum range product-squared κ can vary from 6 to >20 dB, depending upon geometry. The lower limit is the special case of one site very far away, where $\delta = 6$ dB. The upper limit occurs at small α and $R_R \approx L$, which limits operation to regions around the transmitter. Since PBRs typically operate at larger values of α, they can often achieve somewhat less than 12 dB per octave monostatic path loss.

40 And as outlined previously, will trend to 0 as R_R becomes smaller so that $R_T >> R_R$.

Table 6–10 Required Increase, δ, in the Bistatic Maximum Range Product-Squared, κ, When R_R Is Doubled and $L = 50$ km, as a Function of the Receiver Look Angle α

α (°)	R_T (km)	$(R_T+\Delta R_T)$ (km)	δ (dB)
(*a*) $R_R = 50$ km			
2	1.7	50	35.2
10	8.7	51.5	21.5
45	38.3	73.7	11.7
90	70.7	111.8	10.0
135	92.4	139.5	9.6
180	100	150	7.8
(*b*) $R_R = 100$ km			
0	50	150	15.6
10	51.5	151.0	15.4
45	73.7	168.4	13.2
90	111.8	206.2	11.3
135	139.9	238.0	10.6
180	150	250	10.5

Occasionally, the argument is made that the number of terrestrial TV and FM transmitters can be so large in some geographic areas that their combined effective radiated power density becomes constant over a particular region. Hence, as in the case of a satellite transmitter, the PBR can operate with the *6-dB per octave* path loss. This argument assumes that (1) these regions are well identified and of interest for air surveillance, (2) the PBR has the capability to exploit specific transmitters for specific range-azimuth surveillance regions, and (3) multiple transmitters illuminate each of these surveillance regions if the PBR employs multilateration for target location. While such regions may exist, they should be considered very special cases, and not relied upon on for a robust PBR design.

PBRs typically operate in a multistatic, multilateration radar configuration using multiple transmit–receive pairs, all with overlapping coverage. Such a configuration is usually needed to generate acceptable target location accuracies for air surveillance, as detailed in Section 6.6. Thus, optimizing (free-space) coverage in a multistatic system is an interactive exercise in geometry and the bistatic radar range equation. For example, once an air surveillance region is selected and suitable transmitter–illuminator(s) is located, the bistatic receiver(s) can be sited and their common coverage (i.e., the intersection of their individual coverages) estimated or measured. When a specific site is found to truncate the common coverage, it can be reoriented, resited, or even redesigned if necessary, with the aid of equation (6.19).

Specifically, the bistatic maximum range product-squared κ is calculated and a ΔR_R required to reduce the truncation is determined; then ΔR_T as a function of α and L is calculated, which determines δ. If reorienting or resiting the receiver does not sufficiently reduce δ, the system must be redesigned to increase κ. Otherwise, coverage will be reduced or performance within the desired coverage will be degraded, in terms of noisy, fractured, or lost tracks and increased false tracks.

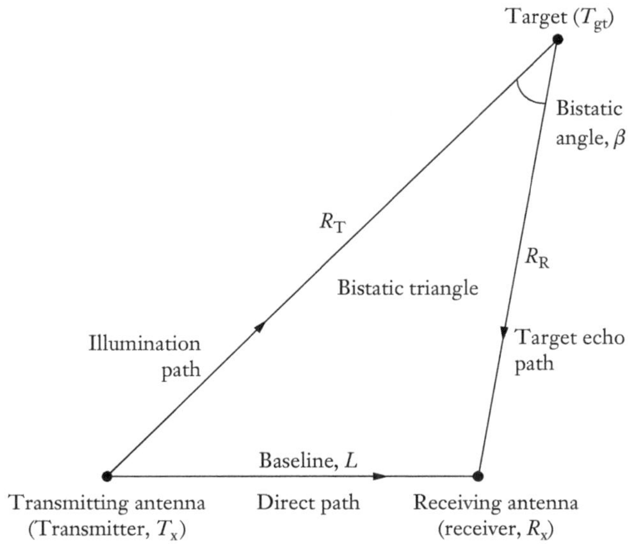

Figure 6–30 Bistatic radar triangle, which defines the bistatic plane. All range measurements are taken in this plane. Direction of arrival (DOA) measurements can be taken as a monostatic radar does, for example, using North and vertical references. Oftentimes it is convenient to reference DOA measurements to the baseline or a perpendicular to the baseline. From [1].

6.6 TARGET LOCATION

6.6.1 Introduction

While estimating target location (or more comprehensively the target state, which includes target velocity and sometimes acceleration) is a straightforward, well-established process for a monostatic radar, it is more complicated for a bistatic or multistatic radar. In nearly all cases the target is located by determining the bistatic radar triangle formed by the transmitter, receiver, and target, as shown in Figure 6–30.

In the standard approach to this problem, the bistatic receiver measures both bistatic range ($R_T + R_R$) and DOA of a target and obtains a separate estimate of the baseline range L. Then explicit equations, for example developed in [113] and expanded in [114, 115], are used to establish the triangle and hence target location with respect to the receiver. Approaches using other measurements have been developed and are also summarized in [113].[41] The receiver can be configured to measure azimuth DOA for 2D target location or azimuth and elevation DOA for 3D target location, depending on the operational requirement.[42]

With one exception, configuring a VHF or UHF PBR to measure target DOA with sufficient accuracy for air surveillance is usually not a viable option. The antenna needed for this task becomes so large that many of the advantages enumerated in Section 6.3.1 (covertness, mobility,

41 The target can also be located without first determining the bistatic triangle by multistatic triangulation, which uses DOA-only data from multiple receive sites and typically the global positioning system (GPS) to locate each receive site. Triangulation is used by the *Space Surveillance System* (*SPASUR*) for satellite location, but because of the enormous aperture sizes (array lengths for *SPASUR*) required for sufficiently accurate DOA measurements at useful ranges, triangulation is seldom considered for other applications.

42 Except for special circumstances, elevation DOA measurements are needed to estimate target altitude.

low cost) are compromised. Thus the PBR is constrained to operate with degraded (or no) DOA estimates, and must find methods to mitigate this limitation. This limitation also applies to pseudomonostatic target location, that is, when the bistatic angle, β, is small and bistatic geometry approaches the monostatic case. Even here, target location is greatly complicated without good DOA estimates—as it would also be in the traditional monostatic case.

The exception is the so-called *Lanterman Tower Excursion*, a fixed-site, urban defense scenario using a very large, building-mounted (or tree-hidden) phased array receive antenna as a PBR stealth trap. This configuration operates in the classic bistatic mode outlined above to generate a long-range, autonomous, 3D air surveillance capability against stealth vehicles. Consequently, it is not confronted with the limitations outlined in this section and does not require further discussion.

For the remaining PBR systems two mitigation approaches have been explored over the past 50 years: parameter estimation and multilateration. The former is typically used with a bistatic (one receiver and one transmitter) configuration; the latter with a multistatic (multiple receivers and/or multiple transmitters) configuration.

6.6.2 Bistatic Configuration

As outlined above, target state estimates are more complicated for a bistatic radar. And, they are vexingly more complicated for a PBR using a single transmitter, a single receiver and a small receive antenna. Such a configuration typically measures bistatic range and/or bistatic doppler with accuracy adequate for air surveillance. But, because of the necessarily small receive antenna, it measures target DOA with poor accuracy, or simply gives up on a DOA measurement altogether. As a result, the fundamental issue becomes converting a sequential set of relatively accurate range and/or doppler measurements into target state estimates either alone or with little help from relatively inaccurate DOA measurements.

6.6.2.1 Patton Approach The first attack on this problem was published by R. B. Patton Jr. in 1960 [116], albeit for a slightly different scenario: determining satellite orbits on a single pass from doppler observations taken by a ground-based CW bistatic radar. Patton reports his solution:

> The method of solution consists of fitting a computed curve to a set of Doppler observations by a least squares technique in which a compatible set of approximations to the six orbital parameters are derived and then improved by repeatedly applying differential corrections until convergence is achieved. Keplerian motion is assumed. The equations of condition are derived from the Taylor expansion about that point which consists of the approximate values of the six orbital parameters...
>
> ...Since there are six unknowns, a minimum of six equations are required for a solution. In practice, sufficient data are available to provide an overdetermined system, thus permitting the least squares solution...
>
> Of primary importance to the success of the method... is the capability of establishing a compatible set of initial approximations which are sufficiently close to the actual values to permit convergence of the computing process. It has been determined that the computation will converge with input consisting of data from a single receiver system, if the base line from the transmitter to the receiver is not excessively short, and the initial approximation... [for the position and velocity components] is moderately accurate. For example, with base lines of the order of 400 miles and either continuous or intermittent observations over time intervals of approximately five minutes, convergence may reasonably be expected when the error in each coordinate of the initial estimate is not in excess of 50 to 75 miles and the velocity components are correct to within 1/2 to 1 mile per second. For a two-receiver system, on the other hand, an accuracy of 150 to 200 miles in each coordinate and 1 to 2 miles per second in each velocity component has been found sufficient to secure convergence....
>
> Clearly, a supporting computation to furnish moderately accurate initial approximations is essential to the successful application of the computing method. Several approaches to this phase of the problem have

been considered, but one in particular is preferred. It involves using an analog computer to fit a computed curve to an observed Doppler frequency–time curve. In the fitting procedure, the boundary values of a second-order differential equation are systematically varied until the computed *S* curve [a typical doppler versus time curve for any passing object] matches the observed *S* curve... [116]

Patton then wrestled with the initial approximations. One parameter, from which satellite altitude z_0 was derived, assumed a circular orbit. Patton observed that highly eccentric orbits significantly degraded that estimate. In such cases,

...The preferred procedure is to assume that $\dot{z}_0 = 0$ and $z_0 = \check{z}$, where $\check{z}$ is an estimate of the altitude based on experience. The former is always an adequate approximation while the latter may usually be estimated to within 100 to 200 miles by merely observing the general characteristics of the S curve. In the event that the original estimate of z_0 proves to be so poor that is prevents convergence, a series of values for $\check{z}$ may be tried without becoming involved in excessive computation [116].

Patton then presented results of his convergent solutions, using field data collected by the *DOPLOC* system along with *Space Track* data from the National Space Surveillance Control Center.[43] Patton reports that

The initial successful solution with field data from the DOPLOC system was achieved for Revolution 9937 of Sputnik III. Measurements were recorded for 28 seconds in the south antenna beam of the system, 7 seconds in the center beam, and 12 seconds in the north beam, with two gaps in the data of 75 seconds each... On the first pass through the computing machine, the computation converged in three iterations to initial position and velocity components that were equivalent to the following orbital parameters

Parameters	*DOPLOC* data	*Space Track* data
a (semimajor axis)	4149 miles	4111 miles
e (eccentricity)	0.0153	0.013
σ (mean anomaly at epoch)	288.04°	257.78
i (inclination)	65.37°	65.06°
Ω (right ascension of the ascending node)	178.24°	178.22°
ω (argument of perigee)	104.62°	137.95°

From this comparison Patton observed that

...there is reasonably good agreement in the values for a, e, i, and Ω, particularly for the latter two. This is characteristic of the single-pass solution when the eccentricity is small and the computational input is limited to Doppler frequency. Since the orbit is almost circular, σ and ω are less significant than the other parameters and likewise, are more difficult for either system to determine accurately... To summarize, when limited to single-pass, single-receiver observations, the DOPLOC system provides an excellent determination of the orientation of the orbital plane, a good determination of the shape of the orbit, and a fair-to-poor determination of the orientation of the ellipse within the orbital plane [116].

43 *DOPLOC* (*DOppler Phase LOCk*) was a ca 1960 U.S. Army ground-based bistatic/multistatic radar designed to detect and track satellites. It is detailed in [117]. For these tests *DOPLOC* used a 50-kW CW transmitter at Fort Sill, OK, and a receiver at Forrest City, AK. (The White Sands Missile Range receiver was inoperative at the time so *DOPLOC* operated bistatically.) The receiver generated three fixed fan beams. As the satellite passed through these beams, it generated a doppler-shifted echo, which was collected by the receiver using a phase-locked, doppler tracking filter. Hence Patton's "S curve." *DOPLOC* was eventually retired in favor of the U.S. Navy's *SPASUR*, which locates satellites by triangulation (Section 2.2).

On an historical note Patton reports that

> ...Computing times are reasonable since convergent solutions have required from 20 to 40 minutes on the ORDVAC with the coding in floating decimal, whereas more modern machines would perform the same computation in 2 to 4 minutes.[44]

This 45-year-old report establishes the analytical precedent for all succeeding work in this area. Now Patton enjoyed some advantages in the satellite scenario that are not available to the air defense scenario, notably, a kinematically well-behaved target and no constraints on time latency. Even so, he suffered the fundamental limitation of any parameter estimation process that depends on integration: the pesky requirement to independently establish the constant of integration (i.e., Patton's initial approximations) with sufficient accuracy to ensure convergence. That he succeeded as well as he did is impressive, but it was not good enough when compared to the brute-force triangulation system, *SPASUR*. All subsequent parameter estimation schemes must contend with this basic requirement, even when using more modern estimation and prediction processes, such as Kalman filters, which incidentally were introduced in 1960.[45]

6.6.2.2 Howland Approach #1 P. E. Howland was next to wrestle with this problem starting in the mid-1990s, and to publish results in the open literature culminating in his elegant 1999 paper [120]. And, like Patton, he invoked both theory and field testing. Howland's system took bistatic doppler and coarse DOA (i.e., azimuth) measurements of air targets illuminated by the narrowband vision carrier of a 543-MHz television transmitter—again from the venerable *Crystal Palace* transmitter site in London. The transmit antenna was elevated 195 m above sea level. Echoes were received at Pershore, England, 150 km away.

The bistatic receiver used a pair of eight-element Yagi-Uda antennas separated about 0.6λ in a phase interferometer configuration for the azimuth measurements. They were elevated 18 m (33λ) above the ground to minimize the horizon multipath null. The ground at this point was approximately 30 m above sea level. Thus, from equation (6.15), the maximum transmit antenna-to-receive antenna LOS would be 86 km, which placed the receive antenna well below the transmitter's LOS horizon. While this separation greatly reduced the direct path interference, it also limited surveillance of lower altitude air targets as will be detailed subsequently.

The essence of Howland's target state estimate scheme lay in two sequential steps. In the first step, doppler and DOA returns were processed over an extended period of time, typically 60 seconds, into profiles for target association. The doppler scheme proceeded as follows. First, two-second samples of the received signals were processed by a two-channel FFT, one for each antenna channel. Approximately 30 sequential FFTs were taken. Next, the strong but stable, 50-Hz harmonics of the transmitter signal were excised using the first few blocks of the FFT data set to generate a noise map. Residual data were processed by a cell-averaging CFAR, which declared targets if they exceeded the noise floor by at least 6 dB and their power was greater than in adjacent doppler bins. Doppler

---

44 *ORDVAC* (*ORDnance Variable Automatic Computer*) was built by the University of Illinois and delivered to the Ballistic Research Laboratories, Aberdeen Proving Ground, MD, in 1951. It used a binary number system, punched tape input and output media, and an electrostatic memory provided by multiple cathode ray tubes. Its logic was developed by the Institute for Advanced Study, Princeton, NJ. It used 2,091 transistors, 915 diodes, and 3,430 vacuum tubes, weighed 3,000 lb and occupied 630 ft^3. Adds were performed in 14 µs; multiplies and divides in 700 µs. With many upgrades, it was still in use in 1961 [118].

45 In a 1963 paper [119], Dawson developed equations for determining the position of a vehicle traveling on any path using doppler-only measurements from a multistatic CW radar consisting of at least four transmit-receive pairs. Dawson's equations were then extensively summarized in [115]. Dawson stated that the doppler measurements could be converted to range-sum increments by an iteration process using "...the first vehicle location as iterative variables." Unfortunately, Dawson did not say where this first vehicle location came from or further elaborate on that process. Thus, the problem of establishing initial conditions remained unsolved.

profiles (plotted as doppler frequency vs FFT number) of each detected target were then extracted by a two-state Kalman filter. Howland reports that azimuth profiles developed in a secondary Kalman filter were not accurate enough to be used for target association.

The second step converted these doppler and DOA profiles into target state (location and velocity) estimates on a standard $x-y$ Cartesian display. Howland used an extended Kalman filter (EKF) for this profile-to-state conversion. It was initialized by an innovative—but tortured—two-stage process: first, the approximate target locations were estimated by a genetic algorithm; second, those estimates were refined by a Levenberg–Marquardt optimization algorithm. At this point the EKF was initialized. The EKF could then run in real time, updated by the latest pair of doppler and DOA measurements, similar to the monostatic radar case. Howland [120] comments:

> Note that, unlike a conventional radar, in which a single tracking phase is used to combine the radar plots, this system requires two distinct tracking phases. The first phase, using the Kalman filter, is used in a conventional manner to associate doppler/bearing plots. The second phase, using the extended Kalman filter, is used to estimate the target's location and velocity from the doppler and bearing. Although it would be possible to use the extended Kalman filter to perform both tasks, the two stages were kept separate to improve the reliability of the system. Without this distinction, the extended Kalman filter would have a tendency to misassociate the doppler/bearing plots whenever the target tracking accuracy was poor, resulting in disastrous positive feedback.

Howland then walked through each step of the processing, starting with a seven-minute sample of the data's mean power spectrum collected by the Pershore receiver. Howland comments:

> Note that at this stage, target echoes are barely visible, and the dominant features are the vision carrier and its harmonics, spaced at multiples of 50 Hz. Fortunately, the carrier and its harmonics are extremely stable, and hence are easily removed in the next stage of processing [using the cell-averaging CFAR algorithm].

After describing the target profile extraction, Howland comes to the crux of his problem. He first develops equations for the nth measurement of doppler shift F_d (n) and bearing $\theta(n)$ in terms of Cartesian coordinates of the target and target's velocity, coherent integration time and transmitter wavelength. He continues:

> To initialise the target track, it is necessary to perform a nonlinear fit of... [these two equations] to a batch of doppler and DOA samples. This nonlinear fit can be expressed as the minimization of the functional formed from the least squares difference of the equations and data...

where the functional is the standard expression that includes a measurement vector containing N measurements of doppler and DOA and a vector containing the corresponding (i.e., previously cited) measurement equations for each time sample. Howland observes that "Consideration of the information content in a batch of data for example, (using the Cramer–Rao lower bound) shows that, for correct initialisation, N should be of the order of 30."

The implication is that since each sample is taken over a 2-sec coherent integration time, 60 seconds of data must be collected for initialization. Howland continues with his initialization scheme:

> It has been found, in practice, that the correct minimisation of eqn. 3 [the equation for bearing θ (n)] using a classical minimisation algorithm will only occur if the algorithm is initialised with an estimate close to the true result. This implies that eqn. 3 is poorly approximated in these algorithms; that there are multiple local minima; or that there is insufficient information in the data set for there to be a clear minimum. Although the latter problem can only be resolved by improving the basic measurements, the first two problems can be addressed by using a minimisation algorithm which assumes nothing about the error surface and is not seduced by local minima. In this work a simple genetic algorithm... is applied to the problem.

Following a description of the genetic algorithm, Howland shows results on a simulated data set using "...a batch of doppler and DOA samples." He found that at generation 50 it calculated the initial target state with enough accuracy to initialize the Levenberg–Marquardt algorithm. (Howland did not report the size of his batch or the time to calculate a generation.) That was, in turn, sufficient to initialize the EKF, which then ran in real time, updated by current doppler and DOA measurements, just as in the monostatic case.

Howland then presents $x-y$ tracks of air targets over Southern England, which matched reasonably well with ground truth data from a secondary surveillance radar (SSR) at Debden. Howland reports that "...It can be seen that one target was tracked well into the English Channel at a range of about 260 km from DERA Pershore and 100 km from Crystal Palace." These values yield an equivalent monostatic maximum range estimate of $(260 \times 100)^{1/2} = 161$ km, equation (6.7). Howland [120] observes that

> ...After allowing for low-level targets near the SSR, the TV-based radar was found to detect nearly all the targets seen by the conventional radar. Of these targets, it was possible to track about one third, with the remaining two thirds either lost in the CFAR and initial Kalman filtering, or having ambiguous or too inaccurate bearing estimates.

Using equation (6.12) with the values $r_R = 260$ km and $h_R = 0.048$ km yields a target altitude $h_t = 3.17$ km $= 10.4$ kft, which is the lowest target altitude visible at the receiver's approximate maximum range of 260 km. Of course, at shorter ranges, lower altitude targets will be visible to the receiver, absent terrain shadowing and multipath nulls.

In addition to the low-altitude target visibility limitation and the detection-to-track loss of targets, the other major issue is the initial time latency in reporting target tracks by the processing scheme: a one-minute delay for collecting about 30, two-second FFT samples plus time required to initialize the EKF with the genetic and Levenberg–Marquardt algorithms. These current limitations fall somewhat short of satisfying the usual air defense surveillance criterion: a clutter-free, real-time display of all threatening air vehicles. In particular, the real-time display of all targets comes up short. Despite these limitations, Howland was the first to document a full solution for converting bistatic doppler and coarse AOA measurements into target state estimates.[46]

6.6.2.3 Howland Approach #2 P. E. Howland developed and tested a second PBR for the North Atlantic Treaty Organization (NATO), and reported results in 2005 [121, 122]. These tests exploited an FM broadcast transmitter located in Lopik, the Netherlands, at 375-m altitude, 50-kW mean effective radiated power and 96.8-MHz frequency. Bistatic range, bistatic doppler, and coarse DOA, that is, azimuth, of air targets of opportunity were measured by a receiver at 75-m altitude and 50 km away at the NATO C3 Agency in The Hague. With this shortened baseline geometry, low-altitude target visibility was greatly improved over Howland's first approach. Howland [121] reports that

> After correlation processing the radar has measurements of the target that are very similar to those of conventional radars: bistatic-range, bearing [i.e., azimuth] and doppler. The accuracy of each of these measurements, however, is quite different. Range and bearing are a factor of ten or so worse than a conventional microwave radar owing to the lower bandwidth of the FM radio signal, whereas doppler is two or three orders of magnitude more accurate owing to the extended integration times [1–2 second] possible with passive radar...

46 Paul Howland must also be commended for his clear, detailed reporting of innovative passive bistatic radar concepts, a refreshing change from proprietary or classified—and thus arcane—concepts that have frustrated the understanding of passive bistatic radars in the past.

The receive antenna used two vertically polarized, half-wave dipoles over a wire-mesh back-plane 1.5 $\lambda \times 1.5\ \lambda$ in size in a phase interferometer configuration, and was oriented to place the transmitted signal in a null. The receive antenna pattern covered a 120° sector looking over the North Sea toward the United Kingdom. The transmitter was located behind the receiver in an *over-the-shoulder* geometry. At a baseline range of 50 km and antenna altitudes of 375 m and 75 m, equation (6.15) shows that the transmitter is well within LOS of the receiver (50 km vs 115 km). Consequently, direct path feed through of the transmitter signal became a major source of RFI—the immediate trade-off for increasing low-altitude target visibility. Howland [121] details this problem:

> The greatest limitation on system performance is the interference received from the transmitter being used to detect aircraft. This unwanted direct signal correlates perfectly with the reference signal and produces range and doppler side lobes that are several orders of magnitude greater than the echoes that are sought. To detect anything but the closest of targets it is necessary to remove this signal, by both angular nulling with the antenna and adaptive echo cancellation in the receiver. However, eventually the dynamic range of the receiver limits the cancellation and so the principal limitation on system performance lies with the analogue-to-digital converter technology.

Howland calculates that this direct path interference is "...up to 90 dB greater than the echo we would expect to see from a 10 m^2 aircraft at a range of 150 km [from the receiver], and illustrates the need for good interference rejection." He then details his development of a two-stage adaptive noise canceller. The first stage is an adaptive M-stage lattice predictor, with prediction order $M = 50$, and the second an adaptive tapped delay line. The first is equivalent to the Gram–Schmidt algorithm and the second to a multiple regression filter.

Howland reports in [122] that this noise canceller suppressed the direct path interference by approximately 75 dB. That suppression coupled with the spatial nulling was adequate to allow target detections 150 km from the receiver. In this case, with a baseline of 50 km and an over-the-shoulder geometry, the equivalent monostatic maximum range, $R_M = (150 + 200)^{1/2} = 173$ km, roughly the same value as Howland's first system.

Howland then proceeds to a principal objective of the work: reducing time latency in (1) the target detection/cross-correlation process and (2) the target association and state estimation process. For the former, a straightforward Fast Fourier Transform of the input data set consisting of 2- to 3-km range resolution cells and 1-Hz doppler resolution cells was found to require excessively long processing times: 25 seconds on one PC to produce a complete correlation surface. That time was reduced to 4.5 s by adding an initial decimation technique that discarded doppler data where targets were not expected. It was later reduced to 0.4 seconds by resurrecting an "old algorithm, FMCW-like processing" running on a Pentium 4, 2.4-GHz PC [122].

The target association/state estimation process was to simultaneously associate doppler, range, and DOA (azimuth) in an EKF, which, if successful, would eliminate the need for the previous, time-consuming track initiation process using genetic and Levenberg–Marquardt algorithms. Howland successfully established target association tracks in range–doppler space, using a basic Kalman filter. However, incorporation of all three measurements in this filter or in the subsequent EKF for target state estimation was not reported. Howland explains [121]:

> At the time of writing, the integration of bearing into the system is very new. We therefore use the simple trigonometrical transformation from bistatic range and bearing into a local Cartesian coordinate scheme for displaying the target on a geographical map... Whilst this serves to quickly show the detections on a map, it effectively discards all the excellent Doppler information provided by the system. We are therefore currently implementing a proper estimation of target location from range, bearing and Doppler by extending the earlier work...

Thus, to this point, conversion of range–doppler target association tracks into target state estimates, either with or without DOA data, has not been accomplished. And again, part of the problem appears to lie in establishing initial conditions necessary for integrating very accurate doppler data into the EKF.

Howland showed similar results in his later briefing of this work [122], suggesting work was still ongoing in..." nonlinear estimation techniques, such as the EKF or particle filter." Also, in both references Howland suggests that a significant change in approach was being considered: expanding the system to work with multiple FM radio stations. Such a configuration of course is a multistatic radar, which can now abandon DOA and use range and possibly the more accurate doppler measurements in a multilateration scheme. This scheme... "Promises to offer far greater target location accuracy and more robust detection." The particle filter is described in the next section.

6.6.3 Multistatic Configuration

Target location measurements by the PBR can be significantly improved by abandoning coarse angular measurements in favor of implementing range and/or doppler multilateration techniques [115, 123, 124].[47] These techniques require a multistatic radar configuration where, at a minimum, multiple transmitters are used with one receiver or multiple receivers are used with one transmitter. Then a target's isorange contours from each transmitter–receiver pair can be combined to produce intersecting contours, which locate the target. Alternatively, isodoppler contours can be used whenever initial conditions can be satisfied. They can also be integrated with isorange contours in an attempt to improve location accuracy. This target location is then processed by a tracking algorithm, for example, an alpha–beta tracker or Kalman filter, to generate target state estimates, just as in a traditional monostatic radar.

Given a favorable multistatic geometry, specifically a favorable *GDOP*, for example where sites are widely separated and the target is located between sites, multilateration can accurately locate targets in three dimensions: range, azimuth, and elevation. Use of a PBR sited around a launch area to track ballistic missile or space shuttle launches is an example of such a geometry. In contrast, when a PBR is used for air surveillance or ballistic missile defense, the GDOP is not nearly as favorable and targets can usually be located in only two dimensions: range and azimuth.

In fact, multilateration systems posses two characteristics unique in the radar world: (1) these range-independent, GDOP-driven location estimates, and (2) ghosts, or false targets that inevitably appear in a multitarget environment. The effects of each will be detailed following a brief review of multistatic systems, which will provide an insight into multilateration location capabilities and limitations.

6.6.3.1 Multistatic System Review Multistatic systems have a long history of development and deployment, dating back to World War II. They were first used in the United States for aerospace range instrumentation, as reported by Scavullo and Paul in their 1965 chapter on "Instruments for Trajectory Measurements" [123]. The earliest (and probably the most notable) system was the *Doppler Velocity And Position System* (*DOVAP*), originally developed by Germany during World War II and subsequently adapted for U.S. range instrumentation by the U.S. Army Ordnance's Ballistic Research Laboratories at Aberdeen Proving Ground, MD. It operated with a ground-based, 36.8-MHz CW interrogator, a transponder on the target—usually a missile under test—and three or more ground-based doppler receivers for each interrogator. The transponder doubled the illuminator frequency, responding at 73.6 MHz.

47 Also called trilateration when three receive sites are used. Trilateration evolved from the surveying technique of measuring sides of triangles, and was first configured using three range-only monostatic radars [115].

`*DOVAP* initiated doppler track at missile launch. Track initialization was performed simply by inputting the missile's launch coordinates. Scavullo and Paul document the scale of *DOVAP*'s operation:

> Chain operation is typical. At the White Sands Missile Range where DOVAP has been a principal electronic trajectory system, some eighty receivers are sited, together with six transmitter stations and four central recording stations. Typically for a long missile trajectory [~150–200 km], as many as twenty receiver stations are operated [123].

However, Scavullo and Paul report that "Signal interruptions lead to the requirement for start points for the [doppler] counting. Bias errors occur if the point at which the object starts to move cannot be related exactly to the DOVAP, but the problem of restarting after an interruption is often insoluble."

So, here again the pesky initialization (and more importantly the reinitialization) problem rises up to plague doppler-only multilateration systems when subjected to intermittent data.

On an historical note, Scavullo and Paul note that *DOVAP*'s

> ...high order of measurement precision has been offset by the delays experienced in postflight data reduction efforts where real-time computation facilities have not been provided. Consequently, it is applied principally when the motion is almost entirely along a radial trajectory away from the station complex as in the case of ICBM and other rocket launchings [123].

Such delays were a major problem then, but not today.

Many variants of *DOVAP* were developed including the *VHF Triangulation Doppler System* (*TRIDOP*), which measured three or more doppler shifts simultaneously. But again, "TRIDOP data cannot produce absolute trajectory information without reference to an absolute fix point obtained independently from other instruments... Momentary signal loss can make all the data useless."

The *Spherical Doppler System* (*SPHERDOP*) used a stable oscillator in the missile's transponder and four rather than three receive sites to improve doppler accuracy. *UHF Doppler* (*UDOP*) was deployed at ballistic missile launch sites to track ballistic missile lift-off and midcourse trajectories.

The VHF *Passive Ranging Doppler System* (*PARADOP*) was the first to operate without a missile transponder, becoming a true radar. Inasmuch as *PARADOP* used omnidirectional antennas, the received SNR was greatly reduced over the transponder-augmented systems. Much of the lost SNR was recovered through the use of a newly developed *phase-locked tracking filter*. The *Microwave Doppler* (*MIDOP*) approximated the *PARADOP* scheme at microwave frequencies in a further attempt to improve accuracy.

Both *Doppler Phase Lock* (*DOPLOC*) and *Microwave Phase Lock* (*MICROLOC*) were used to track ballistic missiles and satellites. *DOPLOC* operated in both a transponder and skin-track mode, and was detailed in Section 6.6.2.1, including the initialization issue. *MICROLOC*, operating at 108 MHz, used phase-locked loops at 12 stations around the earth to track the doppler of incoming radio telemetry signals from satellites. It also used an interferometer antenna to make angular measurements of the satellite's position, similar to *SPASUR*, the survivor of the lot.

The prevalence of—and optimism for—these early doppler instrumentation systems cannot be overstated. For example, a companion *DOPLOC* paper to Patton's paper [116] in the 1960 *IRE Transactions on Military Electronics* by L. G. deBey [117] finished with the following assertion:

> Over the span of time since the first guided missile was fired in this country, there have been often-repeated predictions that one or more of the more sophisticated radars, interferometers or other tracking systems would replace doppler systems in precision tracking applications. Far more sophisticated and complex systems now have been developed, and second-generation systems are in the research and development stage.

> In spite of the technological advancements in these areas, doppler systems continue to be the choice of many missile system designers for obtaining the most needed information concerning flight performance. There is no indication that contemplated developments in these other fields will replace the Doppler systems which will grow out of systems of the current DOPLOC type.

However, a major unsolved problem with these VHF/UHF systems operating with broad antenna beams was multipath, which generated nulls in the antenna patterns. The nulls, in turn, could cause momentary signal dropout and disruption of the doppler count. As reported previously, reinitialization efforts were usually futile. The multipath also caused cyclic phase errors at each receiving site, greatly increasing the error relative to the random component resulting from thermal noise.

In fact, by 1960 the doppler systems had already been replaced by monopulse tracking radars such as the *AN/FPS*-16, which remain the primary instrumentation to this day. Only the satellite triangulation-tracking *SPASUR* survived.

Scavullo and Paul also describe *Sequential Correlation Ranging* (*SECOR*), a range-only multilateration system, also known as *Distance-Measuring Equipment* (*DME*). *SECOR* operated at VHF, then later at UHF, with three or more DME stations separated by relatively long (up to 500 nmi) baselines. Each station sequentially interrogated a transponder carried by a missile, booster, or satellite with a FM/CW signal. Five modulation frequencies were used, which, when demodulated by the DME using a servomechanism that drove resolvers through a gear train, yielded an unambiguous estimate of range. In this sense it operated as a range analogue of a phase interferometer antenna. The three, time-adjusted range estimates were then combined to provide a position estimate.

More complex versions of *SECOR* were developed, including one that used pulse codes to control each of 12 DME stations to cover lower altitude trajectories. Another design extended range beyond 2000 nmi, added doppler measurements and performed multitarget tracking. Finally, Scavullo and Paul observe that

> Each instrument in the SECOR system is capable of fine resolution, and random errors are surprisingly low. However, great care is needed to correlate the uncertainties between stations in any multistation system. Bias errors may be found with more than 100 times the dispersion of the random errors. SECOR has no magic protection from this problem, which also plagues the other long baseline systems [123].

This plague was explained in the coauthor's 1976 *Radar System Analysis* [124]:

> The accuracy obtained in both range and Doppler measurements... is so high that the accuracy of long-baseline trilateration systems becomes limited by knowledge of the shape of the earth and the location of the continents and islands. It was estimated by a scientific panel in 1961 that the best accuracy of a short baseline would be about one part in 10^7, but that the range between more distant points within the continental United States would be in error by about five parts in 10^6. Thus for a baseline of 2000 miles, the error would be about 60 ft. Baselines extending over intercontinental distances [such as *SPASUR*] would be known to an accuracy of 100 to 300 ft, as a standard deviation...

Use of the *Global Positioning System* (GPS) with long-term averaging has now reduced these location errors to a few feet, thus eliminating the problem.

The final section in Scavullo and Paul's chapter [123] covered *angle measurement equipment* (*AME*), also *called interferometers*. Many systems incorporating AME were developed and deployed, including *AZUZA*, *COTAR*, *MINITRACK*, *MIDAS*, *MISTRAM* (*Missile Trajectory Measurement System*), and *SPASUR*. Both *AZUSA* and *MISTRAM* used narrow-beam microwave dishes, which mitigated multipath problems.

MISTRAM requires attention because it represents a third generation of hybrid system development—and a final chapter—in these multistatic instrumentation systems. As reported by Scavullo and Paul, *MISTRAM* was deployed at the Atlantic Missile Range at Cape Kennedy. It was designed to locate a ballistic missile, independently of other instrumentation, soon after launch, track it for more than 1,000 miles and produce accurate position and velocity data in real time. Impact of a ballistic nose-cone was to be computed within 0.1 mr at a downrange distance of 9,000 miles.

MISTRAM used a CW tracking radar that transmitted two X-band signals to a missile-borne beacon. Five slaved receiving antennas configured as an interferometer with baseline length of up to 100,000 ft, also tracked the beacon replies. Range, range-rate, differential range, and DOA were measured using the phase shift between transmitted and received signals. Scavullo and Paul report that "Attainment of ultimate performance objectives is now [circa 1965] contingent upon the eventual solution of microwave transmission problems between receiver and transmitter stations and upon control of other propagation errors [123]."

However, control of the other propagation errors was never accomplished and *MISTRAM* did not achieve its ultimate performance objectives, for the following reasons. Angular accuracy in interferometers becomes inversely proportional to baseline length whenever the measured phase-difference error is found to be independent of baseline length. One contributor to this error is anomalies in the troposphere. At the time, tropospheric models assumed that the anomalies affecting radar had scale lengths comparable to the width of the individual radar antennas, which was approximately 3 m for both *MISTRAM* and its predecessor *AZUSA*. Thus using an interferometer baseline of 30 m for *AZUSA* would not significantly increase tropospheric contributions to the phase-difference error; consequently, angular accuracy would be increased by a factor of 10. Unfortunately, these tropospheric assumptions were in error and *AZUSA*'s angular accuracy was found to be much less than the expected factor of 10.

During the subsequent *MISTRAM* design, engineers concluded that the *AZUSA* baseline had not sufficiently exceeded the scale length of these tropospheric anomalies, but that an increase to 3,000 m for the angle measurement and 30,000 m for the angle rate measurement would do so. But again, angular accuracy was found to be much less than expected.

After further, extensive tropospheric measurements and analysis, which revised the models but did not explain the problem, it was concluded that a unique tropospheric ripple hung over the Cape, causing an increase in phase-difference errors. As a result, another 100 million dollars was spent to place a second *MISTRAM* on one of the downrange islands. Unfortunately, this unique tropospheric difficulty was also reproduced there.

The coauthor's 1963 paper [125] finally offered an explanation for the problem: the errors were not caused by temporal tropospheric fluctuations, but by large-scale, spatial anomalies in the troposphere. These anomalies introduced time-varying phase changes whenever the angular rate of the interferometer-to-target LOS was sufficiently high (>0.1 mr/sec) so that it passed through these anomalies during the measurement period. This, in turn, increased the rate of change of the phase-difference error by factors ranging from 10 to 100. This component of error was found to vary inversely as the ¼ power of both radar aperture width and interferometer baseline length. Thus, interferometers would still overcome some of the mechanical error sources of radar, and could reduce tropospheric errors, but not nearly by the factors advertised.

Improved monostatic radars and optical trackers, which approach and sometimes exceed typical 0.1 mil instrumentation accuracy specifications, are the current instruments of choice for nearly all test ranges. While interferometers have not shown much utility in PBR air defense scenarios, they have for scientific measurements. The *MRR* described in Chapter 7 is an example.

The latest system to appear on the multistatic radar scene is the SAIC-developed, *HDTV-Based Passive Radar*, which was first briefed to the AOC 4th Multinational PCR Conference, held in Syracuse, NY, in October 2005 [126]. The system currently exploits one *HDTV* transmitter, which broadcasts a finite-length pseudonoise (PN) waveform with a useable bandwidth of about 2.5 MHz, equivalent to a monostatic range resolution of about 60 m. Four receivers are deployed within short ranges (~10 km) of the transmitter for range-only multilateration. Doppler is also measured to excise ghosts that inevitably arise in any multilateration system. Section 6.6.3.4.3 summarizes preliminary details of their ghost excision algorithm.

The SAIC system was developed and tested from 2003 to 2005 using in-house funds and a small U.S. Army R&D contract. It is designed to detect and track low-flying aircraft and helicopters as a gap-filler (typically < 1.5 km altitude) for other air surveillance systems, principally air traffic control radar systems. As such, it is designed for an equivalent monostatic range, R_M = 30 km on a 1 m^2 target. Because each receiver must simultaneously view a large surveillance region, typically > 60°, it operates with little azimuth directivity. However the low-altitude-only requirement allows some elevation antenna shaping, yielding an 8-dBi antenna gain.

Because transmitter–receiver separations are small, significant direct path breakthrough occurs. It is reduced with a brute-force, least-squares channel estimator, yielding a 20- to 25-dB noise figure. (SAIC found that performance of a cross-correlator was roughly 40 dB inferior in reducing the noise floor because it did not account for side lobes of the PN waveform in the ambiguity plane.) The cost of this process is to more than double the computational load to 20.5 B floating-point operations per second.

Primitive detection reports from each receiver are sent via landline to a central node, where multilateration, ghost excision, and target contact estimates are made via a 5-node, dual-pentium chip, Xeon cluster. SAIC found that the time latency for these operations ranged from 0.5 seconds for processing 25 targets to 7 sec for 200 targets. The contact data is then passed to an "Adaptive Fusion Tracker" for track generation and merging with other data sources. Total system latency was found to be about 10 sec with an update rate of five seconds. Further details are summarized in Table 6–2.

In tests using the 1-MW ERP, 677-MHz WRC-HD transmitter in Washington, DC, with four receivers located in nearby Northern Virginia, low-altitude, commercial air traffic was detected and tracked, although altitude estimates were found to be noisy and constrained to 500-m bands.[48] Considerable ground traffic was also detected and not always filtered out, especially close-in traffic. Tracking errors were generally less than 50 m. In one instrumented test a small Cessna aircraft was detected out to approximately 40 km at a postdetection SNR of approximately 16 dB. Helicopters were also detected and recognized by their blade modulation out to approximately 20 km [126, 127].

While the system and its test were designed primarily as a gap-filler for air traffic control, implications for air defense are clear. Furthermore, when the eventual conversion of TV broadcasts from conventional analog to high definition, digital format is complete, VHF HDTV transmissions will become available, with their inherent counter-stealth capabilities, as detailed in Section 6.3.

6.6.3.2 Multistatic Location For simple geometries, multistatic location can be established analytically by using simultaneous range-sum measurements from just two bistatic transmit-receive pairs—and knowledge of their baselines. A typical solution using the law of cosines for one transmitter and two receivers configured in the colinear baseline geometry of Figure 6–31, with

48 Because of large geometric dilution of precision (GDOP) factors which are quantified for such multilateration systems in Section 6.6.3.3.

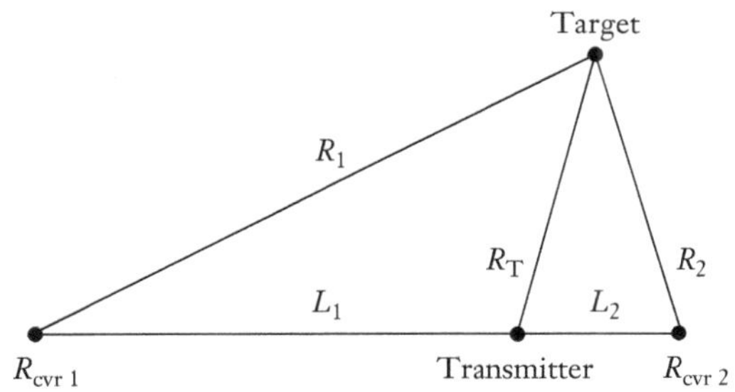

Figure 6–31 Multistatic geometry demonstrating target location in the bistatic plane.

measurements L_1, L_2, K_1 (= $R_T + R_1$) and K_2 (= $R_T + R_2$) shown on the figure is as follows:

$$R_1 = \frac{C^2 - A^2 - B\,(L_1^2 - K_1^2)}{2A + 2BK_1} \tag{6.21}$$

$$\mathrm{R}_2 = \mathrm{R}_1 + \mathrm{A} \tag{6.22}$$

where

$$A = K_2 - K_1$$
$$B = (1 + L_2/L_1)$$
$$C = (L_1 + L_2)$$

and

$$R_T = K_1 - R_1$$

With this *multistatic triangle* (defined by the two receivers and target) established, target bearing from each receiver can be established in the bistatic plane with another application of the law of cosines.

For less compliant geometries, techniques modeling the intersection of isorange contours from each transmit-receive pair may be needed. Also note that a third receive site is required to convert bistatic plane estimates into Cartesian coordinates. Alternatively, isodoppler contours can be used whenever initial conditions can be satisfied. This target location is then processed by a tracking algorithm, for example, an alpha–beta tracker or Kalman filter, to generate target state estimates, just as in a traditional monostatic radar.

6.6.3.3 Multistatic Measurement Accuracy and GDOP This section develops a simple model for estimating the target location errors of a PBR operating in a multilateration mode with two or more FM broadcast transmitters. Errors for both bistatic range and bistatic doppler measurements are developed. The purpose of the multilateration model is to provide a simple procedure for predicting the relationship of target location errors to system geometry and target trajectory so that test results (past and future) may be interpreted and extrapolated to other system configurations and targets. The use of complex models and simulations for specific systems will yield more accurate error predictions, but may not illuminate the factors that control the errors. A generic example will be carried along with the development. It is based upon characteristics of typical PBRs that have been developed and tested over the past ten years.

6.6.3.3.1 Estimation of Range Error The first step in multilateration range error modeling is to express the error in the basic time-delay measurement made by the system. The thermal noise component of time-delay data for an ideal estimator is given by [128] as

$$\sigma_t = \frac{1}{\beta_a \sqrt{2\frac{E}{N_0}}} \tag{6.23}$$

where β_a is the *rms bandwidth* of the signal defined by

$$\beta_a^2 \equiv \frac{\int_{-\infty}^{\infty} (2\pi f)^2 |A(f)|^2 \, df}{\int_{-\infty}^{\infty} |A(f)|^2 \, df} \tag{6.24}$$

and $A(f)$ is the signal spectrum, E/N_0 is the ratio of received signal energy to noise spectral density evaluated over the measurement interval, and equal to the receiver output SNR ratio when a matched filter is used. For a rectangular signal spectrum of width B_r, $\beta_a = \pi B_r/\sqrt{3} = 1.81B_r$, while for a Gaussian spectrum having half-power width B_3, $\beta_a = 2.67B_3$. The typical FM broadcast signal can be represented by a rectangular spectrum of width $B_r \approx 45$ kHz, resulting in $\beta_a = 82 \times 10^3$ rad/sec.

Assume that the generic multilateration system requires a *SNR* of +12 dB = 16, integrated over a period $t_I = 5$ sec, to initiate a track file. At maximum range, the time-delay error $\sigma_t = 2.2 \times 10^{-6}$ sec for each range path, evaluated over the 5-second integration period. This corresponds to the two-path, bistatic range-sum ($R_T + R_R$) error, $\sigma_{rs} = \sigma_t c = 660$ m.

While these errors should be reduced in proportion to the square of the geometric mean range ($R_T R_R$) and by further smoothing in the tracking filter, other error components occur in the receiver/processor that hold the minimum time-delay error to the order of 1.5 or 2 μsec for most target situations. These error components will be discussed later. For the generic example to follow, a 2-μsec error will be assumed.

An important issue in the use of FM broadcast transmitters is the *dead air* (no program modulation.) In this case both the delay error σ_t and the width of the delay resolution cell T_B increase without limit. In general, once a track has been established, such large error spikes, if not occurring too frequently, can be edited out by a nonlinear tracking filter. Depending on the editing procedure, these spikes may extend the tails of the error distribution beyond the level that would be expected from a normally distributed error. The statistics will be similar to those of target glint in conventional radar, which follows the infinite-variance Student distribution with truncation by hardware nonlinearities. Section 6.4.1 quantifies the duration and statistics of this dead air for typical broadcast stations.

6.6.3.3.2 Two-Coordinate Error as a Function of Range Error Because radar range errors normally are much smaller than the product of range and angle errors, range multilateration systems have been used in test range applications to obtain improved tracking data. (Section 6.6.3.1 provides a brief review of these systems.) The primary limitation has been the need to locate three ranging stations such that their LOS to the target maintain angles of intersection as close as possible to 90°, minimizing the geometrical dilution of precision (GDOP) in the three-coordinate output data. Location of stations in space provides such a geometry over wide regions, which of course is the basis of the GPS, using a constellation of some 25 satellites to ensure adequate geometry worldwide.

When target altitudes are small relative to their ranges from the measurement stations and altitude data are not needed, navigation systems based on intersecting range circles from two properly located ground stations can provide accurate position data. Similar systems using range-sum measurements (e.g., *LORAN*) have been used for long-range navigation.

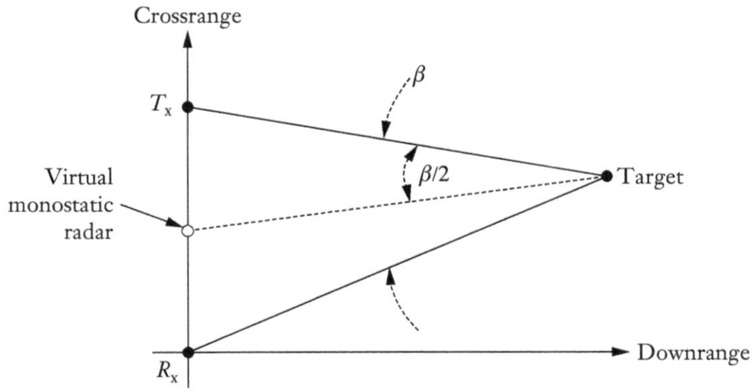

Figure 6–32 Modeling of a bistatic link by a virtual monostatic radar.

A multilateration system measuring bistatic range over two or more bistatic links represents a variant of the range trilateration system discussed in [124, pp. 507–510]. Each bistatic link (from transmitter to target to receiver) can be modeled as a virtual monostatic radar having a path along the bisector of the bistatic angle (Figure 6–32) over which path the range is measured with an error σ_{ri} given by

$$\sigma_{ri} = \frac{\sigma_{rsi}}{2\cos(\beta_i/2)} = \frac{c\sigma_{ti}}{2\cos(\beta_i/2)} \quad i = 1, 2, \ldots \tag{6.25}$$

where σ_{rsi} is the error of the range-sum measurement over the ith bistatic link, σ_{ti} is the error of the time-delay measurement over the ith link, β_i is the bistatic angle of the ith link, and c is the velocity of light.[49]

To obtain measurements in a plane containing the system elements and the target, two transmitters can be used with one receiver (or vice versa), providing two bistatic angles β_1 and β_2 whose bisectors differ by an angle $2\alpha = (\beta_1 + \beta_2)/2$, as shown in Figure 6–33.[50]

The angle 2α between the virtual monostatic paths determines the GDOP factor for the two-coordinate measurement. The geometry of such a two-station multilateration system and its error ellipse, approximated as a parallelogram, is shown in Figure 6–34.

Using the equations of [124, p. 508], expanded for unequal range errors, yields

$$\sigma_{dr} = C_e' \frac{\sqrt{\sigma_{r1}^2 + \sigma_{r2}^2}}{2\sqrt{2}\cos\alpha} \tag{6.26}$$

$$\sigma_{cr} = C_e \frac{\sqrt{\sigma_{r1}^2 + \sigma_{r2}^2}}{2\sqrt{2}\sin\alpha} \tag{6.27}$$

49 This *virtual monostatic radar* model is based on the following property of bistatic geometry: The bisector of the bistatic angle is orthogonal to a tangent of the isorange contour (ellipse) at any point on the isorange contour. Since the tangent is a reasonable approximation to the isorange contour, and hence the bistatic range cell, a virtual monostatic radar lying along the bistatic bisector will generate a nearly identical range cell, increased in width by $1/\cos(\beta/2)$.

50 This geometry is the same as the colinear baseline geometry used in Section 6.6.3.2.

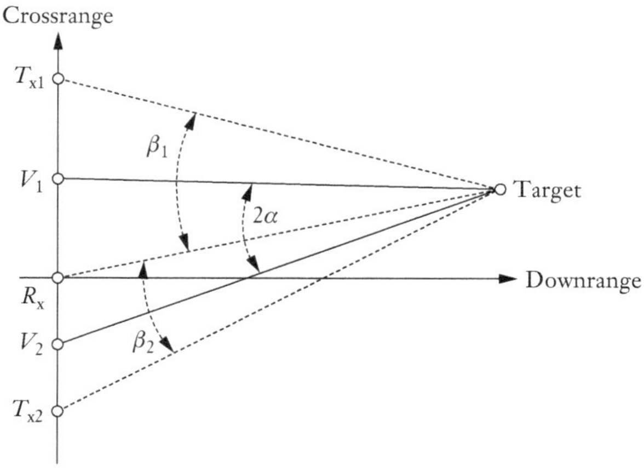

Figure 6–33 Modeling of a two-link multilateration system by virtual monostatic radars.

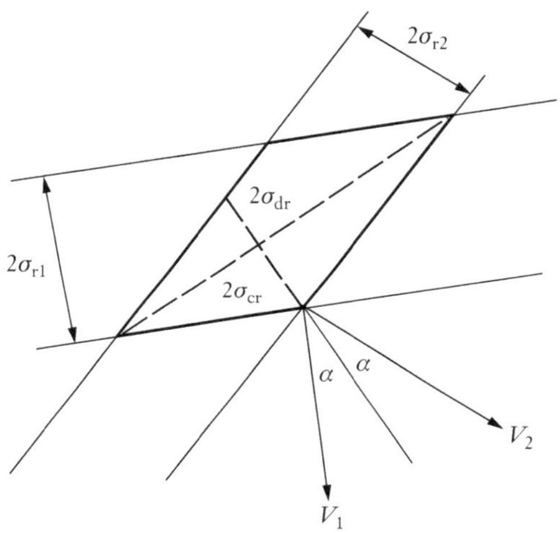

Figure 6–34 Two-station multilateration error ellipse, approximated as a parallelogram, for two virtual monostatic radars, V_1 and V_2, with range errors σ_{r1} and σ_{r2}, respectively, where σ_{dr} is the downrange error and σ_{cr} is the crossrange error.

where σ_{dr} is the downrange error, measured along the bisector of the two virtual monostatic path angles; σ_{cr} is the crossrange error, measured normal to downrange (approximately normal in Fig. 6–34); σ_{ri} are the equivalent monostatic range errors given by equation (6.23); $C_e = \sqrt{2(1-u)}$ and $C_e' = \sqrt{2(1+u)}$ are factors depending on the correlation coefficient u of errors measured along the two paths: $u = 1$ for correlated errors; $u = 0$ for uncorrelated errors. Thus for correlated errors $C_e = 0$ and $C_e' = 2$; for uncorrelated errors $C_e = C_e' = \sqrt{2}$.

For uncorrelated and equal errors over both paths

$$\sigma_{dr} = \frac{\sigma_{r1}}{\sqrt{2}\cos\alpha} \tag{6.28}$$

$$\sigma_{cr} = \frac{\sigma_{r1}}{\sqrt{2}\sin\alpha} \tag{6.29}$$

To illustrate in this simple case the bistatic GDOP factors can be defined as follows:

$$(\mathrm{GDOP}_{\mathrm{dr}})_{e/\mathrm{unc}} = \sigma_{\mathrm{dr}}/\sigma_{r1} = 1/(\sqrt{2}\cos\alpha) \tag{6.30}$$

$$(\mathrm{GDOP}_{\mathrm{cr}})_{e/\mathrm{unc}} = \sigma_{\mathrm{cr}}/\sigma_{r1} = (1/\sqrt{2}\sin\alpha) \tag{6.31}$$

Thus for a GDOP factor of unity (the optimum case) the two paths must intersect at 90° angle, giving $\alpha = 45°$. Hence, the error ellipse becomes a circle of radius σ_{r1}. Note, however, that from (6.23) this error will generally be smaller than the thermal noise component of the range error σ_{rs}. Consequently the bistatic GDOP factors are more commonly defined as the ratios σ_{dr}/σ_{rs} and σ_{cr}/σ_{rs}, where $\sigma_{rs} = (\sigma_{rs1} + \sigma_{rs2})/2$, yielding

$$\mathrm{GDOP}_{\mathrm{dr}} = C_e' \frac{\sqrt{\sigma_{r1}^2 + \sigma_{r2}^2}}{2\sqrt{2}\sigma_{rs}\cos\alpha} \tag{6.32}$$

$$\mathrm{GDOP}_{\mathrm{cr}} = C_e \frac{\sqrt{\sigma_{r1}^2 + \sigma_{r2}^2}}{2\sqrt{2}\sigma_{rs}\sin\alpha} \tag{6.33}$$

The use of additional transmitters for the two-coordinate case provides overdetermination of the target position, reducing the final error value. However, the error will not be significantly less than that obtained using the two transmitters having angular separations farthest from 0° or 180°. The importance of the third transmitter is that it may provide the ability to estimate target altitude, if its virtual monostatic path intersects either of the first two paths at a sufficiently large angle γ in the vertical plane.

6.6.3.3.3 Altitude Measurement Measurement of target altitude presents a more difficult problem to the multilateration system when all transmitters and the receiver are located near the earth's surface. Large angles of intersection (and resulting small GDOP factors for the crossrange error in the vertical plane) can be obtained only for targets at high altitude. Figure 6–35 shows a typical geometry for an aircraft target, for which the intersection angle γ depends on the difference in ranges of the two virtual monostatic radars. In most cases of air vehicle tracking, as will be shown below, this intersection angle is a few degrees or less.

6.6.3.3.4 Example of Predicted Multilateration Tracking Performance Consider now a generic PBR test environment, for which three FM broadcast stations are chosen to provide the best available illumination over a test area, which is located at sea off the east coast of the United States. In each

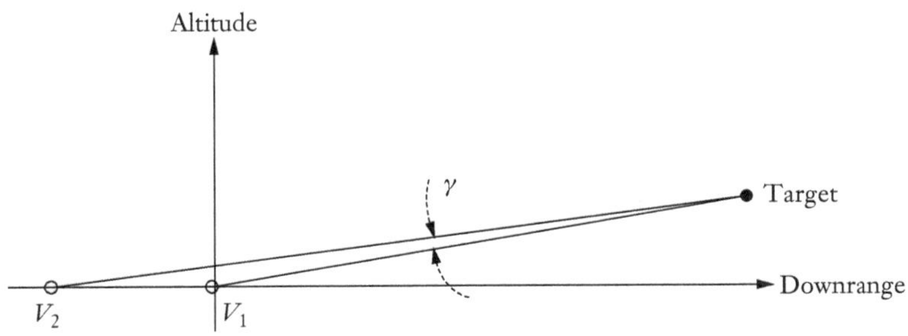

Figure 6–35 Modeling of two-link multilateration elevation measurement by virtual monostatic radars.

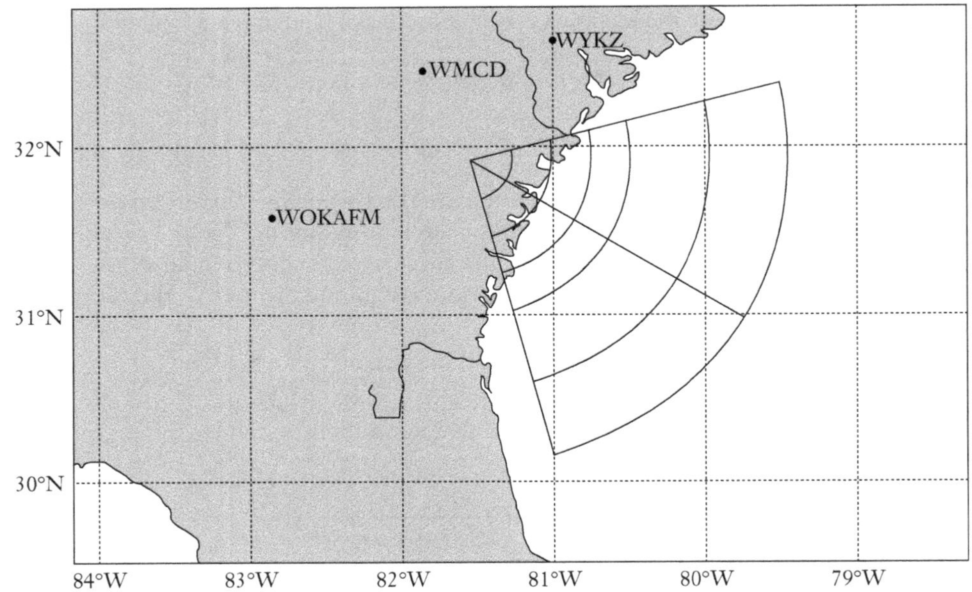

Figure 6–36 Multilateration site configuration and planned coverage.

case, the transmitters are either behind the plane of the receiving array or near 90° to the array broadside. Hence, modeling the system using virtual monostatic radars located along the bisectors of the bistatic angles appears applicable to estimating tracking errors and incidentally, bistatic cross section.

Figure 6–36 shows the locations of transmitters, receiver, and areas in which target tracks are expected. The azimuth coverage fan is marked as a 90° sector with range lines at 25-km intervals to 100 km and at 150 and 200 km. The areas for target tracking accuracy evaluations are between 15 km and 75 km from the receive site.

Figure 6–37 shows an expanded view of the sites, with the broadside axis reoriented to the right of the figure. The receiver is arbitrarily placed 25 km from the origin of the coordinates. A target area is centered 25 km from the receiver. Using the monostatic multilateration model, the virtual monostatic radars are located at points V1, V2, and V3, selected as $(R_T + R_R)/2$, which average the elevation angles over the transmit and receive paths. The small figure at the bottom right shows a projection of the best two sites into the vertical plane for a target at 10-km altitude located 25 km downrange from the receive site.

Assuming that each bistatic link has a range-sum error σ_{rs} = 300 m and that the errors from each link are uncorrelated (as they should be with independent program modulations) then from equation (6.25–6.27) the downrange and crossrange errors, measured in a plane passing through the origin and the target, are for the short-range tracks:

Downrange: σ_{dr} = 117 m (using V2 and V3, bistatic GDOP factor 0.39)
Horizontal crossrange: σ_{crh} = 246 m (using V1 and V3, bistatic GDOP factor 0.82)
Vertical crossrange: σ_{crv} = 3088 m (using V1 and V3, bistatic GDOP factor 10.3).

The virtual monostatic range errors entering the multilateration process for this geometry vary from 155 m to 205 m, slightly greater than half the input range-sum error.

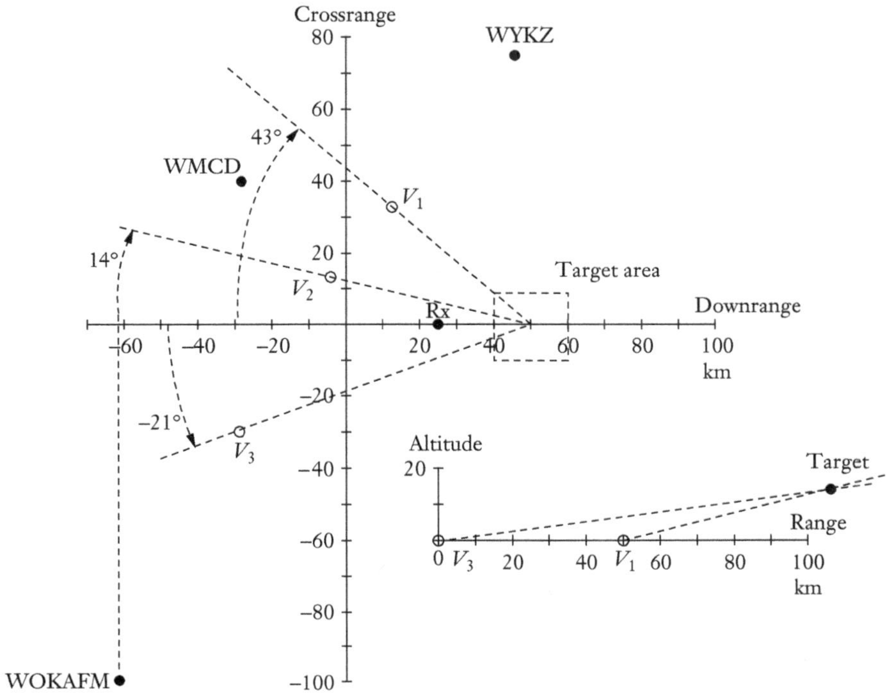

Figure 6–37 Expanded view of the short-range target tracking geometry.

The 3 km vertical-plane error for the short-range target at 10-km altitude, which is based on a maximum intersection angle of 4.8°, precludes useful measurements of target height data. In addition, a lack of target height data causes a bias error in the target's ground-plane coordinates given approximately by $R_R(1 - \cos\theta) = 1{,}000$ m, primarily in the downrange coordinate, where R_R is the receiver-to-target range and θ is the elevation angle from the receiving site. This bias error would be corrected if the altitude in the tracking filter could be forced to the *a priori* altitude of the aircraft, but that is seldom an option in air defense scenarios.

For targets in an area 50 km from the receiving site, Figure 6–38 shows the geometry and the virtual monostatic sites. For these medium-range tracks, the random errors are as follows:

Downrange: $\sigma_{dr} = 112$ m (using V2 and V3, bistatic GDOP factor 0.37)
Horizontal cross range: $\sigma_{crh} = 272$ m (using V1 and V3, bistatic GDOP factor 0.91)
Vertical crossrange: $\sigma_{crv} = 3{,}953$ m (using V1 and V3, bistatic GDOP factor 13.2).

Results are essentially unchanged from the short-range tracking results. For longer tracking ranges, the GDOP factors in the two crossrange coordinates will increase and the downrange coordinate will decrease slightly.

These predictions were subsequently compared to data from a typical multilateration system under similar test conditions. Measurement data for 15 tracks indicated median errors of 287-m downrange and 448-m crossrange (relative to the receiving antenna broadside angle). The median error corresponds to approximately 67% of the rms value, indicating rms errors of 430 m and 670 m, respectively. These measured errors included bias errors, multipath effects, and a component resulting from uncertainty in ground truth data. The maximum track ranges varied from 3.5 km to 61 km, with most in the 40- to 50-km region.

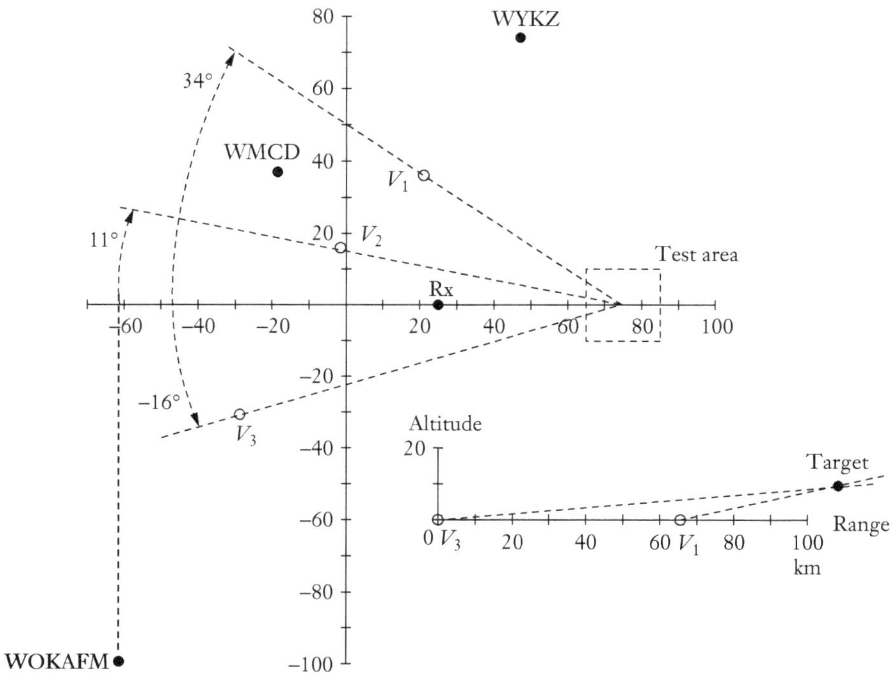

Figure 6–38 Expanded view of the medium-range target tracking geometry.

Even without a detailed analysis of the geometry for each track, it is clear that the model results, which predicted standard deviations of the random error to be 160-m (vs 430 m) downrange and 385-m (vs 670 m) crossrange, are optimistic but not necessarily inconsistent with measured data, which include multipath and ground-truth errors.

6.6.3.3.5 Potential Use of Doppler Data Since a multilateration radar can also measure bistatic doppler with each transmiter–receiver pair, there remains the possibility of obtaining target doppler components in two or three coordinates and integrating them to obtain target position with respect to some initially known target reference point. The thermal noise component of doppler data for an ideal estimator is given by [128] as follows:

$$\sigma_f = \frac{1}{\alpha_a \sqrt{2\dfrac{E}{N_0}}} \tag{6.34}$$

where α_a is the *rms time duration* of the signal:

$$\alpha_a^2 = \frac{\int_{-\infty}^{\infty} (2\pi t)^2 |a(t)|^2 \, dt}{\int_{-\infty}^{\infty} |a(t)|^2 \, dt} \tag{6.35}$$

and $a(t)$ is the signal waveform. Since the transmitted waveform is of indefinite duration, the coherent processing time $t_c \approx 0.5$ sec will determine the precision of doppler data, leading to $\alpha_a \approx t_c$ (for low side lobe weighting in the correlator). For the generic system, the integrated *SNR* is 12 dB over a 5-sec period, leading to an rms doppler error $\sigma_f = 0.35$ Hz, close to the values reported for low *SNR* in the multilateration tests. The corresponding velocity error is $\sigma_v = \sigma_f \times \lambda = 1$ m/sec.

The relationship of this velocity error to its down- and crossrange components is the same as for the range error to its down- and crossrange components. Likewise, the bistatic GDOP factors for doppler measurements with uncorrelated and equal errors are the same:

$$\sigma_{v1} = \frac{\sigma_v}{2\cos(\beta/2)} = \frac{\sigma_f \lambda}{2\cos(\beta/2)} \tag{6.36}$$

$$\sigma_{dv} = \frac{C_e'\sigma_{v1}}{2\cos\alpha} = \frac{\sigma_{v1}}{\sqrt{2}\cos\alpha} = \sigma_{v1}\mathrm{GDOP}_{dv} \tag{6.37}$$

$$\sigma_{cv} = \frac{C_e\sigma_{v1}}{2\sin\alpha} = \frac{\sigma_{v1}}{\sqrt{2}\sin\alpha} = \sigma_{v1}\mathrm{GDOP}_{cv} \tag{6.38}$$

If an initial target position can be established, target position data can be found by integrating the doppler data in three coordinates. Starting with a random error $\sigma_v = 0.5$ m/sec for 5-second smoothing, the random error for 30-second smoothing is

$$\sigma_{vs} = \sigma_v\sqrt{\frac{5}{30}} = 0.5\sqrt{0.167} = 0.22 \text{ m/s}$$

and the range error resulting from integration of velocity over 30 second is

$$\sigma_{rs} = 30 \times 0.22 = 6.6 \text{ m.}$$

Applying typical bistatic GDOP factors predicted for the 50 km generic example (0.37 downrange, 0.91 horizontal crossrange, and 13.2 vertical crossrange) the integrated position error after the 30-second integration would be about 2.4-m downrange, 6.0-m horizontal crossrange, and 87.1 m vertically, relative to the initial position. Thus, if an initial position were known or could be estimated by some method more accurate than by range multilateration, subsequent trajectory points could be established using integrated doppler data with much better accuracy than by using range multilateration.

6.6.3.3.6 Multipath Error The multipath error component has not been considered in this analysis. It is caused by the arrival of (1) direct path transmit signals reflected from surface features lying within the range resolution of the system, and (2) echoes from surface features within the combined range and doppler resolution of the system. The former are likely to be the dominant source of error when the receiver antenna is only a few meters above the earth's surface, since the Fresnel zone defined by the range resolution will encompass a huge region on this surface. This region usually contains many strong reflecting sources (also causing ghost images in broadcast TV reception over the same path).

Historically, multipath has prevented exploitation of systems using wide antenna beamwidths for precision range instrumentation applications. Such systems tested at White Sands Missile Range (then White Sands Proving Ground) included the original *DOVAP* and several subsequent systems such as *COTAR* and *MOPTARS* [129]. *DOVAP* and *COTAR* are described in Section 6.6.3.1. While thermal noise performance predictions for these systems showed great promise, multipath nulls caused signal fading and loss of doppler track. As reported in Section 6.6.3.1, reinitialization of doppler track was nearly always futile. (*MOPTARS* did not integrate doppler to determine range, and hence did not suffer the initialization problem.)

6.6.3.3.7 Dynamic Lag Errors The factor that prevents smoothing of target data over an interval covering the radar's entire FOV is the presence of unpredictable target accelerations. A simple way of modeling dynamic lag errors is by defining the acceleration error constant K_a of the tracking loop, related to the bandwidth β_n and time constant t_o of the loop by [124, p. 306]

$$K_{\mathrm{a}} = 2.5\beta_n^2 = 0.63/t_o^2 \tag{6.39}$$

The resulting dynamic lag error is

$$\varepsilon_{\mathrm{a}} = \frac{a_t}{K_{\mathrm{a}}} = \frac{0.4a_t}{\beta_n^2} = 1.6a_t t_o^2 \tag{6.40}$$

where a_{t} is the target acceleration.

Many radar tracking tests use nonmaneuvering targets, which reduce the lag error to less than the noise error. For the generic example, the time constant will be some multiple of the signal integration time t_{I} = 5 seconds, determined by tuning the filter to the expected target dynamics. A minimum time constant $t_o = t_{\mathrm{I}}$ = 5 seconds generates a lag error of about 400 m on a 1 g target, somewhat larger than the expected thermal noise error. However, in the first configuration of the tracking filter used in these tests, the tracker broke lock when the target aircraft executed a planned turn. Assuming a_{t} = 10 m/sec^2 and t_o = 20 seconds, the lag error would have been

$$\varepsilon_{\mathrm{a}} = 1.6 \times 10 \times 20^2 = 6400 \text{ m}$$

carrying the target beyond the track correlation window.[51] For specific cases, it is appropriate to model this lag error and then to consider as the system error the rss sum of the dynamic lag and the thermal noise error for the tracker configuration actually used.

Unfortunately, Kalman filters are not easily characterized by bandwidths, time constants, or acceleration error coefficients. The model given above can be considered representative of the limitations generally applicable, except in cases where the output data can be fitted to an *a priori* trajectory model with high confidence. These cases seldom apply in air defense and range instrumentation scenarios.

6.6.3.4 Multistatic Ghosts A unique problem arises when using unaided multilateration techniques (i.e., techniques without transponders) in a multitarget environment: false targets or *ghosts*. James Caspers defined ghosts in his bistatic and multistatic radar chapter in Skolnik's first handbook [115], as "...target indications that do not correspond to an actual target." Specifically a false target indication will be generated when isorange contours from two targets intersect at locations where no target exists. To distinguish these ghosts from a second type (called *ghost targets*), these intersections will be called *common ghosts*. In fact, Caspers worried about common ghosts in both active multistatic radars and passive emitter location systems, which have a similar—but much more easily solved—problem. Caspers then cites the two standard remedies for ghost removal: collection of redundant data—the more data the better, and angular resolution—the higher the resolution the better.

Angle resolution schemes are reasonably tractable in high-frequency systems, but much less so in low-frequency systems where very large antennas are required. They might be feasible in

51 These parameters were selected to match observed test results, since specific parameters of the filter were not made available.

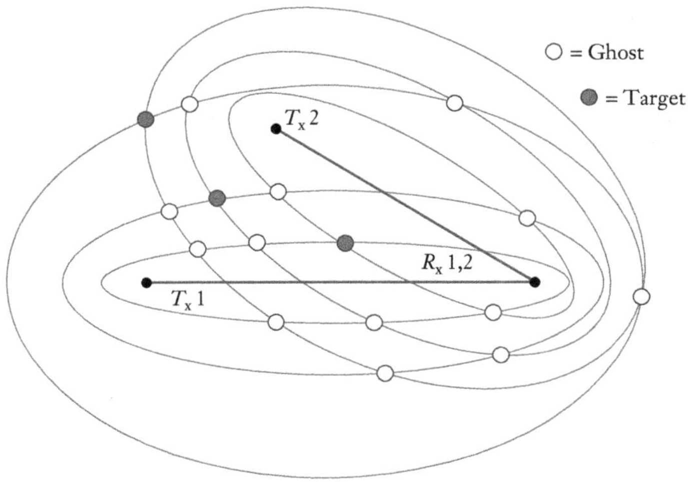

Figure 6–39 Common ghosts showing two transmit–receive pairs (with receivers colocated) detecting three targets. For this geometry, 13 ghosts appear versus the maximum of 15 possible because some ellipses in one set intersect some ellipse in the other set just once, due to differences in ellipse eccentricities.

nonhostile range instrumentation scenarios, but not in hostile air defense scenarios, as outlined earlier. Thus PBRs that use multilateration techniques in a multitarget environment must rely on data redundancy to resolve common ghosts.

6.6.3.4.1 Common Ghosts A simple example will illustrate the common ghost problem. In a range-only bistatic multilateration system using two transmit–receive pairs, each pair generates an isorange contour (an ellipse) for each target it detects. Thus, each pair will generate n confocal ellipses for n detected targets. But, since an ellipse from one set can potentially intersect an ellipse from the other set in two spots, there are a maximum of $2n^2$ possible intersections of these two ellipse sets. Consequently, $2n^2 - n$ common ghosts will be generated, where a common ghost is defined as an intersection where there is no target. Figure 6–39 shows the case for three detected targets. For ten targets, a maximum of 190 ghosts can appear.

A general expression for the maximum number, g, of common ghosts generated by n detected targets and N transmit–receive pairs is

$$g = (2n^2 - n)\,(N^2 - N)/2. \tag{6.41}$$

This expression is developed using the combinatorial expression to calculate the number of all possible pairs of ellipse intersections, reduced by (a) the number of confocal ellipses having no intersections with each other and (b) the number of pairs of ellipse intersections at target locations having more than one such intersection. For $N = 2$, there are no locations with more than two ellipse intersections. Thus $(N^2 - N)/2 = 1$, and $g = 2n^2 - n$.

This analysis assumes that all ellipse sets are measured simultaneously and completely. The former assumption can in theory be satisfied by using a GPS time standard and sending a sequence of time-tagged ellipse sets from each site to a central node, where the central node time-correlates the sets.

The latter assumption requires an adequate SNR for all measurements within a set, which is not always available. Thus, some sets can have missing ellipses, depending a specific target's SNR. In these cases, both target locations and ghost declarations will have probabilities < 1. The analysis also assumes isorange contours of vanishingly small width, which generates perfect resolution. That is, every intersection can be determined precisely and without ambiguity. In real-world implementations the contours will have finite width and less-than-perfect resolution, which for $N > 2$ will generate the second type of ghost, the *ghost target*, as discussed subsequently.

All of these ghost detections will then be processed by the tracking algorithm into a *ghost track*, just like a real target. Ghost tracks will be maintained as long as successive ghost detections satisfy boundary conditions of the tracker. Once a parameter is exceeded, say with an acceleration exceeding prescribed target maneuver limits, the ghost track is dropped. But with enough of these ghost track fragments displayed, the air situation map can become so cluttered as to be unusable. Consequently, the ghosts must be excised.

The brute-force, ghost excision remedy of collecting redundant data is accomplished by adding another transmit–receive pair, for example to the two-pair system in Figure 6–39. This three-pair system then declares a target detection only when three ellipse intersections occur. For example, in an M:N binomial detection logic where $N = 3$, when $M = 3$, a target is declared; when $M = 2$, a ghost is declared. This process significantly reduces the number of common ghosts given adequate spatial diversity of the sites and SNR. Absent such conditions, more sites with more ellipse intersection criteria may then be required to keep the common ghost problem in check.

6.6.3.4.2 Ghost Targets While the foregoing redundancy-plus-binomial detection scheme can usually control the common ghost problem, it arrives with its own baggage: *ghost targets*. A. Vincent Mrstik in his seminal 1978 paper [130] illustrates the ghost target problem and a third remedy to excise these ghosts: improving range resolution, that is, decreasing the size of the range resolution cell. This remedy applies particularly to range-only multilateration systems, since doppler measurements are usually made with very high resolution. His illustration is reproduced in Figure 6–40.

Unfortunately, range resolution is not a *control variable* in the PBR designer's kit. Specifically, it is limited by the available bandwidth of the host transmitter, which for an FM broadcast transmitter is typically 45–50 kHz and a range resolution of about 3 km, roughly the size of that shown in Figure 6–39(*a*). Consequently the designer's options to excise these target ghosts become limited. The severity of the problem is analyzed next.

Mrstic's 1978 paper [130] can be updated to analyze the severity of the ghost target problem for PBRs using multilateration techniques. Mrstic analyzed a range-only multilateration system consisting of N bistatic transmit–receive pairs operating in a multitarget environment. The system used the M-out-of-N (M:N) binomial detection logic, also called coincidence detection, as follows:

> ...a detection is declared when M-out-of-N possible receivers observe a video signal-plus-noise amplitude exceeding a predefined threshold V_T. That is, a first threshold V_T is applied to the output of each radar [i.e., transmit-receive pair], and a second threshold M is applied to a count of the receivers which record a crossing of the first threshold at time delays corresponding to possible target locations.

Mrstic observes that while binomial detection is a noncoherent and thus nonoptimal detection scheme, it is easy to implement with simple processing and insensitive to phase distortions along the multiple receiving paths that can plague coherent schemes.[52]

Mrstic's development for estimating the composite false-alarm rate of the multistatic system, P_{FA}, proceeds as follows. He first defines a multistatic range cell as the intersection of two or more

52 Coherent multistatic operation is treated in [113]. Examples are thinned, random, distorted, and distributed arrays and interferometers with interferometric radars being the principal implementation.

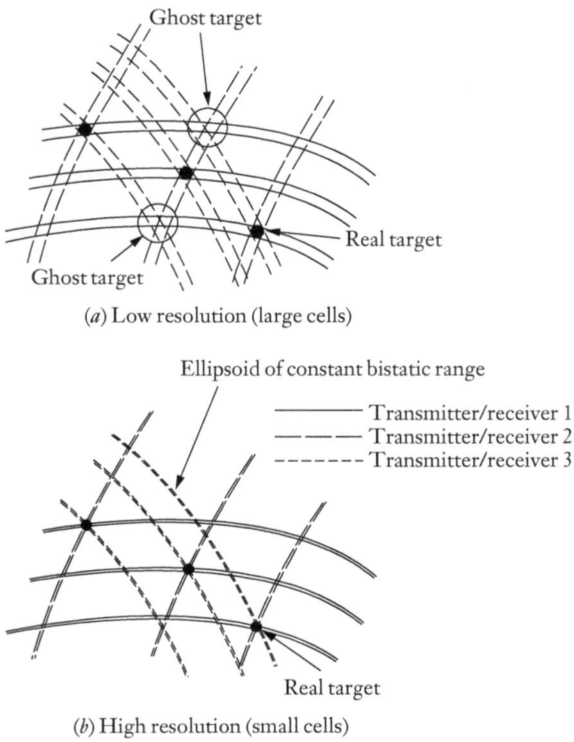

Figure 6–40 Ghost target illustration for a range-only multilateration system with $n = 3$, $N = 3$, from [130]. In this case, Mrstik asserts that decreasing the size of the range resolution can "...resolve many apparent multiple intersections as two or more distinct intersections." Specifically both ghost targets in (*a*) are resolved in (*b*). Note that decreasing resolution does not affect the number of two-ellipse intersections, or common ghosts. Mrstik then cautions: "The feasibility and utility of reducing the processing cell size are limited by the radar range resolution, the available data processing, and the radar measurement accuracy. Making the cells smaller than the basic radar accuracy (including bias errors) would be counterproductive, since real targets as well as ghosts would be lost." © IEEE, 1974

bistatic range ellipses, or ellipsoids in three dimensions. Each ellipse will have a width of the bistatic range resolution, ΔR_B, defined by the signal bandwidth of a specific transmit–receive pair and a specific geometry. Specifically from [113],

$$\Delta R_B = c/2B_n \cos(\beta/2) \tag{6.42}$$

where B_n is noise bandwidth of the bistatic receiver, β is bistatic angle, and c is speed of light.

False alarms in a multistatic range cell are caused by either random noise detections or a target that is detected anywhere on one of the bistatic range ellipses forming the cell under test, but not in that cell.

Mrstic then quantifies false alarms for a single transmit–receive pair and for the multistatic system of N such pairs, under three simplifying assumptions: (1) the SNR is the same at all N receivers; (2) the targets are randomly distributed; and (3) the receivers are widely spaced. The first assumption simplifies the analysis but yields optimistic results. A lower bound to system performance can be set by selecting the lowest SNR in the N-receiver set. The second assumption depends on the scenario and in particular on target density, which will be evaluated subsequently. The third assumption is needed to minimize GDOP, which would occur when the receivers are clustered together—an undesirable and unlikely configuration for multistatic radars in the first place.

The Mrstic analysis also assumes pulsed radar operation, directional antennas and N-receivers operating with one transmitter. These assumptions are easily modified for CW waveforms, near-omni antennas, and any number of transmitters and transmit–receive pairs used by passive multistatic radars.

Mrstic defines the probability that a single transmit–receive pair will falsely declare a first threshold crossing in a cell, P_{fa}, as,

$$P_{fa} = P_n + P_g - P_n P_g \tag{6.43}$$

where P_n is probability of noise crossing the first threshold V_T in the bistatic range ellipse of the cell under test and P_g is probability that a target is detected in the bistatic range ellipse of the cell under test, but not in that cell.
Now

$$P_g = P_e P_d \tag{6.44}$$

where P_d is probability of detecting a target in the cell, given at least one target in the range cell and P_e = probability of having one or more targets in the bistatic range ellipse of the cell under test, but not in that cell.

Thus,

$$P_{fa} = P_n + P_e P_d (1 - P_n). \tag{6.45}$$

The term P_e is related to target density, which can be approximated as the expected number of targets, n, in the surveillance sector of the transmit–receive pair. If the targets are either sparse or uniformly distributed throughout the surveillance sector, P_e will closely approximate the target density. Air traffic control regions near major airports typically have uniformly distributed targets. En route regions typically have sparse targets. World War II bomber formations and dogfights, where many aircraft are clustered together, are counterexamples but not are likely to be encountered in modern air defense scenarios.

The surveillance sector can be defined by the number N_B of contiguous bistatic range ellipses of width ΔR_B bounded by the maximum range oval of Cassini as an outer detection limit. Thus P_e can be approximated as follows:

$$P_e \cong n/N_B. \tag{6.46}$$

In general, the bistatic ellipses and ovals span different areas depending on the baseline L and the equivalent monostatic range R_M, as detailed in [113]. However, in many PBR configurations $R_M > L$, so that both the ellipse set and the oval approximate circles as in the monostatic case. In these cases, N_B can be estimated by evaluating the surveillance sector along the extended baseline, that is, the baseline extended from the receive site out to the oval, which is R_R. That distance is given by Table 6–5 as $(R_M^2 + L^2/4)^{1/2} - L/2$.
Consequently

$$N_B \cong R_R/\Delta R_B \tag{6.47}$$

and

$$P_e \cong n\, \Delta R_B/R_R. \tag{6.48}$$

Also, in this geometry $\beta \rightarrow 0$, so ΔR_B, collapses to the monostatic case:

$$\Delta R_B = c/2B_n \tag{6.49}$$

Thus

$$P_e \cong cn/2\,B_n\,R_R \tag{6.50}$$

Equation (6.47) is somewhat pessimistic in that it uses the smallest ΔR_B over the bistatic surveillance sector, which in turn minimizes P_e and P_{fa}. If the count were made along the perpendicular bisector of the baseline, $\beta > 0$, which will broaden ΔR_B. Although this bisector distance (between baseline and oval) is slightly longer than the distance along the baseline, the change in ΔR_B dominates the count. This in turn will slightly reduce P_e and ultimately the estimated number of target ghosts. The difference is not significant as long as the bistatic ellipses and ovals approximate circles.

The final step in Mrstic's analysis is to calculate the composite false-alarm probability P_{FA} for the full multistatic system, that is, Mrstic's ghost targets, using the cumulative binomial distribution:

$$P_{fa} = \sum_{K=M}^{N} \binom{N}{K} P_{fa}^{K} \left(1 - P_{fa}\right)^{N-K} \tag{6.51}$$

where N = number of transmit–receive pairs in the multistatic net, M = second threshold detection criterion, and P_{fa} = probability that a single transmit–receive pair will falsely declare a first threshold crossing in a cell.

Consider an example using parameters of the PBR in *Howland's Approach #2* (Section 6.6.2.3), extended to operate with multiple transmitters in a multilateration configuration using a binomial detection logic:

	$R_M = 173$ km	Calculated
	$L = 50$ km	Given
	$R_R = 150$ km	Table 6–5
	$\Delta R_B = 3$ km	$B_n = 45$ kHz.
Thus,	$N_B = 50$ cells	Equation (6.47)
	$P_e \cong n/50$	Equation (6.48)

Assume the SNR at each transmit–receive pair is sufficient to generate:

	$P_d = 0.9$	Assumed
	$P_n = 10^{-6}$	Assumed
Thus,	$P_{fa} \cong P_e\,P_d$	Equation (6.45)
	$= 0.9\,P_e$.	

For a typical air surveillance scenario the number of targets, n, can vary from a few to many, say from 1 to 20. Thus P_e, the probability of having one or more targets at the bistatic range of the cell under test, will range from 1/50 to 20/50, or from 0.02 to 0.4, and the probability that a single transmit–receive pair will falsely declare a first threshold crossing in the cell under test, P_{fa}, will range from 0.018 to 0.36. Finally, the ghost target composite false-alarm probability P_{FA} is calculated from equation (6.51) using the following M:N detection criteria: for 3 transmit–receive pairs, 3:3; for 4 pairs: 3:4 and 4:4; and for five pairs, 3:5. The composite probability of detecting true targets, P_D, was also calculated for each M:N criterion by replacing P_{fa} with $P_d = 0.9$ in equation (6.51). Note that common ghosts are automatically excised by requiring $M > 2$ for each criterion. Results are shown in Figure 6–41.

As expected, 3:3 and 4:4 detection criteria show the best ghost target rejection, since all transmit–receive pairs must report a first threshold crossing. That, in turn, severely penalizes

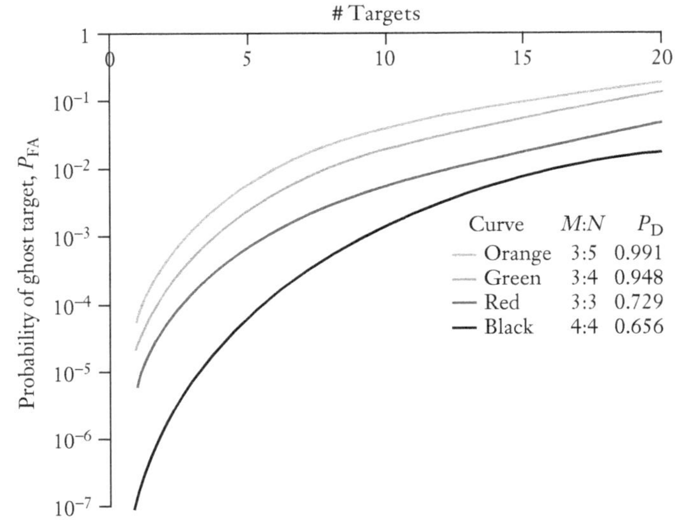

Figure 6–41 Composite ghost target probabilities (and true target detection probabilities) for a range-only multilateration system using *M*:*N* binomial detection logic.

true target detection performance. As the criterion is relaxed, P_D increases; but so does P_{FA}. The net result for most detection criteria is a manageable P_{FA} (10^{-4} to 10^{-2}) for a small number (<5) of targets. But as the number of targets approach 20, P_{FA} exceeds 10^{-1} for acceptable probability of detections (>0.9). This situation can rapidly clutter an air situation display with ghost target tracks. A 3:4 binomial detection logic appears to be the best compromise in these cases.

6.6.3.4.3 Ghost Excision How ghost excision, in particular target ghost excision, is carried out operationally depends primarily on the number of targets observed. In sparse or moderately dense target environments, for example, as shown in Figure 6–39, operator-controlled target association and deghosting is possible. This process is similar to operator-controlled (or -aided) target detection and track initiation in old monostatic search radars displaying target echo data on *plan-position-indicator* (*PPI*) scopes.

In denser target environments, displays can become too cluttered for an operator to sort through, for example, as shown in Figure 6–42. In this case, some type of automatic gating or processing is needed.

The need for automated ghost excision was established shortly after World War II when (a) multiple sensors were used to sort out multiple targets in expected military engagements, and (b) new, digital computers were becoming available for both data processing and data integration. Atkin and Marple were the first to address this problem in the open literature in 1962 [131].

Their problem was to locate targets using angle-only measurements from multiple sensors with overlapping coverage, similar to the jammer triangulation problem. And again, true target locations were established by multiple, that is, >2, angle intersections, while just two intersections were declared ghosts. The major issue was that straightforward testing of all combinations of intersections against a preestablished test tolerance required excessively long times on the available computers. For example, they estimated that 4½ hours would be required to process data from three sensors on 200 targets.

Atkin and Marple were able to reduce this time to an estimated one second by developing a data interrogation algorithm. They describe it as "...an inversion of the normal operating

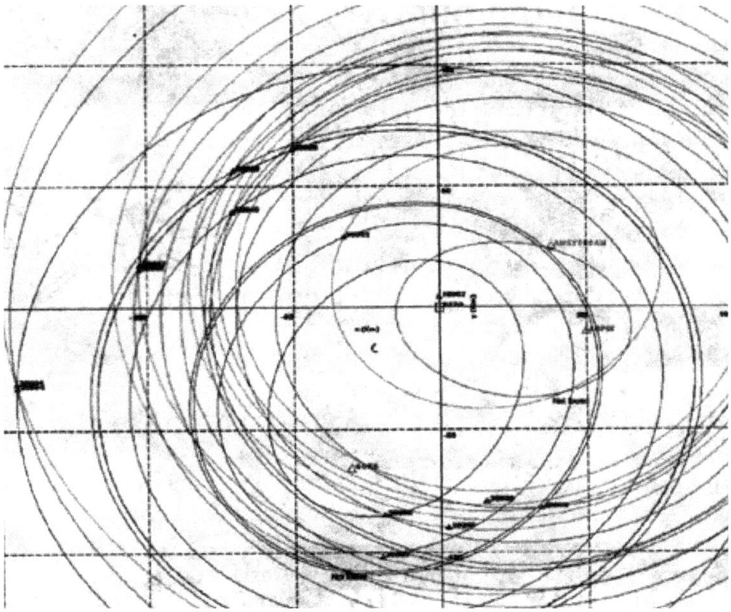

Figure 6–42 Excerpt from Chart 34, Multistatic Data Association, of Howland's *FM-Radio Based Bistatic Radar* brief [122] showing data from his NATO test bed in the Netherlands. Receiver is at (0, 0) and three transmitters are shown in red, blue, and green. Nine targets with nine isorange ellipses per transmit–receive pair are shown (eight for blue). *Ground-truth* target locations are plotted from air traffic control radar data, which also show triplet intersections. All other intersections are ghosts, although some of these intersections appear to be triplets, and thus would become a ghost target. In either case, this level of detail is clearly beyond the capability of an operator to sort through in real time. (© 2005 NC3A.)

philosophy. Usually one computes functions of the data; in a data interrogation one determines if data corresponding to values of the required function is present."

James Caspers provided a tidy description of the process as follows [115]:

> ...The region of interest is divided into cells each of which is defined by a stored set of quantized data. These stored sets of data are then used to interrogate the observations. When the observations and stored data match each other within the limits of a tolerance criterion, the set of observations is assumed to represent a target. Considerable time can sometimes be saved by first dividing the space into relatively few coarse cells. Those cells for which the observed data test affirmatively are then further subdivided.

A functional block diagram of the process is shown in [131] and repeated in [115]. With slight modifications, it can be adapted from intersecting angle strobes to intersecting ellipses.[53]

While formal details of the SAIC-developed *HDTV-Based Passive Radar* deghosting process have not been published, the system is reported to use a sequence of postdetection processing steps.[54] The first step is to form range and range-rate estimates at each range ellipse intersection for each receiver location. Next, the range-rate estimates are compared to measured doppler

53 On an historical note, Atkin and Marple estimate that their algorithm could be implemented on a general purpose processor using NOR logic, operating at a 2-MHz clock rate and requiring 830 transistors, 3,900 diodes and 53,500 bits of memory. Obviously their algorithm is no longer implementation-limited.

54 A. P. Andrews, private communication, January and February 2006.

data from each receiver. Comparisons falling within preestablished error criteria, which include GDOP factors, are declared targets.

Since ground traffic (cars and trucks) in an urban environment was found to be a significant source of both real (but nuisance) and ghost detections, the processing was configured to allow multiple geographic zones with different binomial detection logic. Specifically, since it was less likely that a surface target would present a clear LOS to each of the four receivers in the system, a 4:4 logic was selected in the short-range, low-doppler region. For air targets at long range and higher doppler, a 3:4 logic was selected.

This deghosting scheme was exercised in the 2005 Washington, DC, tests. Target densities reported by one receiver ranged from 50 to 100 detections per 0.5-second coherent dwell; by the central node, 5–25 multilaterated target contacts; and by the tracker, 0–15 tracks. As detailed in Section 6.6.3.4.1, 50 target detections per each of the four receivers will generate $(2n^2 - n)(N^2 - N)/2 = 3 \times 10^4$ ghosts (isorange intersections where no target exists), where n is the number of detected targets per site and N is the number of sites. Thus, the ghost excision via doppler correlation algorithm appeared to excise most of the ghosts, although as reported earlier, the ground traffic was so dense that breakthroughs occurred especially at close range. SAIC also reported that these target densities did not challenge the system's load capacities.

Lockheed Martin's *Silent Sentry*™ PBR [132] suffered from ghost tracks and track fragments during its development.[55] In dense target environments, it now uses up to six transmitters and one receiver [133], along with some type of ghost excision algorithm, possibly an M:N binary detection logic coupled with a data interrogation algorithm. The result is a situation display relatively clear of ghost track fragments at some loss in probability of detection. Unfortunately, details are not available, since nearly all the signal and data-processing algorithms are held proprietary to Lockheed Martin.

However the data processing is accomplished, a penalty for this type of multilateration ghost excision is the need for many transmit and/or receive sites—all with overlapping coverage. Finding an increasing number of transmitter sites becomes more difficult, which limits the regions where the system can operate. The use of many small and inexpensive receive sites operating with one transmitter might offer a more flexible solution.

A new approach to ghost excision, and in fact to target state estimation, was presented as a companion paper [134] to Howland's 2005 IEE paper [121]. Tobias and Lanterman (T/L) attacked the multistatic, multiple target tracking problem by extending the *probability hypothesis density (PHD)* approach developed by Mahler [135] to PBRs. T/L provide the rationale for this approach as follows:

> One of the more elegant aspects of traditional Kalman filtering is the way in which the prior and posterior distributions are characterised by a small set of sufficient statistics that are easily propagated in the Kalman recursion. When target tracking is generalized to the multitarget, multisensor scenario, however, no simple analogous implementation seems to appear.

T/L then define their approach as follows:

> Ronald Mahler introduced the concept of a probability hypothesis density (PHD), which is defined as being any function that, when integrated over any given area, specifies the expected number of targets present in the area. More specifically, the PHD is the factorial moment density found in point process theory... and it provides a straightforward method of estimating the number of targets in a region

---

55 In the special case of a very dense target environment with well-distributed targets, such as will occur near major airports usually during early morning or late afternoon, n can be greater than 30. In this case, $P_{fa} = 0.36$ for $n = 30$ and 0.67 for $n = 40$. *Silent Sentry*™, with performance parameters similar to the Howland system, was field-tested in such an environment, which caused many ghost tracks. The source of such tracks was identified but not quantified at the time.

under observation. We thus expect the PHD to be a useful tool for tracking multiple targets, especially in handling the many ghost targets that arise from noisy bistatic radar measurements.[56] Using probability generating functionals and set calculus, Mahler derives Bayesian time-update and data-update equations that use the PHD, respectively, to perform motion prediction and incorporate sensor observations... This allows the multitarget tracker to incorporate both range and doppler observation information, which we expect to produce better tracking results than using range-only information.

We use the particle filter implementation of the update equations... whereby the PHD is represented by a collection of particles and their corresponding weights. At time-step k, each particle in the filter is a vector of the form $\xi_i = [x_i y_i \dot{x}_i \dot{y}_i]^T$ and has a weight $w_{i,k}$, where $(x_i y_i)$ specify the particle's location and $(\dot{x}_i \dot{y}_i)$ specify its velocity components. As per the defining property of the PHD,

$$\tilde{N} = E\,[\text{no. of targets}] = [N_{k/k}]_{\text{nearest integer}} \tag{6.52}$$

where

$$N_{k/k} = \sum_i w_{i,k} \tag{6.53}$$

Specifically, the PHD is expected to (a) automatically estimate the number of targets, (b) resolve ghost targets, and (c) fuse sensor (i.e., transmit–receive pair) data with no explicit *report-to-track* association needed [136]. T/L describes these traditional association techniques as *soft*, for example, as performed by joint probabilistic data association and *hard*, for example, as performed in multiple frame assignment using Lagrangian relaxation techniques.

T/L then simulated the performance of the PHD particle filter implementation, using three bistatic transmit–receive pairs measuring first range and then range/doppler on two air targets flying in the Washington, DC, area. Three local FM stations served as the transmitter and a variant of the Lockheed Martin's *Silent Sentry*™ was used as the receiver, located 30–50 km from the transmitters. (The variant assumed no receive antenna gain; hence, the system's performance was similar to that of the two Howland systems described earlier.) Adequate target visibility, overlapping coverage, and no multipath were assumed. Calculated SNR ratios ranged from 12.2 dB to 32.5 dB.

In its simplest characterization, the simulation began by independently and randomly assigning the particles' two-dimensional position and velocity components to fall within the FOV of each transmit–receive pair. Particle weights were initialized to 0. These particles were then propagated forward in one-second steps. *Birth particles* with random positions and velocities were added at each time step to model new targets. One new target and hence one birth particle was assumed to appear at each one-second step.

The PHD then assigned (and updated) particle weights $w_{i,,k+1}$ at each time step by incorporating range/doppler observations, calculated probability of detection, Poisson-distributed false alarms, and a single-target likelihood function. Finally, the expected number of targets in the FOV was calculated via equation (6.53). Locations of the $\tilde{N}$ expected targets were found by extracting the $\tilde{N}$ highest peaks from the PHD represented by these weights. T/L initially used an expectation–maximization algorithm for this extraction [134], but they found it computationally inefficient and time consuming, so plan to use a peak-extraction algorithm designed to exploit the integral property of the PHD [136].

Results of the simulation were mixed but encouraging. Two sets of runs were made, first with a false-alarm probability of 10^{-4} and then with 10^{-2}. The first set was initially run with 1,000 particles and 500 birth particles. It showed the range-only tracker performing somewhat better than the range–doppler tracker in that the latter would occasionally fail to detect the second

56 "Noisy bistatic radar measurements" should be considered a generic term. As detailed earlier, principal factors establishing the number of ghosts are number of targets, their density, bistatic resolution cell width, and bistatic coverage area.

target, whereas the former appeared to always detect both targets. Doubling the total number of particles allowed the range–doppler tracker to generate a tighter track over the range-only tracker on both targets. However, it exhibited a time latency of from 1 to 18 seconds to detect the second target, which was ascribed to filtering out ghosts. More particles may reduce this latency. Also, while both trackers were off by a factor of 1 and occasionally 2 in estimating the correct number of targets present in some of the time steps, the range–doppler tracker demonstrated more stability in maintaining a correct estimate once the track was established.

The second, 10^{-2} false-alarm probability set was run with 3,000 total particles and the same SNR; thus, the probability of detection was increased slightly. At these settings the range-only tracker was swamped with false alarms and could not track the two targets. In contrast, the range–doppler tracker was able to track the two targets, but overestimated the number of targets 10%–20% of the time with a maximum overestimate of 4. It underestimated about 2.5% of the time. It also dropped track of the targets on occasion.

T/L also observed that in regions of low SNR within the 80 × 80 km FOV the filter considerably overestimated the number of targets present, which result in ghosts. This effect was ascribed to the birth particle model, which placed these particles in areas of low SNR, and thus low probability of detection, and did not receive any radar data to contradict the placement. One identified solution was to restrict either the FOV or the model to operate in regions of sufficiently high SNR. Other possible solutions are to place birth particles only in absence of propagated particles, and/or to augment the particle state to include target identification [136].

Tobias subsequently reported [137] that an improved method for placement of birth particles was developed, thus removing the need to restrict particles to areas of high SNR. The computationally more intensive method involved using a precomputed grid, the variable spacing of which depended on the range variances in the simulation, and least-squares logic to place birth particles at intersections of the bistatic range ellipse and doppler observations. This method also had the effect of reducing the number of particles needed from a few thousand to a few hundred.

A further concern is the incorporation of multipath effects—inevitable with broad antenna beam widths—into the simulation. Specifically, multipath nulls will exacerbate the ghosting problem by adding more areas of low SNR. Finally, the potential order-of-magnitude improvement in location accuracy afforded by doppler measurements as predicted in Section 6.6.3.3.5 did not materialize and should be rationalized.

Tobias reports [137] that multipath effects were later inserted into the simulation, which showed that "...incorrectness in the sensor likelihood functions caused by the discrepancy between the true Pd and SNR and the Pd and SNR assumed by the PHD filter," unsurprisingly leads to a deterioration in target tracking performance; and further that "in the presence of false alarms, the PHD filter is highly sensitive to the probabilities of detection and signal-to-noise ratios present in the tracking scenario".[57]

T/L characterized these findings and limitations as preliminary and suggested "cleverer [algorithms and] proposal densities," which could improve performance. They also proposed comparing performance of the PHD-based particle filter to other multitarget tracking methods for PBR operation, a necessary step in these types of research and development programs.

In summary, these early results demonstrated, with clear and elegant documentation, a completely new approach to multistatic target state estimation and ghost excision. While the scenario was not particularly stressing, in that three bistatic pairs and just two targets were modeled,[58] the algorithms show potential for converging to useful surveillance tracks with minimum ghosting.

57 M. Tobias, private communication, October 20, 2006.
58 Three targets were subsequently reported [136].

6.7 ELECTRONIC COUNTERMEASURES

6.7.1 Introduction

As outlined in Section 6.3, PBRs are inherently covert. Thus ECM would not be used against a PBR until an adversary (a) identifies PBR operation through visual reconnaissance, infiltration, or compromise, or (b) deduces that a covert air defense net must be operating in the battle area based on unexplained aircraft losses and *a priori* knowledge that PBRs have been deployed. Assuming that one of these events has occurred, it remains to assess the vulnerability of a PBR operating in a military air surveillance scenario, specifically in the hostile air defense scenario that includes various types of noise jamming ECM.

Noise jamming is but one of many techniques in an ECM arsenal. Deception ECM, chaff, and decoys are others. However, noise jamming is usually the first step in a radar vulnerability analysis since it is (1) analytically robust, using well-established link equations; (2) credible, since it is available in most threat inventories; (3) traditionally recognized as a primary method for countering air surveillance radars; and (4) not usually classified. This last reason is nearly unique in the ECM community, since other techniques are often classified [139]. Thus, this section will concentrate on noise jamming techniques.[59]

Assessing the effects of noise jamming on PBRs is divided into two steps: (1) modifying the bistatic radar's geometry-invariant, benchmark range by the effects of noise jamming, and (2) extracting from the modified benchmark range values of receiver-to-target detection range and coverage as a function of the geometry. This process is a direct extension of the method used to evaluate PBR range performance in Section 6.5.

In that section, the *benchmark range* (also known as the equivalent monostatic maximum range or the geometric mean range) was defined as $R_M = (R_T\ R_R)^{1/2}$, where R_T is the range between transmitter and target and R_R is the range between receiver and target. It was calculated using the canonical range equation given in equation (6.8). It represents performance of the bistatic radar when transmitter and receiver are colocated; that is, when the baseline range $L = 0$. This benchmark range will be modified by the *jamming-to-noise ratio* (JNR) to account for the effects of noise jamming and tagged with the subscript n: $(R_M)_n = (R_T R_R)_n{}^{1/2}$.

The second step accounts for the bistatic geometry by defining a fixed range-difference $\Delta = R_T - R_R$, which then can be solved for R_R as a function of Δ and $(R_M)_n$. When the target lies on the extended baseline, for example, the baseline L extended beyond the receive site, $\Delta = L$, and expressions for calculating the maximum and minimum values of R_R, $(R_R)_{max}$, and $(R_R)_{min}$, as well as bistatic coverage given in Table 6–5 apply, with L and Δ used interchangeably in the expressions. Again the subscript *n* will be used for the noise jamming case.

These steps for calculating the detection range of a PBR in the presence of noise jamming are expanded in the following procedure:

1. Establish the PBR's benchmark range $R_M = (R_T R_R)^{1/2}$ using equation (6.8).
2. Calculate the JNR using equation (6.54) and an estimate for the input noise temperature.
3. Divide the benchmark range R_M by $(\text{JNR} + 1)^{1/4}$ to find the benchmark range with noise jamming present $(R_M)_n = (R_T R_R)_n{}^{1/2}$, equation (6.55).

59 That said, VHF/UHF noise jammers are not particularly common in some ECM inventories, simply because there are not many VHF/UHF radars in their opponent's air defense inventory. (In other inventories the reverse should be true, with counters to the Russian *NEBO 55G6U* VHF (180–220 MHz) radar being a prime example.) Thus some development and production of VHF/UHF noise jammers may be required, which would be expected to parallel the development and production of PBRs if that should occur. In short, *if you build it, they will come.*

4. Assume a specific value of the range difference $\Delta = R_T - R_R$ to establish the geometry. Alternatively, use the baseline range L.
5. Use procedures (3) and (4) and expressions in Table 6–5 (also Table 6–13) to find receiver-to-target range and coverage with jamming present $(R_R)_n$.

6.7.2 Noise Jamming Equation

A method of calculating noise jamming effects that is independent of specifics of the receiver/processor is to compare the input noise temperature produced by noise jamming to that of thermal noise. The effective noise temperature of a jammer is given by [140]:

$$T_j = \frac{P_j G_j G_R \lambda^2 F_j^2 F_p^2}{(4\pi)^2 k B_j R_j^2} \tag{6.54}$$

Where P_j = jammer power (W), G_j = jammer antenna gain, G_R = receiver antenna gain, λ = wavelength (m), F_j = pattern-propagation factor for the jammer-to-receiver path, F_p = jammer-to-receiver polarization factor, k = Boltzmann's constant, B_j = jammer bandwidth (Hz), and R_j = jammer-to-receiver range (m).

The advantage of using the noise temperature calculation is that the performance change in a radar can be calculated without making any assumptions as to the processing method. As long as the receiver does not process signals from frequencies beyond the assumed jammer bandwidth, the receiver bandwidth and methods of integration will not change the signal-to-interference ratio (SIR) required for detection. Whatever limitations apply to normal thermal noise will apply also to noise jamming, and the only change will be in the benchmark range, which in turn changes the detection and track initiation ranges.

To calculate the benchmark range with noise jamming present, $(R_M)_n$, a *system input interference temperature* $T_i' = T_i + T_j$ replaces T_i in equation (6.8). Specifically, equation (6.8) is multiplied by $(T_i/T_i')^{1/4} = [T_i/(T_i + T_j)]^{1/4} = (1 + T_j/T_i)^{-1/4}$ to yield $(R_M)_n$. Thus

$$(R_M)_n = (R_M) / (1 + T_j / T_i)^{1/4} = (R_M)/(\text{JNR} + 1)^{1/4} \tag{6.55}$$

where T_j/T_i is the JNR. Benchmark ranges R_M and noise temperatures T_i for air surveillance PBRs are given in Table 6–9 and Section 6.5.5.3, respectively. Thus, it remains to exercise equation (6.54 and 6.55) to calculate the benchmark range with jamming present $(R_M)_n$.

6.7.3 Noise Jamming Configurations

Implicit in equation (6.54) are parameters that describe noise jamming air defense scenarios. First is the effective radiated power of the noise jammer $P_j G_j$. Tactical aircraft typically carry deception ECM internally and noise jammers in pods with self-contained power supplies. Pods might generate an ERP ranging from 100 W to 1 kW [139]. The VHF/UHF airborne jammer antenna cannot be made sharply directional, and hence its pattern floods an azimuth sector of at least 90°, making it unnecessary to know the location of the PBR receiver site. Since the polarization of all PBR-suitable transmitters in any scenario can be determined *a priori*, the jammer can be matched to that polarization. Thus the jammer-to-receiver polarization factor $F_p = 1$.

The jammer bandwidth term B_j defines barrage and spot noise jamming. For bistatic or multistatic PBRs operating with a single broadcast transmitter in a sparse (e.g., rural) RF environment, spot noise jamming can be selected. For example, an FM transmitter would require $B_j \approx 100$ kHz, which is easily implemented since the frequency assignments of all broadcast stations in a given area are exactly known. If a few suitable transmitters were present and it was not clear which was being exploited, each could be so jammed. On the other hand, a multistatic PBR operating with multiple FM transmitters in dense (e.g., urban) RF environments would require many such spots, which could rapidly devolve into 20-MHz barrage noise jamming over the total 88–108 MHz FM broadcast band. Similar arguments apply to U.S. TV transmitters, where an approximately 6-MHz spot noise, and 12-, 42-, and 200-MHz barrage noise bandwidths are needed, depending on the band.

The pattern-propagation factor for the jammer-to-receiver path F_j defines main lobe and side lobe jamming. Since mobile PBR receivers typically use a fixed, broad beam to cover the surveillance sector, jamming from within the surveillance sector will be through the main lobe and $F_j = 1$. An exception is the hypothetical urban defense scenario outlined in Section 6.3.2.3, which exploits a very large, building-mounted (or tree-hidden), 400-element (~26 dBi) phased array receive antenna to generate a 340-km PBR stealth trap. This fixed-site PBR excursion can be considered a major—and costly—extension of early versions of the *Silent Sentry*™ PBR, which used a modest gain, multibeam receiving antenna. For a jammer located outside the main lobe of such an antenna, $F_j^2 \approx -35$ dB might be obtained, which reduces JNR and increases the detection and track initiation ranges.

It is possible that the implementation of adaptive nulling in a receiving array might reduce the side lobe level below −35 dB, but the resulting azimuth null would be broad enough to place a serious gap in the target coverage (especially if the expected multipath reflections of the jamming were spread over a few degrees of azimuth around the actual jammer position). Additional jamming platforms would also extend the sectors lost to nulls from the adaptive array. Thus the −35-dB level appears realistic for most situations.

The jammer-to-receiver range R_j defines stand-off, escort, and self-protection noise jamming. Modern attack aircraft seldom—if ever—carry noise jammers for self-protection. ARMs at microwave frequencies and ground-based emitter locators in all bands are both numerous and too well integrated with air defenses to permit such a tactic. Escort jammers generally fall in the same category, especially with stealthy attack aircraft that employ low observables in the RCS, RF, and IR dimensions: no need to draw RF attention to their operating area. Thus, stand-off noise jammers remain the ECM of choice for support of penetrating aircraft. Stand-off ranges are typically selected to be greater than the expected PBR-supported air defense engagement range, which is approximately 100 km for the mobile PBRs and 300 km for the fixed-site PBR.

Pattern-propagation factor effects for the stand-off jammer-to-receiver path should also be considered. Figure 6–43 shows results of calculating the propagation factor for a 3-m receive antenna height above the surface and a 100-km jammer-to-receiver range plotted against jammer height H_j above a smooth, 4/3 earth radius [140].

The figure shows a value slightly above unity for $H_j = 10$ km. The calculations for jamming effectiveness would be approximately correct (±3 dB) for any jammer altitude between 6 km and 14 km. For $R_j = 300$ km, appropriate jammer altitudes are 22 and 45 km. Table 6–11 summarizes these ECM-controlled parameters.

6.7.4 Benchmark Range Modified by Noise Jamming

Data from Table 6–9 shows that the most capable, mobile PBR currently used for air surveillance exploits FM transmitters with 250-kW ERP. The benchmark range for this system is 120 km.[60]

60 The planned U.S. HD-TV transmitter operating at VHF shows the same performance in a clear environment and will show superior performance against noise jamming, since both its operating band and its instantaneous bandwidth are considerably broader. It becomes the preferred PBR illuminator when deployed in sufficient numbers.

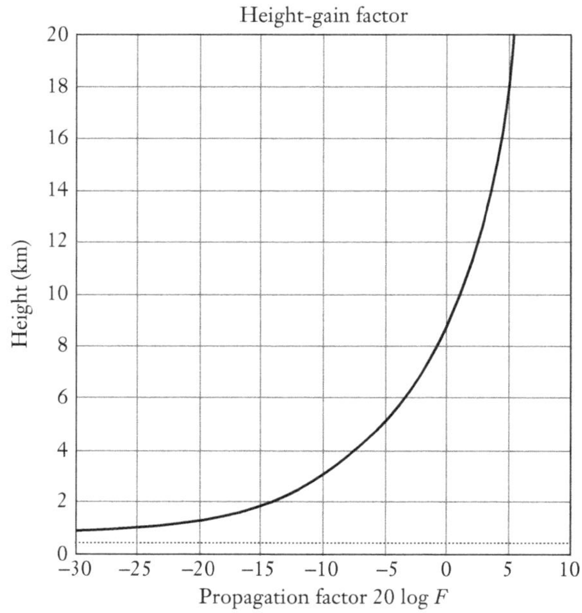

Figure 6–43 Jammer-to-receiver propagation factor for a 3-m receive antenna height and a 100-km jammer [141].

Table 6–11 Noise Jamming Parameters From Equation (6.54)

Parameter	Notation	Value
Effective radiated power (ERP)	$P_j G_j$	100, 1,000 W
Polarization factor	F_p	1
Pattern-propagation factor	F_j	0 dB: main lobe
		−35 dB: side lobe (fixed-site PBR)
FM jamming bandwidth	B_j	100 kHz: spot (one transmitter)
		20 MHz: barrage (many transmitters)
TV jamming bandwidth	B_j	6 MHz: spot
		12, 42, 200 MHz: barrage
Jammer-to-receiver range	R_j	100 km: mobile PBR
		300 km: fixed-site PBR

The benchmark range for the fixed-site PBR excursion using the same transmitter and now a 26-dBi receive antenna is 340 km. The VHF noise temperature T_i for both systems is taken from Section 5.5.3 to be 10^5 K. Resulting noise jamming-modified benchmark ranges $(R_M)_n$ for this system are summarized in Table 6–12.

Spot noise jamming is directed at bistatic and multistatic PBRs operating with one transmitter. If a few transmitters in that scenario must be jammed $(R_M)_n$ is multiplied by $n^{1/4}$, where n is the number of spots required and $T_j/T_i >> 1$. Barrage noise jamming is directed at PBRs operating with multiple transmitters usually over the total broadcast band.

Table 6–12 Noise Jamming-Modified Benchmark Ranges $(R_M)_n$ for Mobile and a Fixed-Site Passive Bistatic Radars Using 250-kW FM Transmitters Operating Against One Stand-Off Noise Jammer*

	Main Lobe (mobile)		Main Lobe/Side Lobe (Fixed-Site)	
Jammer ERP	**Spot Noise**	**Barrage Noise**	**Spot Noise**	**Barrage Noise**
100 W	9.4 km	36	16/122 km	62/320
1,000 W	5.3	20	9/69	35/242

* The jammer is 100 km from the mobile PBR and 300 km from the fixed-site PBR. The mobile PBR has an 8-dBi gain receive antenna; the fixed-site PBR, 26-dBi gain with −35-dB side lobes.

Since mobile PBRs operate with broad fixed beams, main lobe values apply across the PBR surveillance sector. For the fixed-site PBR excursion both main lobe and side lobe values apply, depending on jammer location. Note that barrage noise values for $(R_M)_n$ in the side lobes do not map as $(1{,}000)^{1/4}$. Since $T_j/T_i \rightarrow 1$ in equation (6.55), the full term $(1 + T_j/T_i)^{1/4}$ must be used in the calculation.

6.7.5 Geometry Modified by Noise Jamming

The maximum and minimum detection ranges and coverage for a bistatic radar as a function of its baseline L (or range difference Δ) and benchmark range R_M are given by the formulae in Table 6–5, and reproduced below.

Tables 6–14 and 6–15 summarize calculations for main lobe spot and barrage noise jamming, respectively, of the mobile PBR. Values for $(R_R)_{n\text{-min}}$ are given since that typically sets the receive site spacing requirement for multistatic operation used by mobile PBRs.

It is clear from Table 6–14 that stand-off, spot noise jamming at both 100 and 1,000 W reduces mobile PBR air surveillance coverage to unacceptable levels at all baseline ranges (and range differences) of interest: maximum receiver-to-target detection ranges $(R_R)_{max}$ of < 2 km. Table 6–15 shows similar results for all but the 50 km baseline case operating against a 100-W barrage noise jammer: <11 km. In the 50 km/100 W case while the coverage has been reduced by more than 90%, multistatic operation remains possible at short surveillance ranges: approximately 20 km along the extended baseline, that is, in the forward-threat sector.

Tables 6–16 and 6–17 summarize calculations for side lobe barrage and spot noise jamming, respectively, of the fixed-site PBR. For this excursion with a stand-off barrage noise jammer positioned in the receive antenna's –35-dB side lobes, the ECM outcome is reversed: PBR

Table 6–13 Formulae for Calculating Bistatic Radar Detection Ranges and Coverage, Extracted From Table 5.3*

Case	L (Δ)	Area (of one oval)	$R_{R(max)}$ (on R_x oval)	$R_{R(min)}$ (on R_x oval)
Circle (Monostatic Benchmark)	0	πR_M^2	R_M	R_M
One oval	$<2R_M$	$\sim\pi [R_M^2 - L^4/(64 R_M^2)]$	$(R_M^2 + L^2/4)^{1/2} + L/2$	$(R_M^2 + L^2/4)^{1/2} - L/2$
Two ovals	$>2R_M$	$\sim\pi R_M^2 [R_M^2/L^2]$	$L/2 - (L^2/4 - R_M^2)^{1/2}$	$(R_M^2 + L^2/4)^{1/2} - L/2$
	$>3R_M$	$\sim\pi R_M^2 [R_M^2/L^2]$	$\sim R_M^2/L$	$\sim R_M^2/L$

* Range difference, $\Delta = R_T - R_R$ can be substituted for the baseline range L. $(R_M)_n$ replaces (R_M) for the noise jamming case.

Table 6–14 Summary of Calculations for 100 km Stand-Off, Spot Noise Jamming in the Main Lobe of a Generic, Mobile Passive Bistatic Radar Operating With One 250-kW FM Transmitter

Baseline *L* or Range diff. Δ (km)	Scenario (W)	# ovals	$(R_M)_n$(km)	Full oval area (km²)	$(R_R)_{n\text{-min}}$ (km)
50	Clear	1	120	45,200	98
50	100	2	9.4	10	1.8
50	1,000	2	5.3	1	0.57
120	Clear	1	120	44,500	74
120	100	2	9.4	2	0.74
120	1,000	2	5.3	0.2	0.23

Table 6–15 Summary of Calculations for 100-km Stand-Off, Barrage Noise Jamming in the Main Lobe of a Generic, Mobile Passive Bistatic Radar Operating With Multiple 250-kW FM Transmitters Over the Broadcast Band

Baseline *L* or Range diff. Δ (km)	Scenario (W)	# ovals	$(R_M)_n$ (km)	Full oval area (km²)	$(R_R)_{n\text{-min}}$ (km)
50	Clear	1	120	45,200	98
50	100	1	36	3835	19
50	1,000	2	20	201	7
120	Clear	1	120	44,500	74
120	100	2	36	366	11
120	1,000	2	20	34	3

Table 6–16 Summary of Calculations for 300-km Stand-Off, Barrage Noise Jamming in the −35-dB Side Lobes of a Generic, Fixed-Site Passive Bistatic Radar Operating With Multiple 250-kW FM Transmitters Over the Broadcast Band

Baseline *L* or Range diff. Δ (km)	Scenario (W)	# ovals	$(R_M)_n$ (km)	Full oval area (km²)	$(R_R)_{n\text{-min}}$ (km)
50	Clear	1	340	363,000	316
50	100	1	320	322,000	296
50	1,000	1	242	184,000	218
120	Clear	1	340	363,000	285
120	100	1	320	322,000	266
120	1,000	1	242	184,000	189

performance is diminished only slightly, as shown in Table 6–16. Specifically, long-range surveillance capability is retained in all cases.

As expected, spot noise jamming through a −35-dB side lobe (Table 6–17) shows some reduction of capability, with medium-range (50–100 km) surveillance possible in most cases on the extended baseline in the forward-threat sector.

Table 6–17 Summary of Calculations for 300-km Stand-Off, Spot Noise Jamming in the −35-dB Side Lobes of a Generic, Fixed-site Passive Bistatic Radar Operating With One 250-kW FM Transmitter.

Baseline L or Range diff. Δ (km)	Scenario (W)	# ovals	$(R_M)_n$ (km)	Full oval area (km²)	$(R_R)_{n\text{-min}}$ (km)
50	Clear	1	340	363,000	316
50	100	1	122	47,000	96
50	1,000	1	69	15,000	48
120	Clear	1	340	363,000	285
120	100	1	122	46,000	76
120	1,000	2	69	5,000	31

Finally, main lobe jamming of the fixed-site PBR will occur whenever the main beam of the PBR's receiver scans past the stand-off jammer. In this case the azimuth 3-dB beamwidth is approximately 10°, broadening to approximately 20° at the 0-dBi level. As shown in Table 6–12, the modified benchmark ranges for this case are only slightly greater than noise jamming in the main lobe of the mobile PBR. Thus, no useful surveillance capability is possible against spot noise and an only short range (20–40 km) capability is possible against barrage noise jamming.

6.7.6 ECM–ECCM Responses [142–144]

The mobile PBR starts at a significant disadvantage: 100-W spot noise jamming at 100 km defeats the PBR (Table 6–14). Thus, a first ECCM design response could be to force the ECM adversary into a barrage mode by exploiting multiple transmitters across the full broadcast band (Table 6–15). This action regains short-range surveillance capability against the 100-W jammer, but nothing against the 1,000-W jammer. A second ECCM design response could be to implement either modestly directional PBR receive antennas with –20-dB side lobes or 20-dB direct path jammer cancellation. Either ECCM will increase the modified benchmark range by a factor of $(100)^{1/4} = 3.16$, recovering a medium-range surveillance capability against 100 W barrage noise jamming.

In response, an ECM adversary could spot jam only those transmitters that have been assessed to be suitable for PBR operation in a specific scenario. This type of multiple spot noise jamming is more effective than barrage jamming, depending on the number of spots required. If the adversary deployed more than one such jammer, surveillance coverage would be further reduced, ultimately driving the PBR performance down- to a shortrange capability. Furthermore, if jammer cancellation is used, these multiple spot noise jammers will desensitize a wedge of the surveillance sector near each of the jammer-to-PBR receiver's LOS.

In contrast, the fixed-site PBR with its large receive array antenna starts at a significant advantage: the PBR retains long-range surveillance against 100-W and 1000-W side lobe barrage noise jamming at 300 km (Table 6–16). If the ECM adversary is able to invoke spot noise jamming as described above, PBR performance is reduced to medium ranges (Table 6–17). As a counter, one ECCM design response is to reduce side lobes to −40 dB, or −14 dBi, which is feasible in array antennas but at a penalty of some main beam broadening. This measure will reestablish a formidable surveillance capability. The ECM adversary's response is to reduce the stand-off range or to build bigger jammers.

And so the measure–countermeasure game continues until terminated by ever-increasing costs. Usually the ECCM defender reaches the cost limit first, and thus loses the paper game. Faced with this prospect, the PBR designer of both mobile and fixed sites might consider extending *covert* operation into *war mode* operation. In parallel, the ECM/strike designer could invoke a basic mode of his offense: attack. For example, he could disable all broadcast transmitters in the defended area via a precursor strike. Alternatively, he could attempt a *shoot-look-shoot* doctrine with UAV probes—with the game continuing until assets from one side are either destroyed or exhausted.

6.7.7 Use of a Jammer as the Illuminator

It is sometimes postulated that the jammer itself could provide useful illumination of the target, negating the results derived above. A simple calculation shows that this is not the case for a stand-off jammer, and in fact any type of noise jammer.

Consider the previous scenario with a 1,000-W noise jammer standing off 100 km from the PBR receive site. The PBR in this scenario exploited a 250-kW FM transmitter for a benchmark range R_M = 120 km in a clear environment. If that PBR now exploited the noise jammer as its illuminator with all other terms in the range equation, equation (6.8), remaining constant, its benchmark range would be reduced by a factor of $(250\ \text{kW}/1\ \text{kW})^{1/4} = 4$, to 30 km. Now the baseline range L for this new system is simply the jammer stand-off range of 100 km. Since L/R_M = 100 km/30 km = 3.3, the $L > 3\ R_M$ formulae for the two oval case in Table 6–13 apply, yielding two nearly circular ovals, one around the receiver and one around the jammer, each with average radius of 9 km. The coverage area for one oval is 255 km^2.

While the receiver-centered oval is not useful for air surveillance, targets flying within 9 km of the jammer will be detected by the PBR receiver. However, since a stand-off jammer by definition does not escort the target, the probability of its lying within 9 km of the target would be low, and even if the target were to pass within that distance of the jammer when approaching the defended area it would not stay that close for a prolonged period.

A more likely scenario would be for the jammer to maintain tens of kilometers or greater offset from a penetrating raid as the jammer approached its assigned 100-km stand-off range. Calculation of the aircraft spacing requirements would be part of the mission planning if the PBR system were the intended victim of the jammer. Furthermore, since both escort and self-protection noise jammers are nearly always proscribed in these air defense scenarios, exploiting noise jammers of any type does not appear to be a feasible option for the PBR.

Appendix 6–A
A Review of UHF/VHF Monostatic and Bistatic Radar Cross-Section Data

Little VHF—and no UHF—bistatic RCS data on air targets is available in the open literature. What is available is old, but venerable: Siegel, Crispin, and Newman's horizontally polarized, horizontal plane, 250-MHz monostatic and bistatic "experimental data on a large jet aircraft" (subsequently identified as a *B-47*), published in 1968 [60], reproduced in [40] and Figure 6–A.1. This data becomes the basic source for the benchmark bistatic RCS at VHF.[61]

The plots for this 250-MHz data are averaged over 10° of aspect angle and so are relatively smooth. They exhibit small undulations, a peak at the broadside specular and are relatively flat elsewhere. The monostatic nose, tail, and broadside median RCS values are approximately 80, 125, and 8,000 m^2, respectively. At other aspect angles, they fluctuate about 100 m^2. Most bistatic data are within ±3 dB of the monostatic values, with an occasional excursion to ±6 dB for the $\beta = 135°$ curve. Broadside specular values are identical, except for the $\beta = 135°$ curve, which is about 4 dB less. Excluding the monostatic and bistatic specular peaks, this aspect-independent, nearly constant RCS phenomenon is typical of a target response in its resonance region.

Siegel et al. observes that "...even when the wavelength is relatively large by comparison with body dimensions... the bistatic theorem holds quite well for moderate values of β. Thus, in the 'average' sense we can certainly apply this theorem with reasonable confidence as long as β is not too large [60]."

Siegel et al. used the sufficiently smooth, vanishing wavelength version of the bistatic theorem, which states that the bistatic RCS on the bistatic bisector is equal to the monostatic RCS. These results are not unexpected since resonance scatter should dominate other bistatic effects such as shadowing and loss of retroreflectors on complex targets.

Furthermore, the magnitude of the broadside specular depends on the sine of the incidence angle. For the monostatic case the incidence angle is 90°, and only for the $\beta = 135°$ curve is there any significant reduction: $\sin(90 - 135/2)° = 0.38 = -4$ dB, as shown in the data—probably more coincidence than design.

Extrapolation to other frequencies appears in earlier, horizontally polarized *B-47* monostatic data provided by K. Siegel to M. I. Skolnik for his 1st ed. (1962) of *Introduction to Radar Systems* [Figure 2.17 of (61)]. Four frequencies, each from a different source, are plotted: 150 MHz (Evans Signal Laboratory), 195 MHz (Ohio State University), 330 MHz (University of Michigan), and 600 MHz (Radiation, Inc.). The 195 and 330 MHz data are theoretical; the 150 and 600 MHz data are experimental. Skolnik reports that "The accuracy of theoretical cross sections is claimed to be from 2 to 10 dB. Experimental-measurement accuracies also are of the same order of magnitude. The frequency dependence is seen to be slight."

Unfortunately, security restrictions were even more stringent at that time, so that this data has no calibration on the ordinate, just a "σ in a relative scale" notation, so a direct comparison with the 1968 data cannot be made. In spite of this limitation, all data show virtually the same shape and characteristics as the 250 MHz data cited above [60]: small undulations, a peak at

61 The *B-47* medium jet-bomber has a wing span of 35 m and a length of 33 m, comparable to that of a Boeing 737 jetliner, which incidentally has been among the targets of opportunity for many PBR tests. Thus, their RCS should be similar.

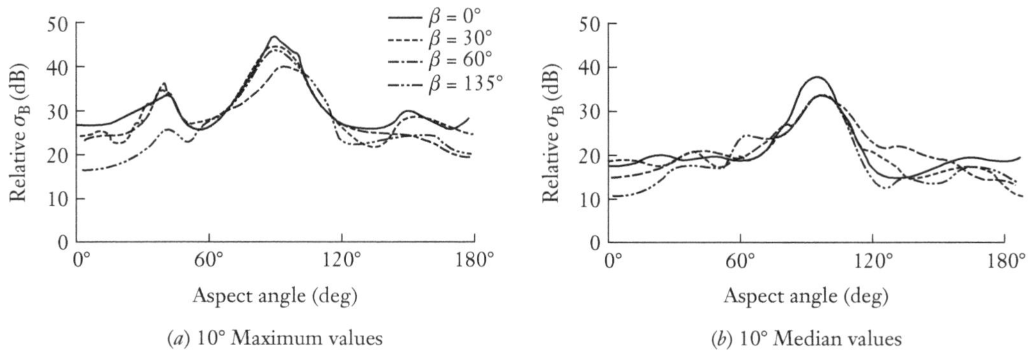

Figure 6–A.1 Relative monostatic and bistatic RCS for a *B-47* jet aircraft at 250-MHz, horizontal polarization, where aspect angle is the angle between the aircraft nose and the bistatic bisector in the horizontal plane. This RCS data is plotted "in decibel-relative scale," clearly because of security restrictions at that time. By correlating other *B-47* plots by Siegel, the ordinate can be changed to dBsm. Median RCS values are typically used in radar range calculations. (Courtesy of Academic Press [60].)

the broadside specular, and relatively flat elsewhere. Hence the same conclusion: the monostatic resonance effect is apparent, extending up to 600 MHz, well into the UHF band. And by modest extrapolation, the bistatic RCS would be expected to exhibit the same effect.

However, earlier monostatic data from Siegel contradicts this resonance finding at 600 MHz. Specifically, Siegel's four-frequency data in the 1st ed. was replaced by UHF (600 and 980 MHz) horizontally polarized, horizontal plane, *B-47* monostatic data in Skolnik's 2nd (1980) edition [55]. This data [62] was taken earlier (1954), but now has the ordinate usefully calibrated in m^2, so can be compared with the 250 MHz data.

None of this 1954 data exhibits the resonance effect. Specifically, the measured, 600 MHz nose, tail, and broadside median monostatic data are 10, 20, and 1,000 m^2, or 8–9 dB uniformly lower than at 250 MHz. Elsewhere the 600 MHz data is 10–20 dB lower than the 250 MHz data. The averaged calculations at 980 MHz generally follow the 600 MHz data, except for a few large, narrow spikes caused by leading and trailing edges of the wing and tail and a drop off near the tail.[62]

Thus, the monostatic and by implication the bistatic resonance effect, which appears at all aspect angles in the 250 MHz data and again in the 150–600 MHz data, is absent in both earlier 600 and 980 MHz data for medium bombers and later 667 MHz data for jetliners. While one is tempted to give a heavier weight to the later data, there is no solid rationale for doing so. Thus, a conservative, upper frequency limit for medium bomber and jetliner resonance scattering will be set at 450 MHz.[63]

One VHF/UHF monostatic data point for smaller aircraft is available in [55] but with no attribution as follows:

> A military propeller aircraft such as the AD-4B has a cross section of about 20 m^2 at L band, but a 100 m^2 cross section at VHF [a 7 dB difference, but with no aspect angle specified]. The longer wavelengths at VHF result in greater cross section than microwaves because the dimensions of the scattering objects are comparable to the wavelength and produce resonance effects.

62 Nor did subsequent compact range data in Skolnik's 3rd ed. [47], now for a Boeing 737 at vertical polarization and scaled to 667 MHz, exhibit the resonance effect.

63 The coauthor admits to a further motivation for selecting 450 MHz: A convenient separation in frequencies occurs at that point: the tuning range of candidate transmitters for PBR exploitation falls entirely above or entirely below 450 MHz, thereby simplifying the subsequent analysis.

The *AD* attack aircraft has a wingspan of 15 m and a length of 11.6 m, less than half the size of the *B-47*. Yet its RCS is similar to the *B-47* at VHF, likely attributed to its large nose cowl and massive propeller. Conventional jet attack aircraft of similar size should generate a smaller RCS. For example, Skolnik provides a 2 m^2 monostatic estimate for a small fighter or four-passenger jet at microwave frequencies in his Table 2.1 [47].[64] Since these jets will also exhibit resonance at VHF, their monostatic (and bistatic) RCS should approach 10 m^2 at VHF.

Skolnik also gives a 20 m^2 monostatic estimate at microwave frequencies for a medium bomber or jetliner. Consequently, their monostatic and bistatic RCS should approach 100 m^2 at VHF, which is consistent with the Siegel et al. 1968 VHF data for the B-47 cited earlier. For all this data, Skolnik cautions that "...it is not usual that the target echo can be adequately described by a single number"—a limitation to be remembered but not avoided when manipulating numbers in the radar range equation.

This subject was revisited by Kuschel in 2002 [39], who reports anechoic chamber measurements of a small (3-m length) "generic missile." RCS, averaged over a ±60° nose-on sector, is plotted versus frequency and shows resonance scattering, fluctuating a few decibels about 1 m^2 between 100 and approximately 450 MHz. From there the RCS rolls off monotonically to 0.08 m^2 at 1 GHz, a 12-dB change, likely due in part to the large, 120° averaging region.

Thus, on the basis of this limited data, the 450 MHz upper frequency limit for resonance scattering of large air targets will be extended to small air targets—pending receipt of more definitive data.

64 Similar values have been cited by Blake and Nathanson in their books, but again without attribution.

Appendix 6–B List of Symbols

Symbol	Meaning
a	Semi-major axis
$a(t)$	Signal waveform
a_r	Radial component of target acceleration
a_t	Target acceleration
A	Target shadow area
$A(f)$	Signal spectrum
A_s	Specular reflection area of target
B	RF input bandwidth
B_3	Gaussian signal spectrum half-power width
B_j	Jammer bandwidth
B_n	Noise bandwidth
B_r	Rectangular signal spectrum width
c	Speed of light
C_e	Cross-range correlation factor
C'_e	Down-range correlation factor
d	Target extent or length
d_β	Distance between center of ogive and baseline
e	Eccentricity
E/N_0	Signal energy-to-noise spectral density ratio
f_R	Receive antenna pattern factor
f_T	Transmit antenna pattern factor
$F_d(n)$	nth measurement of target doppler shift
F_j	Pattern-propagation factor for the jammer-to-receiver path
F_n	Receiver noise factor
F_p	Jammer-to-receiver polarization factor
F_R	Receive pattern propagation factor
F'_R	Receive propagation factor
F_T	Transmit pattern propagation factor
F'_T	Transmit propagation factor
g	Maximum number of common ghosts
G_j	Jammer antenna gain
G_p	Processing gain
G_R	Receiver antenna power gain
G_{Rt}	Gain of receive antenna in direction of transmitter
G_T	Transmitter antenna power gain
h_R	Altitude of receive antenna

h_t	Altitude of target
h_T	Altitude of transmit antenna
H_j	Jammer altitude above a smooth, 4/3 earth
I	Inclination
JNR	Jamming-to-noise ratio
k	Boltzmann's constant
K_a	Acceleration error constant
K_i	Range sum of ith multistatic transmit-receive pair
L	Baseline, or distance between transmitter and receiver
L_i	Baseline of ith multistatic transmit-receive pair
L_M	Monostatic transmit and receive losses
L_{min}	Minimum baseline range
L_r	RF component loss
L_R	Receive system losses (>1) not included in other parameters
L_{sp}	Signal processing loss
L_T	Transmit system losses (>1) not included in other parameters
M	Second threshold detection criterion
n	Number of detected targets
N	Number of transmit-receive pairs
$\tilde{N}$	Expected number of targets
N_B	Number of contiguous bistatic range ellipses of width ΔR_B bounded by a maximum range oval of Cassini
N_o	Noise spectral density
P_d	Probability of detecting a target in the cell, given at least one target in the range cell
P_D	Composite probability of detecting true targets
P_{direct}	Power of direct-path signal at receiver
P_e	Probability of having one or more targets in the bistatic range ellipse of the cell under test, but not in that cell
P_{FA}	Ghost target composite false-alarm probability
P_{fa}	Probability that a single transmit-receive pair will falsely declare a first threshold crossing in a range cell
P_g	Probability that a target is detected in the bistatic range ellipse of the cell under test
P_j	Jammer power
P_n	Probability of noise crossing the first threshold V_T in the bistatic range ellipse of the cell under test
P_T	Transmitter average power output
r_R	Line-of-sight range between receive antenna and target
r_T	Line-of-sight range between transmit antenna and target
r_β	Radius of ogive of constant bistatic angle
R_j	Jammer-to-receiver range

R_M	Equivalent monostatic maximum range, or benchmark range
$(R_M)_n$	Benchmark range with noise jamming present
R_R	Range between receiver and target
$R_{R(av)}$	Radius of a circle with area equal to area of one oval of Cassini
$R_{R(max)}$	Maximum R_R permitted on oval of Cassini
$R_{R(min)}$	Minimum R_T available on oval of Cassini
$(R_R)_n$	Receiver-to-target range with noise jamming present
R_T	Range between transmitter and target
$\mathrm{R_x}$	Receiver
SIR	Signal-to-interference ratio
SNR	Signal-to-noise ratio
$(S/N)_{min}$	Minimum signal-to-noise ratio required for detection
t_c	Coherent processing time or interval
t_I	Integration time to establish a track file
t_o	Time constant of a tracking loop
T_a	Antenna contribution to noise temperature
T_e	Receiver contribution to noise temperature
$\mathrm{T_{gt}}$	Target
T_i	System input noise temperature
T_i'	System input interference temperature
T_j	Effective noise temperature of a jammer
T_m	Temperature of man-made interference
T_o	290 K
T_r	RF component contribution to noise temperature
T_s	Temperature of sky noise interference
$\mathrm{T_x}$	Transmitter
u	Correlation coefficient of errors between virtual monostatic paths
$u(t)$	Point target response of the signal
V	Target radial velocity
$\mathrm{V_i}$	Location of ith virtual monostatic radar
V_T	Predefined, first threshold
z_0	Satellite altitude
α	Receiver look angle
2α	Angle between two virtual monostatic look angles
α_a	rms time duration of the signal
β	Bistatic angle
β_a	rms bandwidth of the signal
β_n	Bandwidth of a tracking loop

γ	Angle between virtual monostatic paths in vertical plane
δ	Required change in κ to maintain adequate detection performance
Δ	Fixed bistatic range difference $(R_T - R_R)$
Δf_d	Target doppler spread, or velocity walk
ΔR_B	Width of bistatic range resolution cell
ΔR_R	Incremental change in R_R
ΔR_T	Incremental change in R_T
ε_a	Dynamic lag error
θ	Elevation of target from receive site
$\theta(n)$	n^{th} measurement of target bearing
θ_F	Angular width of the scattered signal
θ_R	Angle of target measured from the receiver in the bistatic plane
κ	Bistatic maximum range product-squared
λ	Wavelength
ρ	Power utilization factor
σ	Mean anomaly at epoch
σ_B	Bistatic radar cross section
σ_{cr}	Range error measured orthogonal to the bisector of $2a$
σ_{cv}	Velocity error measured orthogonal to the bisector of $2a$
σ_{dr}	Range error measured along the bisector of $2a$
σ_{dv}	Velocity error measured along the bisector of $2a$
σ_f	Thermal noise component of doppler data
σ_F	Forward-scatter radar cross-section
σ_M	Monostatic radar cross section
σ_{ri}	Virtual monostatic range measurement error of ith bistatic link
σ_{rsi}	Bistatic range-sum measurement error of ith bistatic link
σ_t	Thermal noise component of time-delay data
σ_{ti}	Bistatic time-delay measurement error of ith bistatic link
σ_v	Bistatic rms velocity measurement error
σ_{vi}	Virtual monostatic velocity measurement error of ith bistatic link
τ	Target delay
τ_B	Width of bistatic delay resolution cell
ν	Target doppler shift
φ	Incident grazing angle
Φ	Transmitter power density at the target
Φ'	Transmit power spectral density at L_{min}
$\lvert X \rvert^2$	Ambiguity function
ω	Argument of perigee
Ω	Right ascension of the ascending node

References

1. N. J. Willis, *Bistatic Radar*, 2nd ed., Technology Service Corp., Silver Spring, MD, 1995, corrected and reprinted by SciTech Publishing, Inc., Raleigh, NC, 2005.
2. O. E. Rittenbach and W. Fishbein, "Semi-Active Correlation Radar Employing Satellite-Borne Illumination," *IRE Transaction on Military Electronics*, April–July 1960, pp. 268–269.
3. Memorandum, Chief of Naval Research to Chief of Naval Operations, Subject: CW transmit site at Spruce Creek, FL, April 29, 1966.
4. J. Richelson, *Wizards of Langley*, Westview Press, Boulder, CO, 2001.
5. J. Fawcette, "Bistatic Radar May Find a 'Sanctuary' in Space," *Electronic Warfare/Defense Electronics*, January 1978, pp. 84–88.
6. V. Koch and R. Westphal, "A New Approach to a Multistatic Passive Radar Sensor for Air Defense," *IEE Proceedings Radar, Sonar and Navigation*, 1995, pp. 22–28.
7. X. He, M. Cherniakov and T. Zeng, "Signal detectability in SS-BSAR with GNSS non-cooperative transmitter," *IEE Proceedings Radar, Sonar and Navigation*, Vol. 152, No. 3, June 2005, pp. 124–132.
8. M. Cherniakov, D. Nezlin, and K. Kubik, "Air Target Detection via Bistatic Radar Based on LEOS Communication Signals," *IEE Proceedings Radar, Sonar and Navigation*, Vol. 149, No. 1, pp. 33–38, 2002.
9. H. D. Griffiths, A. J. Garnett, C. J. Baker and S. Keaveney, "Bistatic Radar Using Satellite-Bourne Illuminators of Opportunity," *Proceedings of RADAR-92 Conference, Brighton; IEE Conference Publ.*, No. 365, October 12–13, 1992, pp. 276–279.
10. R. F. Bernstein, Jr. and N. J. Willis, consultant, "Bistatic Radar Project," Department of Electrical and Computer Engineering, *U.S. Naval Postgraduate School*, Monterey, CA, 2000 (Approved for public release).
11. N. J. Willis, *Bistatic Radar*, 2nd ed., Technology Service Corp., Silver Spring, MD, 1995.
12. H. D. Griffiths and N. R. W. Long, "Television-Based Bistatic Radar," *IEE Proceedings* Vol. 133, No. 7, Pt. F, December 1986, pp. 649–657.
13. "Passive System Hints at Stealth Detection Silent Sentry—A New Type of Radar," *Aviation Week*, November 30, 1998, pp. 70–71.
14. J. Baniak, G. Baker, A. M. Cunningham and L. Martin, "Silent Sentry™ Passive Surveillance," *Aviation Week and Space Technology*, June 7, 1999.
15. *Jane's Radar and Electronic Warfare Systems*, Land-Based Air Defense Radars, United States. Date Posted: October 22, 2004 James' Information Group, Martin Streetly.
16. P. E. Howland, "Target Tracking Using Television-Based Bistatic Radar," *IEE Proceedings Radar, Sonar and Navigation*, Vol. 146, No 3, June 1999, pp. 166–174.
17. P. E. Howland, D. Maksimiuk and G. Reitsma, "FM Radio Based Bistatic Radar," *IEE Proceedings Radar, Sonar and Navigation*, Vol. 152, No. 3, June 2005.
18. P. E. Howland, "FM-Radio Based Bistatic Radar," *AOC 4th Multinational PCR Conference*, Syracuse, NY, October 6, 2005, pp. 107–115.
19. A. Andrews, "HDTV-Based Passive Radar," *AOC 4th Multinational PCR Conference*, Syracuse, NY, October 6, 2005.
20. N. J. Willis, *Passive Coherent Locators*, Air and Missile Defense Advisory Panel, 2003 National Military Sensing Symposium (MSS), Washington, DC, Session H-9, October 23, 2003.
21. "Assessment of Ultra-Wideband (UWB) Technology," DTIC No. ADB146160, July 13, 1990. (Executive summary of the report was published in *IEEE AES Magazine*, November 1990, pp. 45–49.)
22. C. A. Fowler, "The UWB (Impulse Radar) Caper or 'Punishment of the Innocent'," *IEEE AES Magazine*, December 1992, pp. 3–5.
23. P. E. Howland, "Passive Tracking of Airborne Targets Using Only Doppler and DOA Information," in *IEE Colloquium on Algorithms for Target Tracking* (1995/104), May 1995, Savoy Place, London, UK.
24. J. D. Sahr, "Remote Sensing With Passive Radar at the University of Washington," *IEEE Geoscience and Remote Sensing Society Newsletter*, December 2005, pp. 16–21.
25. M. Cetin and A. D. Lanterman, "Region-Enhanced Passive Radar Imaging," *IEE Proceedings Radar, Sonar and Navigation*, Vol. 152, No. 3, June 2005, pp. 185–194.

26. M. Tobias and A. D. Lanterman, "Probability Hypothesis Density-Based Multitarget Tracking With Bistatic Range and Doppler Observations," *IEE Proceedings Radar, Sonar and Navigation*, Vol. 152, No. 3, June 2005, pp. 195–205.
27. R. P. S. Mahler, "Multitarget Bayes Filtering via First-Order Multitarget Moments," *IEEE Transaction on Aerospace Electronics Systems*, Vol. 39, No. 4, 2003, pp. 1152–1178.
28. C. Meier Hoyuela, A. J. Terzuoli, Jr. and R. P. Wasky, "Determining Possible Receiver Locations for Passive Radar," *IEE Proceedings Radar, Sonar and Navigation*, Vol. 152, No. 3, June 2005, pp. 206–214.
29. http://www.roke.co.uk/sensors/stealth/cell_phone_radar_concept.asp and many news/magazine websites, 2005.
30. D. K. P. Tan, H. Sun, Y. Lu, M. Lesturgie, and H. L. Chan, "Passive Radar Using Global System for Mobile Communication Signal: Theory, Implementation and Measurements," *IEE Proceedings Radar, Sonar and Navigation*, Vol. 152, No. 3, June 2005, pp. 116–123.
31. H. D. Griffiths and C. J. Baker, "Passive Coherent Location Radar Systems. Part 1: Performance Prediction," in *IEE Proceedings Radar, Sonar and Navigation*, Vol. 152, No. 3, June 2005, pp. 153–159.
32. G. W. Preston, "The Theory of Stellar Radar," Rand Corp. Mem. RM-3167-Pr, May 1962.
33. E. A. Lewis et al., "Hyperbolic Direction Finding With Sferics of Transatlantic Origin," *Journal Geophysical Research*, Vol. 5, July 1960, pp. 1879–1905.
34. R. H. Espeland, "Experimental Evaluation of SCORDES Resolution Capabilities," *Naval Ordnance Laboratory*, Rep 652, May 1966.
35. P. E. Howland, "Target tracking using television-based bistatic radar," *IEE Proceedings– Radar, Sonar and Navigation*, Vol. 146, No. 3, June 1999, pp. 166–174.
36. A. D. Bernard, , "Comparison of PCR vs. Traditional Radars," in *Third Multinational Conference on Passive and Covert Radar 2003*, MIT Lincoln Laboratory, October 21, 2003.
37. D. Lynch, Jr., *Introduction to RF Stealth*, SciTech Publishing, Inc., Raleigh, NC, 2004.
38. C. J. David , *Radar and Laser Cross Section Engineering*, 2nd ed., AIAA Education Series, Reston, VA, 2005.
39. H. Kuschel, "VHF/UHF Radar, part 1: Characteristics; part 2: Operational Aspects and Applications," *Electronics and Communications Journal*, Vol. 14, No. 2, April 2002, pp. 61–72, and Vol. 14, No. 3, June 2002, pp. 101–111.
40. N. J. Willis, *Bistatic Radar*, 2nd ed., Technology Service Corp., Silver Spring, MD, 1995.
41. K. M. Siegel et al., "Bistatic Radar Cross Sections of Surfaces of Revolution," *Journal of Applied Physics*, Vol. 26, No. 3, March 1955, pp. 297–305.
42. K. M. Siegel, "Bistatic Radars and Forward Scattering," in *Proceedings of 1958 National Conference on Aerospace and Electronics*, Dayton, Ohio, pp. 286–290.
43. H. D.Griffiths and C. J. Baker, "Passive Coherent Location Radar Systems. Part 1: Performance Prediction," in *IEE Proceedings – Radar, Sonar and Navigation*, June 2005, Vol. 152, No. 3, pp. 153–159.
44. N. J. Willis, , "Bistatic Radar," in M. I. Skolnik (ed.) *Radar Handbook*, 2nd ed., McGraw-Hill Co., New York, NY, 1990, Chapt. 25.
45. *The Washington Post*, "The Advanced Technology Bomber" May 1, 1988 (cited from the IEEE Spectrum).
46. D. K. Barton, *Modern Radar Systems Analysis*, Artech House, Inc., Norwood, MA, 1988.
47. M. I. Skolnik, *Introduction to Radar Systems*, 3rd ed., McGraw-Hill Co., New York, NY, 2001.
48. J. M. Caspers, "Bistatic and Multistatic Radar," in M. I. Skolnik (ed.) *Radar Handbook*, 1st ed., McGraw-Hill Co., New York, NY, 1970, Chapt. 36.
49. M. I. Skolnik, "An Analysis of Bistatic Radar," *IRE Transactions on Aerospace and Navigational Electronics*, March 1961, pp. 19–27.
50. D. K. Barton, "Radar Systems Analysis," Artech House, Inc., Norwood, MA, 1975.
51. R. E. Hiatt et al., "Forward Scattering by Coated Objects Illuminated by Short Wavelength Radar," *Proceedings of IRE*, Vol. 48, September 1960, pp. 1630–1635.
52. F-35 Joint Strike Fighter, October 2006 www.airforce-technology.com/projects/
53. X. Luo, L. M. H. Ulander, J. Askne, G. Smith and P. O. Frölind, , "RFI Suppression in Ultra-Wideband SAR Systems Using LMS Filters in Frequency Domain," *Electronics Letters*, Vol. 37, No. 4, February 2001, pp. 241–243.
54. T. A. Weil, "Transmitters," in M. I. Skolnik (ed.) *Radar Handbook*, McGraw-Hill, New York, 1970, Chapt. 4.

55. Merrill I. Skolnik, *Introduction to Radar Systems*, 2nd ed., McGraw-Hill Co., New York, NY, 1980.
56. H. D. Griffiths and N. R. W. Long, "Television-Based Bistatic Radar," *IEE Proceedings*, Vol. 133, No. 7, Pt. F, December 1986, pp. 649–657.
57. H. D. Griffiths, C. J. Baker, H. Ghaleb, R. Ramakrishnan, and E. Williams, "Measurement and Analysis of Ambiguity Functions of Off-Air Signals for Passive Coherent Location," *Electronics Letters*, Vol. 39, No. 13, June 26, 2003.
58. P. E. Howland, D. Maksimiuk and G. Reitsma, "FM Radio Based Bistatic Radar," *IEE Proceedings - Radar, Sonar and Navigation*, Vol. 152, No. 3, June 2005, pp. 107–115.
59. P. E. Howland, "FM-Radio Based Bistatic Radar, in *AOC 4th Multinational PCR Conference*, Syracuse, New York, 6 October 2005.
60. K. M. Siegel et al., "RCS Calculation of Simple Shapes-Bistatic," in *Methods of Radar Cross-Section Analysis*, Academic Press, New York, 1968, Chapt. 5.
61. M. I. Skolnik, *Introduction to Radar Systems*, 1st ed., McGraw-Hill Co., New York, NY, 1962.
62. C. E. Schensted, J. W. Crispin and K. M. Siegel, *Studies in Radar Cross-Sections XV: Radar Cross-Sections of the B-47 and B-52 Aircraft*, University of Michigan, Radiation Laboratory Report 2260-1-T, August 1954, AD 46 741.
63. P. M. Woodward, *Probability and Information Theory, With Applications to Radar*, Pergamon Press, Oxford, 1953; reprinted by Artech House, Inc., Norwood, MA, 1980.
64. "FM Radio Transmitters", January, 2006. Available at: http://www.bbc.co.uk/reception/.
65. M. A. Ringer, G. J. Frazer and S. J. Anderson, "Waveform Analysis of Transmissions of Opportunity for Passive Radar," in *Fifth International Symposium on Signal Processing and its Applications*, Brisbane, Australia, August 1999, pp. 511–514.
66. H. D. Griffiths, C. J. Baker, H. Ghaleb, R. Ramakrishnan and E. Willman, "Measurement and Analysis of Ambiguity Functions of Off-Air Signals for Passive Coherent Location," *Electronics Letters*, Vol. 39, No. 13, June 26, 2003, pp. 1005–1007.
67. C. J. Baker, H. D. Griffiths and I. Papoutsis, "Passive Coherent Radar systems—Part II: Waveform Properties," Special Issue of *IEE Proceedings Radar, Sonar and Navigation* on Passive Radar Systems, Vol. 152, No. 3, June 2005, pp. 160–169.
68. M. Alard, R. Halbert and R. Lassalle, "Principles of Modulation and Channel Coding for Digital Broadcasting for Mobile Receivers," *EBU Review, Technical*, Vol. 224, 1987, pp. 3–25.
69. D. Poullin, "Passive Detection Using Digital Broadcasters (DAB, DVB) with CODFM Modulation," Special Issue of *IEE Proceedings Radar, Sonar and Navigation* on Passive Radar Systems, Vol. 152, No. 3, June 2005, pp. 143–152.
70. "Cell Phone Networks", January, 2006. Available at: http://www.sitefinder.radio.gov.uk.
71. D. K. P. Tan, H. Sun, Y. Lu, M. Lesturgie and H. L. Chan, "Passive Radar Using Global System for Mobile Communication Signal: Theory, Implementation, and Measurements," Special Issue of *IEE Proceedings Radar, Sonar and Navigation* on Passive Radar Systems, Vol. 152, No. 3, June 2005, pp. 116–123.
72. ETSI EN 300 910, "Digital Cellular Telecommunications System (Phase 2+); Radio Transmission and Reception," GSM 05.05, version 8.5.1, Release 1999, November 2000.
73. B. Walke, *Mobile Radio Networks; Networking, Protocols and Traffic Performance*, Wiley, New York, NY, 1998.
74. B. Walke, P. Seidenberg and M. P. Althoff, *UMTS: the Fundamentals*, Wiley, New York, NY, 2003.
75. Universal Mobile Telecommunication System, http://www.three-g.net/3g_spectrum.html, January 2006.
76. Digital Radio Mondiale, http://www.drm.org, January 2006.
77. M. C. Jackson, "The Geometry of Bistatic Radar Systems," *IEE Proceedings*, Vol. 133, Pt. F, No. 7, December 1986, pp. 604–612.
78. T. Tsao, M. Slamani, P. Varshney, D. Weiner, H. Schwarzlander and S. Borek, "Ambiguity Function for Bistatic Radar," *IEEE Transaction on Aerospace and Electronic Systems*, Vol. AES-33, 1997, pp. 1041–1051.
79. L. V. Blake, *Radar Range Performance Analysis*, Artech House, Inc., Norwood, MA, 1986.
80. D. K. Barton, *Modern Radar Systems Analysis*, Artech House, Inc., Norwood, MA, 1988.
81. H. D. Griffiths and C. J. Baker, "Passive Coherent Radar Systems—Part I: Performance Prediction," Special Issue of *IEE Proceedings Radar, Sonar and Navigation on Passive Radar Systems*, Vol. 152, No. 3, June 2005, pp. 153–159.

82. P. E. Howland, D. Maksimiuk and G. Reitsma, "FM Radio Based Bistatic Radar," Special Issue of *IEE Proceedings Radar, Sonar and Navigation* on Passive Radar Systems, Vol. 152, No. 3, June 2005, pp. 107–115.
83. A. Andrews, "HDTV-Based Passive Radar," *AOC 4th Multinational PCR Conference*, October 6, 2005, Syracuse, New York.
84. P. E. Howland, "FM-Radio Based Bistatic Radar," *AOC 4th Multinational PCR Conference*, Syracuse, New York, October 6, 2005.
85. S. Drabowitch, A. Papiernik, H. D Griffiths, J. Encinas and B. L. Smith, *Modern Antennas*, 2nd ed., Springer, New York, NY, pp. 458–474, 2005, ISBN 1 4020 3216 1.
86. H. D. Griffiths, N. Karavassilis, M. R. Jones and D. E. N. Davies, "Broadband Nulls From a Circular Array," in *Proceedings 4th IEE International Conference on Antennas and Propagation*, Warwick, IEE Conference Publication, NY, No. 248, April 16–19, 1985, pp. 304–308.
87. R. A. Monzingo and T. W Miller, *Introduction to Adaptive Arrays*, Wiley-Interscience, New York, NY, 1980.
88. R. T. Compton Jr., *Adaptive Arrays: Concepts and Performance*, Prentice-Hall, Upper Saddle River, NJ, 1988.
89. A. Farina, *Antenna-Based Signal Processing Techniques for Radar Systems*, Artech House, Inc., Norwood, MA, 1992.
90. T. K. Sarkar, M. C. Wicks, M. Salazar-Palma and R. J. Bonneau, *Smart Antennas*, Wiley, New York, NY, 2003.
91. L. C. Godara, *Smart Antennas*, CRC Press, Boca Raton, FL, 2004.
92. H. D Griffiths, "A Four-Element VHF Adaptive Array Processor," in *Proceedings 2nd IEE International Conference on Antennas and Propagation*, York, IEE Conference Publication No. 195, Part I, April 13–16, 1981, pp. 185–189.
93. D. K. Barton, *Modern Radar Systems Analysis*, Artech House, Inc., Norwood, MA, 1988.
94. M. I. Skolnik, *Introduction to Radar Systems*, 2nd ed., McGraw-Hill Co., New York, NY, 1962.
95. M. I. Skolnik, "An Analysis of Bistatic Radar," *IRE Transactions on Aeorspace and Navigational Electronics*, Vol ANE-8 No. 1, March 1961, pp. 19–27.
96. H. D. Griffiths and C. J. Baker, "Passive Coherent Location Radar Systems. Part 1: Performance Prediction," in *IEE Proceedings - Radar, Sonar and Navigation*, Vol. 152, No. 3, June 2005, pp. 153–159.
97. N. J. Willis, *Bistatic Radar*, 2nd ed., Technology Service Corp., Silver Spring, MD, 1995.
98. J. M. Caspers, "Bistatic and Multistatic Radar," in M. I. Skolnik (ed.) *Radar Handbook*, 1st ed., McGraw-Hill Co., New York, NY, 1970, Chapt. 36.
99. N. J. Willis, "Bistatic Radar," in M. I. Skolnik (ed.) *Radar Handbook*, 2nd ed., McGraw-Hill Co., New York, NY, 1990, Chapt. 25.
100. Dunsmore, M. R. B., "Bistatic Radars," Invited paper, Alta Frequenza Vol. LVIII-N.2, March-April 1989, pp. 53–78.
101. L. V. Blake, *Radar Range-Performance Analysis*, Lexington Books, Lexington, MA, 1980.
102. J. Baniak, G. Baker, A. M. Cunningham and L. Martin, "Silent Sentry™ Passive Surveillance," *Aviation Week and Space Technology*, June 7, 1999, reprint.
103. H. D. Griffiths, N. R. W. Long, "Television-Based Bistatic Radar," *IEE Proceedings*, Vol. 133, No. 7, Pt. F, December 1986, pp. 649–657.
104. P. E. Howland, "Target Tracking Using Television-Based Bistatic Radar," in *IEE Proceedings - Radar, Sonar, and Navigation*, Vol. 146, No. 3, June 1999, pp. 166–174.
105. J. D. Sahr and F. D. Lind, "The Manastash Ridge Radar: A Passive Bistatic Radar for Upper Atmospheric Radio Science," *Radio Science*, Vol. 32, No. 6, November–December 1997, pp. 465–482.
106. R. A. Simpson, "Spacecraft Studies of Planetary Surfaces Using Bistatic Radar," *IEEE Transactions on Geoscience and Remote Sensing*, Vol. 31, No. 2, March 1993, pp. 2345–2358.
107. H. Kuschel, "VHF/UHF Radar, Part 1: Characteristics; part 2: Operational Aspects and Applications," *Electronics and Communications Journal*, Vol. 14, No. 2, April 2002, pp. 61–72, and Vol. 14, No. 3, June 2002, pp. 101–111.
108. G. W. Ewell and S. P. Zehner, "Bistatic Radar Cross Section of Ship Targets," *IEEE Journal Oceanic Engineering*, Vol. OE-5(4), October 1980, pp. 211–215.
109. R. E. Kell, "On the Derivation of Bistatic RCS From Monostatic Measurements," *Proceedings of IEEE*, Vol. 53, August 1965, pp. 983–988.
110. M. I. Skolnik, *Introduction to Radar Systems*, 3rd ed., McGraw-Hill Co., New York, NY, 2001, Chapt. 10.

111. P. E. Howland, "FM-Radio Based Bistatic Radar," in *AOC 4th Multinational PCR Conference*, Syracuse, New York, October 6, 2005.
112. A. Andrews, " HDTV-Based Passive Radar," in *AOC 4th Multinational PCR Conference*, Syracuse, New York, October 6, 2005.
113. M. I. Skolnik, "An Analysis of Bistatic Radar," in *IRE Transactions on Aerospace and Navigational Electronics*, Vol. ANE-8, No. 1, March 1961, pp. 19–27.
114. N. J. Willis, *Bistatic Radar*, 2nd ed., Technology Service Corp., Silver Spring, MD, 1995, corrected and reprinted by SciTech Publishing, Inc., Raleigh, NC, 2005.
115. J. M. Caspers, "Bistatic and Multistatic Radar," in M. I. Skolnik (ed.) *Radar Handbook*, 1st ed., McGraw-Hill Co., New York, NY, 1970, Chapt. 36.
116. R. B. Patton, Jr., "Orbit Determination From Single Pass Doppler Observations," *IRE Transactions on Military Electronics*, April–July, 1960, pp. 336-344.
117. L. G. deBey, "Tracking in Space by DOPLOC," *IRE Transactions on Military Electronics*, April–July 1960, pp. 332–335.
118. "ORDVAC," in *Electronic Computers Within the Ordnance Corps*, ENIAC World Wide Web. Chapt. IV http://ftp.arl.mil.~mike/comphist/6lordnance/chap4.html, May, 2006.
119. C. H. Dawson, "Inactive Doppler Acquisition Systems," *AIEE Trans.*, Vol. 81, Pt. I, January 1963, pp. 568–570.
120. P. E. Howland, "Target Tracking Using Television-Based Bistatic Radar," in *IEE Proceedings—Radar, Sonar and Navigation*, Vol. 146, No. 3, June 1999.
121. P. E. Howland, D. Maksimiuk and G. Reitsma, "FM Radio Based Bistatic Radar," *IEE Proceedings—Radar, Sonar and Navigation*, Vol. 152, No. 3, June 2005.
122. P. E. Howland, "FM-Radio Based Bistatic Radar," in *AOC 4th Multinational PCR Conference*, Syracuse, NY, October 6, 2005.
123. J. Scavullo and F. J. Paul, *Aerospace Ranges: Instrumentation*, D. Van Nostrand Co, Inc., Princeton, NJ, 1965.
124. D. K. Barton, *Radar System Analysis*, Artech House, Inc., Norwood, MA, 1976.
125. D. K. Barton, "Reasons for the Failure of Radio Interferometers to Achieve Their Expected Accuracy," *Proceedings of IEEE*, Vol., 51, No. 4, April 1963, pp. 626–627.
126. A. P. Andrews, "HDTV-Based Passive Radar," in *AOC 4th Multinational PCR Conference*, October 6, 2005, Syracuse, NY.
127. S. C. Carson et al., "Issues in Achieving Noise Limited Performance in a Passive Multi-Static Radar System," in *2006 International Conference on Waveform Diversity and Design*, January 22–27, 2006, Lihue, HI, pp. 103–108.
128. P. M. Woodward, *Probability and Information Theory, With Applications to Radar*, McGraw-Hill Co., New York, NY, 1953; Artech House, Inc., Norwood, MA, 1980, p. 105.
129. W. J. Zable, "MOPTARS—A Multi-Object Phase Tracking and Ranging System," in *3rd IRE Mil-E-Con*, Washington, DC, June 28–July 1, 1959, pp. 187–191.
130. A. V. Mrstik, "Multistatic-Radar Binomial Detection," *IEEE Transactions on Aerospace and Electronic Systems*, Vol. AES-14, No. 1, January 1978.
131 J. Atkin and N. B. Marple, " Information Processing by Data Interrogation," *IRE Transactions on Electronic Computers*, April 1962, pp. 181–187.
132. J. Baniak, G. Baker, A. M. Cunningham L. and Martin, "Silent Sentry™ Passive Surveillance," *Aviation Week and Space Technology*, June 7, 1999.
133. *Jane's Radar and Electronic Warfare Systems*," Land-Based Air Defense Radars, United States. Date Posted: October 22, 2004.
134. M. Tobias and A. D. Lanterman, "Probability Hypothesis Density-Based Multitarget Tracking With Bistatic Range and Doppler Observations," *IEE Proceedings - Radar, Sonar and Navigation*, Vol. 152, No. 3, June 2005.
135. R. P. S. Mahler, "Multitarget Bayes Filtering via First-Order Multitarget Moments," *IEEE Transactions on Aerospace and Electronic Systems*, Vol. 39, No. 4, 2003, pp. 1152–1178.
136. M. Tobias and A. D. Lanterman, "Probability Hypothesis Density-Based Multitarget Tracking With Bistatic Range and Doppler Observations," in *AOC 4th Multinational PCR Conference*, 6 October 2005, Syracuse, NY.

137. M. Tobias, *Probability Hypothesis Densities for Multitarget, Multisensor Tracking with Application to Passive Radar*, PhD thesis, Georgia Institute of Technology, 2006.
138. C. Meier Hoyuela, A. J. Terzuoli, Jr. and R. P. Wasky, "Determining Possible Receiver Locations for Passive Radar," *IEE Proceedings - Radar, Sonar and Navigation*, Vol. 152, No. 3, June 2005.
139. L. Van Brunt, Applied ECM, EW Engineering Inc., Dun Loring, VA, 1978, 3 Vol.
140. D. K. Barton, *Modern Radar System Analysis*, Artech House, Inc., Norwood, MA, 1988, p. 34.
141. D. K. Barton, *Radar System Analysis and Modeling*, Artech House, Inc., Norwood, MA, 2005, Chapt. 6, Program 6.5, "Propagation Factors."
142. J. A. Boyd, et al, *Electronic Countermeasures*, Peninsula Publishing, Los Altos, CA, 1978.
143. M. V. Maksimov, et al, *Radar Anti-Jamming Techniques*, Artech House, Inc., Norwood, MA, 1986.
144. D. C. Schleher, *Introduction to Electronic Warfare*, Artech House, Inc., Norwood, MA, 1986.

7

Ionospheric Measurements[1]

John D. Sahr

7.1 INTRODUCTION

The Earth's upper atmosphere includes both neutral and ionized gases. The ionized gas is called a plasma, and the partially ionized region of the atmosphere is called the *ionosphere*. The ionospheric plasma environment is quite unusual; it is relatively cold, and consists of both molecular (e.g., O_2^+ and NO^+) and atomic ions (e.g., O^+, H^+, and He^+) [1]. When the ionospheric plasma is at rest, radio waves with frequencies above 30 MHz will almost always pass through the ionosphere unimpeded. However when the plasma is disturbed, plasma density irregularities can degrade the ionospheric radio channel at frequencies up to 2 GHz.

Plasma density irregularities frequently occur in the ionosphere at altitudes above 90 km. These field-aligned irregularities (FAI) are ion sound waves (near 100 km) or low-density cavities or bubbles (at higher altitudes) that are created by a variety of plasma instabilities [2, 3]. Scientists study these irregularities for a variety of reasons, ranging from pure scientific inquiry to practical applications.

Scientific investigation of plasma turbulence in the lower ionosphere relies heavily upon VHF and UHF radar remote sensing. The neutral atmosphere is too thin to loft balloons above 40 km, and too thick for durable satellites below about 200 km. The entire *in situ* sounding rocket data set for the lower ionosphere probably does not exceed 2-h duration.

A variety of radio methods are used to study the ionosphere, including techniques for measuring D-region absorption (riometers), bulk total internal reflection (ionosondes), over-the-horizon (OTH) HF radars,[2] and scintillation receivers [4]. Over-the-horizon radars are also discussed in Chapter 4 of this book.

This chapter is especially concerned with the technique of "coherent scatter radar," which refers to the use of moderately sensitive VHF and UHF systems that detect scatter of radio waves from large amplitude fluctuations in the plasma density. This is in contrast to the much more sensitive "incoherent scatter radar" technique, which refers to VHF and UHF systems that detect the scatter of radio waves from the thermal fluctuations in the plasma medium. The underlying radar waveforms and receiver systems can be quite similar, as both targets are volume scatterers

1 This material is based upon work supported principally by the National Science Foundation under Grant No. ATM-0310233.

2 http://superdarn.jhuapl.edu.

with similar spatial and temporal scales. Both radar systems coherently detect the scatter of radio waves from a large volume of incoherent scatterers. From the point of view of the radar performance, the main difference is the level of sensitivity required; the scattering cross section of thermal scatter is 40–70 dB smaller than that of turbulent fluctuations [5]. Later, this chapter will describe the *Manastash Ridge Radar*, which is a passive, bistatic VHF (100 MHz) radar performing a "coherent scatter" function.

A useful performance metric for ionospheric radars is given by the product of the transmitter power and the antenna aperture (possibly normalized by the effective receiver temperature[3]). Enormous systems, such as Arecibo's 430-MHz radar[4] and the Jicamarca 50-MHz radar[5] have power–aperture products on the order of 10 MW-ha, whereas coherent scatter systems have power–aperture products in the range of 10 kW-ha, for example, 50-kW transmitter and 17-dBi antenna, at 50-MHz.[6]

Although the effective power–aperture product of passive bistatic radar systems is insufficient to usefully capture incoherent scatter,[7] there is ample illumination from VHF-FM broadcasts to detect meter scale turbulence. To place this in the context of aerospace systems, the ionosphere can occasionally provide thousands of square meters of scattering cross section at frequencies up to 500 MHz and beyond. VHF and UHF passive bistatic radar systems that can detect commercial aircraft at ranges beyond 100 km will have ample sensitivity to detect scatter from ionospheric turbulence.

After briefly discussing the physics, data will be presented along with a description of some engineering issues for the *Manastash Ridge Radar* system.

7.2 FIELD-ALIGNED IRREGULARITIES

Field-aligned plasma density irregularities in the Earth's ionosphere appear in one basic form in the E region (90–130 km altitude) and in another form in the F region (above 150 km). In all cases, a density depletion or enhancement is formed through a release of free energy provided by an electric field or a density gradient. The density perturbations are *always* very highly aligned with the local magnetic field, because electrons may freely stream along or gyrate tightly around the magnetic field lines. The anisotropy can be very strong: a density depletion with a transverse size of 1 m can extend several kilometers along the magnetic field. Density depletions with a transverse scale size of a few hundred meters are known to extend between distant points of the Northern and Southern hemispheres, which share the same magnetic field line.

It would be very difficult to understand the behavior of the ionosphere without taking into account the Earth's magnetic field. Several good models are available, principally the International Geomagnetic Reference Model (IGRF) and the World Magnetic Model (WMM)[6].[8]

7.2.1 Bragg Scatter

The undisturbed ionosphere is effectively transparent at VHF and UHF frequencies because the greatest plasma frequency of the ionosphere is about 15 MHz. However, very strong backscatter is occasionally observed at VHF and UHF, when the ionosphere becomes roughened by large amplitude density structuring.

---

3 See http://cedarweb.hao.ucar.edu/workshop/tutorials/2004/heinselman04.pdf.
4 See http://www.naic.edu/aisr/sas/sashomeframe.html.
5 See http://jicamarca.ece.cornell.edu/.
6 This resembles the Cornell University Portable Radar Interferometer (CUPRI) system.
7 Dr. Frank Lind, personal communication 2005.
8 http://www.ngdc.noaa.gov/seg/geomag/models.html.

This strong scattering of VHF and UHF radio waves can be explained through the means of Bragg scatter. A monochromatic radio wave with frequency ω_i and wavevector $\vec{k}_i$ is incident upon a region of space whose index of refraction is not smooth, for example, by the presence of sound waves, which perturb the density and hence the index of refraction. A space–time Fourier transform of the region then yields a spectrum of phonons with frequency ω_p and wavevector $\vec{k}_\mathrm{p}$. A scattered radio wave with frequency ω_s and wavevector $\vec{k}_\mathrm{s}$ will occur when the following conditions hold:

$$\omega_\mathrm{i} = \omega_\mathrm{p} + \omega_\mathrm{s} \tag{7.1}$$

$$\vec{k}_\mathrm{i} = \vec{k}_\mathrm{p} + \vec{k}_\mathrm{s} \tag{7.2}$$

For longitudinal density waves (sound waves) the wave phase velocity will be far smaller than that of the electromagnetic waves, so that ω_p is negligible. For the case of backscatter, we have $\vec{k}_\mathrm{s} = -\vec{k}_\mathrm{i}$ in which case we achieve

$$\vec{k}_\mathrm{i} = \vec{k}_\mathrm{p} - \vec{k}_\mathrm{i} \tag{7.3}$$

$$2\vec{k}_\mathrm{i} = \vec{k}_\mathrm{p} \tag{7.4}$$

$$\longrightarrow \frac{1}{2}\lambda_\mathrm{i} = \lambda \tag{7.5}$$

Thus, a 50-MHz radar is sensitive to backscatter from sound waves with wavelength 3 m (not 6 m). The orientation of the wavevectors is illustrated in Figure 7–1.

7.2.2 E-Region Irregularities

Although the radar techniques described apply generally to all field-aligned irregularities, the majority of this chapter will be concerned with E-region irregularities, because only high-latitude E-region irregularities have been (formally) studied using passive bistatic radar techniques [7–11].

A fairly mature theory of meter scale auroral E-region irregularities exists [12–14]. The bulk of the theoretical development involves manipulation of the Navier-Stokes equations for an

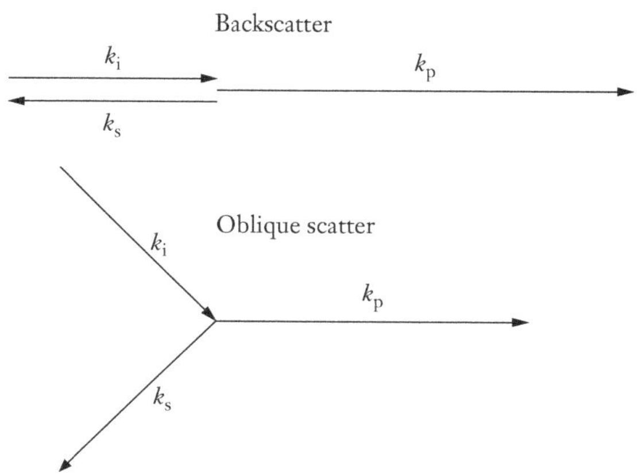

Figure 7–1 Schematic for Bragg scattering. The top case is the common Bragg backscatter condition; the bottom illustrates bistatic scatter. Note that the wavevector of the scattering phonon k_p is smaller for the bistatic geometry than for the monostatic geometry.

electron and ion fluid, with electrostatic coupling between the fields and waves [1, 15, 16]. A variety of radar and sounding rocket observations [3, 17, 18] lead to the inescapable conclusion that these waves are of sufficient amplitude that the nonlinear terms of the Navier–Stokes equations are important, leading to a fully turbulent state. The amplitude of the ion sound waves can approach 10% of the background plasma concentration. This extremely turbulent plasma has resisted detailed theoretical understanding, and since about 1990 computer fluid simulations have begun to provide satisfying models [19, 20] of E-region turbulence. The great majority of the computational simulations consider only the plane perpendicular to $\vec{B}$, because of the tremendous field-aligned anisotropy of meter scale physics.

Despite the complexity of the details, a fairly simple picture emerges. When an electric field is applied to the E region, the electrons perform a Hall drift in the $\vec{E} \times \vec{B}$ direction [15]. In a collisionless plasma, the ions would drift at the same speed, and in the same direction, for no net current. However, in the E region, the ions are collisionally bound to the much denser neutral atmosphere. So, the electrons Hall drift through the ions. If the electric field is large enough, then this Hall drift speed will exceed the local ion sound speed. At this point, the medium erupts into ion sound wave turbulence; it could be described colloquially as a region of continuous "sonic boom" production.

The high-latitude E region has a sound speed that is around 400 m/sec; electric fields greater than about 20 mV/m cause the electron Hall drift to exceed this amount, and ion sound waves ensue [21].

Near the magnetic equator similar physics occur, although the source of the electric field is a bit different, and the threshold for instability is about 10 mV/m, rather than 20 mV/m (because the magnetic field is half as great).

7.2.3 Aspect Angle

As has been mentioned several times, plasma sound wave turbulence is remarkably anisotropic; in particular, the sound waves are restricted very strongly to propagate within a half degree of the plane perpendicular to the local magnetic field; the so-called aspect angle dependence. In contemplation of the Bragg scattering condition of Figure 7–1, a radar will receive scatter from a very restrictive part of the sky, as illustrated in Figure 7–2. This figure was computed for a scattering altitude of 100 km for a bistatic link connecting Oslo, Norway, and Stockholm, Sweden, using the IGRF magnetic field model[9] [22].

The most important thing to take away from the figure is that the scatter occurs within perhaps 0.5° of the exact perpendicular condition, or "zero aspect angle." The figure can also be computed for other altitudes; as a general rule, the aspect angle changes about 1° for every 10 km of altitude, so the figure for 110 km would not look drastically different.

It is possible to be sufficiently far north (or south) of the equator to never achieve perpendicularity for a VHF radar. On the other hand, HF radars can achieve perpendicularity anywhere on the planet because of the significant refraction. The Super Dual Auroral Radar Network (SuperDARN) HF radar system provides substantial coverage of both polar cap regions by tracking weak density irregularities in the F region.[10] SuperDARN is a remarkable network of monostatic radars with a substantial international collaboratory organizational structure that could serve as a model for future growth of the *Manastash Ridge Radar*.

Nearer the Earth's equator, the magnetic field lines are nearly horizontal, so that the perpendicularity condition is achieved near the zenith and to the east and west, in both the E

9 See http://www.ngdc.noaa.gov/IAGA/vmod/igrf.html.
10 http://superdarn.jhuapl.edu/.

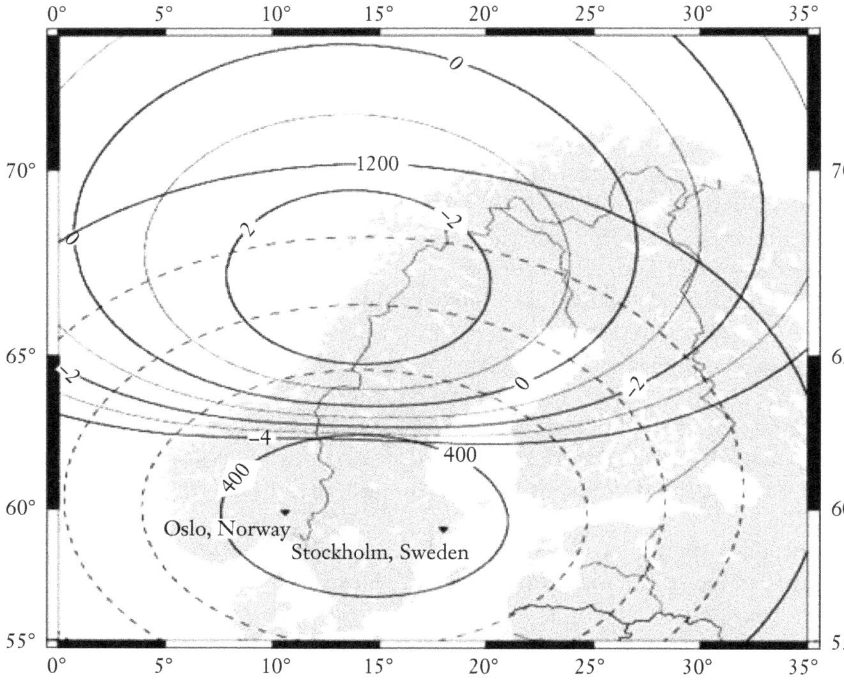

Figure 7–2 Bistatic aspect angle map, for a transmitter–receiver system, is located at Oslo and Stockholm. There are two sets of curves. First there are constant bistatic range ellipses, beginning at 400 km, and increasing by 200 km increments to 1,200 km. There is a second set of ovals to the north labeled in degrees from –4 to +2. These ovals indicate the angle between the disk perpendicular to the magnetic field *B* and the angle bisector of the incident and scattered radio waves, all calculated at a constant scattering altitude of 100 km. The E-region target falls below the horizon at a range of 1,150 km. [Figure courtesy of Dr. M. G. Meyer (mgmeyer@mtu.edu).]

and F regions. A study of the opportunities for passive VHF ionospheric radar at the equator is available [11].

7.2.4 Auroral Target Characteristics

In aerospace radars, the amplitude statistics of targets are sometimes classified into categories known as the Swerling classes [23], which attempt to capture the fading time and (roughly) the number of point-like scatterers in the target; there is also the assumption that the radar itself is a "scanning" system of some kind.

However, the FAI observed by passive bistatic radars are deep, fluctuating targets, not point-like targets, and so a different description is needed. Unfortunately, there is no detailed theory of the irregularities; the best theoretical description of the irregularities has arisen from the computational models, which produce effectively empirical models.

Regarding the overall scattering cross section, several investigators have estimated the volume scattering cross section of meter scale FAI as on the order of 10 m^2/km^3 [3]; others have used incoherent scatter radars to measure the dependence of scattering cross section with aspect angle [6]. Roughly speaking, the scattering cross section decreases by about 10 dB per degree until it fades into the incoherent scatter level by about 10° off perpendicular.

In terms of the temporal behavior, auroral E-region irregularities are essentially ion sound waves, with velocity features on the order of the ion acoustic speed, C_s:

$$C_s = \sqrt{\frac{\gamma_i k_B T_i + \gamma_e k_B T_e}{m_i + m_e}} \tag{7.6}$$

$$\gamma_{i,e} = \text{ratio of specific heats of ions and electrons} \tag{7.7}$$

$$k_B = \text{Boltzmann's constant}, 1.38 \times 10^{-23}\ \text{W/(Hz/K)} \tag{7.8}$$

$$T_{i,e} = \text{ion and electron temperatures, K} \tag{7.9}$$

$$m_{i,e} = \text{ion and electron masses, kg} \tag{7.10}$$

Considerable debate continues over the proper values to use for the specific heats; it is common to use $\gamma_z = 1$ and $\gamma_e = 5/3$. The most common ions (O^+, NO^+) have a mass of approximately 30 AMU. The electron and ion temperatures need not be identical, and are in the vicinity of 300 K. Experimentally, the measured doppler velocities yield sound speeds ranging from about 300 m/sec up to (rarely) 1,000 m/sec. The doppler characteristics of 21,008 auroral echoes have been summarized in Figure 7–3. An earlier summary by Watermann and coworkers is available for 50-MHz data [24].

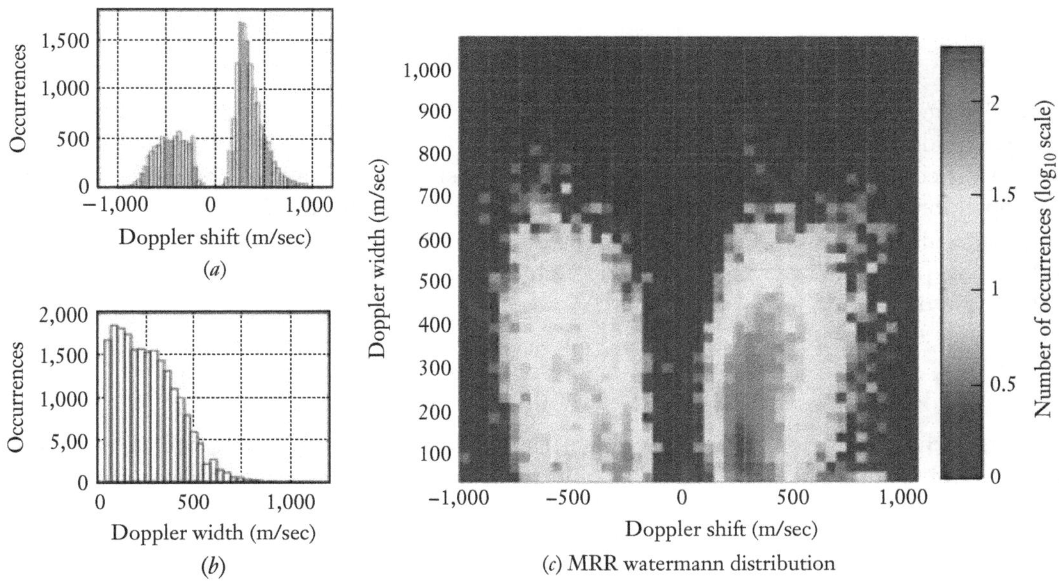

Figure 7–3 Summary statistics of doppler characteristics of high-latitude E-region irregularities. In the upper left panel, 21,008 spectra have been binned by their mean doppler velocity (assuming near-backscatter). The concentration into up- and down-shifted acoustic waves is apparent. In the lower left panel, the doppler width has been binned, showing that most echoes have bandwidth less than 600 m/sec (or 400 Hz). The right panel shows a two-dimensional distribution of the same data in both mean doppler (horizontal) and doppler spread. (This figure and several others were created by Dr. M. G. Meyer (mgmeyer@mtu.edu), and can be found at http://klickitat.ee.washington.edu/mgmeyer/Figures/MRR-Statistics).

When the doppler velocity characteristics are scaled to an operating frequency of 100 MHz, then in near-backscatter conditions the doppler relation

$$f = \frac{2v}{\lambda} \tag{7.11}$$

shows that a radar instrument that is able to recover the full doppler power spectrum of E-region irregularities should accommodate a full doppler spectral width of about 1 kHz. Expressed a bit differently, the correlation time of E-region irregularities may be as short as a few milliseconds and as long as about 100 msec.

The "near-backscatter" approximation is used to compute the doppler velocities. Although the transmitter–receiver baseline is large (about 150 km), all auroral echoes appear at ranges beyond 350 km (700 km round trip), for which the maximum error in doppler velocity computation is about 2.5%.

7.3 DETECTION OF FAI WITH PASSIVE BISTATIC RADAR

In 1997 Sahr and Lind described the basic means for detecting deep fluctuating targets using a distributed coherent receiver [7]. This work contained a mathematical analysis of an estimator of the correlation function of the scattering process. In particular, when an illuminating waveform $x(t)$ results in a scattered signal $y(t)$, an appropriate estimator $Q(r, \tau)$ of the correlation function of range r at lag τ is

$$\hat{Q}_{yx}(r,\tau) = \frac{1}{T}\int_0^T y(t)x^*(t-r)y^*(t-\tau)x(t-\tau-r)\,\mathrm{d}t$$

Here T is the amount of time over which the data are incoherently averaged—often on the order of ten seconds in practice. Sahr and Lind showed that a band-limited, complex Gaussian process $x(t)$ permitted $\hat{Q}_{yx}$ to be a natural estimator, which behaved similarly to the coded long pulse developed for other ionospheric and radar astronomy applications [25, 26].

In particular, Sahr and Lind showed that the estimator $\hat{Q}_{yx}$ achieves the same variance as the coded long pulse, alternating codes, and multipulse codes for deep fluctuating targets whose scattering cross section is a weak function of range. In summary, the result is as follows:

$$\frac{\Delta\hat{Q}(r,\tau)}{Q(r,0)} \approx \sqrt{\frac{\tau_\sigma}{T}} \tag{7.13}$$

Here, τ_σ is the correlation time of the target, and T is the integration time of the observation. Thus, the uncertainty of the correlation estimate scales directly as the number of independent samples of the target, and the random waveform performs as well as any waveform. Sahr and Lind provide complete and rather intricate expressions [7]. A similar analysis has been performed by Ringer et al. using discrete time, rather than continuous time [27].

Sahr and Lind further argued that commercial FM broadcasts had correlation properties similar to that of band-limited Gaussian processes, with $\tau_x \sim 10\ \mu$sec, even though FM broadcasts are not random and are (nearly) constant modulus, and suggested some ways [28–30] for efficiently computing (7.13), which is computationally too expensive to directly evaluate. For a receiver channel sampled at rate B with a number of ranges R to be processed and a number of lags L to be evaluated, the amount of computation required to evaluate equation (7.12) is BRL; for the ionospheric radar, B = 100 kHz, R = 800, and $L \approx 10^5$, for an overall burden of about 10^{13} complex add–multiply operations per second. This is possible but not cheap, and there are much faster ways to do the calculation.

For a waveform that is truly random, there is essentially no range or doppler ambiguity. As a practical matter, finite bandwidth and finite integration time will limit the dynamic range. There are, in addition, some subtle signal processing issues associated with technical differences between the in-phase and quadrature signal developed by digital receivers, and the narrowband "analytic signal" associated with the Hilbert transform [31]. However, these issues can be summarized by noting simply that the *Manastash Ridge Radar* is completely range unambiguous within the solar system (~10^{11} km); and completely velocity unambiguous for even the fastest meteor (±80 km/sec). Very high-speed point targets present an algorithmic challenge associated with range migration, but the underlying radar is nevertheless unambiguous.

The absence of range–doppler ambiguity is one of several potent advantages of passive bistatic radar for observation of ionospheric turbulence.

7.3.1 Computation of the Cross-Ambiguity Function

The usual presentation of range–doppler data is actually a Fourier-transform of (7.12),

$$\chi_{yx}(r,\nu) = \int \hat{Q}_{yx}(r,\tau)\exp(j2\pi\nu\tau)\,d\tau \tag{7.14}$$

which is now the distribution of power in the range–doppler plane. Here $\chi_{yx}(r, \nu)$ is the cross-ambiguity function between signals $y(t)$ and $x(t)$, and becomes the ordinary ambiguity function [23,32] when $y(t)$ is replaced by $x(t)$.

The key to speeding up the computation is to realize that the target bandwidth is much smaller than the transmitter signal bandwidth. In the case of FM broadcasts, most of the transmitter energy is contained within a 100-kHz band, while the E-region signal's doppler content never exceeds 2 kHz. Thus, an intermediate down-sampled signal can be formed,

$$yx(\bar{t};r) = \sum_{n=0}^{50} y(\bar{t}+n)x^*(\bar{t}+n-r) \tag{7.15}$$

This intermediate time series is a noisy but unbiased estimate of the scattering amplitude at range r near time t. For example, rate B (~10^5 Hz for VHF-FM) and range extent R are expressed in time units (e.g., ~8 msec for the ionosphere). The amount of work needed to create these time series is B^2R, about 10^8 add–multiply operations per second, which is straightforward.

The next step is to perform periodogram power spectral estimates for each of these time series. With a little effort it can be seen that the amount of computational effort here is $B'R/\log_2 N$, where $B' = B/50$ is the decimated data rate; thus the periodogram effort is a few million add–multiply operations per second, much less than the initial decimation effort.

A decimation of 50 retains a doppler bandwidth of about ±1,500 m/sec, or about ±Mach 4. If one wished to survey targets whose maximum velocity was Mach 1, then the decimation could be increased to 200 (for 100 kHz raw samples). Note that the computational effort associated with this decimation is not dependent upon the amount of decimation.

The result of this process yields images such as that of Figure 7–4. The *Manastash Ridge Radar* has been in nearly constant operation since 2001, and in its usual mode of operation, it samples the receivers for ten seconds every four minutes, and publishes that data to the web. Those interested can examine the current output and archives of past data by examining the following URL:

http://rrsl.ee.washington.edu/Data

Although not optimized for the detection of aircraft, aircraft signatures frequently occur, as in Figure 7–5.

Sahr and coworkers sometimes refer to this particular organization of the computation of the cross-ambiguity as the Range First algorithm, because the range is selected first, and then the

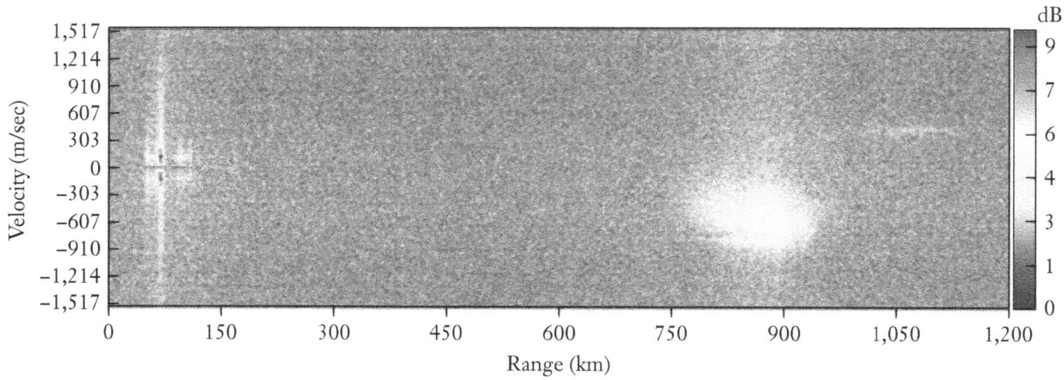

Figure 7–4 Example of a range–doppler cross-ambiguity detection of E-region turbulence. The data were taken from ten seconds of scatter on October 31, 2003, at 081100 UT, using a commercial transmitter at 96.5 MHz. The horizontal axis indicates bistatic slant range from 0 to 1,200 km (in 1.5 km increments), and the vertical axis has been scaled to doppler velocity, ±1,500 m/sec. The scatter at ranges near 70 km is associated with ground clutter; the largest ground clutter signal is not the direct path, but rather scatter from Mt. Rainier, a prominent volcano that stands nearly 3,000 m above the surrounding terrain. The doppler-shifted echo features at 900 and 1,050 km clearly show broad and narrow, up- and down-shifted ionospheric turbulence.

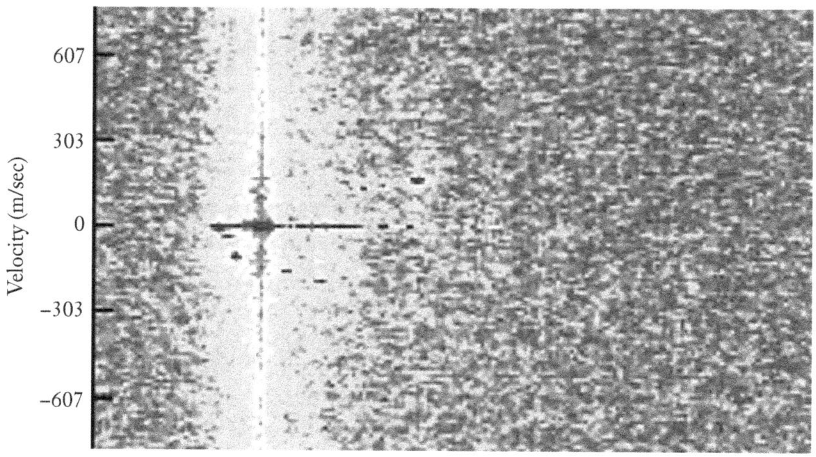

Figure 7–5 The detection signatures of eight separate aircraft can be observed in the region ±303 m/sec and 50–150 km range–doppler plot. The *Manastash Ridge Radar* is not operated in a way to emphasize the detection of aircraft, but they appear quite regularly. Note that they lie near the ground clutter detection and no effort has been made to remove them.

entire doppler power spectrum is computed. As pointed out by Levanon and others [23] the ambiguity function can be manipulated into a Doppler First form, in which a particular doppler shift is first selected and then the content of all ranges is computed at that doppler shift.

The Doppler First algorithm is expensive because it requires lengthy FFT operations to be performed on full rate data. Using the same symbol definitions as above, the work required for the

Doppler First algorithm is on the order of

$$\frac{N}{B'}(2+N)\frac{NB}{B'}/\log\left(\frac{NB}{B'}\right)\sim NBD^2/\log(D)$$

where $D = NB/B' \sim 10^4$, the ratio of the transmitter bandwidth to the desired doppler resolution. Note that this formula does not depend upon the range extent R when $RB'/N << 1$. For FM transmitters, and a desired doppler resolution of 10 Hz, $D \sim 10^4$, the Doppler First algorithm requires the order-10^{17} math operations per second to produce the same result as the Range First algorithm (10^8 math operations per second).

The choice of which algorithm to use depends upon need, and the range and doppler extent of the target. However, for ionospheric applications it is very likely that the Range First algorithm is superior in all cases. Having said that, the Doppler First algorithm could be useful in the cases with very large range extent R, where very few doppler bins N are needed, and could also be helpful in mitigating narrowband jamming.

Sahr and Lind initially performed the cross-ambiguity calculations with 250 kHz data, using a 16-node cluster of computers; Tuan and coworkers created a specialized cross-ambiguity computing engine using field-programmable gate array (FPGA) and a digital signal processor (DSP) chip to perform the computation within a single computer in real time [33]. However, with the increasing speed of computers, and judicious limitation of the bandwidth to 100 kHz, it is now (ca. 2005) possible to perform the cross-ambiguity calculations in near-real time on a conventional work station.

7.3.2 Additional Algorithm Speedups

In anticipation of prosecution of higher bandwidth transmitters (e.g., digital television), as well as processing a large number of transmitter and receiver channels, Morabito investigated an additional means for hastening the computation [30]. The amount of computation associated with the decimation scales as the square of the transmitter bandwidth (higher sample rate and more ranges sampled). So, Morabito artificially split the transmitter and receiver signals into subbands, and processed those subbands separately. Loosely speaking, if the signals are spectrally split into C channels, then the algorithm runs about C times faster. In practice, this ideal speedup is not achieved because of significant data motion, and it should be noted that the algorithm also reduces the range resolution by a factor C.

There are additional opportunities for speeding up the algorithm; a simple one is to channelize the data, and then simply not process all the channels. If a reduction in SNR is tolerable, this can speed up the processing dramatically. Such algorithms may become very useful when digital television (DTV) broadcasts are observed, as their bandwidth is enormous compared to the ionospheric targets.

Significant algorithm bottlenecks arise from data motion through the computer, as opposed to math operations. So it is also desirable to control the precision of the data. The natural data form for the Manastash Ridge Radar is 16-bit, 2's complement, integers. When properly truncated to 8 bits, the loss in SNR is barely detectible and in fact truncation to sign bit only still yields quite useful results [21]; 2-, 3-, or 4-level sampling is routinely used by radio astronomers.

7.3.3 Interferometry

Radar interferometry has been used for several decades to study the size and location of ionospheric turbulence. Meyer has also employed the technique in the MRR passive bistatic radar system [10, 34], with an example shown in Figure 7–6. The principle is a straightforward extension of the original range–doppler method. Now, however, there are several scattering antennas, each

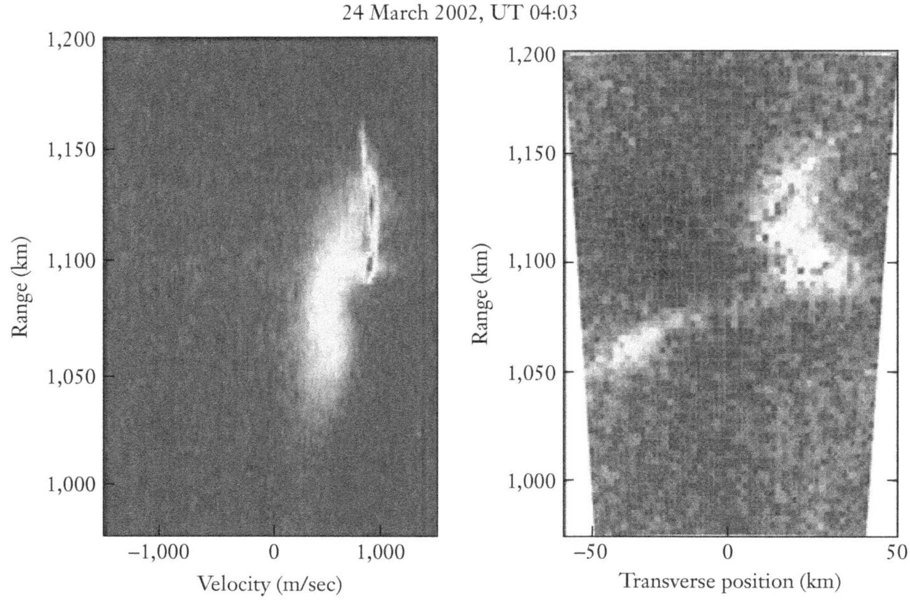

Figure 7–6 Example of passive bistatic radar interferometry. The left panel shows the distribution of power vs. range (vertical axis) and doppler (horizontal axis) from an instance of ionospheric turbulence on March 24, 2002. The right panel shows the same data distributed in range (vertical) and azimuth (horizontal) using the same spatial scale with a two-antenna interferometer as part of the *Manastash Ridge Radar* system. The resolution of the radar is approximately 1.5 km in range, and 2 km in azimuth, at a range of 1,000 km. (This figure was created by Dr. M. Meyer, and can be found at http://rrsl.ee.washington.edu/Data/Interferometer.)

producing a signal $y_n(t)$. Each of these is decimated against the separately sampled illumination $x(t)$ to produce the intermediate signals $yx_n(t, r)$. Upon Fourier transformation to yield $YX_n(f, r)$, the cross-spectrum $C_{nm}(f, r)$ is obtained,

$$C_{nm}(f,r) = \sum YX_n(f,r)YX_m^*(f,r) \tag{7.16}$$

Here the summation is over the several coherent processing intervals. The resulting normalized cross-spectral magnitude indicates the strength of the correlation and the phase indicates direction, as originally introduced by Farley and coworkers [35] and now extended dramatically by Hysell [36].

7.3.4 Frequency-Domain Interferometry

Ionospheric turbulence is extended in range (see Figure 7–4 and Figure 7–6) so that we do not really expect fine range structure. However, meteors and other objects may have much finer structure as a function of range. Kudeki and Stitt introduced a clever technique known as "frequency domain interferometry" (FDI) [37], which permits improved resolution of compact objects in a very clever fashion. In brief, the transmitter is directed to interrogate the target with two slightly different frequencies, and then the phase of the two returns is measured. The mean phase provides a vernier estimate of the target location within a range cell, and the amplitude of the correlation provides an indication of the target range extent.

In passive bistatic radar, of course, the transmitter is uncontrollable. However, variable illumination can be created a posteriori by filtering the transmitter signal $x(t)$ and scattered signal $y(t)$ into subbands. These upper and lower subbands can then be processed in exactly the same manner as multiantenna interferometry (above). Sahr and coworkers have used this technique to resolve meteor trails to about 100 m within the usual 1.5 km range cell.[11].

7.4 SYSTEM ENGINEERING ISSUES

As a "distributed" passive bistatic radar used for ionospheric turbulence studies, the experience gained with the *Manastash Ridge Radar* (MRR) should be a useful guide to others hoping to duplicate or extend its capabilities. By "distributed," we mean that the MRR radar receivers are coherently operated over a large baseline. This stands in contrast to aerospace systems such as the Lockheed Martin Silent Sentry™ [38], in which the reference and scatter signals are determined at a single site.

An array of Silent Sentry-type systems could very well exchange information derived from detection at each site. However, the MRR system exchanges voltage-level data, from which detection is performed among the sites. This is a potential weakness for employing an MRR-type system for aerospace applications, as a relatively large data stream must be transported for real-time, low-latency operation. The MRR is operated in real time, but the latency is long; from two to 15 minutes in practice.

7.4.1 System Structure and Operational Parameters

The gross structure of the *Manastash Ridge Radar* is illustrated in a block diagram in Figure 7–7. A summary of the operating parameters is provided in Table 7–1.

7.4.2 Mitigation of Direct Path Interference

One of the truly great technical successes of the MRR is that it began retriving very high quality range–doppler profiles without needing to perform any significant mitigation of interference on the direct path. The "split receiver" distributed topology proved to be a strong response to the challenge of the direct illumination. This issue is addressed in Section 6.4.2 of this volume.

Examination of Figure 7–4 suggests that the energy of the ground clutter does not extend to the ranges in which the ionospheric scatter occurs. Thus a fair amount of ground clutter can be tolerated without any attempt at mitigation of the direct path.

Having said this, schemes to reduce the direct path and ground clutter interference have been investigated [39]. These have emphasized extension of the "constant modulus algorithm" to clean up the multipath in the reference signal $x(t)$ (see [40]) and space–time-adaptive-processing (STAP) techniques to coherently subtract the ground clutter signal from $y(t)$ (c.f. [41]). Similar techniques are also detailed in Chapter 6, Air Surveillance, and Chapter 11, Adaptive MTI.

7.4.3 Mitigation of Poor Self-Ambiguity

Although common cochannel interference has not been a problem in practice, occasionally the nice ambiguity function properties of the FM broadcast waveform collapse. The effects on range resolution are dramatic, as illustrated in Figure 7–8. A brief study of this issue has been performed by Griffiths [42] and extended by Sahr.[12] A basic result is that FM broadcasters that emphasize music, rather than speech, are more desirable as illuminators. This is because speech tends to have more pauses, which result in low modulation index, narrow bandwidth, and thus extended correlation times.

---

11 Work presented at the URSI General Assembly in New Delhi, India, October 2005.

12 Work presented at the January 2005 North America meeting of URSI in Boulder, CO.

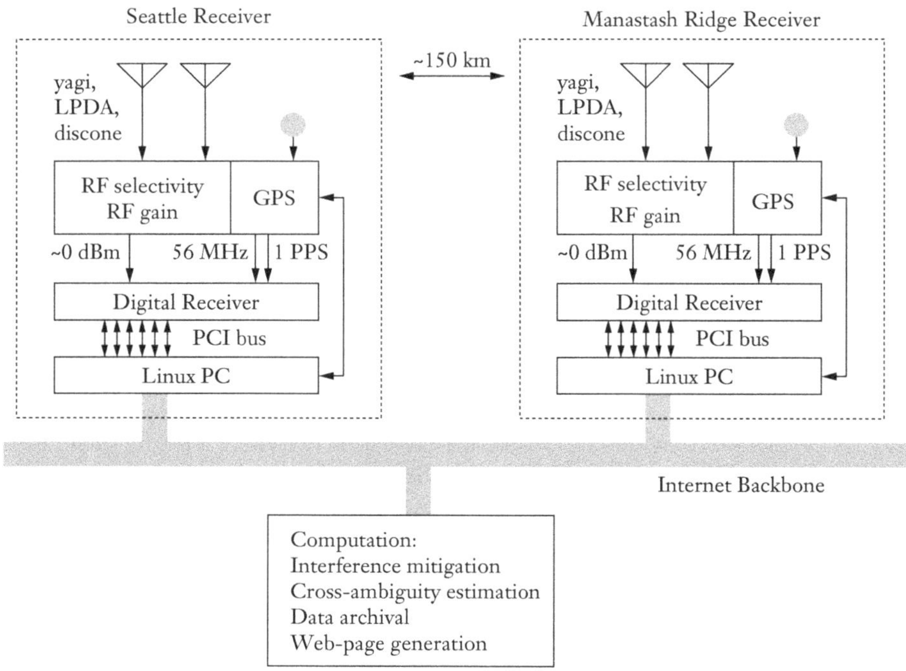

Figure 7–7 The *Manastash Ridge Radar* is composed of several largely identical digital receiver systems. Very simple antennas couple the 100-MHz FM broadcasts into analog bandpass and RF gain blocks. The entire FM spectrum (88–108 MHz) is passed to the digital receivers that operate optimally with 0 dBm input. Note that there is no down conversion; the RF bandwidth is directly sampled at a rate of 56 MHz, which is provided by a GPS receiver. This sample clock maps the FM broadcast spectrum into an unaliased baseband from 4 to 24 MHz. The digital receivers are programmed to deliver 100-kHz samples, in-phase and quadrature, 16 bits. Time series from each receiver are synchronized by a one pulse per second (1PPS) signal also provided by the GPS receiver. The receivers are controlled by a generic Linux computer, which in turn is coordinated by system controlling computers at the University of Washington via conventional internet connections. In the summer of 2005, a third receiver was added near Spokane WA, approximately 300 km further east.

A fairly simple, very effective strategy to mitigate against poor self-ambiguity is employed at MRR. A very simple estimator of the correlation time of the reference signal $x(t)$ is updated every millisecond. If the correlation time exceeds about 50 μsec, then the amplitude of the signal $x(t)$ is set to zero during that block. This has the effect of "turning off" the transmitter for subsequent cross-ambiguity computations. Because periods of poor ambiguity are quite rare, there is no detectable loss in ultimate signal-to-noise ratio in the range–doppler distributions.

Other measures are detailed in Chapter 6.

7.4.4 Mitigation of Jamming

The distributed nature of the *Manastash Ridge Radar* means that the scatter receiver may be exposed to some ground clutter from the intended transmitter, but also to additional transmitter power from other broadcasters. With the barrier of the Cascade Mountains, transmitter frequencies used in Seattle (west of the mountains) may be reused by other broadcasters east of the mountains. This presents both a challenge and an opportunity.

Table 7–1 A Summary of the Basic Features of the *Manastash Ridge Radar.*

Feature	Description
Antennas	Log-periodic dipole array, 6 dBi, H-pol yagi-uda array, 15 dBi, H-pol discone, 4 dBi, V-pol
RF selectivity	Cascaded Lo-pass and Hi-pass filters Helical notch filter for rejection of nearby TX
RF gain	20-dB gain stages 5- and 10-dB attenuators
Down conversion	None; uses direct RF sampling
Digital receiver	Echotek (Mercury) ECDR GC214 TS – two analog (antenna) inputs, four output channels – software selectable channels and bandwidth, 16-bit I&Q
Receiver computer	1*GHz Pentium running Linux (2.4 and 2.6 kernels) 1 GByte RAM, 300 GByte disk
Control computer	Several Linux computers at the UW
Operational modes	
– sample rate	100k samples/sec; occasionally 400k samples/sec
– survey	10 seconds sampled every 4 minutes
– storm	2.5 minutes sampled every 4 minutes
– RFI scan	Sweep each antenna across full band once/day
Cross-ambiguity computation	Decimate 100 kHz time series by a factor of 48 800 ranges (1,200 km; Δr = 1.5 km) 256 doppler bins (±15,00 m/sec, Δv = 12 m/sec) Compute time ≈ real time for 1 channel, 1 CPU Compute latency 4 minutes (due to data transport)
Software	Developed in house; open source Cross-ambiguity calculator in C++ High-level control in Python
Data archival	Retain raw time series during storms Otherwise expire raw time series after ~1 week
Sensitivity	Commercial aircraft (~20 dBsm) at 200 km slant range Large meteor trails out to 950 km (at the horizon) Radar aurora out to 1,150 km (at the horizon)

This design is the second generation of the radar; the original radar had a single RF channel and had an analog down-conversion stage. The authors are in the process of deploying an upgraded system that can simultaneously sample six antennas and process 24 different channels. Note that this design performs "detection" and "estimation" but not "tracking" functions.

The challenge arises simply because of the significant, uncorrelated transmitter power entering the receiver. Such interference can be addressed in two ways. First (and best), the antennas may be oriented to minimize this power in back- or sidelobes. Second, active cancelation techniques may be applied (c.f. references in Chapter 6.).

The opportunity arises when the jamming signal is not too strong. Then it can be used as a source of illumination at the same time, presenting a very efficient reuse of the time series from a particular receiver, since it may be used twice.

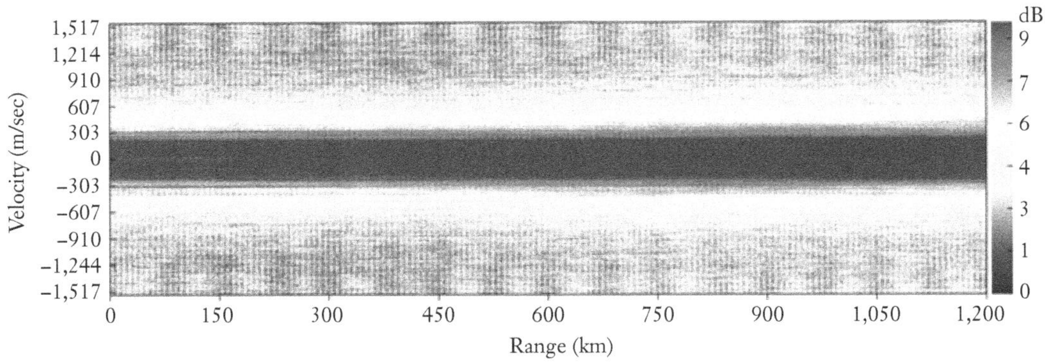

Figure 7–8 An example of poor self-ambiguity from the transmit signal. Occasionally the transmitter will be barely modulated, which results in a long correlation time and subsequent degradation of the range resolution of the passive bistatic radar. This example was derived from ten seconds of data; it is very likely that the period of low modulation lasted less than one second, however.

Occasional sources of interference eliminate even the ground clutter signal, and we find the radar completely desensitized for periods of an hour or so. In general, these periods are infrequent and have not interfered with the geophysical inquiries. Potential sources of this desensitization include the following:

- Atmospheric ducting of distant, cochannel transmitters into the scatter receivers.
- Nearby powerful out-of-band transmitters (our receivers are not terribly selective).

This episodic desensitization should be further investigated, however.

7.4.5 Mitigation of Adjacent Channel Interference

The MRR passive bistatic radar system collects scatter with a "wide open" receiver front end, which is exposed to the full bandwidth of the FM broadcast band and some adjacent spectrum. The RF signal is directly sampled at either 56 or 72 MHz, which preserves the 88–108-MHz spectrum. Then using digital receiver techniques (which perform digital down-conversion, decimation, and filtering), we recover 100-kHz time series for each desired channel.

However, nearby transmitters within the FM band must be accommodated by the digitizer—many bits of the sampler may be consumed by that dynamic range need, and nonlinearities in the sampler, or clock jitter, can radiate numerical noise throughout the spectrum. Thus we have, in some cases, employed fixed-frequency helical notch filters to depress the signal from nearby transmitters by 20–60 dB.

7.4.6 Data Transport

Although real-time operation is not, strictly speaking, necessary for our studies, it is very convenient. Two of the three receivers enjoy connection to very high internet bandwidth; however, the third, at Manastash Ridge, has a network connection of only 500 kbps (60 KBytes/sec). The receiver at Manastash Ridge is capable of producing 400 kBytes/sec per receiver channel, and the receiver has from 4 to 12 channels possible, so the source greatly exceeds the capacity of the channel.

Since the receiver samples at the Manastash Ridge Observatory (MRO) are low SNR and therefore noise-like, they cannot be significantly compressed without loss. As mentioned above, however, lossy compression by truncation of precision works quite well, even to sign bit only—a factor of 16—and we have experimented with some other compression schemes that are even more aggressive [29].

However, the MRR system has a "surveillance" mission primarily, so rather than implementing compression we operate the radar on a restricted schedule. The usual mode is ten seconds of samples each four minutes. All the receiver channels are sampled during that time but only one of them is transported for processing; the others are saved to local disk. If an auroral event occurs, then the radar is operated for 2.5 minutes out of each four minutes, and the much larger data set is transported to the UW over a period of several days. The hard disks at MRO can absorb nearly a week of continuous data before they fill.

Fortunately, the latency challenges for the geophysical experiments are not stringent. There is a plan to increase the network bandwidth up to the observatory, however.

7.4.7 Push vs. Pull Model

There is an interesting design choice that needs to be made for a distributed passive bistatic radar: should the receivers "push" their data to the computation node, or should the computation node "pull" data from the receivers? Strong arguments can be made for both models. Currently MRR uses the "push" model, but will probably transition to the "pull" model, partly to draw upon hypertext transfer protocol (http) software models, which are now quite mature.

7.4.8 Software Issues

For a variety of reasons, the MRR has been implemented on computer and software systems that are nonproprietary:

- Hardware cost. The ionospheric radar community cannot afford high performance, specialized hardware. Data acquisition systems using VME hardware are quite powerful, but would have drastically increased the cost of the MRR system.
- Software flexibility. The ionospheric radar community is fairly capable of manipulating software at low levels. Thus, the computational platform should not prohibit that expertise.
- Sufficiency of open source software. Widespread availability and expertise in the GNU/Linux operating system, programming languages, and related tools were available at no cost.

Thus, the receivers and compute nodes all use a variant of the GNU/Linux operating system;[13] the various software components have been written in open languages, namely C, C++, Python,[14] and Perl.[15]

The cross-ambiguity estimators are written in C++ and available upon request to the author. The reader may also wish to examine the Open Radar Initiative,[16] a fledgling effort to rationalize and share hardware and software designs for ionospheric radars, both active and passive.

7.5 FUTURE PLANS

This chapter has presented some initial results of range–doppler profiles and interferometric observation of ionospheric turbulence using completely passive, bistatic, and coherent passive bistatic radar techniques using commercial FM transmissions as illuminators. The technique has

13 RedHat Fedora, http://www.fedora.redhat.com.

14 See http://www.python.org.

15 See http://www.perl.com.

16 http://www.openradar.org.

proven extremely interesting and valuable, for both geophysical and engineering reasons, and efforts to improve the system are under way.

By the summer of 2007 the *Manastash Ridge Radar* should have expanded to five distributed receivers (from its present three), and Dr. Frank Lind[17] will have created a network of five very capable receivers in New England. Also, Dr. Melissa Meyer[18] is building a two-receiver system in Michigan. These passive receiver systems will have very similar hardware and software, and will provide a significant upgrade in capacity and experience base.

ACKNOWLEDGEMENTS

The author wishes to express his deep gratitude to the National Science Foundation for its decade of support of this work; additional support is gratefully acknowledged from AFOSR and NATO C3I. The Astronomy Department at the University of Washington, the Geology Department at Central Washington University, and the Physics Department at Eastern Washington University have lent us invaluable assistance and laboratory space in field locations. The progress made to date would have been impossible without the tremendous dedication of Dr. Frank Lind and Professor Melissa Meyer, whose doctoral dissertation studies employed the *Manastash Ridge Radar* and greatly extended its capabilities.

References

1. M. C. Kelley, *Ionospheric Plasma Dynamics*, Academic Press, San Diego, CA, 1989.
2. B. G. Fejer and M. C. Kelley. "Ionospheric Irregularities," *Reviews of Geophysics*, Vol. 18, 1980, pp. 401–454.
3. J. D. Sahr and B. G. Fejer, "Auroral Electrojet Plasma Irregularity Theory and Experiment: A Critical Review of Present Understanding and Future Directions," *Journal of Geophysical Research*, Vol. 101, 1996, pp. 26,893–26,909.
4. R. D. Hunsucker, *Radio Techniques for Probing the Terrestrial Ionosphere*, Springer-Verlag, New York, 1991.
5. J. C. Foster, D. Tetenbaum, C. F. del Pozo, J.-P. St.-Maurice, and D. R. Moorcroft, "Aspect Angle Variations in Intensity, Phase Velocity, and Altitude for High-Latitude 34-cm *E* Region Irregularities," *Journal of Geophysical Research*, Vol. 97, 1992, pp. 8601–8617.
6. A. S. Jursa, *Handbook of Geophysics and the Space Environment*, Air Force Geophysics Laboratory, Washington DC, 1985.
7. J. D. Sahr and F. D. Lind, "The Manastash Ridge Radar: A Passive Bistatic Radar for Upper Atmospheric Radio Science," *Radio Science*, Vol. 32, 1997, pp. 2345–2358.
8. J. D. Sahr and F. D. Lind, "Passive Radio Remote Sensing of the Atmosphere Using Transmitters of Opportunity," *Radio Science Bulletin*, Vol. 284, 1998, pp. 4–7.
9. M. G. Meyer, J. D. Sahr, and A. Morabito, "A Statistical Study of Subauroral E-Region Coherent Backscatter Observed Near 100 Mhz With Passive Bistatic Radar", *Journal of Geophysical Research, A*, 109, 2004, p. A07308.
10. M. G. Meyer and J. D. Sahr, "Passive Coherent Scatter Radar Interferometer Implementation, Observations, and Analysis," *Radio Science*, Vol. 39, 2004, RS3009, doi:10.1029/2003RS002985
11. J. D. Sahr and M. G. Meyer, "Opportunities for Passive VHF Radar Studies of Plasma Irregularities in the Equatorial E and F Regions," *Journal of Atmospheric and Solar-Terrestrial Physics*, Vol. 66, No. 17, 2004, pp. 1675–1681.
12. D. T. Farley, "A Plasma Instability Resulting in Field-Aligned Irregularities in the Ionosphere," *Journal of Geophysical Research*, Vol. 68, 1963, pp. 6083–6097.
13. R. N. Sudan, "Unified Theory of Type I and Type II Irregularities in the Equatorial Electrojet," *Journal of Geophysical Research*, Vol. 88, 1983, p. 4853.

17 MIT Haystack Observatory, flind@haystack.mit.edu.
18 Michigan Technological University, mgmeyer@mtu.edu.

14. A. M. Hamza and J.-P. St.-Maurice, "A Self Consistent Fully Turbulent Theory of Auroral *E* Region Irregularities," *Journal of Geophysical Research*, Vol. 98, 1993, p. 11601.
15. D. R. Nicholson, *Introduction to Plasma Theory*, Wiley, 1983.
16. F. F. Chen, *Introduction to Plasma Physics and Controlled Fusion*, Plenum Press, New York, 1984.
17. F. Primdahl, "Ionospheric E-Region Plasma Turbulence and Instabilities Observed In Situ by Sounding Rockets," Paper Presented at *6th Scientific Assembly*, International Association of Geomagnetism and Aeronomy, August, 1989.
18. R. F. Pfaff, J. D. Sahr, J. F. Providakes, W. E. Swartz, D. T. Farley, P. M. Kintner, I. Häggström, Å. Hedberg, H. Opgenoorth, G. Holmgren, D. Wallis, A. McNamara, B. Whalen, A. Yau, S. Watanabe, F. Creutzberg, P. Williams, E. Nielsen, K. Schlegel, and T. R. Robinson, "The E Region Rocket/Radar Instability Study (ERRRIS): Scientific Objectives and Campaign Overview." *Journal of Atmospheric and Terrestrial Physics*, Vol. 54, 1992, pp. 779–808.
19. P. Janhunen, "Three–Dimensional Stabilization Mechanism for the Auroral Farley–Buneman Instability," *Journal of Atmospheric and Terrestrial Physics*, Vol. 54, 1992, p. 1633.
20. M. Oppenheim, "Evidence and Effects of a Wave-Driven Nonlinear Current in the Equatorial Electrojet," *Annales Geophysicae*, Vol. 15, 1997, pp. 899–907.
21. D. T. Farley, "Incoherent Scatter Correlation Function Measurements," *Radio Science*, Vol. 4, 1969, pp. 935–953.
22. N. W. Peddie and E. B. Fabiano, "A Proposed International Geomagnetic Reference Field for 1965–1985," *Journal of Geomagnetism and Geoelectricity*, Vol. 34, 1982, p. 357.
23. N. Levanon, *Radar Principles*, Wiley, New York, 1988.
24. J. Watermann, D. R. McDiarmid, A. G. McNamara, J. A. Koehler, and G. J. Sofko, "Are Type III Radio Aurorae Directly Excited by Electrostatic Ion Cyclotron Waves?" *Journal of Geophysical Research*, Vol. 94, 1989, p. 13479.
25. M. P. Sulzer, "A Radar Technique for High Range Resolution Incoherent Scatter Autocorrelation Function Measurements Utilizing the Full Average Power of Klystron Radars," *Radio Science*, Vol. 21, 1986, pp. 1033–1040.
26. T. Hagfors and W. Kofman, "Mapping of Overspread Targets in Radar Astronomy," *Radio Science*, Vol. 26, 1991, pp. 403–416.
27. M. A. Ringer, G. J. Frazer, and S. J. Anderson, "Waveform Analysis of Transmitters of Opportunity for Passive Bistatic Radar," Technical Report DSTO-TR-0809, DSTO Electronics and Surveillance Research Laboratory, 1999.
28. J. M. Hansen, *A New Radar Technique for Remote Sensing of Atmospheric Irregularities by Passive Observation of the Scattering of Commercial FM Broadcasts*, Master's thesis, University of Washington, Seattle, 1994.
29. P. W. Hall, *Correlative Range–Doppler Detectors and Estimators in Bistatic Radar Using Commercial FM Broadcasts*, Master's thesis, University of Washington, Seattle, 1995.
30. A. N. Morabito, *Improved Computational Performance for Manastash Ridge Radar Processing Through Channelized Data*, Master's thesis, University of Washington, Seattle, 2004.
31. A. D. Whalen, *Detection of Signals in Noise*, Academic Press, New York, 1971.
32. August W. Rihzczek, *Principles of High-Resolution Radar*, McGraw–Hill, New York, 1969.
33. T. Tuan, M. Figueroa, F. Lind, C. Zhou, C. Diorio, and J. D. Sahr, "Spiir: An FPGA-Based Pipelined Correlator for an Ionospheric-Imaging Radar," in *FCCM 00 Symposium on Field-Programmable Custom Computing Machines*, Napa Valley, CA, 2000.
34. M. G. Meyer, *Passive VHF Radar Interferometer Implementation, Observation, and Analysis*, Master's thesis, University of Washington, Seattle, 2003.
35. J. F. Providakes, W. E. Swartz, D. T. Farley, and B. G. Fejer, "First VHF Auroral Interferometer Observations, *Geophysical Research Letters*, Vol. 10, 1983, p. 401.
36. D. L. Hysell, "Imaging Coherent Backscatter Radar Studies of Equatorial Spread F," *Journal of Atmospheric and Terrestrial Physics*, Vol. 61, 1999, pp. 701–716.
37. E. Kudeki and G. R. Stitt, "Frequency Domain Interferometry—A High Resolution Radar Technique for Studies of Atmospheric Turbulence," *Geophysical Research Letters*, Vol. 14, 1987, pp. 198–201.

38. J. Baniak, G. Baker, A. M. Cunningham, and L. Martin, Silent Sentry™ passive surveillance, *Aviation Week and Space Technology*, 7 June 1999.
39. C. C. Zhou, *Application and Extension of Space-Time Adaptive Processing to Passive FM Radar*, Ph.D. thesis, University of Washington, Seattle, 2003.
40. C. R. Johnson, Jr., P. Schniter, I. Fijalkow, and L. Tong. "The Core of FSE-CMA Behavior Theory," in S. Haykin (ed.) *Unsupervised Adaptive Filtering*, Wiley, New York, 2000.
41. S. Haykin, *Adaptive Filter Theory*, 3rd ed., Prentice-Hall, Englewood Cliffs, NJ, 1991.
42. H. D. Griffiths, C. J. Baker, H. Ghaleb, R. Ramakrishnam, and E. Willman, "Measurement and Analysis of Ambiguity Functions of Off-Air Signals for Passive Coherent Location," *Electronics Letters*, Vol. 39, No.13, 2003, pp. 1005–1007.

8

Wind Measurements[1]

Joshua Wurman and Nicholas J. Willis

8.1 INTRODUCTION [1–4]

Accurate knowledge of the full, three-dimensional vector wind field throughout a volume of the atmosphere is useful for aviation, weather forecasting, and research. Specifically, spatially and temporally resolved wind field data can be used to improve (a) prediction of hazardous aviation events, including economically costly wind shifts; (b) general weather forecasting, including weather hazards; (c) mesoscale modeling (modeling of meteorological phenomenon approximately 1–100 km in horizontal extent); (d) media weather presentations; and (e) research into boundary layer fluxes, convective dynamics, and thunderstorm electrification.

A bistatic doppler weather radar network is one method of retrieving these vector wind fields. It consists of one monostatic doppler weather radar and one or more remote bistatic receivers, or *bistatic network receivers* (BNRs), each having a nonscanning, low-gain, passive antenna. A single monostatic radar and single BNR measures the two-dimensional (horizontal) vector wind field by using the *dual-doppler* technique, which is a synthesis of doppler velocity vectors from two separated radars, in this case one from the monostatic radar and one from the BNR. When two or more BNRs hitchhike off a monostatic doppler weather radar, a triple- or *multiple-doppler technique* can be used to retrieve full, three-dimensional vector wind fields. Figure 8–1 is a schematic of a typical network near an airport.

The network has the following advantages when compared to the traditional dual- or multiple-doppler method using two or more monostatic radars: (1) simultaneous measurement of all doppler velocities in individual weather volumes, thus minimizing storm evolution-induced errors; (2) less expensive and simpler installation by using only one transmitter and scanning antenna; (3) hail detection by comparing the intensity of obliquely (bistatically) scattered energy to Rayleigh scattering predictions; and (4) rapid scanning of localized weather, which typically requires multiple monostatic weather radars.

The network has the following disadvantages when compared to the traditional monostatic methods: (1) less sensitivity to weak weather returns due to the low-gain receiving antenna; (2) less

1 Editor's note: This chapter describes a commercial bistatic receiver, for convenience called a bistatic network receiver (BNR). Two or more BNRs hitchhiking off a monostatic doppler weather radar can retrieve full, three-dimensional vector wind fields. BNR is manufactured by Binet Inc., whose principal investigator and developer is Dr. Joshua Wurman. Information for this chapter is extracted and paraphrased when necessary from five publications on BNR [1–5] and a website, www.binet.com, detailing its utility, theory, system characteristics, design trade-offs, test results, and status. Editorial comments are added to provide continuity to the extracted data and clarity between weather and aerospace communities. Dr. Wurman has reviewed and approved this manuscript.

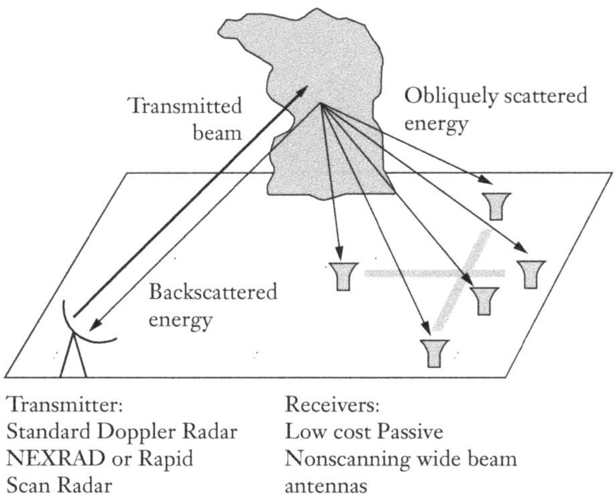

Figure 8–1 Schematic of a typical monostatic/bistatic doppler weather radar network measuring three-dimensional wind vectors near an airport. A single monostatic weather radar transmits a narrow beam and receives backscattered energy from the illuminated weather volume, while five BNRs receive obliquely (bistatically) scattered energy from that volume [1]. © American Meteorological Society.

sensitivity to weather returns near ground level when using horizontal transmit polarization, which are used by many doppler weather radars; (3) more sensitivity to contamination from transmitter side lobes (i.e., direct path breakthrough) and to secondary scattering from weather echoes via transmit antenna side lobe illumination, both due to the broad-beam receiving antenna; and (4) increased measurement errors by about a factor of 2 when operating at large oblique scattering angles. Contamination is acceptable in most situations and can be reduced by using higher-gain antennas when necessary. The lower sensitivity and increased errors can be offset with more BNR sites, a possible solution due to their low cost.

Between 1993 when the first BNR prototype was introduced and 2002, nine sets of BNRs have been sold and deployed at monostatic doppler weather radar sites for research, testing, and operation: one in the United Kingdom, one in Canada, one in Germany, two in Japan, and four in the United States. As of 2006, no additional sales have occurred [www.binet.com].

8.2 EXISTING RADAR METHODS FOR RETRIEVING VECTOR WINDS [2, 3]

Monostatic doppler weather radars measure the radial component of a wind field, that is, components moving toward or away from the radar. Various techniques are then used to estimate three-dimensional vector wind fields using sequential observations from a single radar. Techniques include velocity azimuth display (VAD), volume velocity processing (VVP) and adjoint and tracking radar echoes by correlation (TREC). Both VAD and VVP assume simply varying wind fields, which are suspect in convective storms. Adjoint and TREC apply physical models to sequential observations to infer the complete wind field, which are computationally intensive and depend on predictions of the physical models.

To mitigate these problems, three or more sufficiently separated monostatic weather radars can be used to measure radial velocities within a common volume from different vantage points. Then the full vector precipitation particle (rain drops, hail, etc.) motion field can be retrieved by solving a system of equations to convert the radial velocity measurements into a three-dimensional particle motion field. A correction is then applied for particle fall speed to convert the particle motion field into a wind field.

Alternatively, two sufficiently separated monostatic weather radars can estimate the full vector wind field by solving a system of equations for the horizontal wind field and applying a constraint of mass conservation and a boundary condition to the vertical wind. Combinations, modifications, and improvements to these basic methods have been developed [6].

The major limitation of these multiple monostatic radar techniques is the cost of deploying multiple monostatic radars all with overlapping volume coverage. For example, the construction cost of a single Weather Surveillance Radar-1988 Doppler (WSR-88D), also called Next Generation Radar (NEXRAD), was approximately 3 million dollars, with yearly operating expenses more than 0.3 million dollars [3].[2] As a result, both experiments and operational deployments of such multiple monostatic radars are rare and usually of short duration.

Furthermore, wind field estimates from these multiple monostatic radar networks are degraded due to the nonsimultaneity of observations by radars in the network. Since each radar uses a slow-scanning pencil beam, several minutes can elapse in observing a common volume by two or more radars, thus compromising the estimates, especially in rapidly evolving, severe weather situations [2]. The problem is exacerbated when operational and research radars are used together, since the scan strategy of operational radars cannot be customized, thus complicating coordination and preventing a quick scan of rapidly evolving weather.

Since construction costs of a BNR are estimated to be less than 3% of a typical weather radar [3], it can significantly reduce the cost of such a network. In addition, since BNR uses a wide-beam receiving antenna, it can take simultaneous measurements of the weather volume illuminated by a host monostatic radar, thus improving the wind field estimates.

8.3 SYSTEM THEORY [1, 4]

Radar observations and measurements of atmospheric and weather phenomena started to appear in atmosphere and meteorology journals in the 1950s and 1960s, with texts published by Battan in 1973 [7], Doviak and Zrnic in 1993 [8], and Bringi and Chandrasekar in 2001 [9]. A few bistatic radar articles were interspersed with this early literature, but it was not until the 1990s that detailed analyses and experiments of monostatic/bistatic networks were undertaken [1, 4, 10, 11]. While the lexicon, notation, and coordinate systems for both monostatic and bistatic weather radars often differ from other radar-related disciplines, such as aerospace radars and semiactive missile seekers, the underlying theory, technology, and design issues are virtually identical. Significant differences are emphasized here.

Both *Cartesian* (x–y–z) and *bistatic plane* (plane containing transmitter, receiver, and target) coordinate systems have been developed by the weather radar community, with Cartesian coordinates found to be more practical [4]. Wurman et al. initiated and Satoh and Wurman expanded the development of Cartesian equations for bistatic target state estimates, that is, the location and velocity vector of precipitation echoes.

For the BNR system hitchhiking off a cooperative monostatic weather radar, the target is readily located in three dimensions with respect to the BNR by using (1) monostatic radar coordinates of the target data linked to the receiver: range, azimuth, and elevation angles; (2) the premeasured *baseline range* (range between transmitter and receiver); and (3) a real-time measure of the *time*

2 Current (2006) NEXRAD acquisition and operating costs are estimated to be 10 million dollars and 1 million dollars per year, respectively [Private communication from Dr. Joshua Wurman, October 9, 2006.].

delay, converted to a *range-sum*, $R_T + R_R$, where R_T is the transmitter-to-target range and R_R is the target-to-receiver range.

Wurman et al. [1] and Satoh and Wurman [4] synthesized the velocity vectors into a composite wind field to generate estimates of the full, three-dimensional vector wind field, with Satoh and Wurman [4] evaluating the accuracy of these estimates. These equations are straightforward but exceeding labored, and thus best left in situ.

The *bistatic angle* β (angle between transmitter and receiver with vertex at the target) is often called the *scatter angle* and occasionally the *oblique scattering angle*. And, like other bistatic radars, it mainly governs the accuracy of target state estimates, in this case converted to wind field vectors. The ratio of bistatic-to-monostatic doppler velocity, $[\cos(\beta/2)]^{-1}$, is called the *velocity expansion factor*. Ellipsoids characterize surfaces of constant time delay between transmitter, target, and receiver in all bistatic radar disciplines, but the term *isorange* (and *isodoppler*) *contours* are not used by the weather radar community.

Probably the most significant difference between the weather radar and the aerospace radar communities is the nature of the target: (1) scattering—Rayleigh versus either Mie (resonance) or optical, and (2) size—volume versus discrete. Specifically, Satoh and Wurman [4] assume rain particles are perfect Rayleigh scatterers at typical S-band operating frequencies. Only when large hail is present must Mie scattering formulation be used. In fact, large hail may be indicated in regions where the received power deviates from that predicted by the Rayleigh approximation. The authors continue:

> In the Rayleigh limit, the intensity of the radiation scattered obliquely from a transmitted beam varies with the scattering angle measured relative to the transmitted **E** vector such that $I \propto I_o \sin^2(\theta)$, where I is the intensity of the scattered radiation, I_o is the intensity of the incident radiation, and θ is the angle between the incident **E** vector and the propagation vector of the scattered radiation.

Wurman et al. [1] plots this expression, which shows the scattered radiation approximating a torus, a surface in the shape of a doughnut, but with no hole. The incident radiation enters the torus at a point on the outer circumference. For vertical polarization the torus lies flat with the incident radiation entering perpendicular to the axis of torus rotation. Thus little radiation is scattered vertically. For horizontal polarization the torus sits on the outer circumference, or edge, of the torus. Thus little radiation is scattered horizontally. This latter condition generates a zone, or *notch*, of greatly diminished receiver sensitivity within an arc connecting the transmitter and receiver near the ground, which in turn reduces the capability of a BNR to measure ground winds. Consequently, vertical or circular polarization is desired for such measurements. Unfortunately, many monostatic weather radars are horizontally polarized, requiring multiple BNRs positioned to minimize this notch.[3]

Standard expressions are used for characterizing monostatic and bistatic radar cross sections of volume clutter in the weather radar equation: $V_c\Sigma\sigma_i$, where V_c is the volume (or resolution cell) of the clutter, calculated as the monostatic or bistatic range resolution cell width times the beam area (transmit beam area in the case of BNR) at the clutter range; $\Sigma\sigma_i$ is the average total backscatter cross section of the scatterers per unit volume, also called volume clutter reflectivity, measured in units of m^2/m^3; and σ_i is the cross section of independent scatterers in the cell [12]. For the bistatic case, σ_i is a function of $\sin^2(\theta)$, as outlined previously.

A second formulation of the weather radar equation uses a *radar reflectivity factor*, Z, which can take many forms. One empirically determined version is $Z = ar^b$, where r is the rainfall rate in mm/h and a and b are empirically determined constants. For stratiform rain a widely

3 The 88D (NEXRAD) network may be upgraded for dual-polarization operation, possibly transmitting alternating horizontal and vertical pulses. If so, bistatic receivers can use the vertically polarized pulses to fill in these notches. Also, many newer dual-polarization radars transmit slant polarization, typically at 45°, which does not generate the near-ground notch for bistatic operation [Private communication from Dr. Joshua Wurman, October 9, 2006.].

accepted form is $a = 200$, $b = 1.6$, with Z given in mm^6/m^3. Instead of the rainfall rate r, the intensity of precipitation can be stated as a *dB reflectivity factor*, dBZ = 10 log Z. Thus a rainfall rate, $r = 0.3$ mm/h = 15 dBZ, $r = 1$ mm/h = 23 dBZ, $r = 16$ mm/h = 42 dBZ, and so on [12]. A variation of Z is given in [4] as the *minimum detectable equivalent reflectivity factor*, also in dBZ. It is defined as the minimum weather cross section that can be detected when the signal-to-noise ratio (SNR) = 1, that is when the weather return equals the thermal noise level of the receiver. It is a convenient method of comparing monostatic and bistatic radar sensitivities independent of a specific weather type.

8.4 SYSTEM CHARACTERISTICS AND DESIGN TRADE-OFFS [1–3] (www.binet.com)

A detailed description of the 1993 prototype BNR is available in [3], while technical specifications of the commercial BNR are available at www.binet.com. Wurman et al. [3] report for the prototype:

> ...Many receiver system design requirements were similar to those of traditional Doppler weather radars. There were some additional requirements primarily related to synchronizing both the frequency and timing between the transmitter and remote receiver. Some typical weather radar specifications, particularly in the data acquisition area, were relaxed for this experiment. This first experiment was focused on providing the bistatic multiple-doppler concept rather than retrieving large quantities of research-quality data over broad regions. Many bistatic receiver characteristics, such as filter bandwidths and system noise floor, were similar to those in many Doppler radars, since the system was constructed largely from spare parts from the National Center for Atmospheric Research (NCAR) CP-2 Mile High Radar (MHR), and Electra Doppler Radar (ELDORA) weather radars.
>
> The primary desired system specifications were:

Frequency coherence between transmitter and receiver for doppler velocities:	< one part in 10^9 difference (3 Hz at 3 GHz transmit frequency) (1/6 m/s velocity error) automatically lockable to transmit frequency
Timing coherence between transmitter and receiver for gating and ranging:	< 100 ns drift/jitter during a dwell (64 ms) < 1 μs drift per several minutes automatically locked to transmitter pulses
Data acquisition:	> 200 gates at 1 μs per gate pulse pair algorithm to produce lagged autocovariance 32-64 pulses per dwell at 1 ms per pulse clutter filtering
Transmitter pointing angles:	available at receiver with < 100 ms delay resulting in typically < 1 degree errors
Minimum detectable weather signals:	15 dBZ at 10 km (corresponding to light rain)

> The bistatic network consisted of the NCAR CP-2 weather research radar... as a transmitter and a remote non transmitting receiver as shown in Fig. 2 [Figure 8–2]. The transmitter and remote receiver were linked by telephone so that timing and pointing angle information could be sent to the receiver...

Wurman et al. [3] detailed five major design trade-offs for the BNR. The first was for the receiver system:

> The bistatic dual-Doppler prototype required a new radar receiving system. The use of an existing complete radar system, either a spare or one purchased on the commercial market was considered. There were unknown costs and difficulties involved with modifying spare or discarded equipment that may have been extracted from various laboratories and universities. The data systems in these old systems were likely

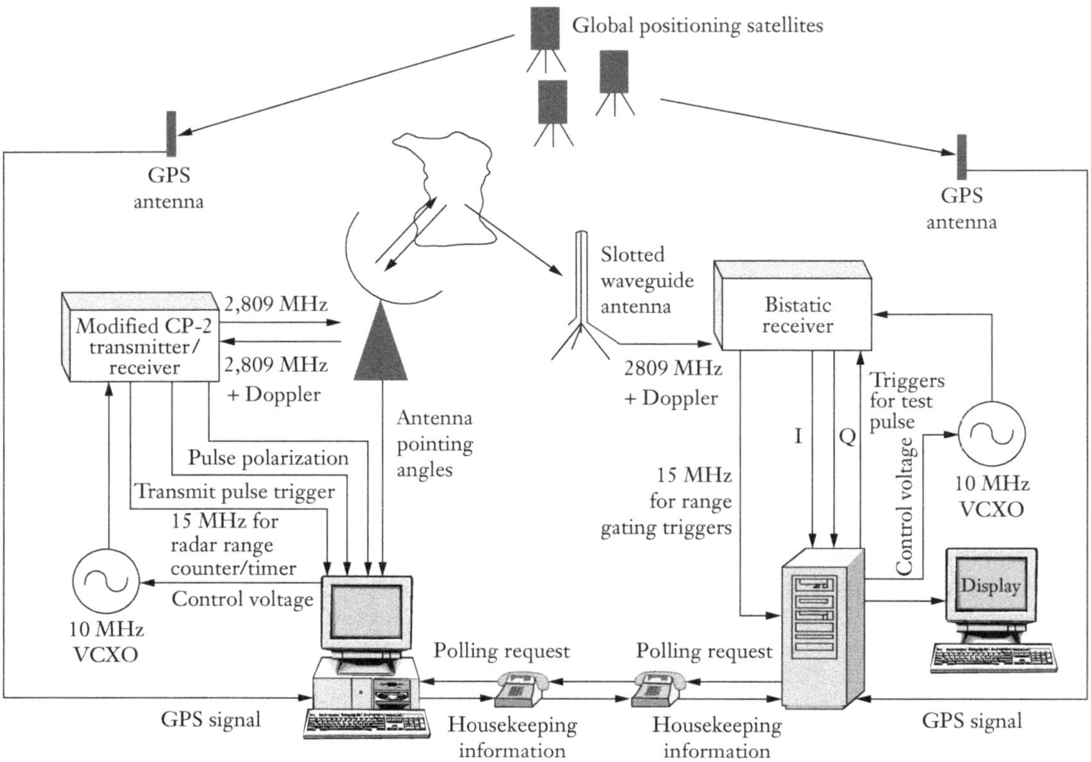

Figure 8–2 Simplified block diagram of the prototype BNR. Synchronization links via telephone lines and Global Positioning Satellite (GPS) are shown. The NCAR CP-2 weather research radar (left) is modified to use frequency sources synchronized to GPS signals. The bistatic site, 10–20-km distant, uses similar frequency generation hardware. A telephone line provides the bistatic site with transmitter antenna pointing angles, pulse transmission times, and other housekeeping information [3]. © IEEE 1994.

to be archaic and difficult to reprogram. New generic systems required modifications to incorporate the unique synchronization hardware and software desired in the bistatic network. Expense precluded the use of a commercially purchased custom designed system.

1) *Receiver:* The new radar receiver (Fig. 4) [Figure 8–3] was constructed from a mix of custom hardware, spares from NCAR radars, and a few commercially purchased parts. Spares were incorporated whenever available, resulting in possibly unusual arrangements such as the three consecutive amplifiers shown...

2) *Data System:* A PC-based data system was constructed (Fig. 5) [Figure 8–3]. Several factors drove the decision to build a PC-based system. Foremost was the availability and low cost of PC data acquisition and other cards. It was desirable to demonstrate that a system with adequate performance could be built using inexpensive components, thus increasing the attractiveness of a bistatic network as an alternative to other methods of retrieving three-dimensional vector windfields. Other factors included the availability of PC programming expertise at NCAR...

The second and third design trade-offs [3] were for the BNR antenna:

One crucial aspect of the prototype system was its use of a low-gain bistatic receiving antenna. This eliminated the need for expensive scanning hardware. [And also avoided the beam scan-on-scan problem [13]. In fact, a high-gain receive antenna with pulse chasing was considered, but rejected [2] because a

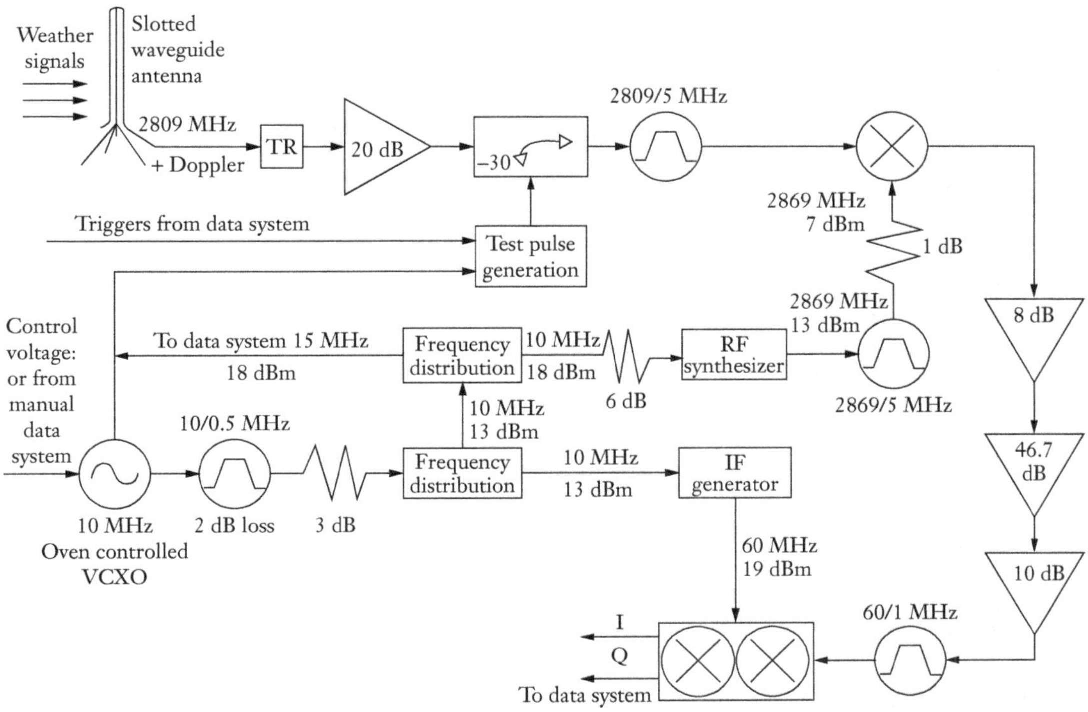

Fig. 4. Block diagram of bistatic receiver. All generated frequencies are locked to the master 10-MHz VCXO which is, in turn, locked to GPS timing signals. A TR tube is used to protect the system from the direct beam of the transmitter.

Figure 8–3 Receiver and processor for the BNR. Note that a simple TR tube is used for direct path excision, just as in a pulsed monostatic radar. (Contrast this method with the complex and expensive cancellation schemes needed to excise a direct path CW signal for passive bistatic radars in Chapter 6.) Also note that a 1990-vintage 486-based PC is sufficient to perform the signal BNR processing [3]. © IEEE 1994.

required electronically scanned, phase array antenna was not available.] Calculations [1] showed that an antenna with a gain of approximately 15 dB would provide the receiver with signals strong enough that the minimum detectable signal (at SNR = 0 dB) would be roughly 15 dBZ, corresponding to light rain, at a range of 20 km.

The initial design employed a standard 10–15-dB gain horn at the receiver. Unfortunately, the bistatic receiver site was on a hill and extremely susceptible to ground clutter contamination. The beam pattern of the horn was very broad, rejecting little of the ground clutter signal. The high site was picked so that it would easily see the direct pulse from the transmitter. This was desirable since the GPS synchronization method was untested. Later tests showed that the transmitter pulses were detectable almost everywhere.

Since the horn rejected little of the gound clutter, an antenna with a smaller beam pattern was needed. Primary design criteria for the new antenna were: a vertically polarized beam, a beam pattern that exhibited a sharp elevation angle sensitivity edge; a broad azimuthal patter, and relatively low sidelobes (<20 dB)...

One of the simplest and most cost-effective ways to achieve these requirements is with a slotted-waveguide antenna... With that in mind, an antenna was constructed in the NCAR design and fabrication laboratory... This 4-m-long slotted-waveguide antenna was manufactured with 58 slots and with a gain of approxiately 13 dB... The antenna exhibited an elevation beam width of 20° and an azimuthial beamwidth of approximately 160°...

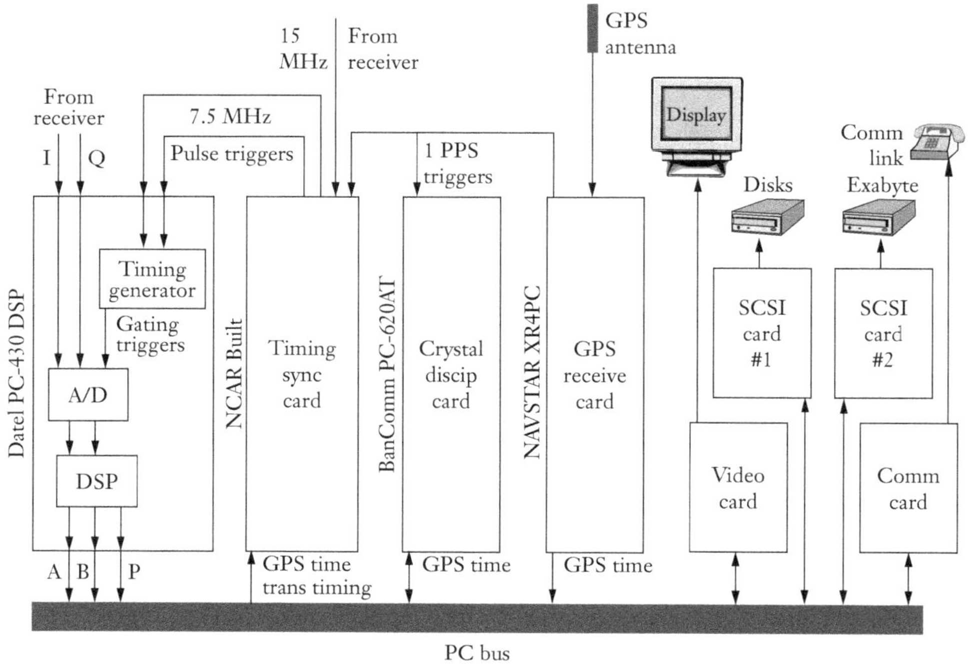

Fig. 5. Block diagram of bistatic receiver data acquisition system. Sampling and signal processing were conducted on PC-430 card. The timing synchronization card provided gating triggers and the gating clock, synchronized to the 10-MHz master VCXO and to GPS timing signals. All cards were contained in a single 486 based PC.

Figure 8–3 *(Continued)*

> Tests with CP-2 transmissions showed that the slotted waveguide antenna significantly reduced the level of ground clutter contamination. This antenna, in conjunction with a simple software clutter filtering scheme, reduced this contamination to acceptable levels...

Figure 8–4 shows the antenna and receiver [2]. A second problem encountered by a wide-beam antenna is signal contamination by multiple and side lobe scattering (i.e., secondary scattering) from weather particles. Wurman et al. [1] outlined the problem:

> The low-gain antennas of the passive receivers cause certain signal contamination problems, usually negligible with transmitting monostatic pencil-beam radars, to become significant. As with monostatic radars, all radiation that arrives at the reciever within a given time period is associated computationally, with a particluar resolution volume. The wide viewing angles of the passive bistatic receivers' antennas allow greater relative levels of undesired radiation to be detected. The scenarios that allow undesired radiation to arrive within the same time period as the desired radiation can be divided into three categories: multiple scattering, first sidelobe coupling, and integrated far sidelobe coupling.
>
> *a. Multiple scattering*
>
> The fractional amount of power scattered per resolution volume can be calculated from the summed cross-sectional area of the scatterers per unit volume and the resolution volume length... Much of this scattered radiation passes completely through intense weather to be detected by the receivers or wasted into space or the ground. A portion, however, is scattered again within the storm or by another storm. Some of this multiply scattered radiation reaches the receiver...

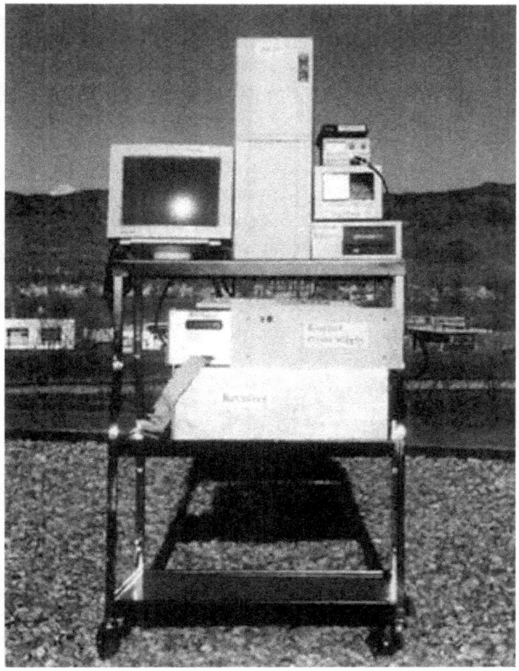

(a) (b)

Figure 8–4 Antenna (*a*) and receiver (*b*) for the BNR. An antenna is mounted on a portable automated mesonet tower (total height 8 m). The waveguide carries a signal to the receiving unit. The receiver uses a 10-MHz oscillator to provide highly stable signals to the receiver and data acquisition system. An RF tuner allows the system to use signals from different transmitters. The modem ingests the CP-2 antenna pointing angle and other information. Data acquisition is conducted on a PC [2]. © American Meteorological Society.

The major effect is to degrade measurements of very sharp reflectivity gradients, such as a high-reflectivity precipitation core with a sharp edge, thus blurring the actual echo boundary. Possible remedies include (1) operating at the longest practical transmitted wavelengths, preferably 0.1 m or even longer; (2) distinguishing between single- and multiple-scattering signals by comparing data from other receivers in the network, for example comparing the measured bistatic reflectivity with a calculated bistatic reflectivity, which is derived from measuring transmitting radar reflectivity and bistatic antenna patterns in the bistatic radar equation and then excising the contaminating vectors, at a cost of measuring wind change around the shear line [2], and using higher-gain receive antennas, at the cost of coverage. Wurman et al. [1] continues:

> *b. First sidelobe coupling*
> Some of the radiation emanating from the first side-lobe of the transmitter will be scattered toward the receiver. It will arrive almost contemporaneously, typically with 1 μs, with radiation scattered from the main beam. As is the case with multiply scattered radiation, these photons will be indistinguishable from those scattering from the main beam. ...the integrated effect will contaminate only those storm regions where reflectivity gradients exceed about 20 dBZ (500 m)$^{-1}$, a very limited region.
>
> *c. Integrated far sidelobe coupling*
> Some of the radiation emanating from all of the transmitter sidelobes will be scattered toward the receiver... Observations with monostatic radars show that the intensity of contamination from this effect

is about 60 dB below the average intensity of radar echoes covering large portions of the sky, effectively limiting the useful dynamic range of monostatic radars. Bistatic receivers... [using] omnidirectional receivers are limited to approximately 30 dB of effective dynamic range. A more typical receiver, employing a 15-dB horn,... would have 45 dB of dynamic range...[with higher-gain antennas providing still greater dynamic range].

The fourth design trade-off [3] was for frequency synchronization:

In a typical weather radar, the transmitted frequency is stable only on short time-scales, typically a dwell time (~100 ms). Drifts of several tens of hertz occur on time scales of several minutes. It was decided that this would unacceptably complicate the task of computing radial velocities at the prototype bistatic site. The drift would have to be frequently detected and corrected for in software, or the hardware at the bistatic site would have to track the varying transmit frequency. While either of these techniques was probably practical, an attempt to generate nearly equal frequencies at both the transmitter and reciever was a simpler option.

Ovenized VCXO's were installed. Laboratory tests showed that the two oscillators maintained frequency coherence to within one part in 10^9 for several days and exhibited very low phase noise. Measurements of the transmitter frequency were taken at the remote bistatic site by detecting sidelobe radiation and the direct transmitted beam. The transmitter frequency maintained frequency coherence with the signals generated at the bistatic reciever to within 2 Hz over the entire test period of one week...

Before the excellent open-loop performance of the oscillators was measured, they were both disciplined with the aid of GPS timing signals. GPS receiving cards provided 1-pulse-per-second (1-pps) signals, locked to GPS time (which is accurate to within 50 ns), to other PC cards. Another card was used to discipline the 10-MHz VCXO. Measurements of the transmitter frequency indicated that it maintained coherence with the bistatic receiver's locally generated frequency to within 1 Hz for weeks. Presumably, this can be extrapolated for much longer time periods since the oscillators were effectively coupled to each other through the GPS link.

The fifth design trade-off [3] was for timing synchronization:

One possible method of synchronization [between transmitter pulses and remote receiver gating] was the construction of a cable between the transmitter and reciever, as had been done in a previous bistatic experiment... Unfortunately, burying a cable between two sites separated by tens of kilometers in an urban area is difficult and costly. In addition, there are unknown complications associated with signal dispersion and the sensitivity of line delays to temperature. This technique was not feasible for this experiment and is generally not likely to be a practical option for most bistatic multiple-Doppler networks.

Two major categories of synchronization techniques were explored. The first employed radiation, pulsed synchronously with the transmitter pulses, purposely or incidentally leaked by the transmitter toward the bistatic reciever. The time of arrival of the pulses could be used to initiate gating at the receiver since the direct path radiation would arrive before any radiation scattered from the transmitter beam. This was an acceptable method of synchronization. Typically, an operator at the bistatic site would manually adjust the gating clock until the gating triggers preceded the arrival of the radiation from the transmitter sidelobes by a few microseconds. The timing offset drifted by less than 1 μs per several hours since the transmitter pulse timing and the receiver gating were both governed by extremely stable 10-MHz oscillators. The few mucroseconds of offset were easily detectable manually and with software.

The second technique involved the use of the previously discussed GPS timing signals... For the PC at the transmitter, a card was constructed that initiated a 15-MHz counter upon the receipt of each 1-pps GPS timing signal. The counter was stopped upon receipt of the next pre-knock pulse from the transmitter indicating that it was about to send a pulse. The counter value was sent to the reciever via the telephone link... A program at the bistatic reciever computed the time of the first transmit pulse that would occur on the next 1-pps epoch. This was used to start the local timing generator which was thus precisely synchronous with the transmitter timing... This was the usual method of synchronization employed during the 1993 experiment.

Wurman et al. [3] report that data processing, in particluar the software, proved to be the most difficult challenge in building the BNR—a typical problem with first and second generation PCs in the early 1990s. Specifically,

> Primary data acquisition was conducted on a PC card… incorporating a two channel 10-MHz 12-bit A/D converter and a single … DSP [digital signal processing] chip. Attempts to integrate the software supplied with the card to function with the PC operating system to provide real-time processing, display and archiving proved problematical. Eventually, a limited data acquisition scheme was developed for the 1993 prototyping experiment. This scheme proved adequate to demonstrate the viability of the prototype, but was difficult to use and has been supplanted with a much improved method…

Specifically, the PC was not fast enough to perform real-time data acquisition. Consequently, data from 32 to 64 transmitted pulses were sampled. Then data collection paused while the DSP executed a pulse-pair algorithm [8]. Pauses lasted roughly 80–130 msec, depending on the number of pulses per beam. I/Q data and power estimates were stored on disk and the sampling process resumed. As a result, grouping of the pulses into beams was not synchronous with the grouping conducted at the CP-2 data system. Because of the data gaps, infinite impulse response clutter filtering schemes could not be applied. A simple mean-removal scheme was used, which was sufficient to suppress ground clutter signals with doppler velocities < 0.5 m/sec. Wurman et al. [3] report that:

> After the conclusion of the 1993 experiment, the data acquisition method was upgraded. The new software allowed continuous collection of data with no dropped pulses [by using a FIFO buffer and dual-port memory.]

Continuous operation also allowed real-time clutter filtering with a four-pole infinite impulse response filter, which provided approximately 35 dB of suppression [2]. Finally,

> Development is continuing on a real-time system, which transmits the raw velocity data to a central site.[4] The data are combined with velocity data from the transmitter to calculate dual-Doppler vectors for every radar gate. Since there is only [one] source of illumination, these calculations can be performed without interpolating the data from the transmitter and receiver(s) to a common grid.

As of 2002, nine sets of BNRs have been deployed at monostatic doppler weather radar sites for research, testing, and operation: one in the United Kingdom, one in Canada, one in Germany, two in Japan, and four in the United States. Table 8–1 lists the technical specifications for such a system. Figure 8–5 shows one antenna installation at the McGill University, Canada site.

8.5 TEST RESULTS [1–5]

Tests of the prototype BNR in a bistatic configuration were conducted in 1993 [1–3]. Wurman et al. [3] summarize the test and results:

> The prototype bistatic dual-Doppler system was tested near Boulder, CO during the summer of 1993. The RAPS-93 [Research Applications Program 1993] weather research field program… was underway to study convective storms in eastern Colorado. In conjuction with this program, the NCAR CP-2 [monostatic] radar was occasionally available for bistatic radar experiments.
>
> The bistatic receiver was deployed so that one of the CP-2/Bistatic dual-Doppler lobes overlapped with one of the CP-2/MHR (Mile High Radar) [a second monostatic radar] dual-Doppler lobes… (A dual-Doppler lobe is that region in which accurate dual-Doppler windfield reconstruction can be conducted.)

4 Specifically in 2006, starting with an analog-to-digital sampling card and sending all the digitized data to a PC for real-time recording and processing [Private communication from Dr. Joshua Wurman, October 9, 2006.].

Table 8–1 Technical Specifications for a Commercially Available BNR

Receiver system	Passive
Size	1 m × 1 m × 1 m
Weight	150 kg
Power	115-V 60-Hz integral gasoline generator or line
Antenna system	Slotted waveguide
Height	4 m
Width	0.5-m triangular tower
Pattern	Area flood
Azimuthal width	50° gaussian
Elevation width	4, 11, 14° flat
Gain	22 dB, 17 dB, 14 dB
Data communication system	900-MHz wireless
Data rate	115.2 kbaud/19.2 kbaud full duplex
Maximum real-time data link range	30-km line of sight
Data system	PC-based PIRAQ processor
Nyquist velocity	±25 m/sec single polarization (vertical) ±12 m/sec dual polarization (vertical receive only)
Recording	Exabyte 8 mm
Number of gates	300
Noise power	−117 dBm
Noise figure	2.1 dB
Dynamic range	67 dB
Bandwidth	0. 74 MHz matched
Minimum detectable dBZ**	
at 30 km	0 dBZ (30-km baseline)
at 100 km	at 100 km* 11 dBZ (100-km baseline)
Dual-doppler baseline	30 km with real-time data link 100 km with record only*
Coverage area	2,700 km² per receiver (30-km baseline) 24,300 km² per receiver (100-km baseline)*
Maximum range	60 km (30-km baseline) 200 km (100-km baseline)*

Key:
* Beyond real time data link range. Data is archived for post processing.
** Using 4°, 22 dB gain antenna.
Source: http://www.binet.com/specs.html, accessed 4/24/06.

In this configuration, convective cells that developed just to the east of Boulder or those that propagated eastward off the mountains could be studied…

Because of the limited area of the bistatic dual-Doppler lobe (~400 km^2) and because of the shared use of the CP-2 transmitter with other researchers, data were collected for a relatively small number of weather events on just a few days.

Figure 8–5 Photograph of a BNR installed as part of the McGill University weather radar network in Canada. Two BNRs are deployed, one approximately 20 km and one approximately 40 km from the monostatic radar. Both dual-doppler and triple-doppler coverage is available. (Photo by Isztar Zawadzki, courtesy of McGill University.)

Vertical-polarization-only transmit pulses were used for the prototype testing experiments so that the problems associated with a low-sensitivity notch in the dual-Doppler lobe [1] would be avoided.

A small and dying convective cell propagated eastward off the mountains over the study area on July 13 producing a weak outflow... [Dual-doppler data showed that] CP-2/Bistatic and CP-2/MHR windfields agree qualitatively, both exhibiting the updraft-downdraft couplet. One major difference in the two vector fields is the higher magnitude of the bistatically retrieved downdraft in the highest reflectivity region. The bistatic system had an enhanced ability to detect the low-level divergence under the cell due to better clutter filtering and shorter range to target...

Throughout most of the storm volume, the dual-Doppler horizontal vector winds calculated using the CP-2/Bistatic and CP-2/MHR couplets agree within 1–2 m/s. Notable differences occur near the ground, mentioned previously, and near the CP-2/Bistatic baseline where bistatic dual-Doppler winds are not retrievable [since bistatic doppler approaches 0 as the target approaches the baseline from any direction].

Reflectivity fields can also be retrieved from the bistatic receiver... It should be noted that the bistatic system was not carefully calibrated so the presented field has had a constant added to it so that the peak reflectivities are in close agreement with those measured by CP-2. Again the most striking feature is that the monostatic (CP-2/MHR) and bistatic fields are in qualitative agreement. While the agreement in peak magnitude has been forced, it is clear the general measured structures are also similar.

Wurman et al. [3] also observed that the bistatic system apparently could not resolve deep minimums of reflectivity or steep gradients. The most likely explanation was that multiple

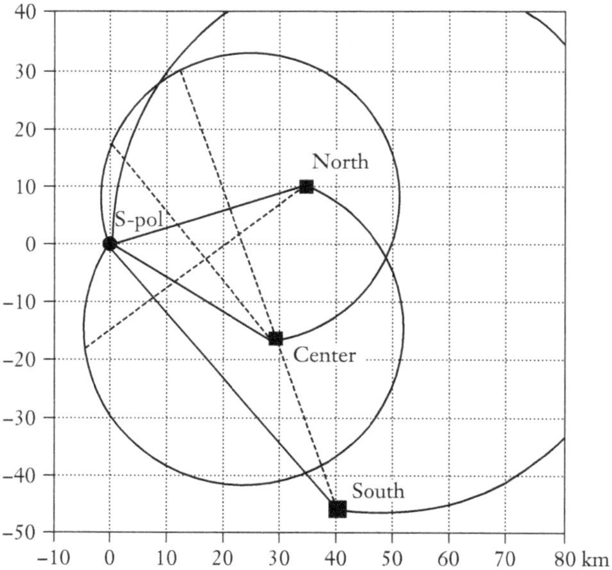

Figure 8–6 Location of the monostatic radar (S-Pol) and three bistatic network receivers (North, Center, South) in a bistatic multiple-doppler network during CASES-97. The arcs indicate the bistatic dual-doppler analysis areas and the dotted lines show the effective bistatic antenna ranges [4]. © American Meteorological Society.

scattering from raindrops into the broad beam receive antenna, described in Section 6.4, contaminated both reflectivity and velocity data, with a higher gain antenna expected to reduce the degradation.

A second set of BNR tests now primarily in a multistatic configuration was conducted during the Cooperative Atmosphere-Surface Exchange Study (CASES) experiment in Kansas in May 1997. Satoh and Wurman [4] report:

> ...In the CASES-97, the National Center for Atmospheric Research's S-band polarized radar (S-Pol) and three bistatic receivers were installed at the locations shown in Fig. 1 [Figure 8–6]. The baseline length between the S-Pol and three bistatic sites (North, Center, and South) was 35.0, 32.9, and 60.9 km, respectively. For the first case study of convective echoes, the North bistatic data is used only because of lack of other bistatic data. For the second case study of stratiform echoes, all data observed by the three bistatic receivers are used to compare with each other in the overdetermined region, and to be composited using a variational method. This study will deal with the [two-dimensional] horizontal wind retrieval using the low-elevation-angle data...

For the two-dimensional case the variational method is a simple average of the horizontal components of the observed wind vectors (u_i, v_i), $i = 1, 2, \ldots n$, weighted by the assumed-constant, but geometry-dependent, variance of the wind vectors, $\alpha_{un} = \alpha_{vn} = (\sin^2\beta)/(\cos\beta + 3)$, where β is the scattering or bistatic angle. Satoh and Wurman [4] continue:

> Figure 12 [Figure 8–7] shows the horizontal distrubution of wind vectors calculated by three pairs of Dual-Doppler synthesis. Each wind vector is calculated in a 1-km-size grid on the cartesian coordinate system... The dual-Doppler analysis region was determined by an enclosed region with the synthesis accuracy (40° < β < 150°)... and the effective bistatic antenna range... [This region is defined by constant

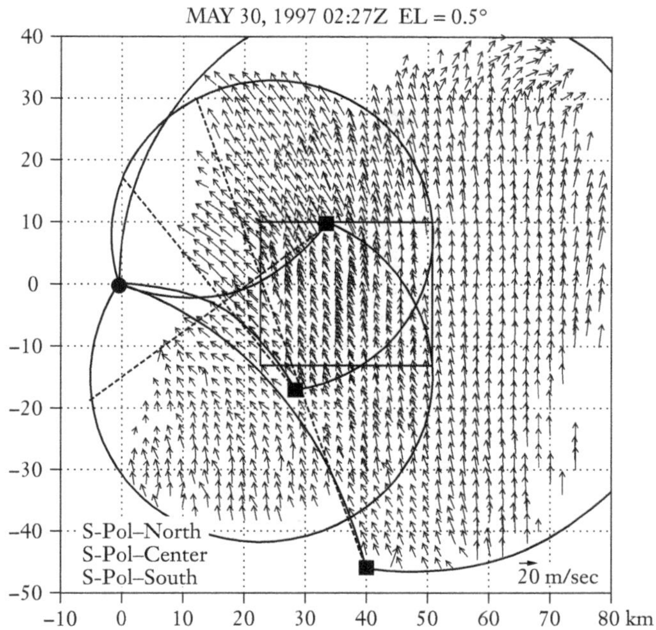

Figure 8–7 Horizontal distribution of wind vectors derived from three pairs of dual-doppler analysis in a stratiform weather system. Doppler data are obtained from a PPI scan at 0.5° in elevation angle of the monostatic radar and three bistatic receivers shown in Figure 8–6. The arcs show bistatic dual-doppler analysis areas bounded by $40° < \beta < 150°$ and the dotted lines show the edge of each bistatic antenna range. The square area shows the overdetermined domain [4]. © American Meteorological Society.

> β contours [13], which are circles of radius $L/(2\sin\beta)$ with center located on the perpendicular bisector the the baseline L, displaced from the baseline by a distance $L/(2\tan\beta)$.]
>
> ...Figure 12 [Figure 8–7] reveals that the reasonable wind vectors were calculated within most of the analysis region. In the northeastern region, however, unrealistic wind vectors appeared. Since the unrealistic vectors derived from the south bistatic data are located a great distance from the bistatic site, the intensity of the received signal is as weak as the noise level... In the overdetermined region around the center of Fig. 12, three kinds of wind vectors appear to roughly correspond with each other...

Satoh and Wurman [4] then calculated the composite wind vectors using the simplified variational method and concluded that:

> ...the composite vectors looked to be realistic with smooth continuity around the boundary between the different dual-Doppler pairs. A large difference among the three kinds of wind vectors appears especially along the baseline. The reason for the inconsistency seems to be the rapid expansion of the bistatic resolution volume, which is a function of β. In fact, the effective range in this case would be $35° < \beta < 120°$... [As outlined in the first tests] ...sidelobe contamination around the strong convective echoes was the main obstacle to accurately retrieving wind fields. In weak stratiform echo cases, [these tests], the minimum detectable reflectivity dominated the determination of the accuracy of the retrieved wind fields.

Friedrich and Hagen [5] observed a limitation in both of the foregoing tests: "...wind velocities measured by the monostatic Doppler radar were used to retrieve both bistatically measured and ...monostatically measured wind vectors, respectively. Therefore, those comparisons were not achieved using two independent wind measurements." Friedrich and Hagen [5] then set out to

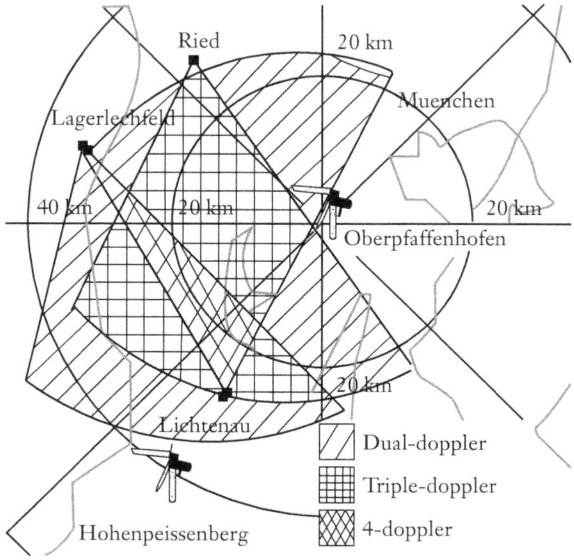

Figure 8–8 Map of the bistatic multiple-doppler radar network at the DLR in OP. The investigation area is restricted by horizontal antenna aperture and range resolution. The DWD C-band monostatic doppler radar on top of Mt. Hohenpeissenberg at 1006 m MSL provides independent measurements of (radial) vector wind data [5]. © American Meteorological Society.

rectify this problem with additional tests designed to compare bistatic multiple-doppler data first with data taken by an independent monostatic doppler radar in the area, and second with in situ flight data taken by a research aircraft. Tests also compared accuracy of each bistatic receiver used in the tests. Specifically [5] reports:

> In this paper measurements of the bistatic multiple-Doppler-radar network operated by the Deutsches Zentrum fur Luft- und Raumfahrt (DLR) on Oberpfaffenhofen (OP) close to Munich in southern Germany, are compared both to those wind measurements achieved by an independent Doppler radar operated by the German Weather Service (DWD) and to wind measurements achieved with the DLR Falcon research aircraft during three stratiform precipitation events...
>
> The DLR bistatic Doppler radar network consists of the monostatic C-band polarimetric diversity Doppler radar system (POLDIRAD...) located ...[in Oberpfaffenhofen] (OP) at 602 m MSL and three bistatic receivers at remote sites each equipped with an antenna and a signal processor... In Fig. 1 [Figure 8–8] the three bistatic receivers at Lichtenau, at Largerlechfeld, and at Ried together with the respective viewing angle of the bistatic antenna are illustrated...
>
> With this configuration, the bistatic radar network covers an area of about 50 km × 50 km horizontally and a height up to 5 km within stratiform precipitation. Horizontal wind fields are determined exactly in the dual-Doppler area (hatched area in Fig. 1) and over-determined in triple-/quadruple-Doppler areas (cross-hatched area in Fig. 1). Due to a limited vertical antenna aperture of 8° oriented close to the ground, the measured wind components are dominated by the horizontal components u and v. As a result, these measurements are used only to determine the horizontal wind vector field directly.

The comparison procedure was carried out by determining the horizontal wind vector based on a least-squares estimation using wind vector data from all available receivers. From this wind vector, a radial wind vector predicted for the independent radar was reconstructed. This prediction was then compared to the actual radial wind vector measured by the independent radar. Error sources that contaminate the measurements were also described, including clutter from ground, biological targets or chaff, system and instrumentation contributions such as antenna motion, integration, aliasing and partial beam shielding, and for bistatic systems, the geometry. Friedrich and Hagen [5] estimated that "For monostatic [weather] radars Doppler velocity measurements have an instrumental error of less than 1 ms^{-1}... Furthermore, errors of about ± 2 ms^{-1} are also associated for traditional monostatic dual-Doppler estimations."

Data comparisons were presented in scatter plots, plan position indicator (PPI) plots, cumulative probability distributions, and data tables, with extensive analysis accompanying each, for two stratiform precipitation events: a low wind shear trough on April 10 and a cold front on June 6, 2001. Also, one in situ flight collected about 1 min of data on April 11, 2001. Results were summarized as follows:

> [1] Comparing direct measurements of the two bistatic receivers, Doppler velocities agree within ± 1 ms^{-1}, which lies within the margin of error for instrumentation accuracy. The Doppler velocity variation is mainly assigned the instrumental error due to the setup of receivers and meteorological conditions...
> [2] The magnitude in Doppler velocity differences ranged mainly in all three case studies between 2 and 3 ms^{-1}...
> [3] The comparison of bistatic Doppler radar measurements with another independent instrument, the in situ flight measurement, confirm those results achieved using a monostatic Doppler radar. Again, differences in Doppler velocities are within ± 2 ms^{-1}.
> [4]...sidelobe contamination has not influenced significantly the radar measurements chosen for this case study since the reflectivity factor was low and on the order of 35 dBZ.

Friedrich and Hagen [5] then concluded:

> Weather events like, for instance, a frontal passage can be well monitored by a bistatic multiple-Doppler radar network as shown in 6 June 2001. One of the great advantages of bistatic network is the real-time display of wind vectors, which allows a rapid and easy detection of signatures in the wind field...

References

1. J. Wurman, S. Heckman, and D. Boccippio, "A Bistatic Multiple-Doppler Radar Network," *Journal of Applied Meterology*, Vol. 32, December 1993, pp. 1802–1814.
2. J. Wurman, "Vector Winds from a Single-Transmitter Bistatic Dual-Doppler Radar Network," *Bulletin of the American Meterological Society*, Vol. 75, No. 6, June 1994, pp. 983–994.
3. J. Wurman, M. Randall, C. L. Frush, E. Loew, and C. L. Holloway, "Design of a Bistatic Dual-Doppler Radar for Retrieving Vector Winds Using One Transmitter and a Remote Low-Gain Passive Receiver" [invited paper], *Proceedings of the IEEE*, Vol. 82, No. 12, December 1994, pp. 1861–1872.
4. S. Satoh and J. Wurman, "Accuracy of Wind Fields Observed by a Bistatic Doppler Radar Network," *Journal of Atmospheric and Oceanic Technology*, Vol. 20, August 2003, pp. 1077–1091.
5. K. Friedrich and M. Hagen, "Evaluation of Wind Vectors Measured by a Bistatic Doppler Radar Network," *Journal of Atmospheric and Oceanic Technology*, Vol. 21, December 2004, pp. 1840–1854.
6. P. P. Neilley, et al., "The Real-Time Analysis and Prediction of Storms Program," in *Proceedings of the 5th International Conference on Aviation Weather Systems*, Boston, MA, 1993, pp. 62–65.
7. L. J. Battan, *Radar Observation of the Atmosphere*, revised edition, University of Chicago Press, Chicago, 1973.
8. R. J. Doviak and D. S. Zrnic, *Doppler Radar and Weather Observations*, 2nd ed., Academic Press, New York, 1993.
9. V. N. Bringi and V. Chandrasekar, *Polarimetric Doppler Weather Radar*, Cambridge University Press, Cambridge, UK, 2001.

10. A. Protat and I. Zawadzki, "A Variational Method for Real-Time Retrieval of Three-Dimensional Wind Field from Multiple-Doppler Bistatic Radar Network Data," *Journal of Atmospheric Oceanic Technology*, Vol. 16, 1999, pp. 432–449.
11. Y. Takaya and M. Nakazato, "Error Estimation of the Synthesized Two-Dimensional Horizontal Velocity in a Bistatic Doppler Radar System," *Journal of Atmospheric Oceanic Technology*, Vol. 19, 2002, pp. 74–79.
12. M. I. Skolnik, *Introduction to Radar Systems*, 2nd ed., McGraw-Hill, New York, 1980.
13. N. J. Willis, *Bistatic Radar*, 2nd ed., corrected and reprinted by SciTech Publishing, Inc., Raleigh, NC. 2005.

PART II

Bistatic Clutter and Signal Processing

9

Clutter

Melvin M. Weiner

9.1 INTRODUCTION AND SUMMARY

9.1.1 Introduction

Bistatic clutter is part of the ground environment that affects the performance of a bistatic radar system. In continuous wave (CW), moving target indication, and pulse doppler radars, bistatic *clutter residue* (clutter return after signal processing) competes as interference with the target signal (see Chapter 11). In synthetic aperture radar, bistatic clutter is a critical part of the target scene (see Chapter 10). In planetary probing radar, bistatic clutter becomes the entire target scene (see Chapter 5). In all cases, the phenomenology of terrain and sea clutter must be characterized, which is the purpose of this chapter.

The radar return of transmitter energy scattered by terrain and sea (called *surface clutter*) from a specific area (called the *clutter cell area*) is a function of the *bistatic angle*. The bistatic angle is the angle subtended at the clutter cell by the lines joining it to the transmitter and receiver that are separated in space. The clutter return before signal processing can be orders of magnitude larger than returns from man-made targets such as aircraft or vehicles in the clutter cell.

Bistatic clutter phenomenology is relevant in applications such as:

a. Air, space, and ground surveillance of moving targets

b. Air and space imaging of fixed targets

c. Electronic countermeasures located in a radar's side lobe that jam via surface scatter through the radar's main lobe, often called *hot clutter*

d. Multipath propagation effects on radars, communication, and telemetry systems

e. Design and evaluation of low side lobe antennas

f. Scientific probing of ocean waves and extra-terrestrial surfaces

Unlike chaff and rain—whose scattering elements are well-defined in terms of geometrical shape, spacing, composition, and electrical properties—terrain and sea surfaces can have a highly irregular profile, and, in the case of terrain, may consist of a nonuniform composite of themes whose electrical properties are dependent on temperature and moisture. For a transmitter and receiver at low grazing angles, the investigation of surface clutter is further complicated because of surface shadowing (incident or reflected waveforms whose lines of sight are obscured by the surface). For this reason, the major emphasis of this chapter is on experimental data of surface clutter with particular attention to an assessment of measurements of the bistatic clutter cross-section per unit area (bistatic surface scattering coefficient).

The system parameters and properties of bistatic surface clutter are discussed in Section 9.2. The database of the bistatic cross section per unit area for terrain and sea scatter is reviewed in Section 9.3.

9.1.2 Summary

1. The clutter cell area for the resolution time-limited case (limited by the receiver resolution time and the transmitter or receiver main lobe beam width) at low grazing angles is more than 30 dB larger in the forward-scattered direction than in the backscattered direction but is less than 10 dB larger for azimuthal angles as large as 135° from the backscattered direction.

2. The surface clutter radar cross section (RCS) is significantly larger in the forward-scattered direction than in the backscattered direction and may nullify any advantages of a possibly enhanced target RCS in the forward-scattered direction, particularly at frequencies greater than 300 MHz.

3. Bistatic radar surveillance at an azimuthal angle of approximately 90° from the backscattered direction offers a potential reduction of surface clutter return compared to that from a monostatic radar, based on measurements of selected vegetative and soil themes.

4. The bistatic surface clutter return at an azimuthal angle as large as 45° from the backscattered direction is not appreciably different from that of the backscattered direction.

5. The definition of clutter cross-section per unit area which is used in determining experimental values may give different values from the definition used in theoretical modeling, depending on the type of averaging performed by the latter. The former definition is useful in characterizing *diffuse scatter* (rays reflected in all directions) from rough *extended surfaces* (surfaces that extend beyond the field of view of the transmitter or receiver) whereas the latter definition was originally used to characterize scattering from nonextended surfaces.

6. Terrain and sea scatter are generally from extended surfaces. Consequently, the number of Fresnel zones subtended by the clutter cell at the receiver or transmitter is usually more than one Fresnel zone. Theoretical models of a clutter cross-section per unit area from extended surfaces should therefore be evaluated for the corresponding Fresnel zone of the clutter cell. However, for zero-mean stochastic processes, the effect of surface roughness on the expected value of the *coherent component* (proportional to the product of the expected values of the scattered field times the expected value of its phase conjugate) of a clutter cross-section per unit area is independent of the number of Fresnel zones subtended by the clutter cell at the receiver.

7. Neither the experimental nor theoretical definitions of a *clutter cross-section per unit area* (scattering coefficient) are inherently independent of the clutter cell area. The coherent contribution to the scattering coefficient in both the main lobe and side lobes of the clutter cell radiation pattern is a function of the clutter cell area. However, if the diffuse *incoherent scatter* (proportional to the variance of the scattered field) in an arbitrary direction is much larger than the coherent scatter in that direction and if the clutter cell area contains a large number of random independent scatterers, then the scattering coefficient is approximately independent of the clutter cell area for scatter in that direction.

8. Although various models (semiempirical, geometrical, and statistical) have been made of surface scatter on the basis of various assumptions about the characteristics of the surface, meaningful results are difficult to achieve except over a relatively narrow range of elevation or out-of-plane angles.
9. The existing open literature database of the bistatic scattering coefficient for terrain and sea scatter is from nine principal investigators over the period 1965–2002 (see Tables 9–1 to 9–4 in Section 9.3.3.1).
10. The most substantial bistatic scattering coefficient data, in terms of number of independent samples, number of terrain and sea themes, and number of data curves, or statistical properties, are given by Cost and Peake, Pidgeon, Domville, Larsen and Heimiller, Ewell and Zehner, Ulaby, and McLaughlin (see Tables 9–1 to 9–4 in Section 9.3.3.1).
11. The Bistatic-Monostatic Equivalence Theorem, when applied to target RCSs or extended surfaces, is generally not valid except for small bistatic angles. However, Willis has successfully used the theorem to interpolate the measured database of Domville (for in-plane scatter from rural land, forest, and urban land) to bistatic angles as large as 135° provided that either the incident or scattering angle was less than 5°.
12. Future efforts in expanding the database of the bistatic scattering coefficient should include measurements as a function of cell area to determine homogeneity of the surface theme and to determine whether the scattering coefficient is predominantly coherent or incoherent.

9.2 SYSTEM PARAMETERS AND PROPERTIES

9.2.1 Main Lobe Clutter Equation

The radar return resulting from transmitter energy scattered by terrain or the sea to a radar receiver is known as surface clutter. The surface clutter area which is within the main lobe of the transmitter and receiver antennas and which may be further restricted—for a pulsed radar by the receiver resolution time to be within a given range delay bin or for a CW-doppler radar by the receiver bandwidth to be within a given doppler frequency bin—is defined as clutter cell area, A_C.

The bistatic surface clutter geometry is shown in Figure 9–1. The transmitter, receiver, and clutter cell are located at points A, B, C, respectively, at separations $\overline{AB} = \rho_{AB}$, $\overline{AC} = \rho$, and $\overline{BC} = r$. The ground reference plane is the plane tangent to the Earth's surface at C. The bistatic angle is $\measuredangle ACB = \beta$, where $\pi - \beta - \gamma$ is the angle subtended at the receiver by the transmitter and clutter cell and γ is the angle subtended at the transmitter by the receiver and clutter cell. The backscattered and forward-scattered directions correspond to $\beta = 0°$ and 180°, respectively. The transmitter and receiver subtend grazing angles ψ_A and ψ_B, respectively, at the clutter cell.

The surface of constant range sum $R = \rho + r$ is an ellipse of revolution about the axis $\overline{AB}$ with the foci at A and B. The clutter return in the absence of range ambiguities, at a time delay R/c, where c is the velocity of wave propagation, is from clutter cells centered along a line determined by the intersection of the ellipsoid of range sum R with the Earth's surface. The time-delay difference, Δ/c, between the indirect signal (clutter return) and the direct signal from the transmitter to receiver, is given by

$$\Delta/c = (R - \rho_{AB})/c \tag{9.1}$$

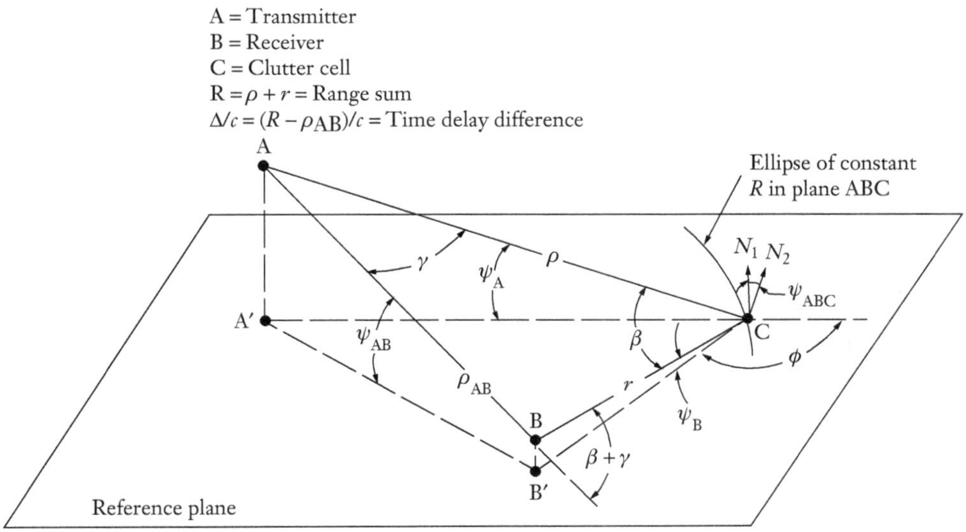

Figure 9–1 Bistatic surface clutter geometry.

The location of clutter cells for a clutter return at a time-delay difference $\Delta/c = (\rho_0 r_0 - \rho_{AB})/c$ corresponding to a main lobe range bin $(\beta_0 + \gamma_0 r_0)$ is shown in Figure 9–2 for the case when the transmitter and receiver are located on the horizon, $(\psi_A = \psi_B = 0)$ the transmitter is on a platform moving with velocity V_A, and the pulse repetition frequency is sufficiently small so that there are no range ambiguities. The main lobe clutter return is from the clutter surface centered along the Δ/c ellipse between points a and b. Points a and b are determined by the common intersection of the transmitter and receiver main lobe azimuthal beamwidths θ_A and θ_B, respectively, with the Δ/c ellipse. The side lobe clutter return is from the cells on the ellipse which are within the line of sight of both the transmitter and receiver but not within the main lobe azimuthal beamwidth of both the transmitter and receiver. In Figure 9–2, these clutter cells correspond to the arcs $\overline{ga}$ and $\overline{be}$.

For a pulse radar, the main lobe clutter return per pulse, at the receiver amplifier and in the absence of range ambiguities, is given by

$$E_C = \int_{A_C} \overbrace{\left(\frac{P_t K \delta t L_\rho G_A}{4\pi\rho^2}\right)}^{\text{Incident energy density}} \overbrace{\left(\frac{L_r G_B \lambda^2}{(4\pi)^2 r^2}\right)}^{\text{Fraction of reflected energy returned to receiver}} \overbrace{\left(\sigma^0\, dA\right)}^{\text{Differential clutter RCS}} \tag{9.2}$$

where E_c is the clutter pulse energy, P_t is the transmitter peak power, and G_A and G_B are the transmitter and receiver antenna power gains, respectively. In the direction of a differential clutter cell area dA, K is the pulse compression ratio, δt is the receiver resolution time $\approx$1/receiver bandwidth, L_ρ and L_r are the one-way propagation losses (excess over free-space loss and including antenna losses) from the transmitter to the clutter cell and clutter cell to receiver, respectively, λ is the radar wavelength, σ^0 is the clutter cross-section per unit area (m^2/ m^2), and A_C is the clutter cell area within the main lobes of the transmitter and receiving antennas.

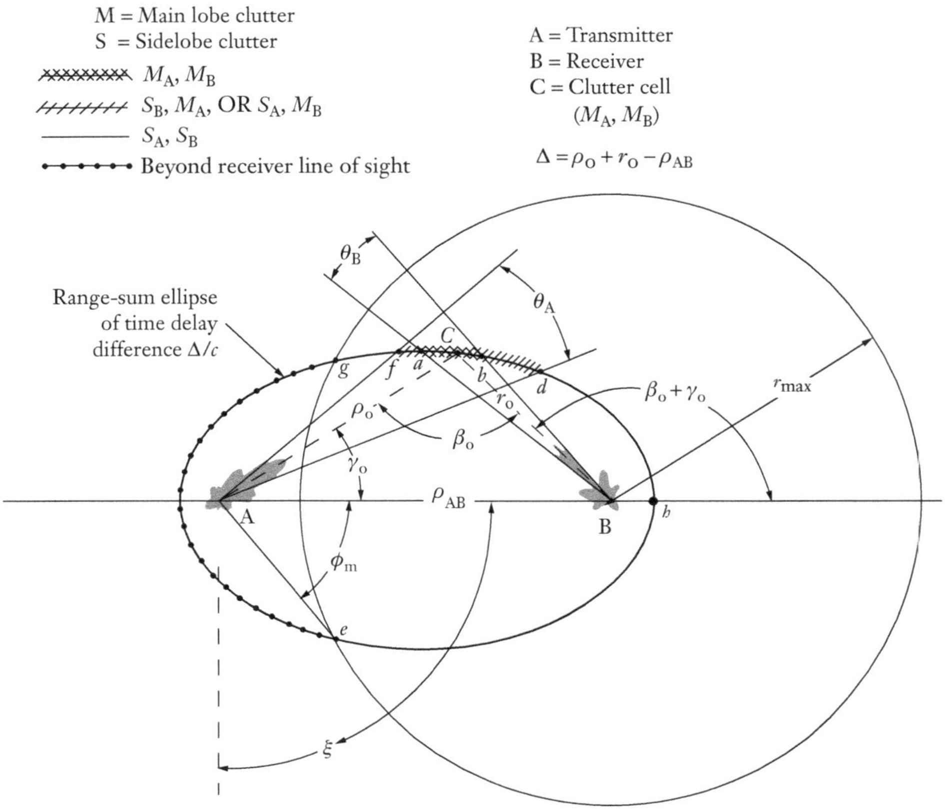

Figure 9–2 Location of clutter cells for range bin ($\beta_0 + \gamma_0$, r_0).

For the ρr product and σ^0 approximately constant over the cell area A_C, equation (9.2) reduces to

$$E_C \approx \left(\frac{P_t K \delta t L_\rho \overline{G_A G_B} \, L_r \lambda^2}{(4\pi)^3 \rho^2 r^2} \right) (\sigma^0 A_c) \tag{9.3}$$

where $\overline{G_A G_B} \equiv \frac{1}{A_C} \int_{A_C} G_A G_B \, dA$ = mean power gain product within the main lobes of the transmitter and, receiver antennas. The factor $\sigma^0 A_C \equiv \sigma_C$ is the main beam clutter RCS.

Equation (9.3) is particularly useful in determining the Improvement Factor required of a doppler processor to detect a weak return from a moving target in the presence of a much stronger quasi-stationary clutter return. For example, for a point target located at the center of the clutter cell and at the peak gains G_{A_0}, G_{B_0} of the transmitter and receiver, respectively, the ratio E_S/E_C, of target return to the main beam clutter return, is given by

$$\frac{E_S}{E_C} = \frac{G_{A_0} G_{B_0}}{\overline{G_A G_B}} \frac{\sigma_t}{\sigma^0 A_C} \tag{9.4}$$

where σ_r is the target RCS. The factor $(G_{A_0}G_{B_0}/\overline{G_A G_B})$ is a function of the transmitter and receiver antenna pattern shapes. The parameters σ_0 and A_C are discussed in Sections 9.4 and 9.2.2, respectively.

9.2.2 Clutter Cell Area

The main lobe clutter cell area may be beamwidth-limited, resolution-time limited, or doppler-frequency limited. The cell area is *beamwidth-limited* when it is the area of the footprint made by the common intersection of the transmitter and receiver main lobes with the clutter surface. The cell area is *resolution-time limited* when, for a pulsed radar, the cell area is restricted by the receiver resolution time to be within a given range delay bin whose size corresponds to an area less than the footprint. The cell area is *doppler-frequency limited* when the cell area is reduced by the video bandwidth to be within a given doppler frequency bin whose size corresponds to an area less than the footprint. In the following discussion of cell area, the notation of Figure 9–1 is utilized.

9.2.2.1 Beamwidth-Limited Case For monostatic radar ($\beta = 0, \rho = r$), the beamwidth-1imited cel1 area is given by [1]

$$A_C = \begin{cases} \dfrac{\pi}{4} D_\theta D_\xi, \text{elliptical beam cross section} & (9.5a) \\ D_\theta D_\xi, \text{rectangular beam cross section} & (9.5b) \end{cases}$$

$$\tan\psi_A > (4\rho/c\delta t)\tan(\xi_A/2),\ \beta = 0,\quad \rho = r \qquad (9.5c)$$

where $D_\theta = 2\rho\ \tan(\theta_A/2)$, $D_\xi = 2\rho\ \tan(\xi_A/2)\csc\psi_A$, θ_A = transmitter main lobe null-to-null azimuthal beamwidth, ξ_A = transmitter main lobe null-to-null elevation beamwidth, ψ_A = beam grazing angle of incidence, and δt = receiver resolution time.

For a pencil beam ($\theta_A = \xi_A << 1$ rad), equation (9.5*a*) reduces to

$$\begin{aligned} &A_C \approx (\pi/4)\rho^2\theta_A{}^2 \csc\psi_A;\ \tan\psi_A > (4\rho/c\delta t)\tan\xi_A/2, \\ &\beta = 0,\quad \rho = r,\quad \theta_A = \xi_A << 1 \text{ rad} \end{aligned} \qquad (9.6)$$

For bistatic radar, the beamwidth-limited cell area has been evaluated for specific antenna pattern functions and for specific geometries by numerical integration techniques [2–4]. For transmitter and receiver pencil beams whose one-way power antenna pattern shapes f_A, f_B, respectively, are approximated in the main lobe region by Gaussian functions $f_A = \exp(-Ax_A^2/2), f_B = \exp(-Bx_B^2/2)$ where x is the angle from the peak gain direction, the beamwidth-limited cell area is given by [5]

$$A_C = 2\pi\rho^2 r^2/C$$

where

$$\begin{aligned} C = {} & A^2 r^4 \sin^2\psi_A + B^2\rho^4 \sin^2\psi_B + AB\,\rho^2\ r^2(\sin^2\psi_A + \sin^2\psi_B \\ & + \cos^2\psi_A \cos^2\psi_B \sin^2\phi)]^{1/2} \end{aligned} \qquad (9.7)$$

ϕ = out-of-plane azimuthal angle between the incident and scattered beams defined in Figure 9–1

Forward scatter is in the elevation plane when $\phi = 0$, whereas backscatter is in the elevation plane when $\phi = 180°$. The constants A and B are related to the pencil-beam half power beamwidths $(\theta_{A,0.5} = \xi_{A,0.5})$, $(\theta_{B,0.5} = \xi_{B,0.5})$, by

$$\begin{aligned} A &= 5.54 / \theta^2_{A,0.5} = 5.54 / \xi^2_{A,0.5} \\ B &= 5.54 / \theta^2_{B,0.5} = 5.54 / \xi^2_{B,0.5} \end{aligned} \tag{9.8}$$

For monostatic radar, equations (9.7) and (9.8) reduce to

$$A_C = (\pi/5.54)\rho^2\theta^2_{A,0.5}\csc\psi_A; A = B,\ \rho = r,\ \psi_A = \psi_B,\ \phi = 180° \tag{9.9}$$

Equation (9.9) should be compared with equation (9.6).

Equation (9.6) is for a beamwidth defined by the nulls of the main lobe, whereas equation (9.9) is for a half-power beamwidth.

9.2.2.2 Resolution-Time Limited Case For monostatic radar, the resolution-time limited cell area for a flat Earth's surface is given by [1]

$$\begin{aligned} A_C &= [2\rho\tan(\theta_A/2)][(c\delta t/2)\sec\psi_A];\ \ c\delta t << 2\rho, \\ &\tan\psi_A \le (4\rho/c\delta t)\ \ \tan(\xi_A/2),\ \ \beta = 0,\ \ \rho = r \end{aligned} \tag{9.10}$$

where δt is the receiver resolution time.

For bistatic radar, the resolution-time limited cell area has been evaluated for specific antenna pattern functions by numerical integration [6]. Algebraic expressions for the resolution-time limited cell area have been determined for small grazing angles on a flat clutter surface at arbitrary bistatic angles and at arbitrary time-delay differences between the indirect and direct signals [7].

For small grazing angles, the cell area is closely approximated by the area in the plane ABC (see Figure 9–1), which is in the field-of-view of the transmitter and receiver azimuthal main beamwidths and which is bounded by the two ellipses of range sums $R + (c\delta t/2)$ and $R - (c\delta t/2)$ both with foci at A and B. For $\Delta/c >> \delta t$, the shape of the clutter cell is a parallelogram whose area is given by equation (9.18) of [7] as

$$A_C = \begin{cases} \rho\theta_A \dfrac{c\delta t}{2}\sec^2\left(\dfrac{\beta}{2}\right), & \theta_A \le (r/\rho)\theta_B << 2 \text{ rad} \\ & \\ r\theta_B \dfrac{c\delta t}{2}\sec^2\left(\dfrac{\beta}{2}\right), & \theta_B < (\rho/r)\theta_A << 2 \text{ rad} \end{cases} \quad ;\ \Delta/c >> \delta t \tag{9.11}$$

for the low grazing angle conditions

$$\begin{aligned} \tan\psi_A &\approx \psi_A \le (2\rho/\delta\rho)\tan(\xi_A/2) \\ \tan\psi_B &\approx \psi_B \le (2r/\delta r)\tan(\xi_B/2) \end{aligned} \tag{9.12}$$

where $\delta\rho = \delta r = c\delta t/2\sec^2(\beta/2) =$ range resolution along $\overline{AC}$ and $\overline{BC}$, respectively, and ξ_A, ξ_B = main lobe elevation beamwidths of the transmitter and receiver, respectively.

It should be noted that the condition $\Delta/c >> \delta t$ is not satisfied for bistatic angles approaching 180°. For monostatic radar ($\beta = 0$, $r = \rho$) and the conditions of equations (9.11) and (9.12), equation (9.12) reduces to equation (9.10). For $\Delta/c = 0$, the shape of the clutter cell is an ellipse

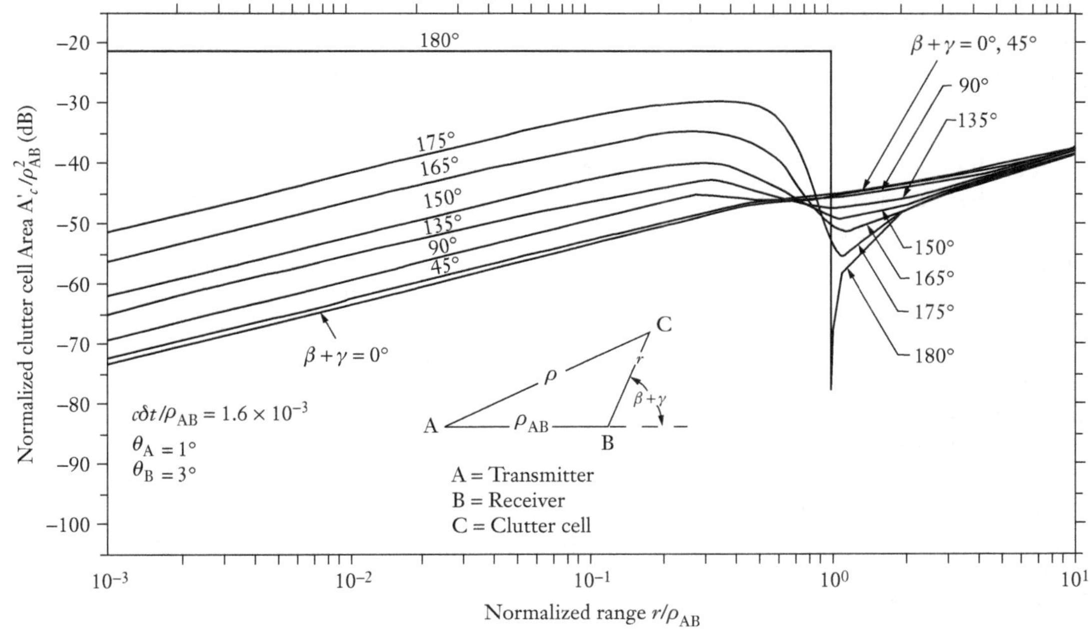

Figure 9–3 Normalized clutter cell area.

truncated by the main beam footprints so that the shape is approximately a hexagon whose area is given by equation (23) of [7]. For arbitrary Δ/c, the cell area may be approximated by quasi-parallelogram and quasi-rhomboidal formulae given by equations (28) and (29) of [7]. Although the cell is a function of the time-delay difference Δ/c and the bistatic angle β, the parameters Δ and γ be expressed as functions of r, β, and ρ_{AB} [7]. The normalized cell area A_C/ρ_{AB}^2 at low grazing angles is plotted in Figure 9–3 as a function of the normalized receiver range r/ρ_{AB} and azimuthal angle $\beta + \gamma$ for a transmitter beamwidth $\theta_A = 1°$, receiver beamwidth $\theta_B = 3°$, and $c\delta t/\rho_{AB} = 1.6 \times 10^{-4}$. The parameter $c\delta t/\rho_{AG} = 1.6 \times 10^{-4}$ corresponds to $\delta t = 10^{-8}$ sec, $\rho_{AB} = 10$ nmi or $\delta t = 10^{-7}$ sec, $\rho_{AB} = 100$ nmi. The clutter cell is more than 30 dB larger in the forward-scattered direction ($\beta + \gamma = 180°$, $r/\rho_{AB} < 1$) than in the backscattered direction ($\beta + \gamma = 0°$) but is only 8 dB larger for azimuthal angles as large as $\beta + \gamma = 135°$. The normalized area $A_C/\rho_{AB}^2 = -22$ dB for $\beta + \gamma = 180°$ corresponds to $A_C = +63$ dBm^2 for $\rho_{AB} = 10$ nmi and to $A_C = +83$ dBm^2 for $\rho_{AB} = 100$ nmi.

9.2.2.3 Doppler-Frequency Limited Case Bistatic scattering measurements of terrain and sea have been made for the case when the cell area is doppler-frequency limited [8, 9]. However, explicit expressions for the cell area is bounded by two iso-dops (lines of constant doppler frequency) and by the azimuthal beam footprints. The doppler-frequency limited cell area has been determined, by numerical iteration techniques, when it is bounded by two iso-dops and by two range-sum ellipses of separation $c\delta t$ [10].

9.2.3 Clutter Radar Cross Section

The clutter RCS, σ_C, is defined in equations (9.3) and (9.4) as

$$\sigma_C = \sigma^0 A_C \tag{9.13}$$

where σ^0 is the clutter cross-section per unit area (also called *scattering coefficient)* and A_C is the clutter cell area. The clutter RCS has the dimensions of area and is often expressed in units of [dBm²] = 10 $\log_{10}$ [area expressed in square meters]. Bistatic measurements of σ^0 are reviewed in Section 9.3.

The cell area A_C is computed from formulae or techniques discussed in the previous section.

In applying the measured values of σ^0 to a particular scenario, some modeling of σ^0 is usually required because system parameters (such as wavelength, polarization, scattering angles, surface roughness, surface theme, or statistics) of the available data may differ from those applicable to the scenario.

The bistatic RCS has been modeled for various surveillance applications including air surveillance by a space-based radar employing a transmitter and a receiver mounted on separate synchronous satellites [11], ground surveillance of moving targets by an airborne transmitter and an airborne receiver [10], and air surveillance of low-altitude targets by an airborne transmitter and ground-based receivers. In [10] and [11], σ^0 was modeled by utilizing monostatic data at an equivalent grazing angle $\psi_{equiv} = (\psi_A + \psi_B)/2$. The clutter cross-section per unit area can be modeled σ^0 by utilizing bistatic data at the actual azimuthal angles but modified by a weighted application of the equivalent grazing angle ψ_{equiv}.

The clutter modeling of M. M. Weiner illustrates the interpolations required in applying measured values of σ^0 to a particular scenario and also illustrates the phenomenology of clutter RCS as a function of the bistatic angle. The 1979 clutter modeling of M. M. Weiner is briefly reviewed in the remainder of this section.

The measured mean values of σ^0 (in decibel), for tall weeds and scrub trees at wavelengths of 3 cm and 23 cm, are interpolated in Figure 9–4 to a wavelength of 10 cm. In Figure 9–4, the azimuthal dependence of the interpolated value of σ^0, for $\psi_A = \psi_B = 10°$ and horizontal polarization of the transmitter and receiver, has a peak value of −2 dB in the forward-scattered

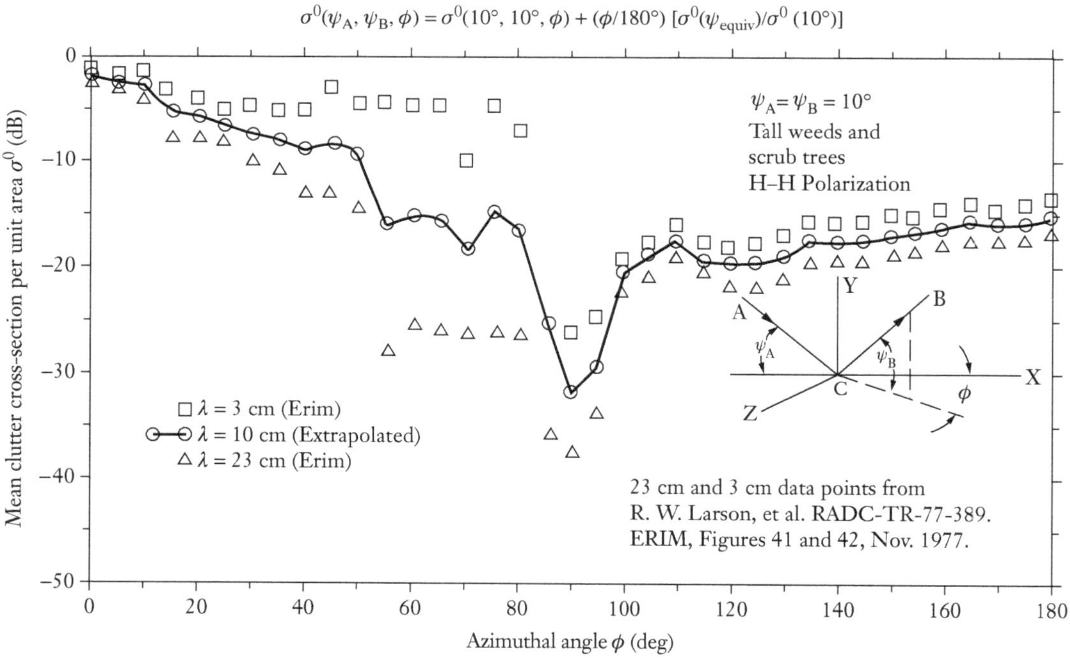

Figure 9–4 Azimuthal dependence of mean clutter cross-section per unit area.

direction ($\phi = 0°$), a minimum value of -32 dB for ($\phi = 90°$), and an intermediate value of -15 dB in the backscattered direction ($\phi = 180°$). Aside from illustrating the phenomology of σ^0 as a function of the azimuthal angle for $\psi_A = \psi_B$, Figure 9–4 also illustrates that the interpolation of monostatic data to a bistatic scenario of arbitrary aziumthal angle at grazing angles $\psi_A = \psi_B$ can introduce errors of approximately ± 15 dB in the mean value of σ^0.

The interpolated curve of Figure 9–4 for $\psi_A = \psi_B = 10°$ may be further interpolated to grazing angles 0°–30° by the addition of the elevation correction term $(\phi/180°)\ \sigma^0\ (\psi_{equiv})/\sigma^0(10°)$ (dB) where $\psi_{equiv} = (\psi_A + \psi_B)/2$ and $\sigma^0\ (\psi_{equiv})/\sigma^0(10°)$ is plotted in Figure 9–5.

The equivalent grazing angle ψ_{equiv} is based on the Bistatic Equivalence Theorem (see Section 9.3.2.4), which, when applied to an extended target such as clutter, is more likely to be valid for scattering angles in the vicinity of the backscattered direction. For this reason, the elevation correction term is weighted by the factor $(\phi/180)$. Figure 9–5 is based on a simple mathematical model by Technology Service Corporation [9] of S-band data for typical monostatic clutter of rural land and wooded areas. The model suggests that the mean value of σ^0 for this surface theme is reduced by 8 dB, at near grazing angles, from its value of $\psi_{equiv} = 10°$.

The clutter RCS, σ_C, is plotted as a function of range r and azimuthal angle ϕ in Figure 9–6 for the clutter models of Figures 9–4 and 9–5, the cell area A_C of Figure 9–3 modified so that A_C is a function of the azimuthal angle ϕ instead of β, and for a transmitter at an elevation $h_A = 25{,}000$ ft, a receiver at an elevation $h_B = 3150$ ft, a receiver resolution time $\delta t = 10^{-7}$ sec, and a baseline separation $\rho_{AB} = 100$ nmi. The clutter RCS is a minimum for the azimuthal angle $\phi = 90°$, and is a maximum for $\phi = 0°$ (elevation plane of forward scatter). The clutter RCS for $\phi = 180°$ (elevation plane of backscatter) is approximately 10 dB larger than for $\phi = 90°$ but is approximately the same for $\phi = 135°$. At a receiver range $r = 30$ nmi, the clutter RCS is approximately 50 dB greater in the forward-scattered direction than in the backscattered direction.

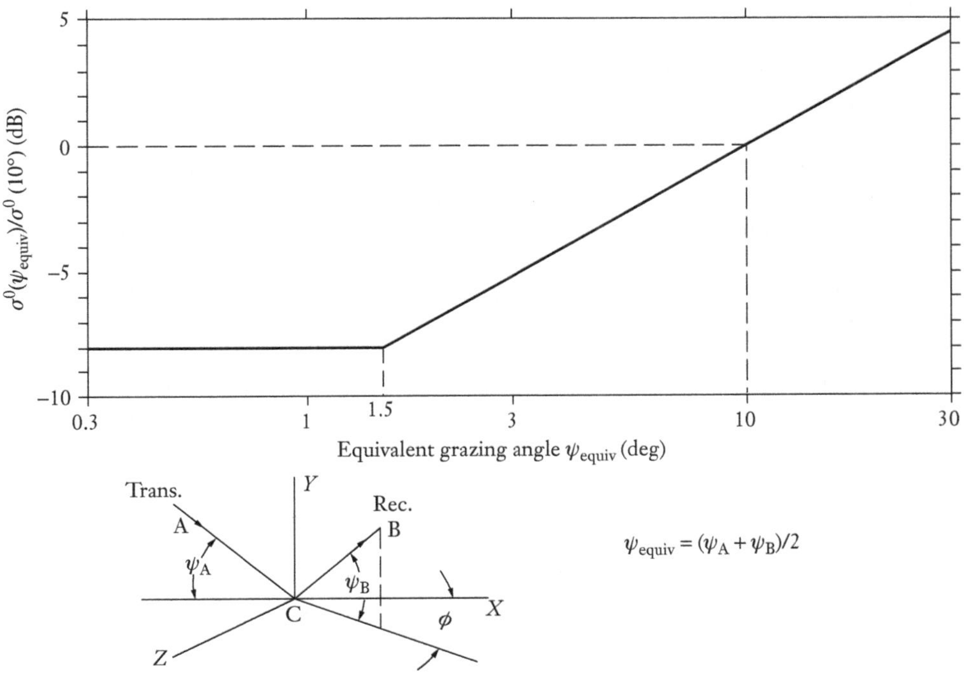

Figure 9–5 Elevation correction factor, clutter cross-section per unit area.

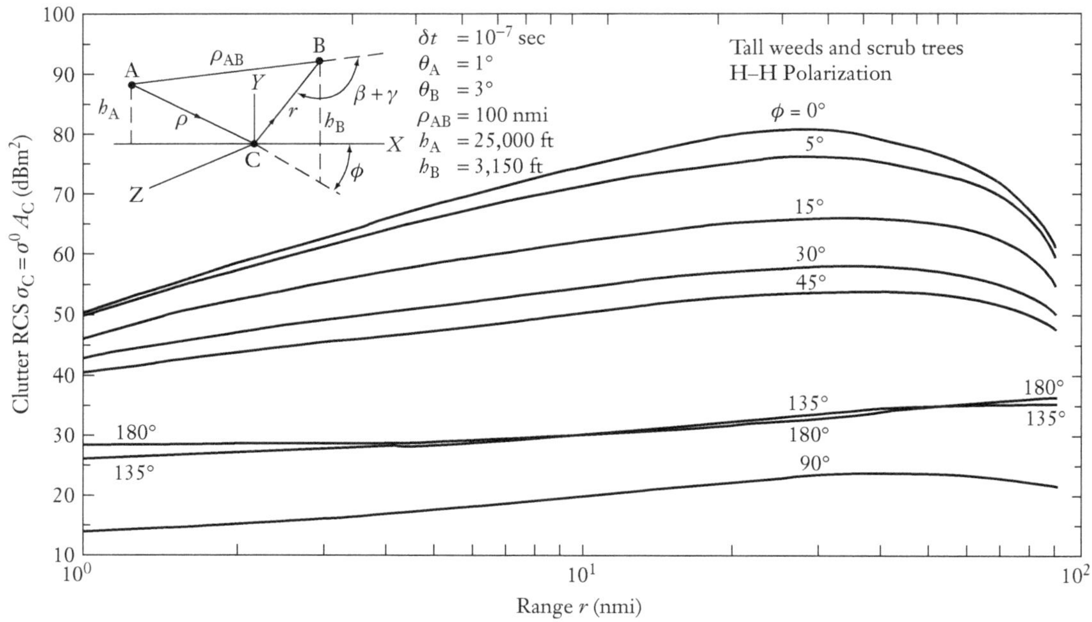

Figure 9–6 Clutter radar cross-section.

9.3 CLUTTER CROSS-SECTION PER UNIT AREA

9.3.1 Definition

9.3.1.1 Nomenclature The surface clutter return, given by equation (9.2), is proportional to the dimensionless parameter σ^0 known as the "clutter cross-section per unit area" (also called "surface scattering coefficient" and "normalized clutter cross-section"). In the older literature, the symbols σ_0 and γ are used instead of σ^0.

With reference to Figure 9–7, it is customary to specify the coefficient σ^0 as $\sigma^0_{pq}(\theta_i, \phi_i; \theta_s, \phi)$, where p, q are the polarization states of the transmitter and receiver antennas, respectively; θ_i, θ_s, are the angles[1] of incidence and scattering respectively measured from the normal N_1, to the mean surface (X–Y plane) of the clutter cell at C; ϕ_i is the azimuthal angle of the clutter cell subtended at the transmitter projection A′ and measured from the X-axis in the X–Y plane; and ϕ is the azimuthal angle of the receiver projection B′ subtended at the clutter cell and measured from the axis A′C. The angles θ_i, θ_s, of Figure 9–7 are related to the grazing angles ψ_A, ψ_B of Figure 9–1 by $\theta_i = (\pi/2) - \psi_A$ and $\theta_s = (\pi/2) - \psi_B$. For linear polarization of the transmitter and receiver, p and q may be either "horizontal" (H) or "vertical" (V) polarization where (H) and (V) denote that the electrical field is parallel to a plane horizontal or vertical, respectively, to the ground.

For homogeneous surface, σ^0 is independent of the azimuthal angle ϕ_i. For such a case, the coefficient σ^0 may be designated by $\sigma^0_{pq}(\theta_i, \theta_s, \phi)$ without loss of generality. In most experiments, a homogeneous surface is assumed even though the surface profile and theme may not be uniform. Nonuniformities are then absorbed into the statistical variations of the measured value of $\sigma^0_{pq}(\theta_i, \theta_s, \phi)$.

1 θ_i, θ_s should not be confused with the azimuthal beamwidth θ_A, θ_B defined in Section 9.2.1.

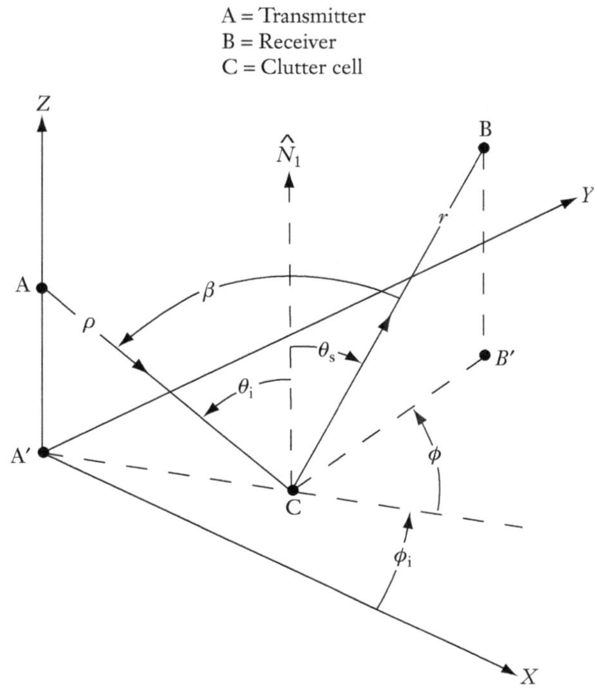

Figure 9–7 Geometry of bistatic surface scattering coefficient.

9.3.1.2 Experimental Definition A definition of $\sigma^0_{pq}(\theta_i, \theta_s, \phi)$, which is used in computing its numerical values from measurement data, may be derived from equation (9.2).

Defining the peak received clutter power $P_C \equiv E_C/K\delta t$, dividing both sides of equation (9.2) by $K\delta t$, and letting $G_A \equiv G_{A_0} f_A$ and $G_B \equiv G_{B_0} f_B$, where f_A, f_B are the power antenna pattern shapes of the transmitter and receiver, respectively, then equation (9.2) reduces to

$$P_{c,q} = \frac{P_{t,p} L_\rho L_r G_{A_0} G_{B_0} \lambda^2}{(4\pi)^3} \int_{A_C} \frac{f_A f_B}{\rho^2 r^2} \sigma^0_{pq} \, dA \tag{9.14}$$

where the subscripts p, q indicate that the transmitted power is of polarization state p and that the received power is of polarization state q. The scattering coefficient σ^0_{pq} is computed from measurement data by treating σ^0_{pq} in equation (9.14) as a constant [12, 13]. Accordingly,

$$\sigma^0_{pq}(\theta_i, \theta_s, \phi) = \frac{P_{c,q}}{P_{t,p}} \frac{1}{K_1 I} \tag{9.15}$$

where

$$K_1 = L_\rho L_r G_{A_0} G_{B_0} \lambda^2 / (4\pi)^3$$

$$I = \int_{A_C} \frac{f_A f_B}{\rho^2 r^2} \sigma^0_{pq} \, dA$$

The scattering coefficient σ^0_{pq} given by equation (9.15) is determined by measuring the ratio $P_{c,q}/P_{t,p}$ and by computing the parameters K_1, I from measured antenna gain patterns and measured propagation losses.

The measured value of $P_{c,q}$ will generally be different for each sampled clutter patch because of random variations in the surface profile from patch to patch. Therefore, σ^0_{pq} is best characterized for a given scattering geometry (θ_i, θ_s, ϕ) by repeated measurements for different clutter patches over an ensemble of several clutter patches, each clutter patch being of cell area A_C. The resulting measurements then can be considered as sample points of the statistical distribution of σ^0_{pq}. From such sample points, the cumulative probability distribution (the probability that σ^0_{pq} is less than a specified value) may be determined. The cumulative probability distribution is useful in estimating clutter false-alarm rates and is sufficient for determining the mean, standard deviation, and median values of σ^0_{pq}. The correlation of σ^0_{pq} with cell area requires additional measurements but such a correlation is useful in determining the homogeneity of the theme and in determining whether the scattering coefficient is predominantly coherent or incoherent (see the following discussion and Section 9.3.2).

In equation (9.14), the received power $P_{c,q}$ is proportional to the sum of scattered powers from elemental areas of the cell area A_C rather than to the magnitude-squared value of the vector sum of the scattered field amplitudes. Therefore, in equation (9.15) one possible interpretation of σ^0_{pq} might be that it is the expected value of the instantaneous scattering coefficient of an elemental scattering element dA averaged over an ensemble of scatterers of random phase within the cell area A_C. This interpretation would suggest that σ^0_{pq} is incoherent in nature and independent of the cell area when the number of scattering elements becomes sufficiently large. In reality, the phases of elemental scatterers from a rough surface are not entirely random but are correlated within some space interval. Therefore, it is useful to express the expected value of the scattering coefficient as the sum of a "coherent" scattering coefficient and an "incoherent" scattering coefficient, the former being proportional to the product of the expected values of the scattered electric field times the expected value of its phase conjugate whereas the latter is proportional to the variance of the scattered electric field [14]. The coherent scattering coefficient is generally a function of the cell area. For example, see equations (9.26) and (9.32) of Section 9.3.2 for scattering in the specular direction. The experimental definition of σ^0_{pq} given by equation (9.15) is useful when incoherent or "diffuse scattering" (rays reflected in all directions) dominates and the clutter cross-section per unit area is independent of cell area.

In equation (9.15), the clutter cell area A_C is generally an extended surface; that is, the cell area generally subtends many Fresnel zones at the transmitter and receiver even though the transmitter and receiver may subtend much less than a Fresnel zone at the cell area. In other words, even though the cell area is in the Fraunhofer far-field of the transmitter and receiver antennas, the transmitter or receiver or both are in the near-field of the clutter cell area.

The extended nature of the clutter surface is illustrated in Figure 9–8 for resolution time- and beamwidth-limited cell areas and compared with a nonextended surface. The number of Fresnel zones $F_{C/A}$ subtended at a field point A by an aperture located at C is simply the path length difference (expressed in the number of half-wavelengths $\lambda/2$) of wavefronts from different points of the aperture and measured at A. In Figure 9–8, it is assumed that the clutter cell at point C is in the far-field of the transmitter and receiver antennas at A and B, respectively. For purposes of illustration, consider ranges $\rho = r = 1$ km, $\lambda = 0.1$ m, bandwidth $\Delta f = 1/\delta t = 3 \times 10^7$ Hz, and antenna aperture $D_A = D_B = 1$ m. For this example, the resolution time- and beamwidth-limited cell area subtend $F_{C/A} = F_{C/B} = 100$ Fresnel zones. Conversely, the number of Fresnel zones subtended by the antenna aperture at the clutter cell is $F_{A/C} \approx D_A^2/\lambda\rho = 1/100$.

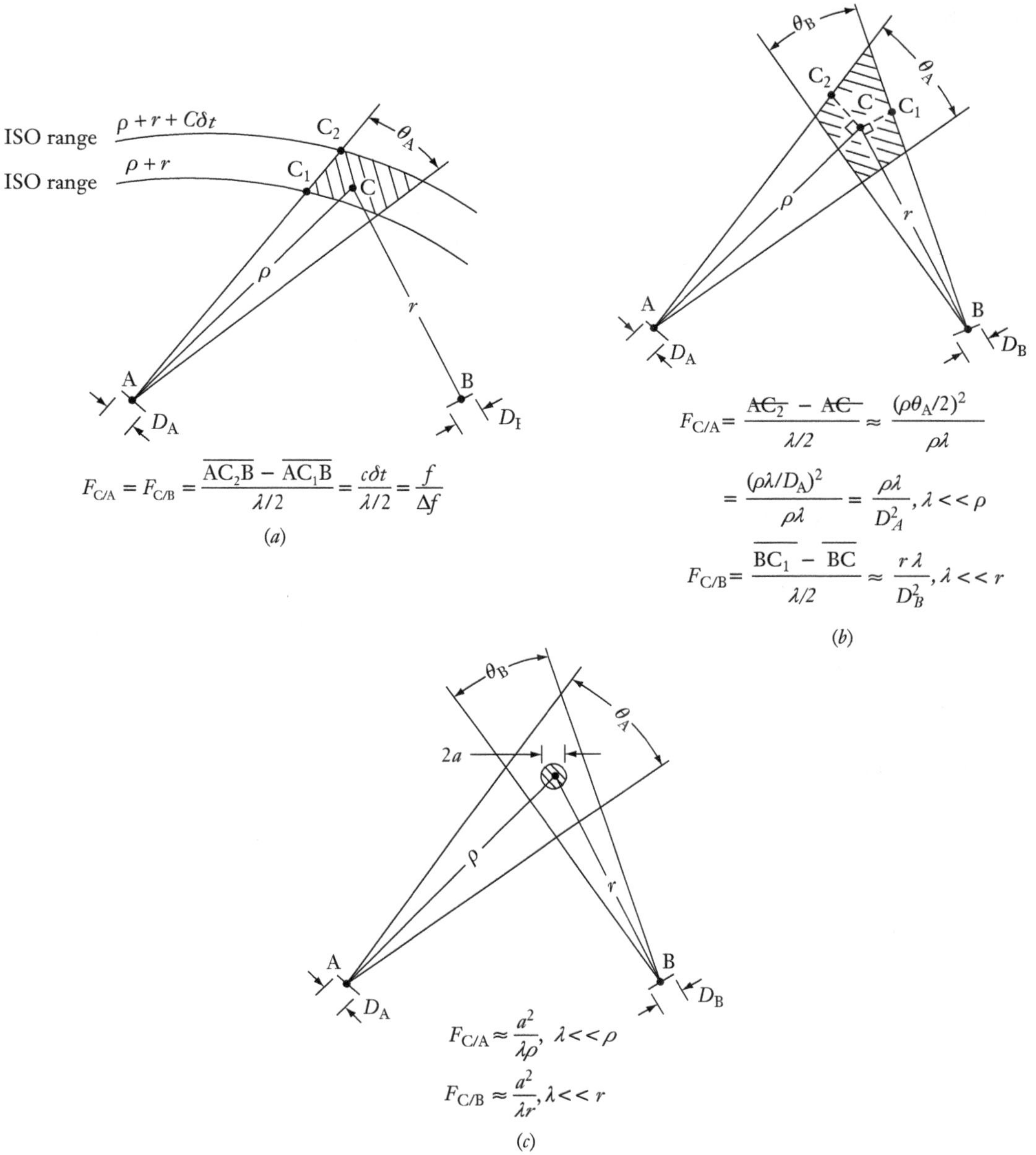

Figure 9–8 Extended nature of clutter cell area. (*a*) Resolution time-limited; (*b*) beamwidth-limited; (*c*) nonextended surface.

For the beamwidth-limited case, $F_{C/A}$ or $F_{C/B}$ *increases* as the range ρ or r is increased. In contrast, the number of Fresnel zones subtended by a nonextended surface decreases as the range ρ or r is increased.

If the range ρ is equal to the Rayleigh range $R = D_A^2/2\lambda$ (the range at which the far-field of the transmitter antenna begins), then the beamwidth-limited cell area subtends only $F_{C/A} = 1/2$ Fresnel zone, which is approximately the minimum number of Fresnel zones that can be subtended by an extended surface.

9.3.1.3 Theoretical Definition For the purposes of theoretical modeling, the RCS per unit area has been defined [15, 16] by first considering the RCS σ_{pq}, of a nonextended surface which is illuminated by a plane wave. The scattering coefficient σ^0_{pq} is then derived by dividing σ_{pq} by the area of the illuminated surface.

With reference to Figure 9–7, the surface at point C is assumed to be in the Fraunhofer far-field of the transmitter and receiver antennas whose polarization states are p and q respectively. The RCS $\sigma^0_{pq}(\theta_i, \theta_s, \phi)$ of dimension (m^2) is defined as [16]

$$\sigma^0_{pq}(\theta_i, \theta_s, \phi) = \lim_{r\to\infty} 4\pi r^2 \frac{\left|\vec{E_{sq}}(r, \theta_i, \theta_s, \phi)\right|^2}{|\vec{E}_{ip}(\theta_i)|^2} \tag{9.16}$$

where

$E_{ip}(\theta_i) = (\vec{E}_0/R_1)\exp(jkR_1) \approx \vec{E}_0 \exp(jk\rho) =$ electric field of incident plane wave front at C with wave number $k = 2\pi/\lambda$

$R_1 = [\rho^2 + x^2_0 + y^2_0]^{1/2}$ = distance of source to a point $Q(x_0, y_0)$ of illuminated surface.

$\vec{E_{sq}}(r, \theta_i, \theta_s, \phi) = (|E_{ip}|/r)\exp(jkr)\vec{U}(r, \theta_i, \theta_s, \phi)$ = electric field of scattered spherical wave measured at B.

$\vec{U}(r, \theta_i, \theta_s, \phi)$ = electric field at B normalized to the spherical wave amplitude $(|E_{ip}|/r)\exp(jkr)$ at B.

It follows from equation (9.16) that

$$\sigma^0_{pq}(\theta_i, \theta_s, \phi) = 4\pi \lim_{r\to\infty} |\vec{U}(r, \theta_i, \theta_s, \phi)|^2 \tag{9.17}$$

The clutter cross section per unit area, σ^0_{pq}, is then defined as [14]

$$\sigma_{pq}(\theta_i, \theta_s, \phi) = \left\langle \sigma_{pq}(\theta_i, \theta_s, \phi) \right\rangle / A_C \tag{9.18}$$

where A_C is the cell area of the illuminated surface and the brackets ⟨ ⟩ refer to some form of spatial or ensemble average. Substituting equation (9.16) into equation (9.18),

$$\sigma^0_{pq}(\theta_i, \theta_s, \phi) = \frac{1}{A_C} \lim_{r\to\infty} 4\pi r^2 \left\langle \left| \frac{E_{sq}}{E_{ip}} \right| \right\rangle^2 \tag{9.19}$$

The scattering amplitude $\vec{U}$ is usually expressed by a diffraction integral of the tangent electric field $\vec{E}_t(Q)$, which propagates from points Q on the illuminated surface to the observation point P which in this case is the receiver at B. Different forms of the diffraction integral for various conditions are given in Appendix 9–A.

The evaluation of equation (9.19) in the limit $r \to \infty$ was originally intended to ensure not only that the transmitted and received electric fields have a $1/r$ dependence (far-radiation zone requirement) but that the received field was in the Fraunhofer far-field of the scattering surface. For a nonextended surface, both of these conditions are satisfied as $r \to \infty$. However, for an extended surface, only the far-radiation zone requirement is satisfied. Furthermore, for an extended surface, the condition that the incident wave is a plane wave (the transmitter is in the Fraunhofer far-field of the surface) is also no longer satisfied.

Depending on the type of averaging which is performed, the theoretically defined scattering coefficient σ^0_{pq} given by equation (9.19) may or may not have characteristics different from those of the experimentally defined coefficient σ^0_{pq} given by equation (9.15). Equation (9.19) is based on the resultant of scattered electric fields, whereas equation (9.15) is based on the resultant of scattered powers. Equation (9.19) is defined for an incident plane wave and for a receiver at infinity whereas equation (9.15) is defined for an incident spherical wave (whose power gain may be nonisotropic) and for a receiver at a finite range. Equation (9.19) was originally applied to a nonextended surface, whereas equation (9.15) is more applicable to diffuse scattering from an extended rough surface.

The clutter cross-section per unit area is closely related to the Bidirectional Reflectance-Distribution Function (BRDF) of optics and radiometry where BRDF is defined for an elemental surface area as [17] the ratio of reflected radiance in the direction toward the viewer to the irradiance in the direction toward a portion of the source. From equation (9.15) or equation (9.19), it can be shown that

$$\sigma^0 = 4\pi \times \text{BRDF} \times \cos\theta_S \tag{9.20}$$

9.3.2 Scattering Models

The surface clutter cross-section per unit area can be found if the scattered electromagnetic fields can be determined. Considerable efforts have been made by many researchers in analytically determining scattering characteristics based on various assumptions about the characteristics of the surface. However, the difficulties in obtaining meaningful results and in determining realistic models are formidable.

Appendix 9–A surveys the theoretical foundations of electromagnetic scattering based on aperture diffraction theory. Section 9.3.3.1 gives references to several reviews of scattering from extended surfaces. In this section, some models of bistatic scattering from extended surfaces are briefly reviewed.

9.3.2.1 Theoretical Foundations Electromagnetic wave scattering from an extended surface may be treated as a problem in aperture diffraction theory if the reflected field at a clear aperture plane of the surface can be determined as a function of the source field which is incident on the surface. The foundation of aperture diffraction theory starts with the vector equivalent Helmoltz–Kirchhoff integral given by equation (A-l) of Appendix 9–A. The source point P_0 corresponds to the transmitter location. The observation point P corresponds to the receiver location. The aperture points Q correspond to the location of clutter cells on the ground. The distance r corresponds to the range from the transmitter to the clutter cell. The distance s corresponds to the range from the clutter cell to the receiver. The field at an observation point P is determined by specifying the electric field $\vec{E}(Q)$ at all points Q of a closed surface $\sum = S_1 + S_2 + S_3$, which encloses the observation point P and which includes points Q on a clear aperture plane S_1 of the scattering surface as shown in Figure A–1 of Appendix 9–A. For the vast majority of problems of interest, an exact evaluation of the Helmholtz–Kirchhoff integral is not feasible because the field $E(\vec{Q})$ is usually not readily expressed in terms of the incident or source field. As a result, a variety of approximation techniques has evolved.

In determining which approximation technique is applicable, it is useful to classify surfaces as "rough" or "smooth," Consider a plane wave incident at a mean angle of incidence θ_i on the mean plane of a surface with height deviations $h(x_0, y_0)$ from the mean plane. Neglecting shadowing effects, the corresponding phase deviation $\beta(x_0, y_0)$ of a reflected wavefront in the mean specularly reflected direction from that of a plane wavefront is

$$\beta(x_0, y_0) = 2(2\pi/\lambda)h(x_0, y_0)\cos\theta_i \tag{9.21}$$

A surface is classified as being "rough" or "smooth" according to the Rayleigh roughness criterion:

$$\begin{aligned} &\beta_{rms} < \pi/2 \text{ rad, smooth} && \text{(a)} \\ &\beta_{rms} < \pi/2 \text{ rad, Rayleigh roughness criterion} && \text{(b)} \\ &\beta_{rms} < \pi/2 \text{ rad, rough} && \text{(c)} \end{aligned} \tag{9.22}$$

where

$$\beta_{rms} = (4\pi/\lambda)\, h_{rms} \cos\,\theta_i = \text{root-mean-squared phase deviation of the reflected wavefront.}$$

9.3.2.2 Smooth Surfaces For smooth surfaces, a well-known approximation technique is the physical optics approximation given by equation (A-7) and conditions A, E, F, H of Appendix A. With reference to Figure A–1, this assumption assumes that (1) the fields at surfaces S_2 and S_3 can be neglected; (2) the tangential reflected electric field $\vec{E_t}(Q)$ at points Q of the clear aperture plane S_1, of the scattering surface is uniformly polarized; (3) the observation point P is located at a distance $\overline{PQ}$, which is large compared to the wavelength and length D of the clear aperture (far radiation zone approximation which is not to be confused with the Fraunhofer far-field diffraction condition); and (4) the tangential electric field $\vec{E_t}(Q)$ can be determined as a function of the incident electric field.

Conditions (2) and (4) are approximately realized for a smooth surface and for an rms surface slope which is small compared to the mean grazing angle of incidence. Shadowing, depolarization, and multiple scattering effects are minimized for these conditions. If the observation point P is confined to directions which are close to the mean specular reflection direction, then the clear aperture plane S_1 at the surface may be chosen orthogonal to the mean specular reflection direction, the phase deviation $\beta(x_0, y_0)$ of the reflected wavefront at S_1, is given by equation (9.22) and Condition (1) is satisfied.

For these conditions, the field $\vec{E}_{sq}$, at an observation point $P(x, y, r)$ where $x = y = 0$ for the specular reflection direction, is given by the Huygens–Fresnel integral of equation (A-9) of Appendix 9–A as

$$\begin{aligned} \vec{E}_{sq}(r, \theta_i, \theta_s, \phi) &= -(j/\lambda r)\cos(\theta_s - \theta_i)|E_{ip}|\hat{t}_0\, r \\ &\cdot \iint_{A_C \cos\theta_i} \exp\,[j\beta(x_0, y_0)\,\exp\,[jk(R_1 + R_2]dx_0\,dy_0 \end{aligned} \tag{9.23}$$

where

$R_1 = [\rho^2 + x^2_0 + y^2_0]^{1/2}$, $R_2 = [r^2 + (x-x_0)^2 + (y-y_0)]^{1/2}$
Γ = Fresnel amplitude reflection coefficient = ratio of reflected to incident electric fields in the specular direction if the surface were perfectly flat and of infinite extent
$\hat{t}_0$ = unit vector tangent to the clear aperture S_1.

The scattering coefficient σ^0 defined by equation (9.19) requires that $4\pi r^2 |E_{sq}|^2$ be evaluated in the limit $r \rightarrow \infty$, where E_{sq} is given by equation (9.23). The results which are obtained depend very much upon whether cell area A is that of a nonextended surface or of an extended surface. The former case corresponds to the Fraunhofer far-field solution of equation (9.23), whereas the latter case corresponds to the Fresnel near-field solution of equation (9.23), which includes a limiting case when the source field can be replaced by its image field (the so-called image solution). Since clutter surfaces are generally extended surfaces, the far-field case can be realized by a clutter surface only if a mask of finite aperture is placed in front of the clutter surface at C. The near-field case is the more appropriate theoretical model for comparison with measurements of scatter from extended surfaces. However, for zero mean stochastic processes, the effect of surface roughness on the expected value of the coherent component of σ^0 is independent of whether the receiver is in the far-field or near-field of the scattering surface [18].

9.3.2.2.1 Far-Field Scatter from Non-Extended Surfaces Consider now the case when the aperture S_1 is limited by a mask of finite aperture placed at S_1. For such a case, when $r \rightarrow \infty$ the integral of equation (9.23) reduces to a Fraunhofer integral, of the form given by equation (A-11) of Appendix A. For specular direction ($\theta_i = \theta_s$ and $x = y = 0$) for an incident plane wavefront, the far-field on-axis intensity is given by

$$\begin{aligned} |E_{sq}(r,\theta_i,\theta_i,0)|^2 &= \frac{|E_{ip}|^2}{(r\lambda)^2}|\Gamma|^2| \iint_{A_C\cos\theta_i} \exp[j\beta(x_0,y_0)]\,dx_0\,dy_0|^2 \\ &= \left[\frac{|E_{ip}|}{r\lambda}A_C\cos\theta_i\right]^2 [|\Gamma|^2 S] \end{aligned} \tag{9.24}$$

where

$$S \equiv \left|(1/A_C\cos\theta_i)A_C \iint_{\cos\theta_i} \exp[j\beta(x_0,y_0)]\,dx_0\,dy_0\right|^2$$

In optics, S is known as the Strehl intensity [19].

The Strehl intensity is plotted in Figure 9–9 (after Reference [19]) for various phase distributions $\beta(x_0, y_0)$ as a function of the normalized rms phase aberration $\beta_{ms}/2\pi$. Please note that the Case II Formula (Marachel approximation) is a poor approximation when the condition $\beta_{ms} << 1$ is not satisfied. Cases I and III correspond to 100% and 0% phase correlation, respectively, between any two points (x_0, y_0) and $(x_0 + \Delta x_0, y_0 + \Delta y_0)$ of the aperture. For a random spatial distribution (x_0, y_0) with a Gaussian probability density of variance $\langle\beta^2\rangle$, S is given by [19]

$$S \equiv \exp\left(-\langle\beta^2\rangle\right) = \exp(-\beta^2_{rms}) \tag{9.25}$$

which is plotted as Case III in Figure 9–9.

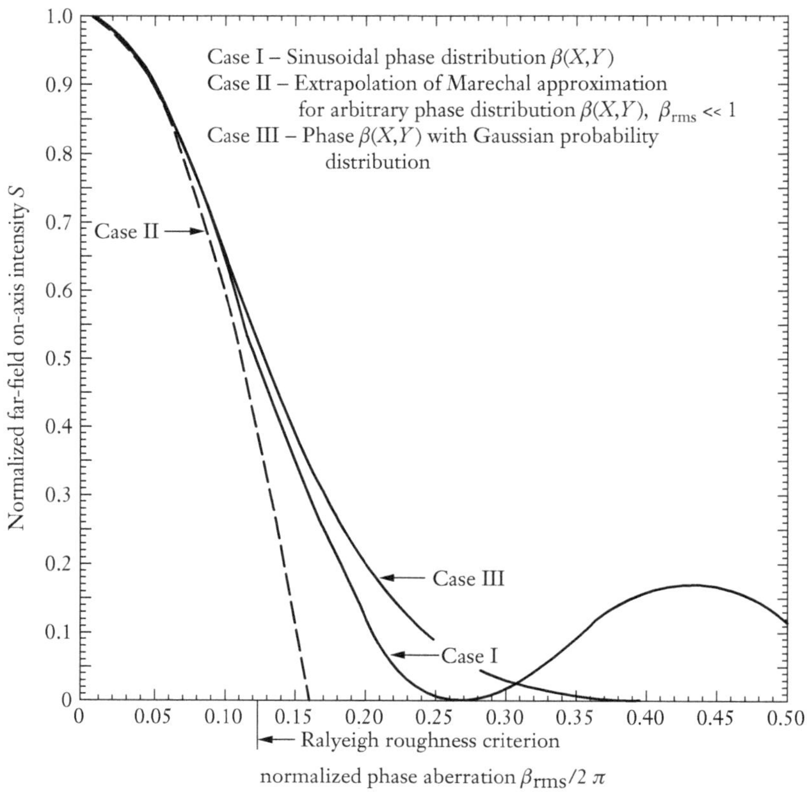

Figure 9–9 Far-field Strehl intensity for uniform intensity and nonuniform phase $\beta(x, y)$ at near field aperture.

In equation (9.24), the far-field on-axis intensity is the product of two factors. The first factor, $[|E_{ip}|A_C \cos\theta_i/r\lambda]^2$, is the far-field on-axis intensity for a perfectly flat surface of perfect reflectivity which radiates as an antenna of aperture area $A_C \cos\theta_i$ with radiated peak intensity $(P_t G_0/4r^2)$ where $P_t = |E_{ip}|^2 A_C \cos\theta_i$ = total radiated power and $G_0 = 4\pi A_C \cos\theta_i/\lambda^2$ = peak antenna gain. The second factor, $|\Gamma|^2 S$, is the fractional amount that the far-field on-axis intensity is reduced by the reflectivity and roughness of the surface. Substituting equation (9.24) into equation (9.19), the far-field scattering coefficient in the specular reflection direction is given by

$$\sigma^0_{pq}(\theta_i, \theta_i, 0) = [4\pi A_C \cos^2\theta_i/\lambda^2]\,[|\Gamma|^2 S], \text{ nonextended surface} \tag{9.26}$$

Coherent scatter from a nonextended surface is therefore characterized by a scattering coefficient which is proportional to (rather than independent of) the cell area A_C. This result will be compared with equation (9.32) for coherent scatter from an extended surface.

The far-field scattering coefficient, in directions near the specular direction, is given by

$$\sigma^0_{pq}(\theta_i, \theta_s, \phi) = [\sigma^0_{pq}(\theta_i, \theta_s, \phi)]_{\text{coherent}} + [\sigma^0_{pq}(\theta_i, \theta_s, \phi)]_{\text{incoherent}} \tag{9.27}$$

where $[\sigma^0]_{\text{coherent}}$ and $[\sigma^0]_{\text{incoherent}}$ are coherent and incoherent scattering coefficients. The coherent scattering coefficient is simply given by

$$[\sigma^0_{pq}(\theta_i, \theta_s, \phi)]_{\text{coherent}} = \sigma^0_{pq}(\theta_i, \theta_i, 0) \times (G/G_0) \tag{9.28}$$

where (G/G_0) is the far-field radiation pattern, normalized to peak gain, of an antenna with aperture area A_C and with uniform intensity and phase. The incoherent scattering coefficient is a function of the autocovariance (coherence or characteristic half-period) length of the surface height variations rather than the entire aperture width of the cell area. The shorter the autocovariance length, the greater is the amount of diffuse scatter at large angles from the specular reflection direction. The fraction of the coherent and incoherent energies within a given solid angle centered about the specular direction are each given by [19] for rectangular and circular apertures.

9.3.2.2.2 Near-Field Scatter From Extended Surfaces Equations (9.24)–(9.18) are applicable to the solution of equation (9.23) for the case corresponding to a transmitter and receiver in the Fraunhofer far-field of a nonextended surface. Now consider the case when the transmitter and receiver are in the Fresnel near-field of a smooth extended surface which is in the far-field of the transmitter and receiver. Equation (9.23) reduces, in the specular direction, to the image solution of a point source above an infinite plane [18]. Introducing a near-field roughness factor W, the image solution in the specularly reflected direction is given by

$$\begin{aligned} P_{\mathrm{c,ql}}/(G_{\beta}\lambda^2/4\pi) &= |E_{\mathrm{sq}}(r,\theta_i,\theta_i,0)|^2 \\ &= \frac{P_{\mathrm{t,p}}L_{\mathrm{p}}L_{\mathrm{r}}G_{\mathrm{A}}}{4\pi} \cdot \frac{|\Gamma|^2 W}{(\rho+r)^2} \end{aligned} \tag{9.29}$$

where $P_{\mathrm{c,q}}$ = received power defined by equation (9.14), $G_B\lambda^2/4\pi$ = receiver aperture area defined by equation (9.14), $P_{\mathrm{t,p}} L_\rho L_{\mathrm{r}} G_A/4\pi$ = transmitter propagation source function defined by equation (9.14), and

$$W = \left| \frac{\int\limits_{A_C}\int\limits_{\cos\theta_i} d_{x_0} d_{y_0} \exp[j\beta(x_0,y_0)] \exp\left[-jk\left(\frac{1}{r}+\frac{1}{s}\right)(x_0^2+y_0^2)/2\right]}{\int\limits_{A_C}\int\limits_{\cos\theta_i} d_{x_0} d_{y_0} \exp\left[-jk\left(\frac{1}{r}+\frac{1}{s}\right)(x_0^2+y_0^2)/2\right]} \right| \tag{9.30}$$

The parameter W is the near-field analog of the far-field Strehl intensity S defined by equation (9.24). For arbitrary but not necessarily random values of $\beta(x_0, |y_0)$ defined on a given patch of the surface, the parameter W is generally not equal to S. However, if the surface profile is a zero mean stochastic process defined for an ensemble of surface patches, each of cell area A_C, then the coherent component of the normalized on-axis intensity is given by [18]

$$\langle W\rangle_{\mathrm{coh}} = \langle S\rangle_{\mathrm{coh}} = \left|\langle \exp[j\beta(x_0,y_0)]\rangle\right|^2 \tag{9.31}$$

where $\langle\ \rangle$ denotes expected value averaged over the ensemble of clutter patches.

Substituting $P_{\mathrm{c,q}}$ given by equation (9.29) into equation (9.15), the scattering coefficient σ^0_{pq} (θ_i, θ_i, 0) corresponding to the near-field image solution of equation (9.23) in the specular direction is given by

$$\sigma^0_{\mathrm{pq}}(\theta_i,\theta_i,0) = \left[\frac{4\pi}{A_C}\frac{\rho^2 r^2}{(\rho+r)^2}\right][|\Gamma|^2 W], \text{ extended surface} \tag{9.32}$$

for the antenna gains approximately constant over the cell area A_C. The factor $[4\pi\rho^2 r^2/A_C(\rho + r)^2]$ may be interpreted as the effective antenna gain of an extended surface illuminated by a point source. Scatter in the specular direction from an extended surface is therefore characterized by

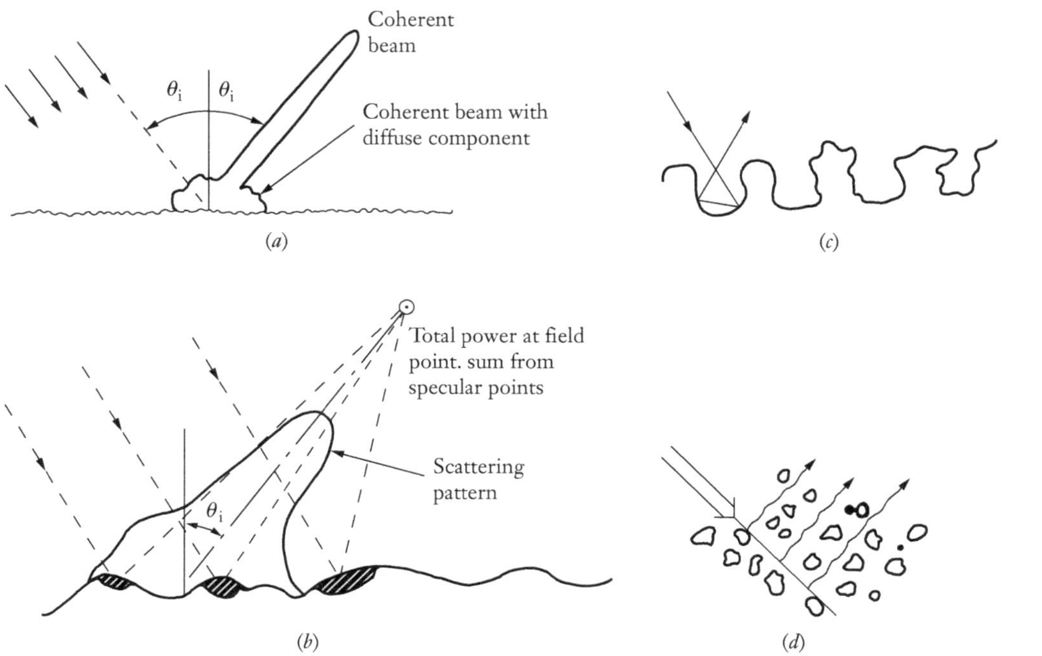

Figure 9–10 Some geometrical scattering models (after Peake and Oliver, 1971).
(*a*) "Slightly rough" surface $h << \lambda$. Scattered beamwidth determined by cell aperture; predominantly coherent.
(*b*) Undulating "rough" surface $h >> \lambda$, slopes << 1. Scattered beam-width ≈ rms surface slope; specular (not coherent).
(*c*) The Lambert type of "rough" surface with large reentrant cavities.
(*d*) The Lommel–Seeliger type of "rough" surface with many independent scatterers.

a scattering coefficient which is a function of the cell area and, unlike that from a nonextended surface [see equation (9.26)], is inversely proportional to the cell area.

A comparison of equation (9.32) with equation (9.26) illustrates the point that the scattering coefficient σ^0 in the specular direction is very much dependent on whether the receiver is in the near field or far-field of the scattering surface. However, for a zero-mean stochastic process, the effect of surface roughness on the expected value of the coherent component of σ^0 is independent of whether the receiver is in the far-field or near-field of the surface.

The radiation pattern of smooth surfaces is illustrated in Figure 9–10*a*. As the surface becomes increasingly rough, the Strehl intensity is reduced and the amount of energy scattered incoherently is increased. At sufficiently large angles from the specular reflection, incoherent scatter predominates. The scattering coefficient is then given primarily by an incoherent scattering coefficient, which is approximately independent of the cell area A_C. Most experiments are designed to measure the incoherent scattering coefficient but occasionally, the coherent scattering coefficient dominates and is mistaken for an incoherent coefficient.

9.3.2.3 Rough Surfaces The physical optics approximation is generally not applicable to rough surfaces. However, one class of rough surfaces for which physical optics is applicable is the case shown in Figure 9–10*b* after Peake and Oliver (W. H. Peake and T. L. Oliver, private communication,

1971) where the surface is gently undulating, that is, the rms slope << 1, as defined in the figure. Each part of this surface is locally smooth and reflects the incident radiation as if it were a tilted flat surface (facet). The total scattered field is the sum of the incoherent reflections from these locally flat areas and the scattered power from a beam centered about the specular reflection direction with a beamwidth approximately equal to the rms slope of the surface. The total power in this incoherent beam is approximately $|\Gamma|^2$ times the power in the incident beam.

Barrick [20] has categorized scattering models of rough surfaces, which do not necessarily have gently undulating slopes, into three classes: geometrical, statistical, and semi-empirical.

The geometrical scattering model employs a surface made up of simple scattering elements arranged randomly over a planar area. Given that the scattering behavior of the individual scattering element is known, the model can be analyzed by multiple scattering techniques. Examples of two such geometrical models are the Lambert rough surface with reentrant cavities of large slopes and the Lommel–Seeliger rough surface with many independent scatters as shown in Figures 9–10*c* and 9–10*d*, respectively. At microwave frequencies, a layer of crushed rock might correspond to a Lambert surface whereas leaves might correspond to a Lommel–Seeliger surface. For both of these surfaces, the scattering is diffuse; that is, there is no strong forward lobe, and the scattering is incoherent and quasi-isotropic.

Some of the types of geometrical scattering models are (a) Twersky surface of randomly spaced, reflecting, conducting hemispheres of equal radii on a perfectly conducting plane; (b) the Peake surface of semi-infinite, lossy, thin dielectric cylinders arranged randomly, but with a predominantly vertical orientation over a flat plane (grassy or wheat type surface); (c) a plane layer of electrically large or small spherical particles; (d) a two-scale rough surface consisting of a fine scale roughness superimposed upon a more coarse-scale roughness; (e) a layer of scatterers with irregular interfaces; (f) a plane dielectric layer with a random dielectric profile; and (g) an inhomogeneous dielectric layer with irregular interfaces. The major problem with the utilization of models of these types is the determination of specific geometrical parameters for correspondence with a real surface.

Statistical models treat the surface height above a reference plane as a random variable and provide results in terms of "averaged" surface properties. Barrick [20] provides a detailed treatment of this type of model. While this type of model provides good insight into the nature of scattering statistics, use of the specific formulas for a real surface requires detailed statistics on the surface parameters, which may be difficult to obtain.

Semi-empirical models contain one or more arbitrary constants which are adjusted to fit empirical data. The model may be based on a geometrical (or statistical) model or may simply be an algebraic expression with one or more arbitrary constants. Examples of semiempirical models are given identically by Barrick [20] and by Peake and Oliver [5] for monostatic scattering and for bistatic scattering (presumably for backscatter in the elevation plane). The most useful semiempirical bistatic models are empirical formulas by principal investigators of the bistatic database, particularly Domville [21–23], who gives not only mean values of the scattering coefficient but also standard deviations. Limitations of semiempirical models are that they offer little insight into the scattering process and the surface properties affecting the scattered fields and they are usually valid only over a relatively narrow range of elevation or azimuthal angles. However, semiempirical models are usually simple and give good agreement with data over a specified range of scattering angles.

9.3.2.4 Scattering Theorems The Reciprocity Theorem and the Bistatic–Monostatic Equivalence Theorems are of interest in modeling bistatic scattering from an extended surface.

An electromagnetic scattering medium, for which the positions of the transmitter and receiver may be interchanged without affecting the received power, is said to be *reciprocal*. Reciprocity holds when the generalized dielectric tensor described by a [3 × 3] matrix is symmetrical about the diagonals of the tensor. Most media are reciprocal. Some exceptions are gyrotropic media such as ferrite materials and the ionosphere. Terrain and sea scattering obey reciprocity. Therefore, the bistatic scattering coefficient satisfies the reciprocity condition.

$$\sigma^0_{pq}(\theta_i, \theta_s, \phi) = \sigma_{pq}(\theta_s, \theta_i, \phi) \tag{9.33}$$

Equation (9.33) is useful in not only reducing by a factor of 2 the amount of bistatic data required to determine the scattered radiation pattern but has also been used to check the validity of experimental data (see [24] and [15]).

There are two variations of the Bistatic–Monostatic Equivalence Theorem which are useful in estimating the bistatic RCS of *nonextended* objects from monostatic measurements. One variation of this theorem by Crispin and Siegal [25] is: for vanishingly small and for sufficiently smooth, perfectly conducting objects, the RCS at a bistatic angle β ($0° < \beta < 180°$) is equal to the monostatic RCS measured on the bistatic angle bisector $\beta/2$. The other variation by Kell [26] is: for small bistatic angle β, the bistatic RCS at a bistatic angle β and frequency f can be approximated by the monostatic RCS measured on the bisector of the bistatic angle and at a frequency $f \cos [\beta/2]$.

In one sense, the latter variation is more relevant to terrain and sea scattering because it applies to both dielectric and finitely conducting bodies as well as perfect conductors. In another sense, the latter variation is more relevant because its limitations to small bistatic angles minimizes the difference between the bistatic scattering coefficient and the monostatic scattering coefficient along the bisector of the bistatic angle. For an *extended* surface, both variations of the theorem, if applicable at all, are applicable to the scattering coefficient and not the RCS because the cell area in the resolution-time limited case increases monotonically with increasing bistatic angle [see equation (9.11)] whereas the scattering coefficient may increase or decrease with increasing bistatic angle. Neither version of the theorem is expected to be relevant in the azimuthal plane to terrain and sea scatter because monostatic scatter is generally independent of azimuthal angle whereas bistatic scatter is a function of azimuthal angle. There is evidence that the Siegal version of the theorem is not applicable for large bistatic angles to terrain and sea scatter in either the azimuthal plane (see the discussion of Figure 9–4 in Section 9.2.3) or the elevation plane (see the discussion of the Childers' analysis in Section 9.3.3.1). Furthermore, none of the semi-empirical formulas given by Barrick (see [20]) for the bistatic scattering coefficient agree with the Bistatic–Monostatic Equivalence Theorem except for small bistatic angles. Therefore, the Bistatic–Monostatic Equivalence Theorem, is generally not valid for target RCS or terrain and sea scatter except for small bistatic angles. However, Willis [27] has successfully used the theorem to interpolate the measured database of Domville (for in-plane scatter from rural land, forest, and urban land) to bistatic angles as large as 135° provided that either the incident or scattering angle was less than 5°.

9.3.3 Database

The bistatic σ^0 database for terrain and sea clutter at microwave and optical frequencies was first published by Weiner in 1980 as [28]. The database included an author index a subject index and an abstract. The full, updated database is reproduced here, with the author and subject data summarized where appropriate.

9.3.3.1 Measurement Programs at Microwave Frequencies The bistatic σ^0 database for terrain and sea clutter at microwave frequencies consists of measurements by primarily nine principal investigators from 10 organizations and reported in the years 1965–2002: (1) Cost and Peake [24, 15]; (2) Pidgeon [30, 31]; (3) Domville [21–23]; (4) Larson and Heimiller [29, 32, 33]; (5) Cornwell and Lancaster [34]; (6) Ewell and Zehner [35, 36]; (7) Ulaby [37], (8) Kochanski [38]; and (9) McLaughlin [39–41]. These measurement programs are summarized in Table 9–1 [(a)–(d)].

Table 9–1(a) Measurement Programs of Bistatic Clutter Cross-Section per Unit Area, σ^0
(a) Year Reported, Organization, Author, Surface Theme

Number	Year Reported	Organization	Reference	Author	Surface Theme
1a	1965	Ohio State University, Antenna Laboratory	[24]	S. T. Cost, W. H. Peake	(1) Sand, smooth (2) Loam
1b	1968		[15]		(3) Foliage, soybean (4) Sand, rough (dry/damp/snow) (5) Loam with stubble (6) Grass (dry/damp/ snow cover
2a	1966	Johns Hopkins University, Applied Physics Lab.	[30]	V. W. Pidgeon	(1) Sea (States 1 ,2, 3)
2b	1967		[31]		(2) Sea (Beaufort wind 5)
3a	1967	G.E.C. Electronics,	[21]	A. R. Domville	(1) Sea, rural land, forest
	1968	Stanmore, England	[22]		
3b	1969		[23]		(2) Forest (3) Semidesert, wet
4a	1977	University of Michigan, Environmental Insitute	[29]	R.W. Larsen, R.C. Heimiller	(1)Flat grass with cement taxiway, dry
4b	1978		[32]	R.C Heimiller, et al.	(2)Tall weeds and scrub trees, dry
4c	1979		[33]	R.W. Larsen, et al.	(3)Snow covered orchard (4)Snow covered weeds and scrub trees
5	1979	Raytheon Co., Equipment Development Labs, Wayland, Mass	[34]	P. E. Cornwell, J. Lancaster	(1) Beach with sand dunes (2) Ocean, sea state 2
6	1982, 1984	Georgia Institute of Technology, EES	[35] [36]	G. W. Ewell, S. P. Zehner	(1) Sea, (0.9, 1.2–1.8 m wave heights)
7	1988	Univ. of Michigan, Dept. of Electrical Engineering and Computer Science	[37]	F. T. Ulaby, et al.	(1) Smooth sand (2) Rough sand (3) Gravel
8	1992	MIT Lincoln Laboratories, Lexinqton, MA	[38]	T. P. Kochanski, et al.	(1) Sea (Sea state 1)
9a	1994	Northeastern University N.U., Boston, MA, University of Massachusetts, Amherst, MA	[39]	D. McLaughlin, et al.	(1) Forested hills
9b	1995		[40]		
9c	2002		[41]		

Table 9–1(b) Measurement Programs of Bistatic Clutter Cross-Section per Unit Area, σ^0
(b) Frequency GHz, Polarization, Scattering Angles, Bistatic Angle

Number	Frequency (GHz)	Polarization*	Scattering Angles (deg)			Bistatic Angle†(deg) β
			ϕ‡	θ_i	θ_s	
1a	10	VV, HH, HV	**0**–145	60–85	60–85	30–170
			0, 180	20–80	0–85	0–165
1b	10	VV, HH, HV	**0–180**	20–85	0–85	0–170
2a	C-band	VV, VH	**180**	87–89.9	0–80	7–89.9
2b	X-band	HH	**180**	82–89	45–78	4–167
3a	X-band	VV	**0**,165,**180**	0–90	0–90	0–180
3b	X-band	HH, VV	**0**,165,**180**	0–90	0–90	0–180
3c	X-band	HH, VV	**0**,165,**180**	0–90	0–90	0–180
4a	1.3, 9.4	HH, HV	**0**–105	50–80	85	75–165
			0–180	70, 75, 80	80	0–160
			0–105	70, 75, 80	70	0–150
4b	1.3, 9.4	HH, HV	**0**–105	50–80	85	5–165
4c	1.3, 9.4	HH, HV	**0**–10	60, 70, 80	60–84	0–164
			0–180	60–80	60–80	0–160
5	9.1	VV	**0**	Unspecified low grazing angles		≈180
6	9.38	VV, HH	95–157	Low grazing angles		23–85
7	35	VV, HH VV, HH VH, HV	**0**–170	66	66	9–132
			0–170	60	60	9–120
			0–90	60	10–80	48–140
8	10	VH, VV	**180**	89.7	50–85	4.7–39.7
9a	S-band	VV, HH	20–70	Low grazing angles		110–160
9b	S-band	VH, HV	20–70	Low grazing angles		110–160
9c	2.71	Fully polarimetric	28–66	Low grazing angles		114–152

*Electric field polarization of transmitter and receiver, respectively. V, vertical; H, horizontal.
† $\beta = \cos^{-1}[\cos\theta_i \cos\theta_s - \sin\theta_i \sin\theta_s \cos\phi]$, where terms are defined on the figure below.
‡ In-plane data are designated by bold type

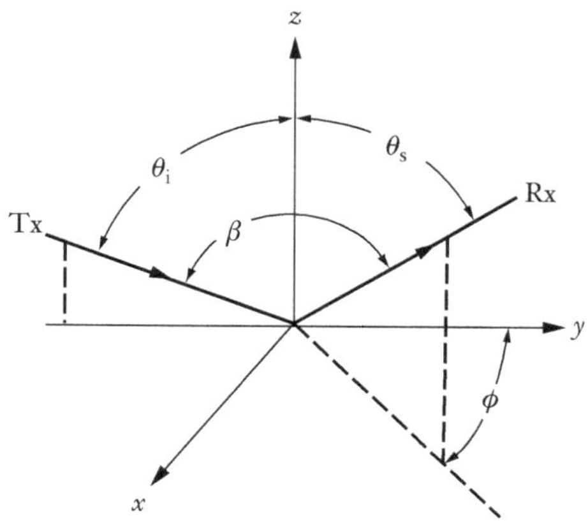

Table 9–1(c) Measurement Programs of Bistatic Clutter Cross-Section per Unit Area, σ^0
(c) Clutter coefficient Statistics, How Measured, Cell Area Limit

Number	Clutter Coefficient Statistics	How Measured	Cell Area Limit
1a 1b	Mean 1–6 independent samples	CW; transmitter and receiver land-based, boom-mounted at 13-ft equidistant ranges from clutter cell; clutter themes in railroad flat-cars; calibrated against measured return from metal sphere; half-power antenna beamwidths of 4°.	Beamwidth
2a	Mean	CW-doppler; land-based transmitter, airborne manual-gain control, receiver with frequency spectrum analyzer	Doppler frequency
2b	Mean	Pulsed, 0.5-μsec pulsewidth; land-based transmitter, 10° beamwidth; airborne receiver, 20° beamwidth	Resolution time
3a, b	Mean	CW; transmitter and receiver in separate aircraft	Beamwidth
3c	Mean	CW; transmitter a few meters above ground; receiver in aircraft with antenna depressed at constant angles	Beamwidth
4	Mean, ≈10–35 independent samples	Pulsed coherent detection, 70- to 90-nsec pulsewidth, 4,000 pulses/sec; airborne transmitter, 12° azimuthal beamwidth; land-based, tower-mounted receiver, 5° beamwidth	Resolution time and receiver beamwidth
5	No reduced data	Pulsed, 1.2-nsec pulsewidth, 25-kHz PRF; helicopter-borne transmitter, 75° beamwidth; land-based, tower-mounted receiver, 1.2° beamwidth; calibrated using average power meter for transmitter power and precision attenuators for received power	Resolution time
6	Standard deviation; bistatic to monostatic ratio	Transmitter, pulsed at constant bearing, land-based, 1° (azimuth) × 3° (elevation) beamwidth; Receiver, land-based, 1.5° beamwidth; calibrated against targets of known RCS	Beamwidth
7	Mean; standard deviation	Swept-frequency polarimeter; transmitter beamwidth 4.2°; receiver beamwidth 1.5°	Transmitter beamwidth
8	Mean; cumulative probability distribution	Breakwater-based transmitter of beamwidth 25°; Airborne receiver of 28 dB gain	Doppler-frequency
9	Mean, median, standard deviation	Ground-based transmitter, 0.9° beamwidth; Ground-based receiver of 6.6° beamwidth	Resolution time and transmitter beamwidth

Table 9–1(d) Measurement Programs of Bistatic Clutter Cross-Section per Unit Area, σ^0
(d) Number of Data Curves, Instrument Error (dB), Author Conclusions

Number	Number of Data Curves	Instrument Error (dB)	Author Conclusions
1a, b	179 in 32 figures	Not stated. Deviation from reciprocity, ±5dB	(1) Pronounced forward lobe in specular direction. (2) Forward lobe less pronounced for rough surfaces. (3) Minimum scatter for azimuthal angle $\phi = 90°$. (4) Backscatter ($\phi = 180°$) larger than for $\phi = 90°$. (5) Forward lobe larger for HH than for VV polarization. (6) Brewster angle effect (VV polarization) observed in specular direction for smooth surfaces. (7) For $\phi = 90°$, scatter larger for VV than HH polarization. (8) HH scatter is small in specular direction but comparable to HH or VV scatter for $\phi = 90°$.
2a	7 in 1 figure	Not stated	(1) Sea state 3 gave 10 dB larger scatter than sea state 1. (2) VH scatter less than VV scatter. VV and HH scatter proportional to sin and $\sin^2$ respectively of transmission depression angle. (3) Scatter relatively independent of receiver depression angles 10°–90° but approximately follows monostatic data for angles 0.2°–3°.
2b	1 in 1 figure	Not stated	(1) Scatter relatively independent of receiver depression angles 12°–55°, transmission depression angles 1°–5°. (2) Scatter increases for transmitter depression angles ≥10°. (3) Scatter follows monostatic data for transmitter depression angles 0°–5°.
3a, b	77 in 4 figures	Not stated	(1) Objective was to obtain a general picture over a wide range of conditions. (2) Measurements first fitted to simple empirical formulae and then combined with other measurements and theory. (3) A computer program is described which, using the data, will calculate clutter spectra for a wide range of conditions.

Table 9–1(d) *(continued)*

Number	Number of Data Curves	Instrument Error (dB)	Author Conclusions
4a, b	[29] 16 in 8 figures [32] 10 in 5 figures	±6.6dB RSS	(1) Scatter larger at 9.4 GHz than at 1.3 GHz for arbitrary ϕ in nonspecular direction. (2) Pronounced forward lobe in specular direction. (3) Minimum scatter for azimuthal angle $\phi = 90°$.
4c	146 in 146 figures	±4 dB RSS (most data)	(1) Snow-covered orchard data compares favorably with forest monostatic data of others. (2) Tall weeds-scrub tree site with and without snow cover displayed little variation in scatter.
5	No reduced data	±2.4 dB, indirect relative to direct signal	(1) Specular reflections, when not masked by beach or ocean, were larger than diffuse reflections. (2) Beach and ocean specular reflections were −18.5 dB and −9.1 dB, respectively, compared to direct signal power.
6a,b	7 in 7 figures	±2 dB	(1) Most clutter amplitude distributions were log normal. (2) Tendency for clutter to be larger for monostatic than bistatic, and for horizontal polarization than vertical.
7	17 in 10 figures	$S/N \geq 10$ dB	(1) For smooth sand, HH clutter is larger than VV in specular direction but smaller than VV and HV outside main lobe. (2) Rough sand clutter and gravel clutter are significantly lower in the specular direction (and their main lobes broader) than for smooth sand clutter.
8	3 in 3 figures	$S/N \geq 30$ dB	(1) Backscatter insensitive over wide range of elevation angles. (2) Mean $\sigma^0 = -45$ dB in good agreement with Domville.
9	15 in 15 figures	12 bit resolution	(1) V-polarized incident fields produce H-polarized weak-scattered fields out of the plane of incidence. (2) H-polarized incident fields produce V- and H-polarized weak scattered fields out of the plane of incidence. (3) Clutter displays prominent polarization angle sensitivity; asymmetric scattering matrix elements; weak but nonzero correlation between co- and cross-polarized components.

The surface themes which have been investigated are nonvegetated terrain (sand, loam, cement taxiway, beach, desert); vegetated farmland (soybean foliage, grass); rural land (tall weeds and scrub trees, tree-tops, tree-covered hills, orchard) forest, urban land, and water (salt, fresh, ocean, sea, beach). Several of the terrain themes are with and without snow cover, damp or dry.

The microwave measurement frequencies are at approximately 1, 3, 5, 9, and 10 GHz in L, C (S, X), and K bands. The antenna polarization states include VV, HH, HV, and VH where the first and second letters denote the polarization of the transmitter and receiver antennas respectively.

The scattering angles range in azimuth from $\phi = 0° - 180°$, in angle of incidence from $\theta_l = 0° - 89.9°$, and in angle of reflectance from $\theta_s = 0° - 89.4°$. The corresponding range of bistatic angles is $\beta = 0°–180°$. Unfortunately, these ranges are for a very limited number of themes and for very limited combinations of ϕ, θ_l, and θ_s. Many of the measurements are in the plane of incidence ($\phi = 0°$ or $180°$, designated "in-plane"). In Table 9–1(b), the angles $\phi = 0°$ and $180°$ are in bold type to emphasize that the measurements are in-plane. For water, the measurements are mostly in the plane of incidence and include backscatter and forward scatter at grazing angles as small as 0.1°. For terrain, there are only three measurement programs with complete azimuthal plane patterns ($\phi = 0°–180°$) for a given θ_l and θ_s (Nos. 1b, 4b, and 4c in Table 9–1(b)). For most out-of-elevation plane measurements of terrain, the grazing angles are larger than 5°. Measurements of scatter in the specular reflection direction exist for both terrain and the sea.

With the exception of data by the more recent measurement programs (Nos. 6–9), only mean values of the scattering coefficient are given because of an insufficient number of independent samples. Programs 6, 7, and 9 also give the standard deviation; 9 gives the median value; and 8 and 9 give the cumulative probability distribution.

Various waveforms and measurement techniques were utilized in obtaining the data. Waveforms included CW, CW-doppler, pulsed coherent and incoherent detection, and CW noise. Locations of the transmitter and receiver included both mounted on land-based stationary booms, land-based transmitter and airborne receiver, airborne transmitter and tower-mounted receiver, and both transmitter and receiver airborne in separate aircraft flying at a constant relative velocity. All the themes that were measured, with the exception of the measurements by Ohio State (No. 1a,b) in which the themes were contained in railroad flat-cars, were naturally occurring themes. Beamwidths of the antennas ranged from 0.9° at S-band (No. 9) to 25° at 1.0 GHz (No. 8).

Several experimental techniques were utilized in defining the cell area A_C. These techniques included the following limitations on cell area: beamwidth, doppler frequency, resolution time, and resolution time and beamwidth.

The number of data curves for each program varied from as few as 3 curves in 3 figures (No. 8) to as many as 179 curves in 32 figures (No. 1a,b).

The instrumentation error, which does not include the statistical spread in the measured data, is not reported by some investigators but when it is reported varies from ±2.4 dB (No. 5) to ±6.6 dB (No. 4a,b).

Table 9–1(d) concludes with a summary of the investigators' conclusions. For the most part, the investigators present their data with few comments and without much effort to relate data curves to various scattering models. However, Domville (No. 3a,b), Ewell and Zehner (No. 6), and Kochanski (No. 9) derive empirical formulas or discuss scattering models which fit the data reasonably well within the ranges of specified scattering angles. Domville's empirical formulae are derived and discussed by Willis [27].

The most substantial bistatic database, in terms of number of independent samples, number of independent samples, number of terrain and sea themes, number of data curves, or statistical properties are by Cost and Peake (No. 1a,b), Pidgeon (No. 2a,b), Domville (No. 3a,b), Larson and Heimiller (No. 4a,b,c), Ewell and Zehner (No. 6), Ulaby (No. 7), and McLaughlin (No. 9).

Selected database curves of the clutter cross section per unit area for the measurement programs of Tables 9–1 to 9–4 are reproduced below with permission.

No. 1a, Cost [24], Ohio State University

Figures 9–11 to 9–14 show the effect of surface roughness in the azimuthal and elevation planes. Figures 9–15 to 9–19 show the effect of antenna polarization in the azimuthal and elevation

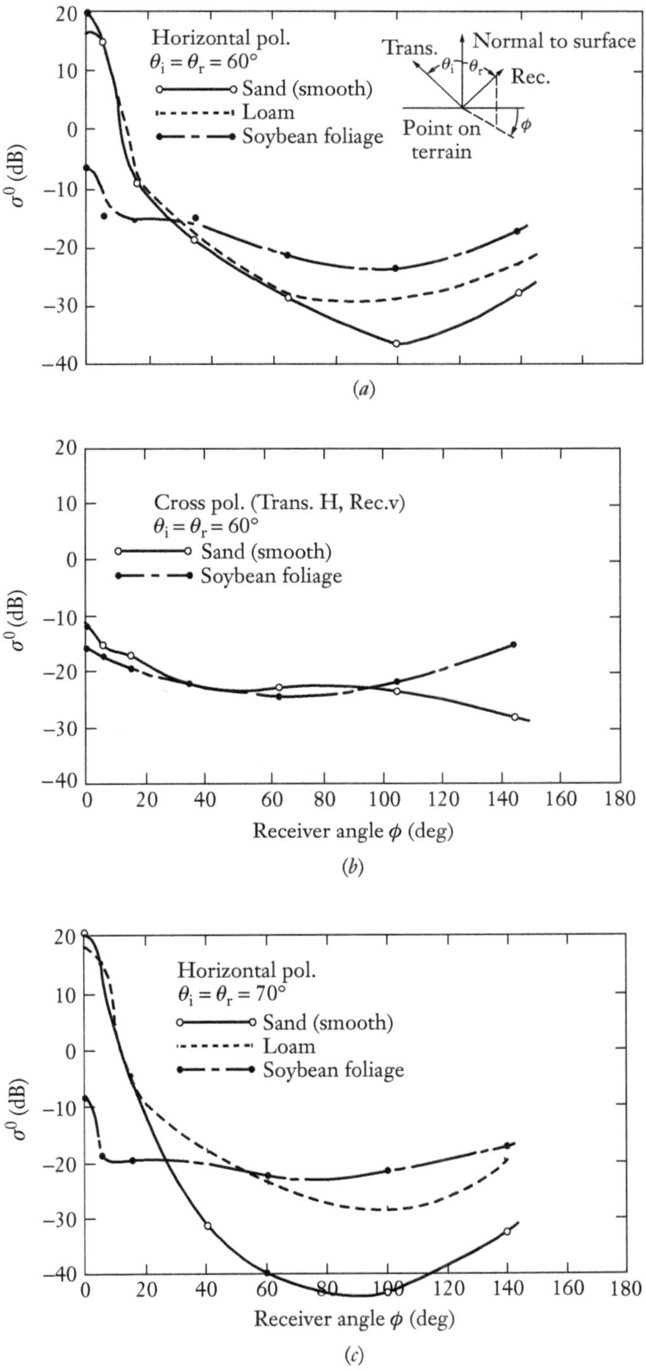

Figure 9–11 Effects of surface roughness on the azimuthal scattering pattern for smooth sand, loam, and soybean foliage (after S. T. Cost, Ohio State University, May 1965 [24]).

(continues)

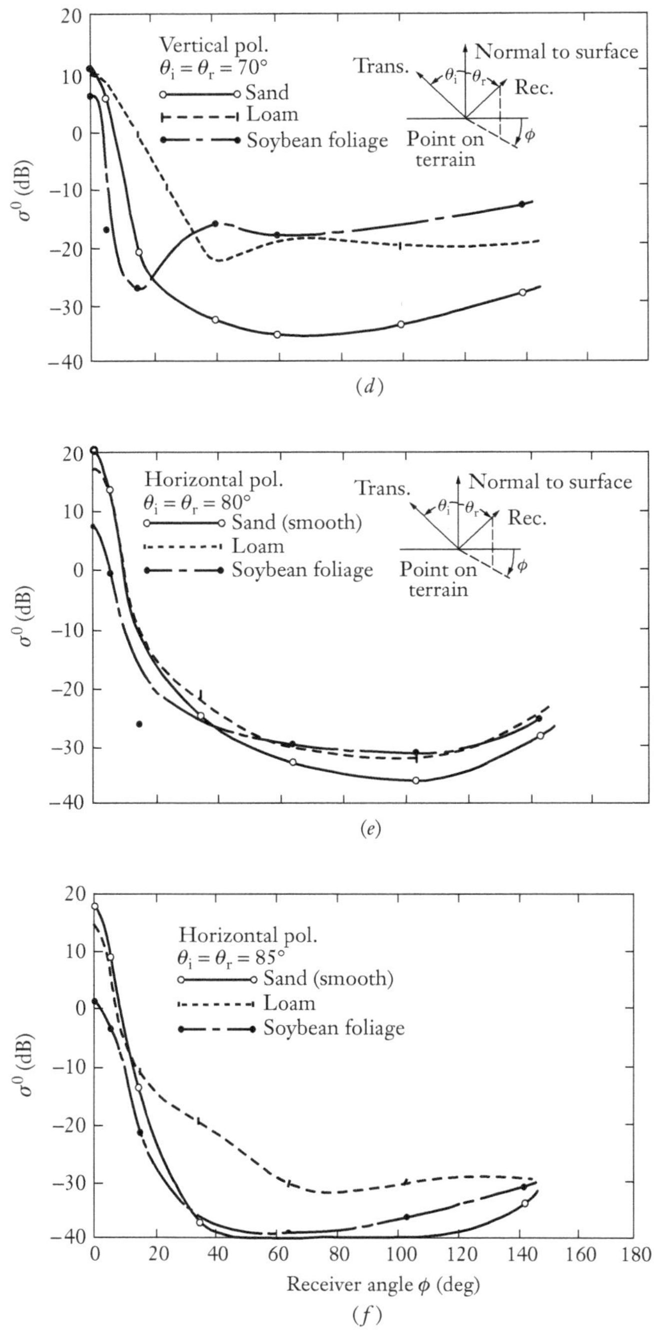

Figure 9–11 Effects of surface roughness on the azimuthal scattering pattern for smooth sand, loam, and soybean foliage (after S. T. Cost, Ohio State University, May 1965 [24]).

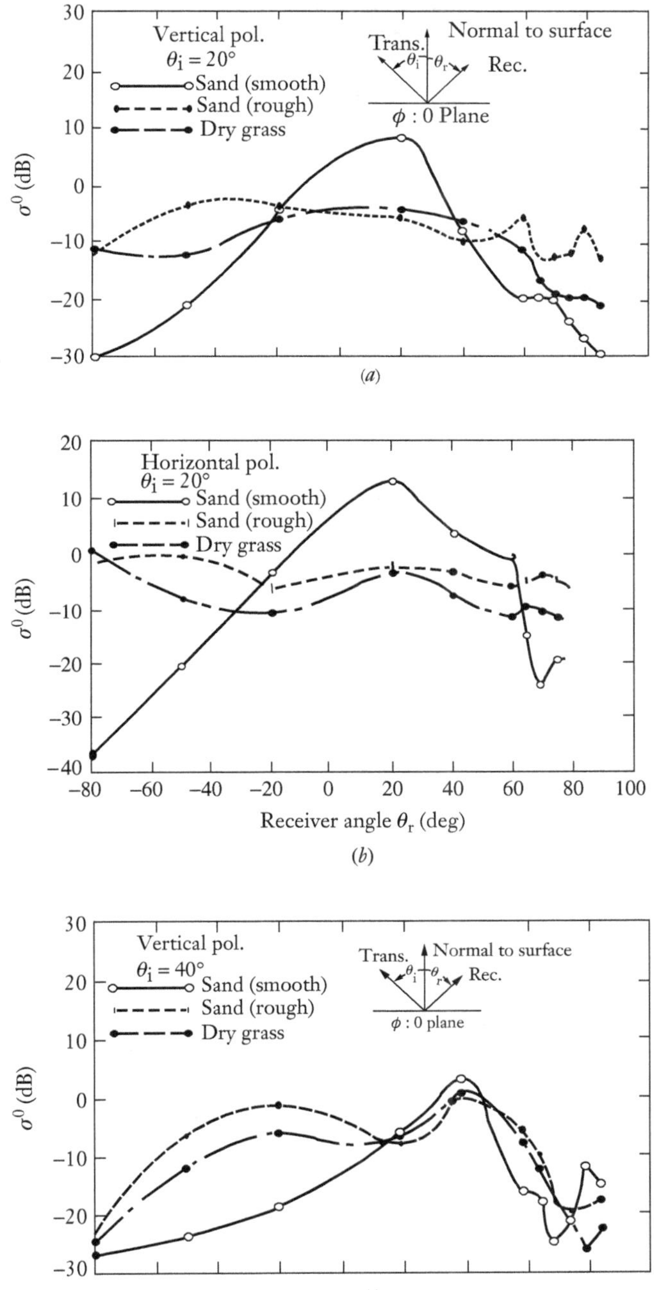

Figure 9–12 Effects of surface roughness in the plane of incidence for smooth sand, dry grass, and rough sand (after S. T. Cost, Ohio State University, May 1965 [24]).

(continues)

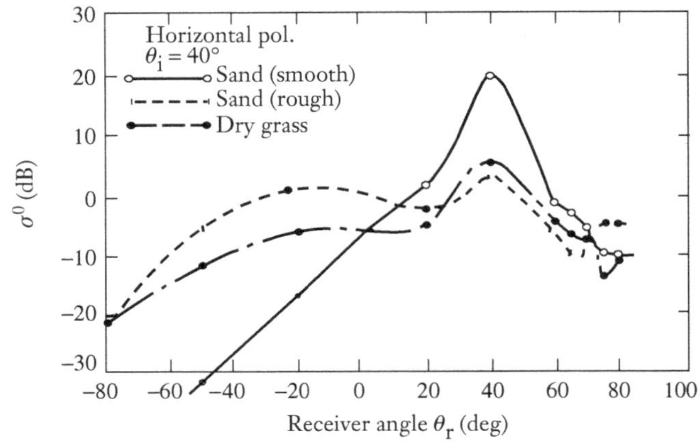

(*d*)

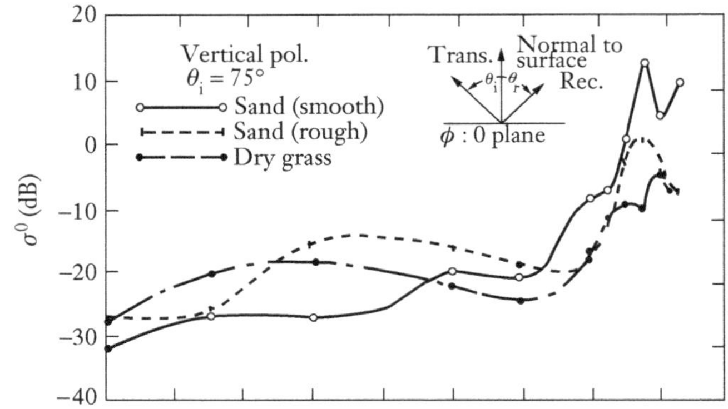

(*e*)

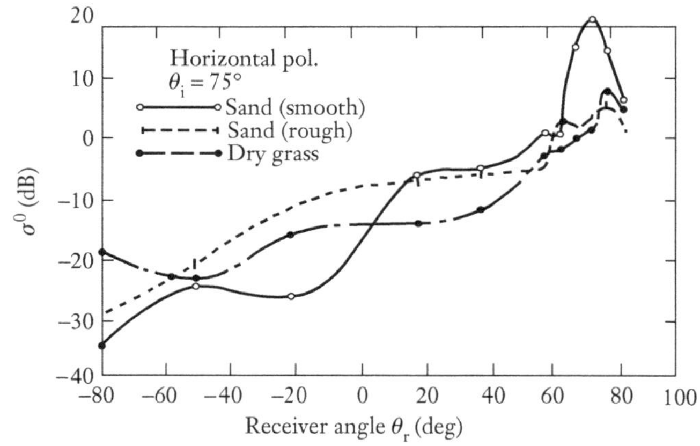

(*f*)

Figure 9–12 Effects of surface roughness in the plane of incidence for smooth sand, dry grass, and rough sand (after S. T. Cost, Ohio State University, May 1965 [24]).

(continues)

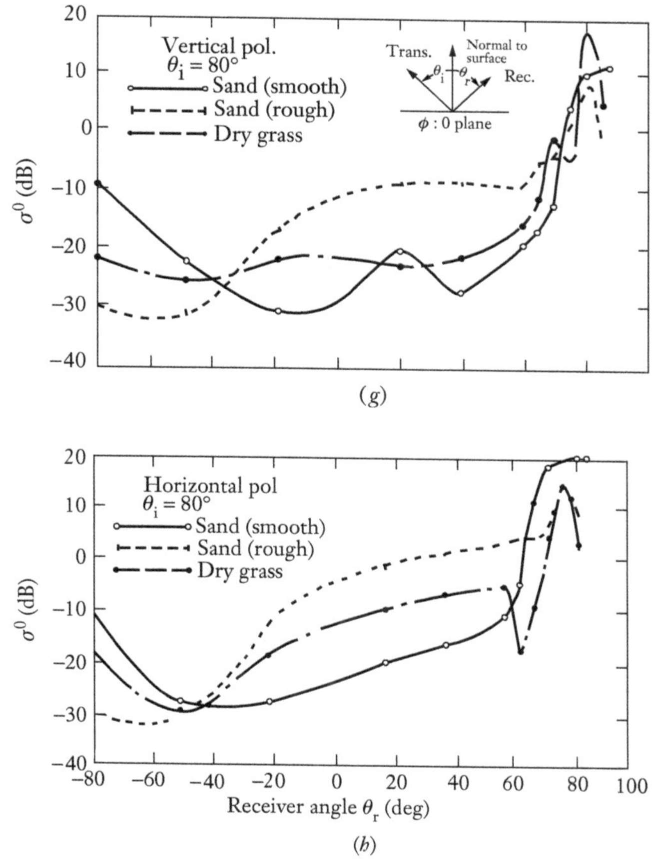

Figure 9–12 Effects of surface roughness in the plane of incidence for smooth sand, dry grass, and rough sand (after S. T. Cost, Ohio State University, May 1965 [24]).

planes. Figure 9–20 shows the effect of angles of incidence on the azimuthal scattering pattern in the elevation plane.

No. 2a,b, Nathanson (after Pidgeon), Johns Hopkins University

Figure 9–21 gives summary curves of sea scatter averaged over many receiver angles.

No. 3a, Domville [21, 22], G. E. C. Electronics

Figures 9–22 to 9–26 give excellent summary curves of data for terrain and sea scatter.

No. 4a,b, Larson [29, 32], Vander Schuur [42], University of Michigan

Figures 9–27 to 9–29 give data in the azimuthal plane for various types of dry vegetated farmland and rural land. Figure 9–30 demonstrates the statistical nature of the scattering coefficient by showing spreads in independent sample data points as much as 35 dB for a given scattering geometry. Figure 9–31 gives some data for rural land with snow cover.

No. 6a,b, Ewell and Zehner [35, 36], Georgia Institute of Technology

Figures 9–32 and 9–33 give the ratio of bistatic to monostatic scattering coefficients for sea clutter at low grazing angles of incidence.

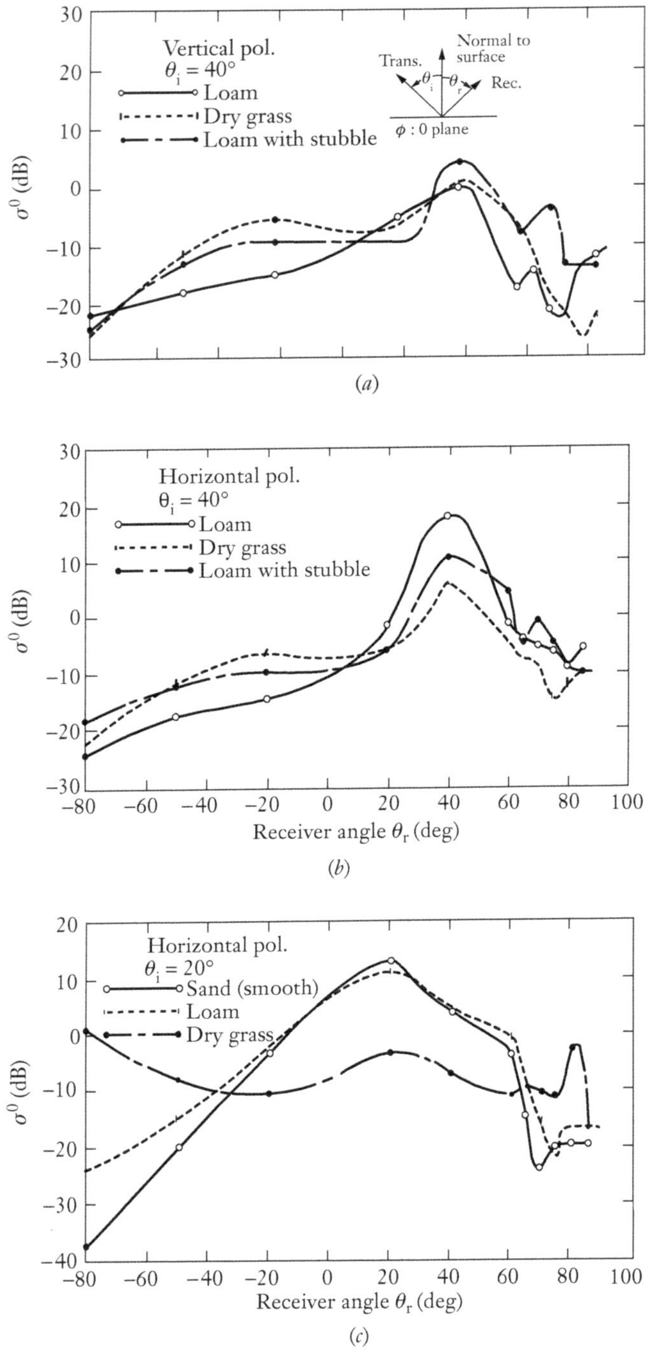

Figure 9–13 Effects of surface roughness in the plane of incidence for loam, dry grass, loam with stubble, smooth sand, and soybean foliage (after S. T. Cost, Ohio State University, May 1965 [24]).

(continues)

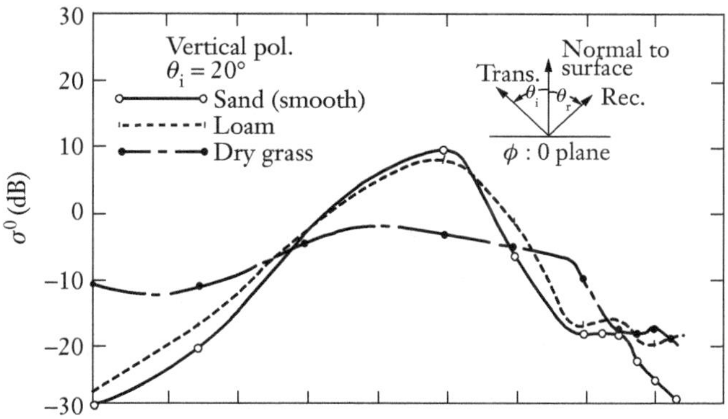

(d)

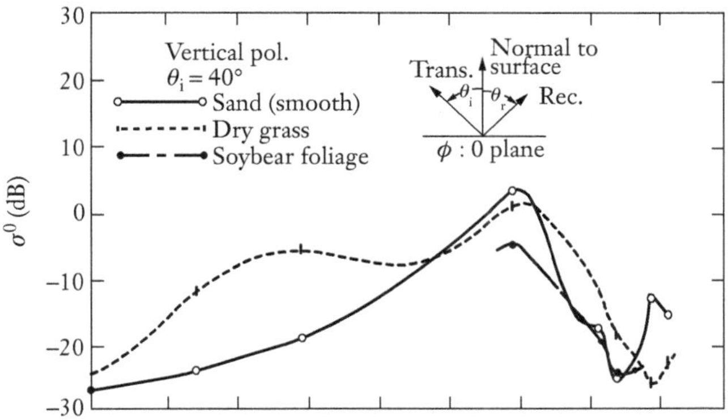

(e)

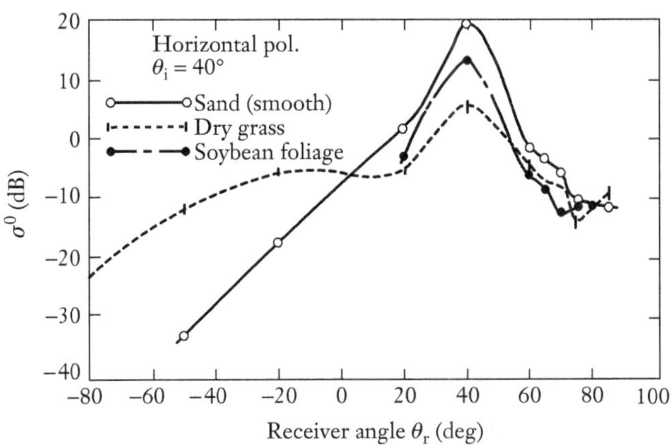

(f)

Figure 9–13 Effects of surface roughness in the plane of incidence for loam, dry grass, loam with stubble, smooth sand, and soybean foliage (after S. T. Cost, Ohio State University, May 1965 [24]).

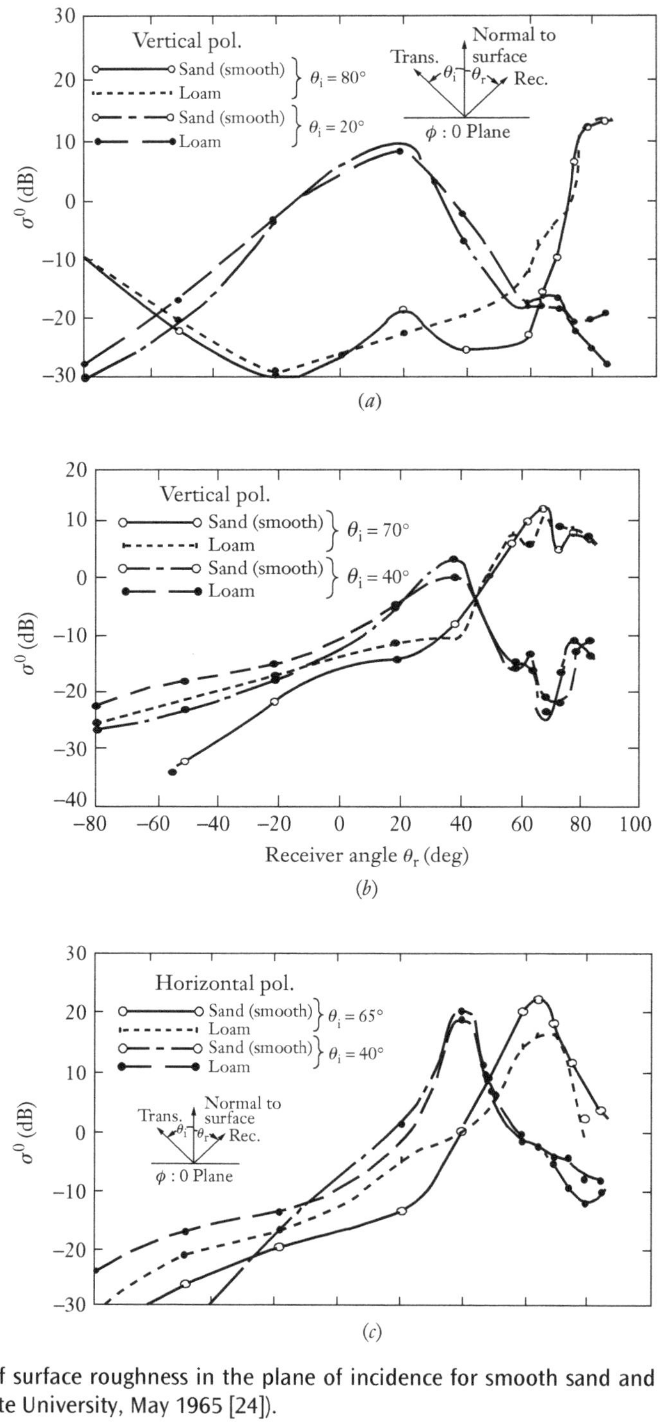

Figure 9–14 Effects of surface roughness in the plane of incidence for smooth sand and loam (after S. T. Cost, Ohio State University, May 1965 [24]).

(continues)

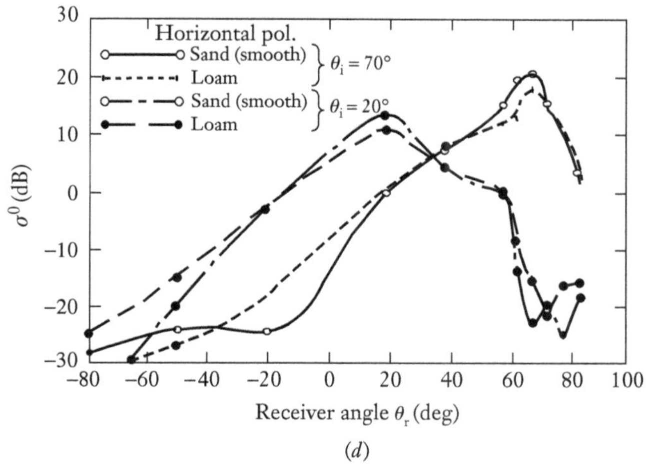

Figure 9–14 Effects of surface roughness in the plane of incidence for smooth sand and loam (after S. T. Cost, Ohio State University, May 1965 [24]).

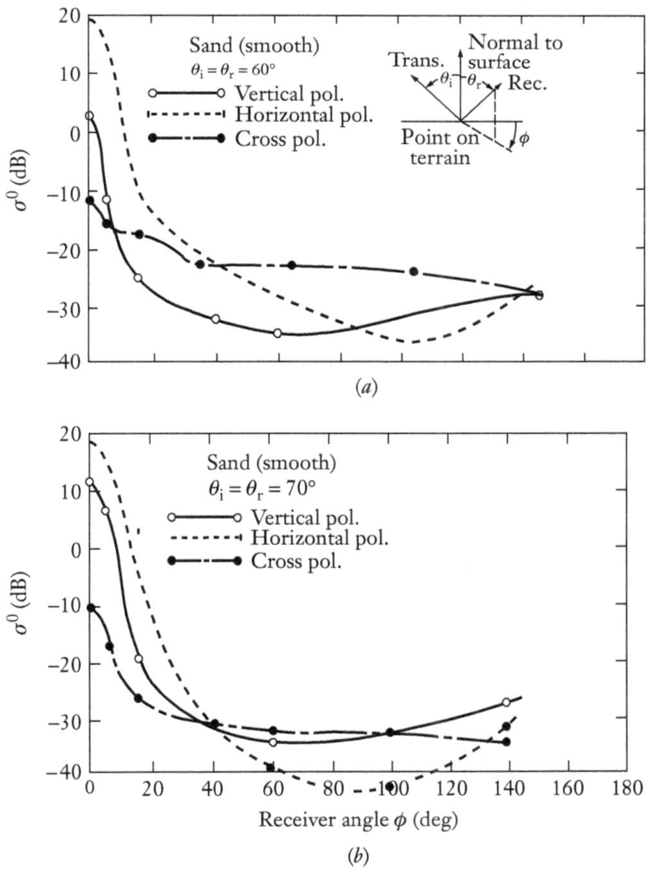

Figure 9–15 Effects of antenna polarization of the azimuthal scattering pattern for smooth sand (after S. T. Cost, Ohio State University, May 1965 [24]).

(continues)

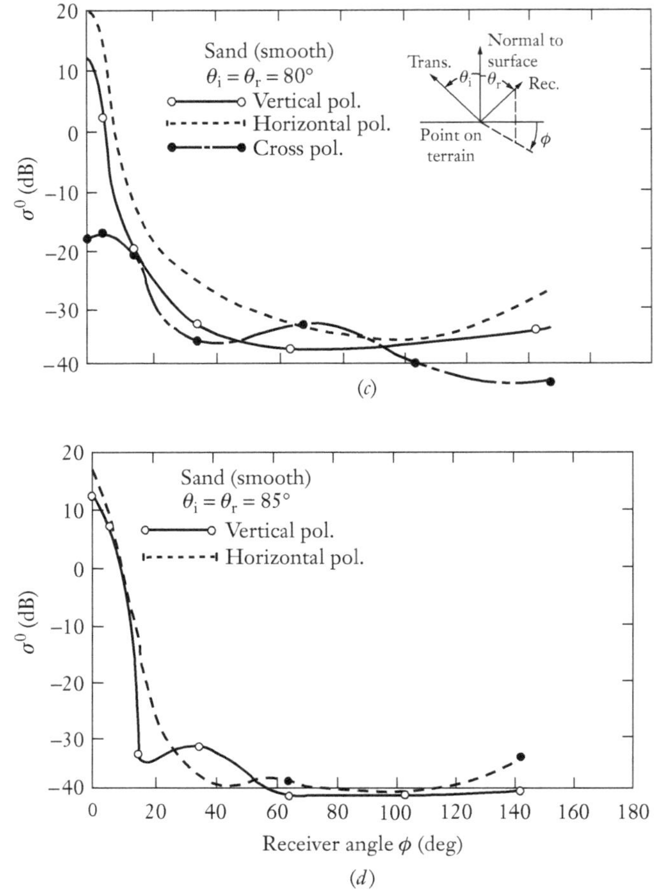

Figure 9–15 Effects of antenna polarization of the azimuthal scattering pattern for smooth sand (after S. T. Cost, Ohio State University, May 1965 [24]).

No. 7, Ulaby et al. [37], University of Michigan

Figures 9–34 to 9–37 are summary curves of the scattering coefficient for smooth sand, rough sand, and gravel, for various polarizations, angles of incidence, and azimuthal angles.

No. 8, Kochanski et al. [38], MIT Lincoln Laboratory

Figure 9–38 gives the scattering coefficient and its cumulative distribution function at a grazing angle of incidence of 0.3°.

No. 9a,b,c McLaughin et al. [39, 40, 41], Northeastern University and University of Massachusetts

Figure 9–39 gives histograms and statistics of the scattering coefficient for forested hills at low grazing angles. Figure 9–40 gives the normalized scattering coefficient as a function of the polarization state of the transmit antenna for copolarized and cross-polarized antennas.

9.3.3.2 Other Measurement Programs In addition to the measurement programs at microwave frequencies, bistatic reflectivity measurements have been made at optical wavelengths by Sauermann [43], at sonic wavelengths by Zornig [44], of buildings by Brambly [45], of airport structures by Brindley [46], of the Venus surface by Pavel'yev [47], and of the Mars

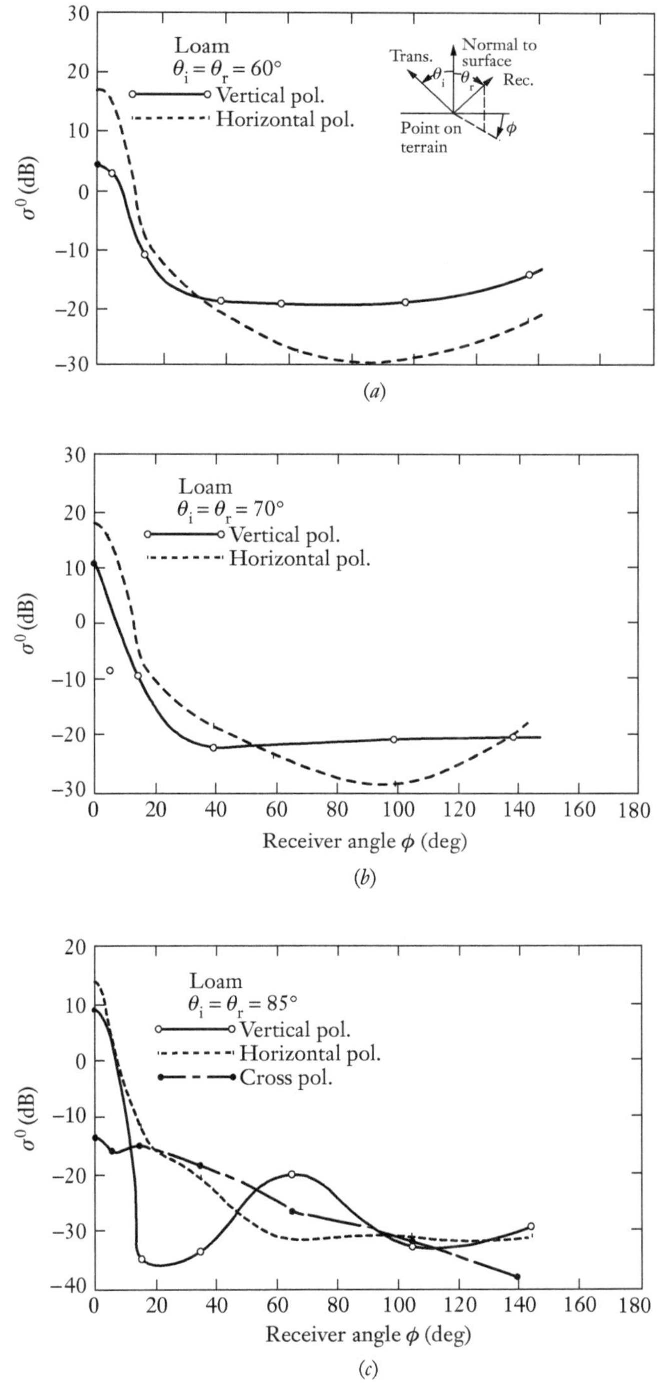

Figure 9–16 Effects of antenna polarization on the azimuthal scattering pattern for loam and soybean foliage (after S. T. Cost, Ohio State University, May 1965 [24]).

(*continues*)

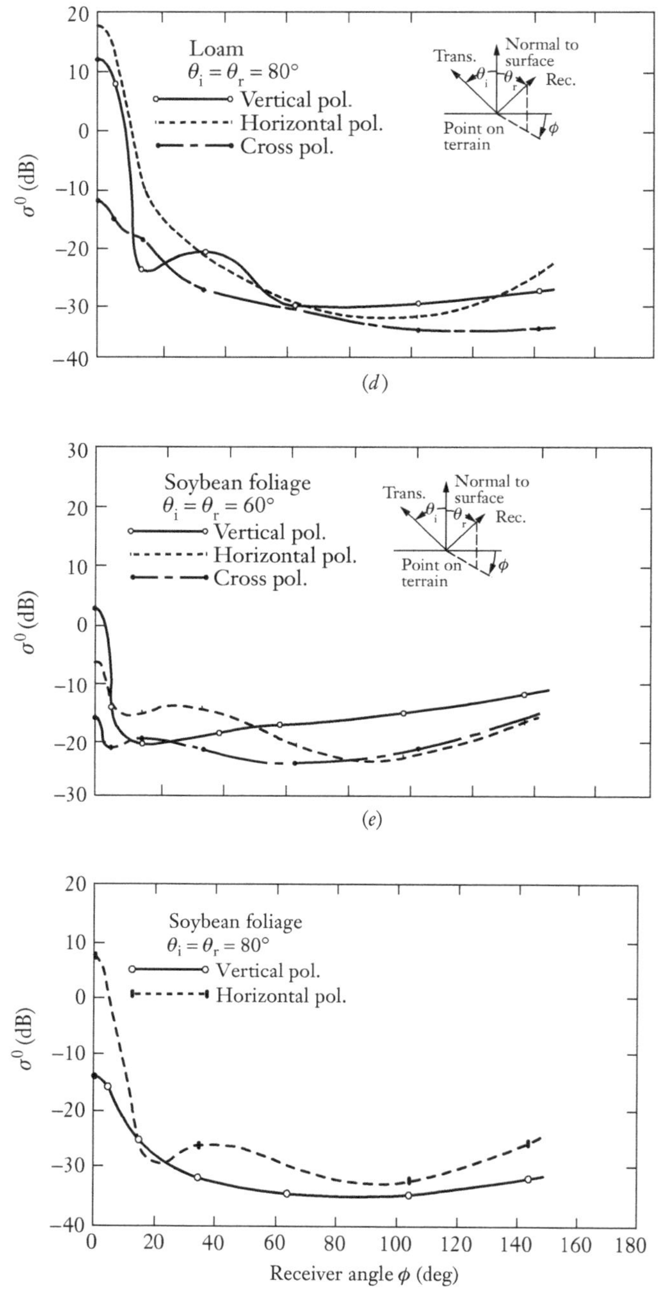

Figure 9–16 Effects of antenna polarization on the azimuthal scattering pattern for loam and soybean foliage (after S. T. Cost, Ohio State University, May 1965 [24]).

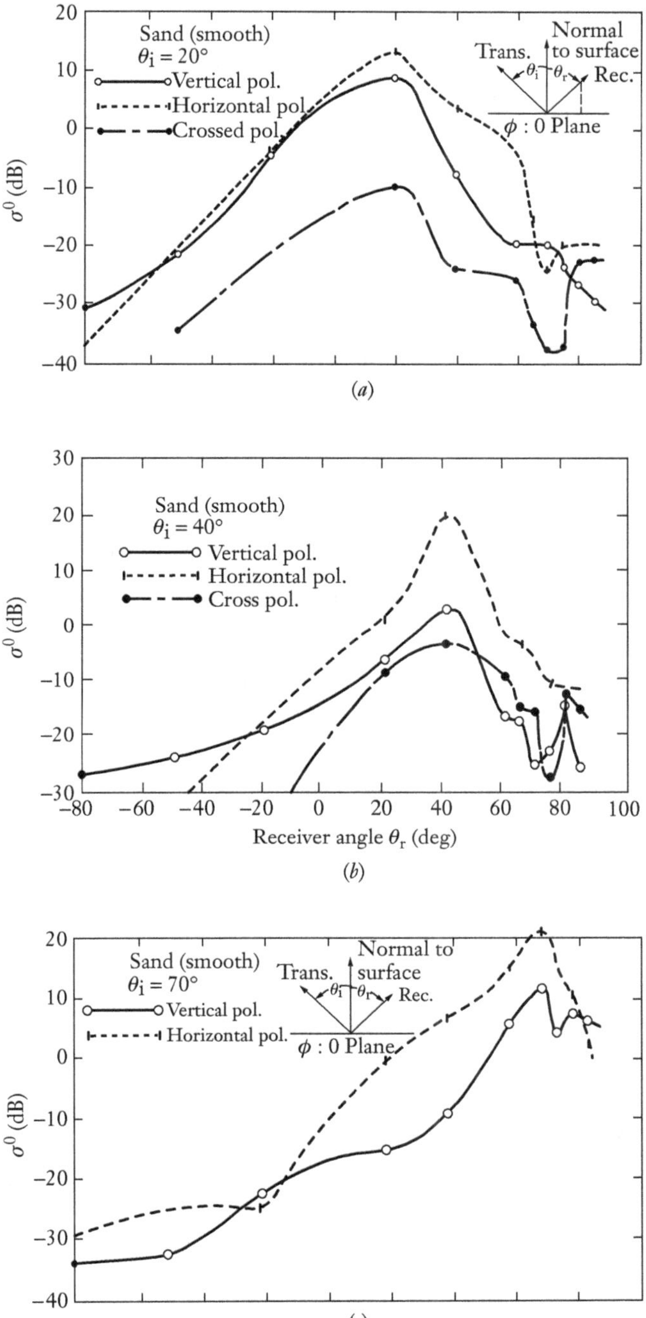

Figure 9–17 Effects of antenna polarization in the plane of incidence for smooth sand, loam, and loam with stubble (after S. T. Cost, Ohio with State University, May 1965 [24]).

(continues)

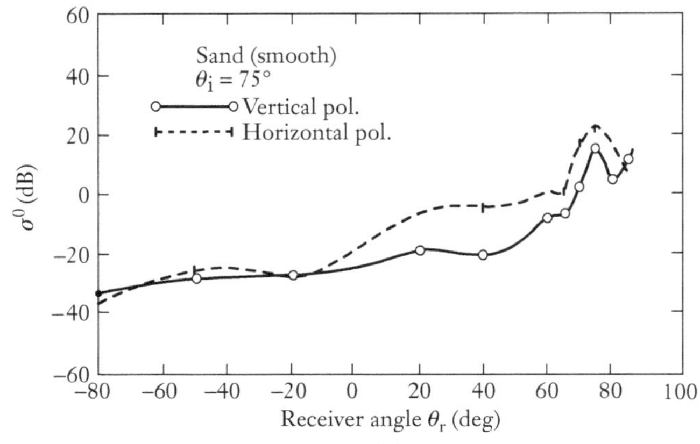

(*d*)

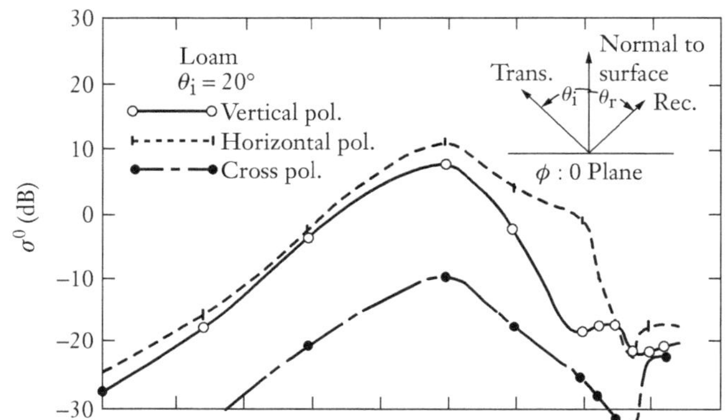

(*e*)

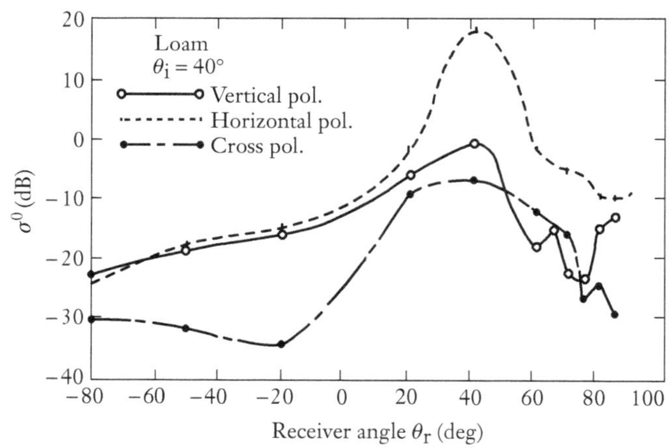

(*f*)

Figure 9–17 Effects of antenna polarization in the plane of incidence for smooth sand, loam, and loam with stubble (after S. T. Cost, Ohio with State University, May 1965 [24]).

(continues)

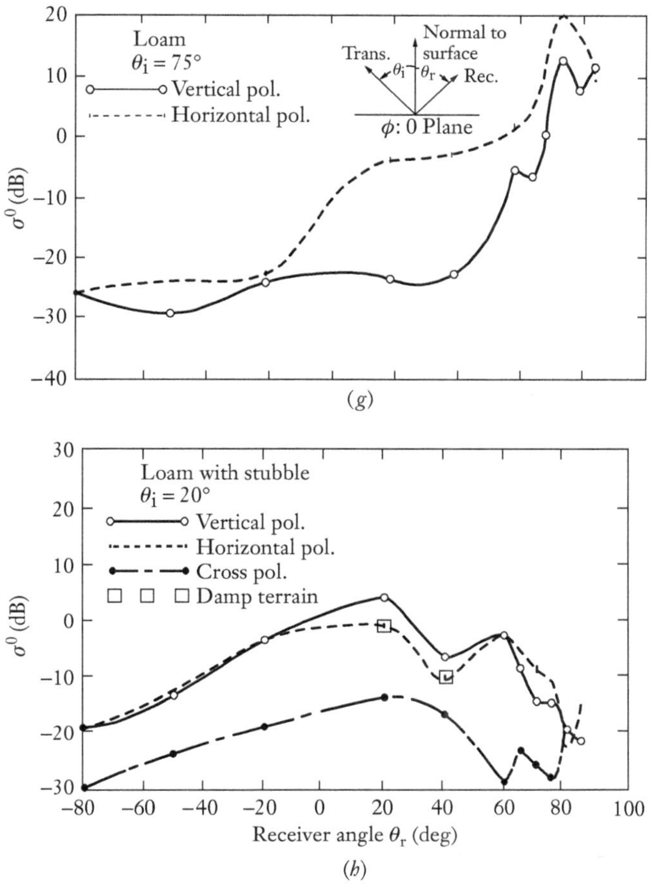

Figure 9–17 Effects of antenna polarization in the plane of incidence for smooth sand, loam, and loam with stubble (after S. T. Cost, Ohio with State University, May 1965 [24]).

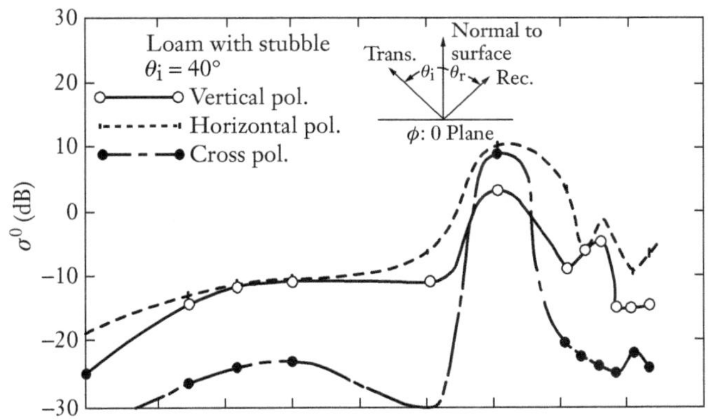

Figure 9–18 Antenna polarization effects in the plane of incidence for loam with stubble and soybean foliage (after S. T. Cost, Ohio State University, May 1965 [24]).

(continues)

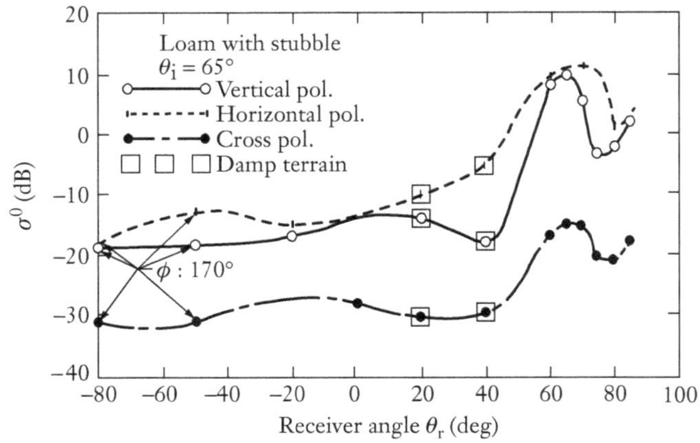

(*b*)

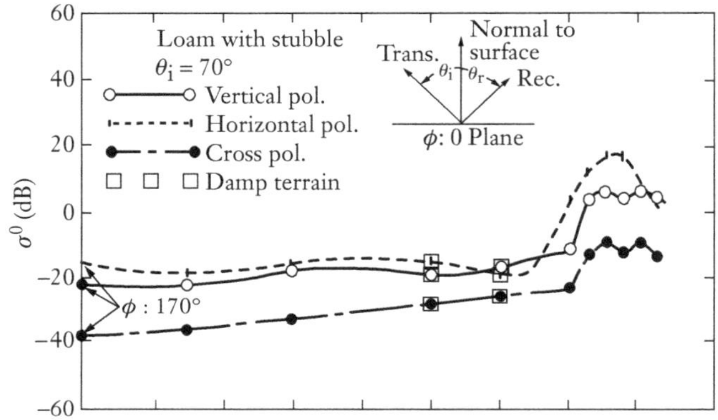

(*c*)

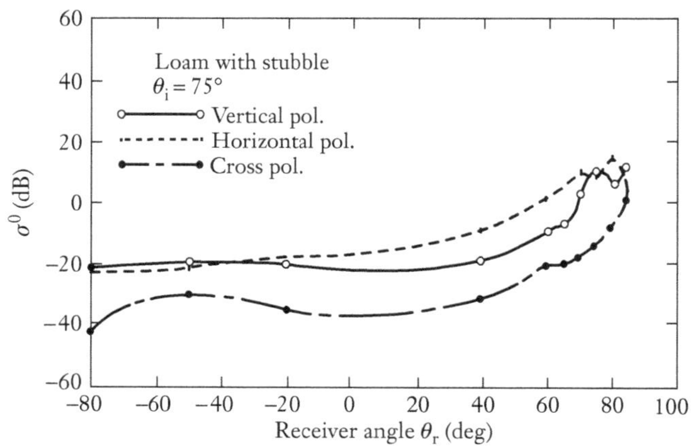

(*d*)

Figure 9–18 Antenna polarization effects in the plane of incidence for loam with stubble and soybean foliage (after S. T. Cost, Ohio State University, May 1965 [24]).

(continues)

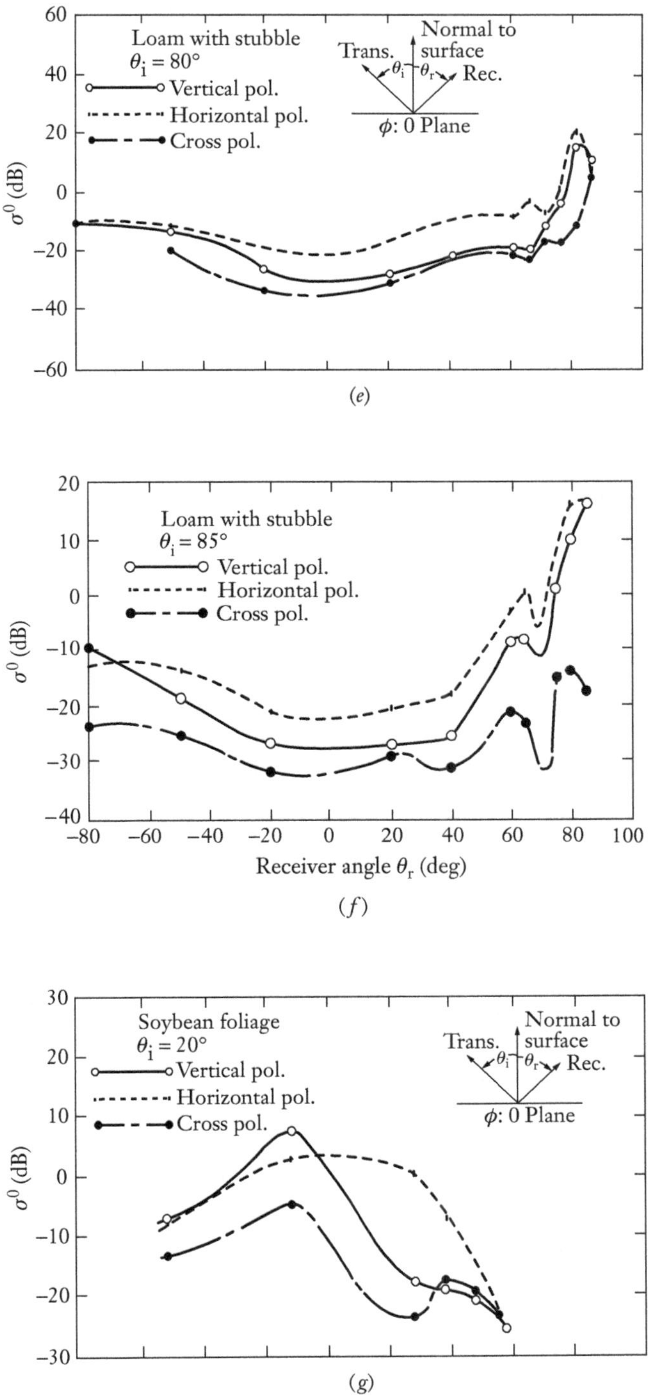

Figure 9–18 Antenna polarization effects in the plane of incidence for loam with stubble and soybean foliage (after S. T. Cost, Ohio State University, May 1965 [24]).

(continues)

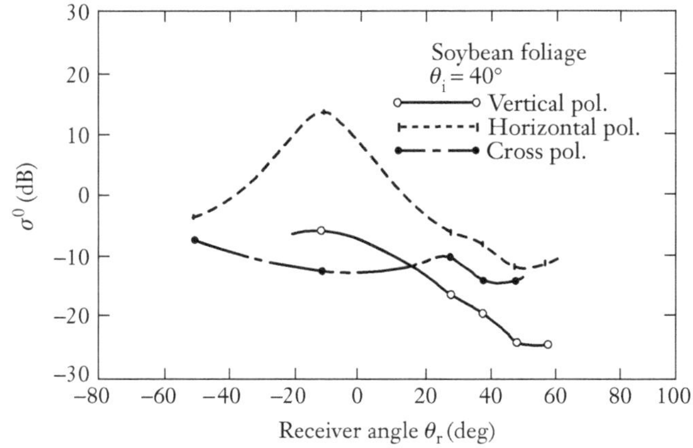

(b)

Figure 9–18 Antenna polarization effects in the plane of incidence for loam with stubble and soybean foliage (after S. T. Cost, Ohio State University, May 1965 [24]).

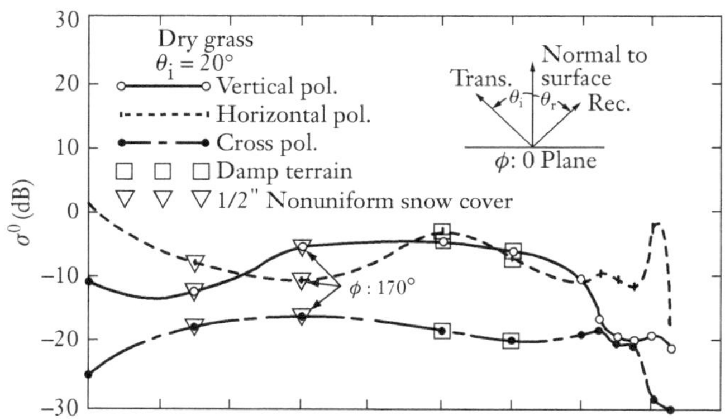

(a)

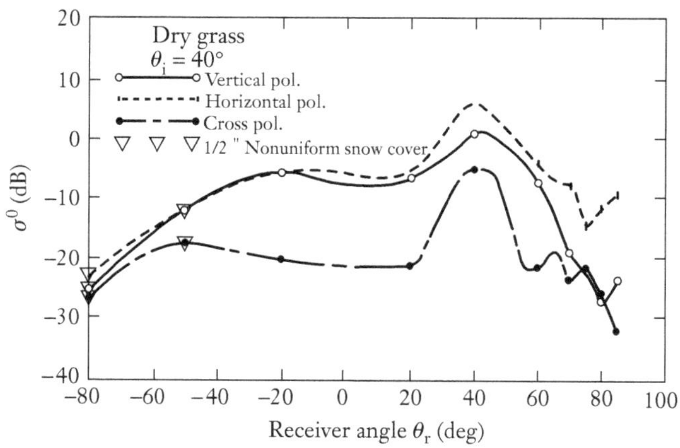

(b)

Figure 9–19 Antenna polarization effects in the plane of incidence for dry grass and rough sand (after S. T. Cost, Ohio State University, May 1965 [24]).

(continues)

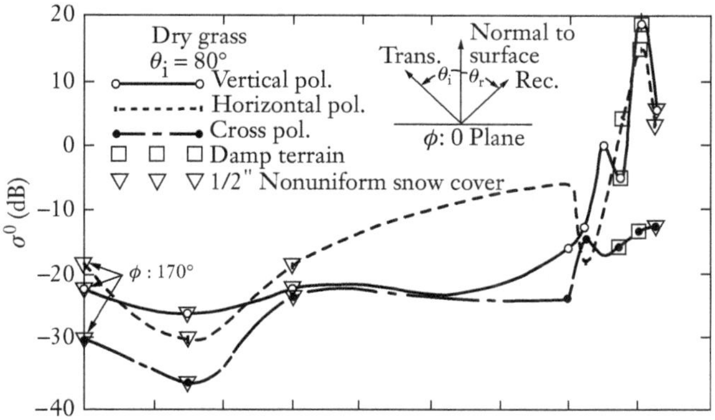

(*c*)

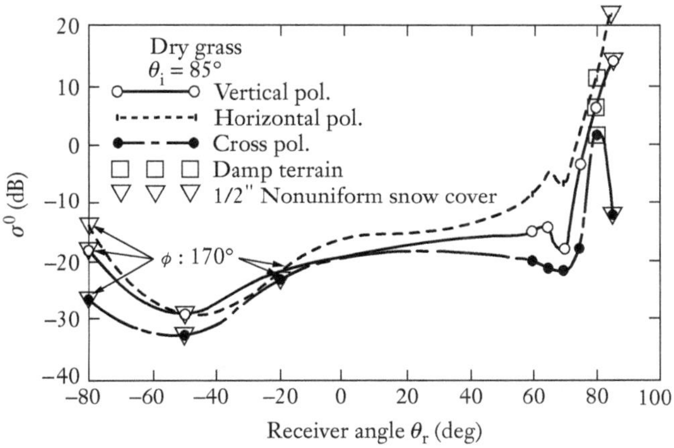

(*d*)

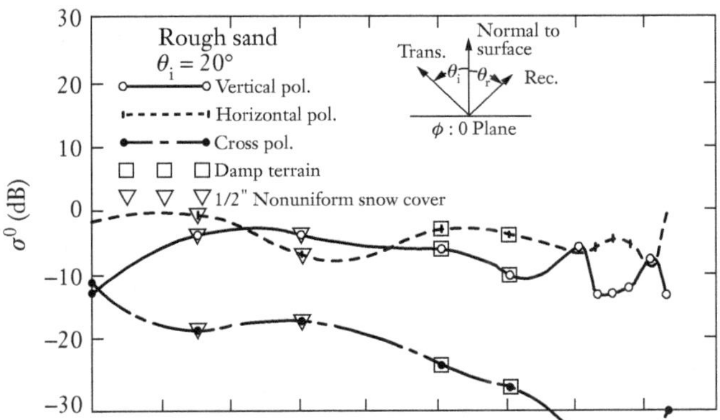

(*e*)

Figure 9–19 Antenna polarization effects in the plane of incidence for dry grass and rough sand (after S. T. Cost, Ohio State University, May 1965 [24]).

(*continues*)

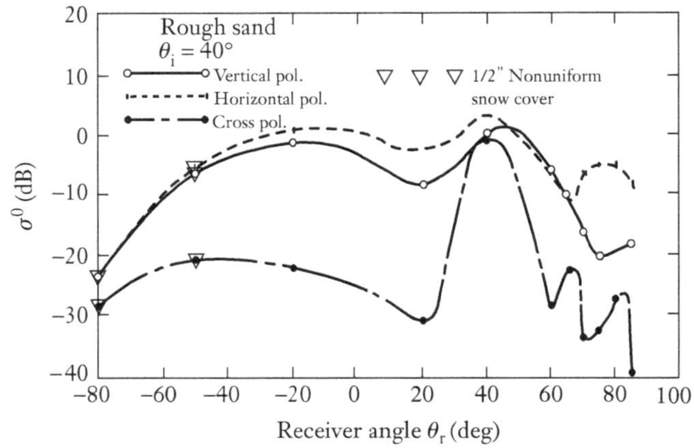

(f)

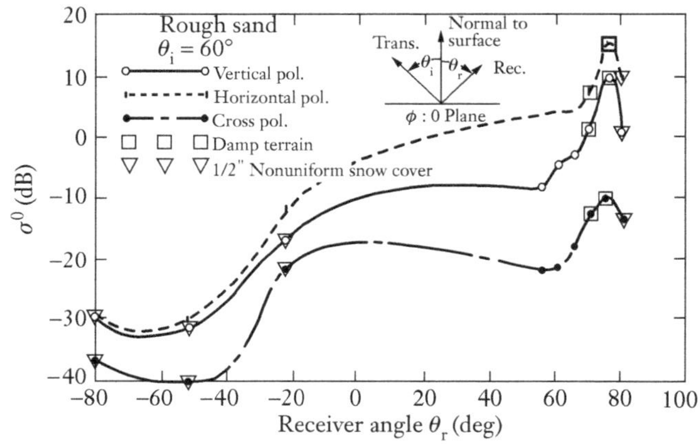

(g)

Figure 9–19 Antenna polarization effects in the plane of incidence for dry grass and rough sand (after S. T. Cost, Ohio State University, May 1965 [24]).

surface by Tang [43]. In each of these measurements, the reflectivity data is expressed in terms of reflected power but not in terms of the cross section per unit area.

Sauermann [44], utilizing laser radiation at 0.68-μm incident on steel and plastic plates of RMS roughness of 1.5–2.5 wavelengths, found that the in-plane polar scattering pattern for grazing angles of incidence of 2°–10° was characterized by three scattering regions—a strong intensity line of diffraction-limited width in the specular direction, a multitude of weaker intensity lines around the forward-scattered direction, and a weak diffuse intensity region at other scattering angles.

Zornig [45], using sound waves at a wavelength of approximately 1 mm incident from below onto a wind-driven water surface of roughness of approximately 5 wavelengths at grazing angles of incidence and reflection of 12° and 17° respectively, found that the scattering strength at azimuthal angles of 0°–180° was dependent on wind direction and surface roughness but weakly dependent

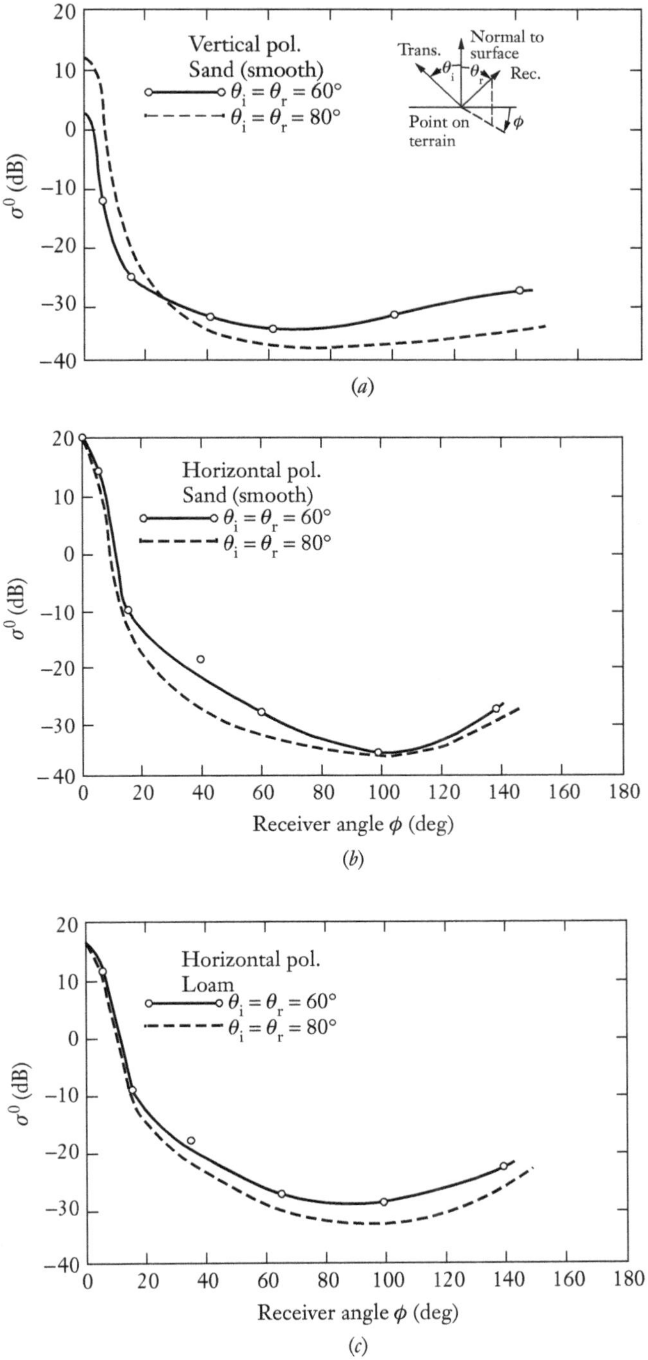

Figure 9–20 Effects of angle of incidence on azimuthal scattering pattern for smooth sand, loam, and soybean foliage (after S. T. Cost, Ohio State University, May 1965 [24]).

(continues)

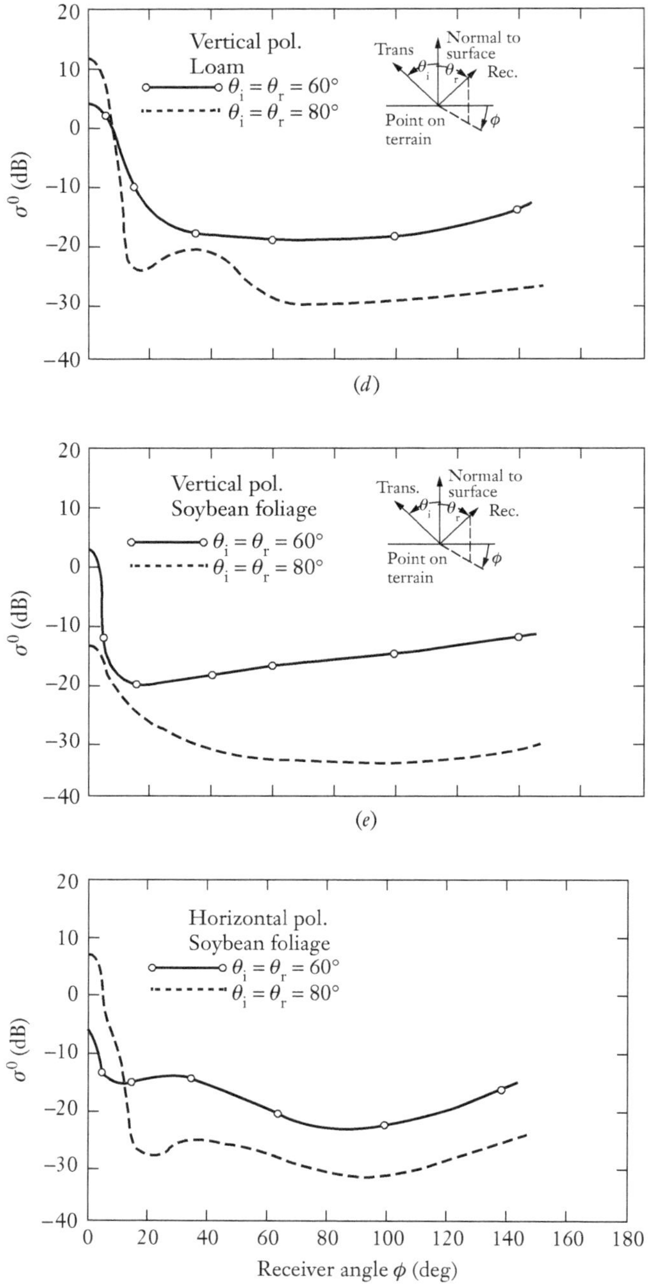

Figure 9–20 Effects of angle of incidence on azimuthal scattering pattern for smooth sand, loam, and soybean foliage (after S. T. Cost, Ohio State University, May 1965 [24]).

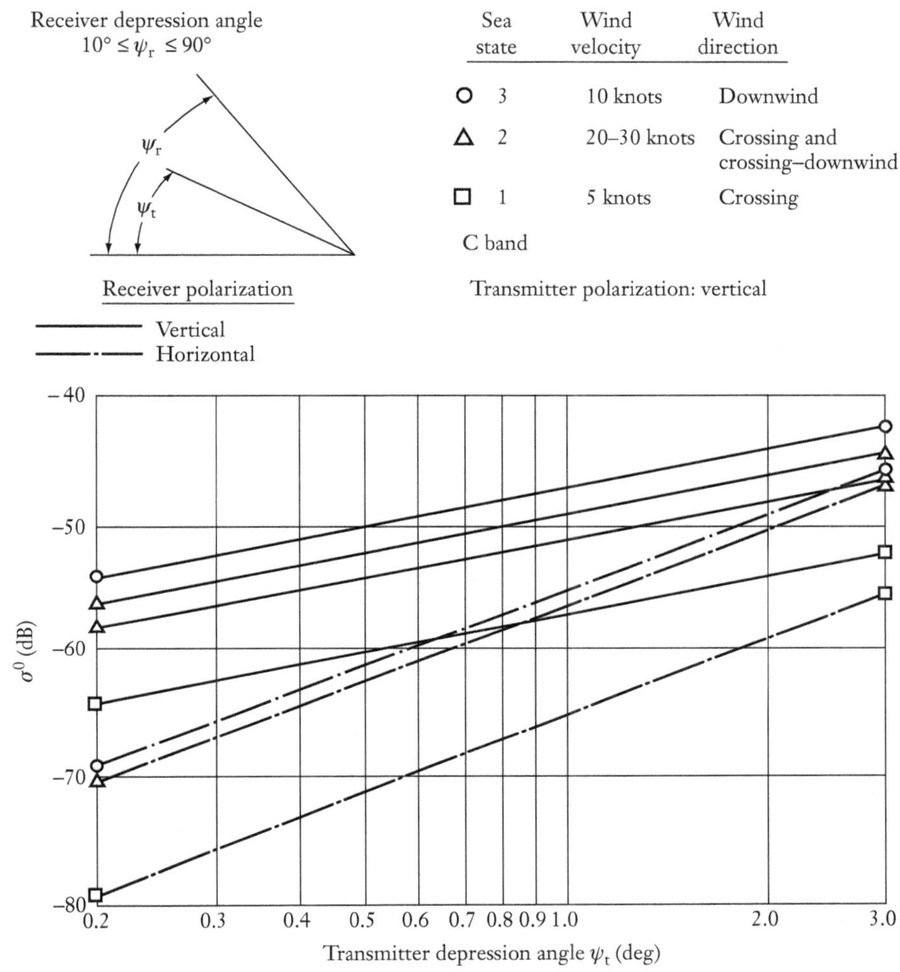

(*a*)

Figure 9–21 Bistatic scattering coefficient of the sea in the plane of incidence at C and X bands (after F. E. Nathanson, Johns Hopkins University data, *Radar Design Principles*, McGraw-Hill, New York, 1969, pp. 88, 89). (*continues*)

on the scalar Rayleigh parameter defined in Section 9.3.2. Zornig concluded that present theory will not predict scattering strength accurately without a knowledge of detected statistics since locally high slopes were generated.

Bramley [46] found that the level of bistatic scatter of pulsed 9.4-GHz radiation incident on tall buildings could be explained reasonably well in terms of physical optics scattering from a number of smooth flat elements and corner reflectors corresponding to visible features of the building. The nonspecular scatter was generally a few decibels stronger for vertical polarization than for horizontal polarization.

Brindley [47] measured the in-plane reflection coefficients of airport structures at grazing angles of 10°–40° with C-band and Ku-band CW radiation of horizontal, vertical, and circular

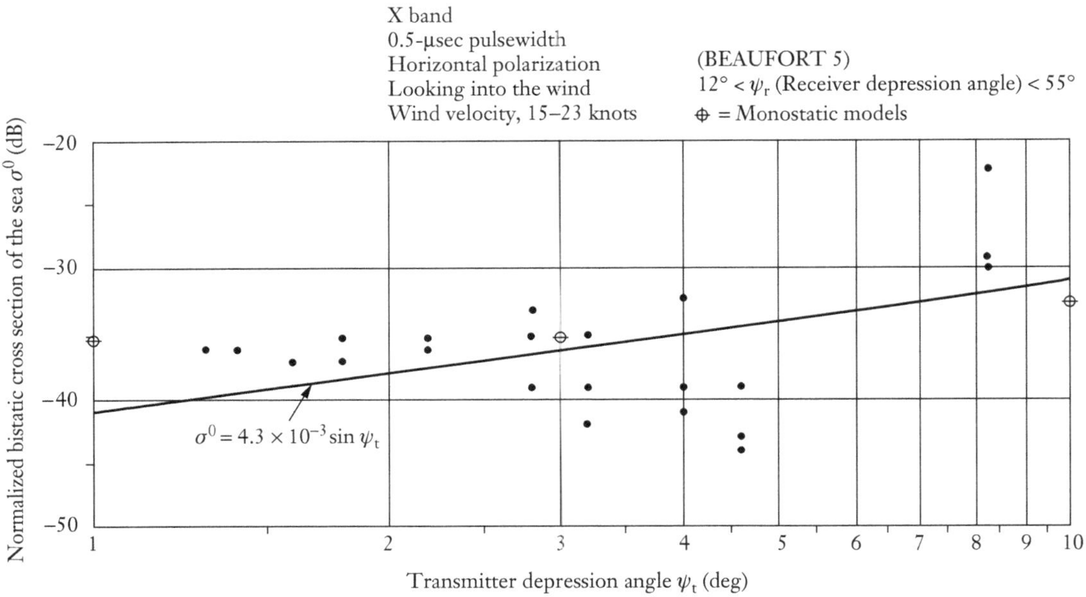

(*b*)

Figure 9–21 Bistatic scattering coefficient of the sea in the plane of incidence at C and X bands (after F. E. Nathanson, Johns Hopkins University data, *Radar Design Principles*, McGraw-Hill, New York, 1969, pp. 88, 89).

polarization. The illuminated spot on the structures was of the order of the size of the first Fresnel zone. Vertical polarization produced stronger specular reflections in every case but weaker nonspecular reflections in more than 80% of the test cases, which is contradictory to the results obtained by Bramley [46].

The data of Pavel'yev [48] and Tang [43] are discussed in Chapter 5 and are illustrative of bistatic reflectivity measurements made by several investigators of planetary surfaces utilizing space satellites, in which either the transmitter or receiver, but not both, are on the satellite. The advantages of these measurements over Earth-based monostatic measurements are the reduction of required transmitter power and access to vast regions of the planet which previously were not within view.

Not included in Tables 9–1 to 9–4 is most of the literature pertaining to ground and sea multipath reflections in the forward-scattered direction because the data is usually expressed as the ratio of indirect signal to direct signal rather than as the scattering coefficient σ^0. An exception is the data of Cornwell [34]. The multipath reflection literature is referenced by Barton [49] and ground multipath measurements are given by Sun [50]. Some older data for asphalt and vegetative surfaces at L-band are given by Cosgriff et al [51].

Also not included in Tables 9–1 to 9–4 are σ^0 bistatic terrain data by Blackwell et al [52] and Daigle et al [53] whose reports were not readily available. In [53], low grazing angle (0°–0.3° incidence, 7°–25° receive), out-of-plane (80°–170°) bistatic measurements were made of desert and mountainous terrain at 5.5 GHz with vertical polarization on transmit, right-hand circular polarization on receive, 160 pulses/sec, 1 μsec pulse width, an airborne receiver, and a ground-based receiver. The cell area was resolution time-limited with large measurement errors because of instrumentation problems. Six data curves in two figures are presented of the mean coefficient.

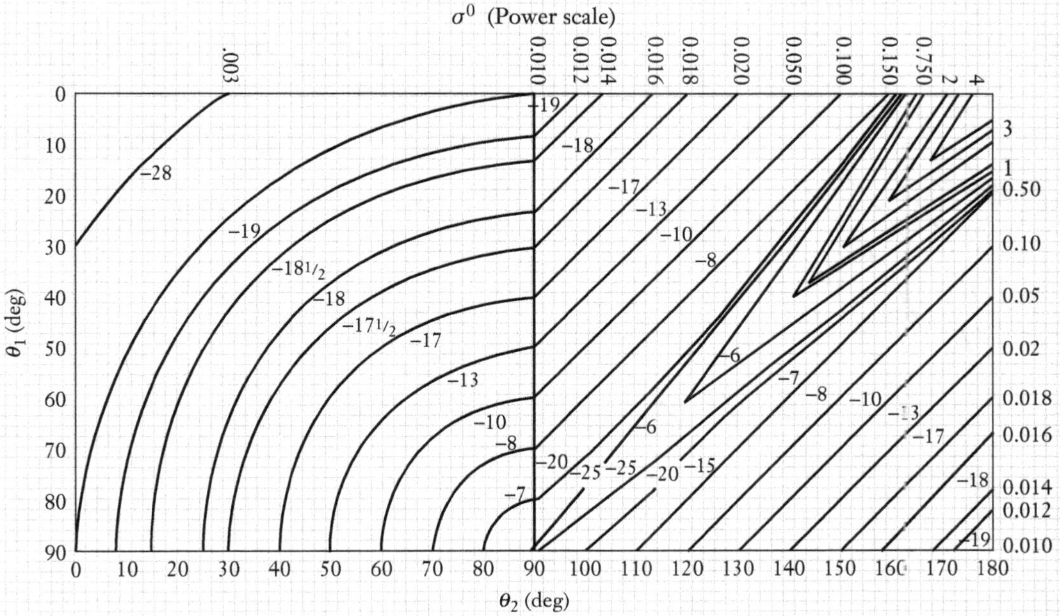

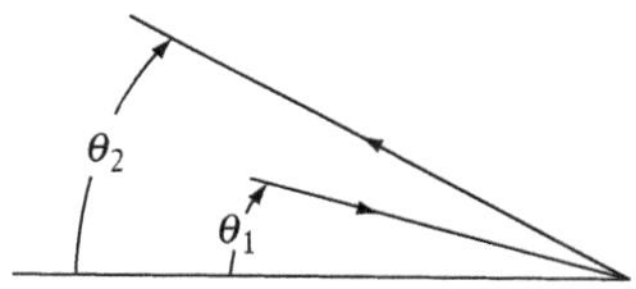

Figure 9–22 Contour plot of X-band scattering coefficient in plane of incidence for rural land. V-V polarization (after A. R. Domville, GEC (Electronics) Ltd., July 1967 [21]).

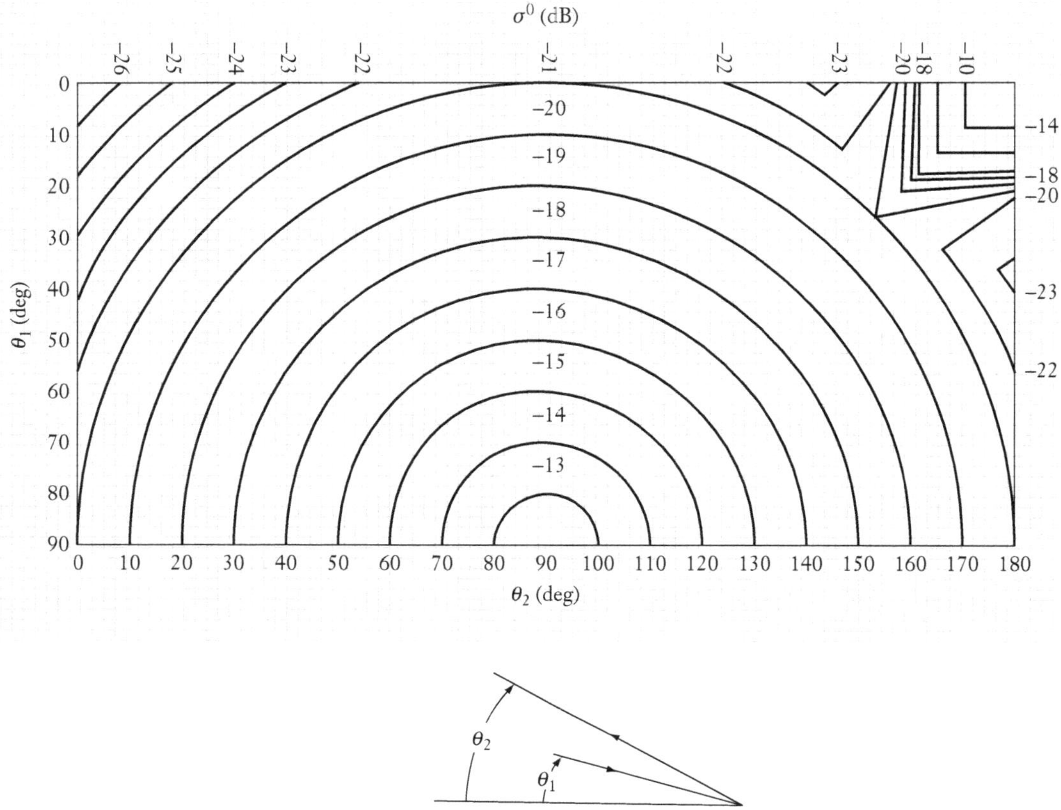

Figure 9–23 Contour plot of X-band scattering coefficient in plane of incidence for forest, V-V polarization (after A. R. Domville, GEC (Electronics) Ltd., July 1967 [21]).

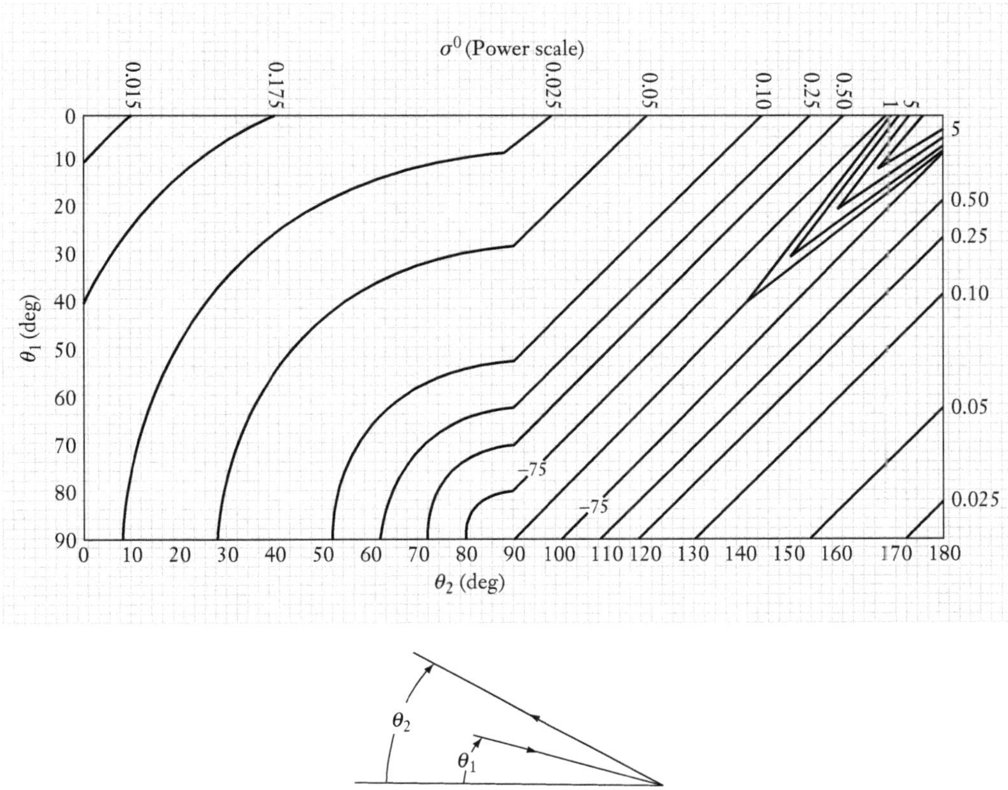

Figure 9–24 Contour plot of X-band scattering coefficient in plane of incidence for urban land. V-V polarization (after A. R. Domville, GEC (Electronics) Ltd., July 1967 [21]).

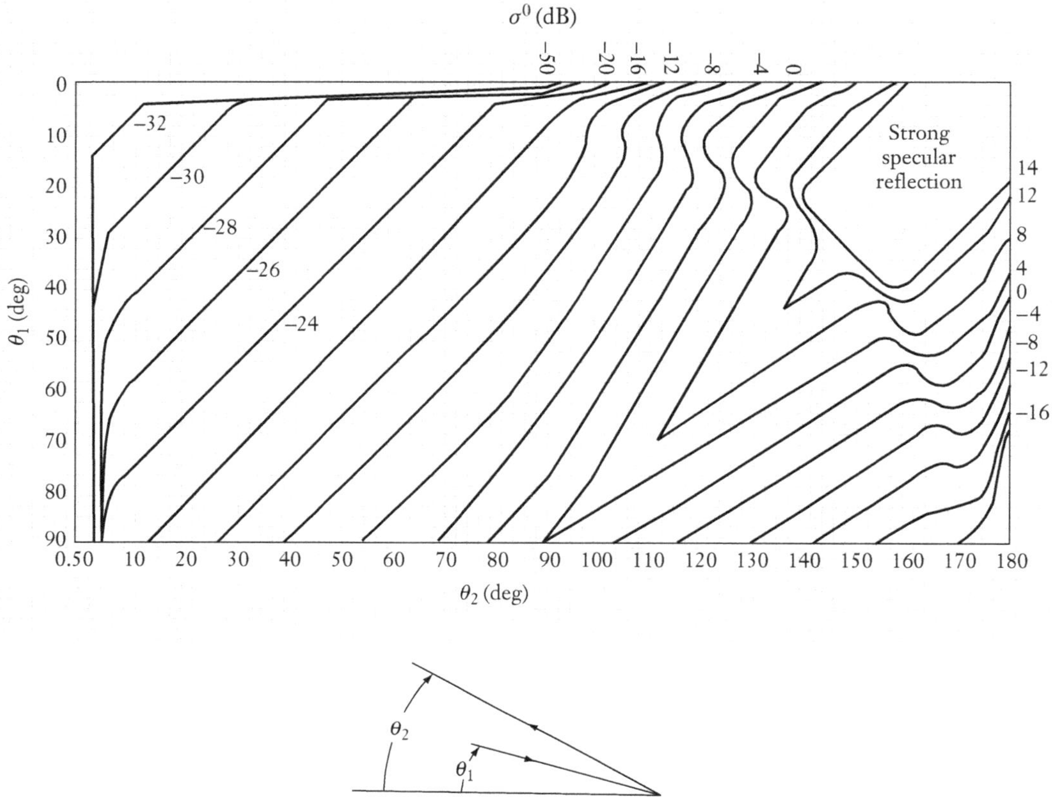

Figure 9–25 Contour plot of X-band scattering coefficient in plane of incidence for the sea, H-H polarization, 20 knot wind (after A. R. Domville, GEC (Electronics) Ltd., July 1968 [22]).

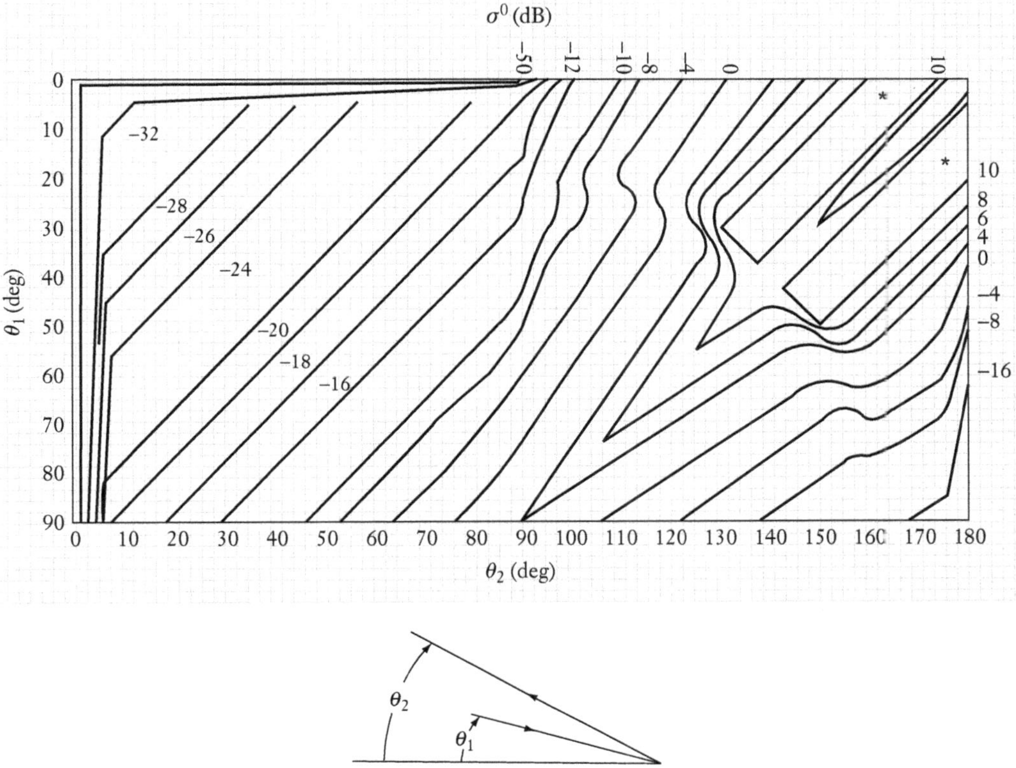

Figure 9–26 Contour plot of X-band scattering coefficient in plane of incidence for the sea, V-V polarization, 20 knot wind (after A. R. Domville, GEC (Electronics) Ltd., July 1968 [22]). *Strong specular reflection

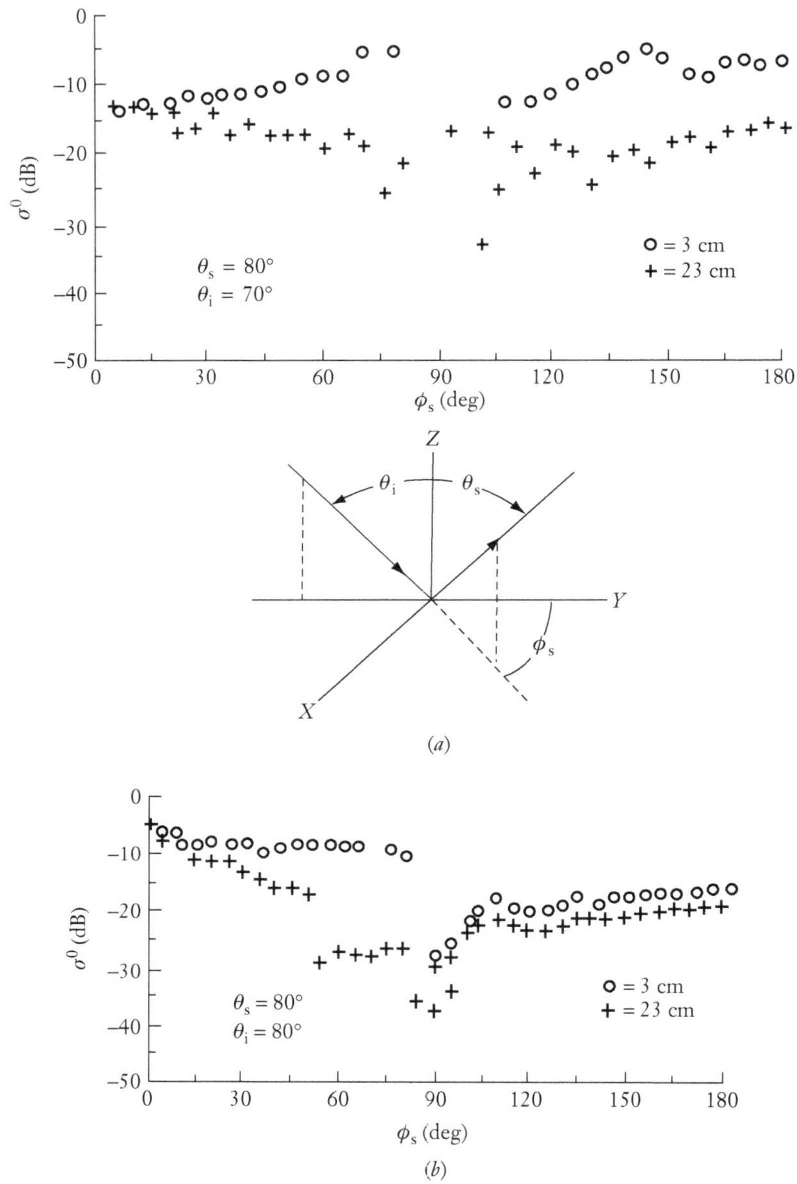

Figure 9–27 Scattering coefficient azimuthal pattern for dry tall weeds and scrub trees, H-H polarization (after R. W. Larson et al., University of Michigan, *IEEE Transactions*, Vol. AP-26, No. 6, November 1978, pp. 801–804 [32]).

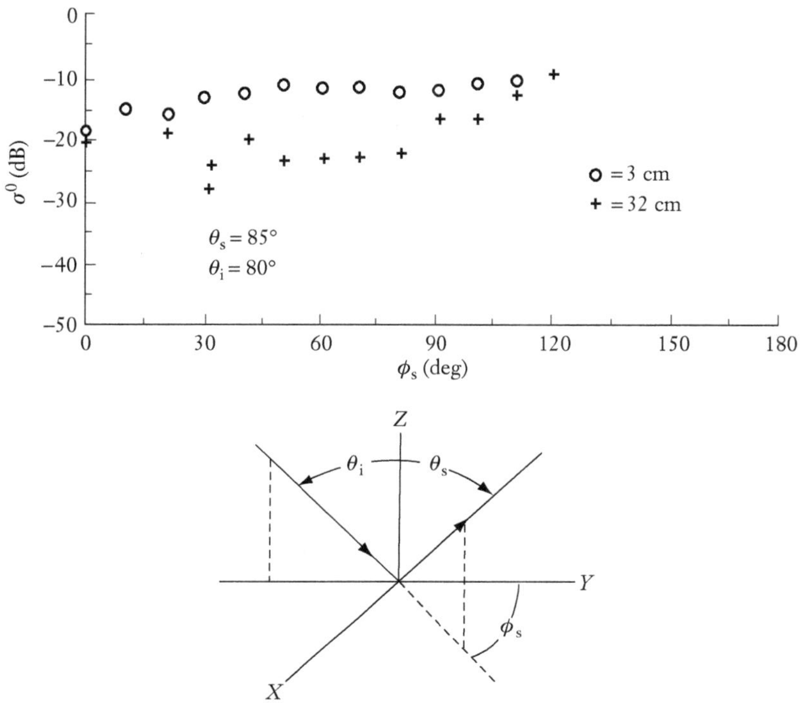

Figure 9–28 Scattering coefficient azimuthal pattern for dry flat grass and cement taxiway, H-H polarization (after R. W. Larson et al., University of Michigan, *IEEE Transactions*, Vol. AP-26, No. 6, November 1978, pp. 801–804 [32]).

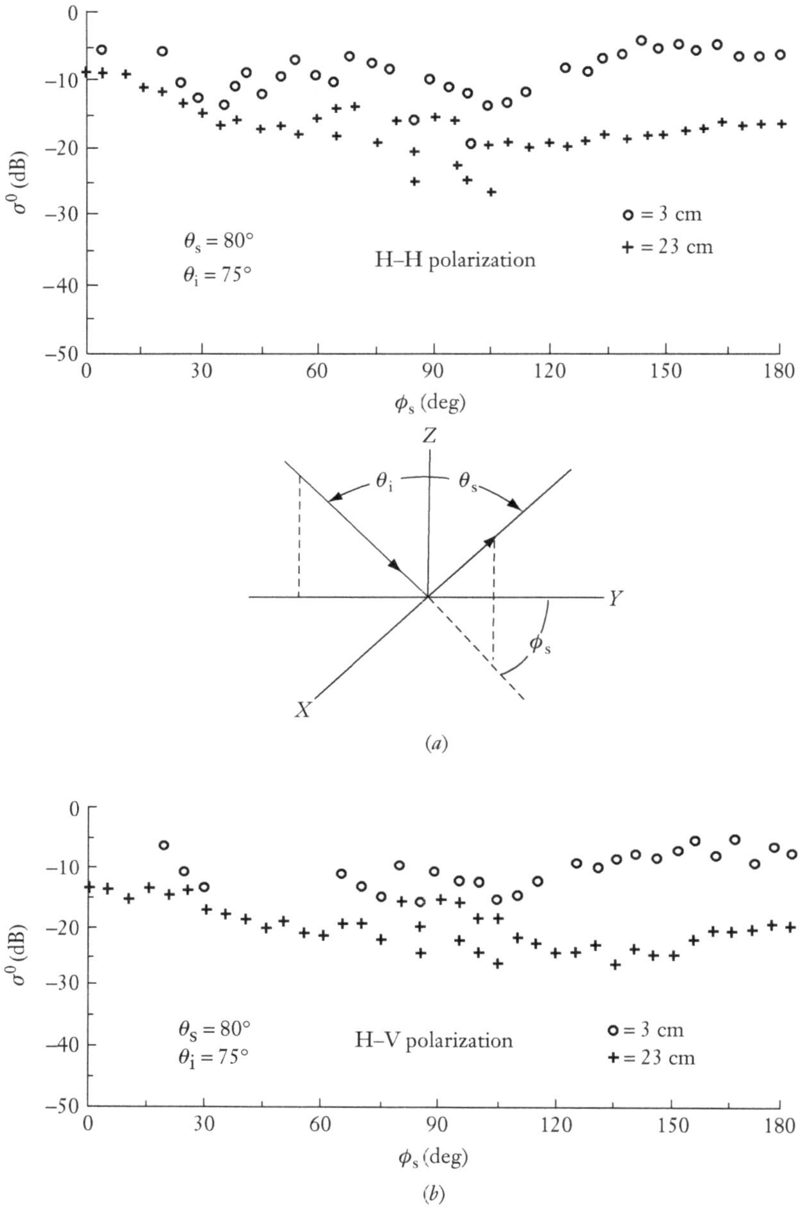

Figure 9–29 Effects of antenna polarization on the azimuthal scattering pattern for dry tall weeds and scrub trees (after R. W. Larson et al., University of Michigan, *IEEE Transactions*, Vol. AP-26, No. 6, pp. 801–804, November 1978 [32]).

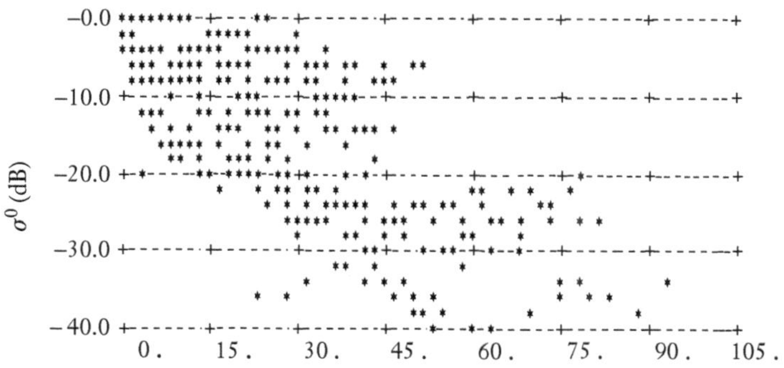

(*a*) All points

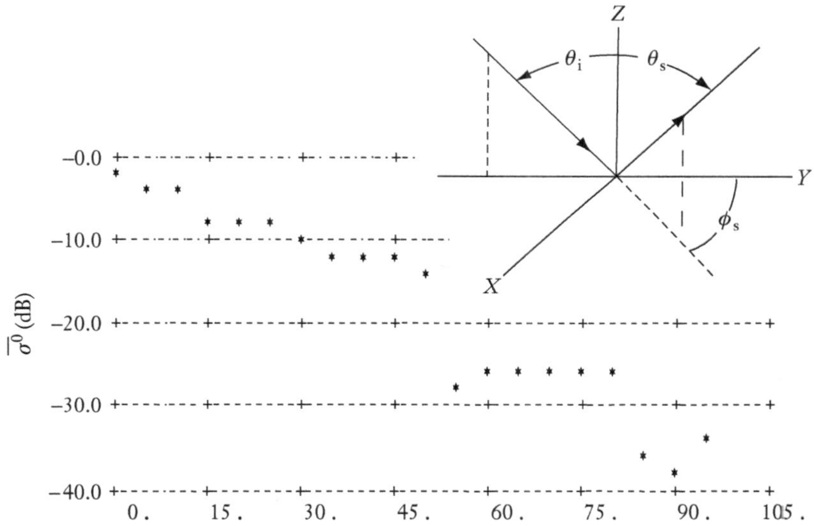

(*b*) Averages

Figure 9–30 Scattering coefficient azimuthal pattern for tail weeds and scrub trees, $\lambda = 23$ cm, H–H polarization, $\theta_s = 80°$, $\theta_i = 60°$. (*a*) All data points; (*b*) mean values averaged over 5° azimuthal intervals (after R. W. Larsen et al., University of Michigan, November 1977 [29]).

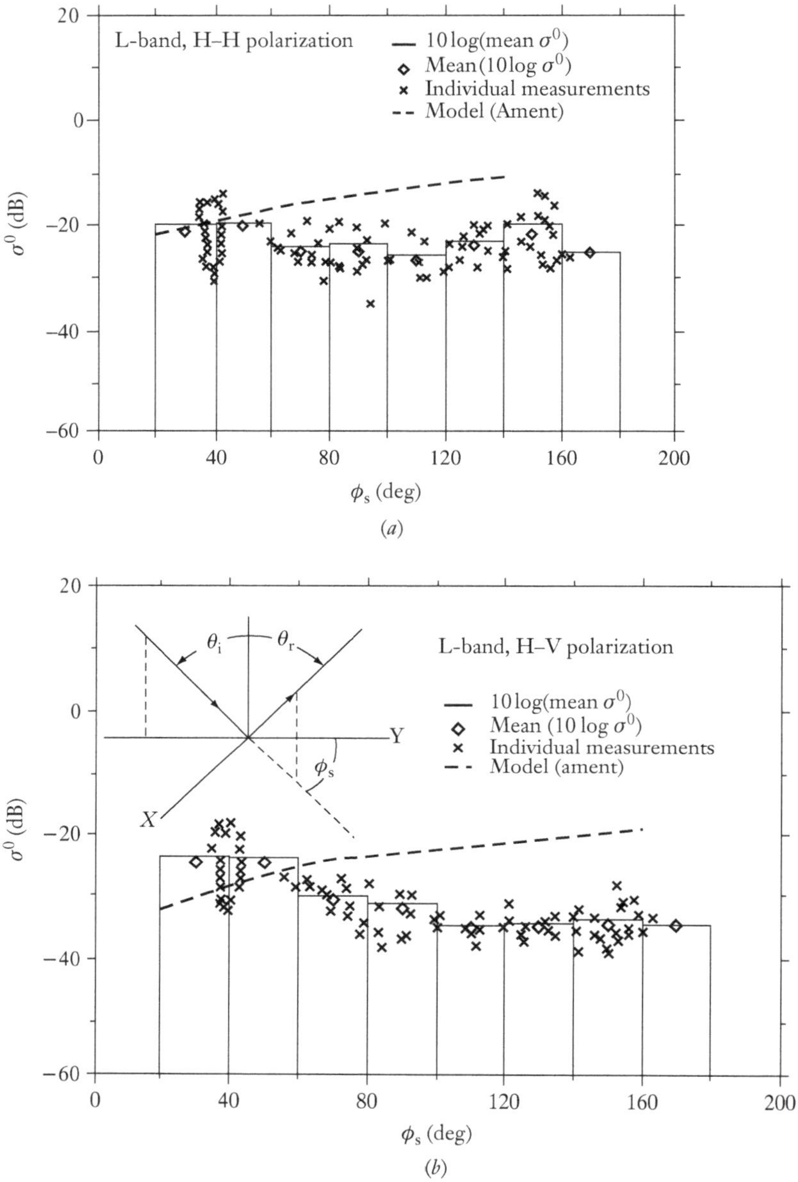

Figure 9–31 Scattering coefficient azimuthal pattern for tree-orchard on wet ground with 6-inch. snow cover; $\theta_s = 80°$, $\theta_i = 60°$ (after Vander Schuur and Tomlinson, University of Michigan data, April 1979 [42]).

(continues)

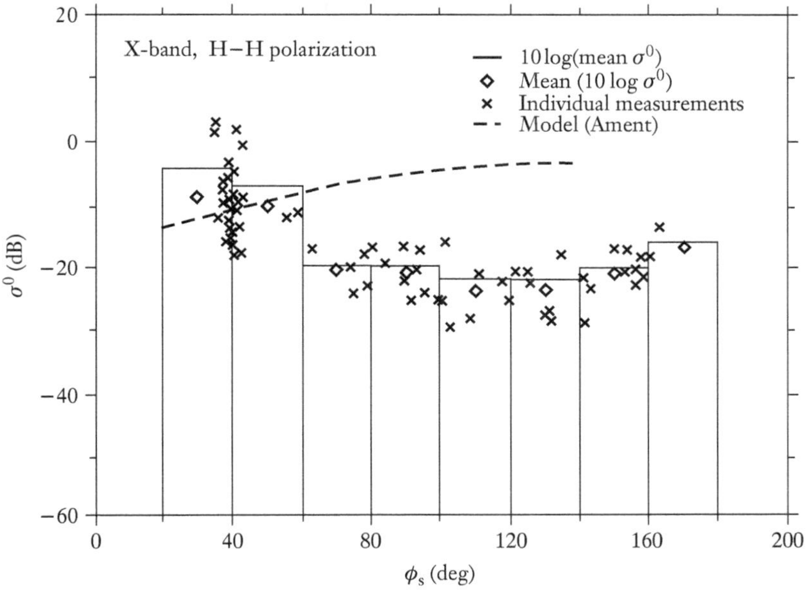

(c)

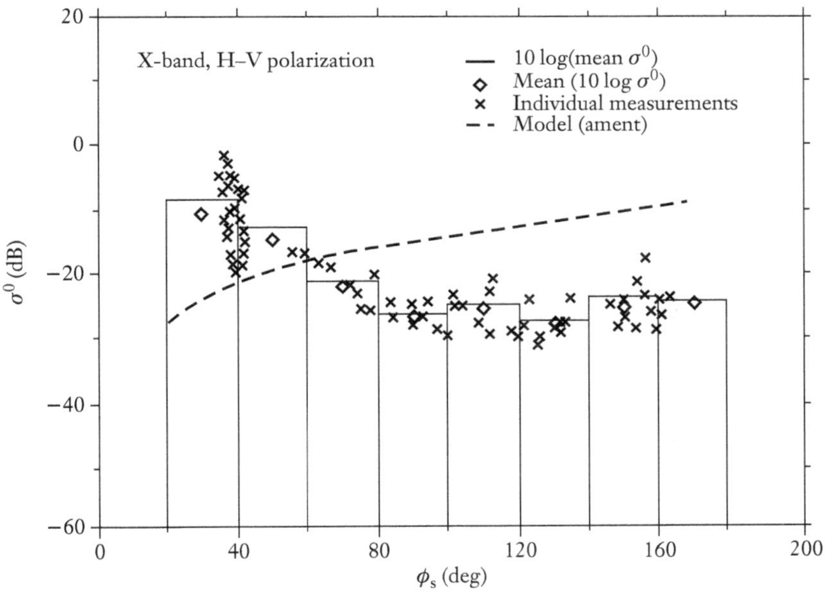

(d)

Figure 9–31 Scattering coefficient azimuthal pattern for tree-orchard on wet ground with 6-inch. snow cover; $\theta_s = 80°$, $\theta_i = 60°$ (after Vander Schuur and Tomlinson, University of Michigan data, April 1979 [42]).

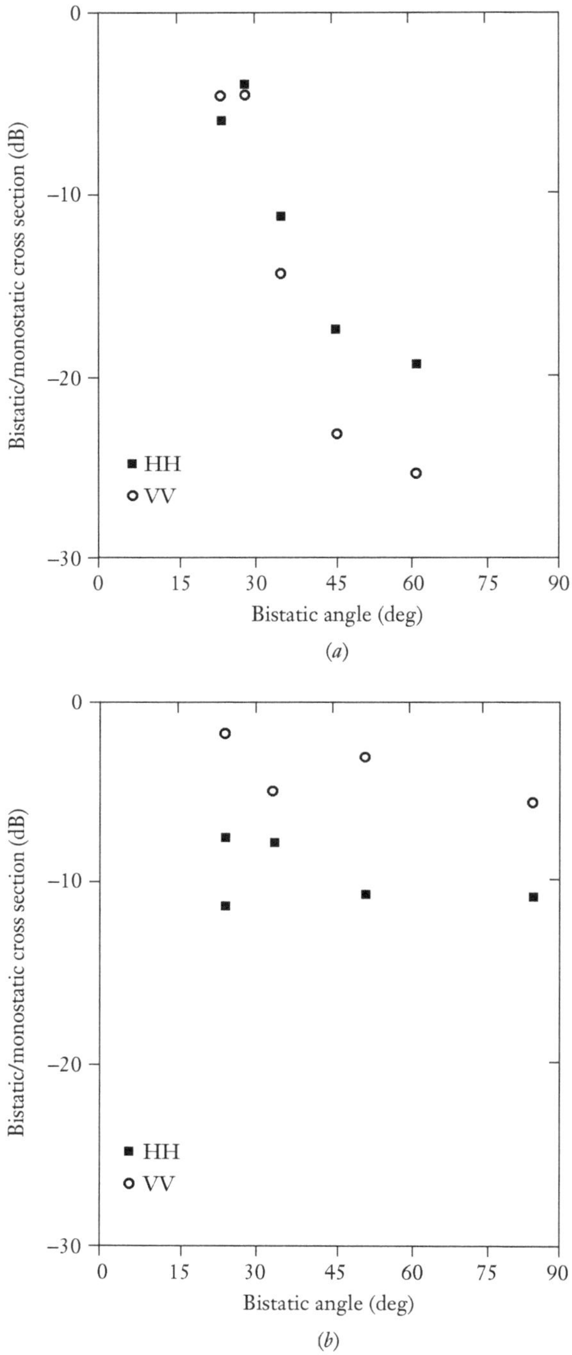

Figure 9–32 Ratio of bistatic to monostatic scattering coefficients (without correction for propagation) for sea clutter at low grazing angles of incidence, H–H and V–V polarizations. (*a*) 1.2–1.8-m wave height, organized sea; (*b*) 1.2–1.8-m wave height, disorganized sea; (*c*) 0.9-m wave height; (*d*) scatter diagram of log normal standard deviations (LSD) for curves (a), (b), (c) (after Ewell and Zehner, 1982 1EE Radar Conference Proceedings, London, pp. 188–192 [35]).

(continues)

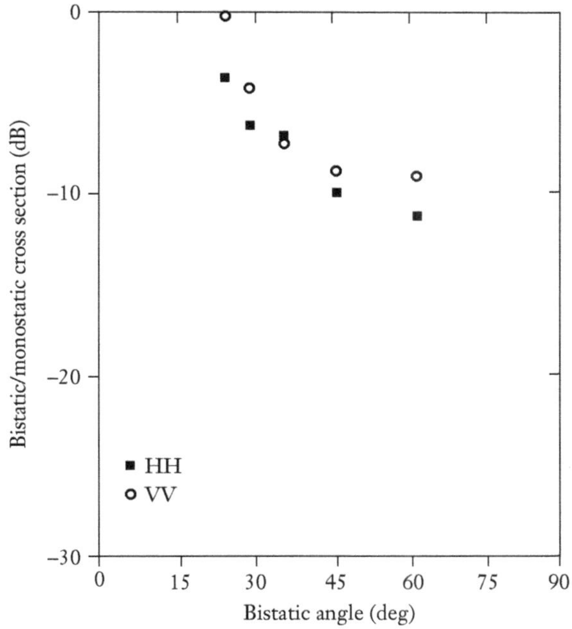

(*c*)

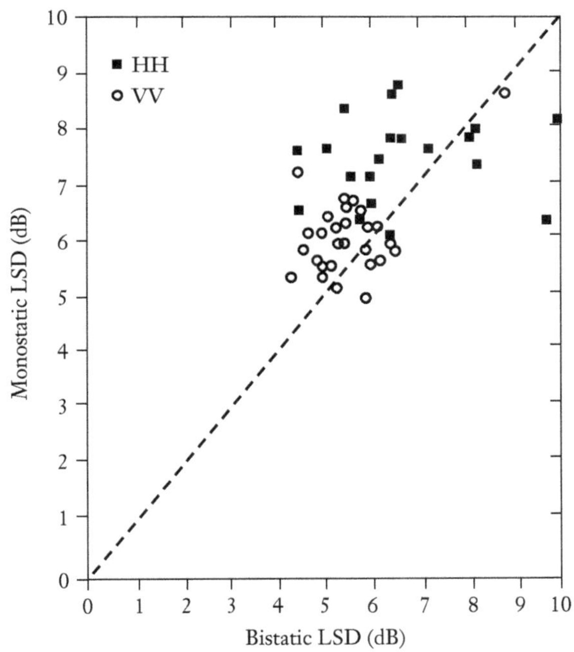

(*d*)

Figure 9–32 Ratio of bistatic to monostatic scattering coefficients (without correction for propagation) for sea clutter at low grazing angles of incidence, H–H and V–V polarizations. (*a*) 1.2–1.8-m wave height, organized sea; (*b*) 1.2–1.8-m wave height, disorganized sea; (*c*) 0.9-m wave height; (*d*) scatter diagram of log normal standard deviations (LSD) for curves (a), (b), (c) (after Ewell and Zehner, 1982 1EE Radar Conference Proceedings, London, pp. 188–192 [35]).

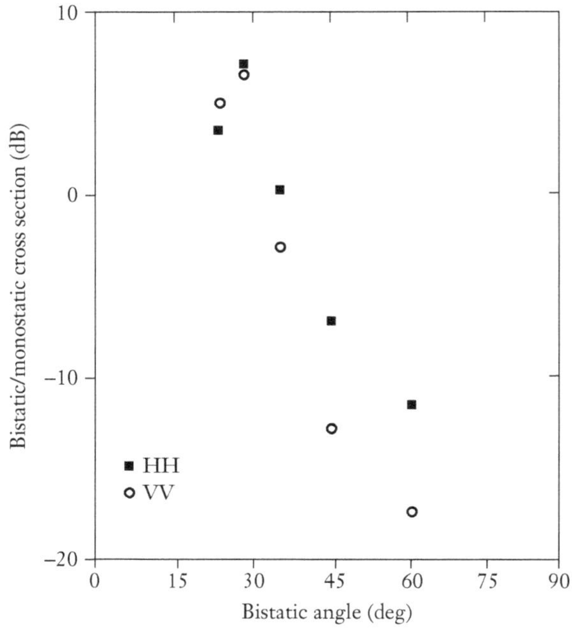

(*a*)

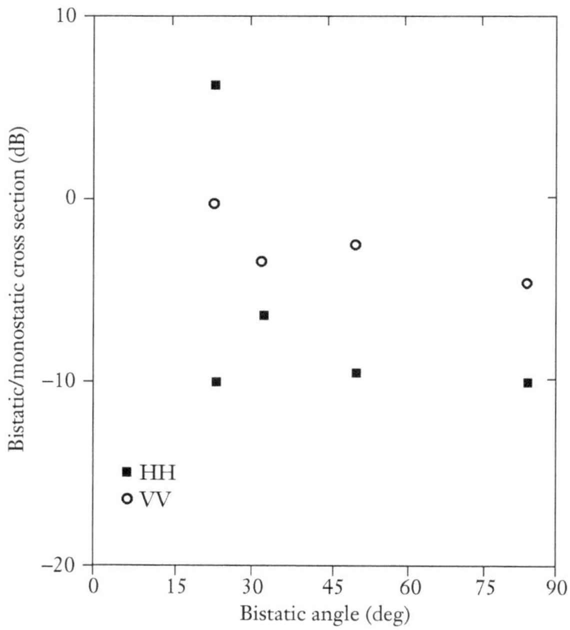

(*b*)

Figure 9–33 Ratio of bistatic to monostatic scattering coefficients (with correction for propagation) for sea clutter at low grazing angles of incidence, H–H and V–V polarizations. (*a*) 1.2- to 1.8-m wave height, organized sea; (*b*) 1.2- to 1.8-m wave height, disorganized sea; (*c*) 0.9-m wave height (after G. P. Ewell, "Techniques of Radar Reflectivity Measurements," in N. C. Currie (ed.) *Bistatic Radar Cross-Section Measurements*, 2nd ed., Artech House, Inc., Norwood, MA, 1984 [36]).

(*continues*)

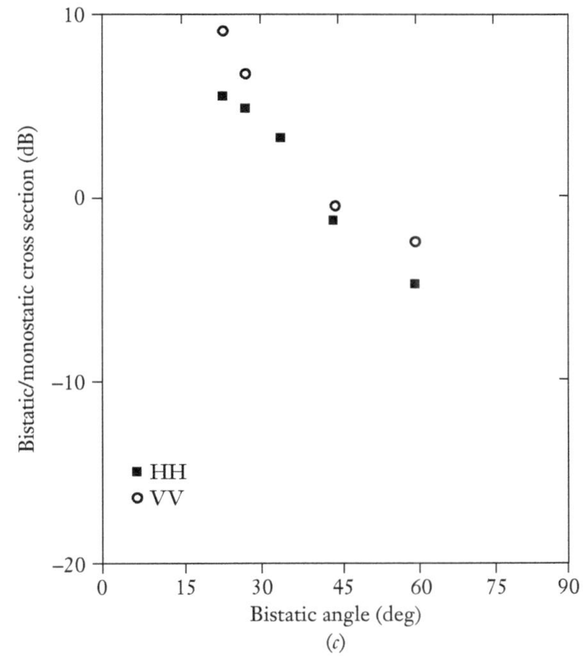

Figure 9–33 Ratio of bistatic to monostatic scattering coefficients (with correction for propagation) for sea clutter at low grazing angles of incidence, H–H and V–V polarizations. (*a*) 1.2- to 1.8-m wave height, organized sea; (*b*) 1.2- to 1.8-m wave height, disorganized sea; (*c*) 0.9-m wave height (after G. P. Ewell, "Techniques of Radar Reflectivity Measurements," in N. C. Currie (ed.) *Bistatic Radar Cross-Section Measurements*, 2nd ed., Artech House, Inc., Norwood, MA, 1984 [36]).

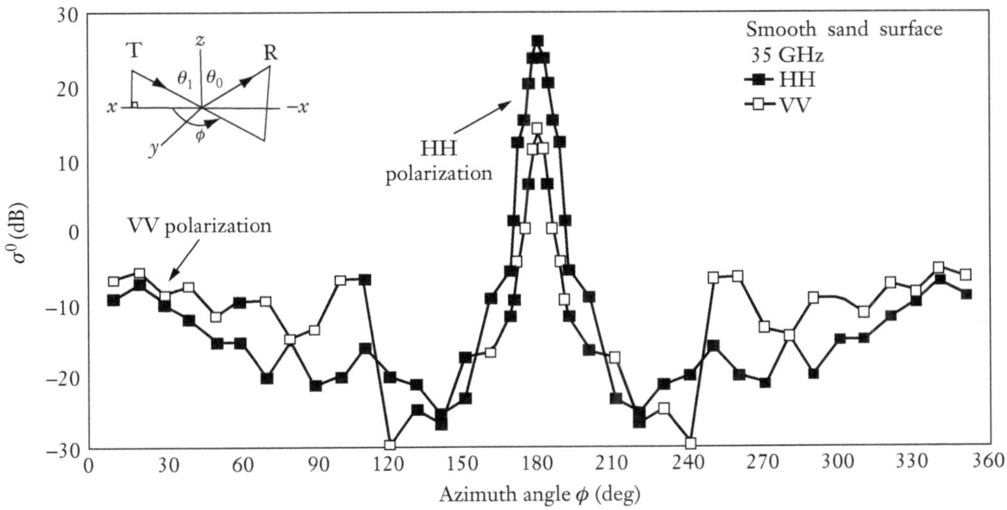

Figure 9–34 Scattering coefficient azimuthal pattern for smooth sand, H–H and V–V polarizations, $\theta_i = \theta_s = 66°$ (after Ulaby et al., *IEEE Transaction*, Vol. GRS-26, No. 3, May 1988, pp. 229–242 [37]).

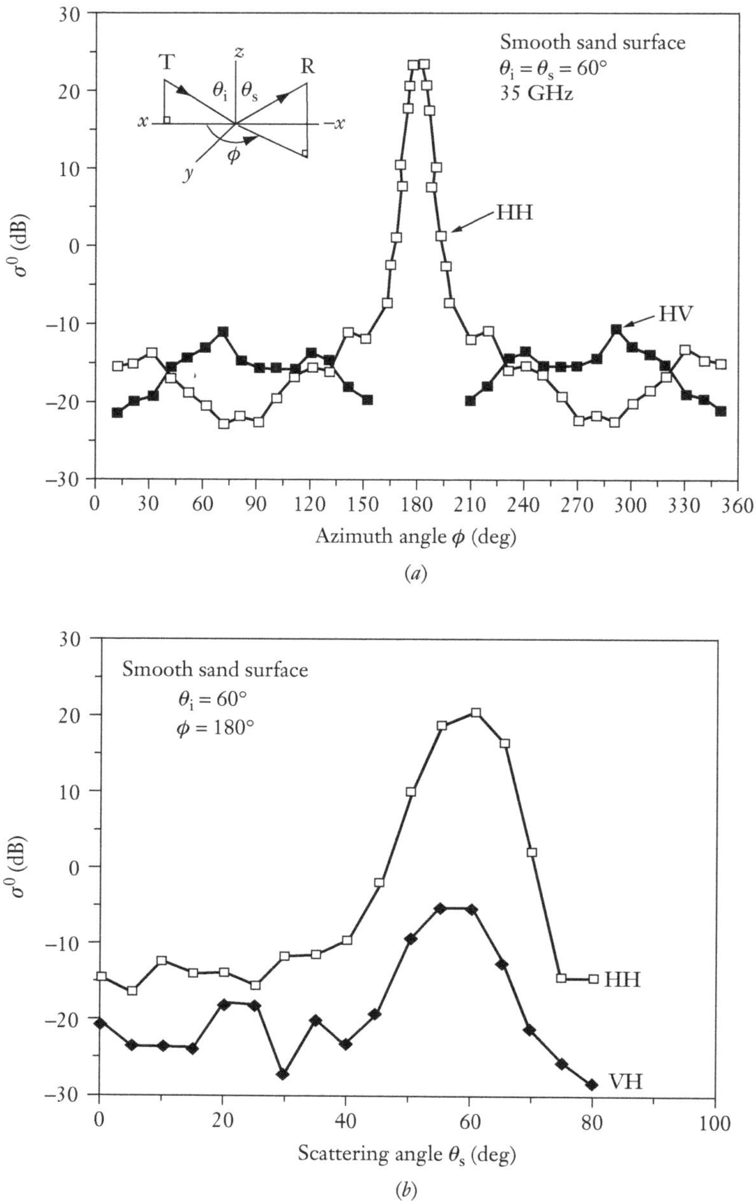

Figure 9–35 Scattering coefficient for smooth sand, H–H, H–V, V–H and V–V polarizations, $\theta_i = 60°$. (*a*) $\theta_s = 60°$, $\phi = 0° - 360°$; (*b*) $\theta_s = 0° - 80°$, $\phi = 180°$; (*c*) $\theta_s = 0° - 80°$, $\phi = 270°$; (*d*) $\theta_s = 0° - 80°$, $\phi = 0° - 360°$ (after Ulaby et al., *IEEE Transaction*, Vol. GRS-26, No. 3, May 1988, pp. 229–242 [37]).

(*continues*)

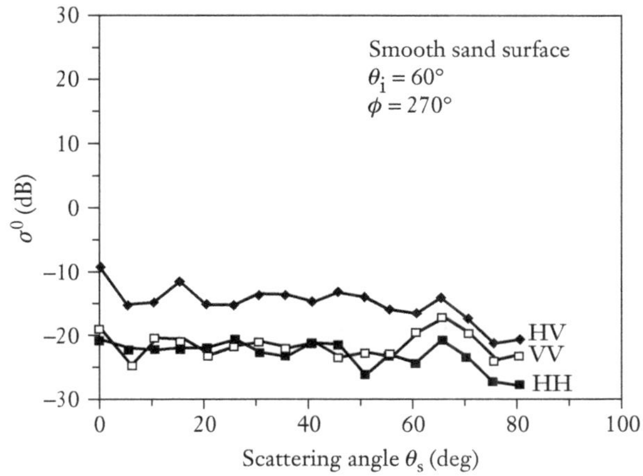

(c)

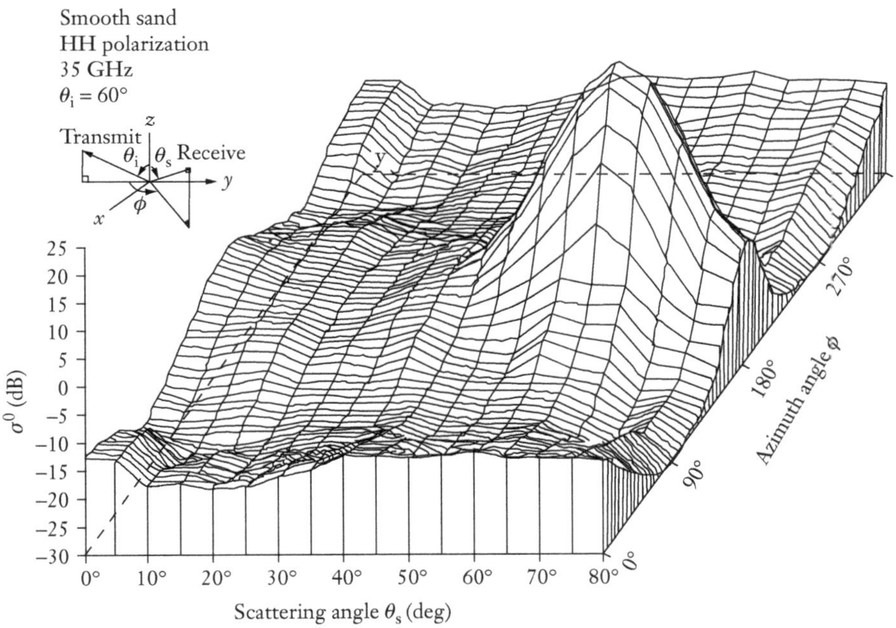

(d)

Figure 9–35 Scattering coefficient for smooth sand, H–H, H–V, V–H and V–V polarizations, $\theta_i = 60°$. (*a*) $\theta_s = 60°$, $\phi = 0° - 360°$; (*b*) $\theta_s = 0° - 80°$, $\phi = 180°$; (*c*) $\theta_s = 0° - 80°$, $\phi = 270°$; (*d*) $\theta_s = 0° - 80°$, $\phi = 0° - 360°$ (after Ulaby et al., *IEEE Transaction*, Vol. GRS-26, No. 3, May 1988, pp. 229–242 [37]).

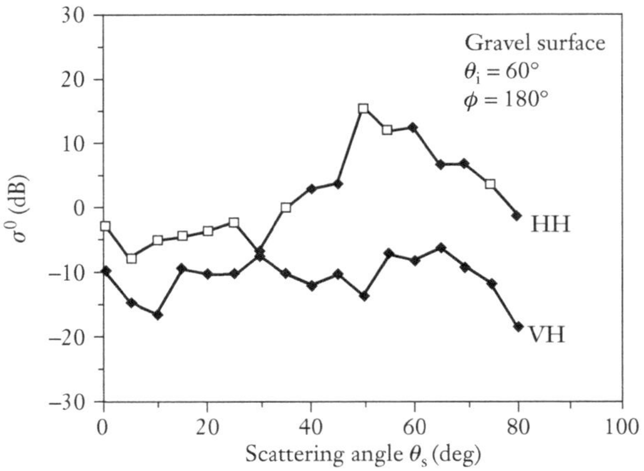

(b)

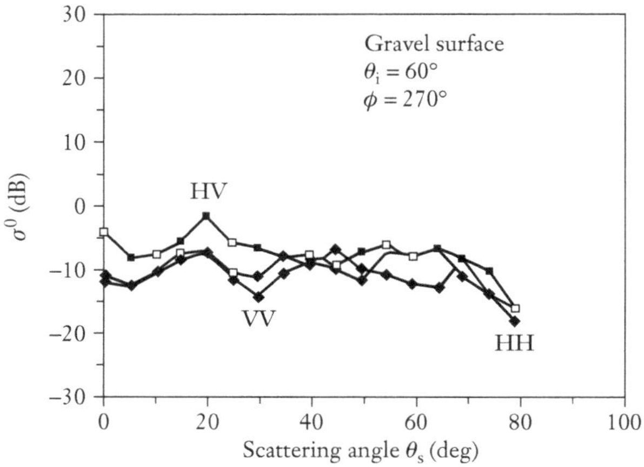

(c)

Figure 9–37 Scattering coefficient for gravel, H–H, H–V, V–H and V–V polarizations, $\theta_i = 60°$. (*a*) $\theta_s = 60°$, $\phi = 0°–360°$; (*b*) $\theta_s = 0°–80°$, $\phi = 180°$; (*c*) $\theta_s = 0°–80°$, $\phi = 270°$ (after Ulaby et al., *IEEE Transaction*, Vol. GRS-26, No. 3, May 1988, pp. 229–242 [37]).

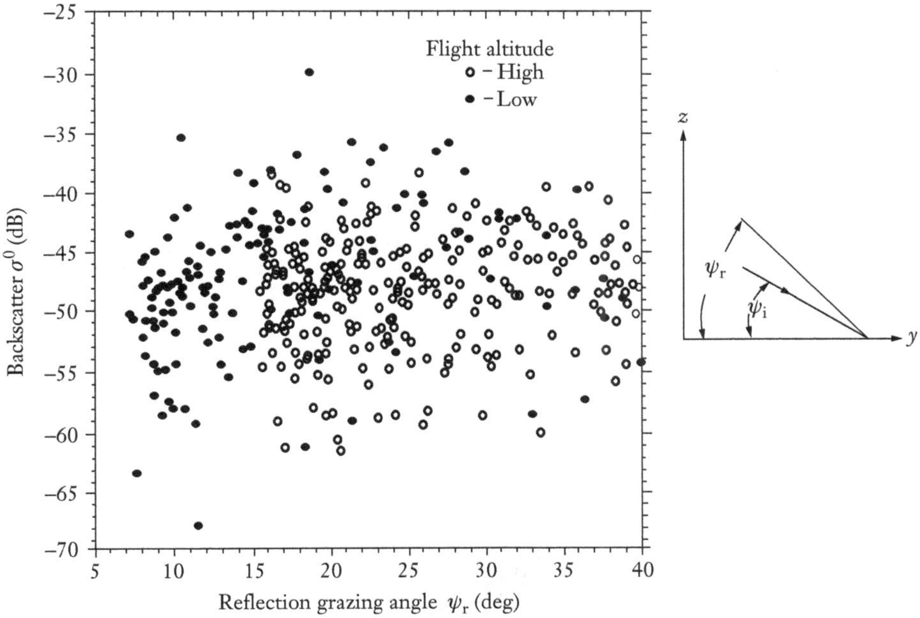

(a)

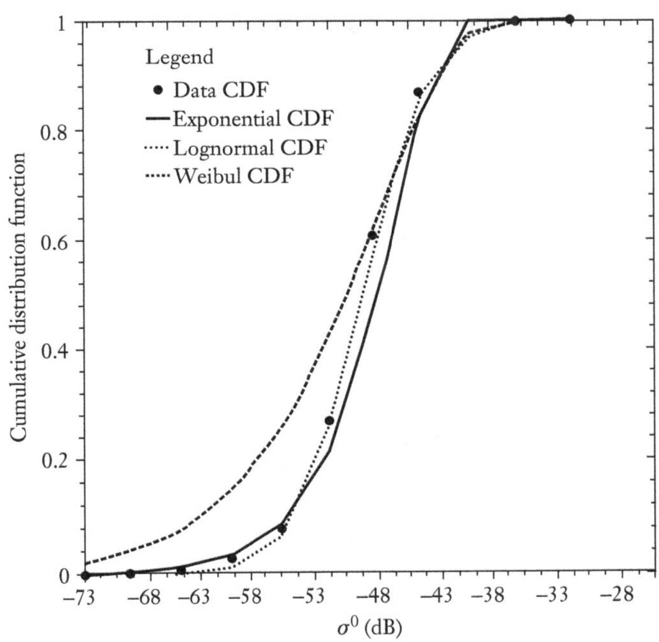

(b)

Figure 9–38 Scattering Coefficient in Plane of Incidence for Sea state 1, V–V and V–H Polarizations, $\psi_i = 0.3°$, $\psi_r = 5°–40°$. (*a*) all data points; (*b*) cumulative distribution function of (*a*). Assuming Statistical Independence of Samples (after Kochanski et al., *IEEE Ocean 92 Conference Proceedings*, Vol. 1, pp. 518–523 [38]).

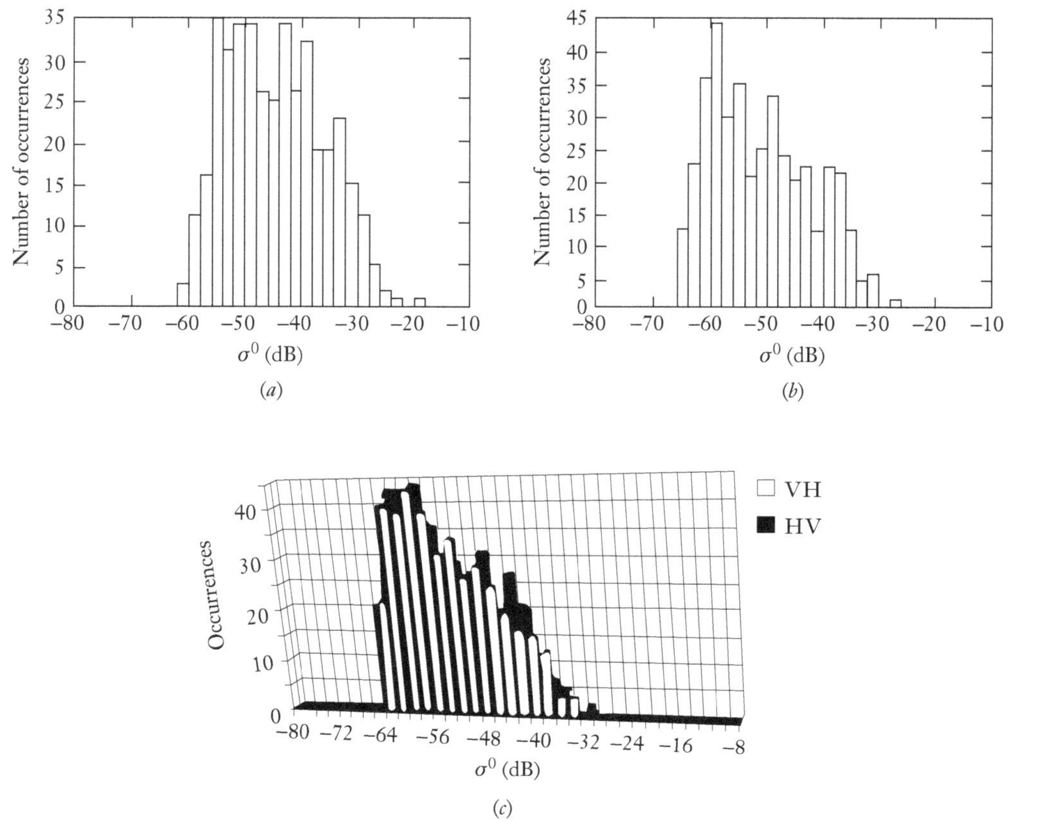

σ^0	VV	VH	HH	HV
	dB	dB	dB	dB
Mean	−36.5	−46.1	−42.5	−46.7
Median	−45.2	−53.7	−52.1	−54.9
Std. dev.	8.6	7.6	8.9	7.7

(d)

Figure 9–39 Scattering coefficient histograms for forested hills at low grazing angles, (averaged over azimuthal angles $\phi = 20–70°$, propagation factor included). (*a*) V–V polarization; (*b*) H–H polarization; (*c*) V–H and H–V polarizations; (*d*) summary of statistics, (*a*)–(*c*) (after McLaughlin et al., *Electronic Letters*, Vol. 30, No. 18, September 1, 1994, pp. 1532–1533, [39]; 31(6):490–491, March 16, 1995 [40]).

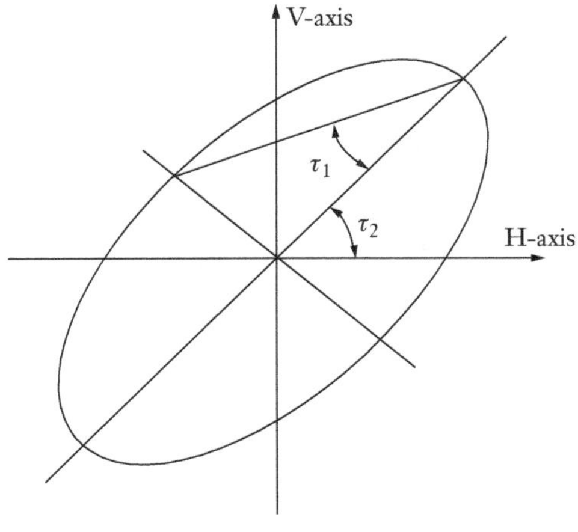

(*a*)

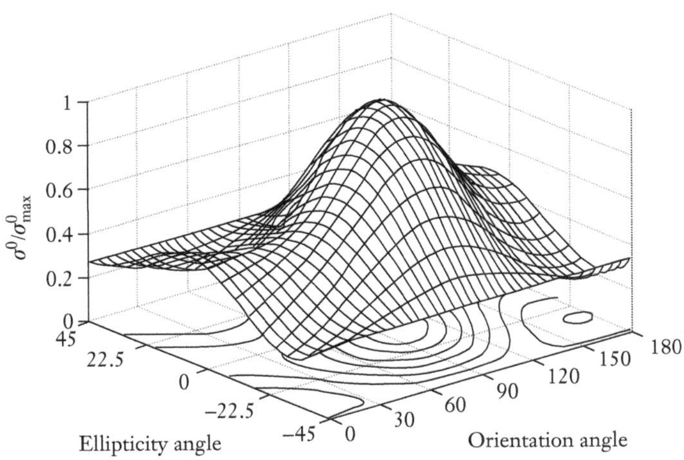

(*b*)

Figure 9–40 Normalized scattering coefficient σ^0/σ^0_{max} as a function of transmit antenna polarization state (forested hills, low grazing angles, azimuthal angle $\phi = 32°$). (*a*) Polarization state defined by ellipticity angle τ_1 and orientation angle τ_2; (*b*) σ^0/σ^0_{max}, Copolarized antennas, $-29.9\text{ dB} = \sigma^0_{HHmax} \leq \sigma^0_{max} \leq \sigma^0_{VVmax} = -29.9\text{ dB}$; (*c*) σ^0/σ^0_{max}, Crosspolarized antennas, $-36.2\text{ dB} = \sigma^0_{HVmax} \leq \sigma^0_{max} \leq \sigma^0_{VHmax} = 32.3\text{ dB}$ (after McLaughlin et al., *IEEE Transactions*, Vol. AP-50, No. 2, February 2002, pp. 101–110 [41]).

(*continues*)

9.3.3.3 Future Data Acquisition The measurement of bistatic scattering coefficients is considerably more difficult than the measurement of monostatic scattering coefficients because bistatic geometry imposes the following constraints:

a. The indirect, scattered signal must be separated from a direct signal which may be orders of magnitude larger.

b. Antenna scan in the azimuthal plane cannot be utilized to obtain a large number of independent samples of the scattering coefficient for a given scattering direction $(\theta_i, \theta_r, \phi)$ because ϕ varies with azimuthal scan.

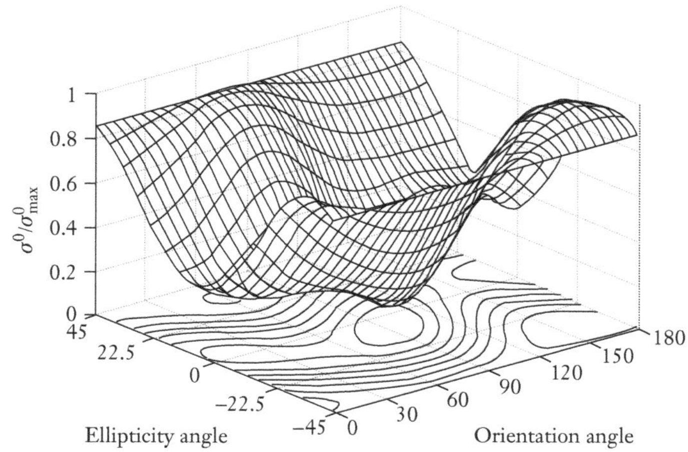

(c)

Figure 9–40 Normalized scattering coefficient σ^0/σ^0_{max} as a function of transmit antenna polarization state (forested hills, low grazing angles, azimuthal angle $\phi = 32°$). (*a*) Polarization state defined by ellipticity angle τ_1 and orientation angle τ_2; (*b*) σ^0/σ^0_{max}, Copolarized antennas, −29.9 dB = $\sigma^0_{HHmax} \leq \sigma^0_{max} \leq \sigma^0_{VVmax}$ = −29.9 dB; (*c*) σ^0/σ^0_{max}, Crosspolarized antennas, −36.2 dB = $\sigma^0_{HVmax} \leq \sigma^0_{max} \leq \sigma^0_{VHmax}$ = 32.3 dB (after McLaughlin et al., *IEEE Transactions*, Vol. AP-50, No. 2, February 2002, pp. 101–110 [41]).

c. The effects of shadowing at low grazing angles of incidence must be determined in two directions rather than one direction.

d. The contribution to the scattering coefficient of coherent scattering, which can be substantial in the forward-scattered direction, must be separated form the incoherent contribution.

e. Calibration of measurement data and calculation of the cell area for specific radiation patterns must be performed over many more scattering angles.

f. An additional antenna at a separate site with support logistics is required.

Prior to 1982 (see Nos. 1–6 in Table 9–1(c) of Section 9.3.3.1), the existing database of terrain and sea scattering coefficients consisted mostly of mean values in the plane of incidence at grazing angles above 5°. The only statistical parameters which had been measured were mean values, some standard deviations, one cumulative probability distribution, a couple of probability density distributions, and no fully polarimetric data. There were almost no measurements of coherent or incoherent spatial correlations with cell size or with the Fresnel number subtended at the receiver by the cell area. Since 1982 (see Nos. 7–10 in Table 9–1(b)), the database includes out-of-plane measurements at low grazing angles of incidence, more statistical data, and fully polarimetric data. Correlation of data with cell size or with Fresnel number is still not available.

The correlation of measured scattering coefficient with cell area is important because it determines the minimum and maximum cell areas for which the theme may be considered homogeneous. The correlation also determines whether the scattering coefficient is predominantly coherent or incoherent. An alternative method for determining the coherent and incoherent contributions to the scattering coefficient consists of measurements at the maxima and minima of the interference pattern between the scattered and direct signals [54]. For a homogeneous theme, the measured scattering coefficient should be independent of cell area for incoherent scattering and strongly dependent on cell area for coherent scattering. For coherent scattering in the far field of the cell area, the scattering coefficient should be proportional to cell area [see equation (9.26)] whereas in the near-field the coefficient should be inversely proportional to cell area [see equation (9.32)].

The design of signal processors, for bistatic radar in which either the receiver or transmitter looks down at terrain or the sea, requires detailed information on the bistatic scattering coefficients and their statistical properties if the false-alarm rate from terrain or sea clutter is to be controlled. For air surveillance radar with either a ground-based receiver or transmitter at ranges of several nautical miles, the information of greatest interest is bistatic scattering coefficients and their statistical distributions at grazing angles less than 5° in both the azimuthal and elevation planes. Furthermore, if the bistatic database is to be transferable with confidence to different radars with different antenna beamwidths or pulse widths, or at different ranges, then the dependence of the scattering coefficient upon cell area must be known.

One method for controlling the cell area is to utilize a mask of given aperture area, located at the theme of interest. The advantages of using a mask to vary cell area are as follows: (a) the receiver and transmitter can then be in the far-field of the cell area so that the number of Fresnel zones subtended by the cell is fixed; (b) several independent samples can be obtained by moving the mask; (c) terrain masking and scattering off abrupt edges of the theme can be controlled; (d) the cell area is more precisely defined; and (e) the transmitter and antenna gains can be made approximately constant over the cell area, thereby simplifying the calculation of the scattering coefficient from measurement data. The requirements for using a mask are as follows: (a) the mask must have much less scattering to the receiver than the clutter cell being measured and (b) the mask must be physically large compared to the clear aperture area, wavelength, and transmitter beam size, to minimize energy diffracted around the edges of the mask. Although the use of masks or limiting apertures are a common technique for optical measurements, it is recognized that considerable design work may be necessary to achieve mask requirements at microwave frequencies.

ACKNOWLEDGEMENTS

Chapter 9 is in large part from M. M. Weiner, "Multistatic Radar Phenomenology, Vol. 1-Terrain and Sea Scatter," MITRE Technical Report MTR-8140A-1, The MITRE Corporation, September 1980, AD BO59351L, as amended and approved for public release; distribution unlimited on July 7, 2003. (Also reissued as RADC-TR-81-75, Vol. 1, Rome Air Development Center, Griffiss Air Force Base, NY 13441, May 1981.) This material is published with permission from The MITRE Corporation.

The work reported in MTR-8140A-1 was sponsored by the U.S. Air Force, Rome Air Development Center under contract F19628-80-C-0001.

The author, Melvin M. Weiner, is indebted to Richard A. Ackley of the Rome Air Development Center for his sponsorship and technical direction. The author is also indebted to Robert I. Millar, Project Leader, for his guidance during the course of the project and J. David Kramer and Chojan S. Teng for helpful discussions.

Some of the figures in this chapter have been published in refereed journals, conference proceedings, or other public domains protected by copyright. Receipt of permission to republish is gratefully acknowledged as follows:

- *Figure 9–13*: M. M. Weiner, "Useful Beam Quality Design Curves for Unstable Resonators," *Optical Engineering*, Vol. 13, No. 2 March/April 1974, pp. 87–91.
- *Figure 9–25*: F. E. Nathanson, *Radar Design Principles*, McGraw-Hill Co., New York, 1969, pp. 88–89.

- *Figures 9–31 to 9–33*: R. W. Larsen, A. L. Maffett, R. C. Heimiller, A. F. Fromm and E. L. Johnson, "Bistatic Clutter Measurements," *IEEE Transactions*, Vol. AP-26, No. 6, November 1978, pp. 801–804.
- *Figure 9–36*: G. W. Ewell and S. P. Zehner, "Bistatic Sea Clutter Return Near Grazing Incidence," in *IEE International Conference Radar-82*, London, Publication No. 216, October 1982, pp. 88–192.
- *Figure 9–37*: G. P. Ewell, "Techniques of Radar Reflectivity Measurements," in N. C. Currie (ed.) *Bistatic Radar Cross-Section Measurements*, 2nd ed., Artech House, Inc., Norwood, MA, 1984.
- *Figures 9–38 to 9–41*: F. T. Ulaby, T. E. Van Deventer, J. R. East, T. F. Haddock and M. E. Coluzzi, "Millimeter-Wave Bistatic Scattering From Ground and Vegetation Targets," *IEEE Transactions*, Vol. GRS-26, No. 3, May 1988, pp. 229–243.
- *Figure 9–42*: T. P. Kochansksi, M. J. Vanderhiil, J. V. Zoloterevsky and T. Fariss, "Low Illumination Angle Bistatic Clutter Measurements at X-Band," in *IEEE International Conference on Oceans-92: Mastering the Oceans Through Technology, Proceedings*, Vol. 1, October 26–29, 1992, pp. 518–523.
- *Figure 9–43*: D. J. Mclaughlin, E. Boltniew, Y. Wu and R. S. Raghaven. "Low Grazing Angle Bistatic NRCS of Forested Clutter," *Electronics Letters*, Vol. 30, No. 18, September 1, 1994, pp. 1532–1533. D. J. McLaughlin, E. Boltniew, R. S. Raghavan and M. J. Sowa, "Crosspolarized Bistatic Clutter Measurements," *Electronics Letters*, Vol. 31, No. 6, March 16, 1995, pp. 490–491.
- *Figure 9–44*: D. J. McLaughlin, Y. Wu, W. G. Stevens, X. Zhang, M. J. Sowa and B. Weijers, "Fully Polarimetric Bistatic Radar Scattering Behavior of Forested Hills," *IEEE Transactions*, Vol. AP-5O, No. 2, February 2002, pp. 101–110.

Appendix 9–A Aperture Diffraction Theory (Survey) (*Sections. 9.3.1 and 9.3.2*)

A.I SUMMARY

Aperture diffraction theory relates the field at an observation point P to the field originating at a sourcefigure point P_0 when the source field and wavefront at points Q of a limiting aperture are modified by the aperture.

Rigorous diffraction theory formulates the field at *P* by either:

1) Integrating the fields that propagate from points on a closed surface which includes the points Q, encloses P, but excludes P_0 or
2) Integrating the fields which propagate from points Q of a plane aperture imbedded in a ground plane.

For suitable small angle approximations, the rigorous diffraction theory reduces to a scalar diffraction theory in which the polarization of the field at the observer is approximately the same as that at of the source point. The scalar diffraction theory, in turn, takes several forms depending on the approximations which are made.

In this Appendix, the various forms of aperture diffraction theory are surveyed by giving an overview followed by a more detailed analysis of the overview. For each form of the theory, the approximations are given.

A.2. OVERVIEW (SEE FIGURE A-1)

I. Rigorous Electromagnetic Theory

(a) Vector Equivalent of Helmholtz and Kirchhoff Integral [55–57] (Integration over closed surface external to sources)

↓

(b) Vector Equivalent of Huygens-Fresnel Integral [56] (Integration over plane aperture imbedded in ground plane)

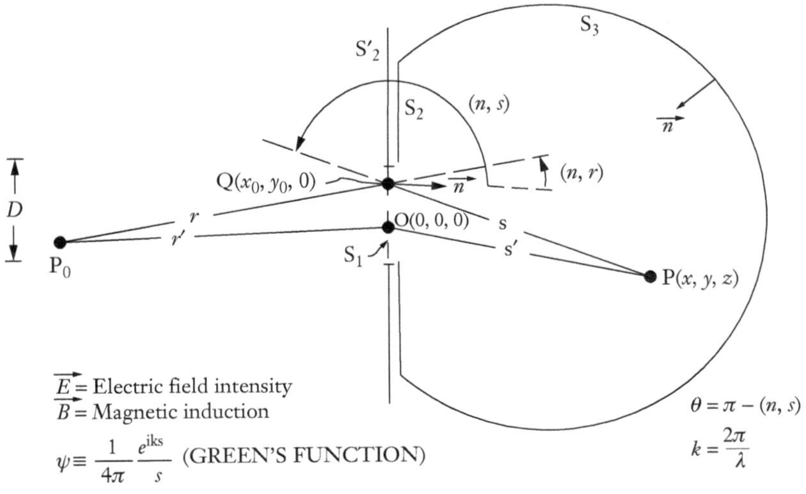

Figure A–1 Aperture diffraction geometry.

II. Scalar Diffraction Theory (Small angle approximations)
From I (a)

(c) Helmholtz and Kirchhoff Integral [55–59] (Source point, arbitrary aperture point, and observation point lie in approximately a straight line)

↓

(d) Fresnel-Kirchhoff Integral [55–59] (Aperture edge fields are neglected)

From I (b)

(e) Huygens-Fresnel Integral [56, 58, 59] (Position vector of observation point makes a small angle with respect to aperture normal)

↓

(f) Fresnel Diffraction Integral [59] $\left\{\frac{s}{8\lambda}\left[\left(\frac{x-x_0}{s}\right)^2+\left(\frac{y-y_0}{s}\right)^2\right]<<l\right\}$

↓

(g) Fraunhoffer Diffraction Integral [55–59] $\left(\frac{x_0^2+y_0^2}{2\lambda s}<<l\right)$

↓

(h) Fourier Integral

A.3 ANALYSIS

1. Rigorous Electromagnetic Theory

(a) Vector Equivalent of Helmholtz and Kirchhoff Integral

$$\begin{aligned}\vec{E}(p) &= \iint_{\Sigma}\left[\vec{E}(Q)\frac{\partial\psi}{\partial n}-\psi\frac{\partial\vec{E}(Q)}{\partial n}\right]dS \equiv \iint_{\Sigma}[\vec{E}(Q)(\vec{n}\cdot\vec{\nabla}\psi)-\psi(\vec{n}\cdot\vec{\nabla})\vec{E}(Q)]dS \\ &= \iint_{\Sigma}\left[\left\{i\frac{2\pi}{\lambda}[\vec{n}\times\vec{B}(Q)]\psi\right\}+\{[\vec{n}\times\vec{E}(Q)]\times\vec{\nabla}\psi\}+\{[\vec{n}\,.\,\vec{E}(Q)]\vec{\nabla}\psi\}\right]dS\end{aligned} \tag{A-1}$$

$\sum = S_1 + S_2 + S_3$ (can be any *closed* surface *external* to the sources)

(b) Vector Equivalent of Huygens–Fresnel Integral

Let S_2' be an infinitely conducting plane screen (Condition A)

By the method of mirror image potentials [56], equation (A-1) reduces to

$$\vec{E}(P) = -2\iint_{S_1}\{[\vec{n}\times\vec{E}(Q)]\times\vec{\nabla}\psi\}dS \tag{A-2}$$

where $\vec{E}(Q)$ is the field in the presence of the screen.

II. Scalar Diffraction Theory (Small Angle Approximations)

(c) Helmholtz and Kirchhoff Integral

Consider the case where $\overline{P_0QP}$ is approximately a straight line and, therefore, satisfies the small angle condition

$$\cos(\overline{P_0Q}, \overline{QP}) \gtrsim 0.9 \qquad \text{(Condition B)}$$

For Condition B, $\vec{E}(P) \approx \hat{t}_0 E(P)$, $\vec{E}(Q) \approx \hat{t}_0 E(Q)$, $\vec{E}(P_0) \approx \hat{t}_0 E(P_0)$, where $\hat{t}_0$ = unit vector parallel to the polarization of $\vec{E}(P_0)$ and is transverse to $\vec{r}$ and $\vec{s}$. For Condition B, equation (A-1) reduces to

$$E(p) = \iint_{\Sigma} \left[E(Q)\frac{\partial \psi}{\partial n} - \psi \frac{\partial E(Q)}{\partial n} \right] \mathrm{d}S \qquad \text{(A-3)}$$

(d) Fresnel-Kirchhoff Integral

High-Frequency Conditions $r >> \lambda$, $s >> \lambda$, $D >> \lambda$ (Conditions C)

Kirchhoff Conditions

$$\left.\begin{aligned} \text{on } S_1: E(0) &\approx E_0 \frac{e^{\mathrm{ikr}}}{r} \\ \text{on } S_2 \text{ and } S_3: E(Q) &\approx 0, \frac{\partial E(Q)}{\partial n} \approx 0 \end{aligned}\right\} \qquad \text{(Condition D)}$$

Conditions D ignore aperture edge effects. It is not clear that Conditions C are sufficient conditions to ignore aperture edge effects. For example, should D be approximately transverse to $\overline{P_0QP}$? Substituting Conditions D into equation (A-3) and noting from Conditions C that $ik >> \frac{1}{r}$, $ik >> \frac{1}{s}$,

$$E(P) = \frac{iE_0}{2\pi} \iint_{S_1} \frac{e^{\mathrm{ik(r+s)}}}{rs} [\cos(n,s) - \cos(n,r] \mathrm{d}S \qquad \text{(A-4)}$$

where cos(n, r), cos (n, s) are subject to Condition B.

For Condition B, [cos (n, s) – cos (n, r)] ≈ –2 cos θ where θ is the angle between $\overline{QP}$ and the aperture normal.

(e) Huygens–Fresnel Integral

Subject only to Condition A, equation (A-2) may be written as

$$\vec{E}(P) = -2\iint_{S_1} \left\{ \left[\vec{n} \times \vec{E}(Q) \right] \right\} \times \hat{s} \frac{e^{\mathrm{iks}}}{4\pi s} \left(ik - \frac{1}{s} \right) \mathrm{d}S \qquad \text{(A-5)}$$

Far (Radiation) Zone Conditions

$$s >> \lambda \quad \text{(Condition E-1)}$$

$$D << s \quad \text{(Condition E-2)}$$

For Conditions E, equation (A-5) reduces to

$$\vec{E}(P) = -(i/\lambda)\frac{1}{s'}\iint_{S_1} \hat{s}' \times \left[\vec{E}_t(Q) \times \hat{z}\right] e^{iks} dS \quad \text{(A-6)}$$

$$\text{(Conditions A, E)}$$

where

$$\vec{E}_t(Q) = \hat{x}E_x(Q) + \hat{y}E_y(Q) = \text{tangential aperture field}$$

$$s = s' - \frac{(xx_0 + yy_0)}{s'} + \frac{x^2{}_0 + y^2{}_0}{2s'} + \frac{(-2xx_0 - 2yy_0 + x^2{}_0 + y^2{}_0)}{R(s')^3} + \cdots$$

Uniform Polarization Condition $\vec{E}_t(Q) = \hat{t}_0\, E_t(0)$ (Condition F)
For Condition F, equation (A-6) reduces to

$$\text{Then } \vec{E}(P) = -(i/\lambda)\frac{e^{iks'}}{s'}\left[\hat{s}' \times (\hat{t}_0 \times \hat{z})\right]\iint_{S_1} E_t(Q) e^{-iks} dS \quad \text{(A-7)}$$

$$\text{(Condition A, E, F)}$$

Small Angle Condition $\sin^2\theta << 1$ (Condition G)
For Condition G, equation (A-6) reduces to

$$\vec{E}(P) = -(i/\lambda)\frac{e^{iks'}}{s'}\cos\theta' \iint_{S_1} \vec{E}_t(Q) e^{-iks} dS \quad \text{(A-8)}$$

$$\text{(Condtion A, E,G)}$$

Source Condition $\vec{E}_t(Q) \approx \begin{cases} \vec{E}_t(Q), & \text{on } S_1 \\ 0 & \text{on } S_2, S_3 \end{cases}$ (Condition H)
where $\vec{E}_0(Q)$ = tangential field at the aperture in the absence of the screen. For Condition H, equation (A-8) reduces to the Huygens–Fresnel integral

$$\vec{E}(P) = -(i/\lambda)\frac{e^{iks'}}{s'}\cos\theta' \iint_{S_1} \vec{E}_t(Q) e^{-iks} dS \quad \text{(A-9)}$$

$$\text{(Condtion A, E, G, H)}$$

(f) Fresnel Diffraction Integral
Fresnel Diffraction Condition

$$k\frac{[-2xx_0 - 2yy_0 + x_0^2 + y_0^2]^2}{8(s')^3} << 2\pi \quad \text{(Condition I)}$$

For Condition I, equation (A-9) reduces to the Fresnel Integral

$$\vec{E}(P) = -(i/\lambda)\frac{e^{iks}}{s'}\cos\theta' \iint_{S_1} \vec{E}_0(Q)\, e^{-i\frac{k}{s'}\left[xx_0 + yy_0 - \frac{(x_0^2+y_0^2)}{2}\right]} dx_0\, dy_0 \tag{A-10}$$

(Conditions A, E, G, H, I)

(g) Fraunhofer Diffraction Integral
Fraunhofer Diffraction Condition

$$\frac{k(x_0^2+y_0^2)}{2s'} << 2\pi \qquad \text{(Condition J)}$$

For Condition J, equation (A-9) reduces to the Fraunhofer Integral

$$\vec{E}(P) = -(i/\lambda)\frac{e^{iks}}{s'}\cos\theta' \iint_{S_1} \vec{E}_0(Q)\, e^{-i\frac{k}{s'}[xx_0 + yy_0]} dx_0\, dy_0 \tag{A-11}$$

(Conditions A, E, G, H, J)

(h) Fourier Integral

$$\text{Let}\quad k'_x = \frac{k}{s'}x = k\sin\theta'\cos\theta'$$

$$k'_y = \frac{k}{s'}y = k\sin\theta'\sin\phi'$$

Then equation (A-11) may be expressed as

$$\vec{E}(P) = -(i/\lambda)\frac{e^{iks}}{s'}\cos\theta'\, \vec{F}_t\left(k'_x, k'_j\right) \tag{A-12}$$

where

$$\vec{F}_t\left(k'_x, k'_y\right) = \iint_{S_1} \vec{E}_0(Q)\, e^{-i\left(k'_x x_0 + k'_y y_0\right)} dx_0\, dy_0 = \text{Fourier transform of } \vec{E}(Q)$$

Appendix 9–B Annotated List of Symbols

Symbol	Annotation	Section of first occurrence
A	Transmitter location point	9.2.1
A_C	Clutter cell area	9.2.1
B	Receiver location point	9.2.1
B	Magnetic field intensity	A.3
BRDF	Bidirectional Reflectance-Distribution Function	9.3.1.3
C	Velocity of wave propagation	9.2.1
C	Clutter cell point	9.2.1
D	Aperture length	A.3
D_A	Transmitter antenna aperture length	9.3.1.2
D_B	Receiver antenna aperture length	9.3.1.2
E	Electric field intensity	A.3
E_C	clutter pulse energy	9.2.1
E_{ip}	Electric field of incident plane wave front	9.2.1.3
E_Q	Electric field at aperture point Q	9.3.1.3
E_{sq}	Electric field of scattered wave front	9.3.1.3
E_S	Target return energy	9.2.1
$E_t(Q)$	Tangential reflected electric field at point Q	9.3.2.1
f_A	Transmitter antenna power pattern shape factor	9.3.1.2
f_B	Receiver antenna power pattern shape factor	9.3.1.2
$F_{C/A}$	Number of Fresnel zones subtended at the transmitter by the clutter cell	9.3.1.2
$F_{C/B}$	Number of Fresnel zones subtended at the receiver by the clutter cell	9.3.1.2
G_A	Transmitter antenna power	9.2.1
G_B	Receiver antenna power	9.2.1
h	Height deviation of earth surface	9.3.2.1
h_A	Altitude of transmitter	9.2.2
h_B	Altitude of receiver	9.2.2
H	Horizontal polarization	9.3.1.1
k	Wavenumber	9.3.1.3
K	Pulse compression ratio	9.2.1
L_P	One-way propagation loss from transmitter to the clutter cell	9.2.1
L_r	One-way propagation loss from clutter cell to the receiver	9.2.1
P	Field observation point	9.3.2.1
P_C	Received peak clutter power	9.3.1.2

$P_{c,q}$	Peak received clutter power for transmitter of polarization state q	9.3.1.2
P_o	Field source point	A.1
P_t	Transmitter peak power	9.2.1
P_t	Total radiated power	9.3.2.2.1
Q	Point on limiting aperture	A.1
r	Distance between receiver and clutter cell	9.2.1
r	Distance from source point to an aperture point	A.3
r	Fresnel amplitude reflection coefficient	9.3.2.1
r'	Distance from source point to aperture origin	A3
R	Range sum	9.2.1
s	Distance from an aperture point to field point	A.2
s'	Distance from aperture origin to field point	A.3
S	Strehl intensity of normalized far-field on-axis beam intensity	9.3.2.2.1
U	Normalized electric field at clutter cell	9.3.1.3
V	Vertical polarization	9.3,1.1
W	Normalized near field on-axis beam intensity	9.3.2.2.2
β	Bistatic angle	9.2.1
β	Phase deviation of reflected wavefront	9.3.2.1
β_{ms}	Root-mean-square phase aberration	9.3.21
γ	Angle subtended at the transmitter by the receiver and clutter cell	9.2.1
Γ	Amplitude of Strehl intensity	9.3.2.1
Δ	Path length difference	9.2.1
$\Delta \mathrm{f}$	Frequency bandwidth	9.3.1.2
δt	Receiver resolution time	9.2.1
θ_A	Transmitter main-lobe null-to-null azimuthal beamwidth	9.2.1
θ_B	Receiver main-lobe null-to-null azimuthal beamwidth	9.2.1
θ_i	Angle of incidence at clutter cell	9.3.1.1
θ_s	Angle of reflection at clutter cell	9.3.1.1
λ	Radar wavelength	9.2.1
ξ_A	Transmitter main lobe null-to-null beamwidth	9.2.2
ρ	Distance between transmitter and clutter cell	9.2.1
ρ_{AB}	Distance between transmitter and receiver	9.2.1
Σ	Any closed surface external to the source	A.3
σ^0	Clutter cross-section per unit area	9.2.1
σ_c	Main beam clutter radar cross section	9.2.1
σ_{pq}	Clutter radar cross-section for polarization states p and q of transmitter and receiver antennas, respectively	9.3.1.3

σ^0_{pq}	Clutter cross-section per unit area for polarization states p and q of transmitter and receiver antennas, respectively	9.3.1.3
ϕ	Out-of-plane azimuthal angle	9.22
ψ	Green's function	A.3
ψ_A	Grazing angle subtended at the clutter cell by the transmitter	9.2.1
ψ_B	Grazing angle subtended at the clutter cell by the receiver	9.2.1
ψ_{equiv}	Equivalent grazing angle	9.2.2

References

1. F. E. Nathanson, *Radar Design Principles*, McGraw-Hill, Co., New York, 1969, pp. 64–65. Please note the typographical errors associated with Eqs. (2-35), (2-36), and Figure 2–5. Please also note that R is used to denote the monostatic range from the transmitter/receiver to the clutter cell and should not be confused with the notation of the present chapter in which $R = p + r =$ range sum.
2. D. E. Barrick, "Normalization of Bistatic Radar Return," Report 1388-13, Ohio State University Foundation, January 15, 1984.
3. S. T. Cost, "Measurements of the Bistatic Echo Area of Terrain at X-Band," Report 1822-2, Antenna Laboratory, Ohio State University, May 1965.
4. W. H. Peake and S. T. Cost, "The Bistatic Echo Area of Terrain at 10 GHz," in *Proceedings* WESCON 1968, Session 22/2, pp. 1–10.
5. W. H. Peake and T. L. Oliver, Private Communication, 1971.
6. R. W. Larson and R. C. Heimiller, "Bistatic Clutter Data Measurement Program," Environmental Research Institute of Michigan, RADC-TR-77-389, November 1977. NTIS, AD-A049037.
7. M. M. Weiner and P. D. Kaplan, "Bistatic Surface Clutter Resolution Area at Small Grazing Angles," Technical Report M82-10, The MITRE Corporation, Bedford, MA, January 24, 1982. Also issued as RADC-TR-82-289, Rome Air Development Center, Nov. 1982, NTIS AD-A1 23660.
8. V. W. Pidgeon, "Bistatic Cross Section of the Sea," *IEEE Transactions*, Vol. AP-14, No. 3, May 1966, pp. 405–406.
9. Technology Service Corporation. Radar Systems Engineering. Notes, lecture 3 of short course, Lexington, MA, May 7–9, 1979, p. 59.
10. J. V. DiFranco and W. L. Rubin, *Radar Detection*, Prentice-Hall, Englewood Cliffs, NJ, 1958, pp. 623–625.
11. D. K. Barton (ed.), *Radars*, Vol. 4, Radar Resolution and Multipath Effect, Artech House, Inc., Norwood, MA, 1975.
12. M. M. Weiner and P. D. Kaplan, "Bistatic Surface Clutter Resolution Area at Small Grazing Angles," Technical Report MTR-3939, The MITRE Corporation, Bedford, MA, April 1980.
13. Technology Service Corporation. Radar Systems Engineering. Notes, lecture 3 of short course, Lexington, MA, May 7–9, 1979, p. 59.
14. P. Beckmann and A. Spizzichino, *The Scattering of Electromagnetic Waves From Rough Surfaces*, Pergamon Press, New York, 1963, pp. 146–151.
15. W. H. Peake and S. T Cost, "The Bistatic Echo Area of Terrain at 10 GHz," *IEEE WESCON*, 1968, Session 22/2, pp. 1–10.
16. J. W. Crispen, Jr. and K. M. Siegal, *Methods of Radar Cross-Section Analysis*, Academic Press, New York, 1968, pp. 5–8, 155.
17. B. K. P. Horn and R. W. Sjoberg, "Calculating the Reflectance Map," *Applied Optics*, Vol. 18, June 1979, pp. 1770–1779.
18. M. L. Boyd and R. L. Deavenport, "Forward and Specular Scattering From a Rough Surface: Theory and Experiment," *Journal of the Acoustical Society of America*, Vol. 53, March 1973, pp. 791–800. See also I. Tolstoy and C. S. Clay, *Ocean Acoustics: Theory and Experiment in Underwater Sound*, McGraw-Hill, Co., New York, 1966, Eq. (4.96).
19. M. M. Weiner, "Useful Beam Quality Design Curves for Unstable Resonators," *Optical Engineering*, Vol. 13, March/April 1974, pp. 87–91.
20. D. E. Barrick, "Rough Surfaces," in G. T. Ruck et al., (eds.) *Radar Cross Section Handbook*, Vol. 2, Plenum Press, New York, 1963, Chapt. 9.
21. A. R. Domville, *The Bistatic Reflection From Land and Sea of X-Band Radio Waves*, Part I. GEC (Electronics) Ltd., Stanmore, England, Memorandum SLM1802, July 1967.
22. A. R. Domville, *The Bistatic Reflection From Land and Sea of X-Band Radio Waves*, Part II. GEC (Electronics) Ltd., Stanmore, England, Memorandum SLM2116, July 1968.
23. A. R. Domville, *The Bistatic Reflection From Land and Sea of X-Band Radio Waves*, Part Il-Supplement. GEC (Electronics) Ltd., Stanmore, England, Memorandum SLM2116 (Supplement), July 1969.
24. S. T. Cost, "Measurements of the Bistatic Echo Area of Terrain at X-Band," Report No. 1822-2, Ohio State University, Antenna Laboratory, May 1965.

25. J. W. Crispin, Jr. and K. M. Siegal, *Methods of Radar Cross-Section Analysis*, Academic Press, New York, 1968, p. 158.
26. R. E. Kell, "On the Derivation of Bistatic RCS From Monostatic Measurement," *Proceedings of the IEEE*, Vol. 53, August 1965, pp. 983–988.
27. N. J. Willis, "In-Plane Ground Clutter," *Bistatic Radar*, Technology Service Corp., Silver Springs, MD, 1995, Sect. 9.1.1, pp. 160–164.
28. M. M. Weiner, "Multistatic Radar Phenomenology, Vol. 1-Terrain and Sea Scatter," MITRE Technical Report MTR-8140A-1, The MITRE Corporation, September 1980, as amended and approved for public release on July 7, 2003. Also available as RADC-TR-81-75, Vol. 1, Rome Air Development Center, Griffiss Air Force Base, NY 13441, May 1981 (as amended on July 7, 2003).
29. R. W. Larsen and R. C. Heimiller, *Bistatic Clutter Data Measurement Program*, Environmental Research Institute of Michigan, RADC-TR-77-389, November 1977, AD-A049037, p. 342.
30. V. W. Pidgeon, "Bistatic Cross Section of the Sea," *IEEE Transactions*, Vol. AP-14, No. 3, May 1966, pp. 405–406.
31. V. W. Pidgeon, "Bistatic Cross Section of the Sea for Beaufort 5 Sea," *Science and Technology*, Vol. 17, 1968, pp. 447–448. Use of Space Systems for Planetary Geology and Geophysics, American Astronautical Society, San Diego.
32. R. W. Larsen, A. L. Maffett, R. C. Heimiller, A. F. Fromm, E. L. Johansen, R. F. Rawson and F. L. Smith, "Bistatic Clutter Measurements," *IEEE Transactions*, Vol. AP-26, No. 6, November 1978, pp. 801–804.
33. R. W. Larsen, A. Maffett, F. Smith, R. C. Heimiller and A. Fromm, "Measurements of Bistatic Clutter Cross Section. Environmental Research Institute of Michigan," Final Technical Report RADC-TR-79-15, May 1979, AD-A071193, p. 229.
34. P. E. Cornwell and J. Lancaster, "Low-Altitude Tracking Over Rough Surfaces II: Experimental and Model Comparisons," *IEEE EASCON-79 Record*, October 1979, pp. 235–248.
35. G. W. Ewell and S. P. Zehner, "Bistatic Sea Clutter Return Near Grazing Incidence," *IEEE International Conference*, Radar 82, London, Publication No. 216, October 1982, pp. 188–192.
36. G. W. Ewell, "Bistatic Radar Cross-Section Measurements," in N. C. Currie (ed.) *Techniques of Radar Reflectivity Measurements*, 2nd ed., Artech House, Inc., Norwood, MA, 1984, Chapt. 7.
37. F. T. Ulaby, T. E. Van Deventer, J. R. East, T. F. Haddock and M. E. Coluzzi, "Millimeter-Wave Bistatic Scattering From Ground and Vegetation Targets," *IEEE Transactions*, Vol. GRS-26, No. 3, May 1988, pp. 229–243.
38. T. P. Kochanski, M. J. Vanderhill, J. V. Zolotarevsky and T. Fariss, "Low Illumination Angle Bistatic Sea Clutter Measurements at X-Band," *IEE International Conference*, Oceans-92:Mastering the Oceans Through Technology, Proceedings, Vol. 1, October 26–29, 1992, pp. 518–523.
39. D. J. McLaughlin, E. Boltniew, Y. Wu and R. S. Raghavan, "Low Grazing Angle Bistatic NCRS of Forested Clutter," *Electronics Letters*, Vol. 30, No. 18, September 1, 1994, pp. 1532–1533.
40. D. J. McLaughlin, E. Boltniew, R. S Raghavan and M. J. Sowa, "Cross-Polarized Bistatic Clutter Measurements," *Electronics Letters*, Vol. 31, No. 6, March 16, 1995, pp. 490–491.
41. D. J. McLaughlin, Y. Wu, W. G. Stevens, X. Zhang, M. J. Sowa and B. Weijers, "Fully Polarimetric Bistatic Radar Scattering Behavior of Forested Hills," *IEEE Transactions*, Vol. AP-5O, No. 2, February 2002, pp. 101–110.
42. R.E. Vander Schurr, P.G. Tomlinson, "Bistatic Clutter Analysis", *Descision-Science Applications, Inc.*, Final Technical Report DSA-106, April 1979, AD-A069385.
43. C. H. Tang, T. I. Boak III and M. D. Gross, "Bistatic Radar Measurements of Electrical Properties of the Martian Surface," *Journal of Geophysical Research*, Vol. 82, No. 28, September 1977, pp. 4305–4315.
44. G. O. Sauermann and P. C. Waterman, "Scattering Modeling: Investigation of Scattering by Rough Surfaces," Technical Report MTR-2762, AFAL-TR-73-334, The MITRE Corp., January 1974, p. 69.
45. J. G. Zornig, P. M. Schultheiss and J. Snyder, "Bistatic Surface Scattering Strength at Short Wavelengths," Report CS-9, Yale University, Department of Engineering and Applied Science, June 1977, AD-A041316, 95 pp.
46. E. N. Bramley and S. M. Cherry, "Investigation of Microwave Scattering by Tall Buildings," *Proceedings of the 1EEE*, Vol. 120, No. 8, August 1973, pp. 833–842.

47. A. E. Brindley, L. C. Calhoun and T. N. Paton, "A Joint Army/Air Force Investigation of Reflection Coefficient at C and Ku Bands for Vertical, Horizontal, and Circular System Polarizations," Final Report TR-76-67, ITT Research Institute, Chicago, July 1976, AD-A031403, p. 160.
48. A. G. Pavel'yev, M. A. Kolosov, O. L. Yakovlev, A. L. Kucher-Yavenkov, Y. E. Molotov and O. Milekhin, "The Study of Venus by Means of the Bistatic Radar Method," *Radio Engineering and Electronic Physics (USSR)*, Vol. 23, October 1978, pp. 1–8.
49. D. K. Barton (ed.) *Radars*, Vol. 4, Radar Resolution and Multipath Effect, Artech House, Inc., Norwood, MA, 1975.
50. D. F. Sun, "Experimental Measurements of Ground Reflection Elevation Multipath Characteristics and Its Effects on a Small Aperture Elevation Tracking Radar," Technical Note 1979-21, MIT Lincoln Laboratory, March 1979. NTIS, AD A077915.
51. R. L. Cosgriff, W. H. Peake and R. C. Taylor, "Terrain Scattering Properties for Sensor System Design," Terrain Handbook II, Bulletin No. 181, Engineering Experiment Station, Ohio State University, May 1960.
52. F. Blackwell and W. Rogers. "Some Measurements of Bistatic Terrain Reflection Coefficient of a Forest Area at X-Band." Memorandum No. 2157, Royal Radar Establishment, Ministry of Aviation, Malvern Worcs, England, January 1965.
53. F. Daigle, K. Wefald and B. Cole, "Bistatic Clutter Experiment," Technical Report BR-6629, Missile Systems Division, Raytheon Co., Bedford, MA, October 1971.
54. C. L. Beard, "Coherent and Incoherent Scattering of Microwaves From the Ocean," *IRE Transactions*, Vol. AP-9, No. 5, September 1961. Please note Fig. 2.
55. S. Silver, *Microwave Antenna Theory and Design, Radiation Laboratory Series*, Vol. 12, McGraw-Hill, Co., New York, 1949, pp. 158–167.
56. J. D. Jackson, *Classical Electrodynamics*, John Wiley, New York, 1962, pp. 280–297.
57. G. W. Stroke, "Diffraction Gratings," in *Handbuch Der Physik, Band XXIX*, Springer-Verlag, New York, 1967, pp. 504–514.
58. A. Sommerfeld, *Optics*, Academic Press, New York, 1954, pp. 195–204.
59. M. Born and E. Wolf, *Principles of Optics*, Pergamon Press, New York, 1964, pp. 375–386.

10

Spotlight Synthetic Aperture Radar

Brian D. Rigling

The crossrange resolution of a monostatic radar is typically limited by the beamwidth of the transmitting and receiving antenna, approximated by [1]

$$\Delta\theta_{\mathrm{B}} \propto \frac{\lambda_{\mathrm{c}}}{D} \text{ (radians)} \tag{10.1}$$

where λ_c is the wavelength of the transmitted waveform and D is the horizontal dimension of the physical aperture. At range R_a, this gives an azimuth resolution of approximately $\rho_a = R_a \Delta\theta_B$. The azimuth resolution is in a direction orthogonal to the range resolution provided by a waveform for some nonzero bandwidth.

The goal of a synthetic aperture radar is to observe a scene over a series of spatial sampling points and coherently combine those samples to synthesize a much larger aperture with finer azimuth resolution. The scene is persistently illuminated by the real beam of the antenna, and each spatial sampling point provides a range-resolved profile. Phase changes between profiles due to doppler are exploited in Synthetic Aperture Radar processing to reconstruct a two- (2-D) or three-dimensional (3-D) representation of the sampled scene. In monostatic SAR, the crossrange resolution is given by [2]

$$\rho_{\mathrm{a}} = \frac{\lambda_{\mathrm{c}}}{4\sin\left(\frac{\Delta\theta}{2}\right)} \tag{10.2}$$

when the scene is viewed at broadside over an angular interval of $\Delta\theta$. This implies that the achievable resolution is ultimately limited by the wavelength of the transmitted waveform, assuming that objects to be resolved are visible over the entire angular interval.

The crossrange resolution of a bistatic radar is similarly limited by the widths of the real beams of the transmit and receive antennas [3, 4]. However, by forming a bistatic synthetic aperture as shown notionally for a bistatic angle of $\beta = 90°$ in Figure 10–1, finer resolution can be achieved [5]. Increasing the angular interval of either the transmit or receive synthetic aperture improves the azimuth resolution but requires the data collection platforms to fly longer trajectories and thus remain coherent with one another for longer periods of time. The achievable amount of coherent integration time is dependent on the ability of each aircraft's inertial navigation system to measure its position and on the level of fidelity of the synchronization link between the transmitter and

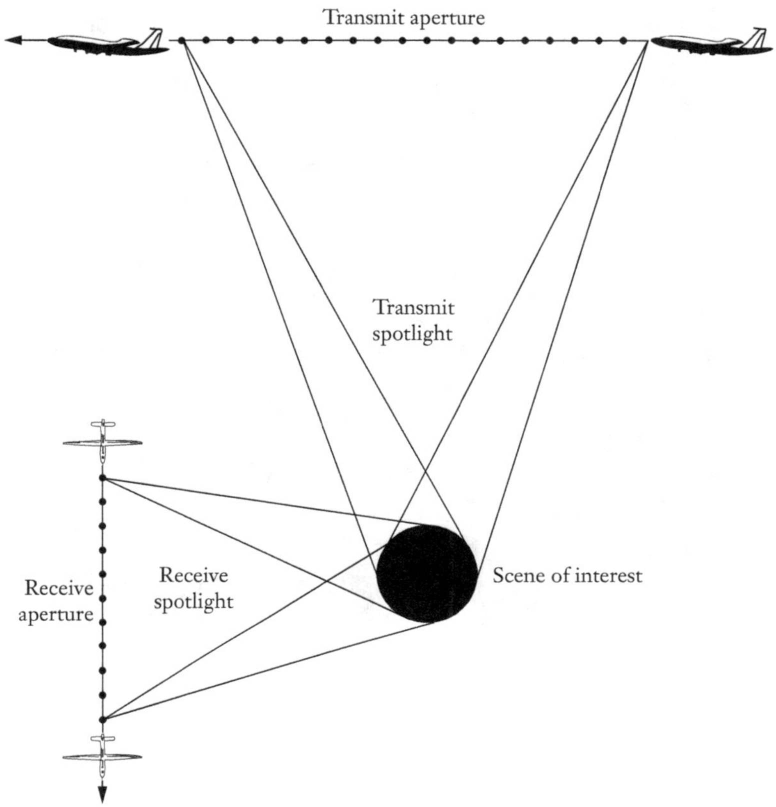

Figure 10–1 Notional spotlight bistatic SAR geometry with a bistatic angle of $\beta = 90°$.

receiver. As of the writing of [6], the limit on coherent time was less than 3 second. This limit has naturally been extended by technological advances, and as will be shown, the achievable coherency can be farther lengthened through use of autofocus in postprocessing.

Synthetic aperture radar is commonly used in air-to-ground targeting, and the operational benefits of doing so bistatically are readily apparent. First, the strike aircraft may function as the bistatic receiver, thereby sensing passively and remaining covert. Second, as the bistatic effective aperture is synthesized by the combined crossrange motion of the transmitter and receiver, a strike aircraft may operate as a receiver in a forward-looking mode. The trajectory of the emitter thus induces all of the target doppler, and allows the receiver to present a lower RCS to defenders, while also executing a more rapid ingress and egress.

Two bistatic SAR programs sponsored by the United States Air Force (USAF) and the Defense Advanced Research Projects Agency (DARPA) successfully illustrated the viability of these concepts in the early 1980s [7]. The COVIN REST (Covert, In-Wx Recce/Strike) program demonstrated the first space-to-air bistatic SAR. Operating at L-band, the Shuttle Imaging Radar served as an illuminator, and a Convair CV-990 acted as the bistatic receiver. Due to the rapid orbital motion of the space shuttle, spotlight illumination was not feasible, and strip map processing was employed instead. As shown in Figure 10–2, a 60-foot resolution image of Sioux City, Iowa was successfully generated. The COVIN REST demonstration thus showed that a SAR image could be formed bistatically through the use of a transmitter safely located in orbit and a passive receiver.

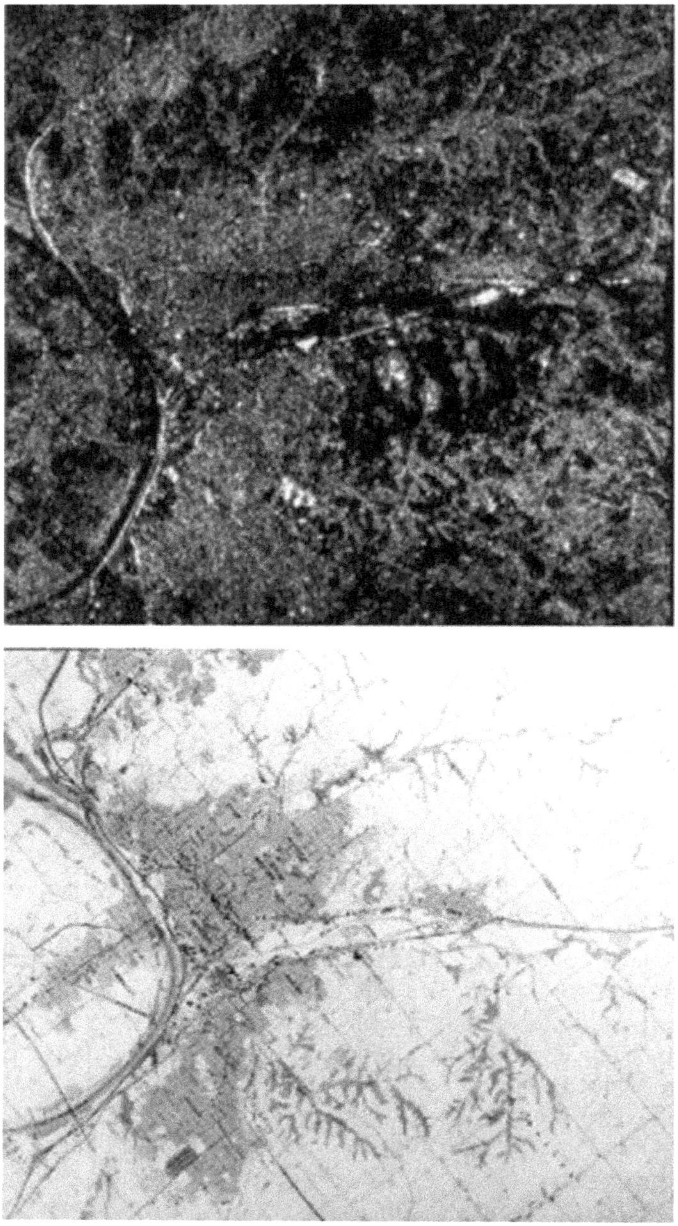

Figure 10–2 Shown on the top is a 60-foot resolution bistatic SAR image of Sioux City, Iowa, generated as a result of the COVIN REST program. At the bottom is a map of the city for ground truth comparison.

The T-BIRD (Tactical Bistatic Radar Demonstration) program exhibited the feasibility of bistatic imaging with a forward-looking receive platform. A C-141 functioning as a transmitter at a 20–30 km slant range was successfully teamed with a receiver onboard a C-130. At a bistatic angle of $\beta = 30°$, the system was able to form a 10-foot resolution image of Demonstration Hill

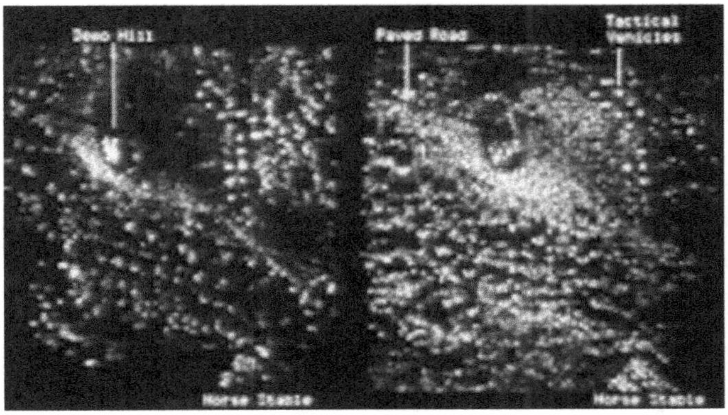

Figure 10–3 On the left is a monostatic SAR image of Demonstration Hill at Fort Huachuca, Arizona, generated as a result of the T-BIRD program. On the right is a 10-foot resolution bistatic ($\beta = 30°$) SAR image of the same area.

within Fort Huachuca, Arizona. For comparison, a monostatic SAR image of the same area was formed simultaneously, as shown in Figure 10–3. Multiple other bistatic SAR data collections are described in the open literature [8–14].

The next few sections describe a generalized bistatic SAR geometry, which is scene-centered rather than platform-centered, and derives expressions for the resolution offered by the collected synthetic aperture data. Methods for phase history data collection are discussed, as well as sampling requirements in frequency and in azimuth.

10.1 BISTATIC PHASE HISTORY DATA

For clarity of exposition, the following discussion assumes the bistatic SAR data collection geometry [5, 15] shown in Figure 10–4; in general, arbitrarily complex geometries may be implemented through straightforward application of the following. The center of the scene to be imaged is located at the origin of coordinates, and the ground plane is the $x-y$ plane. A scatterer within that scene is located at

$$\underline{r}_m = (x_m, y_m, z_m). \tag{10.3}$$

The actual location of the transmitter at a given time τ is

$$\underline{r}_{\mathrm{t}}(\tau) = (x_{\mathrm{t}}(\tau), y_{\mathrm{t}}(\tau), z_{\mathrm{t}}(\tau)). \tag{10.4}$$

The measured transmitter location at time τ is given by

$$\hat{\underline{r}}_{\mathrm{t}}(\tau) = \underline{r}_{\mathrm{t}}(\tau) + \tilde{\underline{r}}_{\mathrm{t}}(\tau), \tag{10.5}$$

where

$$\tilde{\underline{r}}_{\mathrm{t}}(\tau) = (\tilde{x}_{\mathrm{t}}(\tau), \tilde{y}_{\mathrm{t}}(\tau), \tilde{z}_{\mathrm{t}}(\tau)) \tag{10.6}$$

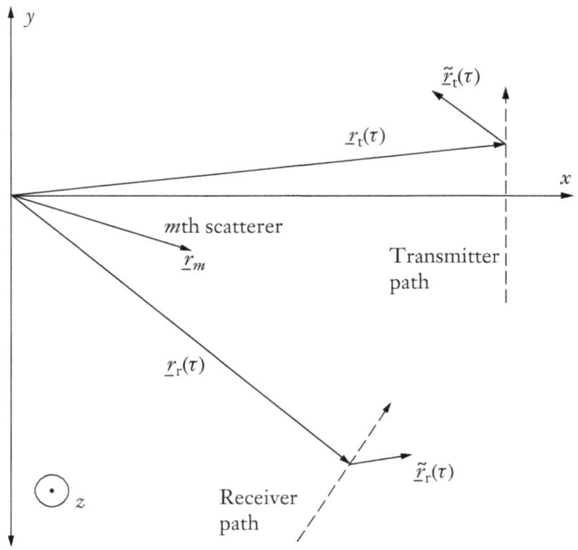

Figure 10–4 Top view of a spotlight bistatic SAR data collection geometry, illustrating vectors from the scene center to the transmitter, receiver, and scatterer locations. The platform motion measurement errors are similarly indicated by vectors originating at the transmitter and receiver positions. The x–y plane is the ground plane.

is the transmitter motion measurement error, due to limitations in the system's inertial navigation. Similarly, the actual location of the receiver at a given time τ is

$$\underline{r}_{\mathrm{r}}(\tau)=(x_{\mathrm{r}}(\tau), y_{\mathrm{r}}(\tau), z_{\mathrm{r}}(\tau)), \tag{10.7}$$

and the measured location of the receiver is

$$\hat{\underline{r}}_{\mathrm{r}}(\tau)=\underline{r}_{\mathrm{r}}(\tau)+\tilde{\underline{r}}_{\mathrm{r}}(\tau), \tag{10.8}$$

where

$$\tilde{\underline{r}}_{\mathrm{r}}(\tau)=(\tilde{x}_{\mathrm{r}}(\tau), \tilde{y}_{\mathrm{r}}(\tau), \tilde{z}_{\mathrm{r}}(\tau)), \tag{10.9}$$

is the receiver location measurement error.

Although linear flight paths are shown in Figure 10–4, the transmitter and receiver may actually take arbitrary nonlinear trajectories. However, a linear flight path is often used, as it is the least taxing on a platform's motion measurement subsystem. To simplify later discussions, the receiving platform is assumed to traverse a path with the same angular progression about the origin as the transmitter (e.g., if the transmitter moves in a clockwise direction, then the receiver also moves in a clockwise direction). This ensures that the samples taken in frequency space, to be described shortly, sweep through a nonoverlapping surface.

As the transmitter moves along its flight path, the radiating antenna periodically transmits pulses of energy in the direction of the scene center. Each transmitted pulse travels from the transmitter to the scene of interest, where it is reflected by scatterers within the area of illumination. This reflected energy disperses in all directions, and some of this energy is observed by the antenna on the receiving platform. The travel time of the pulse from the transmitter to a scatterer to the receiver is assumed to be sufficiently short, with respect to the pulse duration, such that any platform movement within that time period may be neglected. Equivalently, any consequential platform motion is assumed to be compensated for on receive.

The transmitted pulses are assumed to have uniform power over the frequency range $f \in [f_0, f_0 + B]$, where f_0 and B represent the initial frequency and bandwidth of the transmitted pulse, respectively. The *center frequency* or *carrier frequency*

$$f_c = f_0 + \frac{B}{2} \tag{10.10}$$

will often be used in analysis at later points in this text. It is also assumed that scatterers behave as ideal point scatterers within this frequency band. Therefore, the receiver is assumed to observe frequency samples of delayed, band-limited impulses, with uniform power over the frequency band of interest. The delay of each pulse is the travel time from the transmitter to a scatterer to the receiver. The receiver's time of observation is gated such that a pulse reflected by a scatterer at the scene origin will have zero time delay. One may therefore model the received signal in the frequency domain as a complex exponential of some uniform power with a linear phase determined by the relative time delay of the transmitted pulse. The time scale within a pulse is often referred to as *fast time*, while the time scale across pulses is referred to as *slow time*.

At the slow time τ, the actual distance from the transmitter to a scatterer in the scene is

$$d_{tm}(\tau) = \|\underline{r}_t(\tau) - \underline{r}_m(\tau)\|, \tag{10.11}$$

and the actual distance from this scatterer to the receiver is

$$d_{rm}(\tau) = \|\underline{r}_m(\tau) - \underline{r}_r(\tau)\|. \tag{10.12}$$

Also at time τ, the measured distance from the transmitter to the scene origin is

$$d_t(\tau) = \|\underline{r}_t(\tau) + \underline{\tilde{r}}_t(\tau)\|, \tag{10.13}$$

and the measured distance from the receiver to the scene origin is

$$d_r(\tau) = \|\underline{r}_r(\tau) + \underline{\tilde{r}}_r(\tau)\|. \tag{10.14}$$

The time delay of the received pulse, relative to the measured travel time to and from the scene origin, may then be written as

$$t_m(\tau) = \frac{d_{rm}(\tau) - d_r(\tau) + d_{tm}(\tau) - d_t(\tau)}{c} + \tilde{t}(\tau) = \frac{\Delta R_m(\tau)}{c} + \tilde{t}(\tau), \tag{10.15}$$

where c is the speed of light. The quantity $\Delta R_m(\tau)$ is typically known as the *differential range*. In addition to the effects of motion measurement errors, the relative time delay $t_m(\tau)$ may also be subject to a synchronization error $\tilde{t}(\tau)$ between the transmitter and receiver.

Given $\Delta R_m(\tau)$, one may define the linear phase function of the received frequency samples corresponding to a scatterer located at r_m as

$$\Phi_m(f_i, \tau_k) = -2\pi f_i \frac{\Delta R_m(\tau_k)}{c} \qquad i = 1, \ldots, N_f \quad k = 1, \ldots, N_\tau \tag{10.16}$$

where the f_i are the sampled frequencies, and the τ_k are the transmit–receive times. The receiver uniformly samples the frequency band $[f_0, f_0 + B]$, and the transmit–receive times are spaced over the interval $\tau \in [-T/2, T/2]$, where T is the *coherent collection time*. Equation (10.16) defines the *phase history* of a scatterer at position r_m. On the basis of the linear phase function for a single scatterer in equation (10.16), one may represent the total received signal at time τ_k and frequency f_i by the sum of the returns from all the scatterers in the scene; thus

$$S(f_i,\tau_k) = \sum_m A_m e^{j\Phi_m(f_i,\tau_k)} + w(f_i,\tau_k) \tag{10.17}$$

where A_m is a complex constant representing the reflection coefficient of the mth scatterer and $w(f_i, \tau_k)$ is assumed to be circular white Gaussian noise introduced by the receiver. The received SAR data, modeled by equation (10.17), is commonly known as *phase history data*.

Given an imaged scene that is small relative to the platform ranges, a far-field assumption accurately approximates the differential range $\Delta R_m(\tau)$ [16]. This approximation is derived in Appendix A and may be written as

$$\begin{aligned}\Delta R_m(\tau) \approx & -x_m[\cos\phi_\mathrm{t}(\tau)\cos\theta_\mathrm{t}(\tau) + \cos\phi_\mathrm{r}(\tau)\cos\theta_\mathrm{r}(\tau)] \\ & -y_m[\sin\phi_\mathrm{t}(\tau)\cos\theta_\mathrm{t}(\tau) + \sin\phi_\mathrm{r}(\tau)\cos\theta_\mathrm{r}(\tau)] \\ & -z_m[\sin\theta_\mathrm{t}(\tau) + \sin\theta_\mathrm{r}(\tau)]\end{aligned} \tag{10.18}$$

where the contribution of the motion measurement errors, $\tilde{\underline{r}}_\mathrm{t}(\tau)$ and $\tilde{\underline{r}}_\mathrm{r}(\tau)$, has been neglected. The angles ϕ_t and ϕ_r (θ_t and θ_r) denote the measured azimuth (elevation) angles of the transmitter and receiver, with respect to the scene center. The far-field approximation is used in the polar format algorithm for bistatic SAR, described in Section 10.4. Here, the same approximation will be used to relate the dimensions of the data collection geometry to the resulting image resolution. The effects of motion measurement errors are addressed in detail in Section 10.5. They are omitted for now under the assumption that the positions of both platforms are measured perfectly throughout the coherent collection time. It is also assumed that the transmitter and receiver are perfectly synchronized, such that $\tilde{t}(\tau) = 0 \; \forall \, \tau$.

Using equation (10.18), the expression for the recorded data in equation (10.17) may be rewritten as

$$S(f,\tau) \approx \sum_m A_m \exp\left(j\frac{4\pi}{c}[x_m f_x(f,\tau) + y_m f_y(f,\tau) + z_m f_z(f,\tau)]\right) + w(f,\tau) \tag{10.19}$$

where

$$\begin{aligned}f_x(f,\tau) &= \frac{f}{2}\left[\cos\phi_\mathrm{t}(\tau)\cos\theta_\mathrm{t}(\tau) + \cos\phi_\mathrm{r}(\tau)\cos\theta_\mathrm{r}(\tau)\right] \\ f_y(f,\tau) &= \frac{f}{2}\left[\sin\phi_\mathrm{t}(\tau)\cos\theta_\mathrm{t}(\tau) + \sin\phi_\mathrm{r}(\tau)\cos\theta_\mathrm{r}(\tau)\right] \\ f_z(f,\tau) &= \frac{f}{2}\left[\sin\theta_\mathrm{t}(\tau) + \sin\theta_\mathrm{r}(\tau)\right]\end{aligned} \tag{10.20}$$

Section 10.3 presents a few methods by which these frequency domain samples can be acquired. In general, equation (10.19) represents samples of the Fourier transform of the received signal after pulse compression. The *data collection manifold* is defined in frequency space as the surface of

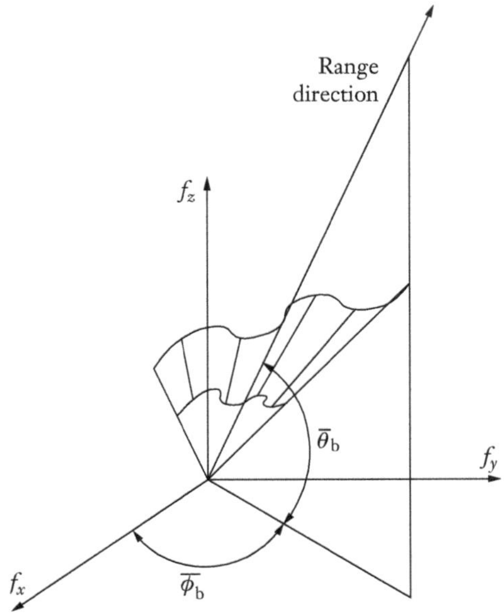

Figure 10–5 Bistatic SAR data collection manifold in 3-D frequency space.

points $\{(f_x(f,\tau), f_y(f,\tau), f_z(f,\tau))\}$ for $f \in [f_0, f_0 + B]$ and $\tau \in [-T/2, T/2]$. This manifold may also be viewed as the surface swept out by the *bistatic SAR line-of-sight*, written in the vector form as

$$\underline{r}_{\mathrm{LOS}}(\tau) = \frac{1}{2}\begin{bmatrix} \cos\phi_{\mathrm{t}}(\tau)\cos\theta_{\mathrm{t}}(\tau) + \cos\phi_{\mathrm{r}}(\tau)\cos\theta_{\mathrm{r}}(\tau) \\ \sin\phi_{\mathrm{t}}(\tau)\cos\theta_{\mathrm{t}}(\tau) + \sin\phi_{\mathrm{r}}(\tau)\cos\theta_{\mathrm{r}}(\tau) \\ \sin\theta_{\mathrm{t}}(\tau) + \sin\theta_{\mathrm{r}}(\tau) \end{bmatrix} \tag{10.21}$$

as shown in Figure 10–5.

It is important to note that this definition for line-of-sight differs from that for the standard bistatic geometry. The standard bistatic line-of-sight is defined as

$$\hat{\underline{r}}_{\mathrm{LOS}}(\tau) = \begin{bmatrix} \cos\phi_{\mathrm{b}}(\tau)\cos\theta_{\mathrm{b}}(\tau) \\ \sin\phi_{\mathrm{b}}(\tau)\cos\theta_{\mathrm{b}}(\tau) \\ \sin\theta_{\mathrm{b}}(\tau) \end{bmatrix} \tag{10.22}$$

where $\phi_{\mathrm{b}}(\tau) = (\phi_{\mathrm{t}}(\tau) + \phi_{\mathrm{r}}(\tau))/2$ and $\theta_{\mathrm{b}}(\tau) = (\theta_{\mathrm{t}}(\tau) + \theta_{\mathrm{r}}(\tau))/2$. In bistatic SAR, the bistatic look angle of an image as a whole, rather than the instantaneous bistatic look angle as a function of slow time, is of greater interest. For future reference, the bistatic look angle of a bistatic SAR image is defined as

$$\overline{\phi}_{\mathrm{b}} = \tan^{-1}\left(\frac{f_y(f,0)}{f_x(f,0)}\right) \tag{10.23}$$

in azimuth, and as

$$\bar{\theta}_b = \tan^{-1}\left(\frac{f_z(f,0)}{\sqrt{f_x(f,0)^2 + f_y(f,0)^2}}\right) \tag{10.24}$$

in elevation.

10.2 IMAGE RESOLUTION

Section 10.4 discusses nonparametric scene reconstruction, or image formation, via matched filtering of phase history data. There, and in [2, 17], it is shown that the resolution of a 2-D scene reconstruction is inversely proportional to the bandwidth of the phase history data in the range and crossrange dimensions. Crossrange resolution improvement can be equivalently derived as fine doppler resolution of stationary objects on the ground, with the doppler induced by the motion of the data collection platforms. By simple extension to three dimensions, the resolution of a 3-D reconstruction in a particular direction is inversely proportional to the bandwidth of the frequency space data collection manifold in that same direction. Thus, without yet considering the process of image formation, one may study how the shape and size of the data collection manifold in Figure 10–5 determine the quality of an image rendered by matched filtering.

Considering the frequency domain measurements of a scatterer in one dimension, written as

$$S(f_u) = A_0\, e^{j\frac{4\pi u_0}{c} f_u}, \tag{10.25}$$

assume that the data are uniformly sampled over the frequency band $f_u \in [f_0, f_0 + B_u]$, such that these samples may be expressed as

$$f_u = f_0 + kB_u/N \qquad k = 0,\ldots,N-1. \tag{10.26}$$

The position of the scatterer is u_0. Taking the discrete Fourier transform of equation (10.25) yields

$$\begin{aligned} P(u) &= \sum_{k=0}^{N-1} e^{j\frac{4\pi u_0}{c}\left(f_0 + \frac{kB_u}{N}\right)} e^{-j\frac{4\pi u}{c}\left(f_0 + \frac{kB_u}{N}\right)} \\ &= e^{-j\frac{4\pi f_0}{c}(u-u_0)} e^{-j\frac{2\pi B_u}{c}\left(\frac{N-1}{N}\right)(u-u_0)} \frac{\sin\left(\frac{2\pi B_u}{c}(u-u_0)\right)}{\sin\left(\frac{2\pi B_u}{cN}(u-u_0)\right)} \end{aligned} \tag{10.27}$$

which allows one to show the relationship between the bandwidth of the sampled data and the corresponding Fourier resolution. From (10.27), two observations may be made. First, the quantity $(u - u_0)$ is unambiguously represented in $P(u)$ only over the range $(u - u_0) \in [-cN/(4B_u), cN/(4B_u)]$. This gives the unambiguous range of the reconstruction in the $\hat{u}$ direction. Second, the first nulls of $P(u)$ occur at $(u - u_0) = \pm c/(2B_u)$. Fourier resolution is typically defined to be one-half the main lobe width [2, 17], implying that the resolution in the $\hat{u}$ direction is $\delta u = c/(2B_u)$. A similar approach is used later to derive Nyquist requirements on the spacing of samples in frequency space, in order to prevent spatial aliasing in the final image.

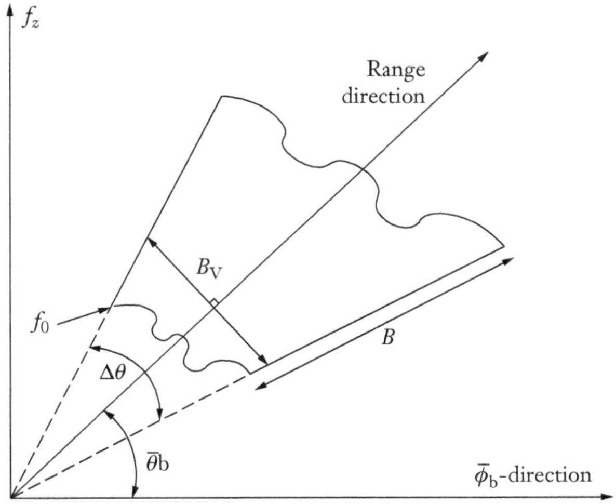

Figure 10.6 Data collection manifold projected in frequency space onto the plane defined by the f_z-axis and the $(\cos\bar{\phi}_b, \sin\bar{\phi}_b)$ direction.

Three-Dimensional Resolution Based on the above discussion, the image resolution supported by $\{S(f,\tau)\}$ in equation (10.17) is most easily analyzed by first assuming that the frequency space data are uniformly sampled in each of three orthogonal directions. One may then compute the bandwidth of the frequency space manifold in each direction, and may thereby approximate the corresponding reconstruction resolution in that direction. While this approach supplies no information about the overall quality of an image reconstruction, it does provide useful information about the Fourier resolution of the system, as defined by the width of the main lobe of the system impulse response. The standard x, y, z-coordinate system is ill-suited for this analysis, as the orientation of the frequency space manifold $\{f_x(f,\tau), f_y(f,\tau), f_z(f,\tau)\}$ is rarely oriented along the standard coordinate axes. As an alternative, three orthogonal directions defined by the geometry of the data collection are examined below.

The first and most obvious direction to consider is the range direction defined by the bistatic look angles as $\hat{\underline{r}} = [\cos\phi_b \cos\theta_b \sin\phi_b \cos\theta_b \sin\theta_b]^T$. As shown in Figure 10–6, the bandwidth of the frequency space manifold in the range direction is equal to the bandwidth of the collected phase histories times the cosine of half the bistatic angle between the transmitter and receiver, defined by

$$\cos\beta = \frac{\underline{r}_t(0)^T \underline{r}_r(0)}{\|\underline{r}_t(0)\|\|\underline{r}_r(0)\|} \tag{10.28}$$

Therefore, the range resolution may be written as

$$\delta r = \frac{c}{2B\cos(\beta/2)} \tag{10.29}$$

The horizontal crossrange direction is next defined to be $\hat{\underline{h}} = [-\sin\bar{\phi}_b, \cos\bar{\phi}_b, 0]^T$, which is orthogonal to the range direction and parallel to the f_x, f_y-plane. As shown in Figure 10–7, the

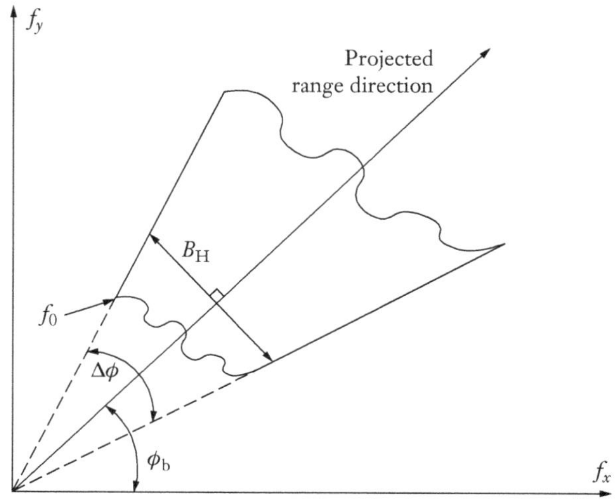

Figure 10.7 Data collection manifold projected in frequency space onto the f_x, f_y-plane.

bandwidth of the frequency space manifold in the horizontal crossrange direction is bounded below as

$$B_{\rm H} \geq 2 f_{\rm c} \sin\left(\frac{\Delta\phi}{2}\right) \cos\bar{\theta}_{\rm b} \cos(\beta/2) \tag{10.30}$$

where

$$\Delta\phi = \max_{\tau} \tan^{-1}\left(\frac{f_y(f,\tau)}{f_x(f,\tau)}\right) - \min_{\tau} \tan^{-1}\left(\frac{f_y(f,\tau)}{f_x(f,\tau)}\right) \tag{10.31}$$

Equation (10.30) implies that the horizontal crossrange resolution is approximately

$$\delta h = \frac{c}{2B_{\rm H}} = \frac{c}{4 f_{\rm c} \sin\left(\frac{\Delta\phi}{2}\right) \cos\bar{\theta}_{\rm b} \cos(\beta/2)} \tag{10.32}$$

By default, the vertical crossrange direction is defined to be $\underline{\hat{v}} = [-\cos\bar{\phi}_{\rm b} \sin\bar{\theta}_{\rm b}, -\sin\bar{\phi}_{\rm b} \sin\bar{\theta}_{\rm b} \cos\bar{\theta}_{\rm b}]^T$, and as shown in Figure 10–6, the bandwidth of the manifold in this direction is bounded below as

$$B_{\rm V} \geq 2 f_{\rm c} \sin\left(\frac{\Delta\theta}{2}\right) \cos(\beta/2) \tag{10.33}$$

where

$$\Delta\theta = \max_{\tau} \tan^{-1}\left(\frac{f_z(f,\tau)}{\sqrt{f_x(f,\tau)^2 + f_y(f,\tau)^2}}\right) - \min_{\tau} \tan^{-1}\left(\frac{f_z(f,\tau)}{\sqrt{f_x(f,\tau)^2 + f_y(f,\tau)^2}}\right) \tag{10.34}$$

This defines the vertical crossrange resolution to be

$$\delta v = \frac{c}{2B_{\mathrm{V}}} = \frac{c}{4f_{\mathrm{c}}\sin\left(\frac{\Delta\theta}{2}\right)\cos(\beta/2)} \tag{10.35}$$

It is important to note that equations (10.29), (10.32), and (10.35) approximate the resolution of an ideal three-dimensional reconstruction using matched filtering, by assuming uniform sampling in each direction. The sparse and non-uniform sampling, which is provided by a nonlinear aperture, will give a misshapen ideal point response, including unfavorably high side lobe levels. Furthermore, the presence of motion measurement errors during data collection, specifically those which give rise to higher order phase errors that may not be completely corrected, can significantly worsen resolution, especially in the crossrange directions [18].

When forming ground map imagery, one is typically most interested in the system resolution in the ground plane, as opposed to the three directions chosen above. The ground range resolution may be defined by projecting the range resolution in equation (10.29) onto the ground plane to yield

$$\delta r_{\mathrm{g}} = \frac{c}{2B\cos(\beta/2)\cos\bar{\theta}_{\mathrm{b}}} \tag{10.36}$$

The horizontal crossrange resolution already lies in the ground plane, but equation (10.32) may be rewritten in terms of physical parameters of the data collection as

$$\delta h_{\mathrm{g}} = \frac{\lambda}{\cos(\beta/2)}\left[\frac{L_{\mathrm{t}}}{R_{\mathrm{t}}}\sin\psi_{\mathrm{t}} + \frac{L_{\mathrm{r}}}{R_{\mathrm{r}}}\sin\psi_{\mathrm{r}}\right]^{-1} \tag{10.37}$$

given the transmitter range, R_{t}; receiver range, R_{τ}; length of the transmitter flight path, L_{t}; length of the receiver flight path, L_{τ}; bistatic angle, β; transmitter squint angle, ψ_{t} ($\pi/2$ for broadside); and receiver squint angle ψ_{τ} ($\pi/2$ for broadside). The wavelength λ is equal to c/f_{c}.

Data Collection Examples In bistatic SAR data collection, the flight paths taken by the transmitter and receiver equally impact the resolution of nonparametric reconstructions. However, most practical applications dictate that the receiving platform, being typically smaller, more maneuverable, and of lower cost, will adjust its trajectory to give the desired performance. The following considers three simple examples of bistatic data collection geometries intended to yield two-dimensional ground images, with resolution in the range and horizontal crossrange directions.

Figure 10–8 shows a bistatic SAR geometry in which the transmitter and receiver fly parallel flight paths in the positive y-direction. Both platforms view the scene at broadside. The parameters for this scenario are given in Table 10–1. The transmit aperture is shown at a long range, and the receiver aperture is shown at a short range. Between these two apertures lies the virtual bistatic aperture, corresponding to the data collection manifold described in the previous section, that is synthesized by the cooperation of the two platforms. The azimuth width of this aperture is 29.1°, thus yielding a horizontal crossrange resolution of δh = 0.3 m, given a center frequency of f_{c} = 1 GHz. The signal bandwidth of B = 500 MHz provides a range resolution of δr = 0.3 m. Figure 10–9 shows the data collection manifold, which results from Figure 10–8, projected onto the f_x, f_y-plane and the imaged ideal point response supported by these data.

Table 10–2 describes a second data collection scenario, shown in Figure 10–10. The transmitter and receiver again traverse parallel flight paths, though they have a squinted view of the scene. The receiving platform is at a longer range, while maintaining the same air speed, thus resulting in a receive aperture with the same length but a smaller angular span. The transmit aperture is shifted

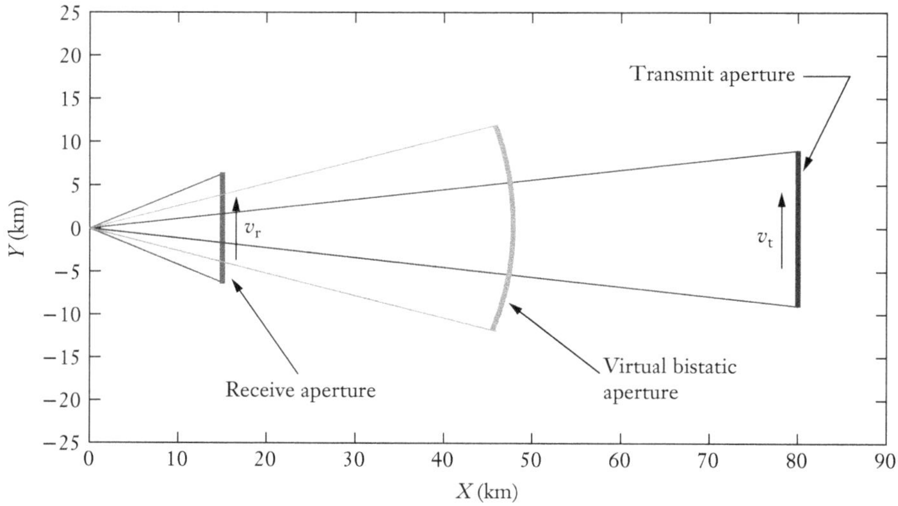

Figure 10–8 Transmit and receive apertures of a bistatic SAR system. Combining the apertures synthesized by the transmitter and receiver forms a virtual bistatic aperture. With a center frequency of 1 GHz, the above system yields a horizontal crossrange resolution of $\delta h = 0.3$ m.

Table 10–1 Bistatic Geometry Parameters Corresponding to Figure 10–8.

$(\bar{x}_t, \bar{y}_t, \bar{z}_t)$	(80 km, 0 km, 10 km)
$(\bar{x}_r, \bar{y}_r, \bar{z}_r)$	(15 km, 0 km, 3 km)
(v_{xt}, v_{yt}, v_{zt})	(0 m/s, 200 m/s, 0 m/s)
(v_{xr}, v_{yr}, v_{zr})	(0 m/s, 140 m/s, 0 m/s)
f_c	1 GHz
B	500 MHz
T_c	90 sec
$(\Delta\phi_t, \Delta\phi_r, \Delta\phi)$	(12.8°, 45.6°, 29.1°)

Table 10–2 Bistatic Geometry Parameters Corresponding to Figure 10–10.

$(\bar{x}_t, \bar{y}_t, \bar{z}_t)$	(80 km, 20 km, 10 km)
$(\bar{x}_r, \bar{y}_r, \bar{z}_r)$	(60 km, 5 km, 8 km)
(v_{xt}, v_{yt}, v_{zt})	(0 m/s, 200 m/s, 0 m/s)
(v_{xr}, v_{yr}, v_{zr})	(0 m/s, 140 m/s, 0 m/s)
f_c	1 GHz
B	500 MHz
T_c	90 sec
$(\Delta\phi_t, \Delta\phi_r, \Delta\phi)$	(12.1°, 11.9°, 12.0°)

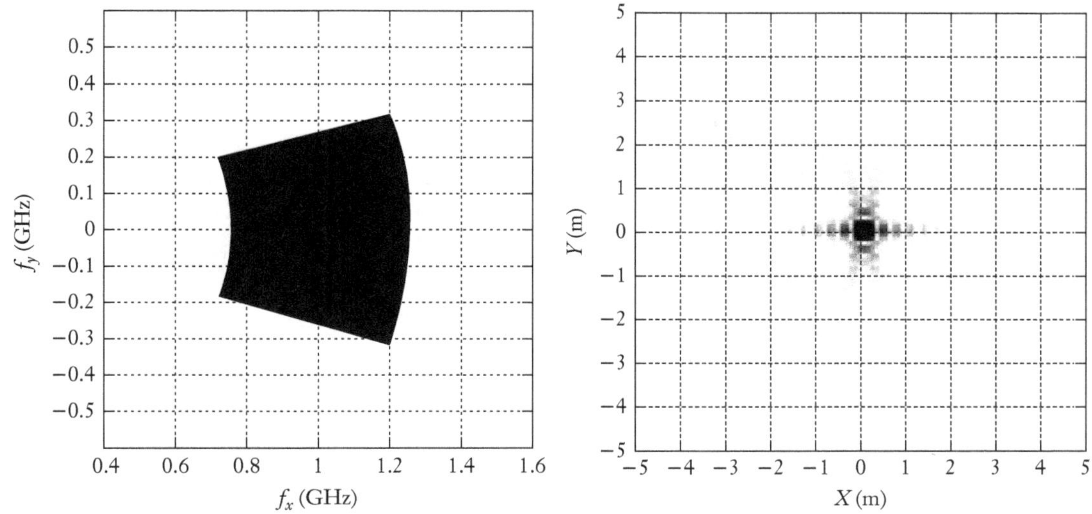

Figure 10–9 The data collection manifold on the left, corresponding to Figure 10–8, supports the ideal point response on the right. The two are related via 2-D Fourier transform.

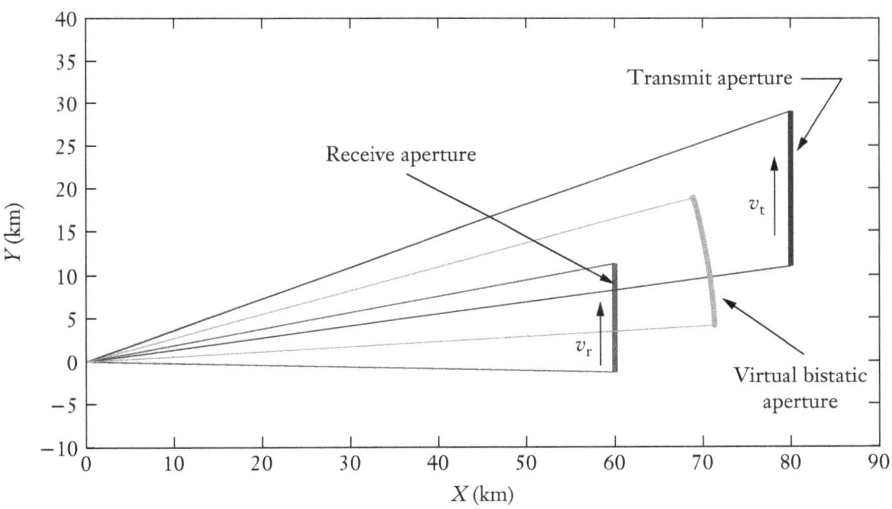

Figure 10–10 Transmit and receive apertures of a bistatic SAR system. Increasing the range of the receiver, while keeping the receive aperture size constant, worsens the azimuth resolution. With a center frequency of 1 GHz, the above system yields a horizontal crossrange resolution of $\delta h = 0.72$ m.

in the y-direction, but this does not result in a significant change in the range. Combining the new receive and transmit apertures yields the virtual bistatic aperture shown in Figure 10–10. This data collection geometry yields a bistatic angular span of $\Delta\phi = 12°$ and a horizontal crossrange resolution of $\delta h = 0.72$ m. Figure 10–11 shows the second data collection manifold, contracted in azimuth relative to the first, and the supported ideal point response with worsened crossrange resolution.

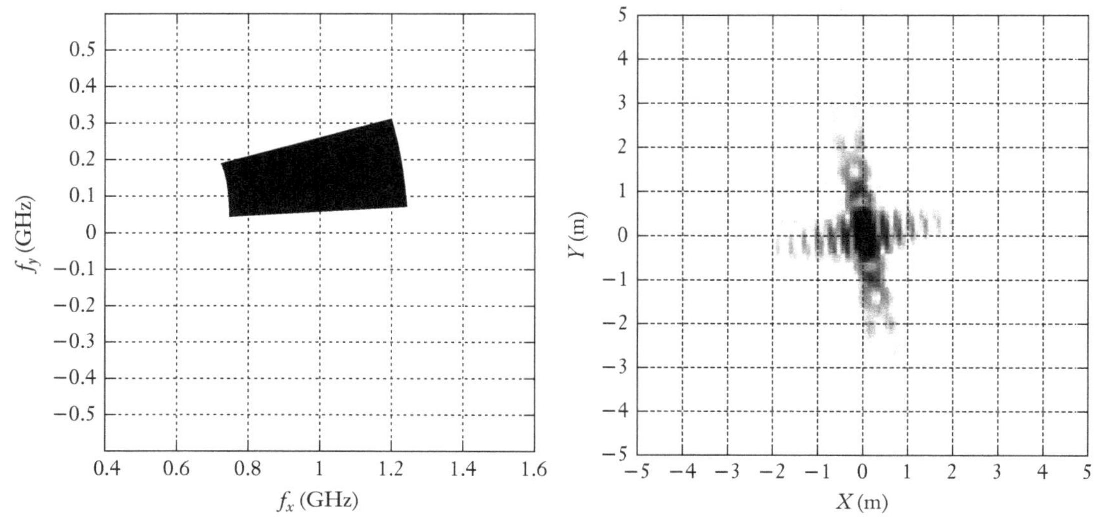

Figure 10–11 The data collection manifold on the left, corresponding to Figure 10–10, supports the ideal point response on the right with worsened crossrange resolution.

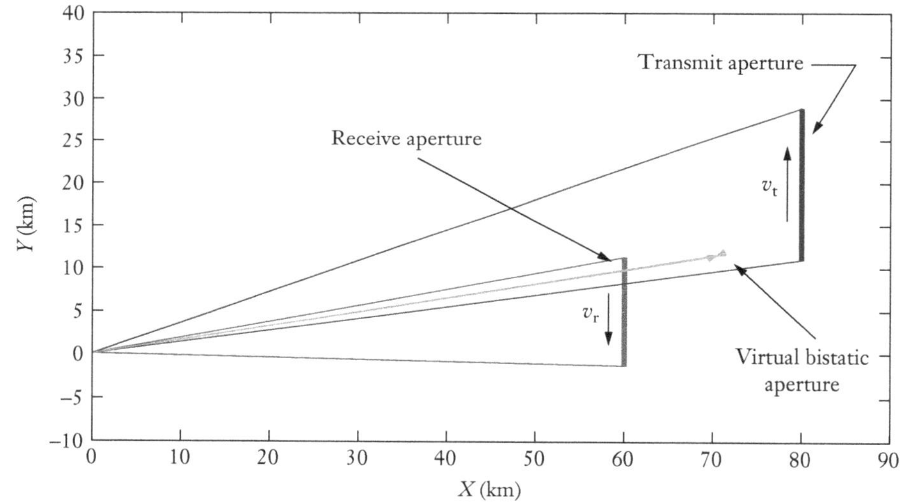

Figure 10–12 Transmit and receive apertures of a bistatic SAR system. The transmitter and receiver traverse their flight paths in opposite directions, thus yielding a significantly smaller bistatic aperture. With a center frequency of 1 GHz, the above system yields a horizontal crossrange resolution of $\delta h = 53$ m.

The last data collection example, described in Table 10–3 and shown in Figure 10–12, demonstrates the impact of the transmitter and receiver taking paths with opposite angular progressions about the scene center. The transmitter moves in a counterclockwise direction, while the receiver moves in a clockwise direction. The resulting bistatic aperture has a very small angular span, $\Delta\phi = 0.2°$ for a horizontal crossrange resolution of $\delta h = 53$ m. This is illustrated by the data collection manifold and corresponding ideal point response in Figure 10–13.

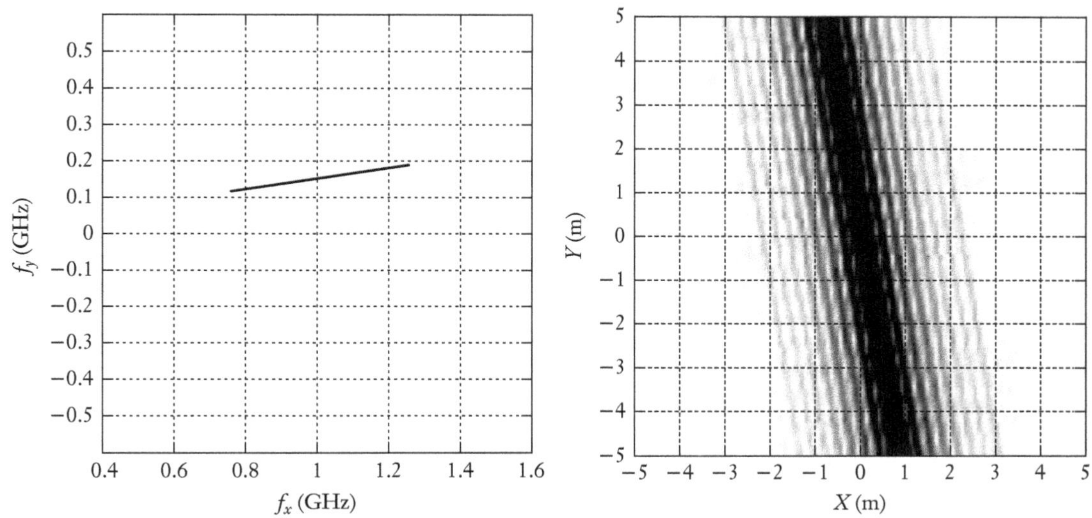

Figure 10–13 The data collection manifold on the left, corresponding to Figure 10–12, provides very limited crossrange bandwidth. The corresponding ideal point response has poor crossrange resolution.

Table 10–3 Bistatic Geometry Parameters Corresponding to Figure 10–12.

$(\bar{x}_t, \bar{y}_t, \bar{z}_t)$	(80 km, 20 km, 10 km)
$(\bar{x}_r, \bar{y}_r, \bar{z}_r)$	(60 km, 5 km, 8 km)
(v_{xt}, v_{yt}, v_{zt})	(0 m/s, 200 m/s, 0 m/s)
(v_{xr}, v_{yr}, v_{zr})	(0 m/s, –140 m/s, 0 m/s)
f_c	1 GHz
B	500 MHz
T_c	90 sec
$(\Delta\phi_t, \Delta\phi_r, \Delta\phi)$	(12.1°, 11.9°, 0.2°)

One can envision scenarios in which improper selection of platform flight paths might lead to a data collection manifold that overlaps itself in azimuth. This overlap can be disruptive to image formation algorithms that project the phase history data onto the f_x, f_y-plane, as discussed in Section 10.4. Differing motion measurement errors within the overlapped portion can lead to loss of coherence. And, even in the cases of perfect motion measurement, slight phase differences due to differing elevation angles or pulse-to-pulse waveform instability can also prevent achieving coherence across the segment of overlapped aperture.

10.3 FREQUENCY SAMPLE DATA COLLECTION

To obtain the frequency samples f_i, at each τ_k in equation (10.17), the transmit and receive hardware may be designed to operate in continuous or discrete modes. Most modern synthetic aperture radar systems transmit discrete pulses, wait an appropriate amount of time for two-way

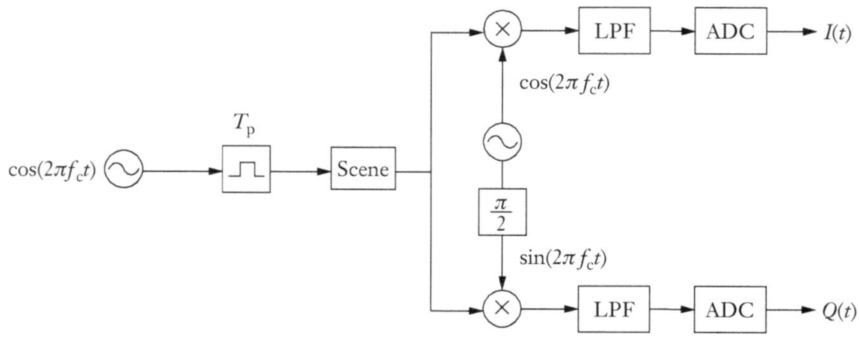

Figure 10–14 Transmit–receive block diagram for a short pulse system.

travel, and then receive and process the returned energy. In such a system, the spacing of the frequency samples f_i is determined by the system's range sampling requirement, to be discussed in Section 10.3.1. Similarly, the pulse repetition frequency (PRF), or the sampling frequency for the slow time samples τ_k, is determined by the system's sampling requirements in azimuth, to be discussed in Section 10.3.2.

This section discusses methods for frequency sample data collection. The use of continuous waveforms, especially in the case of noncooperative bistatic SAR [19–22], is highly relevant to many future system concepts. Nonetheless, the following subsections focus on four examples of pulsed waveform processing: the short pulse, pulsed linear frequency modulation (FM), pulsed linear FM with dechirp-on-receive, and stepped frequency. These techniques may be adapted for processing of continuous waveforms. Segmenting the continuous signal in time allows processing of virtual pulses. The viability of this segmentation approach depends on the autocorrelation properties of the continuous signal [23–27].

Short Pulse The simplest conceivable waveform is a short rectangular pulse of duration T_p modulated at the carrier frequency f_c. The transmitted short pulse waveform is written as

$$s(t) = \mathrm{rect}\left(\frac{\hat{t}}{T_\mathrm{p}}\right)\cos\left(2\pi f_\mathrm{c}\hat{t}\right) \tag{10.38}$$

where

$$\mathrm{rect}(t) = \begin{cases} 1, & |t| \le 0.5 \\ 0, & |t| > 0.5 \end{cases} \tag{10.39}$$

Figure 10–14 shows a block diagram for a short pulse system. There, a short duration tone is used to illuminate the scene, and the returns are demodulated to complex base band, low-pass filtered (LPF), and sampled by analog-to-digital converters (ADC). In classic radar system theory, the range resolution provided by a short pulse, ρ_SP, is given by [1]

$$\rho_\mathrm{SP} = \frac{cT_\mathrm{p}}{2} \tag{10.40}$$

where c is the speed of light. Note that ρ_{SP} is directly proportional to the pulse width. Therefore, to achieve finer range resolution, one must transmit shorter pulses.

For short pulse systems, the frequency samples in equation (10.17) are obtained by sampling the returned signal in fast time and Fourier transforming these data to obtain the corresponding phase history samples. *Fast time* is sampled at the ADC rate of the receiver, and *slow time* is sampled at the PRF of the system. Sampling in slow time in most cases is equivalent to sampling in azimuth. The bandwidth of the frequency domain samples is inversely proportional to the length of the transmitted pulse [1],

$$B \approx \frac{1}{T_{\mathrm{p}}} \tag{10.41}$$

Figure 10–15 shows a time–frequency plot for two short pulse waveforms. A time–frequency image is formed by processing the time series displayed horizontally, which has the power spectrum displayed vertically. Each column of the time–frequency image is computed by windowing a segment of the time series and computing its power spectrum. Thus, the peak of a short pulse time–frequency image appears at the intersection of the time of the pulse's transmission and the associated carrier frequency. The top plot of Figure 10–15 shows a short pulse with a 1 nsec duration and 1 GHz bandwidth, centered at 5 GHz; the bottom plot of Figure 10–15 shows a second pulse with a 4 nsec duration and 250 MHz bandwidth.

Use of a short pulse waveform poses two problems from a hardware perspective. First, as the pulse width shrinks and range resolution becomes finer, the amount of energy transmitted within the pulse width must remain constant to avoid degradation of the received signal-to-noise ratio (SNR). This increases the system's peak power requirement. As peak power is ultimately capped by weight and cost restrictions, this impediment can limit the range or range resolution achievable in a short pulse system, with a shorter pulse providing better resolution but a lower transmit power. Second, in order to satisfy the Nyquist sampling requirements needed to avoid aliasing on receive, the ADC in Figure 10–14 must operate at the bandwidth B of the transmitted pulse, which may be costly if not impossible for very fine range resolution systems.

Linear Frequency Modulation To overcome the peak power requirements of a short pulse system, one may implement a chirped pulse system. Using linear frequency modulation, the transmitted waveform can span the same bandwidth as a short pulse but spreads the transmitted energy over a longer pulse duration. The transmitted linear FM pulse is given by [2]

$$s(t) = \mathrm{rect}\left(\frac{\hat{t}}{T_{\mathrm{p}}}\right)\cos\left(2\pi f_c t + \pi\gamma\hat{t}^2\right) \tag{10.42}$$

where $\hat{t} = t - \tau_k$ represents the within-pulse fast time and $t = \tau_k$ is the slow time of the current pulse. The chirp rate of the pulse is $\gamma = B/T_{\mathrm{p}}$. Figure 10–16 shows a block diagram for transmission and demodulation of a linear FM pulse, which operates in the same fashion as a short pulse waveform with the added requirement of applying a matched filter to the received signal. The matched filter typically consists of a cross-correlation with a copy of the transmitted chirped pulse.

Figure 10–17 shows a time–frequency plot of a linear FM pulse. The simulated pulse has a 40 μsec duration and a chirp rate of γ =25 MHz/μsec, to yield a 1 GHz bandwidth. Chirping the transmitted pulse achieves a significantly greater bandwidth than indicated by the pulse duration, while requiring a lesser peak power.

A matched filter is applied to the received samples upon demodulation. This matched filter compresses the recorded data, such that a copy of the autocorrelation function of equation (10.42)

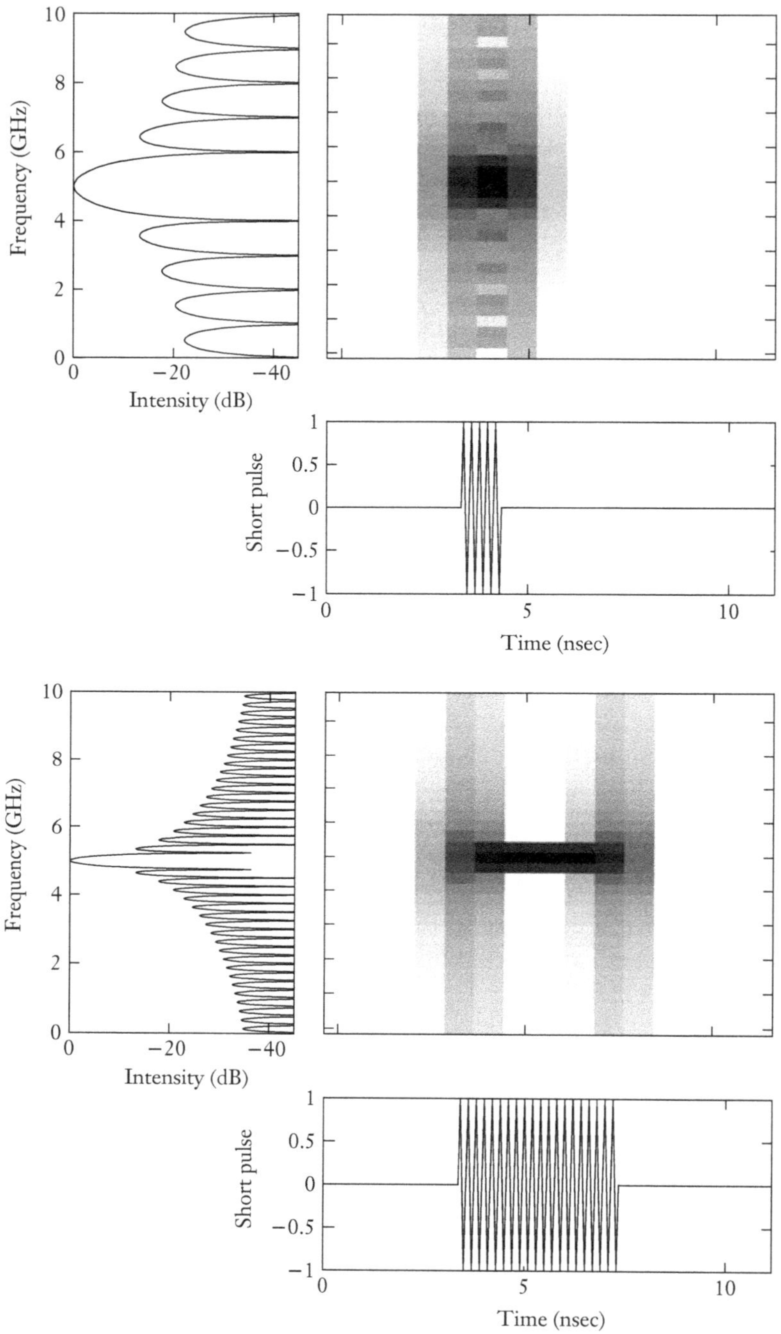

Figure 10–15 Time–frequency plots of two short pulse waveforms, one with a 1 nsec duration (top) and one with a 4 nsec duration (bottom). The bandwidths of the pulses (defined as one-half the main lobe width) are 1 GHz and 250 MHz, respectively, as predicted by equation (10.40).

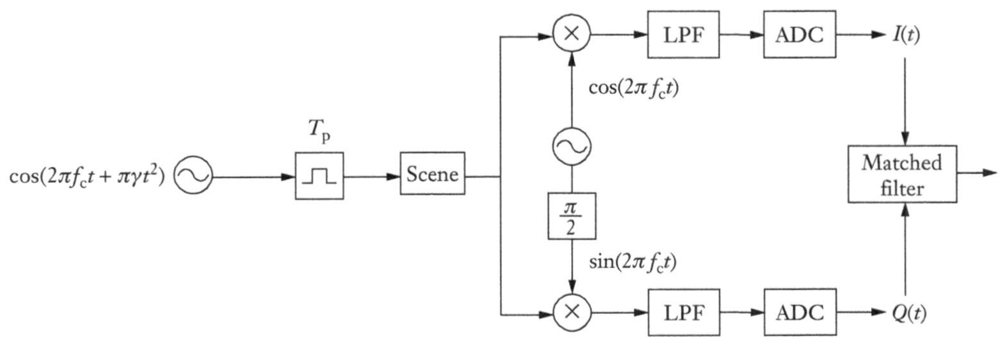

Figure 10–16 Transmit–receive block diagram for a pulsed linear FM system.

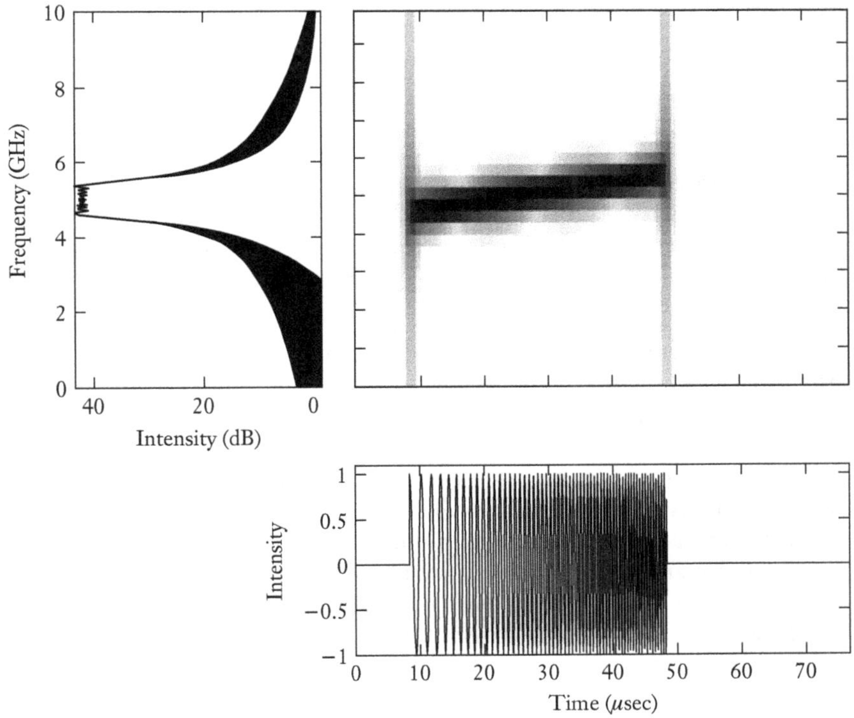

Figure 10–17 Time–frequency plot of a pulsed linear FM waveform, with a 20 nsec duration and 1 GHz bandwidth.

appears at the time delay of each scattered return. The resolution of the compressed data is defined to be equal to half the main lobe width of that autocorrelation function, written as [2]

$$\rho_{\text{LFM}} = \frac{c}{2B} \tag{10.43}$$

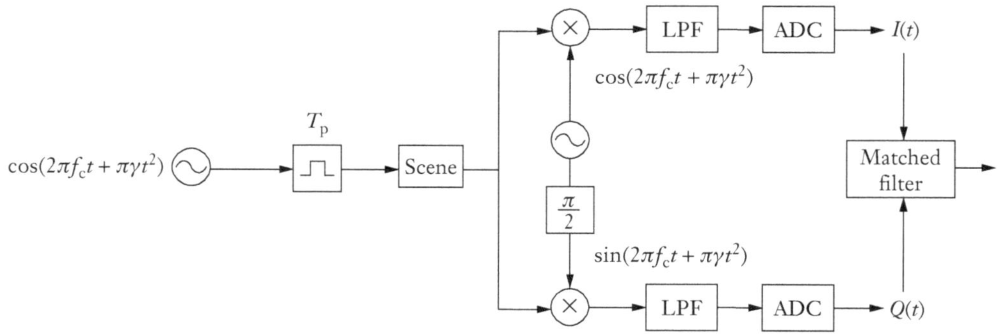

Figure 10–18 Transmit–receive block diagram for a dechirp-on-receive linear FM pulse system.

thus yielding a pulse compression ratio of

$$\text{PCR} = \frac{\rho_{\text{SP}}}{\rho_{\text{LFM}}} = BT_{\text{p}} = \gamma T_{\text{p}}^2 \tag{10.44}$$

The pulse compression ratio indicates the factor of resolution improvement seen by using a linear FM waveform rather than a short pulse of the same duration. Note, however, that the bandwidth of the received linear FM pulse is equal to that of a short pulse with the same resolution, meaning that the use of a linear FM waveform does not ease the Nyquist sampling requirement.

Linear FM Dechirp-on-Receive If one is able to set a minimum and maximum range to be interrogated by the radar, such as the range swath Δr illuminated by the transmitter in SAR, one may reduce the Nyquist sampling requirement for reception of linear FM pulses through use of dechirp-on-receive processing [2, 28], also known as stretch processing [1, 28]. In stretch processing, the received signal is mixed with a reference chirp signal, which has an extended pulse width but the same chirp rate as the transmitted waveform. The output of the mixing operation may then be sampled at a lower rate, to directly yield the desired frequency domain samples of (10.17). However, these samples are corrupted by an artifact commonly known as residual video phase (RVP), which may be largely removed through postprocessing.

Figure 10–18 shows a block diagram for the transmitter and receiver processing of a dechirp-on-receive system. As depicted, the transmitted waveform is given by equation (10.42), and thus the received signal is written as

$$s(t) = A_m \operatorname{rect}\left(\frac{\hat{t} - t_m}{T_{\text{p}}}\right) \cos\left(2\pi f_{\text{c}}(t - t_m) + \pi\gamma(\hat{t} - t_m)^2\right) \tag{10.45}$$

where t_m is the travel time from the transmitter to a scatterer in the scene and then to the receiver. In practice, the received signal will be the superposition of the returns from all of the scattering centers in the scene. Each would have its own time delay, t_m for the mth scatterer, and its own complex reflectivity, A_m. Figure 10–19 shows the timing and bandwidths of the transmitted pulse, the reference pulse on receive, and received pulses from near range, scene center, and far range.

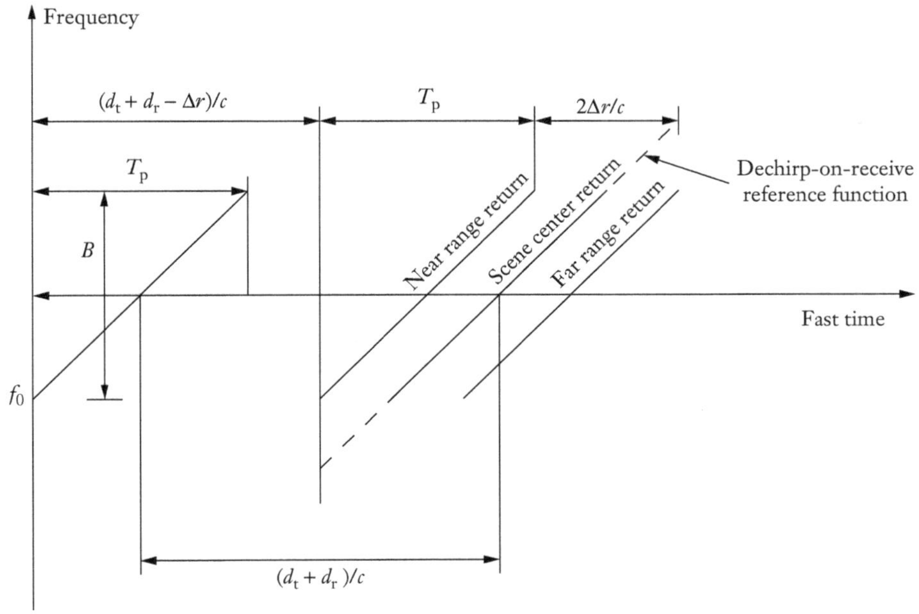

Figure 10–19 Timing of the received signals and the dechirp reference function relative to the pulse duration and ranges of a dechirp-on-receive system.

On receive, the input signal is mixed with a reference signal given by

$$s_I(t) = \cos(2\pi f_c(t - t_{ref}) + \pi\gamma(\hat{t} - t_{ref})^2) \tag{10.46}$$

for the in-phase channel and

$$s_Q(t) = -\sin(2\pi f_c(t - t_{ref}) + \pi\gamma(\hat{t} - t_{ref})^2) \tag{10.47}$$

for the quadrature channel. The time delay

$$t_{ref} = \frac{d_t + d_r}{c} \tag{10.48}$$

is the travel time of a pulse reflected by a scatterer at the scene center. In bistatic SAR data collection, the distances of the scene center from the transmitter and receiver, d_t and d_r, vary with slow time, as defined in (10.13) and (10.14).

After mixing and low-pass filtering, the resulting intermediate frequency [2] signal is

$$\begin{aligned} s_{IF}(t) &= A_m \operatorname{rect}\left(\frac{\hat{t} - t_m}{T_p}\right) e^{-j2\pi f_c(t_m - t_{ref})} e^{j\pi\gamma[(\hat{t}-t_m)^2 - (\hat{t}-t_{ref})^2]} \\ &= A_m \operatorname{rect}\left(\frac{\hat{t} - t_m}{T_p}\right) e^{-j2\pi\gamma\left(\frac{f_c}{\gamma}+\hat{t}-t_{ref}\right)(t_m - t_{ref})} e^{j\pi\gamma(t_m - t_{ref})^2}. \end{aligned} \tag{10.49}$$

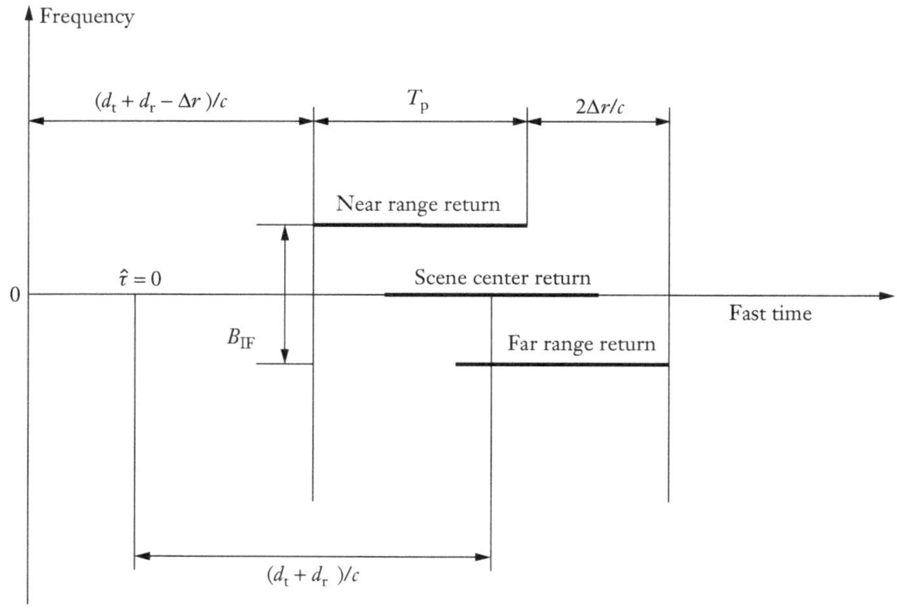

Figure 10–20 Time–frequency representation of $s_{IF}(t)$. Dechirped signals from the near range, scene center, and far range.

The timing and bandwidths of the dechirped signals from the near range, scene center, and far range are shown in Figure 10–20. It is straightforward to see that sampling of s_{IF} at discrete $\hat{t}_i \in [t_{ref} - T_p/2, t_{ref} + T_p/2]$ will yield frequency domain samples at frequencies

$$\begin{aligned} f_i &= \gamma\left(\frac{f_c}{\gamma} + \hat{t}_i - t_{ref}\right) \\ &= f_c + \gamma(\hat{t}_i - t_{ref}) \end{aligned} \tag{10.50}$$

as shown in equation (10.17). Recall that $\hat{t} = t - \tau_k$, thus indicating the timing of these samples in slow time. One may also view the output of dechirp-on-receive domodulation as a collection of tones, where the frequency of each tone is determined by the range of the corresponding scattering center.

The required ADC sampling rate may be found by analyzing the frequencies of the received and dechirped signals in (10.49). The intermediate frequency of the near range returns, meaning $(t_m - t_{ref}) = \Delta r/c$, is $f_{IF,\,near} = -\gamma\Delta r/c$, and likewise, the intermediate frequency of far range returns, meaning $(t_m - t_{ref}) = \Delta r/c$, is $f_{IF,\,far} = \gamma\Delta r/c$. This implies a sampling rate of [2]

$$F_s = B_{IF} = f_{IF,far} - f_{IF,near} = \frac{2\gamma\Delta r}{c} \tag{10.51}$$

on the in-phase and quadrature channels, and a sampling start time of [2]

$$t_{\mathrm{b}} = t_{\mathrm{ref}} - \frac{\Delta r}{c} - \frac{T_{\mathrm{p}}}{2} \tag{10.52}$$

which allows for the range swath width and the transmitted pulse width. The total required sampling time is written as

$$T_{\mathrm{s}} = 2\frac{\Delta r}{c} + T_{\mathrm{p}}. \tag{10.53}$$

One should now note that there is an additional range-dependent phase term $\exp\{j\pi\gamma\,(t_m - t_{\mathrm{ref}})^2\}$ in (10.49), commonly referred to as residual video phase (RVP) [2]. This term must be compensated to avoid range-dependent blurring and distortion in the final SAR image. The RVP term may be viewed as a misalignment in receive time, seen horizontally in Figure 10–20, between the returns from the near range, scene center, and far range. Application of a frequency-dependent delay, to the sampled data, will typically mitigate this effect sufficiently.

To facilitate the following, momentarily neglect sampling in fast time and continue with the continuous time expression in (10.49). Satisfying the Nyquist sampling requirement allows one to work with the continuous and discrete signals interchangeably. Taking the Fourier transform of $s_{\mathrm{IF}}(t)$ with respect to the dummy variable $\eta = \hat{t} - t_{\mathrm{ref}}$ yields

$$\begin{aligned} S_{\mathrm{IF}}(f) &= \int_{-\infty}^{\infty} A_0 \,\mathrm{rect}\left(\frac{\eta + t_{\mathrm{ref}} - t_0}{T_p}\right) e^{-j2\pi\gamma\left(\frac{f_c}{\gamma}+\eta\right)(t_0 - t_{\mathrm{ref}})} \\ &\quad \times e^{j\pi\gamma(t_0 - t_{\mathrm{ref}})^2} e^{-j2\pi f\eta}\, d\eta \\ &= A_0 T_p \,\mathrm{sin}\,c[T_p \pi(\gamma(t_0 - t_{\mathrm{ref}}) + f)]\, e^{-j2\pi(f+f_c)(t_0 - t_{\mathrm{ref}})} \\ &\quad \times e^{-j\pi\gamma(t_0 - t_{\mathrm{ref}})^2} \end{aligned} \tag{10.54}$$

The envelope of this function has its peak at $f = \gamma(t_{\mathrm{ref}} - t_0)$. Thus, if one multiplies (10.54) by $\exp[j\pi f^2/\gamma]$, this will effectively multiply $S_{\mathrm{IF}}(f)$ by $e^{j\pi\gamma(t_0 - t_{\mathrm{ref}})^2}$ at its peak. This approximately cancels the quadratic RVP term, provided T_{p} is large, and also serves to align the returns received from different ranges. Figure 10–21 depicts a block diagram representing the steps necessary for removing residual video phase from the sampled data, where the frequency samples are given by $f_k = F_s k/N$.

10.3.1 Frequency Sampling Requirement

A bistatic SAR system's requirement on frequency sample spacing is determined by the depth of the range swath Δr illuminated by the transmitting platform. Insufficient sampling will alias the received signals, thus causing the near and far portions of the final SAR image to wrap in range.

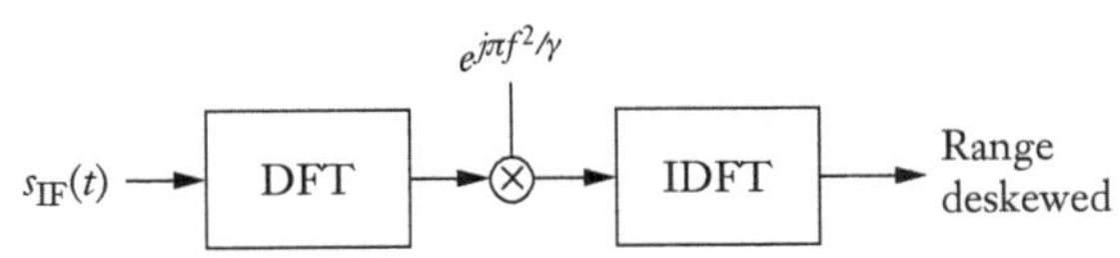

Figure 10–21 Block diagram of removal of residual video phase via range deskew.

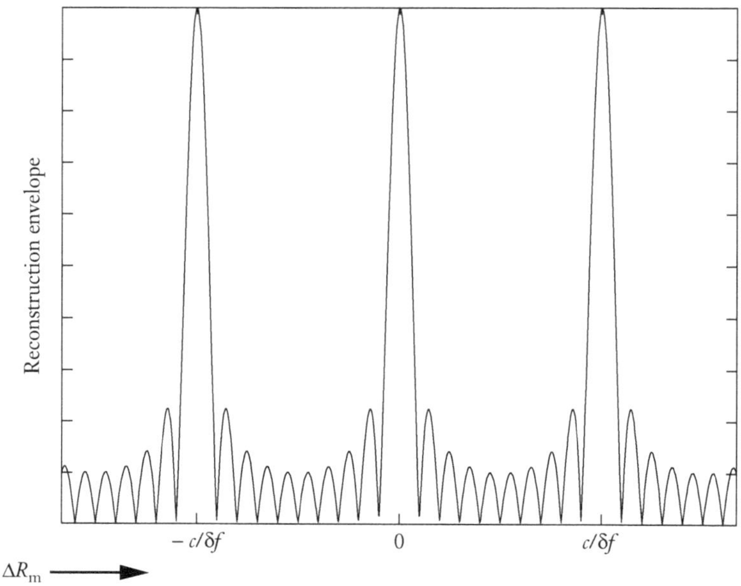

Figure 10–22 Matched filtering reconstruction of a single range profile. The three peaks are spectral replicates of the same scatterer's return.

To illustrate, consider a single sample point in slow time of a scene containing a single-point scatterer. The corresponding frequency samples are written as

$$S(f_i) = A_0\, e^{-j\frac{2\pi f_i}{c}\Delta R_m} \qquad i = 0,\ldots,N-1 \tag{10.55}$$

where ΔR_m is the differential range of the scatterer, and the sampled frequencies are defined as $f_s = f_0 + i\delta f$. Recall from Section 10.1 that the differential range ΔR_m is the difference between the round trip distance to a scatterer in the scene and the round trip distance to a scatterer located at the origin. Matching the filtering of (10.55) to obtain a range profile, as a function of the differential range ΔR, gives

$$\begin{aligned} s(\Delta R) &= \sum_{i=0}^{N-1} A_0\, e^{-j\frac{2\pi f_i}{c}\Delta R_m}\, e^{j\frac{2\pi f_i}{c}\Delta R} \\ &= A_0\, e^{j\frac{2\pi}{c}(\Delta R - \Delta R_m)\left(f_0 + \frac{N\delta f - \delta f}{2}\right)} \frac{\sin\left(\frac{\pi N \delta f}{c}(\Delta R - \Delta R_m)\right)}{\sin\left(\frac{\pi \delta f}{c}(\Delta R - \Delta R_m)\right)}. \end{aligned} \tag{10.56}$$

To find the unambiguous range of this reconstruction, one may examine the periodicity of the envelope of (10.56), shown in Figure 10–22 for $\Delta R_m = 0$. The peaks of $S(\Delta R)$ appear at the zero crossings of the denominator; thus the unambiguous range is

$$\frac{\pi \delta f}{c}(\Delta R - \Delta R_0) \in \left[-\frac{\pi}{2}, \frac{\pi}{2}\right] \;\Rightarrow\; (\Delta R - \Delta R_0) \in \left[-\frac{c}{2\delta f}, \frac{c}{2\delta f}\right]. \tag{10.57}$$

Therefore, given an illuminated range swath of depth Δr, the limit on the maximum allowed frequency sample spacing is defined as

$$\delta f < \frac{c}{2\Delta r}. \tag{10.58}$$

10.3.2 Azimuth Sampling Requirement

The required spacing of a bistatic system's azimuth samples in slow time depends on the range of doppler frequencies observed by the receiving platform. While the ground targets imaged by a synthetic aperture radar are traditionally assumed to be stationary, the data collection platforms are in motion, and it is their motion relative to the targets that induces doppler shift. The doppler frequency of each target is dependent on its position in crossrange. The pulse repetition frequency (PRF) of the transmitter must be greater than or equal to the corresponding doppler bandwidth. As discussed in [6, 29, 30], the doppler frequency observed by the receiver in a bistatic system is equal to the sum of the one-way doppler shifts induced by the transmitter and receiver.

To determine the range of doppler frequencies observed during data collection, the analysis will be divided into separate sections. First, a scene containing only a single scattering center positioned at the origin, will be considered. As the transmitter and receiver traverse their flight paths with respect to this point, the bistatic geometry will naturally change, thereby inducing a change in each platform's line-of-sight velocity to the scene center. As these line-of-sight velocities change, so does the observed doppler frequency. This variation in the observed doppler frequencies of a scatterer at the scene center is referred to as the along-track doppler bandwidth.

Second, at any given point in the data collection the transmit beam is assumed to be illuminating some area of ground with scattering centers positioned throughout. Scatterers that are separated from the origin in the crossrange direction will have a slightly different doppler shift. It is this difference in doppler shifts that allows azimuth resolution to be achieved in SAR imaging. The variation in doppler shifts observed from all of the scatterers in a scene at any given instant is referred to as the spotlighted doppler bandwidth.

The observed doppler frequency of each scatterer in the scene will undergo the same relative variation as a scatterer at the scene center. Thus, the total doppler bandwidth of a scene is equal to the sum of the along-track doppler bandwidth of a scatterer at the scene center and the instantaneous doppler bandwidth of all of the scatterers in the scene. The PRF of the system, corresponding to the azimuth sampling rate, must exceed the total doppler bandwidth of the scene.

Along-Track Doppler Bandwidth From equation (10.16), the phase of the signal returned by a scatterer located at the scene center is given by

$$\Phi(f,\tau) = -2\pi f \frac{d_{\mathrm{r}}(\tau)+d_{\mathrm{t}}(\tau)}{c} \triangleq -2\pi f \frac{d_{\mathrm{rt}}(\tau)}{c}, \qquad f \in [f_0, f_0 + B] \tag{10.59}$$

where the effect of time gating has been omitted. The round-trip distance $d_{\mathrm{rt}}(\tau)$ is equal to the sum of the transmitter and receiver distances to the origin, written as

$$\begin{aligned} d_{\mathrm{rt}}(\tau) &= d_{\mathrm{r}}(\tau) + d_{\mathrm{t}}(\tau) \\ &= \sqrt{x_{\mathrm{r}}(\tau)^2 + y_{\mathrm{r}}(\tau)^2 + z_{\mathrm{r}}(\tau)^2} + \sqrt{x_{\mathrm{t}}(\tau)^2 + y_{\mathrm{t}}(\tau)^2 + z_{\mathrm{t}}(\tau)^2} \end{aligned} \tag{10.60}$$

Focusing on the center frequency f_c, with the corresponding wavelength $\lambda = c/f_c$, the doppler frequency shift observed at the receiver at the slow time τ from a scattering center at the origin is given by

$$\begin{aligned}
f_d(\tau) &= -\frac{1}{\lambda}\frac{\partial d_{rt}(\tau)}{\partial \tau} \\
&= -\frac{1}{\lambda}\left[\frac{x_r(\tau)v_{xr}(\tau)+y_r(\tau)v_{yr}(\tau)+z_r(\tau)v_{zr}(\tau)}{\sqrt{x_r(\tau)^2+y_r(\tau)^2+z_r(\tau)^2}}\right. \\
&\quad \left.+\frac{x_t(\tau)v_{xt}(\tau)+y_t(\tau)v_{yt}(\tau)+z_t(\tau)v_{zt}(\tau)}{\sqrt{x_t(\tau)^2+y_t(\tau)^2+z_t(\tau)^2}}\right] \\
&= -\frac{1}{\lambda}\left[v_{xr}(\tau)\cos\phi_r(\tau)\cos\theta_r(\tau)+v_{xt}(\tau)\cos\phi_t(\tau)\cos\theta_t(\tau)\right. \\
&\quad + v_{yr}(\tau)\sin\phi_r(\tau)\cos\theta_r(\tau)+v_{yt}(\tau)\sin\phi_t(\tau)\cos\theta_t(\tau) \\
&\quad \left.+ v_{zr}(\tau)\sin\theta_r(\tau)+v_{zt}(\tau)\sin\theta_t(\tau)\right]
\end{aligned} \tag{10.61}$$

where $v_{xr}(\tau) = dx_r(\tau)/d\tau$ is the velocity of the receiver in the x-direction, and the platform velocities in each of the other cardinal directions are defined similarly. By introducing velocity vectors for the receiver and transmitter

$$\underline{v}_r(\tau) = \begin{bmatrix} v_{xr}(\tau) \\ v_{yr}(\tau) \\ v_{zr}(\tau) \end{bmatrix} \quad \text{and} \quad \underline{v}_t(\tau) = \begin{bmatrix} v_{xt}(\tau) \\ v_{yt}(\tau) \\ v_{zt}(\tau) \end{bmatrix} \tag{10.62}$$

and normalized position vectors

$$\hat{\underline{r}}_r(\tau) = \begin{bmatrix} \cos\phi_r(\tau)\cos\theta_r(\tau) \\ \sin\phi_r(\tau)\cos\theta_r(\tau) \\ \sin\theta_r(\tau) \end{bmatrix} \quad \text{and} \quad \hat{\underline{r}}_t(\tau) = \begin{bmatrix} \cos\phi_t(\tau)\cos\theta_t(\tau) \\ \sin\phi_t(\tau)\cos\theta_t(\tau) \\ \sin\theta_t(\tau) \end{bmatrix}, \tag{10.63}$$

one may rewrite the instantaneous doppler frequency as

$$f_d^{(1)}(\tau) = -\frac{1}{\lambda}\left[\underline{v}_r(\tau)^T\hat{\underline{r}}_r(\tau)+\underline{v}_t(\tau)^T\hat{\underline{r}}_t(\tau)\right]. \tag{10.64}$$

Given the above expression for the doppler frequency shift, the most general equation for the observed along-track doppler bandwidth is

$$B_d^{(1)} = \max_\tau f_d^{(1)}(\tau) - \min_\tau f_d^{(1)}(\tau) \tag{10.65}$$

where the extrema are found within the sampled range of slow times $\tau \in [-T/2, T/2]$.

By making a few assumptions regarding the flight paths of the transmitter and receiver, a more usable expression for $B_d^{(1)}$ can be derived. First, it is assumed that both platforms traverse straight and level flight paths such that

$$\underline{v}_r(\tau) = \begin{bmatrix} v_{xr} \\ v_{yr} \\ v_{zr} = 0 \end{bmatrix} \quad \text{and} \quad \underline{v}_t(\tau) = \begin{bmatrix} v_{xt} \\ v_{yt} \\ v_{zt} = 0 \end{bmatrix} \tag{10.66}$$

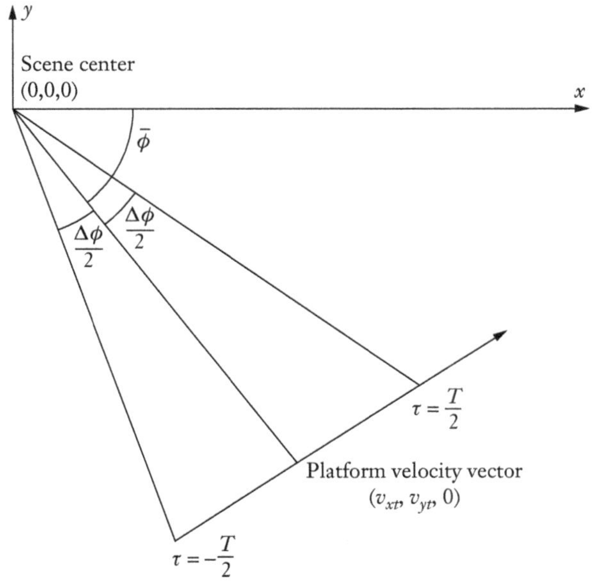

Figure 10–23 Angular span subtended by an antenna platform during data collection.

Second, assume that the flight paths are short enough, and the ground ranges are long enough, such that the elevation angles of the transmitter and receiver may be treated as constants $\theta_t(\tau) = \bar{\theta}_r$ and $\theta_{\mathrm{t}}(\tau) = \bar{\theta}_{\mathrm{t}}$. Third, the transmitter and receiver are both assumed to fly paths, with respect to the scene center, that have the same angular progression in azimuth (i.e., both clockwise or both counterclockwise). Finally, it is assumed that the azimuth angle of the receiver spans $\phi_{\mathrm{r}}(\tau) \in [\bar{\phi}_{\mathrm{r}} - \Delta\phi_{\mathrm{r}}/2, \bar{\phi}_{\mathrm{r}} + \Delta\bar{\phi}_{\mathrm{r}}/2]$ and that the azimuth angle of the transmitter spans $\phi_{\mathrm{t}}(\tau) \in [\bar{\phi}_{\mathrm{t}} - \Delta\phi_{\mathrm{t}}/2, \bar{\phi}_{\mathrm{t}} + \Delta\bar{\phi}_{\mathrm{t}}/2]$. Figure 10–23 shows the geometry of a single such data collection platform. On the basis of these assumptions, one may write the one-way doppler frequency shift induced by the motion of one of the platforms as

$$f_{\mathrm{d}}^{(1)} = \frac{v_x \cos\left(\bar{\phi} \pm \frac{\Delta\phi}{2}\right)\cos\bar{\theta} + v_y \sin\left(\bar{\phi} \pm \frac{\Delta\phi}{2}\right)\cos\bar{\theta}}{\lambda_{\mathrm{c}}}. \tag{10.67}$$

Accordingly, one may rewrite (10.65) as

$$\begin{aligned} B_{\mathrm{d}}^{(1)} = {} & \frac{\cos\bar{\theta}_{\mathrm{t}}}{\lambda}\left[v_{xt}\cos\left(\bar{\phi}_{\mathrm{t}} - \frac{\Delta\phi_{\mathrm{t}}}{2}\right) + v_{yt}\sin\left(\bar{\phi}_{\mathrm{t}} - \frac{\Delta\phi_{\mathrm{t}}}{2}\right)\right] \\ & - \frac{\cos\bar{\theta}_{\mathrm{t}}}{\lambda}\left[v_{xt}\cos\left(\bar{\phi}_{\mathrm{t}} + \frac{\Delta\phi_{\mathrm{t}}}{2}\right) + v_{yt}\sin\left(\bar{\phi}_{\mathrm{t}} + \frac{\Delta\phi_{\mathrm{t}}}{2}\right)\right] \\ & + \frac{\cos\bar{\theta}_{\mathrm{r}}}{\lambda}\left[v_{xr}\cos\left(\bar{\phi}_{\mathrm{r}} - \frac{\Delta\phi_{\mathrm{r}}}{2}\right) + v_{yr}\sin\left(\bar{\phi}_{\mathrm{r}} - \frac{\Delta\phi_{\mathrm{r}}}{2}\right)\right] \\ & - \frac{\cos\bar{\theta}_{\mathrm{r}}}{\lambda}\left[v_{xr}\cos\left(\bar{\phi}_{\mathrm{r}} + \frac{\Delta\phi_{\mathrm{r}}}{2}\right) + v_{yr}\sin\left(\bar{\phi}_{\mathrm{r}} + \frac{\Delta\phi_{\mathrm{r}}}{2}\right)\right] \end{aligned}$$

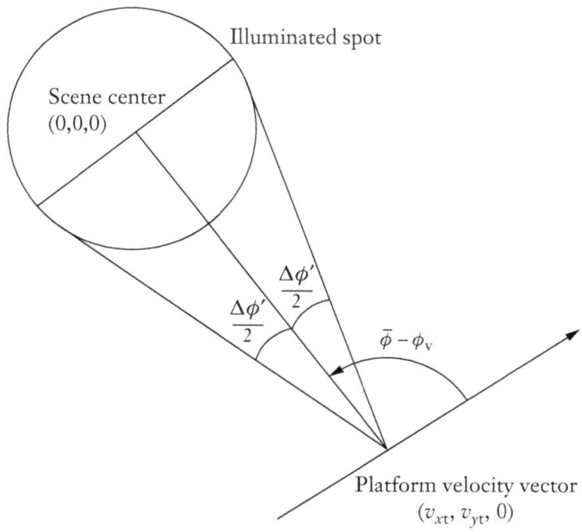

Figure 10–24 Angular span of the illuminated spot, shown with respect to the trajectory of an antenna platform.

$$= \frac{2}{\lambda}\left[v_{yt} \cos\bar{\phi}_t \sin\frac{\Delta\phi_t}{2} \cos\bar{\theta}_t + v_{yr} \cos\bar{\phi}_r \sin\frac{\Delta\phi_r}{2} \cos\bar{\theta}_r \right.$$
$$\left. - v_{xt} \sin\bar{\phi}_t \sin\frac{\Delta\phi_t}{2} \cos\bar{\theta}_t - v_{xr} \sin\bar{\phi}_r \sin\frac{\Delta\phi_r}{2} \cos\bar{\theta}_r \right] \tag{10.68}$$

The above expression gives $B_d^{(1)}$ in terms of only the speed and headings of the two platforms as well as their aspect angles and the spans of their look angles in azimuth. Equation (10.68) defines the breadth of doppler frequencies returned by a stationary scatterer at the scene center, due to the motion of the transmitter and receiver.

Spotlighted Doppler Bandwidth The range of doppler frequencies returned by all of the scatterers within the spotlighted area is now examined. This range of frequencies, at any given time τ, is centered about the along-track doppler frequency given in (10.61). The spotlighted doppler bandwidth is most easily computed by considering the one-way doppler shift induced by each platform and then adding the results. Figure 10–24 shows the position and velocity of a platform with respect to the scene center; the transmitted beam width is represented by $\Delta\phi'$. To compute the one-way doppler bandwidth of this area, calculate the one-way doppler frequencies of points at opposite edges of the spot in crossrange and subtract them. These doppler frequencies are computed as

$$f_d^{(2)} = \frac{v \cos\left(\bar{\phi} - \phi_v \pm \frac{\Delta\phi'}{2}\right) \cos\bar{\theta}}{\lambda_c} \tag{10.69}$$

where $\bar{\theta}$ is the depression angle from the platform to the scene center and $\bar{\phi} - \phi_v$ is the angle separating the line-of-sight to the origin and the platform velocity vector. From (10.69), the spotlighted doppler bandwidth due to the motion of both platforms is

$$
\begin{aligned}
B_{\mathrm{d}}^{(2)} &= \frac{v_{\mathrm{t}}\cos\left(\bar{\phi}_{\mathrm{t}}-\phi_{v\mathrm{t}}-\frac{\Delta\phi_{\mathrm{t}}'}{2}\right)\cos\bar{\theta}_{\mathrm{t}}}{\lambda_{\mathrm{c}}} - \frac{v_{\mathrm{t}}\cos\left(\bar{\phi}_{\mathrm{t}}-\phi_{v\mathrm{t}}+\frac{\Delta\phi_{\mathrm{t}}'}{2}\right)\cos\bar{\theta}_{\mathrm{t}}}{\lambda_{\mathrm{c}}} \\
&+ \frac{v_{\mathrm{r}}\cos\left(\bar{\phi}_{\mathrm{r}}-\phi_{v\mathrm{r}}-\frac{\Delta\phi_{\mathrm{r}}'}{2}\right)\cos\bar{\theta}_{\mathrm{r}}}{\lambda_{\mathrm{c}}} - \frac{v_{\mathrm{r}}\cos\left(\bar{\phi}_{\mathrm{r}}-\phi_{v\mathrm{r}}+\frac{\Delta\phi_{\mathrm{r}}'}{2}\right)\cos\bar{\theta}_{\mathrm{r}}}{\lambda_{\mathrm{c}}} \\
&= \frac{2}{\lambda_{\mathrm{c}}}\Bigg[v_{y\mathrm{t}}\cos\bar{\phi}_{\mathrm{t}}\sin\frac{\Delta\phi_{\mathrm{t}}'}{2}\cos\bar{\theta}_{\mathrm{t}} + v_{y\mathrm{r}}\cos\bar{\phi}_{\mathrm{r}}\sin\frac{\Delta\phi_{\mathrm{r}}'}{2}\cos\bar{\theta}_{\mathrm{r}} \\
&\quad - v_{x\mathrm{t}}\sin\bar{\phi}_{\mathrm{t}}\sin\frac{\Delta\phi_{\mathrm{t}}'}{2}\cos\bar{\theta}_{\mathrm{t}} - v_{x\mathrm{r}}\sin\bar{\phi}_{\mathrm{r}}\sin\frac{\Delta\phi_{\mathrm{r}}'}{2}\cos\bar{\theta}_{\mathrm{r}}\Bigg].
\end{aligned} \tag{10.70}
$$

One may note the similarity between equations (10.65) and (10.70).

Adding the results in (10.65) and (10.70) yields the total observed doppler bandwidth of the data collection, written as

$$
B_{\mathrm{d}} = B_{\mathrm{d}}^{(1)} + B_{\mathrm{d}}^{(2)}. \tag{10.71}
$$

This dictates an azimuth sampling requirement of $PRF = B_{\mathrm{d}}$.

Azimuth Dechirp The two doppler bandwidths that are summed in (10.71) originate from two phenomena: (1) the range of doppler frequencies associated with the scene center through the course of a data collection, and (2) the range of doppler frequencies corresponding to the returns from the entire spotlighted area. The doppler shift associated with the scene center return acts as a carrier frequency about which the spotlighted doppler bandwidth is situated. Given that this carrier frequency should be well known to the receiving platform of the bistatic system at any time $\tau \in [-T/2, T/2]$, the receiver should be able to remove this modulation from each pulse, thus reducing the total observed doppler bandwidth to merely $B_{\mathrm{d}} = B_{\mathrm{d}}^{(2)}$. This operation is commonly known as azimuth dechirp processing. The polar format algorithm [2, 16, 17] for SAR image formation requires that the processed phase history samples be dechirped in azimuth prior to image formation.

The effect of doppler on the sampled signal is well approximated by a slow time-varying doppler shift of the observed frequencies. Taking this time-varying shift into account, one may write the expression for the received signal as

$$
s'(\tau_k,\hat{t}) = s(\tau_k,\hat{t})\exp\{-j2\pi f_{\mathrm{d}}^{(1)}(\tau_k)\hat{t}\} \tag{10.72}
$$

where $f_{\mathrm{d}}^{(1)}$ is functioning as the carrier frequency for $s(\tau_k,\hat{t})$. This doppler phase term manifests itself as an azimuth chirp, meaning a quadratic phase function, in systems employing linear platform trajectories. To provide for linear phase coherence across the synthetic aperture, and to reduce the azimuth sampling requirement, one may choose to remove the carrier by mixing the received signal with

$$
s_{\mathrm{az}}(\tau_k,\hat{t}) = e^{j2\pi f_{\mathrm{d}}(\tau_k)\hat{t}} \tag{10.73}
$$

prior to sampling. Alternative approaches include adjustment of the transmit center frequency or the center frequency of the demodulating signal on receive. In a dechirp-on-receive system,

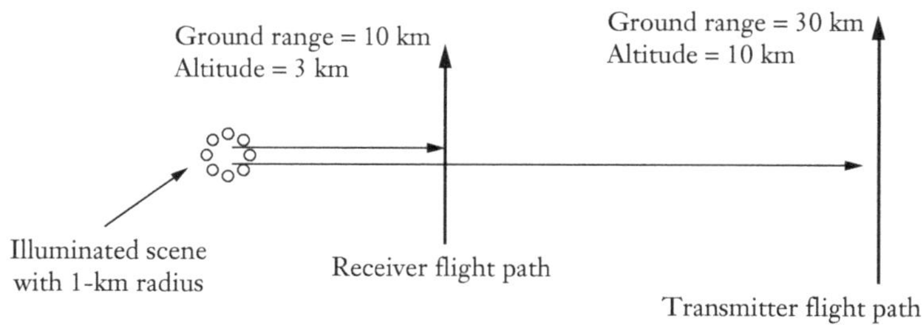

Figure 10–25 Top view of a bistatic system observing a scene with a 1000 m radius. The transmitter speed is 200 m/sec, and the receiver speed is 140 m/sec. The system operates at a center frequency of f_c = 10 GHz (λ_c = 3 cm).

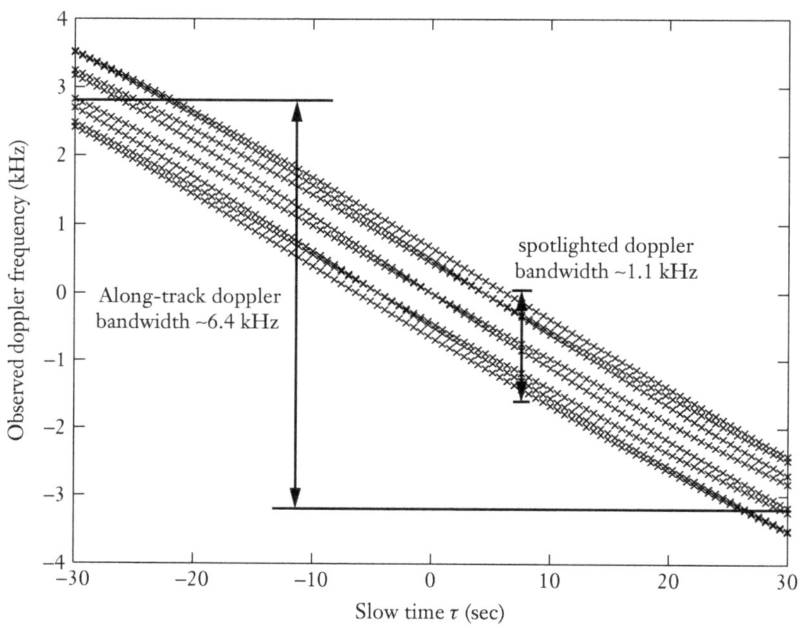

Figure 10–26 Doppler profile for the data collection depicted in Figure 10–25.

this can be accomplished by varying the timing (t_{ref}) of the dechirp signal. Removing the doppler carrier reduces the azimuth sampling requirement to PRF = $B_d^{(2)}$.

To illustrate the above analysis, two simulated data collection examples are presented. Figure 10–25 shows a bistatic data collection geometry for a scene with a 1000 m radius. The transmitter has a ground range of 30 km and an altitude of 10 km, and the receiver has a ground range of 10 km and an altitude of 3 km. The two platforms fly parallel, straight and level trajectories, the transmitter at 200 m/sec and the receiver at 140 m/sec. The system operates at a center frequency of f_c = 10 GHz (λ_c = 3 cm). Over a 60 second data collection interval, the total observed doppler frequency $f_d = f_d^{(1)} + f_d^{(2)}$ was computed at half second intervals for each of eight scatterers on the perimeter of the scene. Figure 10–26 shows the doppler profiles corresponding to returns from the eight scatterers. The along-track doppler bandwidth is approximately 6.4 kHz. The total doppler bandwidth is 7.5 kHz, but application of an azimuth dechirp prior to sampling

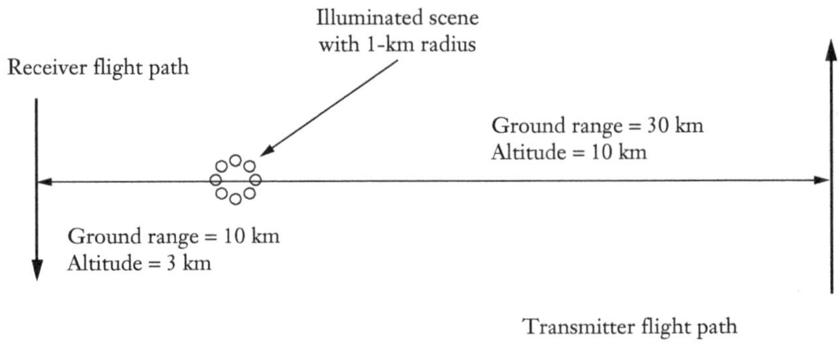

Figure 10–27 Top view of a bistatic system observing a scene with a 1000 m radius. The transmitter speed is 200 m/sec, and the receiver speed is 140 m/sec. The system operates at a center frequency of $f_c = 10$ GHz ($\lambda_c = 3$ cm).

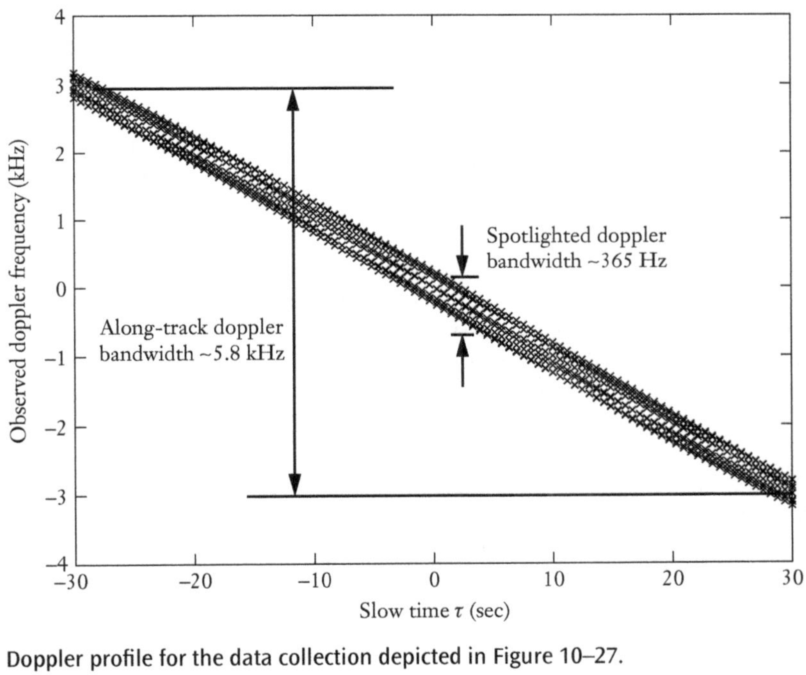

Figure 10–28 Doppler profile for the data collection depicted in Figure 10–27.

would reduce this rate by the along-track doppler bandwidth of approximately 6.4 kHz. The required azimuth sampling rate would then be approximately 1.1 kHz, equal to the spotlighted doppler bandwidth instead of the total bandwidth 7.5 kHz.

In a second simulated example, Figure 10–27 shows a bistatic data collection geometry for a scene with 1000 m radius. The transmitter has a ground range of 30 km and an altitude of 10 km, and the receiver has a ground range of 10 km and an altitude of 3 km. Again, the two platforms fly straight and level trajectories, the transmitter at 200 m/sec and the receiver at 140 m/sec. However, in this example, the receiver is positioned to observe forward scatter from the scene, such that the transmitter and receiver are traveling in opposite directions. The system operates at a center frequency of $f_c = 10$ GHz ($\lambda_c = 3$ cm). Figure 10–28 shows the doppler profiles corresponding to returns from the eight scatterers. The along-track doppler bandwidth is approximately 5.8 kHz, which is nearly identical to the first example. However, the opposite velocities of the transmitter and receiver reduce the spotlighted doppler bandwidth to approximately 365 Hz, which implies a relatively small difference in doppler frequencies for scatterers separated in crossrange and thus relatively poor crossrange resolution.

10.4 BISTATIC SAR IMAGE FORMATION

The previous sections described the bistatic SAR geometry and the process of bistatic SAR data collection. With the collected data, information about the scene that was interrogated will now be reconstructed. This process of nonparametric scene reconstruction is commonly referred to as image formation. The inputs to image formation algorithms typically include the collected phase history data and signal-level information about the transmitted waveform, as well as navigational measurements of the antenna phase centers on the data collection platforms. The resulting image is a top-level view of the observed scene where objects of greater reflectivity appear with greater image intensity.

Multiple image formation processes have been developed for monostatic SAR [2, 17, 31–38]. The mathematically ideal method for image formation is matched filtering (MF). Matched filtering may be implemented more efficiently as the backprojection algorithm (BPA) [32, 33], and thereby offers advantages such as progressive image formation and multiresolution capabilities. However, these theoretically ideal methods come at great computational expense. MF scales with complexity $O(N^4)$, and BPA scales with complexity $O(N^3)$, where N^2 is the number of pixels in the image.

By making explicit assumptions about the collection geometry and the nature of the collected phase history data, SAR systems often use faster algorithms, including the polar format algorithm (PFA) [31], the range migration algorithm (RMA) [34–36], and the chirp scaling algorithm (CSA) [37]. Some recent works [39, 40] detail fast backprojection algorithms that achieve $O(N^2 \log_2 N)$ complexity, but do not correct for range curvature and do not allow progressive image formation. In [41], a fast BPA was presented that gives $O(N^2\sqrt{N})$ complexity, while still correcting for range curvature.

Bistatic SAR systems [19, 21, 42–50], in contrast, have seen much less development in the area of image formation, at least in the open literature. Existing references for bistatic SAR image formation exploit the BPA [51, 52], MF [53], super-resolution [54], direct Fourier inversion [20, 55, 56], nonlinear chirp scaling [57], or special cases of RMA processing [58]. All of these approaches suffer from the high computational costs mentioned above, although a recent fast backprojection algorithm [51] achieves $O(N^2 \log_2 N)$ complexity.

The seminal work by Arikan and Munson [59] provided perhaps the first tomographic formulation of bistatic SAR for a two-dimensional geometry. They arrive at this algorithm through use of an intuitive far-field assumption; they assume that phase history data in frequency space lie along lines defined by the average of the transmitter and receiver look angles. Their algorithm is amenable to polar format processing, thus giving $O(N^2 \log_2 N)$ complexity.

An algorithm presented in [16] extends the work of Arikan and Munson by considering three-dimensional data collection geometries. This permits image formation from arbitrarily configured flight paths and on an arbitrary image plane. A straightforward and rigorous derivation of the differential range approximation that leads to the tomographic formulation is also provided. This derivation serves to define the bistatic look angle that defines downrange and crossrange in a SAR image. This has practical significance in applications that demand autofocus, as is described in [18].

An outline of the remainder of this section is as follows. Continuing with the phase history data model of equation (10.17), the motion measurements are for now assumed to be perfect, yielding an ideal signal model corrupted only by additive white Gaussian noise. This will motivate use of matched filtering for nonparametric scene reconstruction. Section 10.4.1 describes the matched filtering algorithm for formation of images at ground level and at various other altitudes. Next, reorganization of the sums computed in this algorithm, to yield the backprojection algorithm, will improve upon the numerical efficiency of matched filtering, while still yielding a mathematically ideal image reconstruction. To further reduce the computational expense of image formation one may employ far-field approximations as shown in Section 10.4.3. These assumptions linearize the phase of the matched filtering kernel. This allows approximate matched filtering via fast Fourier transforms (FFT) after polar reformatting of the phase history data, a process commonly known as the polar format algorithm. To limit the degree of image degradation introduced by the linearizing approximations, Section 10.4.3 analyzes the error terms in the far-field assumption to determine limits on the size of the imaged scene, when the bistatic PFA is used. The section continues with a brief discussion of two-dimensional interpolation methods for polar reformatting.

10.4.1 Matched Filtering

Given a set of bistatic SAR phase history data $\{S(f_i, \tau_k)\}$, as defined in equation (10.17), one typically wishes to reconstruct an image of the scene of interest. The fact that the uncompensated motion measurement errors in (10.17) are unknown requires one to temporarily assume that they are zero and correct for their effects in postprocessing via autofocus. The image to be formed is defined by a uniformly spaced grid of sample points, or pixels, on a plane in three-dimensional space. By analyzing the image one pixel at a time, one may assume that a scattering center is located at each pixel's grid location. Then, the value of that pixel is computed as the maximum likelihood estimate of the complex reflection coefficient $P(x, y, z)$ at that location, from the measurements $S(f_i, \tau_k)$ for $1 \le i \le N_f$ and $1 \le k \le N_\tau$. Maximum likelihood estimation of this value under the assumption of white Gaussian measurement errors in $\{S(f_i, \tau_k)\}$ results in a least-squares cost function written as

$$\mathcal{J} = \sum_{i=1}^{N_f} \sum_{k=1}^{N_\tau} \left| S(f_i, \tau_k) - P(x, y, z) e^{j\Phi_{xyz}(f_i, \tau_k)} \right|^2 \tag{10.74}$$

where the phase function Φ_{xyz} is defined using equations (10.11)–(10.16) with a scatterer located at $\underline{r}_0 = (x, y, z)$, and with the assumption that the location measurement errors, $\tilde{\underline{r}}_t(\tau)$ and $\tilde{\underline{r}}_r(\tau)$, are zero. Differentiating the cost function $\mathcal{J}$ with respect to the real and imaginary parts of $P(x, y, z)$ gives

$$\nabla_p \mathcal{J} = 2 \sum_{i=1}^{N_f} \sum_{k=1}^{N_\tau} P(x, y, z) - S(f_i, \tau_k)\, e^{-j\Phi_{xyz}(f_i, \tau_k)} \tag{10.75}$$

which, when set equal to zero, allows one to solve for the least-squares estimate of the pixel value, expressed as

$$P(x, y, z) = \frac{1}{N_f N_\tau} \sum_{i=1}^{N_f} \sum_{k=1}^{N_\tau} S(f_i, \tau_k)\, e^{-j\Phi_{xyz}(f_i, \tau_k)} \tag{10.76}$$

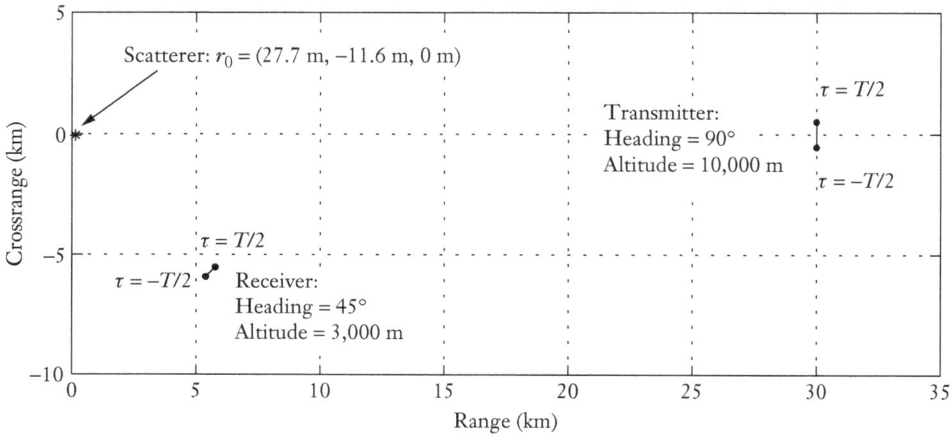

Figure 10–29 Bistatic data collection geometry for MF simulation.

The matched filtering image formation algorithm is given by equation (10.76) for each image pixel centered at location (x, y, z).

The number of operations required to calculate each pixel value scales linearly with the number of frequency samples N_f and linearly with the number of slow time samples N_τ. The number of operations to estimate each pixel value is thus $O(N_f N_\tau)$. Assuming N^2 pixels, and assuming $N_f N_\tau \propto N^2$, then the total number of operations to form an image is $O(N^4)$.

Experimental Study of Layover Although computationally cumbersome, the ideal nature of matched filtering does allow one to explore the nature of bistatic SAR data collection and image formation. More specifically, one may experimentally study the manner in which a three-dimensional scene is projected onto a two-dimensional image by a bistatic SAR system. As an example, equation (10.17) was used to generate phase history data for the bistatic geometry shown in Figure 10–29. It is important to note that the simulated transmitting and receiving platforms traverse straight and level flight paths. In Figures 10–30 and 10–31, MF images have been rendered of a scatterer located at $\underline{r}_0 = (27.7, -11.6, 0)$. Figure 10–30 shows a ground plane image ($z = 0$ m), and Figure 10–31 shows an image formed with the imaging plane at a height of $z = 10$ m.

One can see that the simulated point scatterer is well resolved in both imaging planes. However, the actual location of the scatterer is ambiguous to the MF, in a direction orthogonal to the surface swept out by the bistatic SAR line-of-sight, thus accounting for the translation in the resolved location of the scatterer between imaging planes. For data collection geometries with linear platform trajectories, this ambiguity will motivate one to only consider formation of ground plane images, as images at other altitudes are essentially noninformative.

More complex data collection configurations allow retention of three-dimensional scene information. Section 10.7 poses the possibility of introducing one or more additional receiving antennas, either on the same platform or on separate aircraft, to form a multistatic system. This will allow 3-D surface reconstruction through use of stereo SAR or interferometric processing.

10.4.2 Backprojection Algorithm

While being mathematically ideal, the $O(N^4)$ computational complexity of the MF algorithm prevents it from being implemented in real-time systems. The backprojection algorithm (BPA) [32, 33, 59] is an efficient method for computing (10.76), and does not involve any explicit

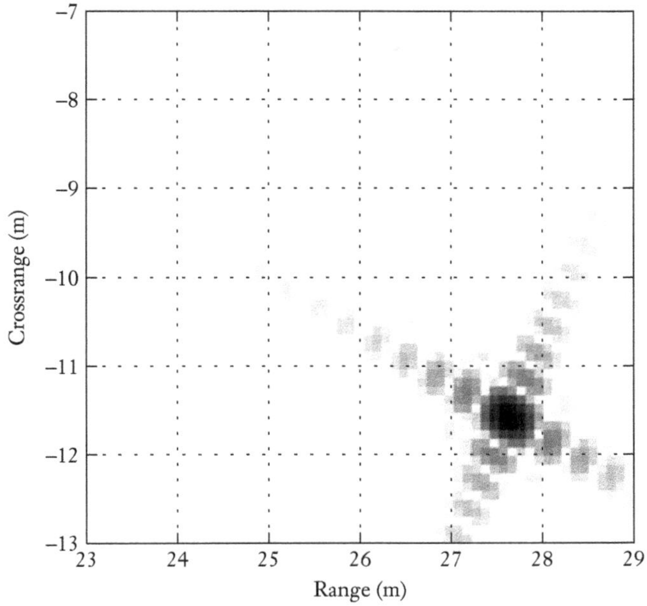

Figure 10–30 Ground plane image (i.e., $z = 0$ m) formed from simulated bistatic SAR data.

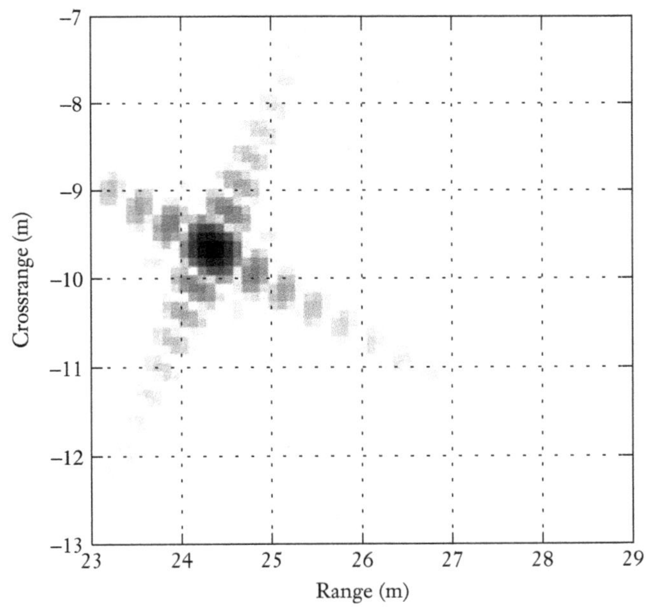

Figure 10–31 Image formed at $z = 10$ m formed from simulated bistatic SAR data.

assumptions that could introduce performance limitations. To illustrate, the sums in (10.76) are reorganized to give

$$P(x,y,z) = \frac{1}{N_f N_\tau} \sum_{k=1}^{N_\tau} \sum_{i=1}^{N_f} S(f_i,\tau_k) \exp\left\{ j2\pi f_i \frac{\Delta R(\tau_k,x,y,z)}{c} \right\} \tag{10.77}$$

In this form, the contribution of each pulse in slow time is added progressively to the estimated pixel values on a pulse-by-pulse basis. The differential range $\Delta R(\tau_k; x, y, z)$ is defined by equations (10.11)–(10.14) for each pixel location (x, y, z) and for each slow time sample τ_k. Substituting discretized sample frequencies

$$f_i = f_0 + \frac{(i-1)B}{N_f} \tag{10.78}$$

into equation (10.77) gives

$$\begin{aligned} P(x,y,z) &= \frac{1}{N_f N_\tau} \sum_{k=1}^{N_\tau} \exp\left\{ j2\pi f_0 \frac{\Delta R(\tau_k, x, y, z)}{c} \right\} \\ &\times \sum_{i=1}^{N_f} S(f_i, \tau_k) \exp\left\{ j2\pi \frac{(i-1)B}{N_f} \frac{\Delta R(\tau_k, x, y, z)}{c} \right\} \\ &= \frac{1}{N_f N_\tau} \sum_{k=1}^{N_\tau} \exp\left\{ j2\pi f_0 \frac{\Delta R(\tau_k, x, y, z)}{c} \right\} \\ &\quad U(\Delta R(\tau_k, x, y, z)) \end{aligned} \tag{10.79}$$

where

$$U(\Delta R) = \sum_{i=0}^{N_f - 1} S(f_i, \tau_k) \exp\left\{ j2\pi \frac{iB}{N_f} \frac{\Delta R}{c} \right\} \tag{10.80}$$

The contribution of the kth pulse to a pixel at (x, y, z) is then the product of $U(\Delta R(\tau_k; x, y, z))$ with the focusing phase $\exp\{j2\pi f_0 \Delta R(\tau_k; x, y, z)/c\}$. One may compute equation (10.80) more efficiently by restricting that calculation to discretized values of the differential range

$$\Delta R_n = \frac{nc}{B} \frac{N_f}{N_{\Delta R}} \qquad n = 0, \ldots, N_{\Delta R} - 1 \tag{10.81}$$

Substituting equation (10.81) into equation (10.80) gives

$$U(\Delta R_n) = \sum_{i=0}^{N_f - 1} S(f_{i+1}, \tau_k) \exp\left\{ j2\pi \frac{in}{N_{\Delta R}} \right\} \tag{10.82}$$

It is then readily apparent that $\{U(\Delta R_n), n = 0, \ldots, N_{\Delta R} - 1\}$ may be computed by FFT for the set of discrete ΔR defined by equation (10.81), provided $N_{\Delta R}$ is a power of 2. In some applications, one may choose to first scale each frequency domain sample by f_i, in order to implement convolution backprojection [33]; although in most high-frequency applications, this convolution step does not have a noticeable impact.

To compute $U(\Delta R(\tau_k; x, y, z))$ for each pixel, one may interpolate $\{U(\Delta R_n), n = 0, \ldots, N_{\Delta R} - 1\}$ with respect to ΔR to obtain estimates of $U(\Delta R(\tau_k; x, y, z))$ for each pixel location. Choosing $N_{\Delta R} >> N_f$ to oversample the FFT of $S(f_i, \tau_k)$ will provide a finer sampling of the ΔR_n. The FFT may be oversampled to a high enough degree that linear interpolation between the sampled ΔR_n will give the necessary phase coherence. This process is repeated for every τ_k in the sum of equation (10.77).

Figure 10–32 compares the impact on BPA image quality of different levels of FFT oversampling prior to linear interpolation. The phase errors and crossrange defocusing, introduced by the linear

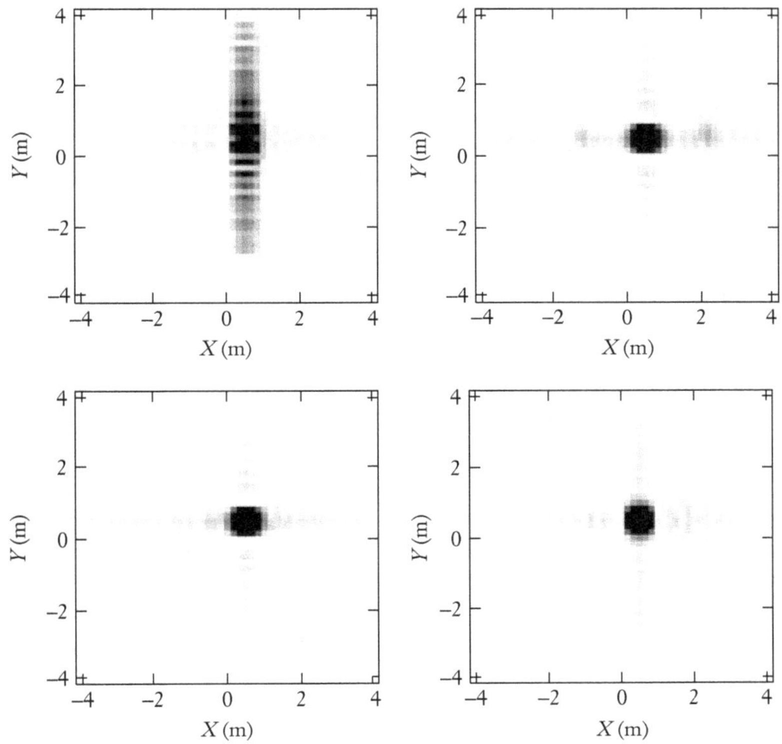

Figure 10–32 The impulse response improves as the amount of FFT oversampling prior to linear interpolation is increased. The above images were formed with oversampling factors of 1 (top left), 2 (top right), 4 (bottom left), and 8 (bottom right).

interpolation with no zero-padding of the FFTs is clearly visible in the upper-left image of Figure 10–32. With eight times oversampling, the image quality in the lower-right image is very good. Detemining the appropriate FFT size for a given application will be subject to parameters of the collected data, including resolution, existing phase errors, and SNR.

Independent of the interpolation method, the contribution of each phase history (indexed by τ_k) to every pixel is computed in $O(N^2)$ operations, and so assuming $N_\tau \propto N$, the backprojection algorithm achieves image formation in $O(N^3)$ operations. Unfortunately, this is not yet of low enough computational complexity to allow real-time implementation. Some variations on the BPA achieve $O(N^2\sqrt{N})$ complexity [41], and others even achieve $O(N^2 \log_2 N)$ complexity [39, 40], but at the expense of using far-field approximations.

10.4.3 Polar Format Algorithm

To further reduce the compuational load of MF image formation, below that of the BPA, one can introduce an approximation of the MF kernel to allow more efficient computation of (10.76). Using the definition of the phase function Φ_{xyz}, one may more explicitly write the kernel of equation (10.76) as

$$\exp\{-j\Phi_{xyz}(f_i,\tau_k)\} = \exp\left\{j2\pi f_i \frac{\Delta R(\tau_k)}{c}\right\} \tag{10.83}$$

and may note its similarity to a Fourier transform kernel. The phase of the MF varies linearly with frequency f_i, which is exploited in the backprojection algorithm, but varies nonlinearly with respect to the slow time τ_k. If the MF kernel could be approximated by a function, the phase of which varies linearly with respect to a downrange spatial frequncy f_x and a crossrange spatial frequency f_y, discrete Fourier transforms (DFTs) could be used to interrrogate the phase history data, thus making available computationally efficient algorithms such as the FFT. The polar format algorithm for monostatic SAR exploits such an approximated matched filter. Development of an analogous method for bistatic SAR is provided below.

In Appendix A, the second-order Taylor series expansion of ΔR is derived for bistatic SAR. Taking the linear terms of this expansion and setting the measurement errors to zero, one may approximate the bistatic differential range at time τ_k with (provided x, y, and z are sufficiently small):

$$\begin{aligned}\Delta R(\tau_k) \approx & -x\cos\phi_{\mathrm{t}}(\tau_k)\cos\theta_{\mathrm{t}}(\tau_k) - x\cos\phi_{\mathrm{r}}(\tau_k)\cos\theta_{\mathrm{r}}(\tau_k)\\ & -y\sin\phi_{\mathrm{t}}(\tau_k)\cos\theta_{\mathrm{t}}(\tau_k) - y\sin\phi_{\mathrm{r}}(\tau_k)\cos\theta_{\mathrm{r}}(\tau_k)\\ & -z\sin\theta_{\mathrm{t}}(\tau_k) - z\sin\theta_{\mathrm{r}}(\tau_k)\end{aligned} \tag{10.84}$$

where the variables ϕ_{t} and ϕ_{r} (θ_{t} and θ_{r}) denote the measured azimuth (elevation) angles of the transmitter and receiver, with respect to the scene center. This approximation is typically referred to as the far-field assumption. If the offset of the scatterer from the scene origin is too large, the above first-order approximation of the differential range will break down, introducing phase errors in the final image. Limits on this offset from the scene origin are derived later in this section.

By inserting the approximation of equation (10.84) in equation (10.76), an approximate matched filter for bistatic image formation may be written as

$$P(x,y,z) = \frac{1}{N_f N_\tau}\sum_{i=1}^{N_f}\sum_{k=1}^{N_\tau} S(f_i,\tau_k)\exp\left(-j\frac{4\pi}{c}[x f_x(f_i,\tau_k) + y f_y(f_i,\tau_k) + z f_z(f_i,\tau_k)]\right) \tag{10.85}$$

where

$$\begin{aligned} f_x(f_i,\tau_k) &= \frac{f_i}{2}[\cos\phi_{\mathrm{t}}(\tau_k)\cos\theta_{\mathrm{t}}(\tau_k) + \cos\phi_{\mathrm{r}}(\tau_k)\cos\theta_{\mathrm{r}}(\tau_k)]\\ f_y(f_i,\tau_k) &= \frac{f_i}{2}[\sin\phi_{\mathrm{t}}(\tau_k)\cos\theta_{\mathrm{t}}(\tau_k) + \sin\phi_{\mathrm{r}}(\tau_k)\cos\theta_{\mathrm{r}}(\tau_k)]\\ f_z(f_i,\tau_k) &= \frac{f_i}{2}[\sin\theta_{\mathrm{t}}(\tau_k) + \sin\theta_{\mathrm{r}}(\tau_k)]\end{aligned} \tag{10.86}$$

are the spatial frequencies sampled in the bistatic data collection. These spatial sample points constitute the data collection manifold described in Section 10.1, and shown conceptually in Figure 10–5. To produce a ground map, the z variable of the approximate matched filter is set to zero to give

$$P(x,y,0) = \frac{1}{N_f N_\tau}\sum_{i=1}^{N_f}\sum_{k=1}^{N_\tau} S(f_i,\tau_k)\exp\left(-j\frac{4\pi}{c}[x f_x(f_i,\tau_k) + y f_y(f_i,\tau_k)]\right) \tag{10.87}$$

where x and y are linear multipliers of the spatial frequencies f_x and f_y.

Note that in equation (10.87) f_x and f_y are not uniformly spaced to allow use of a DFT. However, simple transformations will produce an efficient algorithm, as will be shown below. As an example, Figure 10–33 shows the sampled (f_x, f_y) points for the bistatic geometry

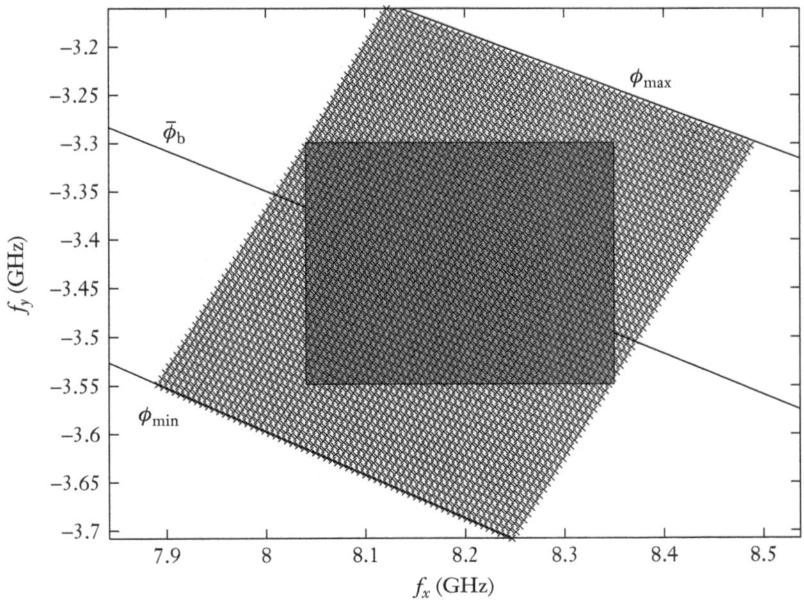

Figure 10–33 Frequency samples projected onto the f_x–f_y plane.

described in Section 10.4.1. In that figure,

$$\phi_{\min} = \arctan\frac{f_y(f_i,\tau_1)}{f_x(f_i,\tau_1)} \tag{10.88}$$

corresponds to the azimuth angle of the first set of frequency samples in slow time, and

$$\phi_{\max} = \arctan\frac{f_y(f_i,\tau_{N_\tau})}{f_x(f_i,\tau_{N_\tau})} \tag{10.89}$$

corresponds to the last set of samples in slow time. The bistatic look angle is defined as

$$\bar{\phi}_b = \arctan\frac{f_y(f_i,\tau_{k_b})}{f_x(f_i,\tau_{k_b})} \tag{10.90}$$

where $k_b = N_\tau/2$ and $\tau_{k_b} \approx 0$. A two-dimensional interpolation of the sampled data onto a rectangular grid, uniformly spaced in f_x and f_y, will allow use of a DFT, but a rectangular grid inscribed in the sampled annulus (as shown by the shaded rectangle in Figure 10–33) may not make efficient use of the collected data.

A rectangular grid rotated by the bistatic look angle $\bar{\phi}_b$, defined by (10.90), will give better coverage of the sampled data. To accomplish this, one may rotate the coordinate system of the collected data by the bistatic look angle $\bar{\phi}_b$ to give

$$\begin{aligned} f_x' &= f_x \cos\bar{\phi}_b + f_y \sin\bar{\phi}_b \\ f_y' &= -f_x \sin\bar{\phi}_b + f_y \cos\bar{\phi}_b \end{aligned} \tag{10.91}$$

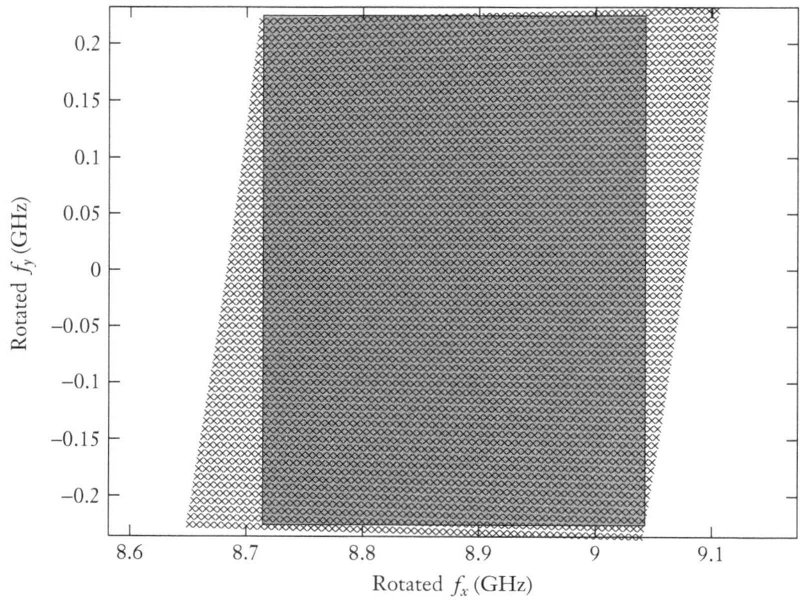

Figure 10–34 Frequency samples projected onto the rotated f_x'–f_y' plane. The shaded rectangle leaves significant areas of the sampled data unused.

and inversely

$$\begin{aligned} f_x &= f_x' \cos\bar{\phi}_b - f_y' \sin\bar{\phi}_b \\ f_y &= f_x' \sin\bar{\phi}_b + f_y' \cos\bar{\phi}_b \end{aligned} \tag{10.92}$$

Correspondingly, the approximate matched filter is evaluated over a rotated coordinate system defined by

$$\begin{aligned} x' &= x\cos\bar{\phi}_b + y\sin\bar{\phi}_b \\ y' &= -x\sin\bar{\phi}_b + y\cos\bar{\phi}_b \end{aligned} \tag{10.93}$$

Equation (10.87) may now be rewritten, using the rotated coordinates, as

$$P(x',y',0) = \frac{1}{N_f N_\tau} \sum_{i=1}^{N_f} \sum_{k=1}^{N_\tau} S(f_i,\tau_k) \exp\left(-j\frac{4\pi}{c}[x'f_x'(f_i,\tau_k) + y'f_y'(f_i,\tau_k)]\right) \tag{10.94}$$

and displaying the sampled data on this rotated coordinate system allows one to inscribe a rectangle that makes better use of the available data, as shown by the shaded area in Figure 10–34.

However, due to the parallelogram shape of the sampled data in frequency space, the inscribed rectangle in Figure 10–34 still does not capture all of the available sample points. Therefore, a grid of nonorthogonal sampling [60] is employed by inscribing a parallelogram into the area of sampled data, as shown in Figure 10–35. The length of the base of the parallelogram, parallel to the rotated f_x-axis, is referred to as B_x, and the height of the parallelogram is referred to as B_y. The slope of the sides of the parallelogram is κ. One may now resample the frequency domain data, by

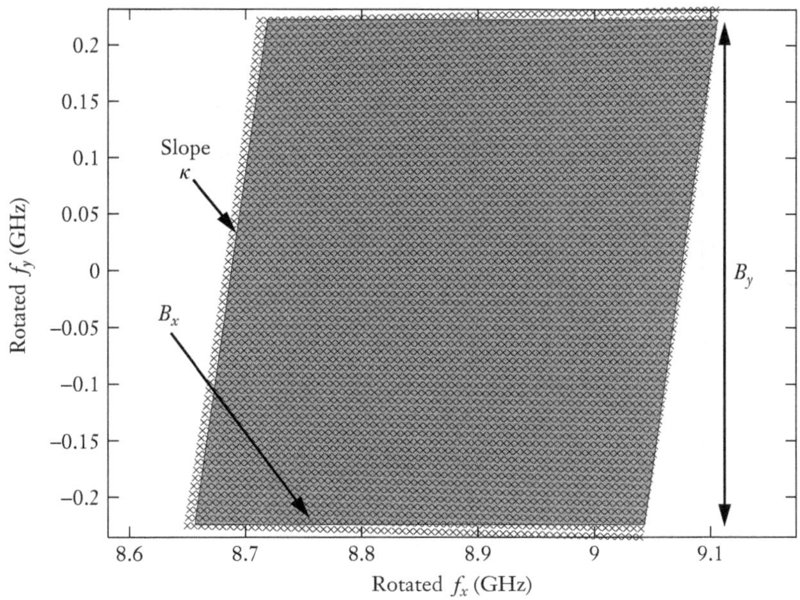

Figure 10–35 Frequency samples projected onto the rotated f_x'–f_y' plane. The shaded parallelogram shows the intended resampling.

two-dimensional interpolation

$$S(f_x', f_y') = \sum_{i=1}^{N_f} \sum_{k=1}^{N_\tau} S(f_i, \tau_k) g(f_x', f_y', f_i, \tau_k) \tag{10.95}$$

onto a uniformly spaced parallelogram grid in $f_x' - f_y'$. This step may be performed through any number of methods [61] in $O(N^2)$ operations. As detailed in [2], 2-D interpolation may be accomplished efficiently and accurately by first performing a series of 1-D interpolations in range, followed by 1-D interpolations in azimuth. While still being $O(N^2)$, the need for accuracy in these computations will often make interpolation the limiting step in the image formation process. While this topic is briefly addressed later, a more thorough discussion regarding the issues surrounding this interpolation step may be found in [2]. The discrete samples of f_x' and f_y' are defined by

$$f_x' = f_0 + \kappa f_y' + B_x \frac{k_x}{N_x} \qquad k_x = 0, \ldots, N_x - 1$$

$$\text{and} \quad f_y' = \frac{-B_y}{2} + B_y \frac{k_y}{N_y} \qquad k_y = 0, \ldots, N_y - 1 \tag{10.96}$$

One typically chooses N_x and N_y to be approximately equal to N_f and N_τ to give a sampling density that is roughly equivalent to what was given in the original f_x–f_y coordinate system. Figure 10–36 compares the footprint of the parallelogram resampled phase history data to the footprint of the raw data set in the original coordinate system.

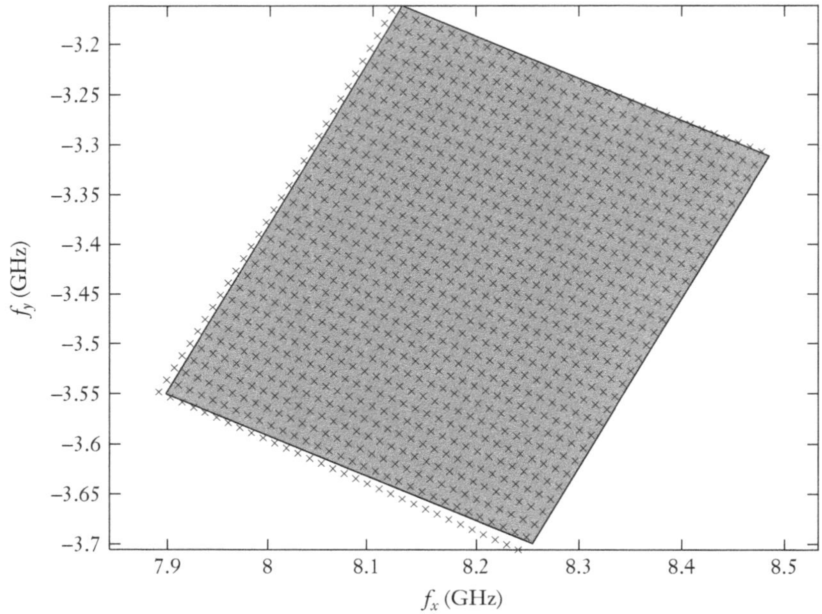

Figure 10–36 The measured data (light shaded region) and inscribed parallelogram region from Figure 10–35 (dark shaded region), shown reformatted to frequency samples on the original f_x–f_y plane as in Figure 10–33. Note the nearly complete data coverage in the f_x–f_y plane of the elliptical annulus of measured data by the inscribed rotated parallelogram.

Rewriting the approximate matched filter equation based on the reformatted sampling gives

$$P(x',y',0)=\frac{1}{N_x N_y}\sum_{k_x=0}^{N_x-1}\sum_{k_y=0}^{N_y-1}S(k_x,k_y)$$
$$\times\exp\left\{-j\frac{4\pi}{c}\left[x'\left(f_0+B_x\frac{k_x}{N_x}+\kappa\left(\frac{-B_y}{2}+B_y\frac{k_y}{N_y}\right)\right)\right.\right.$$
$$\left.\left.+y'\left(\frac{-B_y}{2}+B_y\frac{k_y}{N_y}\right)\right]\right\} \tag{10.97}$$

and allows one to use DFTs to sample $P(x', y', 0)$ on a grid uniformly spaced in x' and y'. The implied discretization of x' and y' is

$$\begin{aligned} x' &= n_x\Delta_x \quad n_x = 0,\ldots,N_x-1 \\ y' &= n_y\Delta_y \quad n_y = 0,\ldots,N_y-1 \end{aligned} \tag{10.98}$$

where Δ_x and Δ_y are the pixel spacings in the x' and y' directions defined by

$$\Delta_x=\frac{c}{2B_x} \quad \Delta_y=\frac{c}{2B_y} \tag{10.99}$$

Substitution of (10.98) and (10.99) into (10.97) yields

$$P(n_x, n_y, 0) = \exp\left\{-j2\pi\left[\frac{n_x}{B_x}\left(f_0 - \kappa\frac{B_y}{2}\right) - \frac{n_y}{2}\right]\right\}$$
$$\times \frac{1}{N_x N_y}\sum_{k_y=0}^{N_y-1}\exp\left\{-j2\pi\frac{n_y k_y}{N_y}\right\}\exp\left\{-j2\pi\frac{n_x k_y}{N_y}\frac{\kappa B_y}{B_x}\right\}$$
$$\times \sum_{k_x=0}^{N_x-1} S(k_x, k_y)\exp\left\{-j2\pi\frac{n_x k_x}{N_x}\right\} \tag{10.100}$$

The approximate matched filter therefore consists of a multiply and accumulate over k_x, followed by a range-dependent phase shift in f_y, and then a multiply and accumulate over k_y. The multiply and accumulate operations over k_x and k_y may be performed by a series of one-dimensional DFTs. This allows the use of FFTs, and thus enables image formation in $O(N^2\log_2 N)$ operations, if $N_x = N_y = N$.

The bistatic PFA process described above is summarized in the flow chart of Figure 10–37. An additional step of weighting the data prior to the final Fourier transform has been added. As in most applications, use of a tapered weighting function [62] in the frequency domain allows control of side lobe levels in the image domain. One may also choose to zero-pad the FFTs to decrease the pixel spacing and give a smoother image. It is typically desirable to rotate the final image from the $x'-y'$ coordinate system into the original x–y coordinate system prior to display, thereby allowing the imaged scene to be registered with respect to a global coordinate system. It is important to note that the bistatic PFA process described above is identical to a monostatic PFA process once the data have been organized in frequency space. This allows the use of specialized processing already developed for monostatic systems [63], including autofocus algorithms, as will be shown in future sections.

Using the simulated phase history data from Section 10.4.1, the above image formation process yields the image of Figure 10–38 in the rotated x'–y' coordinate system, and the image of Figure 10–39 in the global coordinate system. No frequency domain weighting of the data has been used to make the side lobe structure more visible and to facilitate comparison with Figure 10–30.

Scene Size Limits The far-field assumption of equation (10.84) allows formulation of an approximate matched filtering scheme in which ground images may be formed from bistatic SAR data in $O(N^2\log_2 N)$ operations. However, use of this first-order approximation introduces a phase error to images formed using the PFA as compared to using MF. To help quantify this error, the second-order terms of the Taylor expansion of the bistatic differential range are analyzed. This derivation follows [2, Chapter 3] but generalizes it to the bistatic case considered here. For ease of analysis, consider the geometry of Figure 10–40, in which the transmitting and receiving platforms travel parallel paths bisected by the x-axis, and only the worst-case scenario in terms of approximation errors is addressed.

The variables x_t, z_t, x_r, and z_r are assumed to be constant with respect to the slow time τ, while $y_t = v_t\tau$ and $y_r = v_r\tau$ vary linearly with time. Furthermore, it is assumed that the range of the transmitter R_t and the range of the receiver R_r are sufficiently large at the aperture center such that they may be assumed to be constant with respect to the slow time τ. By applying these assumptions to (A.2) and (A.13) from Appendix A and setting the motion measurement errors to

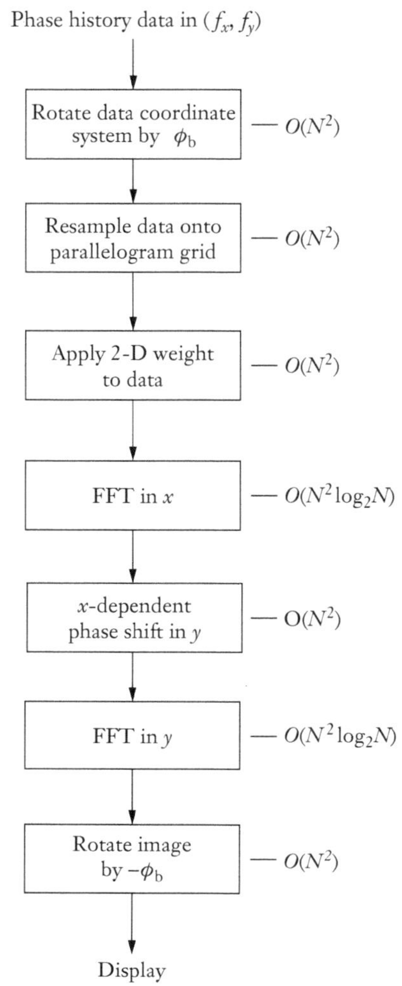

Figure 10–37 Flow chart summarizing the polar format algorithm for bistatic SAR.

zero, one obtains

$$\Delta R_t^{(2)} = \frac{x^2}{2}\left(\frac{v_t^2\tau^2 + z_t^2}{R_t^3}\right) + \frac{y^2}{2}\left(\frac{x_t^2 + z_t^2}{R_t^3}\right) + \frac{z^2}{2}\left(\frac{x_t^2 + v_t^2\tau^2}{R_t^3}\right) - xy\left(\frac{x_t v_t \tau}{R_t^3}\right) - xz\left(\frac{x_t z_t}{R_t^3}\right) - yz\left(\frac{v_t \tau z_t}{R_t^3}\right) \tag{10.101}$$

$$\Delta R_r^{(2)} = \frac{x^2}{2}\left(\frac{v_r^2\tau^2 + z_r^2}{R_r^3}\right) + \frac{y^2}{2}\left(\frac{x_r^2 + z_r^2}{R_r^3}\right) + \frac{z^2}{2}\left(\frac{x_r^2 + v_r^2\tau^2}{R_r^3}\right) - xy\left(\frac{x_r v_r \tau}{R_r^3}\right) - xz\left(\frac{x_r z_r}{R_r^3}\right) - yz\left(\frac{v_r \tau z_r}{R_r^3}\right) \tag{10.102}$$

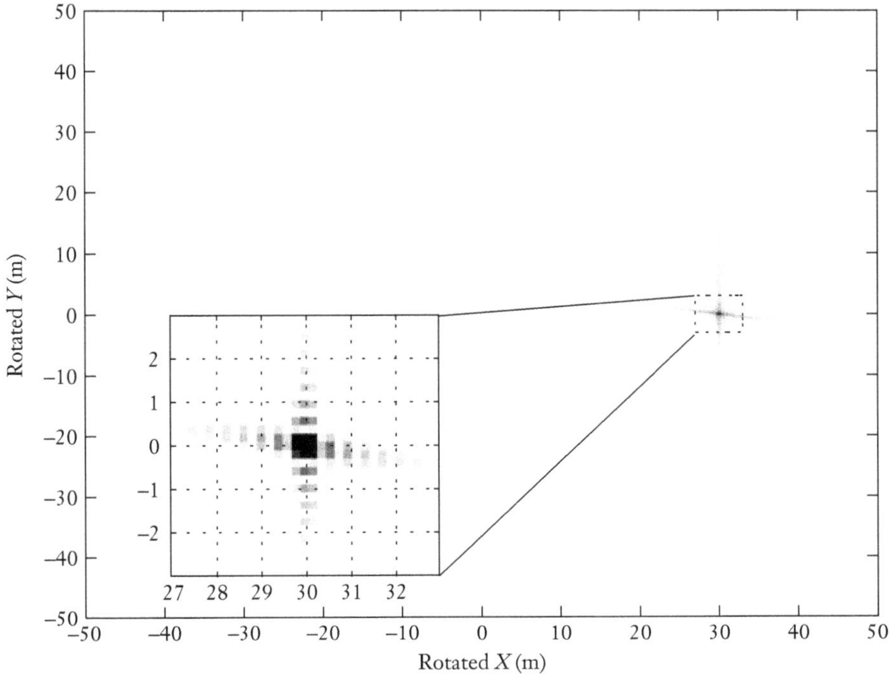

Figure 10–38 Bistatic SAR PFA image, in the coordinate system rotated by the bistatic angle.

It can be seen that each expression above comprises terms that are constant, linear, and quadratic functions of the slow time τ. The constant and linear terms introduce spatially dependent distortions in the final image, but do not cause any blurring or loss of resolution. A spatially dependent distortion causes scatterers that are actually located at (x, y, z) to appear at some $(x + \delta_x, y + \delta_y, z + \delta_z)$, where $\{\delta_x, \delta_y, \delta_z\}$ are functions of the scatterer's actual location. Terms in (10.101) and (10.102) that depend quadratically on time (which are those containing y_t^2 and y_r^2) cause a spatially dependent defocus, or blurring, of scatterers.

The approximation limits for the far-field assumption are determined by bounding the amount of defocus experienced by scatterers on the ground plane [2, 17]. With $z = 0$, only the $x^2y_t^2/(2R_t^3)$ term (with $y_t = v_t\tau$) in (10.101) and the $x^2y_r^2/(2R_r^3)$ term (with $y_r = v_r\tau$) in (10.102) depend quadratically on τ. From equation (10.16), the phase of a received pulse in a bistatic system is defined by

$$\Phi = -2\pi f \frac{\Delta R}{c} \tag{10.103}$$

and thus the defocusing phase error due to range curvature is

$$\Phi_c = \frac{2\pi f_c}{c}\left(\frac{1}{2}\frac{x^2y_t^2}{R_t^3} + \frac{1}{2}\frac{x^2y_r^2}{R_r^3}\right) \tag{10.104}$$

at the center frequency f_c. Analogously to the monostatic case [2, 17], this defocusing phase error is limited to be within $\pm\pi/2$ radians at the maximum crossrange extent of the image $x = \pm r_{max}$

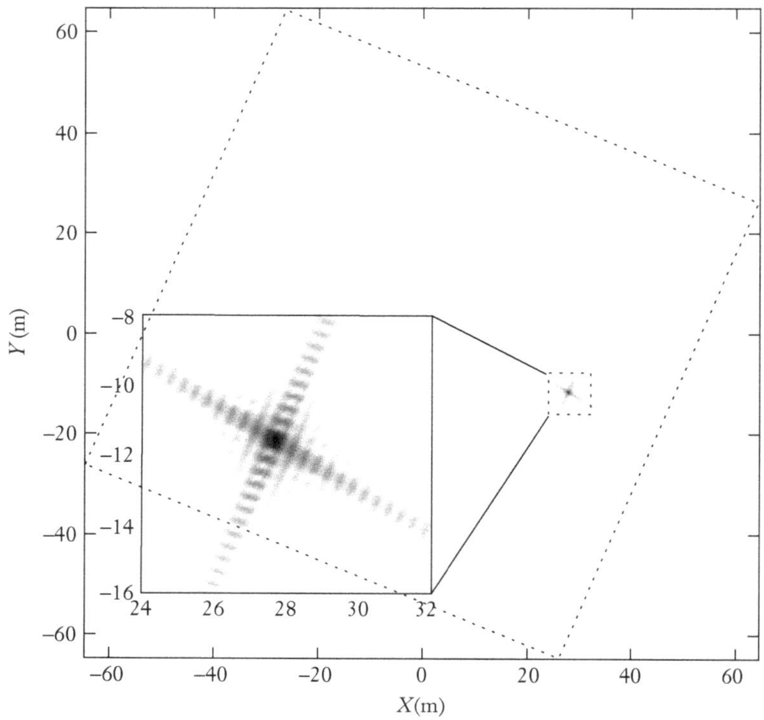

Figure 10–39 Bistatic SAR PFA image in the global coordinate system. The dotted lines mark the boundaries of the rotated image. Differences between the ideal point response shown here and that which is shown in Figure 10–38 are purely due to the graphical display and slight imperfections in the rotation interpolation.

and at the aperture extents $y_t = \pm L_t/2$ and $y_r = \pm L_r/2$, where L_t and L_r are the lengths of the paths traveled by the transmitter and receiver, respectively. By substituting path length dependent expressions for y_t and y_r in (10.104), and replacing c/f_c with the wavelength λ, the inequality $\Phi_c < \pi/2$ becomes

$$\frac{\pi}{2} > \frac{\pi r_{\text{max}}^2}{4\lambda}\left(\frac{L_t^2}{R_t^3} + \frac{L_r^2}{R_r^3}\right) \tag{10.105}$$

Thus, the maximum radius of an imaged scene is limited by

$$r_{\text{max}} < \sqrt{2\lambda}\left(\frac{L_t^2}{R_t^3} + \frac{L_r^2}{R_r^3}\right)^{-1/2} \tag{10.106}$$

where R_t and R_r are typically taken as the ranges to the scene center from the midpoint of each platform's flight path. By way of an example, Figure 10–41 shows the azimuth response of an ideal point scatterer at the scene center, at the maximum scene radius, and at twice the maximum scene radius, when using the polar format algorithm. At the maximum scene radius, the main lobe broadening of an ideal point response is less than 10%.

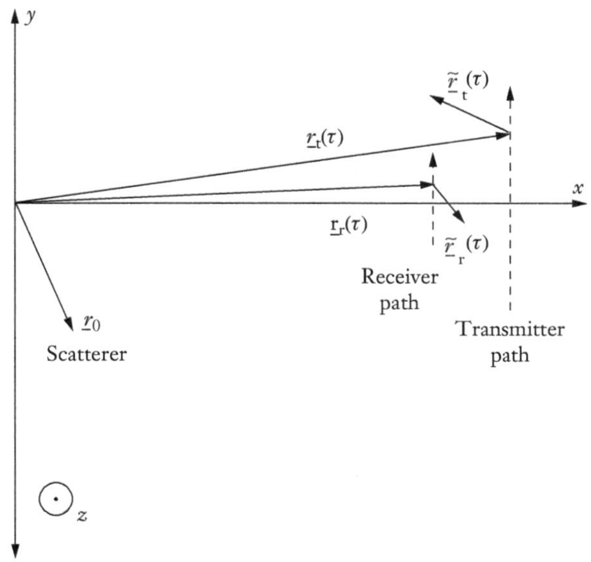

Figure 10–40 Bistatic SAR geometry with the transmitter and receiver traveling nominally parallel paths. Solid arrows indicate vectors from the scene center to the transmitter, receiver, and scatterer positions, and indicate vectors from the ideal transmitter and receiver flight paths (dashed lines) to their measured locations.

Using equation (10.37) and assuming grazing incidence with zero bistatic angle, one may express the crossrange image resolution (for the broadside imaging case depicted in Figure 10–40) as

$$\Delta_y = \frac{\lambda R_t R_r}{L_t R_r + L_r R_t} \tag{10.107}$$

Unlike in the monostatic case, there appears to be no general way of reducing the expression for $r_{\max}$ to a simple function of crossrange resolution. However, specializing to the monostatic case where $L_t = L_r$ and $R_t = R_r$, equations (10.106) and (10.107) become

$$r_{\max} < \sqrt{\lambda\left(\frac{R_t^3}{L_t^2}\right)} \tag{10.108}$$

$$\Delta_y = \frac{\lambda R_t}{2L_t} \tag{10.109}$$

Solving for L_t in (10.109) and substituting into (10.108) yields [2, equation (3.130)]

$$r_{\max} < 2\Delta_y \sqrt{\frac{R_t}{\lambda}} \tag{10.110}$$

where the broadening factor K_a in [2] is assumed to be 1.

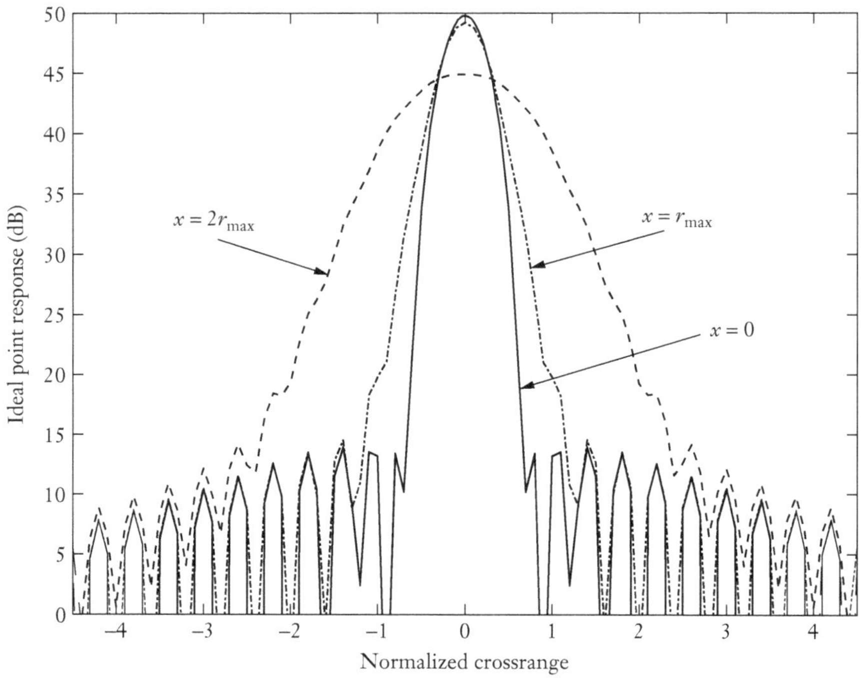

Figure 10–41 The ideal point response of a scatterer at the scene center is unaffected by the far-field approximations. Scatterers at and beyond the maximum scene radius r_{max} begin to exhibit noticeable main lobe broadening.

Polar Reformatting As a critical step in the polar format algorithm, interpolation of the collected phase history data onto a rectangular grid may be accomplished through several methods. The most appropriate option is dictated by the original format of the sampled data in frequency space and the desired format after interpolation. In most cases, sample points should lie on a uniform polar grid, evenly spaced in frequency and in azimuth. However, in some scenarios, such as those posed by multistatic processing, sampled data may be scattered in frequency and azimuth. This would be true if a single receiver were making use of multiple cooperative or noncooperative transmitters. These transmitters would be spatially separated, thus introducing discontinuities in azimuth, and would likely be spectrally separated, thereby leaving gaps in the frequency content.

In general, a parallelogram grid of resampled data is the desired interpolation result. Going from one uniformly sampled grid to another allows use of more well-behaved interpolation techniques. A stable and computationally efficient approach to polar-to-rectangular resampling involves computing a series of one-dimensional interpolations in frequency, along lines of constant azimuth angle, to intersect lines of constant range frequency (f_x). Then, a second series of one-dimensional interpolations achieves the desired uniform sampling in crossrange frequency (f_y). Figure 10–42 shows the initial phase history data in a uniform polar format and a notional parallelogram grid for the resampled data.

By decomposing the two-dimensional interpolation into a series of one-dimensional operations, one may exploit a variety of efficient and accurate interpolation methods. In this case, the initial data set is projected onto the f_x, f_y-plane in an organization defined by sampling along radial

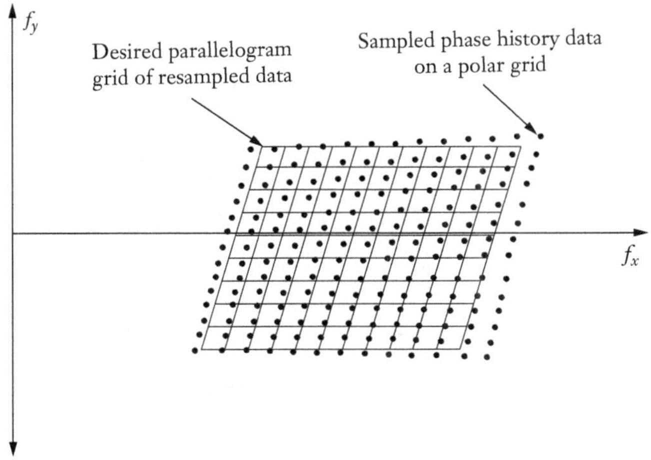

Figure 10–42 Notional sampling grids for the original phase history data and the desired parallelogram grid.

lines of constant azimuth ϕ and along arcs of constant frequency f. Thus, a data set $\{f_x, f_y\}$ that has been rotated by the bistatic SAR look angle $\bar{\phi}_b$, as defined by equation (10.91), has the polar coordinates

$$f' = \sqrt{(f_x')^2 + (f_y')^2} \text{ and } \phi' = \arctan\frac{f_y'}{f_x'} \tag{10.111}$$

The spacing between frequency samples will be uniform for a given azimuth angle, and the spacing between azimuth samples will be approximately uniform, depending on the data collection geometry. This initial data set should be interpolated to the parallelogram grid defined by (10.96).

The first part of the resampling calls for interpolation along the lines of a constant azimuth angle, as shown in Figure 10–43. For a given ϕ', substituting the linear relationship

$$f_y' = f_x' \tan\phi' \tag{10.112}$$

into the expression for f_x' in (10.96) yields

$$f_x' = f_0 + \kappa f_x' \tan\phi' + B_x \frac{k_x}{N_x} \qquad k_x = 0, \ldots, N_x - 1 \tag{10.113}$$

and solving for f_x' gives the desired resampling as a function of k_x, written as

$$f_x' = f\cos\phi' = \frac{f_0 + B_x \frac{k_x}{N_x}}{1 - \kappa\tan\phi'} \tag{10.114}$$

Therefore, for a given azimuth angle, the radially interpolated frequencies as a function of k_x are given by

$$f = \frac{f_0 + B_x \frac{k_x}{N_x}}{(1 - \kappa\tan\phi')\cos\phi'} \tag{10.115}$$

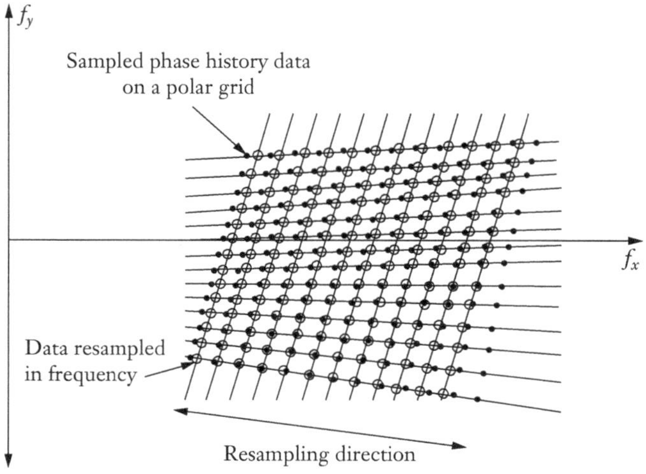

Figure 10–43 Frequency resampling.

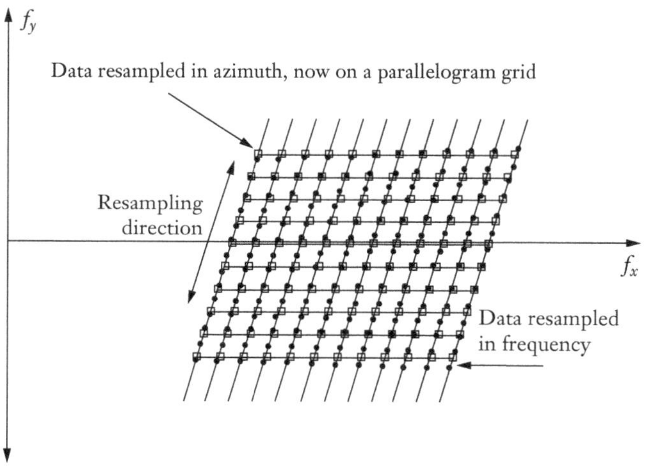

Figure 10–44 Azimuth resampling.

After interpolating along each radial line of constant azimuth, the next step is to resample the data along the lines of constant k_x to achieve samples at the desired f_y', as shown in Figure 10–44. For a given k_x, substituting equation (10.114) into equation (10.112) defines the sampled values of f_y', which resulted from the range frequency interpolation, and the expression for f_y' in (10.96) defines the desired values of crossrange frequency.

10.5 MOTION MEASUREMENT ERRORS (MME6)

The image formation algorithms presented in Section 10.4 are predicated on the assumption that the data collection process, including measurement of platform motion and transmit–receive time synchronization, is ideal. In point of fact, these measurements will never be error free. While synthesizing a bistatic aperture, the transmitter and receiver attempt to traverse predetermined flight paths. The timings of pulse transmissions and receptions are coordinated based on these flight plans. Each platform's motion measurement unit attempts to measure its deviation from ideal at each sample point in azimuth. These measurements are used to adjust the phase of the recorded pulses, such that the return from a scatterer at the scene center will have zero phase.

It is well known from operational monostatic SAR systems that an exact motion measurement is impossible to obtain, and a slight motion measurement error (MME) can have a drastic effect on the quality of the resulting SAR imagery. In bistatic SAR systems, MMEs and interplatform synchronization errors induce phase errors in the azimuth dimension of the phase history data. Very low frequency errors, such as constant and linear terms, will introduce image translations and phase shifts that may affect geolocation accuracy but do not detriment image quality. High-frequency MMEs can lead to degraded image contrast, higher side lobes, and even spurious peaks and paired echoes. Mechanical isolation of the transmitting and receiving antennas, to dampen vibrations, is frequently the only means by which to mitigate high-frequency effects. Phase errors which fall in a middle frequency range often are assumed to take a polynomial form, dominated by a quadratic term. These errors lead to azimuthal smearing of point responses in the final image.

This section seeks to characterize the effects of imperfect motion measurement and platform synchronization on bistatic SAR images, as presented in [18, 47, 64–69]. An understanding of these effects permits one to draw limits on the achievable coherency of a bistatic system, such as the maximum of 3 seconds of coherency referenced at the close of [6]. These limits in turn determine practical ranges and resolutions of operation. The focus of this discussion is primarily on low-frequency errors that permit expression in a polynomial form. The results of this analysis support the manner in which the bistatic polar format algorithm was designed to fit within the framework of legacy monostatic SAR image formation processors. Specifically, one finds that the phase history data rotation, which is prescribed for the bistatic PFA, effectively places all of the phase error effects along a single dimension. This allows application of monostatic autofocus algorithms between the range and azimuth compressions. This analysis also leads to guidelines for the precision of motion compensation necessary to obtain usable imagery from the bistatic SAR phase history data.

An outline of this section is as follows. Section 10.5.1 revises the earlier model for bistatic SAR phase history data collection and image formation to include the effects of imperfect motion measurement. Section 10.5.2 describes a typical model for low-frequency motion measurement errors and analyzes their effects on the collected phase history data. Section 10.5.3 demonstrates the effect of MMEs on bistatic SAR images. Bounds on the allowable MMEs are also derived, based on limiting the degree of quadratic phase error.

10.5.1 Understanding Image Formation Processing

From Section 10.4, one may adapt equation (10.17) to explicitly note the presence of MMEs, written as

$$S(f_i,\tau_k)=\sum_m A_m e^{j\hat{\Phi}_m(f_i,\tau_k)}+w(f_i,\tau_k) \tag{10.116}$$

where

$$\hat{\Phi}_m(f_i,\tau_k)=-2\pi f_i\left(\frac{\Delta\hat{R}_m(\tau_k)}{c}+\tilde{t}(\tau_k)\right) \qquad i=1,\ldots,N_f \quad k=1,\ldots,N_\tau \tag{10.117}$$

is the measured phase of the mth scatterer in the scene. The variable $\Delta R_m(\tau_k)$ represents the measured differential range to the mth scatterer, and $\tilde{t}(\tau_k)$ is the transmit–receive timing error on the kth pulse.

Taylor expansion of the measured bistatic differential range in equation (10.117) gives the far-field approximation (see Appendix A)

$$\begin{aligned}\Delta\hat{R}_m(\tau_k) \approx &-(x_m+\tilde{x}_{\mathrm{t}})\cos\phi_{\mathrm{t}}(\tau_k)\cos\theta_{\mathrm{t}}(\tau_k)-(x_m+\tilde{x}_{\mathrm{r}})\cos\phi_{\mathrm{r}}(\tau_k)\cos\theta_{\mathrm{r}}(\tau_k\\ &-(y_m+\tilde{y}_{\mathrm{t}})\sin\phi_{\mathrm{t}}(\tau_k)\cos\theta_{\mathrm{t}}(\tau_k)-(y_m+\tilde{y}_{\mathrm{r}})\sin\phi_{\mathrm{r}}(\tau_k)\cos\theta_{\mathrm{r}}(\tau_k)\\ &-(z_m+\tilde{z}_{\mathrm{t}})\sin\theta_{\mathrm{t}}(\tau_k)-(z_m+\tilde{z}_{\mathrm{r}})\sin\theta_{\mathrm{r}}(\tau_k)\end{aligned} \tag{10.118}$$

where the variables ϕ_{t} and ϕ_{r} (θ_{t} and θ_{r}) denote the azimuth (elevation) angles of the transmitter and receiver with respect to the scene center. This approximation is separated into two terms. The first term

$$\begin{aligned}\Delta R_m(\tau_k) \approx &-x_m\cos\phi_{\mathrm{t}}(\tau_k)\cos\theta_{\mathrm{t}}(\tau_k)-x_m\cos\phi_{\mathrm{r}}(\tau_k)\cos\theta_{\mathrm{r}}(\tau_k\\ &-y_m\sin\phi_{\mathrm{t}}(\tau_k)\cos\theta_{\mathrm{t}}(\tau_k)-y_m\sin\phi_{\mathrm{r}}(\tau_k)\cos\theta_{\mathrm{r}}(\tau_k)\\ &-z_m\sin\theta_{\mathrm{t}}(\tau_k)-z_m\sin\theta_{\mathrm{r}}(\tau_k)\end{aligned} \tag{10.119}$$

represents the true differential range function, and gives the ideal phase function

$$\Phi_m(f_i,\tau_k)=-2\pi f_i\frac{\Delta R_m(\tau_k)}{c} \qquad i=1,\ldots,N_f \quad k=1,\ldots,N_\tau \tag{10.120}$$

identical to equation (10.16), in which the platform MMEs are assumed to be zero. The second term

$$\begin{aligned}\Delta\tilde{R}_m(\tau_k) \approx &-\tilde{x}_{\mathrm{t}}(\tau_k)\cos\phi_{\mathrm{t}}(\tau_k)\cos\theta_{\mathrm{t}}(\tau_k)-\tilde{x}_{\mathrm{r}}(\tau_k)\cos\phi_{\mathrm{r}}(\tau_k)\cos\theta_{\mathrm{r}}(\tau_k)\\ &-\tilde{y}_{\mathrm{t}}(\tau_k)\sin\phi_{\mathrm{t}}(\tau_k)\cos\theta_{\mathrm{t}}(\tau_k)-\tilde{y}_{\mathrm{r}}(\tau_k)\sin\phi_{\mathrm{r}}(\tau_k)\cos\theta_{\mathrm{r}}(\tau_k)\\ &-\tilde{z}_{\mathrm{t}}(\tau_k)\sin\theta_{\mathrm{t}}(\tau_k)-\tilde{z}_{\mathrm{r}}(\tau_k)\sin\theta_{\mathrm{r}}(\tau_k)\end{aligned} \tag{10.121}$$

represents the error in the measured differential range—due to mismeasurements of the transmitter $(\tilde{x}_{\mathrm{t}}(\tau_k), \tilde{y}_{\mathrm{t}}(\tau_k), \tilde{z}_{\mathrm{t}}(\tau_k))$ and receiver $(\tilde{x}_{\mathrm{r}}(\tau_k), \tilde{y}_{\mathrm{r}}(\tau_k), \tilde{z}_{\mathrm{r}}(\tau_k))$ trajectories—and contributes to the phase error due to MMEs, written as

$$\tilde{\Phi}_m(f_i,\tau_k)=-2\pi f_i\left(\frac{\Delta\tilde{R}_m(\tau_k)}{c}+\tilde{t}(\tau_k)\right) \qquad i=1,\ldots,N_f, \qquad k=1,\ldots, \tag{10.122}$$

Note that equation (10.121) and equation (10.122) are independent of the scatterer location (x_m, y_m, z_m), thus implying that the effects of platform motion measurement errors in the final image will be seen primarily as spatially independent. This is true as long as the scene size limitations imposed by the far-field approximation (see equation (10.106)) are met. In most cases, the azimuth phase errors expressed by $\tilde{\Phi}_m(f_i, \tau_k)$ may also be treated as constants with respect to frequency f_i, provided the system is of a low percent bandwidth (i.e., B/f_c is small).

It is important to note that antenna platform MMEs and synchronization errors are only two contributors to the slow-time (τ_k) dependent phase error $\tilde{\Phi}_m(f_i, \tau_k)$. Atmospheric effects, transmit waveform instability, receiver demodulation instability, and uncompensated target motion may also add to $\tilde{\Phi}_m$.

To analyze the effects of MMEs on bistatic SAR imagery, one must first understand how an image formation algorithm interprets the collected phase history data to form images. Here, some of the steps taken in the previous sections are briefly retraced. Based on the phase history model defined by equations (10.11)–(10.17), the location $\underline{r}_m$ of each scatterer in the observed scene is encoded in the phase of the observed phase histories via each differential range function $\Delta\hat{R}_m(\tau)$, which is approximated by equation (10.118). As in Section 10.4, to form SAR images, one typically assumes that the MMEs are zero and that all of the scatterers in a scene lie on a uniform grid of pixel locations. One may then use a matched filtering algorithm to estimate the reflectivity of a scatterer at each (x, y, z) pixel location, written similar to equation (10.76) as

$$P(x,y,z) = \frac{1}{N_f N_\tau} \sum_{i=1}^{N_f} \sum_{k=1}^{N_\tau} S(f_i, \tau_k) e^{-j\hat{\Phi}_{xyz}(f_i,\tau_k)} \tag{10.123}$$

which implicitly assumes that the measured phase functions $\hat{\Phi}_m(f_i, \tau_k)$ are error free. The phase function

$$\Phi_{xyz}(f_i, \tau_k) = \frac{2\pi f_i}{c} \Delta R_{xyz}(\tau_k) \tag{10.124}$$

represents an ideal phase linearly dependent on the differential range of a scatterer located at (x, y, z). To form a complete image, equation (10.123) is computed for every element of the assumed grid of (x, y, z) locations on the image plane. A peak will appear in the image when the phase of the matched filter matches the phase of an observed point scatterer; this occurs when $\Delta R_{xyz}(\tau_k) \approx \Delta R_m(\tau_k)$.

This begs the question: how is scatterer position information encoded in the bistatic differential range function? By understanding that process, the effects of MMEs on the encoded information may then be studied. To explore the manner in which information is recorded by the differential range function, further approximations are applied to the differential range expression in equation (10.119). First, assume that the ground range $r_t = \sqrt{x_t(\tau)^2 + y_t(\tau)^2}$ and the slant range $R_t = \sqrt{x_t(\tau)^2 + y_t(\tau)^2 + z_t(\tau)^2}$ of the transmitter are sufficiently large such that they may be treated as constants with respect to slow time τ. Next, the same assumptions about the ground range and slant range of the receiver, $r_r = \sqrt{x_r(\tau)^2 + y_r(\tau)^2}$ and $R_r = \sqrt{x_r(\tau)^2 + y_r(\tau)^2 + z_r(\tau)^2}$, are made. Finally, the transmitter and receiver are assumed to traverse linear flight paths at constant velocities. This allows one to make the series of substitutions shown in Table 10–4. Thus, the approximated differential range in equation (10.119) may be further approximated by a linear function of the slow time,

$$\begin{aligned}
\Delta R(\tau) \approx & -x\left(\frac{x_t + v_{xt}\tau}{r_t}\right)\left(\frac{r_t}{R_t}\right) - y\left(\frac{y_t + v_{yt}\tau}{r_t}\right)\left(\frac{r_t}{R_t}\right) - z\left(\frac{z_t + v_{zt}\tau}{R_t}\right) \\
& - x\left(\frac{x_r + v_{xr}\tau}{r_r}\right)\left(\frac{r_r}{R_r}\right) - y\left(\frac{y_r + v_{yr}\tau}{r_r}\right)\left(\frac{r_r}{R_r}\right) - z\left(\frac{z_r + v_{zr}\tau}{R_r}\right) \\
= & \left[-x\left(\frac{x_t}{R_t} + \frac{x_r}{R_r}\right) - y\left(\frac{y_t}{R_t} + \frac{y_r}{R_r}\right) - z\left(\frac{z_t}{R_t} + \frac{z_r}{R_r}\right)\right] \\
& + \left[-x\left(\frac{v_{xt}}{R_t} + \frac{v_{xr}}{R_r}\right) - y\left(\frac{v_{yt}}{R_t} + \frac{v_{yr}}{R_r}\right) - z\left(\frac{v_{zt}}{R_t} + \frac{v_{zr}}{R_r}\right)\right]\tau \\
\triangleq & \ \beta_0 + \beta_1 \tau.
\end{aligned} \tag{10.125}$$

The locations of the transmitter and receiver at their aperture midpoints are given by (x_t, y_t, z_t) and (x_r, y_r, z_r), and the transmitter and receiver velocity vectors are (v_{xt}, v_{yt}, v_{zt}) and (v_{xr}, v_{yr}, v_{zr}), respectively.

Table 10–4 Linear Approximations for Transmitting Platform Trigonometric Expressions.

Trigonometric expression	Linear approximation
$\cos\phi_t(\tau)$	$\frac{x_t + v_{xt}\tau}{r_t}$
$\sin\phi_t(\tau)$	$\frac{y_t + v_{yt}\tau}{r_t}$
$\cos\theta_t(\tau)$	$\frac{r_t}{R_t}$
$\sin\theta_t(\tau)$	$\frac{z_t + v_{zt}\tau}{R_t}$

Note: Expressions for receiving platform substitutions are obtained by replacing $(\cdot)_t$ with $(\cdot)_r$.

Equation (10.125) describes a transformation which relates the linear approximation of the differential range $\Delta R(\tau) = \beta_0 + \beta_1\tau$ to the actual location of a scatterer in the scene, which is written as

$$\begin{bmatrix}\beta_0\\ \beta_1\end{bmatrix} = \begin{bmatrix} -\frac{x_t}{R_t} - \frac{x_r}{R_r} & -\frac{y_t}{R_t} - \frac{y_r}{R_r} & -\frac{z_t}{R_t} - \frac{z_r}{R_r} \\ -\frac{v_{xt}}{R_t} - \frac{v_{xr}}{R_r} & -\frac{v_{yt}}{R_t} - \frac{v_{yr}}{R_r} & -\frac{v_{zt}}{R_t} - \frac{v_{zr}}{R_r} \end{bmatrix} \begin{bmatrix} x\\ y\\ z\end{bmatrix} \tag{10.126}$$

The following analysis will focus on scatterers located on the ground plane ($z = 0$), giving

$$\begin{bmatrix}\beta_0\\ \beta_1\end{bmatrix} = \begin{bmatrix} -\frac{x_t}{R_t} - \frac{x_r}{R_r} & -\frac{y_t}{R_t} - \frac{y_r}{R_r} \\ -\frac{v_{xt}}{R_t} - \frac{v_{xr}}{R_r} & -\frac{v_{yt}}{R_t} - \frac{v_{yr}}{R_r} \end{bmatrix} \begin{bmatrix} x\\ y\end{bmatrix} \triangleq Q \begin{bmatrix} x\\ y\end{bmatrix} \tag{10.127}$$

Therefore, given a measured differential range function for a single scatterer on the ground, one may linearly approximate $\Delta R(\tau)$ and then use

$$\begin{bmatrix} x\\ y\end{bmatrix} = Q^{-1}\begin{bmatrix}\beta_0\\ \beta_1\end{bmatrix} \tag{10.128}$$

to estimate the (x, y) location of the peak that scatterer's response, in a ground plane image.

Provided there are no MMEs, the (x, y) location of a scatterer in the ground plane is encoded in the linear portion of the differential range function with slope β_1 and intercept β_0 via the linear mapping Q. Note that β_0 and β_1 are analogous to range–doppler measurements of a stationary scatterer by two moving bistatic platforms.

10.5.2 Characterizing Motion Measurement Errors

The above derivations assume that there are no MMEs. To implement an actual bistatic SAR system, MMEs must be considered. Errors in measuring the motion of an airborne surveillance platform have frequently been characterized [2, 17, 70, 71] by expressions in the form

$$\underline{\tilde{r}}(\tau) = \underline{\tilde{g}}(\tau) + \underline{\tilde{h}}(\tau) \tag{10.129}$$

where $\underline{\tilde{g}}(\tau)$ represents low-frequency errors and $\underline{\tilde{h}}(\tau)$ represents high-frequency errors. The low-frequency component of the errors is often modeled as a polynomial, whereas the high-frequency errors are more difficult to model and are most often described by their spectral characteristics or RMS values. The high-frequency errors are usually much smaller in magnitude than the low-frequency errors, and typically consist of noise-like elements, and sinusoidal components with periods much shorter than the aperture duration. Errors not due to motion measurement, such as propagation phenomena and atmospheric turbulence, also contribute to the high-frequency errors. The following will focus on the dominant low-frequency elements of the unmeasured platform motion, and will model them with a quadratic polynomial of the form

$$\underline{\tilde{g}}(\tau)=\begin{bmatrix}\tilde{x}(\tau)\\ \tilde{y}(\tau)\\ \tilde{z}(\tau)\end{bmatrix}=\begin{bmatrix}\tilde{p}_x\\ \tilde{p}_y\\ \tilde{p}_z\end{bmatrix}+\begin{bmatrix}\tilde{v}_x\\ \tilde{v}_y\\ \tilde{v}_z\end{bmatrix}\tau+\begin{bmatrix}\tilde{a}_x\\ \tilde{a}_y\\ \tilde{a}_z\end{bmatrix}\frac{\tau^2}{2} \tag{10.130}$$

where $[\tilde{p}_x \;\; \tilde{p}_y \;\; \tilde{p}_z]^T$ represents the error in measuring a platform's location at the aperture midpoint, and $[\tilde{v}_x \;\; \tilde{v}_y \;\; \tilde{v}_z]^T$ and $[\tilde{a}_x \;\; \tilde{a}_y \;\; \tilde{a}_z]^T$ represent the errors in measuring a platform's velocity and acceleration.

By applying the same approximations from Table 10–4 to $\Delta\tilde{R}(\tau)$ in equation (10.121), the error introduced into the observed differential range can be approximated as

$$\begin{aligned}\Delta\tilde{R}(\tau)\approx{}&-\tilde{x}_{\mathrm{t}}(\tau)\frac{x_{\mathrm{t}}+v_{x\mathrm{t}}\tau}{R_{\mathrm{t}}}-\tilde{x}_{\mathrm{r}}(\tau)\frac{x_{\mathrm{r}}+v_{x\mathrm{r}}\tau}{R_{\mathrm{r}}}\\ &-\tilde{y}_{\mathrm{t}}(\tau)\frac{y_{\mathrm{t}}+v_{y\mathrm{t}}\tau}{R_{\mathrm{t}}}-\tilde{y}_{\mathrm{r}}(\tau)\frac{y_{\mathrm{r}}+v_{y\mathrm{r}}\tau}{R_{\mathrm{r}}}\\ &-\tilde{z}_{\mathrm{t}}(\tau)\frac{z_{\mathrm{t}}+v_{z\mathrm{t}}\tau}{R_{\mathrm{t}}}-\tilde{z}_{\mathrm{r}}(\tau)\frac{z_{\mathrm{r}}+v_{z\mathrm{r}}\tau}{R_{\mathrm{r}}}+c\tilde{t}(\tau)\end{aligned} \tag{10.131}$$

where the contribution of the transmit–receive timing error has been included here, instead of analyzing its effect as a separate error. Using equation (10.131), the effect of motion measurement errors on the observed differential range may be characterized by substituting in quadratic polynomial expressions for the MMEs $\{\tilde{x}_{\mathrm{t}}(\tau),\ \tilde{x}_{\mathrm{r}}(\tau),\ \tilde{y}_{\mathrm{t}}(\tau),\ \tilde{y}_{\mathrm{r}}(\tau),\ \tilde{z}_{\mathrm{t}}(\tau),\ \tilde{z}_{\mathrm{r}}(\tau)\}$. These polynomials are of the form given in equation (10.130). For ease of analysis, it is also assumed that the time synchronization error between the transmit and receive platforms is of the form $\tilde{t}(\tau)=\tilde{t}_0+\tilde{t}_1\tau+\tilde{t}_2\tau^2+\tilde{t}_3\tau^3$. The result is a cubic polynomial expression for equation (10.131), written as

$$\Delta\tilde{R}(\tau)\approx\tilde{\beta}_0+\tilde{\beta}_1\tau+\tilde{\beta}_2\tau^2+\tilde{\beta}_3\tau^3 \tag{10.132}$$

Table10–5 shows the errors in the observed differential range induced by transmitter MMEs and time synchronization errors. The errors induced by receiver MMEs are identical. The results in Table 10–5 allow one to study the effect of MMEs on bistatic SAR images.

10.5.3 MME Effects on Bistatic SAR Imagery

Using the polynomial approximation for $\Delta\tilde{R}(\tau)$ in equation (10.132), one may approximate the observed differential range as

$$\begin{aligned}\Delta\hat{R}(\tau)&=\Delta R(\tau)+\Delta\tilde{R}(\tau)\\ &\approx\hat{\beta}_0+\hat{\beta}_1\tau+\hat{\beta}_2\tau^2+\hat{\beta}_3\tau^3\\ &\triangleq(\beta_0+\tilde{\beta}_0)+(\beta_1+\tilde{\beta}_1)\tau+\tilde{\beta}_2\tau^2+\tilde{\beta}_3\tau^3\end{aligned} \tag{10.133}$$

Table 10–5 Examples of Transmitter MME Induced Differential Range Errors

	Coefficients of $\Delta\tilde{R}(\tau)$			
MME	$\tilde{\beta}_0$	$\tilde{\beta}_1$	$\tilde{\beta}_2$	$\tilde{\beta}_3$
$\tilde{t}(\tau)$	$c\tilde{t}_0$	$c\tilde{t}_1$	$c\tilde{t}_2$	$c\tilde{t}_3$
$\tilde{x}_t(\tau) = \tilde{p}_{x\tau}$	$-\frac{\tilde{p}_{xt}x_t}{R_t}$	$-\frac{\tilde{p}_{xt}v_{xt}}{R_t}$	—	—
$\tilde{x}_t(\tau) = \tilde{v}_{xt}\mathrm{t}$	—	$-\frac{\tilde{v}_{xt}x_t}{R_t}$	$-\frac{\tilde{v}_{xt}v_{xt}}{R_t}$	—
$\tilde{x}_t(\tau) = \frac{1}{2}\tilde{a}_{xt}\tau^2$	—	—	$-\frac{\tilde{a}_{xt}x_t}{2R_t}$	$-\frac{\tilde{a}_{xt}v_{xt}}{2R_t}$
$\tilde{y}_t(\tau) = \tilde{p}_{yt}$	$-\frac{\tilde{p}_{yt}y_t}{R_t}$	$-\frac{\tilde{p}_{yt}v_{yt}}{R_t}$	—	—
$\tilde{y}_t(\tau) = \tilde{v}_{yt}\tau$	—	$-\frac{\tilde{v}_{yt}y_t}{R_t}$	$-\frac{\tilde{v}_{yt}v_{yt}}{R_t}$	—
$\tilde{y}_t(\tau) = \frac{1}{2}\tilde{a}_{yt}\tau^2$	—	—	$-\frac{\tilde{a}_{yt}y_t}{2R_t}$	$-\frac{\tilde{a}_{yt}v_{yt}}{2R_t}$
$\tilde{z}_t(\tau) = \tilde{p}_{zt}$	$-\frac{\tilde{p}_{zt}z_t}{R_t}$	$-\frac{\tilde{p}_{zt}v_{zt}}{R_t}$	—	—
$\tilde{z}_t(\tau) = \tilde{v}_{zt}\tau$	—	$-\frac{\tilde{v}_{zt}z_t}{R_t}$	$-\frac{\tilde{v}_{zt}v_{zt}}{R_t}$	—
$\tilde{z}_t(\tau) = \frac{1}{2}\tilde{a}_{zt}\tau^2$	—	—	$\frac{-\tilde{a}_{zt}z_t}{2R_t}$	$\frac{-\tilde{a}_{zt}v_{zt}}{2R_t}$

Note: Expressions for receiving platform errors are obtained by replacing $(\cdot)_t$ with $(\cdot)_r$.

The impact of differential range errors on bistatic SAR images may be determined by using the results of Section 10.5.1 to study equation (10.133). The image domain effects of constant and linear differential range errors are straightforward to analyze. By inserting the observed $\hat{\beta}_0, \hat{\beta}_1$ into (10.128) to get

$$\begin{aligned}
\begin{bmatrix} \hat{x} \\ \hat{y} \end{bmatrix} &= Q^{-1}\begin{bmatrix} \hat{\beta}_0 \\ \hat{\beta}_1 \end{bmatrix} \\
&= Q^{-1}\begin{bmatrix} \beta_0 + \tilde{\beta}_0 \\ \beta_1 + \tilde{\beta}_1 \end{bmatrix} \\
&= \begin{bmatrix} x \\ y \end{bmatrix} + Q^{-1}\begin{bmatrix} \tilde{\beta}_0 \\ \tilde{\beta}_1 \end{bmatrix} \\
&\triangleq \begin{bmatrix} x \\ y \end{bmatrix} + \begin{bmatrix} \tilde{x} \\ \tilde{y} \end{bmatrix}
\end{aligned} \tag{10.134}$$

it is seen that the dominant effect of constant and linear differential range errors is a spatially invariant shift in the imaged location of any scatterer in the scene, such that a scatterer located at (x, y) in the scene will translate into a peak in image intensity at $(\hat{x}, \hat{y}) = (x, y) + (\tilde{x}, \tilde{y})$.

For terms in the observed differential range function that are quadratic or cubic in τ, the effect of the errors is analyzed indirectly. To perform this analysis, make use of the results from Section 10.5.1, which indicated that image formation algorithms extract the scatterer location information $\Delta\hat{R}(\tau)$ that is encoded in $\Delta\hat{R}(\tau)$ and $\Delta\hat{R}'(\tau)$, at every point in the aperture via matched filtering. This is done under the assumption that the observed differential range function $\Delta\tilde{R}(\tau)$ is well modeled as linear. Quadratic or cubic terms in $\Delta\tilde{R}(\tau)$ are therefore approximated with a piecewise linear function of the form

$$\Delta\tilde{R}(\tau) \approx \beta_{0,i} + \beta_{1,i}\tau, \qquad \tau_i - \frac{\delta_\tau}{2} \le \tau < \tau_i + \frac{\delta_\tau}{2} \tag{10.135}$$

$$\beta_{0,i} = \Delta\tilde{R}(\tau_i) \tag{10.136}$$

$$\beta_{1,i} = \frac{\partial \Delta\tilde{R}(\tau_i)}{\partial \tau} \tag{10.137}$$

where δ_τ is the width of each piecewise linear interval, and the τ_i are the centers of those intervals. Each interval in slow time serves to define a subset of the synthesized aperture, commonly referred to as a subaperture. This piecewise linear decomposition serves as the basis for the map drift autofocus algorithm [72] for monostatic SAR. A series of Q_i matrices may then be obtained by computing the elements of the matrix Q based on the locations of the transmitter and receiver at each of the time interval midpoints τ_i. The observed differential range function is now interpreted by computing the location

$$\begin{bmatrix} \hat{x}_i \\ \hat{y}_i \end{bmatrix} = Q_i^{-1}\begin{bmatrix} \hat{\beta}_{0,i} \\ \hat{\beta}_{1,i} \end{bmatrix} = Q_i^{-1}\begin{bmatrix} \beta_{0,i} + \tilde{\beta}_{0,i} \\ \beta_{1,i} + \tilde{\beta}_{1,i} \end{bmatrix} = \begin{bmatrix} x_i \\ y_i \end{bmatrix} + Q_i^{-1}\begin{bmatrix} \tilde{\beta}_{0,i} \\ \tilde{\beta}_{1,i} \end{bmatrix} \tag{10.138}$$

that each linear subinterval indicates. The resulting $\{(\hat{x}_i, \hat{y}_i)\}$ are the approximate locations of the broadened matched filter responses resulting from each subaperture, which will be added by superposition in the final image. Without MMEs, these responses add coherently to form a sharpened response, but in the presence of errors in $\Delta\hat{R}(\tau)$ that are of quadratic or higher order, a smeared response will appear. One may also note that the primary effect of the MMEs, modeled by $Q_i^{-1}[\tilde{\beta}_{0,i}\ \ \tilde{\beta}_{1,i}]^T$, is independent of the location (x_i, y_i) of the observed scatterer. These defocusing effects are therefore considered spatially invariant.

To illustrate the above approach, consider a bistatic scenario where, at their aperture midpoints, the transmitter range is 31 km, the receiver range is 8.5 km, and the transmitter and receiver have an angular separation of 45° in azimuth. Both platforms view the scene at broadside, and the total data collection time is approximately T = 6 sec. The measured motion of the transmitter includes an uncompensated acceleration of 0.005 m/sec^2 in the x direction, which gives an observed differential range function with the quadratic component shown in Figure 10–45. Compute $\{(\tilde{\beta}_{0,i}\ \ \tilde{\beta}_{1,i})|i = 1,2,3\}$ from equations (10.136)–(10.137) to approximate the quadratic component of the differential range error with three line segments, as shown in Figure 10–45. Then, compute the image domain locations $\{(\hat{x}_i, \hat{y}_i)|i = 1,2,3\}$ from equation (10.138). These three points are plotted in Figure 10–46. The image shown in Figure 10–47 was formed from the phase histories resulting from inserting the observed $\Delta\hat{R}(\tau)$ into equations (10.16) and (10.17). It may be observed that a line segment connecting the points $\{(\hat{x}_i, \hat{y}_i)|i = 1,2,3\}$ accurately predicts the direction of main lobe broadening that appears in the final image.

To understand why the defocus seen in Figure 10–47 is predominantly in a single direction, one may analyze the inverses of the Q_i matrices. From equations (10.118), (10.127), and

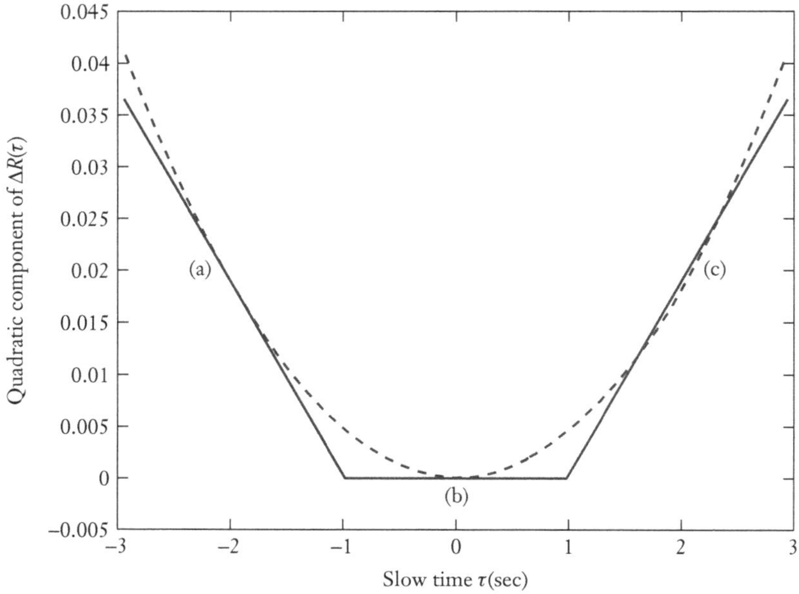

Figure 10–45 Quadratic component of the differential range function induced by an uncompensated 0.005 m/sec^2 acceleration of the transmitter in the x direction.

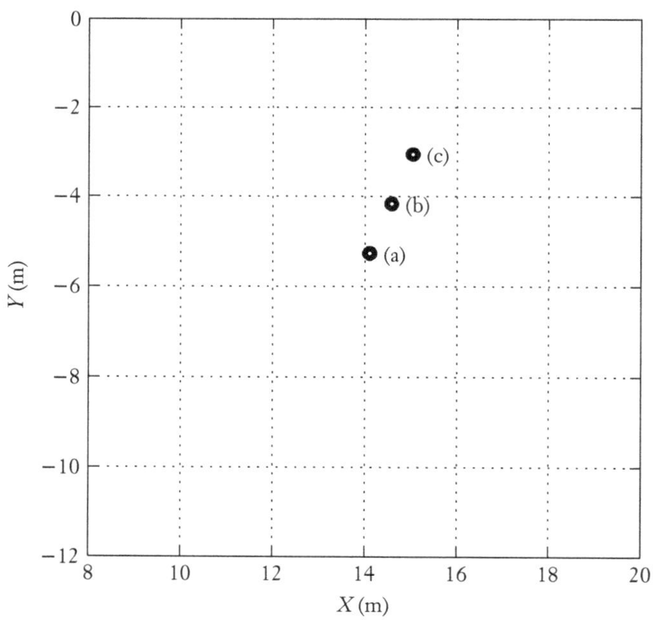

Figure 10–46 Approximate image domain locations corresponding to the three line segments used to approximate $\Delta R(\tau)$ in Figure 10–45.

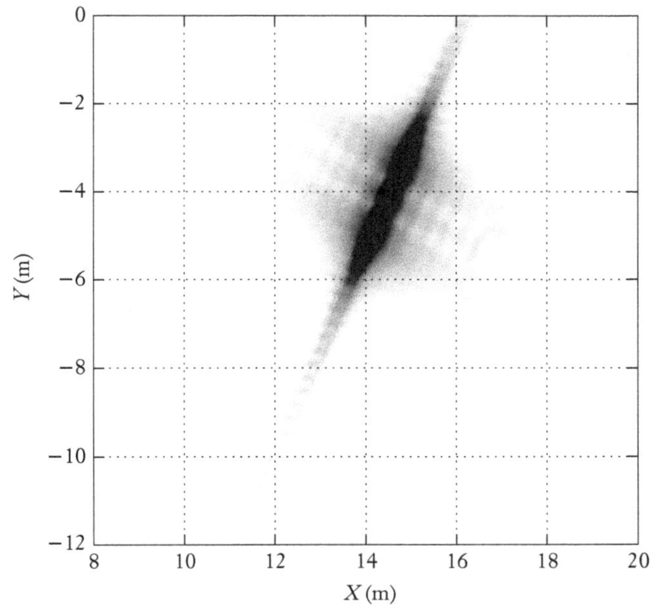

Figure 10–47 Ground plane image of the smeared response caused by an uncompensated 0.005 m/sec² acceleration of the transmit platform in the *x* direction. Without this MME, a focused response would appear instead.

(10.128), one may write

$$Q_i^{-1} = \frac{1}{|Q_i|}\begin{bmatrix} -\frac{v_{yt}}{R_t} - \frac{v_{yr}}{R_r} & \frac{y_t(\tau_i)}{R_t} + \frac{y_r(\tau_i)}{R_r} \\ \frac{v_{xt}}{R_t} + \frac{v_{xr}}{R_r} & -\frac{x_t(\tau_i)}{R_t} - \frac{x_r(\tau_i)}{R_r} \end{bmatrix}$$

$$= \frac{1}{|Q_i|}\begin{bmatrix} -\frac{v_{yt}}{R_t} - \frac{v_{yr}}{R_r} & \sin\phi_t(\tau_i)\cos\theta_t(\tau_i) + \sin\phi_r(\tau_i)\cos\theta_r(\tau_i) \\ \frac{v_{xt}}{R_t} + \frac{v_{xr}}{R_r} & -\cos\phi_t(\tau_i)\cos\theta_t(\tau_i) - \cos\phi_r(\tau_i)\cos\theta_r(\tau_i) \end{bmatrix} \tag{10.139}$$

where $\phi_t(\tau_i)$ and $\phi_r(\tau_i)$ ($\theta_t(\tau_i)$ and $\theta_r(\tau_i)$) are the azimuth (elevation) angles of the transmitter and receiver at the subaperture midpoints. From equation (10.139), the following observations are made. Since R_t and R_r are typically on the order of several kilometers and the platform speeds

$$v_t = \sqrt{v_{xt}^2 + v_{yt}^2} \quad \text{and} \quad v_r = \sqrt{v_{xr}^2 + v_{yr}^2} \tag{10.140}$$

are typically a few hundred meters per second at most, the norm of the first column of Q_i^{-1} will be much smaller than the norm of the second column. And, furthermore, the vector $[\sin\phi_t(\tau_i)\cos\theta_t(\tau_i) + \sin\phi_r(\tau_i)\cos\theta_r(\tau_i) \; -\cos\phi_t(\tau_i)\cos\theta_t(\tau_i) - \cos\phi_r(\tau_i)\cos\theta_t(\tau_i)]^T$ does not vary significantly across the aperture.

Thus, terms of $\Delta\tilde{R}(\tau)$ that are of quadratic or higher order, which therefore have a changing $\beta_{1,i}$, will cause a smearing in the image that is primarily in the direction of

$$\underline{v}_{\text{defocus}} = \begin{bmatrix} \sin\bar{\phi}_t\cos\bar{\theta}_t + \sin\bar{\phi}_r\cos\bar{\theta}_r \\ -\cos\bar{\phi}_t\cos\bar{\theta}_t - \cos\bar{\phi}_r\cos\bar{\theta}_r \end{bmatrix} \tag{10.141}$$

Recall that $\bar{\phi}_t$ and $\bar{\phi}_r$ ($\bar{\theta}_t$ and $\bar{\theta}_r$) are the azimuth (elevation) angles of the transmitter and receiver at the midpoint of their respective flight paths. It is significant to note that $\underline{v}_{\text{defocus}}$ is orthogonal to the

Table 10–6 Quadratic Phase Errors (QPE) Corresponding to Transmit Platform Velocity and Acceleration MMEs, and Maximum Allowable MME to Prevent image Defocus.

MME	QPE	Allowable MME
$\tilde{x}_t(\tau) = \tilde{v}_{xt}\tau$	$\dfrac{\pi\tilde{v}_{xt}v_{xt}T^2}{2\lambda R_t}$	$\tilde{v}_{xt} < \dfrac{\lambda R_t}{4\tilde{v}_{xt}T^2}$
$\tilde{y}_t(\tau) = \tilde{v}_{xt}\tau$	$\dfrac{\pi\tilde{v}_{yt}v_{yt}T^2}{2\lambda R_t}$	$\tilde{v}_{yt} < \dfrac{\lambda R_t}{4\tilde{v}_{yt}T^2}$
$\tilde{z}_t(\tau) = \tilde{v}_{xt}\tau$	$\dfrac{\pi\tilde{v}_{zt}v_{zt}T^2}{2\lambda R_t}$	$\tilde{v}_{zt} < \dfrac{\lambda R_t}{4\tilde{v}_{zt}T^2}$
$\tilde{x}_t(\tau) = \frac{1}{2}\tilde{a}_{xt}\tau^2$	$\dfrac{\pi\tilde{a}_{xt}x_tT^2}{2\lambda R_t}$	$\tilde{a}_{xt} < \dfrac{\lambda}{2T^2\cos(\phi_t)\cos(\theta_t)}$
$\tilde{y}_t(\tau) = \frac{1}{2}\tilde{a}_{yt}\tau^2$	$\dfrac{\pi\tilde{a}_{yt}y_tT^2}{2\lambda R_t}$	$\tilde{a}_{yt} < \dfrac{\lambda}{2T^2\sin(\phi_t)\sin(\theta_t)}$
$\tilde{z}_t(\tau) = \frac{1}{2}\tilde{a}_{zt}\tau^2$	$\dfrac{\pi\tilde{a}_{zt}z_tT^2}{2\lambda R_t}$	$\tilde{a}_{zt} < \dfrac{\lambda}{2T^2\sin(\theta_t)}$

Note: Expressions for receiving platform errors are obtained by replacing $(\cdot)_t$ with $(\cdot)_r$.

direction of the bistatic look angle $[\cos\bar{\phi}_t\cos\bar{\theta}_t + \cos\bar{\phi}_r\cos\bar{\theta}_r \ \sin\bar{\phi}_t\cos\bar{\theta}_t + \sin\bar{\phi}_r\cos\bar{\theta}_r]^T$ indicated by equation (10.23) and used as the angle of reference frame rotation in the bistatic PFA.

10.5.4 Maximum Allowable Motion Measurement Errors

Next system specifications are derived that will limit the degree of defocus in the final image. In monostatic SAR systems, one typically chooses to limit the magnitude of quadratic phase error (QPE) contributed by any one motion measurement error to be less than $\pi/4$ radians [2]. As a bistatic system is being considered, which is dependent on two antenna platforms, the QPEs are limited to be less than $\pi/8$ radians for each platform. In practice, this allotment may be budgeted based on real-world requirements such as dictated by platform cost and complexity tradeoffs. To determine the QPE caused by each form of motion measurement error, insert the assumed $\Delta\tilde{R}(\tau)$ into equation (10.122) to obtain an expression for the phase error $\tilde{\Phi}$. The quadratic component of $\tilde{\Phi}$ is then limited to be less than $\pi/8$ radians to arrive at the transmitter specifications given in Table 10–6.

Recall that the length of the total data collection time interval is represented by T. The motion measurement specifications for the receiving platform are obtained from Table 10–6 by replacing subscripts "t" with subscripts "r". Note that the elements in Table 10–6 generalize the monostatic SAR case, and agree with the expressions in [2, Table 5.7] for the size of quadratic phase errors from major sources, if one sets the broadening factor K_a to unity and takes into account a rotation of coordinate systems.

Higher order phase errors are less dominant in SAR images, are more difficult to analyze, and have consequently been given less treatment in the SAR literature. Typically for monostatic SAR, the allowable levels of high-frequency vibrations are determined based on image quality metrics such as the peak side lobe ratio (PSLR) and the integrated side lobe ratio (ISLR) [2]. In [2, Table 5.12], the allowable sinusoidal motion is specified through a PSLR-dependent limitation on the sinusoid amplitude $A_s \le \lambda/(2\pi)\sqrt{\text{PSLR}}$, and the allowable wide band vibration level is specified with an ISLR-dependent limitation on the RMS vibration value $\sigma_v \le \lambda/(2\pi)\sqrt{\text{ISLR}}$ These limits may easily be extended to bistatic SAR by simply splitting the RMS values of both between the transmitter and receiver. Dividing the requirements equally gives a sinusoidal requirement $A_{t/r} \le \lambda/(2\pi)\sqrt{\text{PSLR}/2}$ and an RMS vibration requirement $\sigma_{t/r} \le \lambda/(2\pi)\sqrt{\text{ISLR}/2}$. The allowable motion may be divided unequally provided that the total RMS value does not exceed that which is specified by [2].

Many monostatic SAR system designs specify overall motion measurement error limits in terms of the allowable line-of-sight acceleration and along-track velocity errors, rather than in terms of each translational degree of freedom in 3-D. By examining Table 10–6, one may write the total quadratic phase error coefficient due to uncompensated platform velocity as

$$\text{QPE}_{\text{v}} = -\frac{\tilde{v}_{x\text{t}} v_{x\text{t}}}{R_{\text{t}}} - \frac{\tilde{v}_{y\text{t}} v_{y\text{t}}}{R_{\text{t}}} - \frac{\tilde{v}_{z\text{t}} v_{z\text{t}}}{R_{\text{t}}} - \frac{\tilde{v}_{x\text{r}} v_{x\text{r}}}{R_{\text{r}}} - \frac{\tilde{v}_{y\text{r}} v_{y\text{r}}}{R_{\text{r}}} - \frac{\tilde{v}_{z\text{r}} v_{z\text{r}}}{R_{\text{r}}} \tag{10.142}$$

and the total quadratic phase error coefficient due to uncompensated platform acceleration as

$$\text{QPE}_{\text{a}} = -\frac{\tilde{a}_{x\text{t}} x_{\text{t}}}{2R_{\text{t}}} - \frac{\tilde{a}_{y\text{t}} y_{\text{t}}}{2R_{\text{t}}} - \frac{\tilde{a}_{z\text{t}} z_{\text{t}}}{2R_{\text{t}}} - \frac{\tilde{a}_{x\text{r}} x_{\text{r}}}{2R_{\text{r}}} - \frac{\tilde{a}_{y\text{r}} y_{\text{r}}}{2R_{\text{r}}} - \frac{\tilde{a}_{z\text{r}} z_{\text{r}}}{2R_{\text{r}}} \tag{10.143}$$

The velocity error and acceleration error of each platform may then be expressed in a vector form, such that

$$\underline{\tilde{v}}_{\text{t}} = \begin{bmatrix} \tilde{v}_{x\text{t}} \\ \tilde{v}_{y\text{t}} \\ \tilde{v}_{z\text{t}} \end{bmatrix} \text{ and } \underline{\tilde{v}}_{\text{r}} = \begin{bmatrix} \tilde{v}_{x\text{r}} \\ \tilde{v}_{y\text{r}} \\ \tilde{v}_{z\text{r}} \end{bmatrix} \tag{10.144}$$

and

$$\underline{\tilde{a}}_{\text{t}} = \begin{bmatrix} \tilde{a}_{x\text{t}} \\ \tilde{a}_{y\text{t}} \\ \tilde{a}_{z\text{t}} \end{bmatrix} \text{ and } \underline{\tilde{a}}_{\text{r}} = \begin{bmatrix} \tilde{a}_{x\text{r}} \\ \tilde{a}_{y\text{r}} \\ \tilde{a}_{z\text{r}} \end{bmatrix} \tag{10.145}$$

Equations (10.142) and (10.143) may then be re-written as

$$\begin{aligned} \text{QPE}_{\text{v}} &= -\frac{\underline{v}_{\text{t}}^{T} \underline{\tilde{v}}_{\text{t}}}{R_{\text{t}}} - \frac{\underline{v}_{\text{r}}^{T} \underline{\tilde{v}}_{\text{r}}}{R_{\text{r}}} \\ &= -\frac{\tilde{v}_{\text{t},AT} v_{\text{t}}}{R_{\text{t}}} - \frac{\tilde{v}_{\text{r},AT} v_{\text{r}}}{R_{\text{r}}} \end{aligned} \tag{10.146}$$

and

$$\begin{aligned} \text{QPE}_{\text{a}} &= -\frac{\underline{r}_{\text{t}}^{T} \underline{\tilde{a}}_{\text{t}}}{2R_{\text{t}}} - \frac{\underline{r}_{\text{r}}^{T} \underline{\tilde{a}}_{\text{r}}}{2R_{\text{r}}} \\ &= -\frac{\tilde{a}_{\text{t,LOS}}}{2} - \frac{\tilde{a}_{\text{r,LOS}}}{2} \end{aligned} \tag{10.147}$$

where $\tilde{v}_{\text{r, AT}}$ and $\tilde{v}_{\text{r, AT}}$ are the along-track (AT) velocity measurement errors of the transmitter and receiver, computed by projecting $\underline{\tilde{v}}_{\text{t}}$ and $\underline{\tilde{v}}_{\text{r}}$ onto the transmitter and receiver velocity vectors

Table 10–7 Quadratic Phase Errors (QPE) corresponding to Transmit Platform Along-Track Velocity and Line-of-Sight Acceleration MMEs, and Maximum Allowable MME to Prevent Image Defocus.

MME	QPE	Allowable MME
$\tilde{v}_{\mathrm{t,AT}}\tau$	$\frac{\pi \tilde{v}_{\mathrm{t,AT}} v_{\mathrm{t}} T^2}{2\lambda R_{\mathrm{t}}}$	$\tilde{v}_{\mathrm{t,AT}} < \frac{\lambda R_{\mathrm{t}}}{4 v_{\mathrm{t}} T^2}$
$\frac{1}{2}\tilde{a}_{\mathrm{t,LOS}}\,\tau^2$	$\frac{\pi \tilde{a}_{\mathrm{t,LOS}} T^2}{4\lambda}$	$\tilde{a}_{\mathrm{t,LOS}} < \frac{\lambda}{2T^2}$

Note: Expressions for receiving platform errors are obtained by replacing $(\cdot)_{\mathrm{t}}$ with $(\cdot)_{\mathrm{r}}$.

$\underline{v}_{\mathrm{t}}$ and $\underline{v}_{\mathrm{r}}$, respectively. The platform speeds are denoted by $v_{\mathrm{t}} = ||\underline{v}_{\mathrm{t}}||$ and $v_{\mathrm{r}} = ||\underline{v}_{\mathrm{t}}||$. The variables $\tilde{a}_{\mathrm{t,LOS}}$ and $\tilde{a}_{\mathrm{r,LOS}}$ are the line-of-sight (LOS) acceleration measurement errors, computed by projecting $\underline{\tilde{a}}_{\mathrm{t}}$ and $\underline{\tilde{a}}_{\mathrm{r}}$ onto the platform line-of-sight directions $\underline{r}_{\mathrm{t}}/R_{\mathrm{t}}$ and $\underline{r}_{\mathrm{r}}/R_{\mathrm{t}}$. Recall that $\underline{r}_{\mathrm{t}}$ is the transmitter position vector and $\underline{r}_{\mathrm{r}}$ is the receiver position vector.

Equations (10.146) and (10.147) allow new MME requirements to be written, as alternatives to those given in Table 10–6. Limiting the peak quadratic phase error due to along-track velocity errors to be less than $\pi/4$ gives

$$\mathrm{QPE_v}\frac{2\pi}{\lambda}\left(\frac{T}{2}\right)^2 < \frac{\pi}{4} \Rightarrow \mathrm{QPE_v} < \frac{\lambda}{2T^2} \tag{10.148}$$

and similarly limiting the peak quadratic phase error due to line-of-sight acceleration errors yields

$$\mathrm{QPE_a}\frac{2\pi}{\lambda}\left(\frac{T}{2}\right)^2 < \frac{\pi}{4} \Rightarrow \mathrm{QPE_a} < \frac{\lambda}{2T^2} \tag{10.149}$$

Allocating the transmitter and receiver each a contribution of $\pi/8$ to both the velocity and acceleration errors yields the motion measurement requirements given in Table 10–7.

In the special monostatic case, where $\tilde{v}_{\mathrm{t,AT}} = \tilde{v}_{\mathrm{r,AT}} = \tilde{v}_{\mathrm{AT}}$ and $\tilde{a}_{\mathrm{t,LOS}} = \tilde{a}_{\mathrm{r,LOS}} = \tilde{a}_{\mathrm{LOS}}$, these limits may be further simplified to give

$$\begin{aligned} \mathrm{QPE_v} &= 2\cdot\frac{\tilde{v}_{\mathrm{AT}} v_{\mathrm{a}}}{R_{\mathrm{a}}} < \frac{\lambda}{2T^2} \\ \tilde{v}_{\mathrm{AT}} &< \frac{R_{\mathrm{a}}\lambda}{4 v_{\mathrm{a}} T^2} \\ \tilde{v}_{\mathrm{AT}} &< \frac{\delta h^2 v_{\mathrm{a}} \sin^2\bar{\phi}_{\mathrm{a}}}{\lambda R_{\mathrm{a}}} \end{aligned} \tag{10.150}$$

and

$$\begin{aligned} \mathrm{QPE_a} &= 2\cdot\frac{\tilde{a}_{\mathrm{LOS}}}{2} < \frac{\lambda}{2T^2} \\ \tilde{a}_{\mathrm{LOS}} &< \frac{\lambda}{2T^2} \\ \tilde{a}_{\mathrm{LOS}} &< \frac{2\delta h^2 v_{\mathrm{a}}^2 \sin^2\bar{\phi}_{\mathrm{a}}}{\lambda R_{\mathrm{a}}^2} \end{aligned} \tag{10.151}$$

Table 10–8 Key System Parameters for a Fictitious Monostatic SAR

Parameter	Value	Units
Range (R_a)	100	km
Velocity (v_a)	150	m/s
Wavelength (λ)	3	cm
Azimuth resolution (δh)	0.3048	m
Squint angle ($\bar{\phi}_a$)	0	degrees
Max QPE velocity (QPE_v)	20π	radians
Max QPE acceleration (QPE_a)	20π	radians

where the azimuth resolution is approximated by

$$\delta h = \frac{\lambda R_a}{2 v_a T \sin \bar{\phi}_a} \tag{10.152}$$

The platform squint angle is represented by $\bar{\phi}_a$, with $\bar{\phi}_a = 90°$ indicating that the scene is viewed at broadside, and the slant range and velocity of the monostatic antenna are denoted by R_a and v_a. Allowing for differences in notation, equations (10.150) and (10.151) match those found in [2, Table 5.12].

The above analysis lends support to a claim that bistatic SAR systems, while more complex in their construction, benefit from greater degrees of freedom in their design. As a simple example, consider equations (10.146) and (10.147), which give the quadratic phase error due to uncompensated along-track velocities and line-of-sight accelerations. In developing a SAR system, a single specification is typically set for the allowable amount of quadratic phase error due to uncompensated motion prior to autofocus. Monostatic systems must meet that requirement by accurately measuring the motion of a single antenna phase center. Bistatic systems are required to track the position of two antenna, which is a more complex requirement but also a more flexible one. This error budget may be split between the transmitter and the receiver in whatever way yields the best performance.

A simple design example can illustrate the utility of this flexibility. Consider a tactical aircraft equipped with a monostatic SAR with the performance parameters given in Table 10–8, including a 20π quadratic phase error allotment for uncompensated velocity and acceleration errors individually. A requirement of 20π is used under the assumption that the system is equipped with an autofocus algorithm capable of correcting a total of 40π of quadratic phase error. To achieve the stated goal of a 1-foot azimuth resolution, the data collection platform's inertial navigation unit must achieve a velocity measurement sensitivity of 37 cm/sec and an acceleration sensitivity of 1.1 mm/sec^2.

If one were to employ this platform as the transmitting end of a bistatic system, its contribution to the total system quadratic phase error would be cut in half, but its allowable contribution would also be halved. Thus, if the receiving antenna was placed coincident with the transmitter and was given the same motion sensing capability, the bistatic system would achieve the same performance as the original monostatic system.

However, a passive and low-cost receiver may operate at a significantly shorter range while remaining at a low level of tactical risk. With the receiver at a closer range, the same azimuth resolution may be achieved with a shorter integration time. This shortened integration time eases the motion sensing requirement placed on both platforms, as shown in Figures 10–48 and 10–49. This assumes that the receiver views the scene at broadside with a bistatic angle of $\beta = 0°$. Due

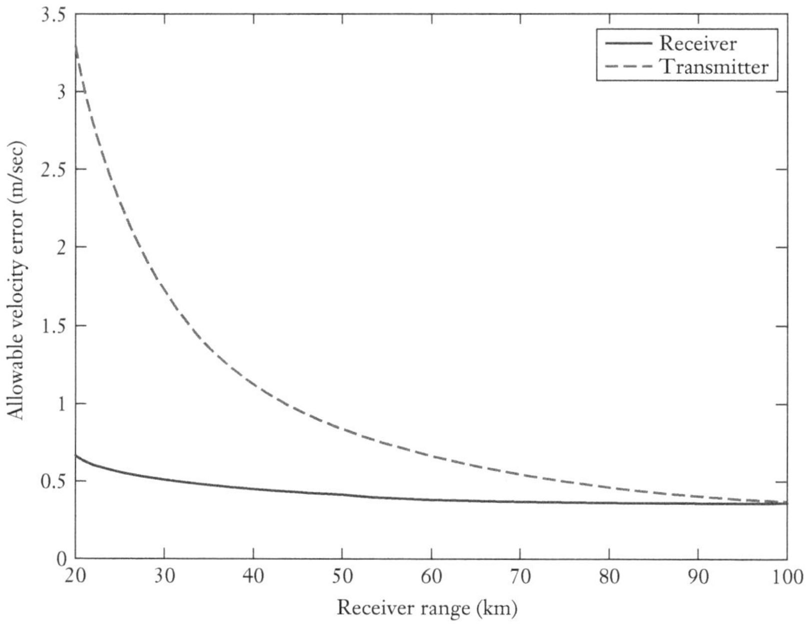

Figure 10–48 Velocity sensitivity requirement for a bistatic SAR consisting of an existing monostatic platform paired with a low-cost receiver.

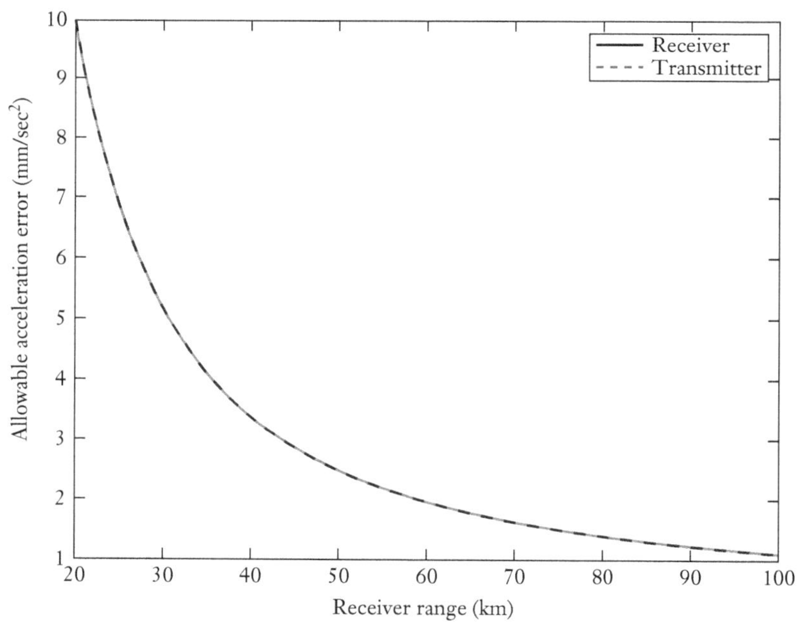

Figure 10–49 Acceleration sensitivity requirement for a bistatic SAR consisting of an existing monostatic platform paired with a low-cost receiver.

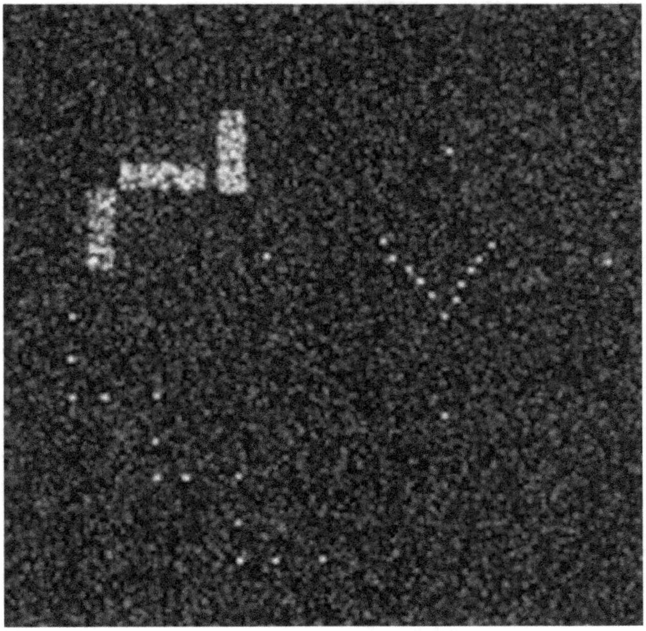

Figure 10–50 Focused SAR image of a simulated scene.

to the range-dependence of the velocity sensitivity, the receiver benefits more from operating at a short range. The acceleration sensitivity is range independent and thus benefits both platforms equally. A lessened motion compensation requirement allows use of a lower quality and cheaper inertial navigation unit, thereby supporting the low-cost development of receive-only platforms.

10.6 AUTOFOCUS

Coherently forming a focused monostatic SAR image requires precise knowledge of the local oscillator phase throughout the duration of a data collection. Inaccuracies in this phase of knowledge can be introduced through hardware instabilities [67, 68, 73], though the high quality of modern RF oscillators can make them a negligible contributor of phase errors. A much more frequent source of phase instabilities is a lack of accurate line-of-sight distance measurements from the antenna phase center to the scene to be imaged. This problem is exacerbated in bistatic SAR where the positions of both the transmitting and receiving platforms must be tracked with subwavelength relative accuracy, and in addition, the timing between the local oscillators of the two systems must be synchronized to within an acceptable level of relative drift.

As explained in Section 10.5, imprecise knowledge of the transmit and receive platform locations results in a miscalculation of the reference time delays used by the receiving platform during data capture. Errors in these time delays manifest themselves in the recorded data as slow time-varying phase errors, which may cause image shift, distortion, and defocus. The same effect is perpetrated by timing drift between the transmitter and receiver and by atmospheric variation over long distances, which causes a variation in the speed of light over the sampling interval.

Low-frequency phase errors that permit representation by low-order polynomials are usually dominant. Higher frequency phase errors, which are more often represented spectrally, result in higher peak side lobes, increased integrated side lobes, and multiplicative noise. Figure 10–50

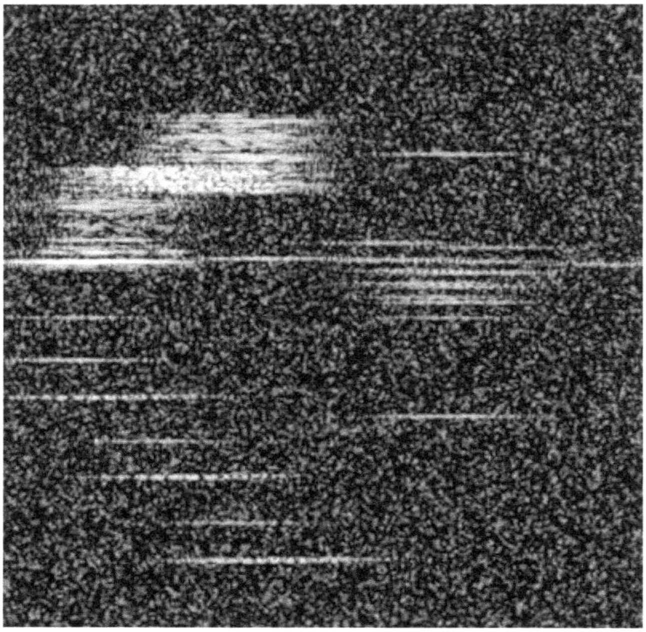

Figure 10–51 The SAR image from Figure 10–50 with a 20 π quadratic phase error. Point responses are defocused and image contrast is reduced.

shows an ideally focused image of a simulated SAR scene. The image was formed via the polar format algorithm, and may therefore be viewed as a monostatic or bistatic example, as monostatic and bistatic data sets are not mathematically distinguishable after polar reformatting. This observation will be exploited in the application of monostatic autofocus algorithms to bistatic imagery. Figures 10–51 and 10–52 show the same image corrupted by a low-frequency quadratic phase error and a high-frequency sinusoidal phase error, respectively. In order to limit the degree of defocus, performance requirements are imposed on the inertial navigation systems of the data collection platforms and on the synchronization between them. For high-resolution and/or long-range imaging, these requirements may be very stringent, as discussed in Section 10.5.

Automatic focusing (autofocus) algorithms can alleviate these requirements by correcting for slow time-varying phase errors as part of the image formation processing or postprocessing. Figure 10–53 illustrates a successful experimental demonstration of this concept [8, 9], with the left figure showing a bistatic SAR image with an uncompensated slow time phase error and the right showing the same image after autofocus. As in monostatic SAR, a reliable autofocus algorithm with the ability to correct a quadratic phase error (QPE) with up to $\Phi_{\max}$ radians of center-to-edge variation will increase the allowable maximum MMEs in the third column of Table 10–6, which are based on an allowance of $\pi/8$ radians of phase per error source, by a factor of $\frac{8}{\pi}\Phi_{\max}$. For the example system analyzed in Figures 10–48 and 10–49, it was assumed that the image processing included autofocus capable of correcting a 40π quadratic phase. This allocates 10π of quadratic phase each to the line-of-sight acceleration and along-track velocity errors of the transmitter and receiver, resulting in respective maximum allowable velocity and acceleration errors of 37 cm/sec and 1.1 mm/sec, for a monostatic geometry. Without autofocus, these requirements would be tightened by a factor of 80, yielding a maximum allowable velocity error of 4.6 mm/sec and a maximum allowable acceleration error of 13.8 μm/sec.

Figure 10–52 The SAR image from Figure 10–50 corrupted by a six-cycle sinusoidal phase error. This produces higher peak side lobes in the form of paired echoes.

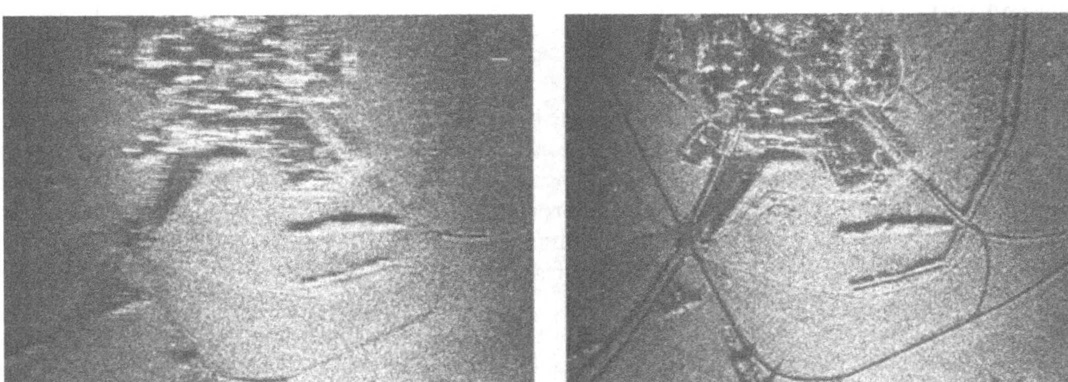

Figure 10–53 Bistatic SAR image before autofocus (left)—the image requires autofocus due to increased levels of phase noise. There are two primary causes of this phase noise: errors in the motion sensor data from the receiver platform; additional phase noise introduced by the use of two radar systems. Bistatic SAR image after autofocusing (right) using an image-based implementation of the phase gradient autfocus. (© QinetiQ Ltd [2003]. All rights reserved.)

To correct bistatic SAR image defocusing in postprocessing, one may in general take advantage of the fact that most of the defocus caused by MMEs is in a single direction. This is done by rotating the imaging coordinate system such that the crossrange dimension of the image is in the $[\sin\bar{\phi}_\mathrm{t}\cos\bar{\theta}_\mathrm{t} + \sin\bar{\phi}_\mathrm{r}\cos\bar{\theta}_\mathrm{r} \;\; -\cos\bar{\phi}_\mathrm{t}\cos\bar{\theta}_\mathrm{t} - \cos\bar{\phi}_\mathrm{r}\cos\bar{\theta}_\mathrm{r}]^T$ direction of equation (10.141). It is not

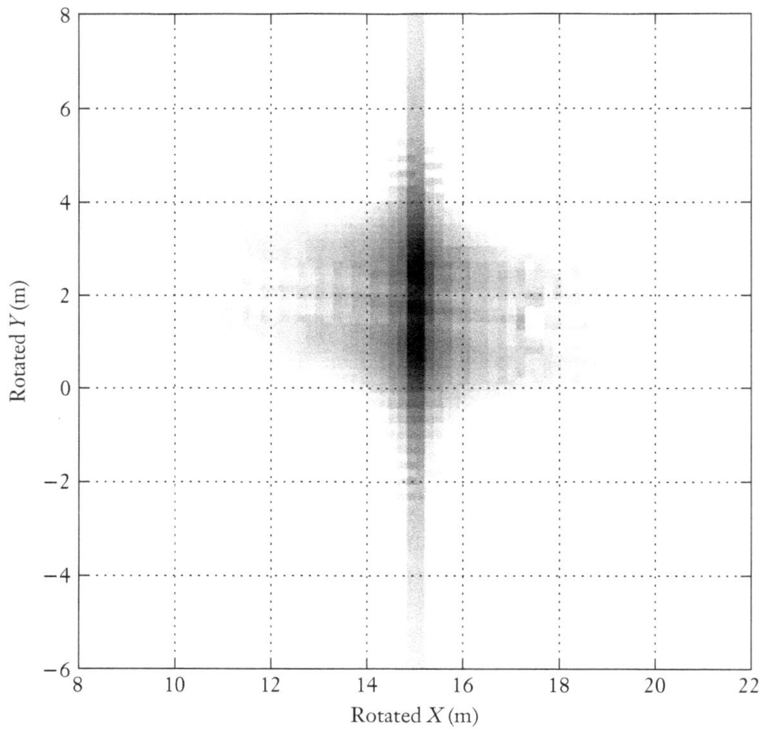

Figure 10–54 Image with MME effects rotated by the bistatic look angle ϕ_b such that the majority of the defocus is in the crossrange (y) direction.

by coincidence that this rotation is precisely the same as that prescribed in (10.91) for the bistatic polar format algorithm. After this rotation, one can then make use of any of a number of autofocus algorithms that have been developed for monostatic SAR (e.g., [2, 17, 72, 74–76]) to correct the phase errors induced by the MMEs. This approach to bistatic SAR image formation processing makes the algorithm highly amenable to existing monostatic SAR software architectures.

The above process is illustrated by way of an example. Figure 10–54 shows the defocused response of Figure 10–47 rotated by the bistatic look angle $\bar{\phi}_b$ of equation (10.23), thus placing the smearing primarily in the rotated crossrange (y') direction. In application, one would not actually have to perform this rotation, as this is the state of the SAR data prior to the final rotation of the bistatic PFA (see Figure 10–37). The (monostatic) phase gradient algorithm [74, 75] is applied to this smeared response to obtain the focused response shown in Figure 10–55. One would then apply the final step of the bistatic PFA, rotating the focused image back to the absolute coordinate system.

10.6.1 Overview

Autofocus algorithms are bootstrapping techniques for correcting slow time phase errors in postprocessing. These algorithms seek to estimate the phase errors present in the phase history data without outside information. The source of the phase corruption (e.g., motion measurement errors, atmosphere, etc.) is transparent to the algorithms' implementation and application, and scene content, that which is to be focused, is also unknown, although simple assumptions about the presence of ideal point returns are sometimes made.

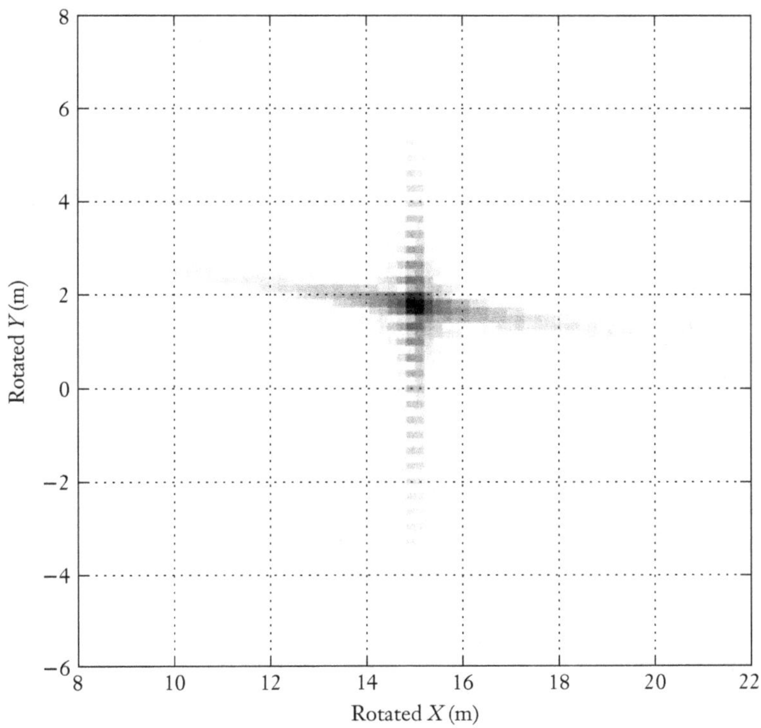

Figure 10–55 Image with MME effects after application of phase gradient algorithm autofocus.

Autofocus is usually applied as an intermediate step in the image formation process, often occurring directly after range compression. Ubiquitously, the assumed model for the range-compressed phase history data is

$$\hat{s}(n_x, k_y) = s(n_x, k_y)\exp\{j\tilde{\phi}(k_y)\} \tag{10.153}$$

where $s(n_x, k_y)$ is the ideal range-compressed data that have been corrupted by the phase error $\hat{\phi}(k_y)$, which is constant with respect to range. Crossrange frequencies are indexed by k_y after polar reformatting, and range cells are indexed by n_x after range compression. Equation (10.153) may be viewed as an intermediate result obtained from equation (10.100), after the first sum has been performed for each k_y and after the range-dependent phase shift has been applied. As slow time-varying phase errors do not prevent pulse compression, energy is still well localized in the range direction. This allows one to select range gates containing a higher level of energy on which to operate. Range gates with a lower amount of total energy are less likely to contain strong responses in the final image. The responses of strong scatterers in the scene inevitably drive the success of phase estimation techniques.

Given that proper design of a bistatic SAR processor allows use of monostatic autofocus algorithms, the content of this section is merely a review of the state-of-the-art of those algorithms. At present, there are three main families of autofocus algorithms which attempt to estimate and remove phase errors: subaperture-based methods, point-based methods, and image quality methods.

The subaperture-based autofocus (SBA) approaches (see, e.g., map drift [72], multiple aperture map drift [2]) are characteristically robust with respect to SAR image content, signal-to-noise ratio (SNR), and the presence of large, low-order phase errors; however, these algorithms become unreliable when faced with phase errors characterized by higher order polynomials or high-frequency sinusoids.

Point-based autofocus (PBA) approaches (see, e.g., phase gradient algorithm [17, 74, 75, 77], prominent point processing [2]) are capable of correcting higher order phase errors, but are restrained by a need for point responses in the imaged scene upon which to operate. Effectively detecting and isolating point scatterers to process levies a requirement on the minimum acceptable SNR and the maximum allowable phase error.

Taking into consideration the strengths and weaknesses offered by each of these classical approaches, researchers have developed autofocus algorithm hybridizations [78]. Hybrid algorithms seek to combine the benefits of several techniques in order to achieve overall performance greater than that of any of them alone.

There also exist image quality metrics, namely entropy, contrast, and sharpness, that may be interpreted to measure the degree of relative focus between two images. These measures may be used to quantify whether an estimated phase correction (from an SBA or PBA) significantly improves the image quality, or direct numeric optimization of these metrics can be used to estimate phase error corrections. In more recent developments, several authors have proposed autofocus algorithms based on optimization of image quality metrics (IQM) [79–82], such as sharpness, contrast, and entropy. The performance of these methods is hampered by the ill-conditioned nature of the numerical optimization problem. The parameterized cost surface is often subject to many local minima, especially in cases of poor signal-to-noise ratio. These autofocus algorithms therefore require robust, and costly, initialization prior to the numerical optimization step.

The remainder of this section discusses each of the three families of autofocus algorithms in greater detail.

10.6.2 Subaperture-Based Autofocus (SBA)

10.6.2.1 SBA Overview The basis of subaperture-based autofocus is an assumption that the phase error function to be estimated ($\tilde{\phi}(k_y)$) may be represented by a polynomial of a degree low enough to permit a succinct piecewise-linear approximation. The effects of these segments of linear phase are then easily analyzed. A linear phase in the azimuth dimension of the phase history data results in an azimuth translation of the image. Assuming a piecewise-linear phase function, one may choose to correspondingly segment the phase histories into subapertures. Individually compressing each of these subapertures in azimuth (i.e., doing the azimuth FFT) results in a spoiled-resolution version of the full image that has been translated by an amount proportional to the linear phase over that segment. The amount of resolution spoiling is inversely proportional to the size of the segment that was processed relative to the full aperture size. If the subaperture is one-third the length of the full aperture, the azimuth resolution cell is three times wider.

Recall that Figure 10–51 is defocused by a quadratic phase error function in the azimuth dimension of the corresponding phase history data. Dividing the range-compressed phase histories into two equal segments results in two subapertures with identical but reversed phase errors. Applying the final azimuth Fourier transform to the first subaperture results in Figure 10–56*a*, which shows a lesser degree of defocus than Figure 10–51 but is shifted to the left. Similarly, compressing the second subaperture results in Figure 10–56*b*, which is identical to the first subaperture image but is shifted to the right. This effect may be most easily understood by approximating the quadratic phase error with two linear segments as shown in Figure 10–57. The negatively sloped linear phase in the first half of the aperture causes the corresponding image to

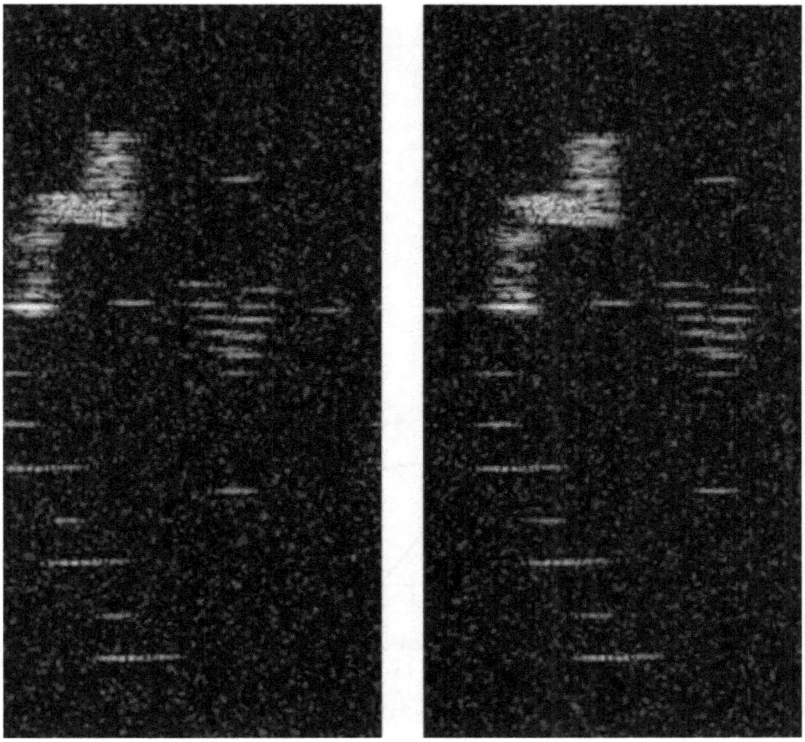

(a) (b)

Figure 10–56 Subaperture images each formed from half of the phase histories used to form the image of Figure 10–51. The first half of the quadratic phase error causes (*a*) to be shifted to the left, while the second half of the quadratic phase error causes (*b*) to be shifted to the right.

be translated to the left, and the positively sloped linear phase in the second half of the aperture causes a shift to the right.

Measuring the difference in the translations seen in the two subaperture images provides an estimate of the difference between the slopes of the two linear segments. Given an estimate of this difference, one may compute a quadratic phase correction to be applied to the range-compressed phase history data. Repeating this process—forming two subaperture images, measuring their relative shifts, and computing a new quadratic phase correction—will iteratively reduce the phase error residual to eventually produce a focused image.

This process may be adjusted to accommodate different degrees of polynomial phase errors by simply changing the number of subapertures that are formed through division of the full aperture. Images are formed from all of the phase history data subsets, and an ensemble of measured interimage shifts is used to compute a polynomial phase correction. The maximum degree of phase error that may be corrected is then theoretically equal to the number of processed subapertures. However, with each additional subdivision, the amount of signal data per subaperture is reduced, thereby diminishing the SNR of each subaperture image and degrading the quality of the interimage shift measurements. Sub-aperture based autofocus is usually only applied to low-order polynomial phase errors (e.g., fifth degree or less) for this reason.

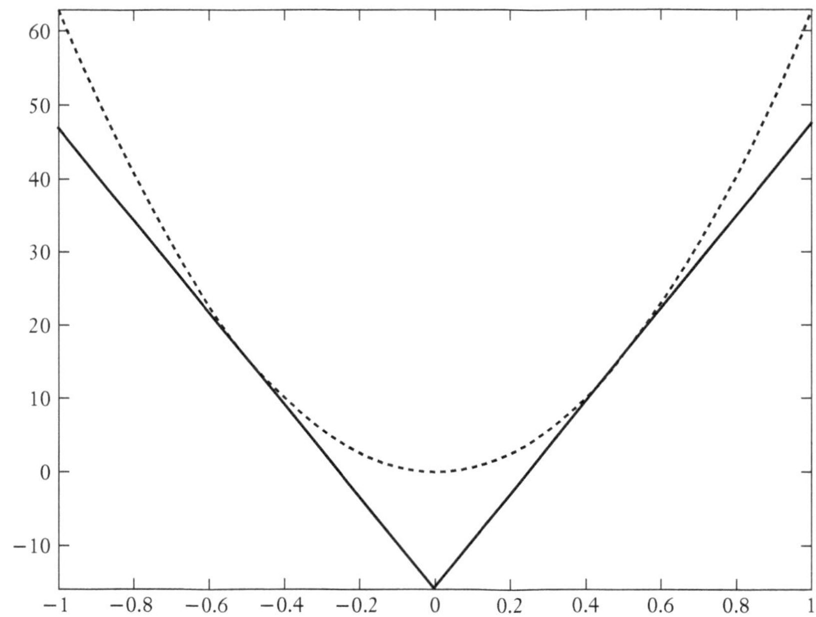

Figure 10–57 The simplest linear approximation of a quadratic function is two piecewise-linear segments. *Note:* the azimuth-dependent variable (*x*-axis) has been normalized to have extents [–1, 1].

10.6.2.2 SBA Theory The mathematical theory necessary for practical implementation of a subaperture-based autofocus algorithm is now developed. This requires development of analytic expressions for the shift of a subaperture image due to a given phase error function. In addition, as only relative translations between subapertures will be measured, a means for relating these shifts back to a phase error estimate is needed.

Derivation of the SBA algorithm begins under the assumption that range-compressed phase history data in the form of equation (10.153) are provided. The phase error $\tilde{\phi}(k_y)$ has an assumed polynomial degree of N_a, and thus the phase histories are divided into N_a subapertures of equal width W_a. If W_a does not divide evenly into the number of azimuth samples N_y, any remainder is discarded, though even division is preferred. The *p*th subaperture of data may then be represented as

$$\hat{s}_p(n_x,k_y)=s(n_x,k_y)\exp\{j\tilde{\phi}(k_y)\} \qquad k_y\in[k_p-W_a/2,k_p+W_a/2] \tag{10.154}$$

where k_p is the midpoint of that subaperture. A first-order Taylor series approximation of $\tilde{\phi}(k_y)$ over the *p*th subaperture is written as

$$\tilde{\phi}_p(k_y)\approx\tilde{\phi}(k_p)+\tilde{\phi}'(k_p)(k_y-k_p) \qquad k_y\in[k_p-W_a/2,k_p+W_a/2] \tag{10.155}$$

where $\tilde{\phi}'(k_y)$ denotes the first derivative of $\tilde{\phi}(k_y)$ with respect to k_y and evaluated at $k_y = k_p$. Only the linear portion of the approximated phase will result in an image translation. To determine the

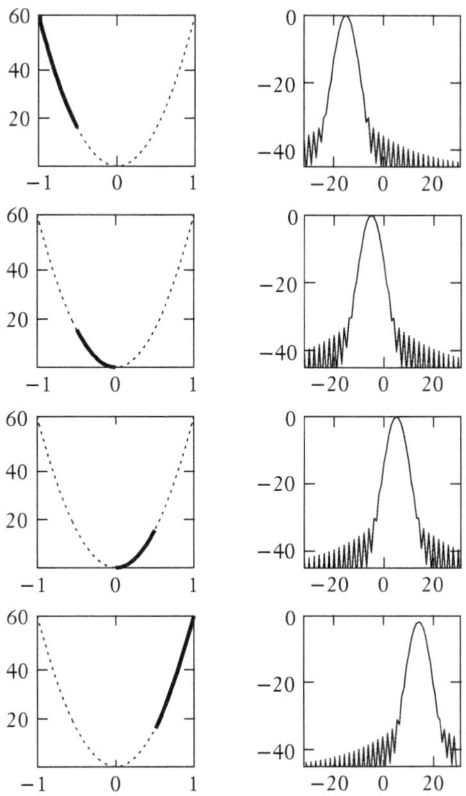

Figure 10–58 The defocus due to a phase error can be understood as the effect of a subaperture dependent translation. Subaperture images formed from different aperture segments are translated according to the segment of the phase error present in that subaperture, here shown as 1-D cuts in crossrange for a fixed range. *Note:* the azimuth-dependent variable (*x*-axis) has been normalized to have extents [–1, 1].

amount of image shift, equation (10.155) may be rewritten in the form

$$\tilde{\phi}_p(k_y) \approx \frac{2\pi k_y}{W_{\mathrm{DFT}}}\left(\tilde{\phi}'(k_p)\frac{W_{\mathrm{DFT}}}{2\pi}\right) + \tilde{\phi}(k_p) - \tilde{\phi}'(k_p)k_p \tag{10.156}$$

where W_{DFT} is the number of points in the discrete Fourier transform that is computed during azimuth compression of a subaperture. The time shifting property of the Fourier transform then implies that the pth subaperture image will be translated by a number of samples equal to

$$n_p = \tilde{\phi}'(k_p)\frac{W_{\mathrm{DFT}}}{2\pi} \tag{10.157}$$

provided that a DFT of length W_{DFT} is used for azimuth compression. A DFT of length $W_{\mathrm{DFT}} > W_{\mathrm{a}}$ is often employed to permit use of a fast Fourier transform.

Now consider two images drawn from different subapertures of the phase history data. As shown in Figure 10–58, each segment of the phase error function determines the shift of the

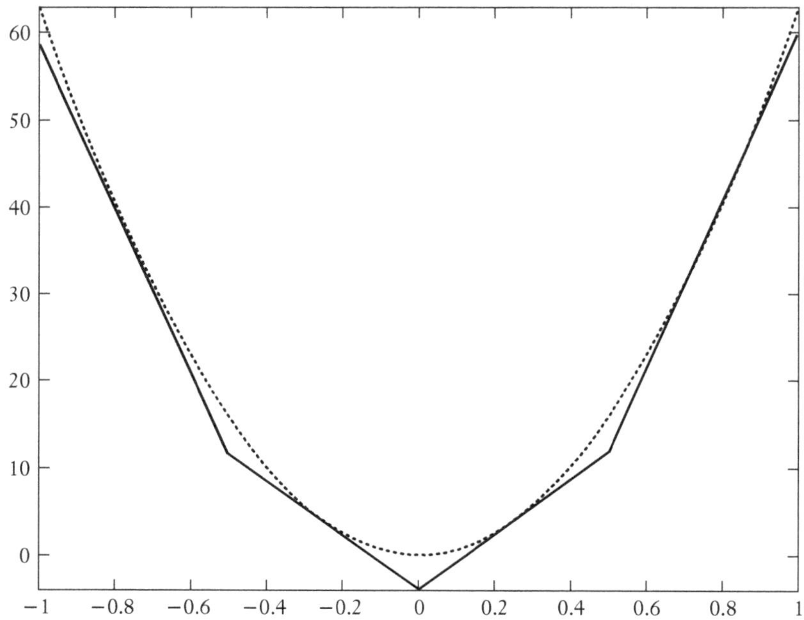

Figure 10–59 The subaperture translations observed in Figure 10–58 are proportional to the slope of the phase error function at each subaperture midpoint. *Note:* the azimuth-dependent variable (*x*-axis) has been normalized to have extents [–1, 1].

corresponding subaperture image. The amount of this shift is proportional to the slope of the phase error at each subaperture midpoint, as illustrated by Figure 10–59. Unfortunately, the azimuth shift of any given subaperture is not directly accessible. One must instead measure the translation of subaperture images relative to one another. Ideally, the measured translation of the pth image relative to the qth image would be (from equation (10.157))

$$n_{p,q} = \left(\tilde{\phi}'(k_p) - \tilde{\phi}'(k_q)\right)\frac{W_{\text{DFT}}}{2\pi} \tag{10.158}$$

An ensemble of these measurements may be used to compute a phase error estimate. By exploiting the assumed polynomial form of the phase error function, this calculation can be formulated as the solution of a straightforward linear least-squares problem.

Assuming a polynomial form of degree N_a, the phase error can be expressed as

$$\tilde{\phi}(k_y) = \sum_{l=0}^{N_a} a_l k_y^l \tag{10.159}$$

and has first derivative

$$\tilde{\phi}'(k_y) = \sum_{l=1}^{N_a} l a_l k_y^{l-1} \tag{10.160}$$

Thus, the difference of first derivatives in equation (10.158) is ideally

$$\begin{aligned}\tilde{\phi}'(k_p)-\tilde{\phi}'(k_q) &= \sum_{l=1}^{N_a} la_l k_p^{l-1} - \sum_{l=1}^{N_a} la_l k_q^{l-1} \\ &= \sum_{l=2}^{N_a} la_l \left(k_p^{l-1}-k_q^{l-1}\right) \\ &= \left[2\left(k_p^1-k_q^1\right)\cdots N_a\left(k_p^{N_a-1}-k_q^{N_a-1}\right)\right]\begin{bmatrix} a_2 \\ \vdots \\ a_{N_a}\end{bmatrix} \\ &\triangleq \underline{\delta}_{p,q}\cdot \underline{a}\end{aligned} \tag{10.161}$$

where the a_1 term has been dropped due to $k_p^{1-1} - k_q^{1-1} = 0$. In practice, the number of subapertures may be different than (but must be greater than or equal to) the degree of the phase error. This implies that N_a in equation (10.161) may be replaced by an integer less than the number of subapertures.

Combining the results in equations (10.158) and (10.161) to yield

$$n_{p,q} = \underline{\delta}_{p,q}\cdot \underline{a}\frac{W_{\text{DFT}}}{2\pi} \tag{10.162}$$

The observed relative shift between images p and q is directly related to the vector of polynomial coefficients $\underline{a}$, which are to be estimated. Measuring the relative translation of several pairs of subaperture images to obtain an ensemble of measurements, for example $\{n_{1,3}, n_{2,3}, n_{4,3}, n_{5,3}\}$ for a five-subaperture case, allows one to construct a matrix equation,

$$\underline{n} = \begin{bmatrix} n_{1,3} \\ n_{2,3} \\ n_{4,3} \\ n_{5,3}\end{bmatrix} = \frac{W_{\text{DFT}}}{2\pi}\begin{bmatrix} \underline{\delta}_{1,3}^T \\ \underline{\delta}_{2,3}^T \\ \underline{\delta}_{4,3}^T \\ \underline{\delta}_{5,3}^T\end{bmatrix}\underline{a} = \Delta\underline{a} \tag{10.163}$$

that may then be solved via least-squares

$$\underline{a} = (\Delta^T\Delta)^{-1}\Delta^T\underline{n} \tag{10.164}$$

The number of rows in Δ is equal to the number of relative shifts that were measured. The number of measured shifts should be greater than or equal to the number of quadratic and higher order polynomial coefficients to be estimated. In addition, each subaperture should be included in at least one measurement. Beyond these general guidelines, other specific measurement variations may yield better or worse performance on a case-by-case basis.

The estimated polynomial coefficients $\underline{a}$ may then be input to equation (10.159) to compute an estimate of the phase error function

$$\hat{\phi}(k_y) = \sum_{l=2}^{N_a} a_l k_y^l \tag{10.165}$$

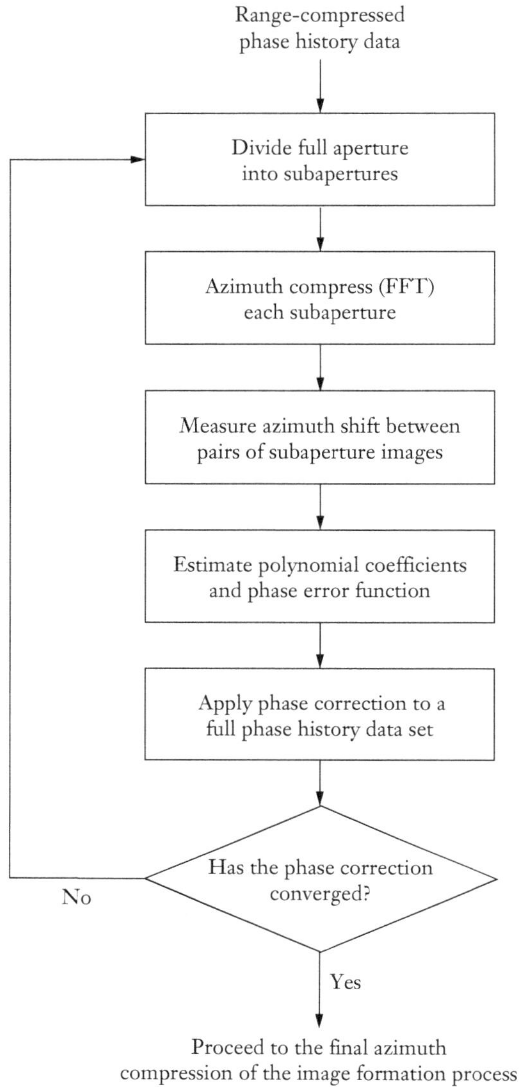

Figure 10–60 Flow chart for an iterative implementation of the SBA algorithm.

As the constant and linear terms, corresponding to a_0 and a_1, do not contribute to image defocus, they may be safely omitted from the sum. The SBA algorithm then applies this result to the full aperture phase history in equation (10.153) in the form of a phase correction,

$$\begin{aligned}\hat{s}_{\mathrm{AF}}(n_x, k_y) &= \hat{s}(n_x, k_y)\exp\{-j\hat{\phi}(k_y)\}\\ &= s(n_x, k_y)\exp\{j\tilde{\phi}(k_y) - j\hat{\phi}(k_y)\}.\end{aligned} \tag{10.166}$$

10.6.2.3 SBA Algorithm Subaperture-based autofocus is most often implemented as an iterative algorithm as shown in Figure 10–60. Although in some hybrid autofocus approaches, SBA may

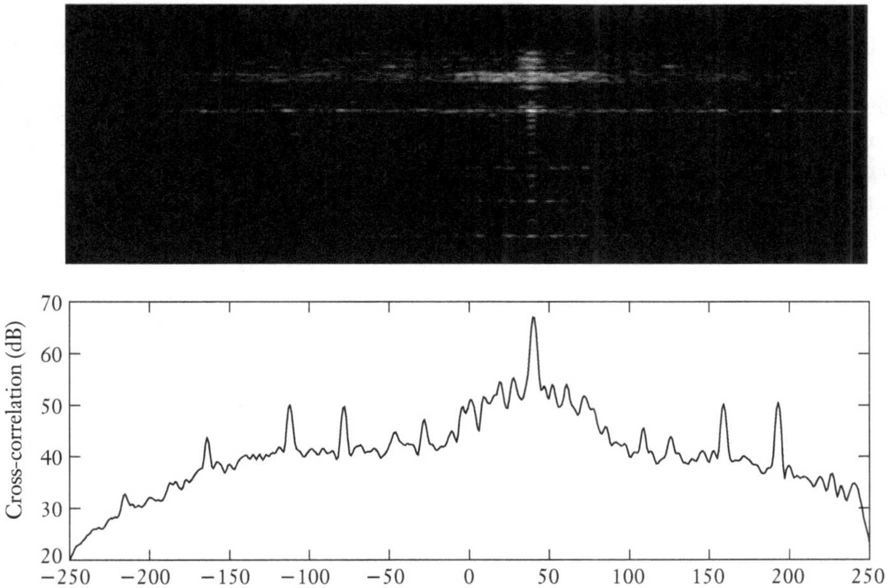

Figure 10–61 Relative subaperture image shifts may be computed by cross-correlating range gates (top), and then summing over range gates to produce an aggregate cross-correlation function (bottom). The index of the peak of this function is the estimated image translation.

be used noniteratively as a prefocus in preparation for a higher order autofocus algorithm such as PBA or IQM optimization, such methods would apply SBA a fixed, small number of times in order to remove any dominant lower order phase errors that would hinder less robust algorithms. The input to the SBA process is range-compressed phase history data. As mentioned in the introduction of this section, autofocus may be inserted as an intermediate step in the polar format image formation algorithm, taking place just before the final azimuth compression. Other image formation algorithms may require one to begin with a fully formed image, and then remove the azimuth compression.

Given range-compressed phase history data, and an assumed phase error polynomial degree N_a, one first divides the azimuth span of data into a number of subapertures greater than or equal to the assumed polynomial order, $N_a \geq N_p$. If the number of azimuth samples N_y is not evenly divisible by N_a, any remaining samples may be discarded. Discarding azimuth samples at either end of the phase history record necessitates that the estimated phase correction will be extrapolated to extend over the unsupported azimuth samples. As the SBA algorithm is meant to operate on lower order polynomial phase errors, discarding a few samples is not overly detrimental.

After segmenting the data, each subaperture is compressed in azimuth with a DFT (or FFT) of equal length W_{DFT}. Choosing W_{DFT} to be a power of 2 permits use of the FFT, which can greatly speed the algorithm's progress. The compressed subaperture images are then detected to yield intensity-only images. These intensity images are shifted in azimuth relative to one another depending on the slope of the phase error over each subaperture. Pairing these images and finding their relative shifts gives a relative measure of the phase error slope corresponding to each. The most common method for calculating the relative shift of an image pair is a one-dimensional cross-correlation of the range gates of both images. The Figure 10–61(top) shows this result when

Figures 10–56*a* and 10–56*b* are used as input. Note that each image was compressed in azimuth with a DFT of length W_{DFT} = 256. Figure 10–61(bottom) is the aggregate cross-correlation function computed by summing over all range gates.

The desired shift measurement is given by the index of the peak value of the aggregate cross-correlation. In Figure 10–61, the value is $n_{2,1}$ = 40. Given an initial number of azimuth samples of N_y = 256 divided into two apertures with midpoints k_1 = 64 and k_2 = 192, the corresponding row vector from equation (10.161) is a single element

$$\delta_{2,1} = 2 \cdot \left(k_2^1 - k_1^1\right) = 2(192 - 64) = 256 \tag{10.167}$$

Applications of the SBA algorithm involving more than two subapertures require several measurements of image shifts in order to fill out the $\underline{n}$ vector and Δ matrix in equation (10.163). Once completed, the estimated phase error coefficients are computed using equation (10.164), and the phase correction is computed using equation (10.165). Application of this correction, as in equation (10.166), will reduce the amount of residual phase error thus producing a better focused image. Following through with the above example, the computed quadratic phase error coefficient is

$$a_2 = \frac{2\pi}{W_{DFT}\delta_{2,1}} \cdot n_{2,1} = \frac{2\pi}{256 \cdot 256} \cdot 40 = 0.0038 \tag{10.168}$$

which may be verified as correct by computing the azimuth extent of the phase error

$$\hat{\phi}\left(k_y = \frac{N}{2}\right) = a_2\left(\frac{N}{2}\right)^2 = 0.0038 \cdot \left(\frac{256}{2}\right)^2 \approx 62 \tag{10.169}$$

This estimate compares well with the actual phase error shown in Figure 10–57, which has a maximum center-to-edge value of 62.8.

At the end of each SBA iteration, evaluation of a convergence criteria takes place. If convergence has been reached, the algorithm terminates, and the final azimuth image compression is performed. Otherwise, another iteration commences. Variations on the SBA algorithm may allow the number of subapertures to change between iterations depending on progress toward convergence or some set regimen. Measuring algorithm convergence may also be an area of contention between different implementations. Some may utilize a fixed number of iterations or may use the amplitude of the phase correction as an indicator. A more computationally intensive approach is to compute the final azimuth compression after each iteration. The resulting image can then be quantitatively evaluated through use of an image quality metric such as entropy, sharpness, or contrast (see Section 10.6.4).

10.6.3 Point-Based Autofocus

10.6.3.1 PBA Overview Optical images of space are frequently defocused due to uncompensated effects of the atmosphere. Volumes of turbulence and other inhomogeneities can function as lenses that bend the incoming phase front and prevent focusing in the image plane. Though the exact nature of the defocusing effect is unknown, focus can be achieved by exploiting known point sources in the far field, namely stars. A parametric representation of the blurring function can be adopted and estimated by determining the function parameters that result in stars appearing as point sources. Point-based autofocus algorithms operate on principles similar to those exploited by focusing techniques for astronomical observation.

It is well known that ideal or near-ideal point responses are commonly found in SAR images. Many geometric configurations can result in an isotropic far-field return, including corners,

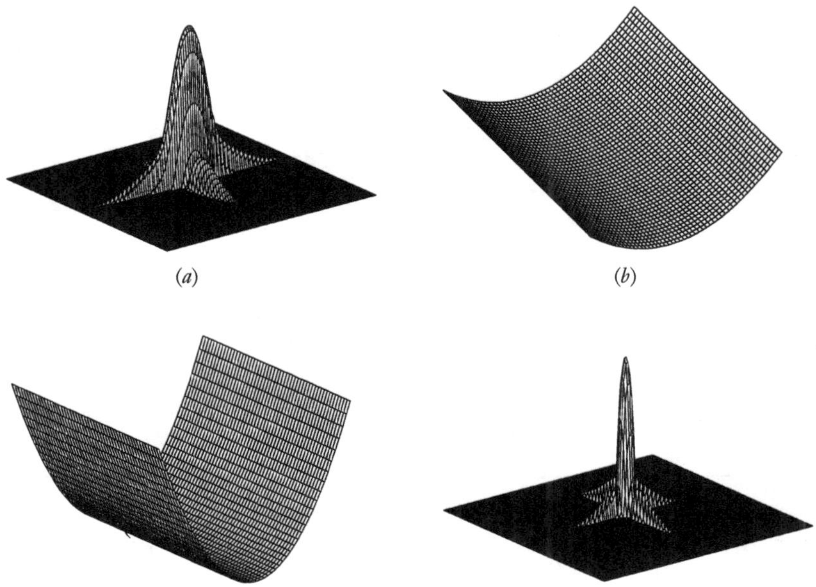

(*a*) (*b*) (*c*) (*d*)

Figure 10–62 A known isolated point scatterer with defocus (*a*) may be easily focused by taking the 2-D FFT of the image and extracting the phase function (*b*), removing the linear components via linear regression to leave the nonlinear phase error (*c*), and applying the result as a phase correction to give a focused image (*d*).

edges, and other distinct surface discontinuities. Moreover, to be a useful point source for a SAR autofocus algorithm, the scattering center need only be point-like in the azimuth dimension; the scattering response may be distributed in the range direction. If such a reflector can be isolated in range and azimuth, PBA algorithms are able to iteratively apply corrections to the phase history data until a focused point appears. This would be straightforward to accomplish if it was known *a priori* that a given data set contained a single isolated point scatterer. Then, one could simply extract the phase function of the phase history data, perform a linear regression and subtraction to leave only nonlinear terms, and then apply the residual as a phase correction. Figure 10–62 illustrates this concept with a phase error that is invariant in the range dimension, as is the case in SAR autofocus.

Without the benefit of this prior knowledge, PBA algorithms are required to extemporaneously find, isolate, and process point responses in SAR images. The ability of a given implementation to find and isolate such responses most often determines the effectiveness of the overall algorithm. Choices to be made in this process include what metrics by which to measure how point-like a given image response is, what size window function should be used to isolate said response, and how many responses should be extracted. Measurements for how isotropic a response is are typically computed in the phase history domain due to the defocusing effect of phase errors in the image domain. Common metrics [83, 84] might take into account amplitude flatness and phase linearity, which are measures commonly applied to the spectral features of digital filters, as well as amplitude contrast.

Point-like responses are typically extracted by centering each response in a given range gate within a one-dimensional window in azimuth. The size of the window determines the severity of the defocus that may be corrected. More severe blurring will require a wider window function,

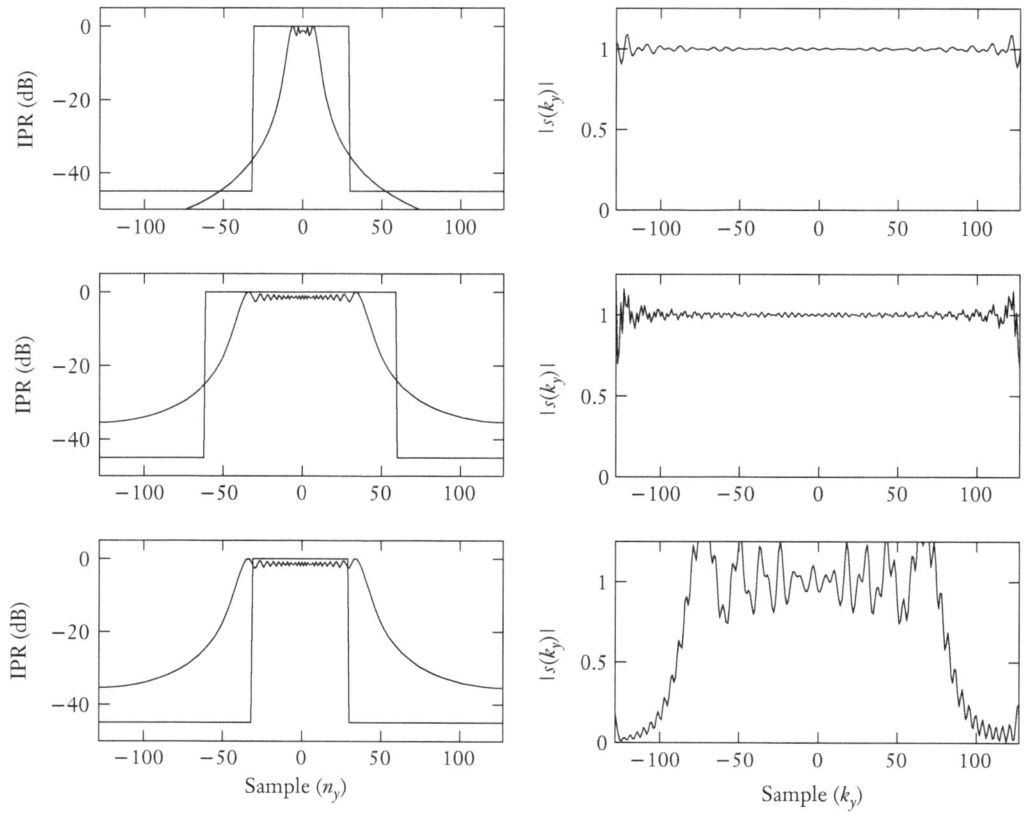

Figure 10–63 It is desirable to match the size of the PBA windowing function to the expected degree of defocus. The left column shows the defocused ideal point response (IPR) and an associated windowing function. The right column shows the resulting phase history data. It is unlikely that a good phase estimate will be extracted from a poorly windowed data set, such as that shown in the bottom two plots.

but will also allow for less isolation of point responses and will admit more noise into the phase estimation process. Choosing a window that is too small may prevent any successful autofocus if the blurred response is excessively truncated by the window. Figure 10–63 illustrates the effect of varied window sizes on different degrees of defocus. Given that the degree of phase error is unknown, one must strike a balance between allowing for greater defocus and providing more strict point response isolation. Some PBA implementations may even choose to use a variable window size [2], which is wide at the first iteration and becomes iteratively narrower as image focus is presumed to improve. Yet other approaches may seek to estimate the appropriate window size for each iteration.

The number of point responses to be isolated is also a subject of contention. Extracting more responses will provide more averaging in phase error estimates, while also increasing the possibility of admitting less favorable scattering centers. As strong point returns are often of greater image intensity, a common approach is to compute the total energy of each compressed range gate (i.e., all of the crossrange cells at a given range) as part of the algorithm initialization. Then, only a select few of the highest energy range gates are processed. One may then azimuth compress the selected range gates and select the strongest point return from each. Other algorithms [85, 86] choose to select multiple scatterers from each range gate in order to greater utilize the available data.

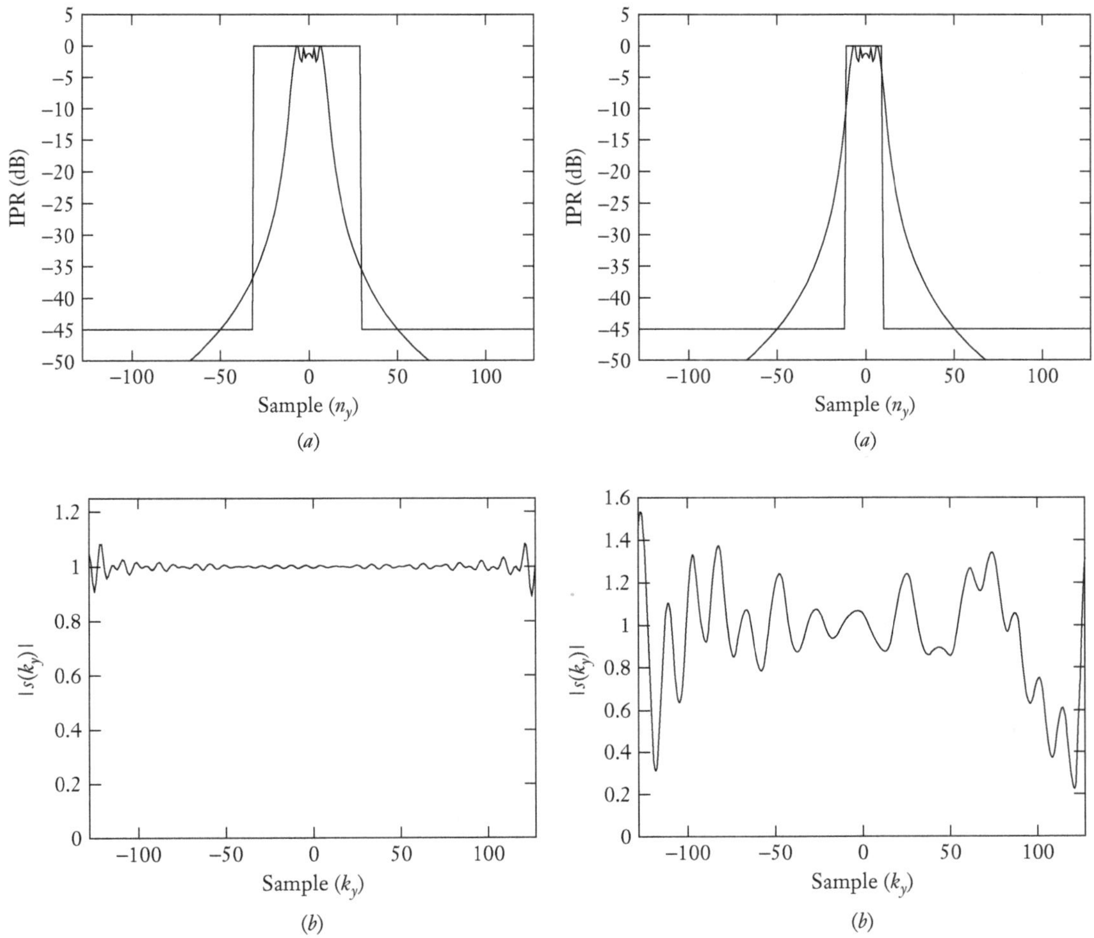

Figure 10–64 The left and right columns show two examples of PBA processing of the same phase error, but with different window widths. Selected point-like responses are first windowed (*a*), and then inverse-Fourier transformed to yield the signal data of a single point (*b*). A phase gradient (*c*) is then computed and integrated to yield a phase correction (*d*). Use of an appropriately sized window function produces a higher quality focused response (*e*); the dashed line shows the original defocused response.

10.6.3.2 PBA Theory The theoretical development of the basic PBA algorithm again begins with equation (10.153), and the following focuses on exploiting a single range gate, with a single value of n_x that will be omitted for brevity. In practice, many range gates undergo the same processing with their results being combined at the output. The first step in most PBA algorithms is to azimuth compress (i.e., Fourier transform with respect to k_y) the selected range gates, window the strongest response in each gate, and then inverse Fourier transform. The result,

$$\begin{aligned} \hat{s}^{(1)}(k_y) &= \hat{s}(n_x, k_y) * h(k_y) \\ &\approx A\exp\{j\omega_y k_y\}\exp\{j\tilde{\phi}(k_y)\} \end{aligned} \tag{10.170}$$

is a phase history record for an isolated point response, with an azimuth location proportional to w_y and with some distortion due to convolution with the transform of the windowing function, $h(k_y)$.

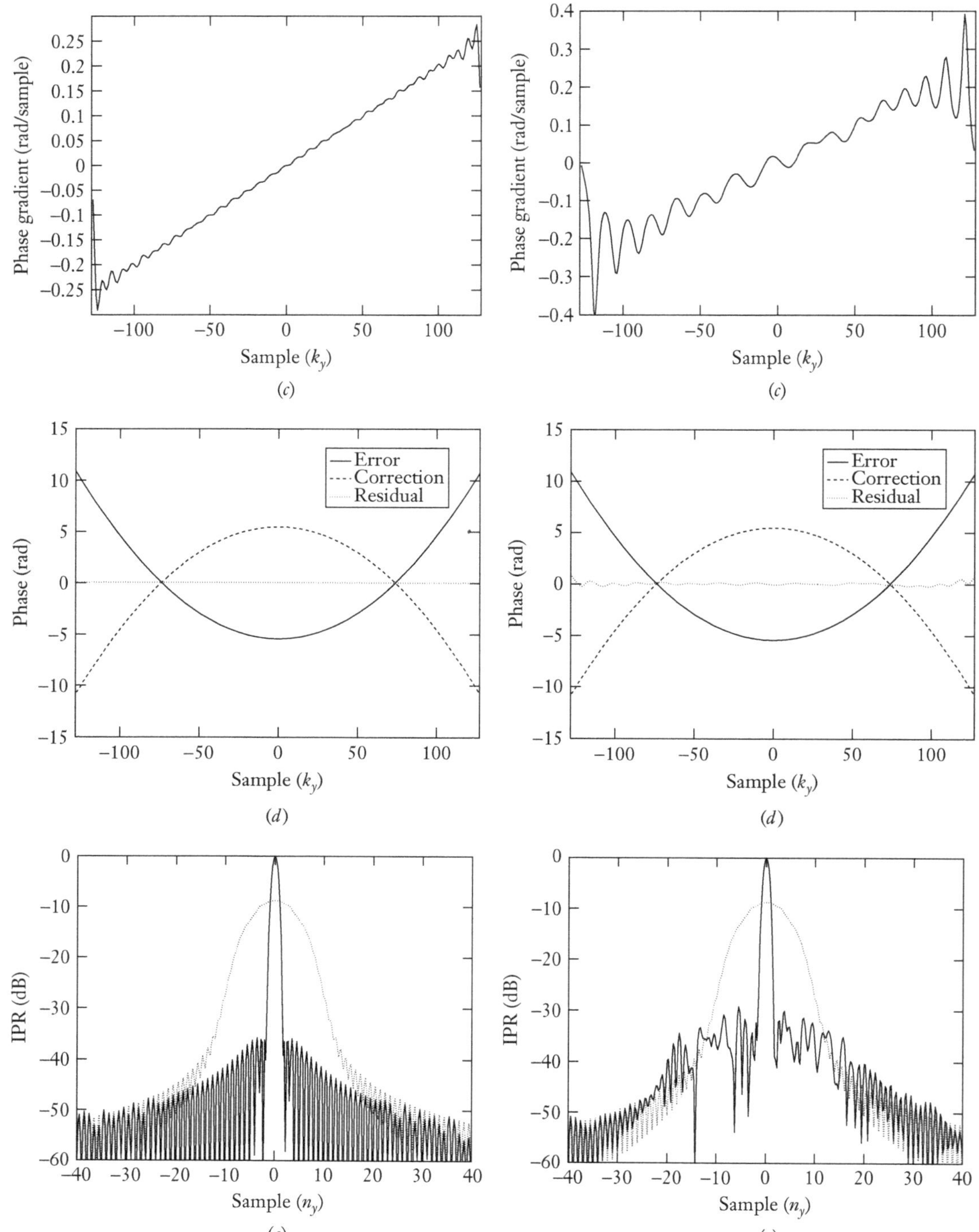

Figure 10–64 *(Continued)*

Providing that the width of the window function was set appropriately, the resulting distortion may be safely neglected. The left and right columns of Figure 10–64*a* and 10–64*b* shows selection of a sufficient and insufficient window size and the effect each has on the signal data. A sufficiently wide window allows the ripple in Figure 10–64*b* (left) to be ignored. A significant phase artifact accompanies the much larger amplitude ripple in Figure 10–64*b* (right).

One might at this point directly extract the error $\tilde{\phi}(k_y)$ by computing the unwrapped phase of equation (10.170) and removing the linear components via linear regression. However, it has been found that consistently better results are obtained by first computing a phase gradient [17, 74, 77], which may then be processed to yield a phase correction. Unwrapping of the phase gradient is more stable in particular, as the maximum sample-to-sample phase step is reduced. Constant phase terms, which are lost in the gradient computation, are not of interest. The most common method for calculation of the phase gradient is a simple first difference of phase, represented by

$$\begin{aligned}\hat{s}^{(2)}(k_y) &= \hat{s}^{(1)}(k_y)\cdot(\hat{s}^{(1)}(k_y-1))^* \quad k_y = 1,\ldots,N_y-1 \\ &= |A|^2 \exp\{j\omega_y\}\exp\{j\tilde{\phi}(k_y) - j\tilde{\phi}(k_y-1)\}\end{aligned} \tag{10.171}$$

where $(\cdot)^*$ denotes complex conjugation. From equation (10.171), the PBA process next extracts the phase gradient function and unwraps it. It is also common to compute and subtract off the mean of the phase gradient function, yielding

$$\tilde{\phi}'(k_y) = \tilde{\phi}(k_y) - \tilde{\phi}(k_y - 1) + \tilde{\phi}_0 \tag{10.172}$$

where $\tilde{\phi}_0$ is equal to the difference between w_y and the mean of the phase gradient. This step effectively cancels the linear phase shift due to the azimuth position of the extracted point response. The desired phase correction should have no linear components, and thus a zero-mean derivative, in order to prevent translation of the final image. The left and right columns of Figure 10–64*c* show the phase gradients resulting from each windowing selection. Note the high-frequency phase ripple in the right plot due to the additional clipping caused by the narrow window in Figure 10–64*a* (right).

Numerically integrating the phase gradient produces an estimated phase correction,

$$\hat{\phi}(k_y) = \begin{cases} 0, & k_y = 0 \\ -\sum_{l_y=1}^{k_y} \tilde{\phi}'(l_y), & 1 \le k_y < N_y \end{cases} \tag{10.173}$$

plotted in Figure 10–64*d* (left and right) along with the initial phase error and the residual. The original and corrected impulse responses are shown in Figure 10–64*e* (left and right). In Figure 10–64*e*, the IPRs are computed with a –35 dB Taylor weight and are twice oversampled, whereas Figure 10–64*a* uses uniform weighting and no oversampling. The degraded side lobes of Figure 10–64*c* (right) illustrate the deleterious effect of the phase ripple introduced by the narrow windowing function in Figure 10–64*a* (right).

10.6.3.3 PBA Algorithm Figure 10–65 shows an iterative implementation of the basic point-based autofocus algorithm. Many variations on this basic form have been developed with varied levels of success. As is the case with such things, it is the fine details that differentiate one algorithm from another and that can significantly affect performance. As mentioned in the overview, the method by which point-like responses are selected for processing is a key discriminator. Beyond that, most algorithms choose to repeat this selection step with each iteration of the algorithm, and may choose to progressively alter the selection criteria. This can be motivated by the fact that initial selections of point-like responses may be very poor due to the phase error itself. After some focusing, true ideal point scatterers will have more intense responses and will be more likely to be chosen.

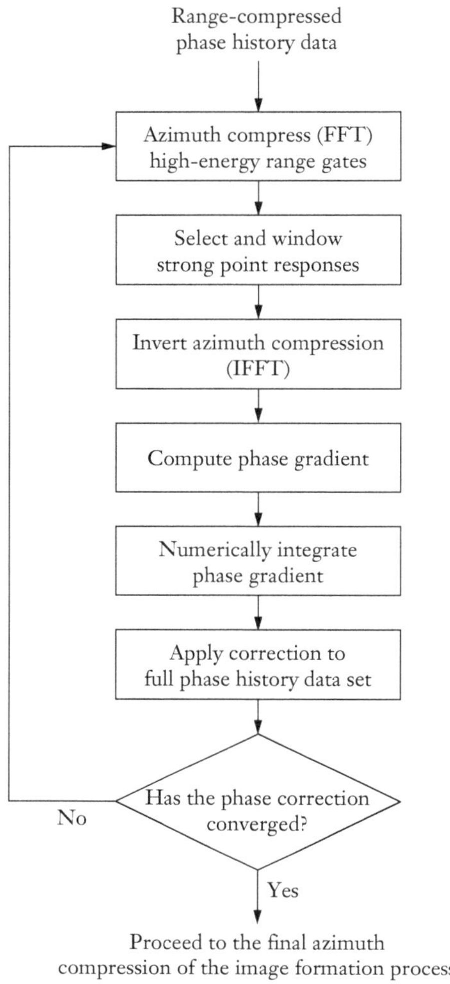

Figure 10–65 Flow chart for an iterative implementation of the PBA algorithm.

A second area of variation between PBA algorithms is the means by which multiple range gates are handled. Though the analysis in the previous section considered only a single scatterer in a single range gate, practical implementations will simultaneously process multiple scatterers in numerous range gates. All of the extracted responses are processed as described above, and the results are aggregated prior to application of a phase correction. Necessarily, the phase corrections estimated from all of the processed scatterers are combined through some form of averaging, a simple arithmetic mean of the phase correction functions being the most simplistic. Other algorithms have suggested using various criteria to weight the average, such as the amplitude contrast metric suggested in [83], in order to emphasize corrections of potentially higher quality. Averaging phase correction estimates from multiple point scatterers produces a higher quality autofocus solution.

Lastly, some variations on the PBA algorithms may choose to implement an additional smoothing step just prior to application of the phase correction. One of the principle advantages of point-based autofocus is that no parametric assumptions are required, thus allowing high-frequency phase errors to be compensated. However, one may improve the robustness of PBA by constraining it with a parametric requirement, enforced just before phase correction. Given a nonparametric phase estimate, consisting of N_y samples, a parametric solution may be determined through two common approaches: polynomial regression or low-pass filtering. As low-frequency phase error terms are typically dominant, the intention of both is to suppress high-frequency phase artifacts that may have originated in the PBA processing or that are due to measurement noise, and that do not require correction in the signal data. Use of polynomial regression implies that the desired phase correction is obtained by fitting a low-order polynomial to the nonparametric phase estimate. The correct degree of polynomial is typically predictable through some analysis of the motion compensation design. Similarly, if the envelope of the motion compensation residual is predicted analytically, a low-pass filter can be used to suppress components of the autofocus correction that lie outside those frequency boundaries.

10.6.4 Image Quality Autofocus (IQM)

10.6.4.1 IQM Overview Many methods for measuring SAR image quality are commonly used. Popular metrics include the ideal point response (IPR) 3-dB width, 30-dB width, peak side lobe ratio (PSLR), and integrated side lobe ratio (ISLR) [2]. These quantities tend to be subjective evaluators of image quality as they are biased by the quality of the "ideal" point chosen for processing. In practice, to measure the IPR statistics of a SAR system, one must provide a calibrated ideal point reflector in the scene. Analyzing the image of such a reflector will give a more reliable measurement of system performance. However, the presence of a strong ideal point scatterer in the scene may also allow artificially superior performance to be achieved, especially if a point-based autofocus algorithm is employed.

More objective measures for image quality, which do not rely on analysis of an ideal point scatterer, include image contrast, sharpness, and entropy. By definition, contrast metrics (see, e.g., [82, 87]) evaluate the difference in brightness between the light and dark areas of an image. A well-focused image will have brighter bright areas and darker dark areas, as phase errors tend to smear energy from bright regions into dark ones.

Sharpness metrics typically employ a function of the image intensity that will emphasize bright areas and de-emphasize dark areas. Parseval's theorem states that the 2-D sum over image intensity (see equation (10.179) below) will not change as a function of the azimuth phase error. However, a 2-D sum over a higher order power of image intensity, such as amplitude cubed or intensity squared, will provide sensitivity to the defocusing effects of azimuth phase errors.

The entropy metric may also be classified as a higher order function of image intensity. In the physical sciences, entropy is classically used to quantify energy dispersion. Viewing the image defocusing effect of phase errors as a type of energy dispersion implies that, in general, image entropy will increase as phase errors worsen.

Figure 10–66 illustrates the behavior of three image quality metrics as a function of the quadratic phase error amplitude. Figures 10–66*a*, 10–66*b*, and 10–66*c* show a point scatterer with 0 (0°), 5π (900°), and 10π (1800°) center-to-edge quadratic phase errors, respectively. Clearly, the image contrast decreases as the phase error worsens, and one may qualitatively observe a greater degree of energy dispersion and a weakening of the peak point response. Figures 10–66*d*, 10–66*e*, and 10–66*f* plot the image entropy, sharpness, and contrast as a function of the center-to-edge phase error. The image entropy increases as the center-to-edge phase error increases, while the sharpness and contrast decrease. If the image only contains a single point return, that point response must have high SNR, as shown in Figure 10–66, in order to successfully drive the image quality metric. However, in general, SAR images will contain many returns of lower intensity, with an aggregate power sufficient to drive an IQM optimization.

The general trends of the plots shown in Figure 10–66 suggest that these metrics are good indicators of the degree of image focus. As such, they may be used to complement the SBA and

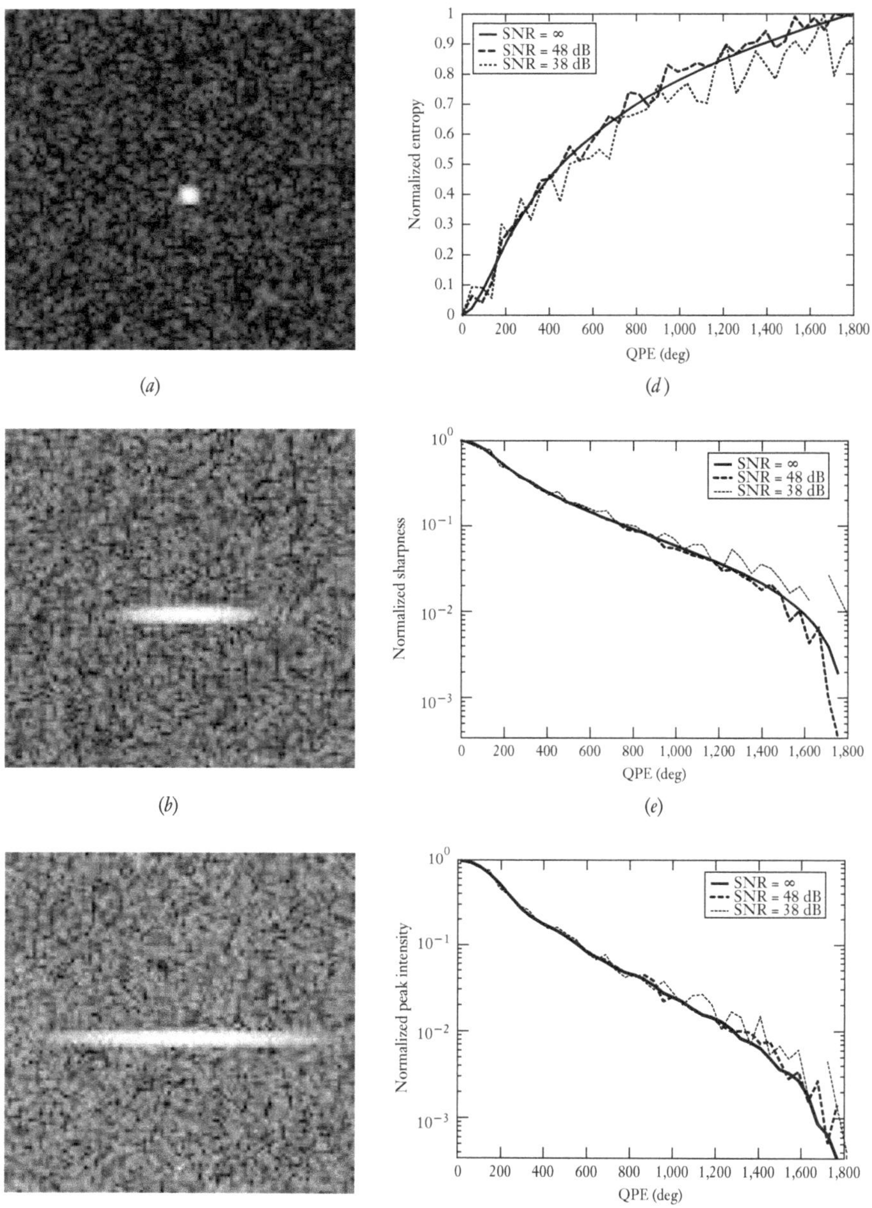

Figure 10–66 The behavior of image quality metrics as a function of the quadratic phase error amplitude. The left column shows images with (*a*) 0 (0°), (*b*) 5π (900°), and (*c*) 10π (1800°) of the quadratic phase error. The right column plots examples of (*d*) entropy, (*e*) sharpness, and (*f*) contrast image quality metrics.

PBA algorithms that were discussed earlier. With each iteration of these algorithms, an IQM may be computed to quantify the degree of focus that has been achieved. One may even design stopping conditions based on these measures, such as terminating iteration when the IQM fails to significantly improve.

A more direct application of an IQM is to autofocus an image by numerically optimizing that quantity. This requires a parametric definition of the phase correction function. Then, steepest descent/ascent algorithms may be employed to find the parameter values that minimize the image entropy, or maximize the image sharpness or contrast. The performance of these approaches can be highly dependent on the image content, the signal-to-noise ratio, and the initial conditions of the search. Image content determines the overall behavior of the IQM function, as scenes with bright localized returns will tend to have greater sensitivity to defocus. As seen in Figures 10–66*d*, 10–66*e*, and 10–66*f*, the smoothness of the IQM cost functions is degraded by a reduced SNR, and proper initialization will help avoid convergence to a local minimum.

10.6.4.2 IQM Theory Measures of contrast, sharpness, and entropy have been known to take many forms, and each expression will tend to have different characteristics depending on scene content and the phase error to be corrected. Three contrast metrics in particular are suggested in [82]:

1. the maximum value of the image amplitude

$$C_1 = \max_{n_x, n_y} |P(n_x, n_y)|, \tag{10.174}$$

2. the ratio of the standard deviation to the mean of the image amplitude

$$C_2 = \frac{\sqrt{A\{[|P(n_x, n_y)| - A\{|P(n_x, n_y)|\}]^2\}}}{A\{|P(n_x, n_y)|\}} \tag{10.175}$$

3. the ratio of the standard deviation to the mean of the image intensity

$$C_2 = \frac{\sqrt{A\{[|P(n_x, n_y)|^2 - A\{|P(n_x, n_y)|^2\}]^2\}}}{A\{|P(n_x, n_y)|^2\}} \tag{10.176}$$

In these definitions, A{·} is the arithmetic mean operator

$$A\{Y(n_x, n_y)\} = \frac{1}{N_x N_y} \sum_{n_x=0}^{N_x-1} \sum_{n_y=0}^{N_y-1} Y(n_x, n_y) \tag{10.177}$$

Sharpness metrics tend to take on a more general form, and typically operate on the image intensity

$$I(n_x, n_y) = |P(n_x, n_y)|^2 \tag{10.178}$$

Parseval's theorem states that the total image energy, computed as the 2-D sum over intensity

$$E = \sum_{n_x=0}^{N_x-1} \sum_{n_y=0}^{N_y-1} I(n_x, n_y) \tag{10.179}$$

will remain constant, independent of any defocusing due to phase errors in the phase history domain. However, computing this same sum over the image intensity taken to some power not equal to 1 will provide sensitivity to image defocus. Generally, these IQMs may be expressed as

$$S_p = \sum_{n_x=0}^{N_x-1} \sum_{n_y=0}^{N_y-1} I(n_x, n_y)^p \qquad p \neq 1 \tag{10.180}$$

Entropy metrics can be viewed as a variation on the sharpness metric wherein multiple powers of the image intensity are employed. Image entropy is usually computed as

$$H = \sum_{n_x=0}^{N_x-1}\sum_{n_y=0}^{N_y-1} I(n_x, n_y)\log I(n_x, n_y) \tag{10.181}$$

All of the above functions share certain commonalities in that they are functions of the image amplitude or intensity and they require a sum or average to be computed over all image pixels.

The authors of [81] have most clearly illustrated the process of autofocus through generalized image quality metric optimization. Following the derivation given in [81], these algorithms typically begin with range-compressed phase history data represented by equation (10.153). An inverse Fourier transform of equation (10.153) with respect to the slow time dimension k_y yields a phase error-corrupted image, written as

$$P_{\mathrm{d}}(n_x, n_y) = F_{k_y}^{-1}\{\hat{s}(n_x, k_y)\} \tag{10.182}$$

Note that the use of an *inverse* Fourier transform, instead of a *forward* Fourier transform, effectively just frequency-reverses the result of crossrange compression, and this approach is used here in order to remain consistent with the work of [81]. Application of a phase correction $\hat{\phi}(k_y)$ gives a corrected signal history

$$\hat{s}_{\mathrm{AF}}(n_x, k_y) = \hat{s}(n_x, k_y)e^{-j\hat{\phi}(k_y)} \tag{10.183}$$

and a corresponding corrected image

$$P(n_x, n_y) = F_{k_y}^{-1}\{\hat{s}_{\mathrm{AF}}(n_x, k_y)\} \tag{10.184}$$

As described above, several metrics designed to measure the quality of an image have been shown to correlate well with the amount of uncorrected phase error that is present in the phase history data. In keeping with the notation of [81], these metrics are generally computed as

$$S_{\Gamma} = \sum_{n_x, n_y} w(n_x)\Gamma[I(n_x, n_y)] \tag{10.185}$$

where $\Gamma(\cdot)$is an arbitrary nonlinear function and $I(\cdot)$ is the intensity image. The function $w(x)$ applies a range-dependent weighting, which is typically chosen to emphasize selected range gates, such as those with a greater share of the total image energy. Comparison with equation (10.181) shows that for the entropy metric

$$\Gamma[I(n_x, n_y)] = I(n_x, n_y)\log I(n_x, n_y) \tag{10.186}$$

and equation (10.180) similarly implies that for sharpness metrics

$$\Gamma[I(n_x, n_y)] = I(n_x, n_y)^p \qquad p \neq 1 \tag{10.187}$$

In [82], equivalent expressions for contrast measures were derived that may be computed in forms analogous to equation (10.187).

Iterative numerical optimization of S_{Γ} will yield a focused image. This process is facilitated by the availability of analytic expressions for the gradient of the generalized sharpness metric S_{Γ}. From [81], the gradient of a generalized IQM with respect to each sample of the phase correction function is given by

$$\frac{\partial S_\Gamma}{\partial \hat{\phi}(k_y)} = \sum_{n_x, n_y} w(n_x) \frac{\partial \Gamma[I(n_x, n_y)]}{\partial I(n_x, n_y)} \frac{\partial I(n_x, n_y)}{\partial \hat{\phi}(k_y)}$$

$$= \frac{2}{N_y} \sum_{n_x} w(n_x) \cdot \Im\left\{ \hat{s}_{\mathrm{AF}}(n_x, k_y) \left(P(n_x, n_y) \frac{\partial \Gamma[I(n_x, n_y)]}{\partial I(n_x, n_y)} \right)^* \right\} \tag{10.188}$$

where $\Im(\cdot)$ indicates taking the imaginary part of the argument. As discussed in the next subsection, this expression may be exploited in algorithm implementations that use parametric representations for $\hat{\phi}(k_y)$. Note that the partial derivative of $\Gamma[I]$ with respect to I is straightforward to derive from equations (10.186) and (10.187).

10.6.4.3 IQM Algorithm Implementation of IQM autofocus algorithms is relatively straightforward. To allow for numerical optimization, one must first choose a parametric representation for the phase correction function. The simplest approach is to allow each sample of $\hat{\phi}(k_y)$ to be an independent parameter, such that $\hat{\phi}(k_y) = a_{k_y} = a_m$. In that case, the gradient expression in equation (10.188) is directly employed in steepest descent/ascent algorithms. However, an expansion in terms of an orthogonal basis set (e.g., Legendre polynomials), written as

$$\hat{\phi}(k_y) = \sum_{m=1}^{M} a_m L_m(k_y) \tag{10.189}$$

is most common. The authors of [74] provide an alternative expression for the gradient of S_Γ when the phase correction is represented by a parametric expression in the form of equation (10.189). The partial derivative of a generalized sharpness metric with respect to each coefficient of this expansion may be computed as

$$\frac{\partial S_\Gamma}{\partial a_m} = \sum_{k_y} L_m(k_y) \frac{\partial S_\Gamma}{\partial \hat{\phi}(k_y)} \tag{10.190}$$

Autofocus is then achieved by iteratively optimizing S_Γ over the search space $\{a_m | m = 1, \ldots, M\}$. The optimal phase correction is thus defined by equation (10.189), given coefficients that satisfy

$$\{\hat{a}_m\} = \arg \max_{\{a_m\}} S_\Gamma \tag{10.191}$$

In practice, this multidimensional optimization is an ill-conditioned problem. Depending on the image content and SNR, the entropy cost surface may be highly irregular and may contain many local minima.

Often, to avoid convergence to a nonglobal minimum, one must map out the cost surface by gridding the parameter space with initial guesses. The computational complexity of this initialization is $O(N^M)$, where M is the number of parameters to be initialized and N is the number of sample points tested for each parameter. As an example, consider initialization of the search space defined by a second-order polynomial, with a desired $\pm 40\pi$ pull-in phase and a sample spacing of π, or by a fourth-order polynomial with the same requirements. This implies that the number of parameters is $M = 1$ for the second-order polynomial and $M = 3$ for the fourth-order polynomial, as constant and linear terms are neglected, and that the number of sample points is $N = 81$ for both. In comparing the two scenarios, it is seen that the complexity required for the second-order polynomial is $O(N^M = 81^1 = 81)$ and for the fourth-order polynomial is $O(N^M = 81^3 = 531,441)$, a drastic difference. In addition, a good initial guess does not ensure fast and accurate convergence on a cost surface that may be ill-behaved.

More recent implementations of IQM optimization [88] have employed hierarchical methods. Starting with a low-order phase correction, optimizing this correction, and then iterating with higher order phase corrections has been shown to improve the robustness of IQM autofocus algorithms to low SNR and varied scene content. This behavior is driven by the prevalence of low-order phase errors, primarily quadratic and cubic, in most measured SAR images. By first correcting the lower order errors, the effective SNR of the image is improved enough to gain sensitivity to the side lobe degradation caused by higher order phase errors. Hierarchical approaches also tend to reduce optimization time, without the use of a costly, brute force initialization.

10.7 THREE-DIMENSIONAL SURFACE RECONSTRUCTION

Assuming that bistatic SAR image formation may be accomplished successfully, one might consider the possibilities offered by the availability of multiple images of the same scene, but at different aspect angles. In point of fact, using multiple receiving platforms with a single transmitting platform to form a multistatic system and thus multiple bistatic images has been suggested [55, 56, 89, 90]. Each bistatic image would be a 2-D representation of the 3-D scene of interest. The data collection geometry associated with each image determines the manner in which the 3-D world is projected into the resulting image. With the availability of images from multiple bistatic look angles, three-dimensional reconstruction of surfaces within the imaged scene may be possible.

In monostatic SAR one may increase the dimensionality of one's data collection manifold in frequency space by employing multiple synthetic apertures. This has been accomplished through two different methods: interferometric SAR [91–93] and stereo SAR [94, 95]. In interferometric SAR, two receiving antennas, separated by a known baseline distance, are placed on the same platform. The spatial diversity offered by the two antennas effectively creates a monostatic data collection manifold consisting of two parallel planes. A point scatterer projects slightly differently onto the two planes, thus disambiguating its 3-D position. Stereo SAR involves deploying two monostatic SAR platforms to interrogate and form images of the same scene of interest, but from different aspect angles. Measuring the layover differences between the two images allows extraction of 3-D information.

As shown in [96], analogous techniques may be applied in bistatic SAR. Two antennas may be placed on a single receiving platform to allow bistatic interferometric processing [97]. One may also deploy multiple transmitters or receivers to form a multistatic system, thereby allowing use of stereo SAR techniques [98]. The stereo SAR approach is more attractive in the bistatic arena than in the monostatic. In deploying a second monostatic SAR platform, the cost of the assets at risk will often be doubled. However, as bistatic SAR systems may cast relatively low-cost unmanned aerial vehicles in the roles of receiving platforms, fielding one or more additional receivers may come at reduced cost. This is especially true in the case where all of the receiving platforms are operating passively.

As will be shown, monostatic stereo techniques can be modified for application to images created by a multistatic system, incorporating multiple transmitting and/or receiving platforms, and interferometric techniques may be applied to a pair of bistatic SAR images, formed by a receiver equipped with two or more antenna.

In a bistatic SAR geometry involving a single transmitter and receiver, the planar data collection manifold that results from strictly straight and level flight paths fails to unambiguously encode three-dimensional scene information. Design of nonlinear receiver flight paths, to warp the data collection manifold, would cause point scatterers to be projected onto a nonplanar surface. If this projection is sufficiently nonlinear, unambiguous recording and reconstruction of 3-D scatterer position information is possible.

However, this approach presents serious yet surmountable practical implementation issues, motion compensation being at the forefront. The out-of-plane maneuvers required to record high-fidelity 3-D information could severely tax the capabilities of the receiving platform's inertial navigation unit, thus introducing uncompensated azimuth phase errors and image defocus. In addition, removal of these errors in postprocessing is complicated by the flight path nonlinearities that were intentionally added. Scatterers that do not lie within the image formation plane will be defocused by these nonlinearities, and the degree of defocus will depend on the out-of-plane height of each individual scatterer. The spatial dependence of these defocusing effects will hamper the application of traditional monostatic SAR autofocus algorithms.

This section develops the theory necessary for three-dimensional surface reconstruction using bistatic SAR interferometry and bistatic stereo SAR. An outline of the remainder of this section is as follows. Section 10.7.1 briefly reviews the previous model for the bistatic phase history data collection and bistatic SAR image formation. Based on these models, the means by which three-dimensional point scatterers are projected into a two-dimensional bistatic SAR image is reintroduced. This projection is commonly known as layover. Section 10.7.2 derives an interferometric processing algorithm for height estimation from a pair of registered bistatic SAR images, which were formed by two closely spaced apertures. Section 10.7.3 develops an algorithm that estimates the height of surfaces in a scene, based on the difference in layover between two bistatic images formed at different look angles.

10.7.1 Bistatic Phase History Data

As in earlier sections, the bistatic SAR data collection geometry shown in Figure 10–4 is considered. The center of the scene to be imaged is located at the origin of coordinates, and the ground plane is the x–y plane. A scatterer within that scene is located at $\underline{r}_m = (x_m, y_m, z_m)$. At a given time τ, the location of the transmitter is $\underline{r}_t(\tau) = (x_t(\tau), y_t(\tau), z_t(\tau))$, and the location of the receiver is $\underline{r}_r(\tau) = (x_r(\tau), y_r(\tau), z_r(\tau))$.

As the transmitter and receiver traverse their respective respective flight paths, they interrogate the scene at a discrete number of instances in slow time, given by $\tau_i \in [-T/2, T/2]$. The transmitter radiates a pulse at the time of each slow time sample, and the receiver observes the returns from the scene. The received phase history data span the frequency range $f \in [f_0, f_0 + B]$, where f_0 and B represent the initial frequency and bandwidth of the transmitted pulses, respectively. The collected data may be represented by

$$S(f_i,\tau_k)=\sum_m A_m \exp\left\{-j2\pi f_i \frac{\Delta R_m(\tau_k;x_m,y_m,z_m)}{c}\right\}+w(f_i,\tau_k) \tag{10.192}$$

where

$$\begin{aligned}\Delta R_m(\tau_k;x_m,y_m,z_m)\approx&\sqrt{(x_t(\tau_k)-x_m)^2+(y_t(\tau_k)-y_m)^2+(z_t(\tau_k)-z_m)^2}\\&-\sqrt{(x_t(\tau_k))^2+(y_t(\tau_k))^2+(z_t(\tau_k))^2}\\&+\sqrt{(x_r(\tau_k)-x_m)^2+(y_r(\tau_k)-y_m)^2+(z_r(\tau_k)-z_m)^2}\\&-\sqrt{(x_r(\tau_k))^2+(y_r(\tau_k))^2+(z_r(\tau_k))^2}\end{aligned} \tag{10.193}$$

and the corresponding matched filter for image formation is

$$P(x,y)=\frac{1}{N_f N_\tau}\sum_{i=1}^{N_f}\sum_{k=1}^{N_\tau} S(f_i,\tau_k)\exp\left\{-j\frac{2\pi f_i}{c}\Delta R(\tau_k;x,y)\right\} \tag{10.194}$$

where

$$\begin{aligned}\Delta R_m(\tau_k;x,y) \approx & \sqrt{(x_t(\tau_k)-x)^2+(y_t(\tau_k)-y)^2+z_t(\tau_k)^2} \\ & -\sqrt{(x_t(\tau_k))^2+(y_t(\tau_k))^2+(z_t(\tau_k))^2} \\ & +\sqrt{(x_r(\tau_k)-x)^2+(y_r(\tau_k)-y)^2+z_r(\tau_k)^2} \\ & -\sqrt{(x_r(\tau_k))^2+(y_r(\tau_k))^2+(z_r(\tau_k))^2}\end{aligned} \tag{10.195}$$

to form a ground plane image ($z = 0$). Recall that N_τ is the number of samples in azimuth or slow time, N_f is the number of frequencies sampled at each sampling point in azimuth, and A_m is the complex-valued reflectivity of a scatterer in the scene.

From equations (10.192) and (10.194), the following observations are made. The (x, y, z) location of each scatterer is encoded in the observed differential range $\Delta R_m(\tau)$ of that scatterer. However, the matched filter in equation (10.194) used in image formation assumes that all scatterers lie on the ground plane, thus implying that a projection takes place in the image formation process.

To determine the nature of this projection, the motion measurement errors are all assumed to be zero, and the same set of far-field approximations used in Section 10.5.1 are employed. The substitutions given in Table 10–4 allow one to approximate the differential range function in equation (10.118) with a linear function of slow time,

$$\begin{aligned}\Delta R_m(\tau_k) \approx & -x_m\frac{\bar{x}_t+v_{xt}\tau}{r_t}\frac{r_t}{R_t}-y_m\frac{\bar{y}_t+v_{yt}\tau}{r_t}\frac{r_t}{R_t}-z_m\frac{\bar{z}_t+v_{zt}\tau_k}{R_t} \\ & -x_m\frac{\bar{x}_r+v_{xr}\tau}{r_r}\frac{r_r}{R_r}-y_m\frac{\bar{y}_r+v_{yr}\tau}{r_r}\frac{r_r}{R_r}-z_m\frac{\bar{z}_r+v_{zr}\tau_k}{R_r} \\ = & -\left[x_m\left(\frac{\bar{x}_t}{R_t}+\frac{\bar{x}_r}{R_r}\right)+y_m\left(\frac{\bar{y}_t}{R_t}+\frac{\bar{y}_r}{R_r}\right)+z_m\left(\frac{\bar{z}_t}{R_t}+\frac{\bar{z}_r}{R_r}\right)\right] \\ & -\left[x_m\left(\frac{v_{xt}}{R_t}+\frac{v_{xr}}{R_r}\right)+y_m\left(\frac{v_{yt}}{R_t}+\frac{v_{yr}}{R_r}\right)+z_m\left(\frac{v_{zt}}{R_t}+\frac{v_{zr}}{R_r}\right)\right]\tau_k \\ \triangleq & \ \beta_0+\beta_1\tau_k\end{aligned} \tag{10.196}$$

Equation (10.196) describes a transformation that relates the linear approximation of the differential range $\Delta R(\tau) = \beta_0 + \beta_1\tau$ to the actual location of a scatterer in the scene. This transformation may be written as

$$\begin{bmatrix}\beta_0\\ \beta_1\end{bmatrix}=\begin{bmatrix}-\frac{\bar{x}_t}{R_t}-\frac{\bar{x}_r}{R_r} & -\frac{\bar{y}_t}{R_t}-\frac{\bar{y}_r}{R_r} & -\frac{\bar{z}_t}{R_t}-\frac{\bar{z}_r}{R_r}\\ -\frac{v_{xt}}{R_t}-\frac{v_{xr}}{R_r} & -\frac{v_{yt}}{R_t}-\frac{v_{yr}}{R_r} & -\frac{v_{zt}}{R_t}-\frac{v_{zr}}{R_r}\end{bmatrix}\begin{bmatrix}x_m\\ y_m\\ z_m\end{bmatrix}\triangleq Q\begin{bmatrix}x_m\\ y_m\\ z_m\end{bmatrix} \tag{10.197}$$

In contrast, the image formation process in equations (10.194) and (10.195) implicitly assumes that scatterers are located on the ground plane ($z = 0$), and thus defines a transformation

$$\begin{bmatrix} \hat{x}_m \\ \hat{y}_m \end{bmatrix} = \begin{bmatrix} -\frac{\bar{x}_t}{R_t} - \frac{\bar{x}_r}{R_r} & -\frac{\bar{y}_t}{R_t} - \frac{\bar{y}_r}{R_r} \\ -\frac{v_{xt}}{R_t} - \frac{v_{xr}}{R_r} & -\frac{v_{yt}}{R_t} - \frac{v_{yr}}{R_r} \end{bmatrix}^{-1} \begin{bmatrix} \beta_0 \\ \beta_1 \end{bmatrix} \triangleq P \begin{bmatrix} \beta_0 \\ \beta_1 \end{bmatrix} \tag{10.198}$$

relating the linear approximation of the differential range to an (x, y) pixel location. It is at this approximate pixel location $(\hat{x}_m, \hat{y}_m)$ that a point response will appear, corresponding to the original scatterer located at (x_m, y_m, z_m). Thus, using equations (10.197) and (10.198), the linear operation that projects three-dimensional point scatterers into a 2-D image is defined as

$$\begin{bmatrix} \hat{x}_m \\ \hat{y}_m \end{bmatrix} = PQ \begin{bmatrix} x_m \\ y_m \\ z_m \end{bmatrix} \tag{10.199}$$

Note that if $\bar{z}_t = \bar{z}_r = v_{zt} = v_{zr} = 0$ then $PQ = [I \quad 0]$ and scatterers project vertically onto a ground plane image.

10.7.2 Bistatic Interferometric SAR

10.7.2.1 Theory Now, consider a bistatic interferometric system involving a transmitting platform and a single receiving platform equipped with two physically separated receiving antennas. This allows formation of two coherent bistatic SAR images, one from the phase history data from each receiving antenna. The differences between these two images, caused by the physical separation of the receiving antennas, allows height estimation to be accomplished interferometrically. Given the geometric configuration of the bistatic system, one can compute the projection matrices from equations (10.197)–(10.199) for the two bistatic apertures; then, the ideal layover in each image due to a scatterer height of 1 m is

$$\begin{bmatrix} \Delta x_{ml} \\ \Delta y_{ml} \end{bmatrix} = P_l Q_l \begin{bmatrix} 0 \\ 0 \\ 1 \end{bmatrix} \qquad l = 1, 2 \tag{10.200}$$

Accordingly, the primary effect of layover is a small change in the differential range,

$$\Delta \bar{R}_l = \left[-\Delta x_{ml} \left(\frac{\bar{x}_t}{R_t} + \frac{\bar{x}_{rl}}{R_{rl}} \right) - \Delta y_{ml} \left(\frac{\bar{y}_t}{R_t} + \frac{\bar{y}_{rl}}{R_{rl}} \right) \right] \qquad l = 1, 2 \tag{10.201}$$

that imparts a shift in the imaged location of a scatterer and a shift in the phase of the data collected by each receiver

$$\begin{aligned} S_l(f_i, \tau_k) &= S(f_i, \tau_k) \exp\left\{ -j \frac{2\pi f_i}{c} \Delta \bar{R}_l \right\} \\ &\approx S(f_i, \tau_k) \exp\left\{ -j \frac{2\pi f_c}{c} \Delta \bar{R}_l \right\} \end{aligned} \tag{10.202}$$

which is approximately constant across frequencies for narrow band systems, with center frequency $f_c = f_0 + B/2$. Thus, two images of the same scene formed from receivers with slightly different positions can be represented by

$$s_1(x, y) = s(x, y)\exp\left\{-j\frac{2\pi f_c}{c}\Delta\bar{R}_1\right\} \tag{10.203}$$

$$s_2(x, y) = s(x, y)\exp\left\{-j\frac{2\pi f_c}{c}\Delta\bar{R}_2\right\} \tag{10.204}$$

where $s(x, y)$ is the image of a point scatterer with zero height but shifted in x and y due to layover.

Computing the product of the first image with the complex conjugate of the second image yields

$$\begin{aligned} C(x, y) &= s_1(x, y)\overline{s_2(x, y)} \\ &\approx |s(x, y)|^2 \exp\left\{-j\frac{2\pi}{\lambda}\left(\Delta\bar{R}_1 - \Delta\bar{R}_2\right)\right\} \end{aligned} \tag{10.205}$$

where $\lambda_c = c/f_c$. The phase of $C(x, y)$ is

$$\begin{aligned} \angle C(x, y) &\approx -\frac{2\pi}{\lambda}\left(\Delta\bar{R}_1 - \Delta\bar{R}_2\right) \\ &\approx \frac{2\pi z_m}{\lambda_c}\left[(\Delta x_{m1} - \Delta x_{m2})\left(\frac{\bar{x}_t}{R_t} + \frac{\bar{x}_{r1}}{R_{r1}}\right) + (\Delta y_{m1} - \Delta y_{m2})\left(\frac{\bar{y}_t}{R_t} + \frac{\bar{y}_{r1}}{R_{r1}}\right)\right] \end{aligned} \tag{10.206}$$

where it is assumed that $x_{r1} \approx x_{r2}$, $y_{r1} \approx y_{r2}$, and $R_{r1} \approx R_{r2}$. It is then apparent that the height of the dominant scatterer in an imaged pixel may be estimated as

$$z_m = \frac{\lambda_c \angle C(x, y)}{2\pi}\left[(\Delta x_{m1} - \Delta x_{m2})\left(\frac{\bar{x}_t}{R_t} + \frac{\bar{x}_{r1}}{R_{r1}}\right) + (\Delta y_{m1} - \Delta y_{m2})\left(\frac{\bar{y}_t}{R_t} + \frac{\bar{y}_{r1}}{R_{r1}}\right)\right]^{-1}. \tag{10.207}$$

Thus, interferometric height estimates may be computed from two bistatic SAR images by computing the complex product of equation (10.205), taking the phase of this product and inserting it into equation (10.207). The maximum and minimum unambiguous heights may be calculated by substituting $\angle C(x, y) = \pm\pi$ into equation (10.207). Note that the above result is derived based on a point scatterer assumption. Scintillation and decorrelation effects, which are not modeled here, are a significant source of height estimation inaccuracy. A thorough discussion of these topics may be found in [2, 99, 100].

10.7.2.2 Algorithm Interferometric SAR processing begins by forming two bistatic SAR images, one from each of the receive antennas. These images may in general suffer from differing geolocation errors, thus requiring image registration to be applied in postprocessing. The complex product of equation (10.205) is computed, and the phase resulting from this product is used in equation (10.207) to compute a height estimate for a scatterer that has laid over into that pixel. Given a height estimate, the true (x, y) location of a scatterer may also be estimated by removing the shift due to layover using equations (10.199) and (10.200). Specifically, if a scatterer is observed in the first image at $[\hat{x}_{m1} \quad \hat{y}_{m1}]$ and has estimated height $\hat{z}_{12}^{(m)}$, then the true (x, y) location of that scatterer may then be computed as

$$\begin{bmatrix} \hat{x}_{12}^{(m)} \\ \hat{y}_{12}^{(m)} \end{bmatrix} = \begin{bmatrix} \hat{x}_{m1} \\ \hat{y}_{m1} \end{bmatrix} - P_1 Q_1 \begin{bmatrix} 0 \\ 0 \\ \hat{z}_{12}^{(m)} \end{bmatrix} \tag{10.208}$$

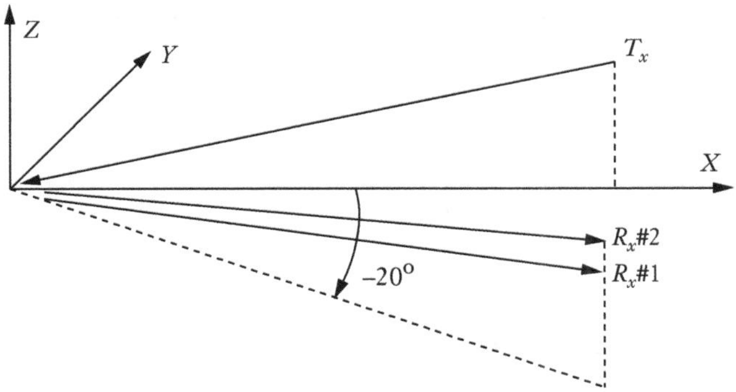

Figure 10–67 Simulated bistatic interferometric SAR geometry. The parameters used to generate this example are given in Table 10–9.

This calculation is commonly known as orthorectification. Equation (10.208) represents a bistatic generalization of corresponding monostatic orthorectification equations found in, for example, [2, 17].

To demonstrate, two images were generated of a scene illuminated by a transmitter with a flight path orthogonal to the positive x-axis and with its midpoint centered on the x-axis. As shown in Figure 10–67, a simulated receiving platform was located at an azimuth angle of ϕ_r = –20° and observed the scene at broadside with two antennas. The two receiving antennas were given a vertical separation of 10 m. Figures 10–68*a* and 10–68*b* show the two bistatic images formed from each receiving antenna; their magnitudes are nearly identical and differences in their phases encode the scatterer height. All of the scattering centers shown were equal in amplitude, and they were scaled prior to addition of white Gaussian noise in order to yield a peak signal-to-noise ratio (SNR) of 30 dB. The above algorithm was used to coherently process the two images in Figures 10–68*a* and 10–68*b* to yield height estimates for each point scatterer in the scene. The parameters used in this example are given in Table 10–9. Fusing these height estimates with one of the original images yields the 3-D reconstruction shown in Figure 10–68*c*. The true locations of the simulated point scatterers are shown in Figure 10–68*d*. Note that Figure 10–68*c* gives height estimates only at the point scatterer locations, due to the pixel-by-pixel processing of IFSAR, and does not give the surface effect that will be seen in stereo processing.

10.7.3 Stereo Height Estimation from Two Bistatic Apertures

10.7.3.1 Theory A system wherein two receiving platforms, each with a single antenna, are teamed with a transmitting platform is now considered. The first receiving platform traverses a linear flight path with $(\bar{x}_{r1}, \bar{y}_{r1}, \bar{z}_{r1})$ as its midpoint and $(v_{xr1}, v_{yr1}, v_{zr1})$ as its velocity vector. Similarly, the flight path of the second receiving platform is defined by $(\bar{x}_{r2}, \bar{y}_{r2}, \bar{z}_{r2})$ and $(v_{xr2}, v_{yr2}, v_{zr2})$. Based on these parameters, projection matrices for the first and second bistatic apertures are defined as

$$Q_i = \begin{bmatrix} \frac{\bar{x}_t}{R_t} + \frac{\bar{x}_{ri}}{R_{ri}} & \frac{\bar{y}_t}{R_t} + \frac{\bar{y}_{ri}}{R_{ri}} & \frac{\bar{z}_t}{R_t} + \frac{\bar{z}_{ri}}{R_{ri}} \\ \frac{v_{xt}}{R_t} + \frac{v_{xri}}{R_{ri}} & \frac{v_{yt}}{R_t} + \frac{v_{yri}}{R_{ri}} & \frac{v_{zt}}{R_t} + \frac{v_{zri}}{R_{ri}} \end{bmatrix} \qquad i = 1, 2 \tag{10.209}$$

and

$$P_i = \begin{bmatrix} \frac{\bar{x}_t}{R_t} + \frac{\bar{x}_{ri}}{R_{ri}} & \frac{\bar{y}_t}{R_t} + \frac{\bar{y}_{ri}}{R_{ri}} \\ \frac{v_{xt}}{R_t} + \frac{v_{xri}}{R_{ri}} & \frac{v_{yt}}{R_t} + \frac{v_{yri}}{R_{ri}} \end{bmatrix}^{-1} \qquad i = 1, 2 \tag{10.210}$$

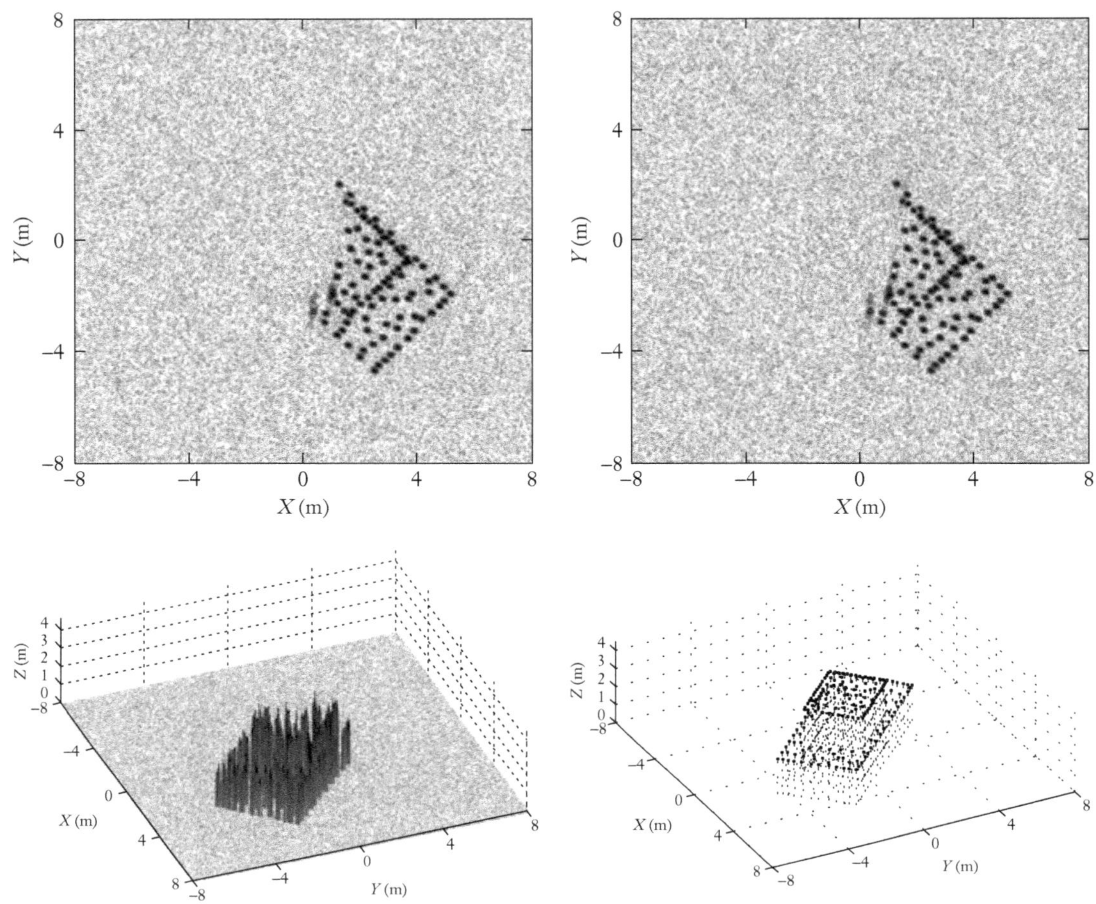

Figure 10–68 The transmitter aperture is centered at $\phi_t = 0°$. (*a*) Image formed by a receiver with aperture centered at $\phi_{rt} = -20°$. (*b*) An image with an identical envelope formed by a second receiver, displaced 10 m vertically from the first. (*c*) Topographical image formed interferometrically from the two bistatic images. (*d*) True locations of the simulated point scatterers. The parameters used to generate this example are given in Table 10–9.

Using equations (10.209) and (10.210), the locations of the *m*th scatterer in the scene as observed by the first and second receivers, respectively, are

$$\begin{bmatrix} \hat{x}_{m1} \\ \hat{y}_{m1} \end{bmatrix} = P_1 Q_1 \begin{bmatrix} x_m \\ y_m \\ z_m \end{bmatrix} + \begin{bmatrix} \tilde{x}_1 \\ \tilde{y}_1 \end{bmatrix} \tag{10.211}$$

and

$$\begin{bmatrix} \hat{x}_{m2} \\ \hat{y}_{m2} \end{bmatrix} = P_2 Q_2 \begin{bmatrix} x_m \\ y_m \\ z_m \end{bmatrix} + \begin{bmatrix} \tilde{x}_2 \\ \tilde{y}_2 \end{bmatrix} \tag{10.212}$$

where the vectors $[\tilde{x}_1\ \tilde{y}_1]^T$ and $[\tilde{x}_2\ \tilde{y}_2]^T$ represent shifts in the imaged scene due to uncompensated motion measurement and geolocation errors. Note that a scatterer at a nonzero height will project differently into the two 2-D images formed from the two bistatic apertures, due to the difference in their layover geometries and their individual platform motion measurement errors. By measuring the difference in this projection, an estimated height for that scatterer may be calculated. The layover offset of the *m*th scatterer is represented with

$$\begin{aligned}\underline{\hat{u}}_m &= \begin{bmatrix}\hat{x}_{m1}\\ \hat{y}_{m1}\end{bmatrix} - \begin{bmatrix}\hat{x}_{m2}\\ \hat{y}_{m2}\end{bmatrix}\\ &= P_1Q_1\begin{bmatrix}x_m\\ y_m\\ z_m\end{bmatrix} - P_2Q_2\begin{bmatrix}x_m\\ y_m\\ z_m\end{bmatrix} + \begin{bmatrix}\tilde{x}_1\\ \tilde{y}_1\end{bmatrix} - \begin{bmatrix}\tilde{x}_2\\ \tilde{y}_2\end{bmatrix}\\ &= (P_1Q_1 - P_2Q_2)\begin{bmatrix}0\\ 0\\ z_m\end{bmatrix} + \underline{\tilde{u}}_{12}\\ &\triangleq \underline{u}_z z_m + \underline{\tilde{u}}_{12}\end{aligned} \tag{10.213}$$

where $\underline{\tilde{u}}_{12}$ is the registration error induced between the two images due to their differing motion measurement errors. Furthermore, coherency of the image phases is not assumed for the derivation below, and in practice, it would be difficult, but not impossible, to maintain image phase coherence across two different platforms.

From equation (10.213), one may compute a height estimate for the *m*th scattering center as

$$\hat{z}_{12}^{(m)} = \frac{\underline{u}_z^T \underline{\hat{u}}_m}{\|\underline{u}_z\|^2} = \frac{\underline{u}_z^T(\underline{u}_z z_m + \underline{\tilde{u}}_{12})}{\|\underline{u}_z\|^2} = z_m + \tilde{z}_{12} \tag{10.214}$$

The lack of perfect geolocation prevents one from recovering an absolute estimate of z_m. However, all of the scatterers in the scene will suffer from the same height estimate offset $\tilde{z}_{12}$. This implies that relative height measurements, between two scattering centers, may still be meaningful. Given a point of reference with a known height, the offset $\tilde{z}_{12}$ may be removed to yield absolute height estimates for each scatterer in the scene.

10.7.3.2 Algorithm An algorithm for performing multistatic stereo SAR height estimation is now described. This implementation faces two impediments. First, it is possible that multiple scatterers of different heights may lay over into the same image location, thus complicating one's ability to make accurate observations. Second, scatterers observed in the two images must be correctly associated, and adding to this difficulty, some scatterers may be visible in one image but not the other. To circumvent these complications, the following standard assumptions are made.

(1) Scatterers lie on a surface $h(x, y)$ that is smoothly varying, such that multiple scatterers do not lay over into the same image location.
(2) This surface is locally flat.
(3) Multiple scatterers on a locally flat area of this surface may be correctly associated by cross-correlation of subimages extracted from the two bistatic images.

Though integral to stereo height estimation algorithms, these assumptions do not always hold in practice, and when violated, some performance loss may result.

The algorithm for multistatic SAR stereo height estimation is a standard stereo matching technique based on cross-correlation, which is briefly described below. Two bistatic SAR images are formed on the same pixel grid. Given these two images, a constant false alarm rate detector [101] is applied to the first image to obtain the pixel locations of strong scatterer returns. A peak finding algorithm is then applied, such that out of a cluster of adjacent detections only the pixel with the largest magnitude is returned. This effectively reduces the complexity of the algorithm by limiting the number of detected pixels corresponding to each point scatterer response in the image.

The algorithm now iterates through the list of peak detections and attempt to find the shift in location between each detection in the first image and the corresponding return in the second image. To do this, a square subimage is extracted, centered about the current peak detection, from the first image, and a larger square subimage is extracted, centered about the same pixel location, from the second image. The subimage taken from the first image should be large enough to possess distinct features, which will sufficiently differentiate it from other areas of the scene. The size of the second subimage is determined by the size of the first subimage, the maximum shift expected due to layover differences, and the maximum shift expected due to geolocation errors. The two-dimensional cross-correlation of the magnitudes of the two subimages is now computed. Extracting the location of the peak cross-correlation gives the relative shift between the two subimages. After determining relative shifts for all of the peak detections, height estimates are computed using equation (10.214) and orthorectified using equation (10.208).

The following two examples demonstrate the above algorithm. In the first example, two images are generated of a scene illuminated by a transmitter with a flight path centered on the positive x-axis and two receivers located at azimuth angles of $\phi_{r1} = -20°$ and $\phi_{r2} = 5°$, as shown in Figure 10–69. The elevation angles of all three platforms are approximately 20°, and all three platforms view the scene at broadside. The scene contains 190 point scatterers, configured in the shape of a target. As in the interferometric examples, the amplitudes of all of the scattering centers were set equal, and they were scaled before adding white Gaussian noise in order to yield a peak SNR of 30 dB. The two original images both with square 4-inch resolution cells are shown in Figures 10–70*a* and 10–70*b*. Note that a lateral separation (in x–y) was maintained between scatterers at different heights. This prevents ambiguities due to overlapping layover, thus artificially satisfying assumption (1) above.

The above algorithm is applied twice, alternating which image is used as the first image and using 1 m squared subimages, thus obtaining two sets of height estimates $\{\hat{z}_{12}^{(m)}\}$ and $\{\hat{z}_{21}^{(m)}\}$. Using equation (10.208), the subimages of the first image were then shifted to their true locations, as indicated by the $\{\hat{z}_{12}^{(m)}\}$, and the subimages of the second image were shifted as indicated by the $\{\hat{z}_{21}^{(m)}\}$. The magnitudes of these two height registered images were then averaged and fused with the extracted height estimates. The resulting 3-D reconstruction is shown in Figure 10–70*c*. The 3-D target reconstruction is fairly accurate, except in areas corresponding to the shadow boundaries of each image. Shadow boundaries were simulated by assigning, on a pulse-by-pulse basis, an amplitude of 1 to scatterers with azimuth angles

$$\phi_m = \arctan\left(\frac{y_m}{x_m}\right) \tag{10.215}$$

within 90° of the transmitter and receiver look angles, and by assigning an amplitude of 0 to all other scattering centers.

In the second example, the same data collection geometry was implemented, but the simulated scattering centers were arrayed in a tighter configuration in x–y. The second set of example images is shown in Figures 10–71*a* and 10–71*b*, where geometry and scatterer configuration used

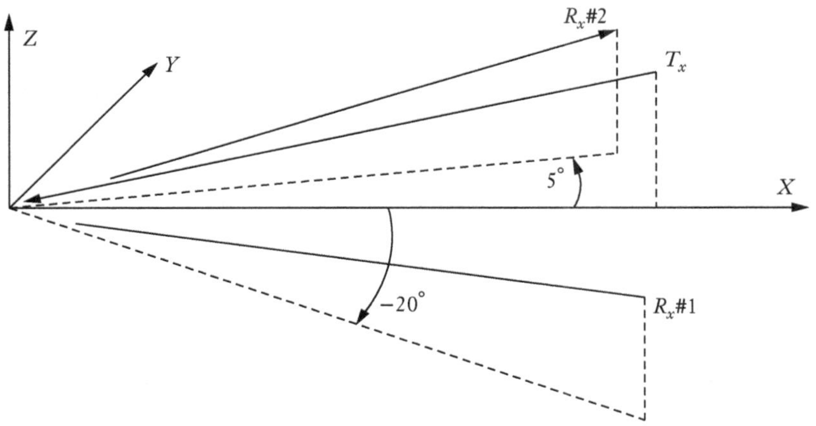

Figure 10–69 Simulated bistatic stereo SAR geometry.

Table 10–9 Bistatic geometry parameters corresponding to Figure 10–68.

$\Delta x_{m1} - \Delta x_{m2}$	–0.613 cm
$\Delta y_{m1} - \Delta y_{m2}$	0.0133 cm
$(\bar{x}_t, \bar{y}_t, \bar{z}_t)$	(30 km, 0 km, 10 km)
$(\bar{x}_{r1}, \bar{y}_{r1}, \bar{z}_{r1})$	(7.5 km, –2.7 km, 3 km)
$(\bar{x}_{r2}, \bar{y}_{r2}, \bar{z}_{r2})$	(7.5 km, –2.7 km, 3.01 km)
λ_c	2.64 cm
σ_ψ	0.081

for the image shown in Figure 10–71*a* is identical to that which was used for Figure 10–68*a*. Stereo processing of these images gives the 3-D reconstruction shown in Figure 10–71*c*. One may observe that regions containing scatterers laid over from both simulated heights are subject to greater variance in their height estimates. The accuracy of this reconstruction is lower than that of Figure 10–70, but the 3-D shape of the object is still recognizable.

Both stereo [95] and interferometric processing have been demonstrated in monostatic SAR systems, with interferometric processing being of greater emphasis in recent research [102–104]. Implementation of these techniques in bistatic systems should be similarly feasible. However, stereo SAR may face greater challenges in terms of phenomenology. Stereo techniques operate under the assumption that the same scene returns are observable in two images viewed from different aspect angles. In a bistatic scenario, this implies that the same scattering mechanisms will be illuminated by a transmitting platform, observed by both receiving platforms, and will permit registration between the two bistatic images. Interferometric processing does not make such stringent phenomenological assumptions.

10.8 PROPERTIES OF BISTATIC SAR IMAGES

Although bistatic SAR images have broadly similar appearance to their monostatic counterparts, they have some particular properties that are of interest and importance. Relatively few examples of bistatic SAR imagery have been reported in the open literature, though since 2004 some results

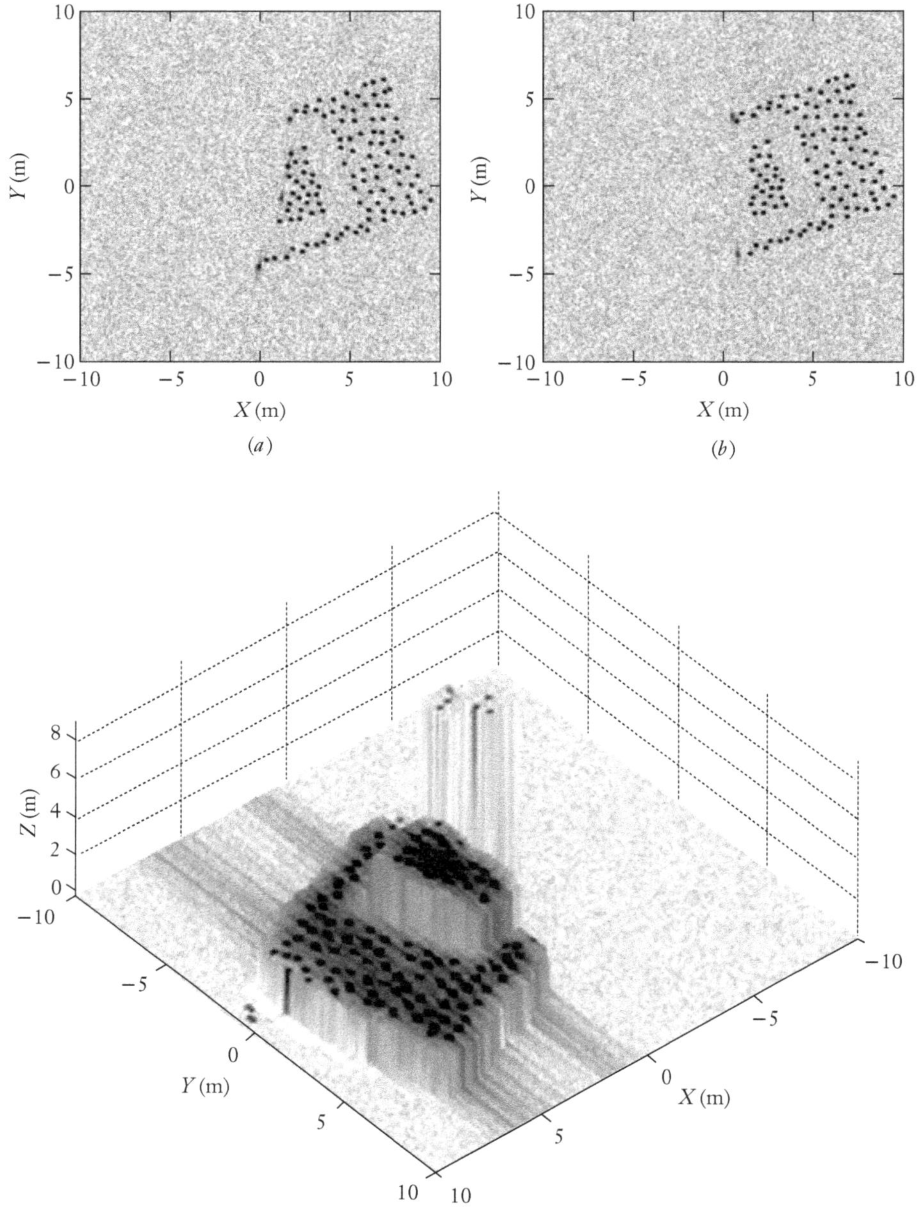

Figure 10–70 The transmitter aperture is centered at $\phi_t = 0°$. (*a*) Image formed by a receiver with aperture centered at $\phi_{r1} = -20°$. (*b*) Image formed by a receiver with aperture centered at $\phi_{r2} = 5°$. (*c*) 3-D surface reconstruction formed from (*a*) and (*b*). Lateral separation between scatterers at different heights prevents ambiguous height estimates.

have started to appear, notably in the Proceedings of the Biannual European *EUSAR* series of conferences [105].

These properties are illustrated below using some bistatic SAR imagery from the British QinetiQ organization. The radar data were acquired in a series of trials in 2003, using the

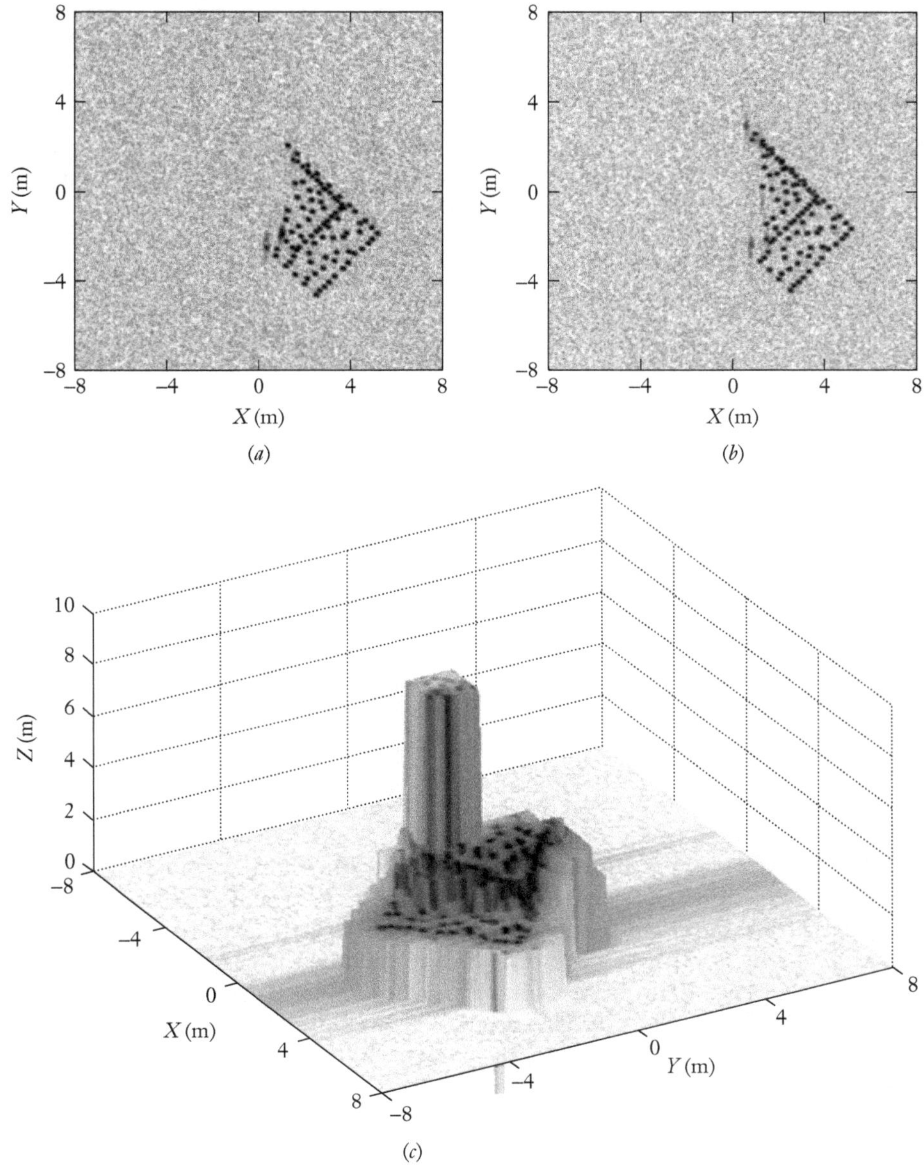

Figure 10–71 The transmitter aperture is centered at $\phi_t = 0°$. (*a*) Image formed by a receiver with aperture centered at $\phi_{r1} = -20°$. (*b*) Image formed by a receiver with aperture centered at $\phi_{r2} = 5°$. (*c*) 3-D surface reconstruction formed from (*a*) and (*b*). Tightly spaced scatterers lay over onto the same image regions, thus hindering stereo height estimation.

Enhanced Surveillance Radar (ESR) carried by a *BAC-111* aircraft and operating at X-band and a receiver carried by the ADAS helicopter platform. The arrangement is shown schematically in Figure 10–72*a* and an example image is shown in Figure 10–72*b*. The target scene in this case is a village in southwest England. Further information on the trials and on the SAR imagery can be found in references [106] and [107].

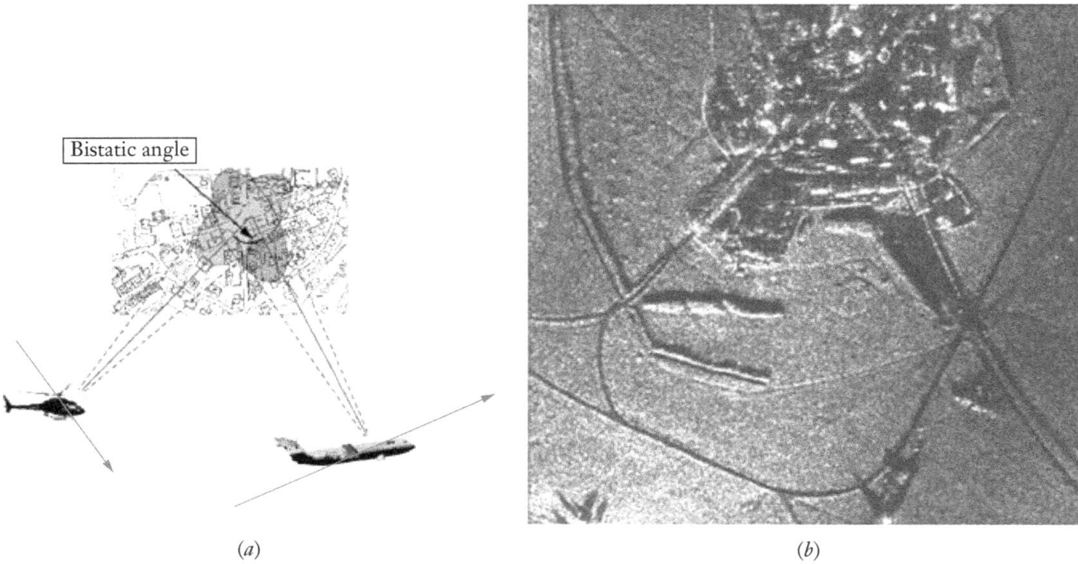

Figure 10–72. *(a)* Bistatic geometry, showing aircraft-borne transmitter, helicopter-borne receiver, and target scene; *(b)* bistatic image. ()

First, inspection of the target at the bottom left-hand corner of Figure 10–72*b* shows that the target has two shadows: one behind the target from the direction of the transmitter, and one behind the target from the direction of the receiver. In cases where the target shadow is used to infer the target shape [108], this property provides important additional information which may improve target classification performance.

Second, the bistatic radar signatures of targets will be different than monostatic signatures, as already discussed in Section 2.3, Chapter 2. Targets with dihedral and trihedral features will show strong monostatic backscatter which will not be present in the equivalent bistatic imagery. This will be particularly true of dihedrals and trihedrals formed by the vertical walls of buildings and horizontal ground, and is evident from a comparison of Figures 10–73*a* and 10–73*b*.

Although this effect is clearly evident in monostatic and bistatic SAR imagery of target scenes with buildings, there is some suggestion that it might be more generally true. In other words, bistatic clutter may be less ‘spiky’ than the equivalent monostatic clutter; that is, there is a higher probability of a higher-amplitude clutter echo, and hence a longer ‘tail’ to the clutter pdf. If so, this would give bistatic radars a performance advantage over monostatic radars, allowing a lower false-alarm rate for a given probability of detection.

This has been investigated [106, 107] by fitting the compound K-distribution (a widely used statistical clutter model [109]) to monostatic and bistatic image data. A random variable X is said to have a K distribution with parameters b and ν if its probability density function is given by

$$f(x;b,\nu)=\begin{cases} 0, & x\leq 0 \\ \frac{2b}{\Gamma(1+\nu)}\left(\frac{bx}{2}\right)^{\nu+1}K_\nu(bx), & x>0 \end{cases} \tag{10.215}$$

where Γ is the gamma function, K_ν is the modified Bessel function of order ν, ν is the shape parameter, and b is the scale parameter. Low values of ν (<1) indicate spiky clutter. This two-parameter family

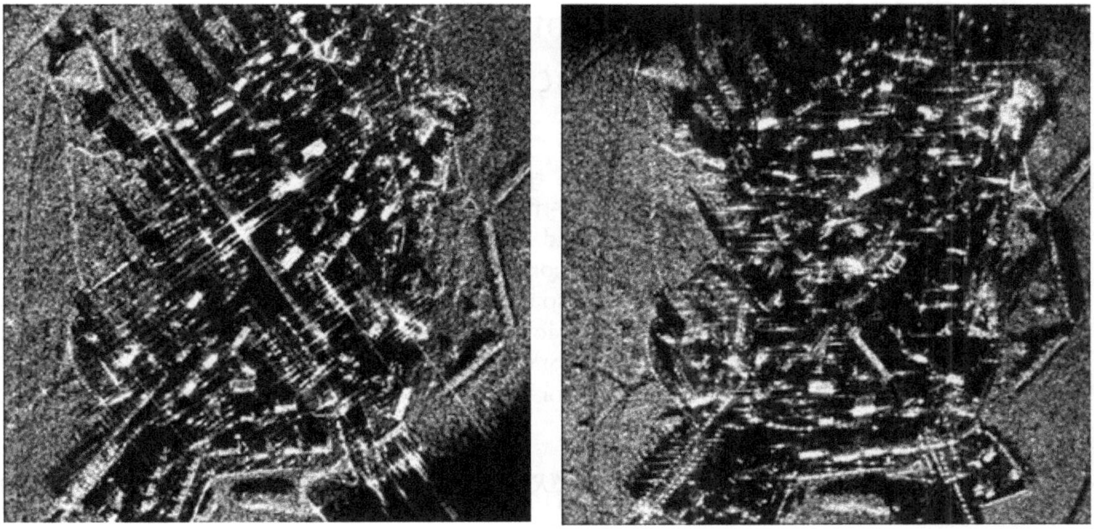

(*a*) (*b*)

Figure 10–73 (*a*) Monostatic and (*b*) bistatic image of a village target scene. (© QinetiQ Ltd (2003). All rights reserved.)

Table 10–10. Values of Shape Parameter ν and Goodness of Fit for Three Field Regions of Monostatic and Bistatic SAR Imagery [106, 108].

	Monostatic		**Bistatic**	
	Fit	ν	**Fit**	ν
Field 1	16%	4.27	80%	10.01
Field 2	72%	6.02	58%	6.25
Field 3	46%	4.10	44%	5.64

of distributions, denoted $X \sim K(b, \nu)$, was originally introduced by Jakeman and Pusey [110] to model the statistical properties of the magnitude (envelope) of the electric field scattered by the sea, and has since been applied to many other kinds of radar clutter.

Table 10–10 presents the results of fitting the *K*-distribution to three field regions of the bistatic imagery and the equivalent monostatic imagery and shows the values of the shape parameter ν and the goodness of fit in each case. The results are not totally convincing, but all suggest that the bistatic clutter is indeed less spiky. Similar results have been obtained from monostatic and bistatic imagery of London [111] using the European Space Agency's ASAR synthetic aperture radar carried by the ENVISAT satellite and a fixed, ground-based bistatic receiver [112].

Although it would be wrong to draw firm conclusions on the basis of such limited experiments and data sets, this nevertheless suggests that it is a promising area for future research.

Appendix A: Taylor Expansion of Differential Range for a Bistatic Geometry

The bistatic differential range $\Delta R(\tau)$ measures the distance from the transmitter to a scatterer to the receiver relative to the measured round-trip distance to the scene center. The first-order terms of the Taylor expansion of $\Delta R(\tau)$ are useful for linearly approximating the phase of scatterer returns. This linearizing approximation, more commonly known as the far-field approximation, is used in bistatic SAR image formation and is also useful for performance analysis. Placing bounds on the second-order terms of the Taylor expansion provides limitations on the use of the far-field assumption. In this appendix, the first- and second-order Taylor expansion terms are derived.

From Section 10–1, the differential range for a bistatic geometry is computed, with the time dependence suppressed, as

$$\Delta R = \Delta R_{\mathrm{t}} + \Delta R_{\mathrm{r}} \tag{A.1}$$

where $\Delta R_{\mathrm{t}} = d_{\mathrm{t}0} - d_{\mathrm{t}}$ is the differential range of the transmitter, and $\Delta R_{\mathrm{r}} = d_{\mathrm{r}0} - d_{\mathrm{r}}$ is the differential range of the receiver. The distance from the transmitter to a point on the ground (x, y, z) is

$$d_{\mathrm{t}0} = \sqrt{(x_{\mathrm{t}} - x)^2 + (y_{\mathrm{t}} - y)^2 + (z_{\mathrm{t}} - z)^2} \tag{A.2}$$

and the measured distance from the transmitter to the scene center is

$$d_{\mathrm{t}} = \sqrt{(x_{\mathrm{t}} + \tilde{x}_{\mathrm{t}})^2 + (y_{\mathrm{t}} + \tilde{y}_{\mathrm{t}})^2 + (z_{\mathrm{t}} + \tilde{z}_{\mathrm{t}})^2} \tag{A.3}$$

Similarly, the distance from the receiver to the same point on the ground is

$$d_{\mathrm{r}0} = \sqrt{(x_{\mathrm{r}} - x)^2 + (y_{\mathrm{r}} - y)^2 + (z_{\mathrm{r}} - z)^2} \tag{A.4}$$

and the measured distance from the receiver to the scene center is

$$d_{\mathrm{r}} = \sqrt{(x_{\mathrm{r}} + \tilde{x}_{\mathrm{r}})^2 + (y_{\mathrm{r}} + \tilde{y}_{\mathrm{r}})^2 + (z_{\mathrm{r}} + \tilde{z}_{\mathrm{r}})^2} \tag{A.5}$$

To obtain the Taylor expansion of ΔR_{t}, one may expand $d_{\mathrm{t}0}$ with respect to the ground location about the point $(x, y, z) = (0, 0, 0)$ and expand d_{t} with respect to the measurement error about the point $(\tilde{x}_{\mathrm{t}}, \tilde{y}_{\mathrm{t}}, \tilde{z}_{\mathrm{t}}) = (0,0,0)$. The difference between the constant terms of these expansions is zero, implying the $\Delta R_{\mathrm{t}}^{(0)}$ is zero. The first-order terms of ΔR_{t} are

$$\begin{aligned}\Delta R_{\mathrm{t}}^{(1)} &= \left.\frac{\partial d_{\mathrm{t}0}}{\partial x}\right|_{x=0} x + \left.\frac{\partial d_{\mathrm{t}0}}{\partial y}\right|_{y=0} y + \left.\frac{\partial d_{\mathrm{t}0}}{\partial z}\right|_{z=0} z \\ &\quad - \left.\frac{\partial d_{\mathrm{t}}}{\partial \tilde{x}_{\mathrm{t}}}\right|_{\tilde{x}_{\mathrm{t}}=0} \tilde{x}_a - \left.\frac{\partial d_{\mathrm{t}}}{\partial \tilde{y}_{\mathrm{t}}}\right|_{\tilde{y}_{\mathrm{t}}=0} \tilde{y}_{\mathrm{t}} - \left.\frac{\partial d_{\mathrm{t}}}{\partial \tilde{z}_{\mathrm{t}}}\right|_{\tilde{z}_{\mathrm{t}}=0} \tilde{z}_{\mathrm{t}} \\ &= \frac{-x_{\mathrm{t}}(x + \tilde{x}_{\mathrm{t}}) - y_{\mathrm{t}}(y + \tilde{y}_{\mathrm{t}}) - z_{\mathrm{t}}(z + \tilde{z}_{\mathrm{t}})}{\sqrt{x_{\mathrm{t}}^2 + y_{\mathrm{t}}^2 + z_{\mathrm{t}}^2}}\end{aligned} \tag{A.6}$$

By expressing the transmit antenna location (x_t, y_t, z_t) in polar coordinates (R_t, ϕ_t, θ_t), where $x_t = R_t \cos\phi_t \cos\theta_t$, $y_t = R_t \sin\phi_t \sin\theta_t$, and $z_t = R_t \sin\theta_t$, (A.6) can be expressed more simply as

$$\Delta R_t^{(1)} = -(x+\tilde{x}_t)\cos\phi_t\cos\theta_t - (y+\tilde{y}_t)\sin\phi_t\cos\theta_t - (z+\tilde{z}_t)\sin\theta_t \tag{A.7}$$

Similarly, the first-order terms of ΔR_r may derived to yield

$$\Delta R_r^{(1)} = -(x+\tilde{x}_r)\cos\phi_r\cos\theta_r - (y+\tilde{y}_r)\sin\phi_r\cos\theta_r - (z+\tilde{z}_r)\sin\theta_r \tag{A.8}$$

Given the block diagonal form of the Hessian matrix of ΔR_t, one may write the second-order Taylor terms of the transmitter differential range as

$$\Delta R_t^{(2)} = d_{t0}^{(2)} - d_t^{(2)} \tag{A.9}$$

where

$$\begin{aligned} d_{t0}^{(2)} = {} & \left.\frac{\partial^2 d_{t0}}{\partial x^2}\right|_{x=0}\frac{x^2}{2} + \left.\frac{\partial^2 d_{t0}}{\partial x\partial y}\right|_{x,y=0}\frac{xy}{2} \\ & + \left.\frac{\partial^2 d_{t0}}{\partial x\partial z}\right|_{x,z=0}\frac{xz}{2} + \left.\frac{\partial^2 d_{t0}}{\partial y^2}\right|_{y=0}\frac{y^2}{2} \\ & + \left.\frac{\partial^2 d_{t0}}{\partial y\partial x}\right|_{y,x=0}\frac{yx}{2} + \left.\frac{\partial^2 d_{t0}}{\partial y\partial z}\right|_{y,z=0}\frac{yz}{2} \\ & + \left.\frac{\partial^2 d_{t0}}{\partial z^2}\right|_{z=0}\frac{z^2}{2} + \left.\frac{\partial^2 d_{t0}}{\partial z\partial x}\right|_{z,x=0}\frac{zx}{2} \\ & + \left.\frac{\partial^2 d_{t0}}{\partial z\partial y}\right|_{z,y=0}\frac{zy}{2} \end{aligned} \tag{A.10}$$

and

$$\begin{aligned} d_t^{(2)} = {} & \left.\frac{\partial^2 d_t}{\partial \tilde{x}_t^2}\right|_{\tilde{x}_t=0}\frac{\tilde{x}_t^2}{2} + \left.\frac{\partial^2 d_t}{\partial \tilde{x}_t\partial \tilde{y}_t}\right|_{\tilde{x},\tilde{y}_t=0}\frac{\tilde{x}_t\tilde{y}_t}{2} \\ & + \left.\frac{\partial^2 d_t}{\partial \tilde{x}_t\partial \tilde{z}_t}\right|_{\tilde{x}_t,\tilde{z}_t=0}\frac{\tilde{x}_t\tilde{z}_t}{2} + \left.\frac{\partial^2 d_t}{\partial \tilde{y}_t^2}\right|_{\tilde{y}_t=0}\frac{\tilde{y}_t^2}{2} \\ & + \left.\frac{\partial^2 d_t}{\partial \tilde{y}_t\partial \tilde{x}_t}\right|_{\tilde{y}_t,\tilde{x}_t=0}\frac{\tilde{y}_t\tilde{x}_t}{2} + \left.\frac{\partial^2 d_t}{\partial \tilde{y}_t\partial \tilde{z}_t}\right|_{\tilde{y}_t,\tilde{z}_t=0}\frac{\tilde{y}_t\tilde{z}_t}{2} \\ & + \left.\frac{\partial^2 d_t}{\partial \tilde{z}_t^2}\right|_{z=0}\frac{\tilde{z}_t^2}{2} + \left.\frac{\partial^2 d_t}{\partial \tilde{z}_t\partial \tilde{x}_t}\right|_{\tilde{z}_t,\tilde{x}_t=0}\frac{\tilde{z}_t\tilde{x}_t}{2} \\ & + \left.\frac{\partial^2 d_t}{\partial \tilde{z}_t\partial \tilde{y}_t}\right|_{\tilde{z}_t,\tilde{y}_t=0}\frac{\tilde{z}_t\tilde{y}_t}{2} \end{aligned} \tag{A.11}$$

Evaluating equations (A.10) and (A.11) yields

$$\begin{aligned}\Delta R_{\mathrm{t}}^{(2)} = {} & \frac{x^2}{2}\left(\frac{y_{\mathrm{t}}^2+z_{\mathrm{t}}^2}{R_{\mathrm{t}}^3}\right)+\frac{y^2}{2}\left(\frac{x_{\mathrm{t}}^2+z_{\mathrm{t}}^2}{R_{\mathrm{t}}^3}\right)+\frac{z^2}{2}\left(\frac{x_{\mathrm{t}}^2+y_{\mathrm{t}}^2}{R_{\mathrm{t}}^3}\right) \\ & -xy\left(\frac{x_{\mathrm{t}}y_{\mathrm{t}}}{R_{\mathrm{t}}^3}\right)-xz\left(\frac{x_{\mathrm{t}}z_{\mathrm{t}}}{R_{\mathrm{t}}^3}\right)-yz\left(\frac{y_{\mathrm{t}}z_{\mathrm{t}}}{R_{\mathrm{t}}^3}\right) \\ & -\frac{\tilde{x}_{\mathrm{t}}^2}{2}\left(\frac{y_{\mathrm{t}}^2+z_{\mathrm{t}}^2}{R_{\mathrm{t}}^3}\right)-\frac{\tilde{y}_{\mathrm{t}}^2}{2}\left(\frac{x_{\mathrm{t}}^2+z_{\mathrm{t}}^2}{R_{\mathrm{t}}^3}\right)-\frac{\tilde{z}_{\mathrm{t}}^2}{2}\left(\frac{x_{\mathrm{t}}^2+y_{\mathrm{t}}^2}{R_{\mathrm{t}}^3}\right) \\ & +\tilde{x}_{\mathrm{t}}\tilde{y}_{\mathrm{t}}\left(\frac{x_{\mathrm{t}}y_{\mathrm{t}}}{R_{\mathrm{t}}^3}\right)+\tilde{x}_{\mathrm{t}}\tilde{z}_{\mathrm{t}}\left(\frac{x_{\mathrm{t}}z_{\mathrm{t}}}{R_{\mathrm{t}}^3}\right)+\tilde{y}_{\mathrm{t}}\tilde{z}_{\mathrm{t}}\left(\frac{y_{\mathrm{t}}z_{\mathrm{t}}}{R_{\mathrm{t}}^3}\right)\end{aligned} \tag{A.12}$$

and analogously, the second-order Taylor terms of the receiver differential range may be expressed as

$$\begin{aligned}\Delta R_{\mathrm{r}}^{(2)} = {} & \frac{x^2}{2}\left(\frac{y_{\mathrm{r}}^2+z_{\mathrm{r}}^2}{R_{\mathrm{r}}^3}\right)+\frac{y^2}{2}\left(\frac{x_{\mathrm{r}}^2+z_{\mathrm{r}}^2}{R_{\mathrm{r}}^3}\right)+\frac{z^2}{2}\left(\frac{x_{\mathrm{r}}^2+y_{\mathrm{r}}^2}{R_{\mathrm{r}}^3}\right) \\ & -xy\left(\frac{x_{\mathrm{r}}y_{\mathrm{r}}}{R_{\mathrm{r}}^3}\right)-xz\left(\frac{x_{\mathrm{r}}z_{\mathrm{r}}}{R_{\mathrm{r}}^3}\right)-yz\left(\frac{y_{\mathrm{r}}z_{\mathrm{r}}}{R_{\mathrm{r}}^3}\right) \\ & -\frac{\tilde{x}_{\mathrm{r}}^2}{2}\left(\frac{y_{\mathrm{r}}^2+z_{\mathrm{r}}^2}{R_{\mathrm{r}}^3}\right)-\frac{\tilde{y}_{\mathrm{r}}^2}{2}\left(\frac{x_{\mathrm{r}}^2+z_{\mathrm{r}}^2}{R_{\mathrm{r}}^3}\right)-\frac{\tilde{z}_{\mathrm{r}}^2}{2}\left(\frac{x_{\mathrm{r}}^2+y_{\mathrm{r}}^2}{R_{\mathrm{r}}^3}\right) \\ & +\tilde{x}_{\mathrm{r}}\tilde{y}_{\mathrm{r}}\left(\frac{x_{\mathrm{r}}y_{\mathrm{r}}}{R_{\mathrm{r}}^3}\right)+\tilde{x}_{\mathrm{r}}\tilde{z}_{\mathrm{r}}\left(\frac{x_{\mathrm{r}}z_{\mathrm{r}}}{R_{\mathrm{r}}^3}\right)+\tilde{y}_{\mathrm{r}}\tilde{z}_{\mathrm{r}}\left(\frac{y_{\mathrm{r}}z_{\mathrm{r}}}{R_{\mathrm{r}}^3}\right)\end{aligned} \tag{A.13}$$

REFERENCES

1. M. I. Skolnik, *Introduction to Radar Systems*, McGraw-Hill, Co., New York, NY, 1980.
2. W. G. Carrara, R. S. Goodman, and R. M. Majewski, *Spotlight Synthetic Aperture Radar: Signal Processing Algorithms*, Artech House, Inc., Norwood, MA, 1995.
3. M. C. Jackson, "The Geometry of Bistatic Radar Systems," *IEE Proceedings*, Vol. 133, No. 7, December 1986, pp. 604–612.
4. L. R. Moyer, C. J. Morgan, and D. A. Rugger, An Exact Expression for Resolution Cell Area in Special Case of Bistatic Radar Systems, *IEEE Transactions on Aerospace and Electronic Systems*, 25(4): July 1989, 584–587.
5. I. Walterscheid, A. R. Brenner, and J. H. G. Ender, Results on Bistatic Synthetic Aperture Radar, *Electronics Letters*, 40(19):1224–1225, September 2004.
6. N. J. Willis, Bistatic Radar, Technology Service Corporation, Silver Spring, MD, 1995.
7. F. Johnson, Synthetic Aperture Radar (SAR) Heritage: An Air Force Perspective. June, 2003,
8. G. Yates, A. M. Horne, A. P. Blake, R. Middleton, and D. B. Andre, "Bistatic SAR Image Formation," in *Proceedings of EUSAR 2004, 5th European Conference on Synthetic Aperture Radar*, Ulm, Germany, May 2004, Vol. 2, pp. 581–584.
9. G. Yates and A. Blake, "Bistatic SAR Image Utility," in *Proceedings of EUSAR 2006, 6th European Conference on Synthetic Aperture Radar*, Dresden, Germany, May 2006.
10. P. Dubois-Fernandez, H. Cantalloube, O. Ruault du Plessis, M. Wendler, B. Vaizan, C. Coulombeix, D. Heuze, and G. Krieger, "Analysis of Bistatic Scattering Behavior of Natural Surfaces," in *Proceedings of EUSAR 2004, 5th European Conference on Synthetic Aperture Radar*, Ulm, Germany, May 2004, Vol. 2, pp. 573–576.

11. M. Wendler, G. Krieger, M. R. Cassola, R. Horn, Bernd Gabler, P. Dubois-Fernandez, B. Vaizan, O. Ruault du Plessis, and H. Cantalloube, "Analysis of Bistatic Airborne SAR Data," in *Proceedings of EUSAR 2004, 5th European Conference on Synthetic Aperture Radar*, Ulm, Germany, May 2004, Vol. 2 pp. 571–572.
12. I. Walterscheid, A. R. Brenner, and J. H. G. Ender, Geometry and System Aspects for a Bistatic Airborne SAR-Experiment, in *Proceedings of EUSAR 2004, 5th European Conference on Synthetic Aperture Radar*, Ulm, Germany, May 2004, Vol. 2, pp. 567–570.
13. J. Palmer, M. Martorella, B. Littleton, and J. Homer, Emulated Bistatic Synthetic Aperture Imaging: Real World Data Analysis--Pt I--ISAR, in *Proceedings of EUSAR 2006, 6th European Conference on Synthetic Aperture Radar*, Dresden, Germany, May 2006.
14. J. Palmer, B. Littleton, J. Homer, and M. Martorella, Emulated Bistatic Synthetic Aperture Imaging: Real World Data Analysis--Pt II--SAR, in *Proceedings of EUSAR 2006, 6th European Conference on Synthetic Aperture Radar*, Dresden, Germany, May 2006.
15. O. Loffeld, H. Nies, U. Gebhardt, V. Peters, S. Knedlik, and W. Wiechert, Bistatic SAR—Some Reflections on Rocca's Smile, in *Proceedings of EUSAR 2004, 5th European Conference on Synthetic Aperture Radar*, Ulm, Germany, May 2004, Vol. 1, pp. 379–384.
16. B. D. Rigling and R. L. Moses, Polar Format Algorithm for Bistatic SAR, *IEEE Transactions on Aerospace and Electronic Systems*, 40(4):1147–1159, October 2004.
17. C. V. Jakowatz, D.E. Wahl, and P. H. Eichel, *Spotlight-Mode Synthetic Aperture Radar: A Signal Processing Approach*, Kluwer Academic Publishers, Boston, MA, 1996
18. B. D. Rigling and R. L. Moses, Motion Measurement Errors and Autofocus in Bistatic SAR, *IEEE Transactions on Image Processing*, 15(4):1008–1016, April 2006.
19. C. Mikhail, K. Kurt, and N. David, Bistatic Synthetic Aperture Radar with Non-Cooperative LEOS Based Transmitter, *IEEE 2000 International Geoscience and Remote Sensing Symposium Proceedings*, 2:861–862, July 2000.
20. M. Soumekh, Wide-Bandwidth Continuous-Wave Monostatic/Bistatic Synthetic Aperture Radar Imaging, *International Conference on Image Processing*, 3:361–365, October 1998.
21. L. Cazzani, et al., A Ground Based Parasitic SAR Experiment, *IEEE IGARSS*, 3:525–1527, 1999.
22. W. Keller and K. Kubik, Bistatic SAR Using GPS Signals Reflected at the Sea-Surface, in *Proceedings of EUSAR 2004, 5th European Conference on Synthetic Aperture Radar*, Ulm, Germany, May 2004, Vol. 2, pp. 779–782.
23. H. D. Griffiths and N. R. W. Long, Television-based bistatic radar, *IEE Proceedings*, 133(7):649–657, December1986.
24. H. D. Griffiths, C. J. Baker, H. Ghaleb, R. Ramakrishnan, and E. Willman, Measurement and Analysis of Ambiguity Functions of Off-Air Signals for Passive Coherent Location, *Electronics Letters*, 39(13):1005–1007, June 2003.
25. C. J. Baker, H. D. Griffiths, and I. Papoutsis, Passive Coherent Radar Systems--Part II: Waveform Properties, *IEE Proceedings of Radar, Sonar and Navigation*, 152(3):160–169, June 2005.
26. H. D. Griffiths, A. J. Garnett, C. J. Baker, and S. Keaveney. Bistatic Radar Using Satellite-Borne Illuminators of Opportunity, in *Proceedings RADAR-92 Conference*, October 1992, pp. 276–279.
27. M. A. Ringer and G. J. Frazer, Waveform Analysis of Transmissions of Opportunity for Passive Radar, in *Fifth International Symposium on Signal Processing and its Applications*, August 1999, pp. 511–514.
28. W. J. Caputi, Stretch: A Time-Transformation Technique. *IEEE Transactions on Aerospace and Electronic Systems*, 7(2):269–278, March 1971.
29. F. He, D. Liang and J. Liu, Doppler Properties of Bistatic Spaceborne SAR, in *Proceedings of EUSAR 2004, 5th European Conference on Synthetic Aperture Radar*, Ulm, Germany, May, 2004, Vol. 2, pp. 783–786.
30. G. Krieger, N. Gebert, and A. Moreira, Digital Beamforming and Non-Uniform Displaced Phase Centre Sampling in Bi- and Multistatic SAR, in *Proceedings of EUSAR 2004, 5th European Conference on Synthetic Aperture Radar*, Ulm, Germany, May 2004, Vol. 2, pp. 563–566.
31. J. L. Walker, Range-Doppler Imaging of Rotating Objects, *IEEE Transactions on Aerospace and Electronic Systems*, 16:23–52, January, 1980.
32. D. C. Munson, J. D. O'Brien, and W. K. Jenkins, A Tomographic Formulation of Spotlight-Mode Synthetic Aperture Radar, *Proceedings of the IEEE*, 72:917–925, August 1983.

33. M. D. Desai and W. K. Jenkins, Convolutional Backprojection Image Reconstruction for Spotlight Mode Synthetic Aperture Radar, *IEEE Transactions on Image Processing*, 1:505–517, October 1992.
34. F. Rocca, C. Cafforio and C. Prati, Synthetic Aperture Radar: A New Application for Wave Equation Techniques, *Geophysical Prospecting*, Vol. 37, No. 7, October 1989, pp. 809–830.
35. C. Cafforio, C. Prati, and F. Rocca. SAR Data Focusing Using Seismic Migration Technique, *IEEE Transaction on Aerospace and Electronic Systems*, 27:194–206, March 1991.
36. C. Cafforio, C. Prati, and F. Rocca, Full Resolution Focusing of SEASAT SAR Images in the Frequency-Wave Number Domain, in *Proceedings of the 8th EARSeL Symposium*, March 1988, pp. 336–355.
37. R. K. Raney, H. Runge, R. Bamler, I. G. Cumming, and F. H. Wong, Precision SAR Processing Using Chirp Scaling, *IEEE Transactions on Geoscience and Remote Sensing*, 32:786–799, 1994.
38. D. L. Mensa. *High Resolution Radar Cross-Section Imaging*, Artech House, Inc., Norwood, MA, 1991.
39. S. Basu and Y. Bresler, $O(N^2 \log_2 N)$ filtered backprojection reconstruction algorithm for tomography, *IEEE Transactions on Image Processing*, 9(10):1760–1773, October 2000.
40. S. Basu and Y. Bresler, Error analysis and performance optimization of fast hierarchical backprojection algorithms, *IEEE Transactions on Image Processing*, 10(7):1103–1117, July 2001.
41. A. F. Yegulalp, Fast Backprojection Algorithm for Synthetic Aperture Radar, in *Proceedings of the IEEE Radar Conference*, April 1999, pp. 60–65.
42. H. M. Jones, Predicted Properties of Bistatic Satellite Images, in *IEEE National Radar Conference*, April 1993 pp. 286–291.
43. G. P. Cardillo, On the Use of the Gradient to Determine Bistatic SAR Resolution, *Antennas and Propagation Society International Symposium*, 2:1032–1035, May 1990.
44. Y. Wu and D. C. Munson, Wide-Angle ISAR Passive Imaging Using Smoothed Pseudo Wigner-Ville Distribution, in *IEEE Radar Conference Proceedings*, May 2001, pp. 363–368.
45. Y. Wu, D. C. Munson, and A. D. Lanterman, Multistatic Passive Radar Imaging of Aircraft: A Feasibility Study Using FISC, in *Proceedings of URSI National Radio Science Mtg.*, January 2002.
46. J. Detlefsen, Application of Multistatic Radar Principles to Short-Range Imaging, *IEE Proceedings*, 133(7):658–663, December 1986.
47. A. M. Horne and G. Yates, Bistatic Synthetic Aperture Radar, in *IEE Radar 2002*, October 2002, pp. 6–10.
48. M. Lowe, Algorithms for High Resolution Bistatic SAR, in *IEE Radar 2002*, October 2002, pp. 512–515.
49. J. Sanz-Marcos and J. J. Mallorqui, A Bistatic SAR Simulator and Processor, in *Proceedings of EUSAR 2004, 5th European Conference on Synthetic Aperture Radar*, Ulm, Germany, May 2004, Vol. 2, pp. 585–588.
50. J. Palmer, M. Martorella, and I. D. Longstaff, Airborne ISAR Imaging Using the Emulated Bistatic Radar System, in *Proceedings of EUSAR 2004, 5th European Conference on Synthetic Aperture Radar*, Ulm, Germany, May 2004, Vol. 1, pp. 375–378.
51. Y. Ding and D. C. Munson, A Fast Backprojection Algorithm for Bistatic SAR Imaging. *International Conference on Image Processing*, 2:449–452, 2002.
52. C. Hu, T. Zeng, T. Long, and C. Juan, Fast Backprojection Algorithm for Bistatic SAR with Parallel Trajectory, in *Proceedings of EUSAR 2006, 6th European Conference on Synthetic Aperture Radar*, Dresden, Germany, May 2006.
53. A. D. M. Garvin and M. R. Inggs, Use of Synthetic Aperture and Stepped-Frequency Continuous Wave Processing to Obtain Radar Images, in *South African Symposium on Communications and Signal Processing 1991*, 1991, pp. 32–35.
54. I. Erer, S. Kent, M. Kartal, and S. Kargin, Superresolution Bistatic Spotlight SAR Imaging Using Image Fusion, in *Proceedings of EUSAR 2004, 5th European Conference on Synthetic Aperture Radar*, Ulm, Germany, May 2004, Vol. 2, pp. 787–790.
55. M. Soumekh, Bistatic Synthetic Aperture Radar Inversion with Application in Dynamic Object Imaging, *IEEE Transactions on Signal Processing*, 39:2044–2055, September 1991.
56. M. Soumekh, Multistatic Echo Imaging in Remote Sensing and Diagnostic Medicine, in *Sixth Multidimensional Signal Processing Workshop*, 1989, p. 200.
57. Y. L. Neo, F. Wong, and I. Cumming, An Efficient Non-Linear Chirp Scaling Method of Focusing Bistatic SAR Images, in *Proceedings of EUSAR 2006, 6th European Conference on Synthetic Aperture Radar*, Dresden, Germany, May 2006.

58. J. H. G. Ender, A Step to Bistatic SAR Processing, in *Proceedings of EUSAR 2004, 5th European Conference on Synthetic Aperture Radar*, Ulm, Germany, May 2004, Vol. 1, pp. 359–364.
59. O. Arikan and D. C. Munson, A Tomographic Formulation of Bistatic Synthetic Aperture Radar, in *Proceedings ComCon'88*, October, 1988, p. 418.
60. D. E. Dudgeon and R. M. Mersereau, *Multidimensional Digital Signal Processing*, Prentice-Hall, Englewood Cliffs, NJ, 1984.
61. C. F. Gerald and P. O. Wheatley, *Applied Numerical Analysis*, Addison-Wesley, Reading, MA, 1994.
62. F. J. Harris, On the Use of Windows for Harmonic Analysis with the Discrete Fourier Transform, *Proceedings of the IEEE*, 66:51–83, January 1978.
63. D. D'Aria, A. Monti Guarnieri, and F. Rocca, Precision Bistatic Processing with a Standard SAR Processor, in *Proceedings of EUSAR 2004, 5th European Conference on Synthetic Aperture Radar*, Ulm, Germany, May 2004, Vol. 1, pp. 385–388.
64. M. Weiβ, Time and Frequency Synchronisation Aspects for Bistatic SAR Systems, in *Proceedings of EUSAR 2004, 5th European Conference on Synthetic Aperture Radar*, Ulm, Germany, May 2004, Vol. 1, pp. 395–398.
65. M. Younis, R. Metzig, G. Krieger, and R. Klein, Performance Prediction and Verification for Bistatic SAR Synchronization Link, in *Proceedings of EUSAR 2006, 6th European Conference on Synthetic Aperture Radar*, Dresden, Germany, May 2006.
66. T. Lei, T. Zeng, and C. Hu, Effect of Time and Frequency Synchronization Errors on Bistatic SAR Image Formation, in *Proceedings of EUSAR 2006, 6th European Conference on Synthetic Aperture Radar*, Dresden, Germany, May 2006.
67. C. Gierull, C. Pike, and F. Paquet, Mitigation of Phase Noise in Bistatic SAR Systems with Extremely Large Synthetic Apertures, in *Proceedings of EUSAR 2006, 6th European Conference on Synthetic Aperture Radar*, Dresden, Germany, May 2006.
68. W. Wang, X. Liang, C. Ding, Y. Wang, and W. Hong, A Phase Synchronization Approach for Bistatic SAR Systems, in *Proceedings of EUSAR 2006, 6th European Conference on Synthetic Aperture Radar*, Dresden, Germany, May 2006.
69. H. Nies, O. Loffeld, and K. Natroshvili, Analysis and Focusing of Bistatic Airborne SAR Data, in *Proceedings of EUSAR 2006, 6th European Conference on Synthetic Aperture Radar*, Dresden, Germany, May 2006.
70. D. Blacknell, et al., Geometric Accuracy in Airborne SAR Images, *IEEE Transactions on Aerospace and Electronic Systems*, 25:241–257, March 1989.
71. T. A. Kennedy, Strapdown Inertial Measurement Units for Motion Compensation for Synthetic Aperture Radars, *IEEE AES Magazine*, October 1988.
72. C. E. Mancill and J. M. Swiger, A Map Drift Autofocus Technique for Correcting Higher Order SAR Phase Errors (U), in *27th Annual Tri-Service Radar Symposium Record*, June, 1981, pp. 391–400.
73. S. Zhang and R. Yang, Analysis of Oscillator Phase Noise Effects on Bistatic SAR, in *Proceedings of EUSAR 2006, 6th European Conference on Synthetic Aperture Radar*, Dresden, Germany, May 2006.
74. P. H. Eichel, D. C. Ghiglia, and C. V. Jakowatz, Jr., Speckle Processing Method for Synthetic Aperture Radar Phase Correction. *Optics Letters*, 14(1), January 1989.
75. D. E. Wahl, et al., Phase Gradient Autofocus—A Robust Tool for High Resolution SAR Phase Correction, *IEEE Transactions on Aerospace and Electronic Systems*, 30:827–834, July 1994.
76. W. D. Brown and D. C. Ghiglia, Some Methods for Reducing Propagation-Induced Phase Errors in Coherent Imaging Systems. I. Formalism, *Journal of the Optical Society of America A*, 5:924–942, June 1988.
77. P. H. Eichel and C. V. Jakowatz, Jr., Phase-Gradient Algorithm as an Optimal Estimator of the Phase Derivative. *Optics Letters*, 14(28):1101–1109, October 1989.
78. B. Rigling, et al., Improved SAR Image Auto-focus Through Algorithm Hybridization, in *EUSAR 2002 Proceedings*, June 2002.
79. A. F. Yegulalp, Minimum Entropy Auto-Focus, in *Seventh Annual Adaptive Sensor Array Processing Workshop at Lincoln Laboratories*, March 1999, pp. 60–65.
80. R. G. Paxman and J. C. Marron, Aberration Correction of Speckled Imagery with an Image-Sharpness Criterion, in *Statistical Optics, Proceedings of the Meeting, Society of Photo-Optical Instrumentation Engineers*, August 1988.

81. J. R. Fienup and J. J. Miller, Aberration Correction by Maximizing Generalized Sharpness Metrics, *Journal of Optical Society of America A*, 20(4):609–620, April 2003.
82. F. Berizzi and G. Corsini, Autofocus of Inverse Synthetic Aperture Radar Images Using Contrast Optimization. *IEEE Transactions on Aerospace and Electronic Systems*, 32(3):1185–1191, July 1996.
83. E. Yadin, SAR Auto-Focus Viewed as Adaptive Beam-Forming on Prominent Scatterers, in *Proceedings of the National Radar Conference*, 1994, pp. 138–1994.
84. R. Levy-Nathansolm and E. Yadin, Performance Evaluation of a Minimum ISLR SAR Auto-Focusing Algorithm, in *EUSAR 2000, 3rd European Conference on Synthetic Aperture Radar*, Munich, Germany, May 2000.
85. H. L. Chan and T. S. Yeo, Noniterative Quality Phase-Gradient Autofocus (QPGA) Algorithm for Spotlight SAR Imagery. *IEEE Transactions on Geoscience and Remote Sensing*, 36(5):1531–1539, September 1998.
86. H. L. Chan and T. S. Yeo, Comments on "Noniterative Quality Phase-Gradient Autofocus (QPGA) Algorithm for Spotlight SAR Imagery," *IEEE Transactions on Geoscience and Remote Sensing*, 40(11):2517, November 2002.
87. F. Berizzi, G. Corsini, M. Diani, and M. Veltroni, Autofocus of Wide Azimuth Angle SAR Images by Contrast Optimisation, *Geoscience and Remote Sensing Symposium*, 1996. International, 2:1230–1232, May 1996.
88. B. D. Rigling, "Multistage Entropy Minimization for SAR Image Autofocus," in E. G. Zelnio (ed.) *Algorithms for Synthetic Aperture Radar Imagery XIII*, *Proceedings of SPIE*, Vol. 6237, April 2006.
89. A. Moccia and D. Robustelli, Bistatic and Interferometric SAR Evolution of the Italian COSMO-SkyMed Constellation, in *Proceedings of EUSAR 2004, 5th European Conference on Synthetic Aperture Radar*, Ulm, Germany, May 2004, Vol. 1, pp. 371–374.
90. G. Krieger, H. Fiedler, and A. Moreira, Bi- and Multi-Static SAR: Potentials and Challenges, in *Proceedings of EUSAR 2004, 5th European Conference on Synthetic Aperture Radar*, Ulm, Germany, May, 2004, Vol. 1, pp. 365–370.
91. L. C. Graham, Synthetic Interferometer Radar for Topographic Mapping. *Proceedings of the IEEE*, 62:763–768, June 1974.
92. H. Zebker and R. Goldstein, Topographic Mapping from Interferometric SAR Observations. *Journal of Geophysical Research*, 91:4993–4999, 1986.
93. A. K. Gabriel and R. M. Goldstein, Crossed Orbit Interferometry: Theory and Experimental Results from SIR-B. *International Journal of Remote Sensing*, 9:857–872, 1988.
94. M. D. Desai, Spotlight Mode SAR Stereo Technique for Height Computation. *IEEE Transactions on Image Processing*, 6:1400–1411, 1997.
95. C. V. Jakowatz, Jr., D. E. Wahl, and P. A. Thompson, Three-Dimensional SAR Imaging Using Cross-Track Coherent Stereo Collections, *Thirty-First Asilomar Conference on Signals, Systems & Computers*, 2:1199–1203, 1997.
96. B. D. Rigling and R. L. Moses, "Three-dimensional Surface Reconstruction from Multi-Static SAR Images," in E.G. Zelnio (ed.) *Algorithms for Synthetic Aperture Radar Imagery X*, *Proceedings of SPIE*, Vol. 5095, April 2003, pp. 75–85.
97. H. Cantalloube, M. Wendler, V. Giroux, P. Dubois-Fernandez, and G. Krieger, Challenges in SAR Processing for Airborne Bistatic Acquisitions, in *Proceedings of EUSAR 2004, 5th European Conference on Synthetic Aperture Radar*, Ulm, Germany, May 2004, Vol. 2, pp. 577–580.
98. J. Palmer, B. Littleton, and J. Homer, SAR Generation Through Bistatic Radar Emulation, in *Proceedings of EUSAR 2004, 5th European Conference on Synthetic Aperture Radar*, Ulm, Germany, May 2004, Vol. 2, pp. 803–806.
99. E. Rodriguez and J. M. Martin, Theory and Design of Interferometric Synthetic Aperture Radars, *IEE Proceedings-F*, 139(2):147–159, April 1992.
100. D. Just and R. Bamler, Phase Statistics of Interferograms with Applications to Synthetic Aperture Radar, *Applied Optics*, 33(20):4361–4368, July 1994.
101. G. Minkler and J. Minkler, *CFAR: The Principles of Automatic Radar Detection in Clutter*, Magellan Book, Baltimore, MD, 1990.

102. C. D. Austin and R. L. Moses, "IFSAR Image Reconstruction with Multiple Scattering Centers in a Resolution Cell," in E. G. Zelnio (ed.) *Algorithms for Synthetic Aperture Radar Imagery XIII, Proceedings of SPIE*, Vol. 6237, April 2006.

103. R. L. Moses and C.D. Austin, "IFSAR Processing for 3D Target Reconstruction," in E.G. Zelnio (ed.) *Algorithms for Synthetic Aperture Radar Imagery XII, Proceedings of SPIE*, Vol. 5808, March, 2005.

104. R. L. Moses, P. Adams, and T. Biddlecome, "Three-Dimensional Target Visualization from Wide-Angle IFSAR Data," in E. G. Zelnio (ed.) *Algorithms for Synthetic Aperture Radar Imagery XII, Proceedings of SPIE*, Vol. 5808, March 2005.

105. *Proceedings 5th European Conference on Synthetic Aperture Radar EUSAR 2004*, Ulm, Germany, May 2004; *Proc. 6th European Conference on Synthetic Aperture Radar EUSAR 2006*, Dresden, Germany, May 2006.

106. G. A. Yates, *Bistatic Synthetic Aperture Radar*, Ph. D. thesis, University College London, January 2005.

107. G. A. Yates, A. M. Horne, A. P. Blake, and R. Middleton, "Bistatic SAR Image Formation", *IEE Proceedings—Radar, Sonar and Navigation*, Vol. 153, No. 3, June 2006, pp. 208–213.

108. R. Bolter and F. Leberl, "Shape-From-Shadow Building Reconstruction From Multiple View SAR Images", *Proc. 24th Workshop of Austrian Association of Pattern Recognition*, Vol. 142, 2000, pp. 199–206.

109. K. D. Ward, R. J. A. Tough, and S. Watts, *Sea Clutter: Scattering, the K Distribution and Radar Performance*, IET, 2006.

110. E. Jakeman and P. N. Pusey, "A Model for Non-Rayleigh Sea Echo," *IEEE Transactions on Antennas and Propagation*, Vol. AP-24, No. 6, 1976, pp. 806–814.

111. M. Oudinot, *Clutter Study in the Case of Bistatic Radars*, M. Sc. dissertation, University College London, 2006.

112. H. D. Griffiths, C. J. Baker, J. Baubert, N. Kitchen, and M. Treagust, "Bistatic Radar Using Spaceborne Illuminator of Opportunity", in *Proceedings on the RADAR 2002 Conference*, Edinburgh, October 15–17, 2002, IEE Conf. Publ. No. 490, pp. 1–5.

11

Adaptive Moving Target Indication

William L. Melvin

11.1 OVERVIEW

A mechanism to suppress doppler-spread ground clutter reflections is essential to the effective operation of an aerospace bistatic moving target indication (MTI) radar. Bistatic space–time adaptive processing (STAP) techniques combine spatial and temporal radar samples in a finite impulse response (FIR) filter to cancel stationary clutter returns while providing maximal gain on target. This strategy holds the potential to maximize output signal-to-interference-plus-noise ratio (SINR), thereby increasing the probability of detection for a fixed false alarm rate [1]. This chapter describes adaptive bistatic clutter cancellation methods based on STAP.

As defined in [2], a significant distance separates transmitter and receiver sites in bistatic radar. For example, the distance between transmitter and receiver may be comparable to the distance between transmitter and target. Simply providing separate transmitter and receiver locations—as in a typical continuous wave (CW) radar system—does not by itself constitute a bistatic configuration. Rather, in a somewhat qualitative manner, we identify the bistatic configuration by either significant transmitter–receiver separation and/or the occurrence of phenomenological change relative to strictly monostatic conditions. As we describe, the bistatic configuration presents unique challenges to STAP implementation.

Despite the fact that nearly all modern radar systems are monostatic [2–4], the research community continues to show interest in bistatic radar development for at least the following reasons:

- A desire to overcome space loss in systems limited by power or aperture. Examples include radar mounted on unmanned aerial vehicles (UAVs) and space-based moving target indication radar [5, 6]. Moreover, cost considerations may favor bistatic operation [3].
- The capability for silent radar operation [2]. With a separated transmitter, the receiver is less vulnerable to location and attack.
- If the separated transmitter is also remote from the battle area, it is less vulnerable to attack.
- Given diminishing radar spectrum, merging radar operation with available signals of opportunity in a cooperative or noncooperative manner is highly desirable.

- Vast and continuing improvements in signal processing and sensor technology enable bistatic radar system deployment. Digital beamforming (DBF) provides an effective mechanism for pulse chasing and enables STAP insertion into the radar system. *Bistatic space–time adaptive processing* supports target detection in clutter-limited environments [7–18].

Advantages of bistatic STAP include a response tailored to the unknown, dynamic characteristics of the operating environment; the potential for slow moving target detection is based on the STAP's ability to precisely place deep, narrow nulls at closely spaced frequencies occupied by clutter. However, bistatic STAP is not a direct application of general STAP theory [7, 19, 20]. With this in mind, we discuss the principles underlying adaptive bistatic clutter cancellation methods within this chapter. We focus specifically on those bistatic systems where both transmitter and receiver are in motion. Thus, methods described herein can be applied to aerospace bistatic radar configurations.

Since the motions of individual platforms generate a highly complex, nonstationary clutter response, coping with this nonstationary clutter behavior is an integral aspect of bistatic STAP methods. The spread of the ground clutter response in both spatial and doppler domains limits MTI performance. Specifically, the stronger clutter response overwhelms the weaker target signal. Suppressing the ground clutter response to enhance the target signal is the purpose of bistatic STAP. Adaptive processing methods apply data-dependent weightings to capture perceived characteristics of the dynamic clutter response, thereby enhancing the processor's output SINR. As probability of detection is a monotonic function of SINR for a fixed false alarm rate [1], the adaptive processor holds the potential to vastly improve bistatic radar detection performance. Bistatic STAP implementation, however, is not without its challenges.

The field of bistatic STAP is relatively new; early papers include [8–12]. Due to the nonstationary nature of bistatic ground clutter, effective STAP methods require novel modification to traditional, monostatic approaches. Significant progress has been made in this regard through the intervening years. The remainder of this chapter describes these recent advances to the field of bistatic radar.

11.1.1 Organization

The remainder of this chapter is organized as follows:

- Section 11.1 (Notation; Variable Names; Acronyms). The remainder of this section provides notation, principal variables, and acronyms. The use of basic linear algebraic notation and operations is needed to discuss bistatic STAP. The notation described is standard in the signal processing literature.
- Section 11.2 (Bistatic Moving Target Indication). This section introduces the objectives of bistatic MTI radar and performance metrics. It also describes the various types of applicable MTI processing. The reader should note the use of signal vector notation, as well as the form of the space–time filter applied to the observation vector. While the MTI processor in terms of a general weight vector and corresponding FIR filter is formulated, only one weight vector is optimal from the perspective of maximum probability of detection. A simple derivation of this optimal weight vector is provided, thereby forming the basis for STAP. Finally, practical STAP application is highlighted and requirements for effective operation are provided. At this stage, covariance estimation errors and briefly examine their impact on performance are described.
- Section 11.3 (Bistatic Clutter Angle–doppler Response). Bistatic clutter fundamentally limits moving target detection. This section presents a space–time bistatic clutter model and corresponding clutter covariance matrix. Characteristics of the clutter response over range and identify nonstationary behavior are also examined.

- Section 11.4 (Adaptive Bistatic Clutter Cancellation Methods). After establishing the need for bistatic STAP early in this section, a number of bistatic STAP techniques available to the radar designer are described in detail. The different methods fall into three major categories: localized processing, time-varying weights, and data warping methods. Each of these categories includes multiple techniques. The benefits and shortcomings of the approaches are then qualitatively described.
- Section 11.5 (Bistatic STAP Performance Characteristics). Using results from Section 11.3, the detection performance of the bistatic STAP algorithms given in Section 11.4 is quantified. This section presents a general taxonomy describing four typical bistatic radar operating modes: pseudomonostatic, forward adjunct, hybrid and 90° bistatic. This discussion is used to motivate the utility of several of the bistatic STAP approaches.
- Section 11.6 (Summary).

11.1.2 Notation

This section provides a guide to the notation used throughout the remainder of the chapter. In general, a boldface, lowercase variable indicates a vector quantity; a boldface, upper case variable indicates a matrix; a variable with a caret (^) is an estimate of an unknown quantity. Superscripts T or H applied to a vector or matrix denote the transpose or Hermitian (conjugate) transpose operations. The notation $\mathbf{a} \sim CN(\boldsymbol{\mu}_a, \mathbf{R}_a)$ indicates that $\mathbf{a}$ is complex normal (Gaussian) with mean $\boldsymbol{\mu}_a$ and covariance matrix $\mathbf{R}_a$. The notation $\mathbf{x} \in C^{M\times 1}$ indicates $\mathbf{x}$ is an element of the set of complex vectors of dimension $M \times 1$, while $\mathbf{X} \in C^{M\times N}$ indicates a matrix belonging to the set of complex matrices of dimension $M \times N$. Also, $[\mathbf{x}]_m$ or x_m is the mth element of $\mathbf{x}$, $[\mathbf{X}]_{m,n}$ or $x_{m,n}$ is the mnth element of $\mathbf{X}$, $(f_m)_{m=1}^{M}$ is a length-M column vector with elements $f_1, f_2, \ldots, f_M$, $\mathbf{x}_k$ or $\mathbf{X}_k$ represent the kth realization of the given vector or matrix, and $\mathbf{x}_k(n)$ is the nth sample of the kth realization (e.g., $\mathbf{x}_k(n)$ might represent the vector of voltages measured at M antenna elements for the kth range and nth pulse).

The statistical expectation operator is denoted by $E[\cdot]$. The covariance matrix of $\mathbf{a}$ is thus $\mathbf{R}_a = E[(\mathbf{a}-\boldsymbol{\mu}_a)(\mathbf{a}-\boldsymbol{\mu}_a)^H] = E[\mathbf{a}\mathbf{a}^H] - \boldsymbol{\mu}_a\boldsymbol{\mu}_a^H$.

The following linear algebraic operations are used: inner product, outer product, Kronecker (tensor) product, and Schur (Hadamard) product. The inner product describes the output of an FIR filter. Given the weight vector $\mathbf{w} \in C^{P\times 1}$ and data vector $\mathbf{x} \in C^{P\times 1}$, the inner product defines the peak output of a P-tap FIR filter,

$$\mathbf{w}^H\mathbf{x} = \begin{bmatrix} w_1^* & w_2^* & \cdots & w_P^* \end{bmatrix} \begin{bmatrix} x_1 \\ x_2 \\ \vdots \\ x_P \end{bmatrix} = \sum_{m=1}^{P} [\mathbf{w}]_m^* [\mathbf{x}]_m \tag{11.1}$$

where $[\mathbf{w}]_m^* = w_m^*$ is the conjugate of the mth element of $\mathbf{w}$. Optimization problems typically involve the expected value of the outer product; the outer product of the data vector is the $P \times P$ matrix

$$\mathbf{x}\mathbf{x}^H = \begin{bmatrix} x_1x_1^* & x_1x_2^* & \cdots & x_1x_P^* \\ x_2x_1^* & x_2x_2^* & \cdots & x_2x_P^* \\ \vdots & \vdots & & \vdots \\ x_Px_1^* & x_Px_2^* & \cdots & x_Px_P^* \end{bmatrix} \tag{11.2}$$

The Kronecker product conveniently describes a collection of multidimensional observations. For example, given vectors $\mathbf{s}_s \in C^{M\times 1}$ and $\mathbf{s}_t \in C^{N\times 1}$, the Kronecker product is

$$\mathbf{s}_t \otimes \mathbf{s}_s = [[\mathbf{s}_t]_1 \mathbf{s}_s^T \quad [\mathbf{s}_t]_2 \mathbf{s}_s^T \quad \cdots \quad [\mathbf{s}_t]_N \mathbf{s}_s^T]^T \in C^{NM\times 1} \tag{11.3}$$

where $[\mathbf{s}_t]_m$ is the mth element of $\mathbf{s}_t$. Lastly, the Schur product is tantamount to element-wise multiplication of two vectors of the same length. For example, given vector $\mathbf{s}_s \in C^{M\times 1}$ and the window function (taper) $\mathbf{a}_s \in C^{M\times 1}$, the Schur product is

$$\mathbf{a}_s \odot \mathbf{s}_s = [[\mathbf{a}_s]_1[\mathbf{s}_s]_1 \quad [\mathbf{a}_s]_2[\mathbf{s}_s]_2 \quad \cdots \quad [\mathbf{a}_s]_M[\mathbf{s}_s]_M]^T \tag{11.4}$$

The $P \times P$ identity matrix is given as $\mathbf{I_P}$ · $\mathbf{1}_p$ is a p-length vector of 1s. And, $\mathbf{0}_p$ is a p-length vector of 0s.

11.1.3 Variable Names

The following lists many of the variable names found within this chapter:

ϕ_T, θ_T = azimuth and elevation angle in the transmit true north (radar) coordinate system;
ϕ_R, θ_R = azimuth and elevation angle in the receiver true north (radar) coordinate system;
ϕ_c = receive array cone angle;
f_B = bistatic doppler (Hz);
ω_B = bistatic doppler (rads/sec);
$\tilde{\omega}_B$ = normalized bistatic doppler (rads)
$\tilde{\omega}_s$ = normalized spatial frequency (rads);
R_T = range from transmitter to target or clutter;
R_R = range from receiver to target or clutter;
$\mathbf{k_T}$ = wavenumber vector, transmitter radar coordinate system;
$\mathbf{k_R}$ = wavenumber vector, receiver radar coordinate system;
$\mathbf{k_{a/T}}$ = wavenumber vector, transmit array coordinate system;
$\mathbf{k_{a/R}}$ = wavenumber vector, receive array coordinate system;
f_0 = center frequency (Hz);
ω_0 = center frequency (rads/sec);
λ_0 = center wavelength (m);
M = number of channels;
N = number of pulses;
L = number of available range bins;
T = pulse repetition interval (sec);
c = velocity of propagation (speed of light, m/sec);
$\mathbf{v_{p/T}}$ = transmit platform velocity vector (true north);
$\mathbf{v_{p/R}}$ = receive platform velocity vector (true north);
$\mathbf{s}_s$ = spatial steering vector;
$\mathbf{v}_s$ = hypothesized spatial steering vector;
$\mathbf{s}_t$ = temporal steering vector;
$\mathbf{v}_t$ = hypothesized temporal steering vector;
$\mathbf{s}_{s\text{-}t}$ = space-time steering vector;
$\mathbf{v}_{s\text{-}t}$ = hypothesized space-time steering vector;
$\mathbf{x}_{s,k}(n)$ = spatial data snapshot, kth range cell, nth pulse;
$\mathbf{x}_{t,k}(m)$ = temporal data snapshot, kth range cell, mth channel;
$\mathbf{x}_k$ = space-time data snapshot, kth range cell;

$\mathbf{c}_k$ = clutter space-time snapshot, kth range cell;
$\mathbf{n}_k$ = uncorrelated noise space-time sanpshot, kth range cell;
$\mathbf{t}_k$ = target space-time snapshot, kth range cell;
$\mathbf{R}_k$ = null-hypothesis covariance matrix, kth range cell;
$\hat{\mathbf{R}}_k$ = null-hypothesis covariance estimate, kth range cell;
$\mathbf{w}_\mathrm{k}$ = space-time weight vector;
$\hat{\mathbf{w}}_k$ = adaptive space-time weight vector;
$L_{\mathrm{s},1}$ = clairvoyant SINR loss;
$L_{\mathrm{s},2}$ = adaptive SINR loss;
P_d = probability of detection;
P_f = probability of false alarm;
σ_s^2 = single channel, single pulse signal power;
σ_n^2 = single channel, single pulse noise power;
σ_c^2 = single channel, single pulse clutter power;
P_s = output signal power;
P_n = output noise power;

11.1.4 Acronyms

Commonly used acronyms in this chapter include the following:
ADC—angle–doppler compensation;
A^2DC—adaptive angle–doppler compensation;
AMTI—airborne moving target indication;
CNR—clutter-to-noise ratio;
CPI—coherent processing interval;
CUT—cell under test;
CW—continuous wave;
DBF—digital beamformer;
DBU—derivative-based updating;
DoF—degrees of freedom;
EFA—extended factored algorithm;
FFT—fast Fourier transform;
FIR—finite impulse response;
GMTI—ground moving target indication;
HODW—higher order doppler warping;
JDO—joint domain optimal;
MCARM—Multichannel Airborne Radar Measurements;
MDV—minimum detectable velocity;
MTI—moving target indication;
MVDR—minimum variance distortionless response;
PRI—pulse repetition interval;
PSD—power spectral density;
RCS—radar cross section;
RD-STAP—reduced-dimension space–time adaptive processing;
RMB—Reed, Mallett, and Brennan;
RR-STAP—reduced-rank space–time adaptive processing;
Rx—receiver;
SAR—synthetic aperture radar;
SINR—signal-to-interference plus noise ratio;

SNR—signal-to-noise ratio;
STAP—space–time adaptive processing;
SVD—singular value decomposition;
TVW—time-varying weights;
Tx—transmitter;
UDSF—useable doppler space fraction;
ULA—uniform linear array.

11.2 BISTATIC MOVING TARGET INDICATION

11.2.1 Goals

The purpose of MTI radar, as the name aptly suggests, is to detect moving targets. There are two primary MTI modes: ground moving target indication (GMTI), the detection of vehicles and other surface targets; airborne moving target indication (AMTI), the detection of targets in motion above the Earth's surface. Usually the two modes are mutually exclusive. In each case, a file describing target detections and corresponding ranges, angles (or position), and possibly velocity, is the principal output. The MTI radar system then uses this information to track targets of interest.

In both bistatic MTI modes, the presence of clutter signals and receiver noise degrade target detection. The confluence of broadened two-way antenna gain and the generally range-varying nature of stationary clutter returns—those reflections from the Earth's surface—complicate clutter mitigation procedures and, thus, adversely impact detection performance potential in the absence of ameliorating techniques. Bistatic adaptive clutter cancellation, or bistatic STAP, enables effective detection performance potential. STAP is an important *adaptive clutter cancellation* technique for both GMTI and AMTI radar systems [7, 19, 20].

The objective of bistatic SAR, on the other hand, is to estimate the bistatic reflectivity of stationary scatterers, including terrain and man-made objects [21]. An RF image of the scene is the radar data product. For this reason, SAR is often called *imaging* radar. Chapter 10 discusses bistatic SAR. SAR does not directly apply to moving target detection, though efforts have been made in this regard [22].

11.2.2 MTI Methods

Discriminating a target signal from clutter and noise is the purpose of MTI radar operating with a pulsed waveform. The measurement degrees of freedom (DoFs), filtering strategy, characteristics of the clutter environment, and the radar system design determine the effectiveness of the bistatic MTI radar.

The bistatic radar range equation is the starting point for any MTI analysis and defines the signal-to-noise ratio (SNR) for a target of a particular radar cross section (RCS) at a specified bistatic range. The single channel, single pulse SNR at the *m*th subarray is given as

$$\mathrm{SNR}_m^{(1,1)}(\phi_\mathrm{T}, \theta_\mathrm{T}, \phi_\mathrm{R}, \theta_\mathrm{R}) = \left(\frac{P_\mathrm{T} G_\mathrm{T}(\phi_\mathrm{T}, \theta_\mathrm{T}) F_\mathrm{T}}{4\pi R_\mathrm{T}^2 L_\mathrm{T}}\right)\left(\frac{\sigma_\mathrm{P}(\beta) F_\mathrm{R}}{4\pi R_\mathrm{R}^2}\right)\left(\frac{\lambda_0^2 g_{\mathrm{R}/m}(\phi_\mathrm{R}, \theta_\mathrm{R})}{4\pi k_\mathrm{B} T_\mathrm{sys} B_\mathrm{n} L_\mathrm{R}}\right) G_\mathrm{sp} \tag{11.5}$$

The peak transmit power is P_T, G_T (ϕ_T, θ_T) is the transmit gain at the specified azimuth and elevation angles, R_T and R_R represent one-way ranges from transmitter to target and target to receiver, F_T and F_R are transmit and receive propagation losses, L_T and L_R are transmit and receive RF losses, σ_P (β) is the target RCS at bistatic angle β (see Figure 11–4), $g_{\mathrm{R}/m}$ (ϕ_R, θ_R) is the *m*th subarray receive gain, k_B is Boltzmann's constant, T_{sys} is the system temperature (including both noise figure and standard temperature terms), B_n is the receiver bandwidth, and G_{sp} is the pulse compression gain [2].

In addition to providing adequate SNR, the radar system must provide a mechanism to suppress clutter and jamming, collectively known as interference. The output SINR strongly influences the corresponding probability of detection and is given by

$$\text{SINR}(\phi_T, \theta_T, \phi_R, \theta_R; f_B) = \text{SNR}(\phi_T, \theta_T, \phi_R, \theta_R) \cdot L_{s,1}(\phi_T, \theta_T, \phi_R, \theta_R; f_B) \cdot L_{s,2}(\phi_T, \theta_T, \phi_R, \theta_R; f_B) \quad (11.6)$$

where f_B is bistatic doppler frequency in hertz, SNR is the integrated signal-to-noise ratio (involving the coherent combination of all pulse and channel data), $L_{s,1}$ is known as *clairvoyant* SINR loss, and $L_{s,2}$ is *adaptive* SINR loss. Both $L_{s,1}$ and $L_{s,2}$ are bound between unity (no loss) and zero (full loss). Clairvoyant SINR loss represents the degradation in output SINR due to the presence of interference (colored noise, such as ground clutter) at the sensor's input presuming perfect statistical knowledge of the interfering source. Such clairvoyance is not available in practice, and so the real-world implementation must substitute an appropriate estimate; errors between that estimate and the actual statistical behavior lead to adaptive loss. If the system does not require estimated quantities, $L_{s,2}$ equals unity, but generally the $L_{s,1}$ term greatly increases, thus offsetting any advantage. The next section provides formal definitions for these loss factors.

As seen from equation (11.6), SINR can never exceed SNR. Colored noise suppresses detection performance potential over the unambiguous doppler interval. A complete bistatic MTI radar system design must manage all three factors on the right-hand side of equation (11.6). This design must consider the antenna, algorithm selection, and characteristics of the operating environment in addition to the traditional power–aperture concerns of the radar engineer.

Consider a bistatic receive antenna consisting of M channels[1]. A receiver and analog-to-digital converter are present at each channel. The multiple channels provide spatial DoFs and access to directional information characterizing a propagating electromagnetic wavefront. Moreover, the receiver collects N pulses transmitted with a pulse repetition interval (PRI) of T. The M channels constitute the spatial aperture, whilst the N pulses define the coherent processing interval (CPI); the PRI defines a slow-time measurement domain. We further presume the availability of L fast-time, or bistatic range samples, after conversion to complex baseband and pulse compression operations. We thus define the space–time snapshot for the kth range as $\mathbf{x}_k \in C^{MN \times 1}$. The mnth element of $\mathbf{x}_k$ is the complex baseband and pulse compressed voltage measured for the mth channel at the nth pulse and kth range. The data from element to element in $\mathbf{x}_k$ are coherent, a feature requiring stable phase characteristics of the local oscillator at the receive site.

Two models characterize $\mathbf{x}_k$,

$$\begin{aligned} H_0 &: \mathbf{x}_k = \mathbf{c}_k + \mathbf{n}_k \\ H_1 &: \mathbf{x}_k = \mathbf{t}_k + \mathbf{c}_k + \mathbf{n}_k \end{aligned} \quad (11.7)$$

where H_0 is known as the null hypothesis and corresponds to target absence, and H_1 is the alternative hypothesis of target presence. The components $\mathbf{c}_k$, $\mathbf{n}_k$, and $\mathbf{t}_k$ represent clutter, receiver noise, and target contributions, respectively. The MTI processor must combine the elements of $\mathbf{x}_k$ and then judge the corresponding model—null or alternative hypothesis—leading to a particular output

1 A spatial channel is also known as a subarray. Generally, multiple antenna elements and an analog beamforming network comprise a subarray.

observation. Since ambiguity exists between the two models, a statistical procedure is needed to provide a rational decision mechanism.

The likelihood ratio test is the optimal strategy to combine and threshold the multiple spatial and slow-time samples for a particular range realization [1, 7]. For the purposes of discussion, the detector takes the form

$$f_\psi\left(\mathbf{w}_k^H \mathbf{x}_k\right) \underset{H_0}{\overset{H_1}{\gtrless}} \upsilon_T \tag{11.8}$$

where $f_\psi(\cdot)$ is a nonlinear operator, $\mathbf{w}_k \in C^{MN \times 1}$ is a weight vector defining a linear, FIR filter, and υ_T is a threshold set according to the output conditional probability density function characterizing the null-hypothesis case and meeting the specified false alarm rate [1, 23]. Different linear weightings are possible and include the class of data-independent (deterministic) and data-dependent (adaptive) formulations. The choice of the weight vector strongly influences the overall detection performance. Specifically, the weight vector leading to maximum output SINR minimizes the amount of overlap between output null and alternative hypothesis conditional probability density functions, thereby maximizing the probability of detection for a fixed false alarm rate [1]. We describe the maximum SINR weight vector subsequently.

The measurement DoFs are critical to achieving superior performance. In general, radar systems may measure the angle of arrival, doppler frequency, range delay, amplitude, and polarization. The first three measurements couple to time difference of arrival: the angle of arrival relates to the spatial variation of time delay across a multichannel array, doppler frequency results from the varying time delay from pulse to pulse, and range information results from the summed time delay from transmitter to target and then target to receiver. Accessing the different measurement DoFs requires specific hardware, such as multiple spatial channels to ascertain angle information or polarimetric channels to measure the polarization of the receive signal.

This chapter's interest lies in exploiting spatial and slow-time DoFs to suppress ground clutter returns[2]. These measurement domains and matched filtering operations are considered a cornerstone to the discussion on STAP.

11.2.2.1 *Doppler Processing* Doppler processing involves testing for different linear phase slopes across a temporal aperture. This phase slope and the corresponding matched filter accomplishing this phase slope testing procedure are defined.

Suppose the initial bistatic range to a target of interest is $R_T(0) + R_R(0)$ and the target remains in a single resolution cell over the CPI. If the target moves with constant velocity over the dwell, the range variation is

$$R_\Sigma(n) = R_T(n) + R_R(n) = (R_T(0) + n\Delta R_T) + (R_R(0) + n\Delta R_R) \tag{11.9}$$

where ΔR_T and ΔR_R represent the pulse-to-pulse change in range between the target and either transmitter or receiver, and $n = 0, 1, \ldots, N-1$ is the pulse number. $R_\Sigma(0)$ is uniformly distributed over the range resolution cell, since the precise target starting range is unknown. The range variation from the initial pulse is the critical factor. The corresponding time delay

2 Angle of arrival is the Fourier transform of space, whereas doppler frequency is the Fourier transform over the pulse domain.

is $\Delta t(n) = R_\Sigma(n)/c$, where c is the velocity of propagation (nominally the speed of light) and the resulting signal phase is

$$\psi_B(n) = \Delta t(n) \cdot 2\pi f_0 = \frac{2\pi}{\lambda_0} R_\Sigma(n) \tag{11.10}$$

f_0 is center frequency of the transmit waveform, λ_0 is the corresponding wavelength, and $f_0 \lambda_0 = c$. The time derivative of phase yields radian frequency,

$$\omega_B = \frac{d}{dt}\psi_B(n) = \frac{2\pi}{\lambda_0}\left(\frac{\Delta R_T}{T} + \frac{\Delta R_R}{T}\right) \tag{11.11}$$

The resulting bistatic doppler frequency is just equation (11.11) scaled by the inverse of 2π, or

$$f_B = \frac{v_{r/T} + v_{r/R}}{\lambda_0} \tag{11.12}$$

$v_{r/T} = \Delta R_T/T$ and $v_{r/R} = \Delta R_R/T$ are radial velocity components between transmitter and target and target and receiver.

Equations (11.9) through (11.12) summarize the bistatic doppler frequency as the slope of the time-varying phase function across the temporal aperture resulting from motion between transmitter, target, and receiver. The use of the fast Fourier transform (FFT), synonymous with doppler processing, tests for different phase slopes across the collection of multipulse data.

It is convenient to introduce steering vector notation at this point as a more general framework for doppler processing. Using the first pulse as a reference, the complex baseband voltages measured at the mth channel over N pulses and corresponding to a point source are

$$\mathbf{x}_{t,k}(m) = \alpha_{s,k}(m)\left[1 \quad \exp(j\tilde{\omega}_B) \quad \exp(j2\tilde{\omega}_B) \quad \cdots \quad \exp(j(N-2)\tilde{\omega}_B) \quad \exp(j(N-1)\tilde{\omega}_B)\right]^T \tag{11.13}$$

$\alpha_{s,k}(m)$ is a complex, random scalar measured at the mth channel. It exhibits spatial correlation consistent with the target direction of arrival (a matter to be discussed in the next section) and $\sigma_s^2 = E[|\alpha_{s,k}(m)|^2]$ is the single channel, single pulse signal power. Also, $\tilde{\omega}_B = 2\pi f_B T$ is the normalized doppler frequency. Moreover, we identify $\mathbf{x}_{t,k}(m)$ as the temporal snapshot, whilst the column vector in equation (11.13), viz.,

$$\mathbf{s}_t(\tilde{\omega}_B) = \left[1 \quad \exp(j\tilde{\omega}_B) \quad \exp(j2\tilde{\omega}_B) \quad \cdots \quad \exp(j(N-2)\tilde{\omega}_B) \quad \exp(j(N-1)\tilde{\omega}_B)\right]^T \tag{11.14}$$

is known as a temporal steering vector. The linear phase variation among the elements of equation (11.14) indicates the doppler frequency. The FFT approach to doppler processing projects a collection of temporal steering vectors against the data to test for a particular doppler frequency in a numerically efficient manner. Mathematically, this projection operation is

$$y_k(m, \tilde{\omega}_B) = s_t^H(\tilde{\omega}_B)\mathbf{x}_{t,k}(m) \tag{11.15}$$

Defining a real temporal weighting, such as a Hamming or Chebychev weighting (sometimes called apodization), by $\mathbf{a}_t \in R^{N \times 1}$, the weights corresponding to the doppler filter generally appear as

$$\mathbf{w}_{t,k} = \mathbf{a}_t \odot \mathbf{s}_t(\tilde{\omega}_B) \tag{11.16}$$

so that $y_k(m, \tilde{\omega}_B) = \mathbf{w}_{t,k}^H(\tilde{\omega}_B)\, \mathbf{x}_{t,k}(m)$.

The elements of the temporal steering vector define a matched filter. The matched filter maximizes the output SNR. To see this, express SNR as

$$\text{SNR} = \frac{P_s}{P_n} = \frac{E[|\mathbf{w}_{t,k}^H \mathbf{x}_{t,k}(m)|^2]}{E[\mathbf{w}_{t,k}^H \mathbf{n}_{t,k}(m)\mathbf{n}_{t,k}^H(m)\mathbf{w}_{t,k}]} = \frac{\sigma_s^2 |\mathbf{w}_{t,k}^H \mathbf{s}_{t,k}(\tilde{\omega}_B)|}{\sigma_n^2 \mathbf{w}_{t,k}^H \mathbf{w}_{t,k}} \tag{11.17}$$

where P_s and P_n are output signal and noise powers, respectively, and $\mathbf{n}_{t,k} \sim CN(\mathbf{0}, \sigma_n^2 \mathbf{I}_N)$ is the temporal noise snapshot with input noise variance of σ_n^2. Using the Schwarz inequality, (11.17) is expressed as

$$\text{SNR} \le \frac{(\mathbf{w}_{t,k}^H \mathbf{w}_{t,k})(\mathbf{s}_t^H(\tilde{\omega}_B)\mathbf{s}_t(\tilde{\omega}_B))}{\mathbf{w}_{t,k}^H \mathbf{w}_{t,k}} \tag{11.18}$$

thereby indicating that the optimal choice for the weight vector—i.e., the one maximizing SNR—is $\mathbf{w}_{t,k/opt} = \mu \mathbf{s}_t(\tilde{\omega}_B)$, for arbitrary scalar μ. The apodization $\mathbf{a}_t$ controls doppler side lobe levels at the expense of degraded SNR and main lobe width, since a mismatch occurs according to equation (11.18).

11.2.2.2 Digital Beamforming Using an array of spatial sensors, the radar determines the direction of a propagating electromagnetic wave by effectively measuring time difference of arrival as a phase variation across the aperture [23]. Specifically, the array digitizes the voltage outputs of M channels, thereby gathering spatial information embodied in the relative phase of each of the measured signals. A collection of antenna elements, previously defined as a subarray, feeds each receiver channel. For the purpose of this discussion, we assume the following conditions: the far-field approximation applies, indicating a planar, propagating wave; the signals are narrowband, suggesting the signal bandwidth is a very small fraction of the signal carrier frequency; M equal-size subarrays comprise the overall antenna array; the array configuration exhibits uniform spacing with horizontal, linear orientation.

Figure 11–1 depicts the corresponding scenario of a propagating plane wave impinging on a uniform linear array (ULA). The angle ϕ_c identified in Figure 11–1 is known as a cone angle. A ULA does not uniquely measure azimuth and elevation, generically given as ϕ and θ in the figure; rather, it senses the cone angle. In this case, the cone angle defines a conical ambiguity surface centered on the horizontal axis with the apex at the origin. The unit vector, $\mathbf{k}(\phi, \theta)$, points orthogonal to the propagating wavefront and is sometimes referred to as the wavenumber vector[3].

The signal angle of arrival, frequency, and physical location of a given subarray influence the varying phase shift that the propagating wave induces among each of the channels. Let $\tau_s(m)$ represent the delay between the time it takes the wavefront to arrive at a suitably designated reference point and the mth subarray. The corresponding phase shift at the mth channel is then

$$\gamma_s(m) = \tau_s(m)\omega_0 \tag{11.19}$$

3 Often, the wavenumber vector has a scaling, such as $2\pi/\lambda_0$.

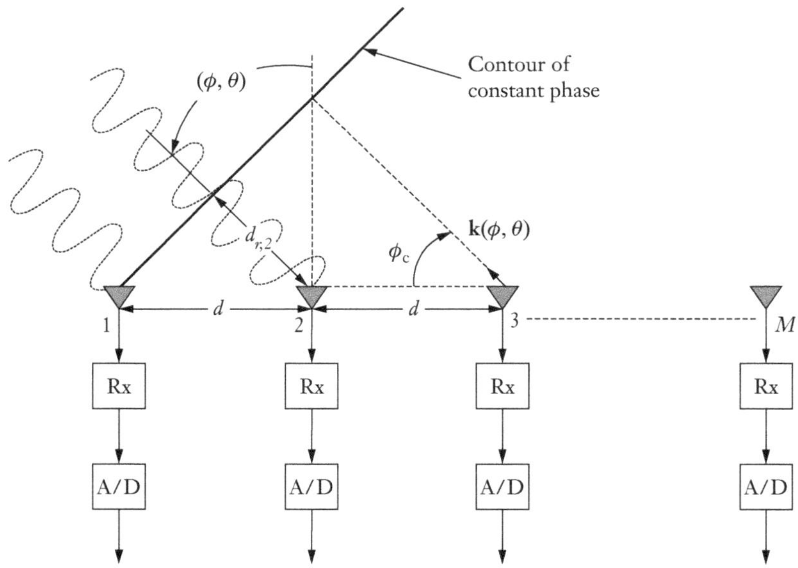

Figure 11–1 Propagating plane wave impinging on a uniform linear array comprised of *M* channels (Rx denotes "receiver" and A/D indicates "analog-to-digital converter").

where the time delay is given by $\tau_s(m) = d_{r,m}/c$. $d_{r,m}$ is the distance the wavefront must travel after impinging on the reference point[4]. From the geometry in Figure 11–1,

$$d_{r,m} = (m - 1)\, d \cos \phi_c \tag{11.20}$$

with the first channel as the reference. The phase shift at the mth channel follows as

$$\gamma_s(m) = (m-1)\frac{\omega_0}{c} d \cos\phi_c = (m-1)\frac{2\pi}{\lambda_0} d \cos\phi_c \tag{11.21}$$

Taking the spatial derivative of equation (11.21) and then normalizing by the sensor spacing yields a normalized spatial frequency

$$\tilde{\omega}_s = 2\pi \frac{d}{\lambda_0} \cos\phi_c \tag{11.22}$$

As seen, the characteristics of the spatially sampled signal are mathematically similar to the temporal sampling case of the prior section.

Using the first channel as a reference, the complex baseband voltages measured at the nth pulse over M channels and corresponding to a point source are

$$\mathbf{x}_{\mathbf{s,k}}(n) = \alpha_{\mathrm{t,k}}(n)\left[1 \quad \exp(j\tilde{\omega}_s) \quad \exp(j2\tilde{\omega}_s) \quad \cdots \quad \exp(j(M-2)\tilde{\omega}_s) \quad \exp(j(M-1)\tilde{\omega}_s)\right]^T \tag{11.23}$$

4 The reference point is arbitrary. For simplicity, the first channel to the far left in Figure 11–1 has been chosen.

$\alpha_{t,k}(n)$ is a complex, random scalar measured at the nth pulse. It exhibits temporal correlation consistent with the target bistatic doppler frequency and $\sigma_s^2 = E[|\alpha_{t,k}(n)|^2]$ is the single channel, single pulse signal power, consistent with our earlier discussion. We identify $\mathbf{x}_{s,k}(n)$ as the spatial snapshot, whilst the column vector in equation (11.23) is the spatial steering vector:

$$\mathbf{s}_s(\tilde{\omega}_s) = [1 \quad \exp(j\tilde{\omega}_s) \quad \exp(j2\tilde{\omega}_s) \quad \cdots \quad \exp(j(M-2)\tilde{\omega}_s) \quad \exp(j(M-1)\tilde{\omega}_s)]^T \tag{11.24}$$

The linear phase variation among the elements of equation (11.24) indicates the spatial frequency.

Following an argument similar to that in the prior section, it may be shown that the spatial matched filter weights are

$$\mathbf{w}_{s,k} = \mu\mathbf{s}_s(\tilde{\omega}_s) \tag{11.25}$$

As in the temporal case, a real weighting, $\mathbf{a}_s \in R^{M\times 1}$, controls side lobe levels at the expense of SNR and main lobe width.

11.2.2.3 Space–Time Adaptive Processing (STAP) Doppler and spatial beamforming methods are components of STAP. Both doppler and spatial beamformers are matched filters—or close approximations thereof—and thus maximize the signal power relative to the uncorrelated noise component. In contrast, STAP attempts to maximize output SINR through the use of a data-dependent weighting. It is used when colored noise, such as ground clutter reflection, is present [7, 19, 20].

We first consider the optimal weights in the maximum SINR sense. Maximizing the output SINR is a key objective of the space–time processor. From equation (11.7), the output SINR is the ratio of signal-to-interference-plus-noise power at the filter output:

$$\text{SINR} = \frac{E[\mathbf{w}_k^H \mathbf{t}_k \mathbf{t}_k^H \mathbf{w}_k]}{E[\mathbf{w}_k^H \mathbf{x}_{k/H_0} \mathbf{x}_{k/H_0}^H \mathbf{w}_k]} = \frac{\sigma_s^2 |\mathbf{w}_k^H \mathbf{s}_{s\text{-}t}(\tilde{\omega}_B, \tilde{\omega}_s)|^2}{\mathbf{w}_k^H \mathbf{R}_k \mathbf{w}_k} \tag{11.26}$$

We assume $\mathbf{t}_k = \alpha_{s\text{-}t,k}\mathbf{s}_{s\text{-}t}(\tilde{\omega}_B, \tilde{\omega}_s)$, where $\mathbf{s}_{s\text{-}t}(\tilde{\omega}_B, \tilde{\omega}_s) = \mathbf{s}_t(\tilde{\omega}_B) \otimes \mathbf{s}_s(\tilde{\omega}_s)$ is the space–time steering vector. Also, we consider the colored Gaussian case, wherein $\mathbf{x}_{k/H_0} \sim CN(\mathbf{0}, \mathbf{R}_k)$, and $\mathbf{R}_k = \mathrm{E}[\mathbf{x}_{k/H_0} \mathbf{x}_{k/H_0}^H]$ is the interference covariance matrix. $\alpha_{s\text{-}t,k}$ is a complex RMS voltage and $\sigma_s^2 = E[|\alpha_{s\text{-}t,k}|^2]$ (i.e., this is the previously noted single channel, single pulse signal power). Further note that the Kronecker product indicates the data organization of $\mathbf{x}_k$: the first M elements correspond to the spatial snapshot for the first received pulse, the second M correspond to the spatial snapshot for the second pulse, and so forth, i.e., $\mathbf{x}_k = (\mathbf{x}_{s,k}(n))_{n=1}^N$.

The *optimal weight vector* maximizes the output SINR and takes the form $\mathbf{w}_k = \beta_w \mathbf{R}_k^{-1}\mathbf{s}_{s\text{-}t}(\tilde{\omega}_B, \tilde{\omega}_s)$, for arbitrary scalar β_w [1]. To see this, express equation (11.26) as

$$\text{SINR} = \sigma_s^2 \frac{|\hat{\mathbf{w}}_k^H \hat{\mathbf{s}}|^2}{\hat{\mathbf{w}}_k^H \hat{\mathbf{w}}_k} \leq \sigma_s^2 \frac{\hat{\mathbf{w}}_k^H \hat{\mathbf{w}}_k \hat{\mathbf{s}}^H \hat{\mathbf{s}}}{\hat{\mathbf{w}}_k^H \hat{\mathbf{w}}_k} \tag{11.27}$$

where $\mathbf{w}_k = \mathbf{R}_k^{-1/2}\hat{\mathbf{w}}_k$ and $\mathbf{s}_{s\text{-}t}(\tilde{\omega}_B, \tilde{\omega}_s) = \mathbf{R}_k^{1/2}\hat{\mathbf{s}}$. For the covariance matrices of interest, $\mathbf{R}_k = \mathbf{R}_k^{1/2}\mathbf{R}_k^{1/2}$. By choosing $\hat{\mathbf{w}} = \hat{\mathbf{s}}$, equation (11.27) achieves the upper bound. Substituting the prior expressions gives

$$\hat{\mathbf{w}}_k = \mathbf{R}_k^{1/2}\mathbf{w}_k = \hat{\mathbf{s}} = \mathbf{R}_k^{-1/2}\mathbf{s}_{s\text{-}t}(\tilde{\omega}_B, \tilde{\omega}_s) \Rightarrow \mathbf{w}_k = \mathbf{R}_k^{-1}\mathbf{s}_{s\text{-}t}(\tilde{\omega}_B, \tilde{\omega}_s) \tag{11.28}$$

Scaling the weight vector by β_w does not alter the output SINR.

In practice, neither $\mathbf{R}_k$ nor $\mathbf{s}_{s\text{-}t}(\tilde{\omega}_B, \tilde{\omega}_s)$ is known with certainty. STAP attempts to approximate the optimal filter response by replacing $\mathbf{R}_k$ with the estimate $\hat{\mathbf{R}}_k$, and $\mathbf{s}_{s\text{-}t}(\tilde{\omega}_B, \tilde{\omega}_s)$ with a surrogate space–time steering vector $\mathbf{v}_{s\text{-}t}(\tilde{\omega}'_B, \tilde{\omega}'_s)$; the prime over the frequency variables suggests a mismatch between actual and hypothesized target doppler and spatial frequencies. The corresponding STAP weight vector is $\hat{\mathbf{w}}_k = \hat{\beta}_w \hat{\mathbf{R}}_k^{-1} \mathbf{v}_{s\text{-}t}(\tilde{\omega}'_B, \tilde{\omega}'_s)$.

The most common approach to estimate $\hat{\mathbf{R}}_k$ is given by the expression

$$\hat{\mathbf{R}}_k = \frac{1}{K} \sum_{m=1}^{K} \mathbf{x}_m \mathbf{x}_m^H \tag{11.29}$$

The data vectors $\mathbf{x}_m$ used in equation (11.29) are known as *secondary* or *training* data, and K describes the number of vectors included in the computation. When the training data are independent and identically distributed (iid), equation (11.29) is a maximum likelihood estimate for $\mathbf{R}_k$ [24]. Moreover, it is shown in [24] that choosing $K = 2NM - 3$ leads to 3-dB average loss between adaptive and optimal filters, under the assumption that $\mathbf{v}_{s\text{-}t}(\tilde{\omega}'_B, \tilde{\omega}'_s) = \mathbf{s}_{s\text{-}t}(\tilde{\omega}_B, \tilde{\omega}_s)$.

This result is known as the RMB rule after its inventors. When the data are non-iid, the loss naturally increases.

The difference between $\mathbf{v}_{s\text{-}t}(\tilde{\omega}'_B, \tilde{\omega}'_s)$ and $\mathbf{s}_{s\text{-}t}(\tilde{\omega}_B, \tilde{\omega}_s)$ is known as steering vector mismatch and leads to degradation of the matched filtering operation. Sources of mismatch include straddling (hypothesized values for $\tilde{\omega}'_B$ and $\tilde{\omega}'_s$ straddle the true target spatial and doppler frequencies) and array amplitude and phase errors. Array design and calibration rectifies the latter issue, whilst increasing the number of test frequencies resolves the former.

Covariance estimation errors are significantly more consequential than steering vector mismatch. Covariance estimation error occurs under the following situations:

- The secondary data are iid, but K is finite. The RMB rule accurately captures the consequent performance degradation. This situation is ideal.
- The secondary data are non-iid as a result of sensor-induced nonstationarity. In this case, the covariance structure varies over range in a relatively deterministic manner given precise knowledge of the platform state vectors, receive antenna orientation, and terrain height. This situation is symptomatic of bistatic radar operation.
- Secondary data appear heterogeneous as a result of cultural variations in clutter type [25]. Land–sea interfaces, shadowing due to mountainous terrain, clutter discretes, etc., are all examples of clutter heterogeneity. Monostatic and bistatic radar systems must contend with heterogeneous clutter. Sensor-induced nonstationarity further exacerbates clutter heterogeneity.

Bistatic STAP concerns mitigating covariance estimation errors corresponding to sensor-induced nonstationarity.

11.2.2.3.1 Reduced-Dimension STAP As described in [7, 19, 20], reduced-dimension STAP (RD-STAP) applies a linear transformation to project the space–time data vector $\mathbf{x}_k$ into a lower dimensional subspace. The transformed data vector is

$$\tilde{\mathbf{x}}_k = \mathbf{T}^H \mathbf{x}_k; \quad \mathbf{T} \in C^{NM \times J} \tag{11.30}$$

where $J << NM$ and $\tilde{\mathbf{x}}_k$ has dimension $J \times 1$. The computational burden associated with matrix inversion drops from $O(N^3M^3)$ to $O(J^3)$, and nominal sample support decreases from $2NM$ to $2J$ in accord with the RMB rule.

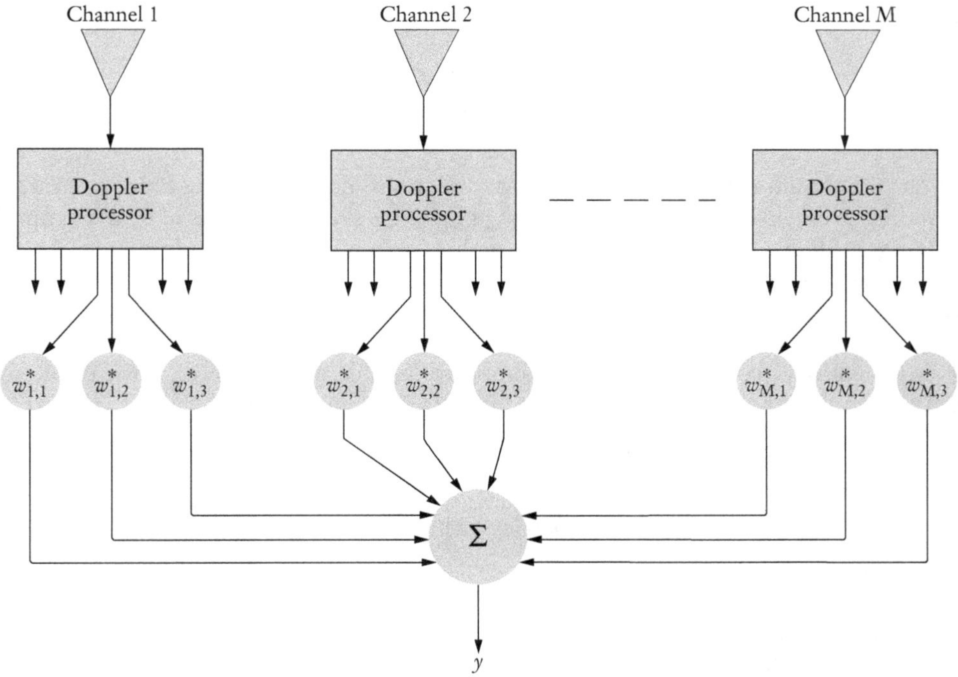

Figure 11–2 Block diagram of the EFA RD-STAP method. In this diagram, $w^*_{m,n}$ is the weighting applied to the *m*th channel and adjacent doppler filter *n*. (After [20], © 2004 IEEE.)

The $\mathcal{J} \times \mathcal{J}$ null-hypothesis covariance matrix corresponding to equation (11.30) is

$$\tilde{\mathbf{R}}_k = E\left[\tilde{\mathbf{x}}_{k/\mathbf{H}_0}\tilde{\mathbf{x}}^H_{k/\mathbf{H}_0}\right] = \mathbf{T}^H\mathbf{R}_k\mathbf{T} \tag{11.31}$$

Applying the same transformation to the space–time steering vector gives $\tilde{\mathbf{s}} = \mathbf{T}^H \mathbf{s}_{s\text{-}t}(\tilde{\omega}_B, \tilde{\omega}_s)$. The corresponding optimal weight vector is $\hat{\tilde{\mathbf{w}}}_k = \tilde{\beta}\hat{\tilde{\mathbf{R}}}_k^{-1}\tilde{\mathbf{s}}$, for arbitrary scalar $\tilde{\beta}_w$. The adaptive solution involves calculating $\tilde{\mathbf{R}}_k$ from equation (11.29) using the transformed training data set $\{\mathbf{T}^H\mathbf{x}_m\}^K_{m=1}$, replacing $\tilde{\mathbf{s}}$ with the hypothesized steering vector $\tilde{\mathbf{v}}$, and then forming and applying the adaptive weight vector $\hat{\tilde{\mathbf{w}}}_k = \hat{\tilde{\mathbf{R}}}_k^{-1}\tilde{\mathbf{v}}$.

A variety of choices for $\mathbf{T}$ are possible [26]. Naturally, the best choices provide an effective combination of DoFs to mitigate interference while minimizing computational burden and the requisite number of training samples for covariance estimation. A method of interest is a multibin, postdoppler, element space method known as the extended factored algorithm (EFA) [27]. Figure 11–2 provides a block diagram of the EFA method; the RD-STAP adapts over all available spatial DoFs and several adjacent doppler filters. EFA is used in subsequent sections of the chapter.

11.2.2.3.2 Reduced-Rank STAP Reduced-rank STAP (RR-STAP) is a weight calculation technique and generally involves manipulating eigenvalues and eigenvectors of the sample covariance matrix. Details of RR-STAP are given in [7, 19]. As an example, the optimal weight vector is expressible as [28]

$$\mathbf{w}_{\mathrm{k}} = \frac{1}{\lambda_{\mathrm{n}}}\left[\mathbf{s}_{\text{s-t}}\left(\tilde{\omega}_{\mathrm{B}},\tilde{\omega}_{\mathrm{s}}\right) - \sum_{m=1}^{NM}\frac{\lambda_{k,m}-\lambda_{n}}{\lambda_{k,m}}\alpha_{k,m}\mathbf{q}_{\mathrm{k,m}}\right] \tag{11.32}$$

where λ_{n} is a minimum eigenvalue characterizing the noise floor, $\lambda_{k,m}$ is the mth eigenvalue of $\mathbf{R}_{\mathrm{k}}$ corresponding to eigenvector $\mathbf{q}_{\mathrm{k,m}}$, and $\alpha_{k,m} = \mathbf{q}_{\mathrm{k,m}}^{H}\mathbf{s}_{\text{s-t}}(\tilde{\omega}_{\mathrm{B}},\tilde{\omega}_{\mathrm{s}})$. Define the rank of $\mathbf{R}_{\mathrm{k}}$ as the number of dominant eigenvalues, r_{λ}. The noise floor eigenvalues do not affect the result in equation (11.32), letting us replace the upper limit on the sum from NM to r_{λ}, where ideally $r_{\lambda} << NM$; this is the reduced-rank formulation. In the adaptive case, the RMB rule becomes roughly $2r_{\lambda}$ versus $2NM$ for the standard formulation.

The computational cost of the eigendecomposition, challenges in rank estimation, and the potential for significant clutter rank (i.e., number of dominant eigenvalues) leads to the preference of the RD-STAP methods in the previous section.

11.2.3 Performance Metrics

Important bistatic radar MTI detection metrics include the following:

- Probability of detection (P_{d});
- Probability of false alarm (P_{f});
- Signal-to-interference-plus-noise ratio (SINR);
- Clairvoyant SINR loss ($L_{\mathrm{s,1}}$);
- Adaptive SINR loss ($L_{\mathrm{s,2}}$).

These metrics are interrelated. For example, P_{d} is a monotonic function of SINR for a fixed P_{f} in the Gaussian case [1, 23], and SINR = SNR $\cdot\, L_{\mathrm{s,1}} \cdot L_{\mathrm{s,2}}$. Other important metrics include the following: minimum detectable velocity (MDV), the lowest velocity relative to stationary clutter wherein reliable detection is achieved, useable doppler space fraction (UDSF), the percentage of the available doppler space where detection meets or exceeds specifications, and area coverage rate. These latter metrics follow directly from detection performance characteristics (P_{d}, SINR, SINR loss). SINR loss is the preferred approach to characterize algorithm capability and compare various methods.

11.2.3.1 Probabilities of Detection and False Alarm An optimum detection statistic for the space–time combiner $y_k = \mathbf{w}_{\mathrm{k}}^{H}\mathbf{x}_{\mathrm{k}}$, under the Gaussian assumption $\mathbf{x}_{\mathrm{k/H_0}} \sim \mathrm{CN}(\mathbf{0}, \mathbf{R}_{\mathrm{k}})$ previously noted, follows from the likelihood ratio test. The decision statistic is of the form of equation (11.8) and takes the form [1, 29]

$$|y_k| \underset{H_0}{\overset{H_1}{\gtrless}} v_{\mathrm{T}} \tag{11.33}$$

The performance of equation (11.33) is given by

$$\begin{aligned} P_{\mathrm{f}} &= \exp\left(\frac{-\beta_{\mathrm{T}}^{2}}{2}\right); \\ P_{\mathrm{d}} &= \int_{\beta_{\mathrm{T}}}^{\infty} u \ \exp\left(\frac{-\left(u^{2}+\alpha_{1}^{2}\right)}{2}\right) I_{0}\left(\alpha_{1}u\right)\mathrm{d}u \end{aligned} \tag{11.34}$$

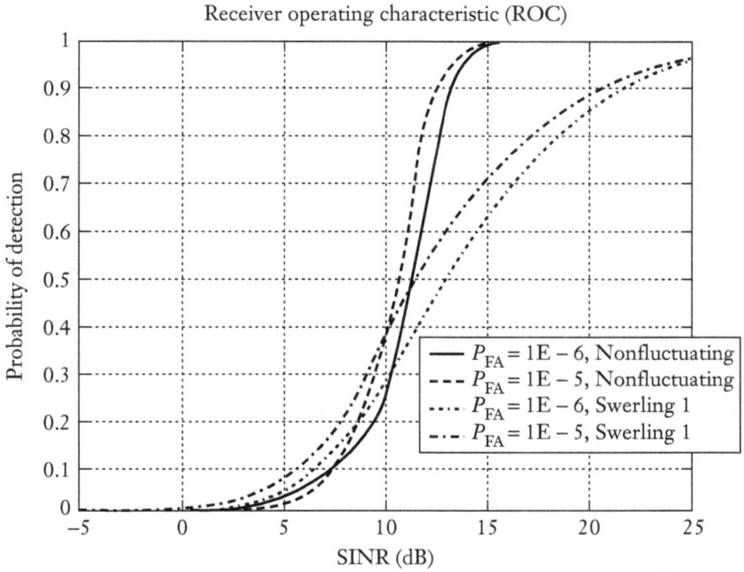

Figure 11–3 Receiver operating characteristics for nonfluctuating and fluctuating targets. (After [20], © 2004 IEEE.)

where P_f is the probability of false alarm, P_d is the probability of detection, $\beta_T = v_T/\sqrt{\mathbf{w}_k^H \mathbf{R}_k \mathbf{w}_k}$ is a normalized detection threshold, $I_0(\cdot)$ is the modified zero-order Bessel function of the first kind, and α_1 is equal to the square root of the peak output SINR:

$$\alpha_1^2 = 2 \times \text{SINR} = 2 \times \frac{\sigma_s^2 |\mathbf{w}_k^H \mathbf{s}_{s\text{-}t}(\tilde{\omega}_B, \tilde{\omega}_s)|^2}{\mathbf{w}_k^H \mathbf{R}_k \mathbf{w}_k} \tag{11.35}$$

Figure 11–3 characterizes equation (11.34), as well as the case of a Swerling 1 fluctuating target [4], and is known as a receiver operating characteristic (ROC) curve. Note that (11.34) is a monotonic function of α_1, and hence α_1^2; as Figure 11–3 shows, maximizing SINR maximizes the probability of detection for a fixed false alarm rate. Equation (11.34) and Figure 11–3 apply for any choice of $\mathbf{w}_k$ whether the interference is colored or uncorrelated Gaussian. The choice of weight vector leading to maximal output SINR thus likewise maximizes P_d for a fixed P_f; this weight vector is given in equation (11.28).

11.2.3.2 SNR, SINR, and SINR Loss Bistatic SNR is given by equation (11.5) for the single channel, single pulse case. After integrating over space and time, the output SNR is

$$\text{SNR} = \frac{\sigma_s^2}{\sigma_n^2} NM \tag{11.36}$$

Output SINR is given by the expression in equation (11.26). Note that when $\mathbf{R}_k = \sigma_n^2 \mathbf{I}_{NM}$ (i.e., a scaled $NM \times NM$ identity matrix, indicating the interference is uncorrelated receiver noise only), equation (11.26) reduces to equation (11.36) when using the optimal weight vector.

SINR loss captures the degradation in SINR as a result of nonideal effects. Each SINR loss is bound between unity (no loss) and zero (complete loss), as previously noted. Clairvoyant SINR loss is given by

$$L_{s,1}(\tilde{\omega}_B, \tilde{\omega}_s) = \frac{\mathrm{SINR}|_{\mathbf{w}_k}}{\mathrm{SNR}|_{\mathbf{w}_k = \mathbf{s}_{s\text{-}t}(\tilde{\omega}_B, \tilde{\omega}_s)}} = \frac{\mathbf{s}_{s\text{-}t}^H(\tilde{\omega}_B, \tilde{\omega}_s)\mathbf{R}_k^{-1}\mathbf{s}_{s\text{-}t}(\tilde{\omega}_B, \tilde{\omega}_s)}{\sigma_n^{-2}NM} \quad (11.37)$$

for the case of a constant amplitude target over the coherent dwell when $\mathbf{w}_k$ is given by equation (11.28) for the numerator calculation. Factors influencing $L_{s,1}$ $(\tilde{\omega}_B, \tilde{\omega}_s)$ include characteristics of the radar system design and properties of the clutter environment, as well as the particular STAP algorithm. A second commonly used SINR loss term is the adaptive loss [24]. This loss term is a random variable and is given as the ratio of the output SINR values for adaptive and optimal filter implementations:

$$L_{s,2}(\tilde{\omega}_B, \tilde{\omega}_s) = \frac{|\mathbf{v}_{s\text{-}t}^H(\tilde{\omega}'_B, \tilde{\omega}'_s)\hat{\mathbf{R}}_k^{-1}\mathbf{s}_{s\text{-}t}(\tilde{\omega}_B, \tilde{\omega}_s)|^2}{\mathbf{v}_{s\text{-}t}^H(\tilde{\omega}'_B, \tilde{\omega}'_s)\hat{\mathbf{R}}_k^{-1}\mathbf{R}_k\hat{\mathbf{R}}_k^{-1}\mathbf{v}_{s\text{-}t}(\tilde{\omega}'_B, \tilde{\omega}'_s)} \Bigg/ (\mathbf{s}_{s\text{-}t}^H(\tilde{\omega}_B, \tilde{\omega}_s)\mathbf{R}_k^{-1}\mathbf{s}_{s\text{-}t}(\tilde{\omega}_B, \tilde{\omega}_s)), \quad (11.38)$$

with $\mathbf{v}_{s\text{-}t}(\tilde{\omega}'_B, \tilde{\omega}'_s)$ the hypothesized space–time steering vector. Both covariance matrix estimation errors and array manifold errors adversely affect equation (11.38). In the bistatic case, $\hat{\mathbf{R}}_k$ may prove an especially erroneous estimate of $\mathbf{R}_k$ due to clutter nonstationarity; $L_{s,2}$ $(\tilde{\omega}_B, \tilde{\omega}_s)$ captures the corresponding degradation.

11.2.4 Adaptation of Monostatic MTI Methods to Bistatic Radar

STAP tailors the null depth, location, and width of a space–time FIR filter to the characteristics of the interference environment. Generally, STAP implementation requires training data to arrive at an estimate of the unknown null-hypothesis covariance matrix. Degraded estimates adversely affect the filter response and output SINR.

Factors affecting the fidelity of the covariance estimate are twofold: clutter heterogeneity and clutter nonstationarity. The former is a result of clutter cultural variation, whilst the latter results from sensor collection geometry. In monostatic radar, mild variation in clutter angle–doppler response occurs over the range dimension [7]; this mainly occurs at near ranges where the array normal and velocity vector coalign (e.g., a forward-looking array). Some loss in performance results from this nonstationary clutter behavior, but partial compensation is possible [30, 31]. Bistatic radar collection geometries can lead to much more significant angle–doppler clutter variation over range.

Bistatic STAP methods are a class of algorithmic techniques distinct from their monostatic counterparts. Minimizing losses associated with clutter nonstationarity—and characterized by $L_{s,2}$ $(\tilde{\omega}_B, \tilde{\omega}_s)$—is the primary objective of bistatic STAP algorithm design. The various bistatic STAP methods generally fall into one of three classes:

- *Localized processing*. Train locally in an attempt at mitigating the variation of the clutter response to arrive at a suitable estimate of the clutter covariance matrix.
- *Time-varying weights*. Allow the weight vector to vary linearly over the range dimension to approximately track the changing clutter angle–doppler response.
- *Data warping*. Apply a range-varying modulation to align the space–time clutter data to a reference, thereby partially homogenizing the training set. The processor applies data warping prior to traditional STAP training and application steps.

Further discussion is given in Section 11.4.

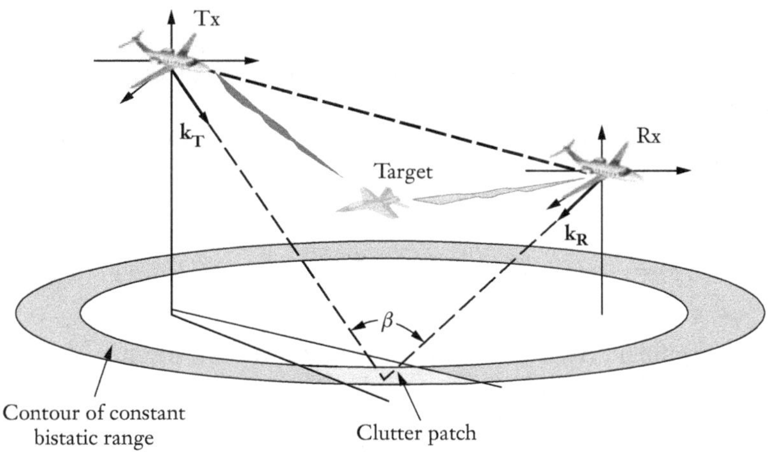

Figure 11–4 Depiction of bistatic space–time clutter patch and simulation model.

11.3 BISTATIC CLUTTER ANGLE–DOPPLER RESPONSE

In this section, we examine the bistatic clutter angle–doppler response and its range variation. An understanding of the space–time clutter model provides insight into failure mechanisms, the required STAP response, and ameliorating solutions.

Figure 11–4 shows a transmitter and receiver oriented in a bistatic configuration. The ground clutter response is a result of the coherent sum of the bistatic space–time responses of individual patches on the Earth's surface corresponding to the particular bistatic range of interest. In this figure two unit vectors are shown, $\mathbf{k}_T$ and $\mathbf{k}_R$, pointing from the transmitter and receiver to a particular clutter patch. It is common to employ a North-referenced coordinate system to orient each unit vector. The transmit and receive antennas mounted on each platform generally employ a separate coordinate system, with spatial orientation between transmit antenna and clutter patch given by the unit vector $\mathbf{k}_{a/T}$, and between receive antenna and clutter patch by $\mathbf{k}_{a/R}$.

The velocity of the transmit platform projected onto the pth clutter patch is

$$v_{r/T}(p) = \mathbf{k}_T(p) \cdot \mathbf{v}_{p/T} \tag{11.39}$$

where $\mathbf{v}_{p/T}$ is the transmit platform velocity vector in the true North coordinate system. The receiver radial velocity at the clutter patch is similarly given by

$$v_{r/R} = \mathbf{k}_R(p) \cdot \mathbf{v}_{p/R} \tag{11.40}$$

where $\mathbf{v}_{p/R}$ denotes the receive platform velocity vector. The bistatic doppler frequency of the pth patch then follows from equation (11.12) as

$$f_B(p) = \frac{v_{r/T}(p) + v_{r/R}(p)}{\lambda_0} \tag{11.41}$$

The unit vectors in each array coordinate system pointing to the pth patch follow from $\mathbf{k}_T(p)$ and $\mathbf{k}_R(p)$ as $\mathbf{k}_{a/T}(p) = \mathbf{M}^{-1}_{r2a/T}\,\mathbf{k}_T(p)$ and $\mathbf{k}_{a/R}(p) = \mathbf{M}^{-1}_{r2a/R}\,\mathbf{k}_R(p)$, where $\mathbf{M}_{r2a/T}$ and $\mathbf{M}_{r2a/R}$ are linear transformations mapping the unit vectors $\mathbf{k}_{a/T}$ and $\mathbf{k}_{a/R}$ into the radar true North coordinate system based on the known transmit and receive array normal pointing directions [32]. We use both $\mathbf{k}_{a/T}$ and $\mathbf{k}_{a/R}$ to calculate antenna gains and receive spatial frequency (see Section 11.2.2.2). The transmit gain to patch p is a function of elevation and azimuth from the transmit array normal and given by $G_T(\mathbf{k}_{a/T}(p))$. Similarly, the receive subarray gain is $g_R(\mathbf{k}_{a/R}(p))$ for the azimuth and elevation angle embodied by $\mathbf{k}_{a/R}(p)$. The scattering from the clutter patch gives rise to a spatial frequency across the array of the form of equation (11.22). To generalize the spatial frequency calculation, we express equation (11.19) as

$$\gamma_s(m,p) = \tau_s(m,p)\omega_0 = \left(\frac{\mathbf{k}_{a/R}(p)\cdot\mathbf{d}_s(m)}{c}\right)\omega_0 = \frac{2\pi}{\lambda_0}\mathbf{k}_{a/R}(p)\cdot\mathbf{d}_s(m) \tag{11.42}$$

In this case, the argument denotes the mth channel for the pth patch and $\mathbf{d}_s(m)$ is the position vector defining the location of the mth channel's phase center. The x-, y- and z-components of $\mathbf{d}_s(m)$ are given in the array coordinate system; in the case of a ULA, only the x-component is nonzero, leading to the form for spatial frequency in equation (11.22). Otherwise, the spatial steering vector to the pth patch is expressed as

$$\mathbf{s}_s(\mathbf{k}_{a/R}(p)) = \left[\exp\left(\frac{2\pi}{\lambda_0}\mathbf{k}_{a/R}(p)\cdot\mathbf{d}_s(1)\right)\ \exp\left(\frac{2\pi}{\lambda_0}\mathbf{k}_{a/R}(p)\cdot\mathbf{d}_s(2)\right)\ \cdots\ \exp\left(\frac{2\pi}{\lambda_0}\mathbf{k}_{a/R}(p)\cdot\mathbf{d}_s(M)\right)\right]^T \tag{11.43}$$

The single pulse, single channel clutter-to-noise ratio (CNR) of the pth patch, kth bistatic range, and jth bistatic range ambiguity is

$$\mathrm{CNR}_{k,j}^{(1,1)}(p;\mathbf{k}_{a/T}(p,k,j),\mathbf{k}_{a/R}(p,k,j)) = \left(\frac{P_T G_T(\mathbf{k}_{a/T}(p,k,j))F_T}{4\pi R_T^2 L_T}\right)\left(\frac{\sigma_p F_R}{4\pi R_R^2}\right)\left(\frac{\lambda^2 g_R(\mathbf{k}_{a/R}(p,k,j))}{4\pi k T_{sys} B_n L_R}\right)G_{sp} \tag{11.44}$$

where the subarray gains appear matched over all M channels, σ_p is the clutter patch RCS (and, implicitly, is a function of $\mathbf{k}_{a/T}(p,k,j)$ and $\mathbf{k}_{a/R}(p,k,j)$), and the remaining terms were otherwise given after introducing equation (11.5). For a complex envelope abiding by the Rayleigh clutter model (Gaussian voltage), the clutter patch subarray voltage for a single pulse is

$$\nu(p,k,j) = \eta_{p,k,j}\sqrt{\mathrm{CNR}^{(1,1)}\sigma_n^2} \tag{11.45}$$

where $\eta_{p,k,j} \sim CN(0,1)$ (see Section 11.1.2) and a constant CNR at each of the M subarrays is assumed. Note that

$$E[|\nu(p,k,j)|^2] = \sigma_c^2(p,k,j) \tag{11.46}$$

is the clutter patch power. The space–time clutter response is then

$$\tilde{\mathbf{c}}(p,k,j) = \eta_{p,k,j}\,\mathbf{s}_{s\text{-}t}(p,k,j) \odot \mathbf{t}_{s\text{-}t}(p,k,j) \tag{11.47}$$

In this case, the space–time steering vector to the clutter patch is

$$\mathbf{s}_{s\text{-}t}(\phi_R,\theta_R\,|\,\phi_T,\theta_T,r) = \mathbf{s}_t(f_B(p,k,j)) \otimes \mathbf{s}_s(\mathbf{k}_{a/R}(p,k,j)) \tag{11.48}$$

where $f_B(p, k, j)$ is the clutter patch doppler frequency and r indicates the bistatic range to the clutter patch. Equation (11.48) explicitly indicates the angle–doppler coupling of the stationary clutter response. The vector $\mathbf{t}_{s\text{-}t}(p, k, j)$ is stochastic, with correlation matrix $\mathbf{T}_{s\text{-}t}(p, k, j)$, and describes decorrelation across the space–time aperture due to intrinsic clutter motion (ICM) [19] and bandwidth dispersion [33].

The bistatic clutter response for range sum $R_\Sigma = R_T + R_R$ is the continuous integral of the voltage response of all scatterers comprising the range resolution cell. The clutter snapshot, from the receiver perspective, is given by

$$\mathbf{c}(R_\Sigma) = \int_{R_\Sigma-\Delta_R}^{R_\Sigma+\Delta_R} \int_0^{2\pi} \sigma_c(\phi_R, \theta_R | \phi_T, \theta_T, r)\eta(\phi_R, \theta_R | \phi_T, \theta_T, r)\mathbf{s}_{s\text{-}t}(\phi_R, \theta_R | \phi_T, \theta_T, r)d\phi_R dr \quad (11.49)$$

where Δ_R defines the bounds of the range resolution cell, $\sigma_c^2(\phi_R, \theta_R | \phi_T, \theta_T, r)$ is the elemental scatterer power at the specified bistatic range and receiver azimuth, $\eta(\phi_R, \theta_R | \phi_T, \theta_T, r) \sim CN(0, 1)$, $\mathbf{s}_{s\text{-}t}(\phi_R, \theta_R | \phi_T, \theta_T, r)$ is the clutter patch space–time steering vector, dr is incremental range, and ϕ_T, θ_T, and θ_R are the transmit angles and receiver elevation angle corresponding to the point on the Earth's surface at the specified range and receiver azimuth, ϕ_R, where $d\phi_R$ is incremental azimuth angle. The prior discussion may be employed to arrive at a discrete approximation to equation (11.49)—a useful approach for numerical analysis—by applying the following steps:

- Divide the Earth's surface into pixels corresponding to a fraction of a radar resolution cell and identify all points commonly within both transmitter and receiver horizons.
- Calculate the transmit and receive gains at each pixel.
- Calculate the pixel bistatic doppler frequency, and the azimuth and elevation angles in the receive antenna coordinate system.
- Determine the bistatic range to each pixel.
- Calculate the pixel snapshot as in equation (11.49) by multiplying the square root of the pixel power by a Gaussian random variate of zero mean and unit variance and the appropriate space–time steering vector.
- Sum all snapshots from a common bistatic range, including all range fold-over terms.

The corresponding discretized clutter model is

$$\mathbf{c}_k = \sum_{j=0}^{N_a} \sum_{p=1}^{N_c} \nu(p,k,j)\mathbf{s}_{s\text{-}t}(f_B(p,k,j), \mathbf{k}_{a/R}(p,k,j)) \odot \mathbf{t}_{s\text{-}t}(p,k,j) \quad (11.50)$$

for N_c patches and N_a range ambiguities. Space–time decorrelation not shown in equation (11.49) is incorporated through the use of the stochastic vector $\mathbf{t}_{s\text{-}t}(p, k, j)$. This model accommodates array errors by applying the angle-dependent array error vector to the corresponding spatial steering vector using the Hadamard product. It is further assumed that each pixel is statistically independent. The accuracy of the model improves for diminishing pixel size, approaching equation (11.49) in the limit.

The clutter space–time covariance matrix follows from equation (11.50) by forming the individual patch covariance matrices and summing over all patches within the bistatic range cell. The covariance matrix corresponding to equation (11.47) is

$$\tilde{\mathbf{C}}(p,k,j) = E[\tilde{\mathbf{c}}(p,k,j)\tilde{\mathbf{c}}^H(p,k,j)] = \sigma_c^2(p,k,j)\mathbf{s}_{s\text{-}t}(p,k,j)\mathbf{s}_{s\text{-}t}^H(p,k,j) \odot \mathbf{T}_{s\text{-}t}(p,k,j) \quad (11.51)$$

Using equation (11.51), and noting the assumed statistical independence of each clutter patch, the following expression is arrived at for the clutter covariance matrix at the kth bistatic range:

$$C_k = \sum_{j=0}^{N_a} \sum_{p=1}^{N_c} \tilde{C}(p,k,j) \tag{11.52}$$

The availability of the clutter covariance matrix in a simulation environment is critical to radar system and algorithm performance assessment.

11.3.1 Numerical Examples of Angle–Doppler Nonstationarity

This section considers the radar system parameters given in Table 11–1.

Figures 11–5 and 11–6 show angle–doppler responses at selected range bins for the bistatic radar geometries shown in Figures 11–7 and 11–8. The minimum variance distortionless response (MVDR) spectrum [34] for range bins 1, 100, 200 is shown, and also for the sum of the covariances for all ranges (denoted as the asymptotic response). The MVDR spectrum shows superresolution properties, clearly identifying the clutter angle–doppler locus, and is given as [34]

$$P(\tilde{\omega}_B, \tilde{\omega}_s) = \frac{1}{\mathbf{s}_{\text{s-t}}^H(\tilde{\omega}_B, \tilde{\omega}_s)\mathbf{R}_k^{-1}\mathbf{s}_{\text{s-t}}(\tilde{\omega}_B, \tilde{\omega}_s)} \tag{11.53}$$

$\mathbf{R}_k$ is the clutter-plus-noise covariance matrix for the kth range. It takes the form $\mathbf{R}_k = \mathbf{C}_k + \sigma_n^2\mathbf{I}_{NM}$. $P(\tilde{\omega}_B, \tilde{\omega}_s)$ is an estimate of the output power of the minimum variance beamformer, exhibiting peaks at the angle–doppler locations of the ground clutter response. The range-varying nature of the clutter response is illustrated through Figures 11–5 and 11–6.

For case A—shown in Figure 11–7—we note the peak clutter response at a receive cone angle (see Figure 11–1) in the vicinity of 60° and –600 Hz for range bin 1; the peak moves to a receive cone angle just above 50° and centered on –400 Hz for range bin 100, and to 50° and –300 Hz for range bin 200. The scan angle of –36.87° in Figure 11–7 corresponds to a cone angle of just over 53° (measured from the velocity vector) to swath center, and consistent with the response for range bin 100 nearing this point (range bin 125 is swath center). The asymptotic response is the MVDR spectrum for the sum of the clutter-plus-noise covariance matrices for range bins 1–250,

Table 11–1 Radar Parameters for Numerical Analysis (After [32], © IEEE)

Parameter	Value	Parameter	Value
Center frequency	5.3 GHz	Transmit array	1.2 m azimuth, 0.5 m elevation
Uncompressed pulse	21.33 μ sec	Receive array	0.6 m azimuth, 0.5 m elevation
Compressed pulse	0.33 μ sec	Element spacing (transmit and receive)	0.58 λ azimuth, 0.74 λ elevation
Peak transmit power	10 kW	Transmit weighting	25 dB Taylor, both azimuth and elevation
PRF	1,400 Hz	Receive channels, M	18
Receiver noise figure	2.6 dB	Pulses, N	16
Total RF loss	9.5 dB	Clutter reflectivity	–15 dB

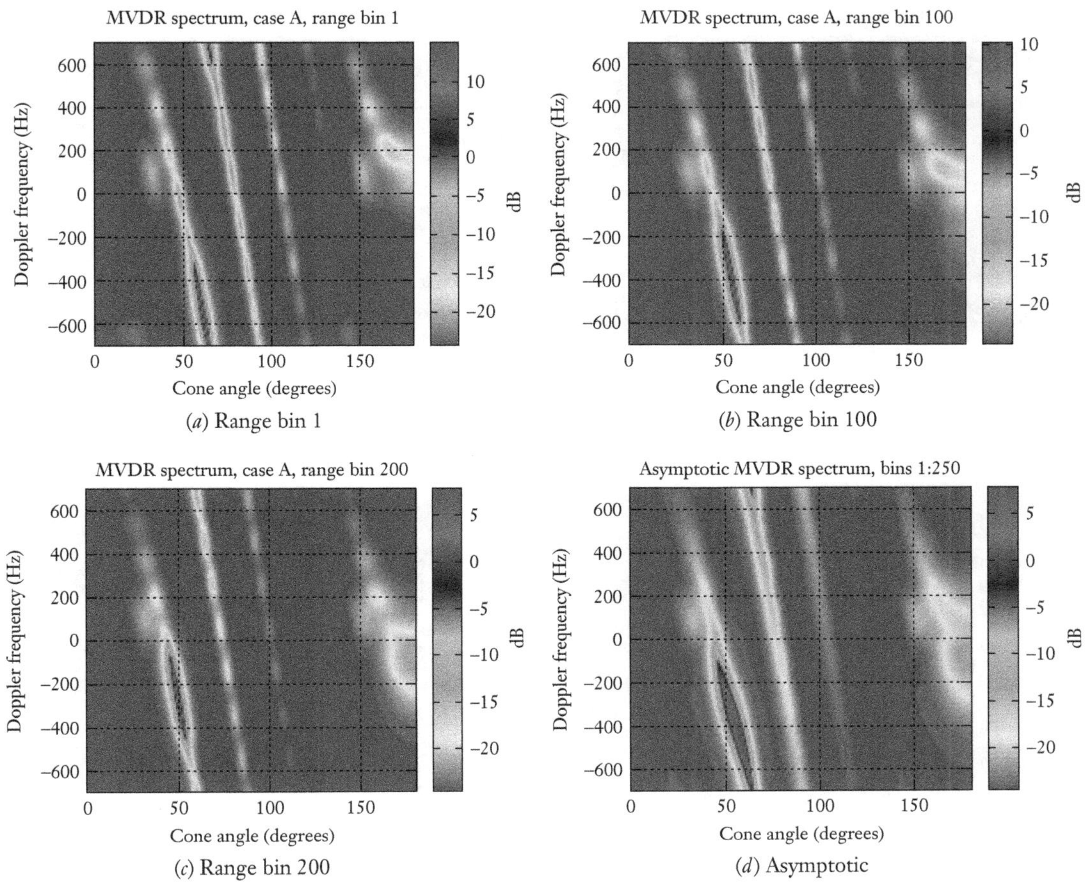

Figure 11–5 MVDR spectra at near, middle, and far range bins relative to scene center for case A. Asymptotic spectrum showing effect of nonstationarity is also shown. (After [32], © IEEE.)

normalized by the number of ranges; the asymptotic response captures the average characteristics of the clutter data comprising the range interval used in the summation. The corresponding STAP response will tend to the inverse of the spectrum shown in Figure 11–6*d*, thus appearing mismatched to any particular range of interest.

The various MVDR spectra for case B are shown in Figure 11–6. Case B shows substantially greater variation over range than case A. At range bin 1, the clutter response is fairly broad over angle, with the peak appearing near an angle of 75° and –375 Hz. The clutter response is more peaked at range bin 100, appearing near 90° cone angle and 0 Hz, whilst at range bin 200, the peak clutter response is near 100° and 200 Hz. The asymptotic response exhibits the walk-off of the peak response over range, as well as the more broadly illuminated ground response at nearer ranges. The response at range bin 100 is consistent with expectations from the bistatic geometry in Figure 11–8, where range bin 100 occurs near swath center (range bin 125 is at swath center).

To assess the performance impact of the nonstationary clutter response on traditional STAP, a commonly used postdoppler STAP method known as the extended factored algorithm (EFA) is applied [27]. EFA is an RD-STAP method discussed in Section 11.2.2.3.1 (see Figure 11–2).

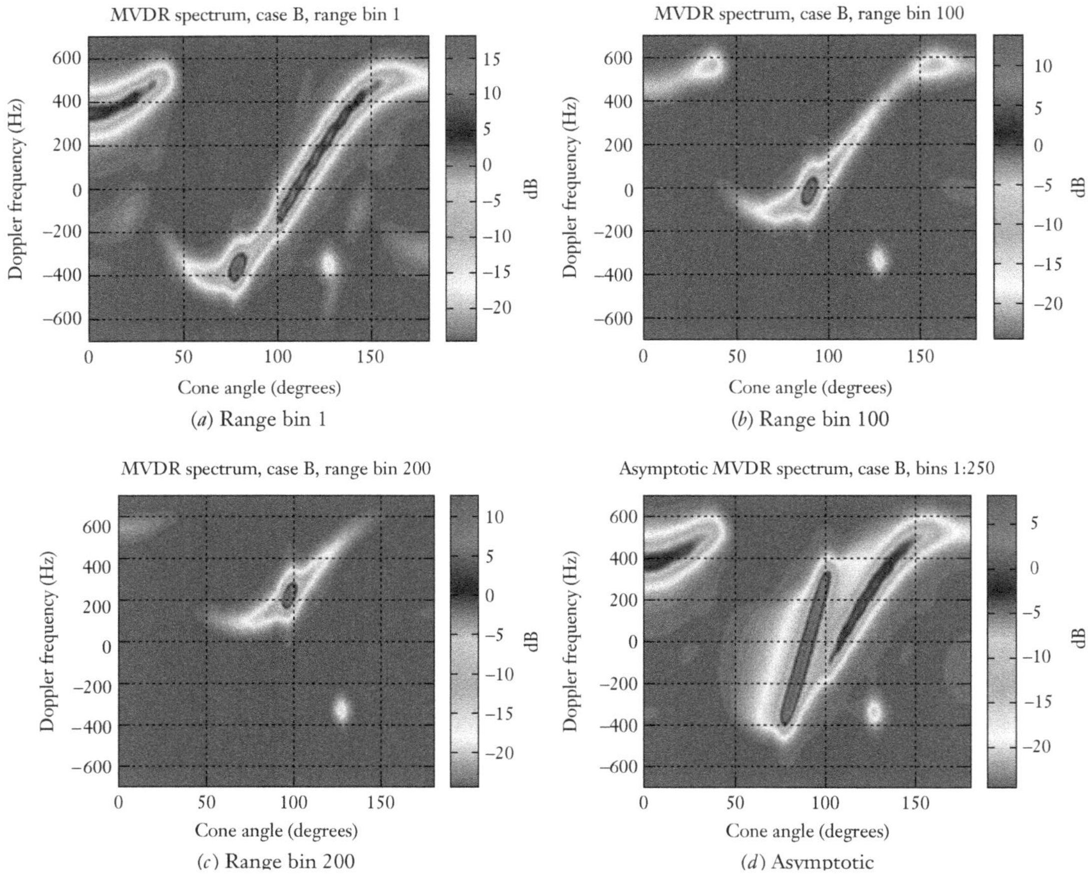

Figure 11–6 MVDR spectra at near, middle, and far range bins relative to scene center for case B. Asymptotic spectrum showing effect of nonstationarity is also shown. (After [32], © IEEE.)

The implementation uses all $M = 18$ spatial DoFs and three adjacent Hanning-weighted doppler filters. Figure 11–9 shows the adaptive SINR loss, $L_{s,2}$, for case A at each range bin (denoted by the various lines), with the covariance matrix estimate calculated via a formula similar to equation (11.29) but in the postdoppler transform domain; all 250 available data vectors are used to calculate the postdoppler covariance matrix. The generic form of the linear transformation from space–time to space–doppler is given in Section 11.2.2.3.1, with further detail in [26]. For the 54 adaptive DoFs (18 channels by 3 adjacent doppler filters), the RMB loss in a homogeneous environment amounts to -1.03 dB [24]. As seen from Figure 11–9, the losses are substantial in this instance, slightly exceeding –20 dB; losses in the clutter-free regions approach the RMB loss. The extensive loss is a result of the clutter nonstationarity over the training window. In this case, the filter response tends to the characteristics of the data comprising the training window (Figure 11–5*d*) versus the required instantaneous response (Figures 11–5*a* to 11–5*c*). Incorrect filter null width, depth, and placement is the principal mechanism for failure. The corresponding mismatch for any particular range bin leads to ineffective filter behavior relative to the optimal filter performance.

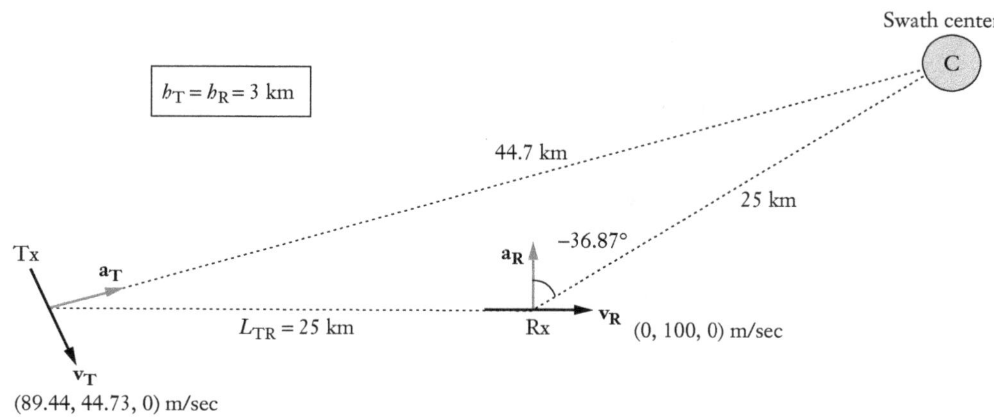

Figure 11–7 Case A, intermediate bistatic angle geometry. (After [32], © IEEE.)

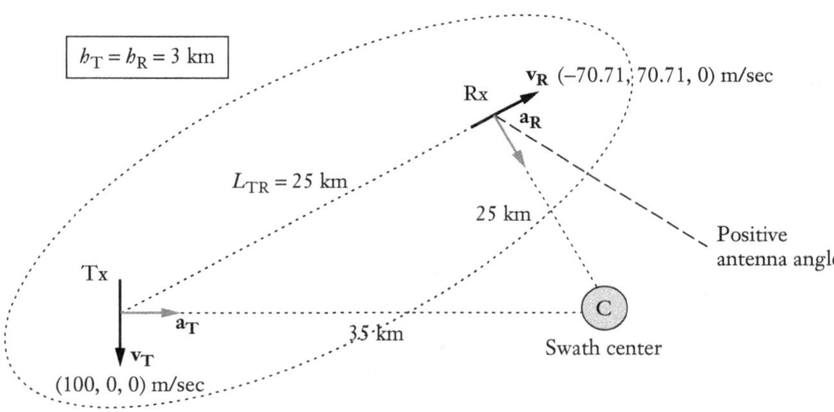

Figure 11–8 Case B, wide bistatic angle geometry. (After [32], © IEEE.)

Figure 11–10 shows adaptive losses for case B for the EFA STAP approach with training data taken over all 250 available space–time vectors. Losses shown in the figure correspond to ranges 50–250; the first 50 bins are excluded to avoid the "ground bounce" region near the radar. The losses are again excessive relative to the optimal as a result of the clutter nonstationarity depicted in Figure 11–6.

11.4 ADAPTIVE BISTATIC CLUTTER CANCELLATION METHODS

The goal of bistatic STAP is to maintain the adaptive losses, $L_{s,2}$, at values more consistent with the RMB loss. Table 11–2 summarizes various bistatic STAP methods and their rationale. As mentioned previously, the approaches given in Table 11–2 generally fall into one of three philosophies that are discussed next: localized training, time-varying weights, and data warping.

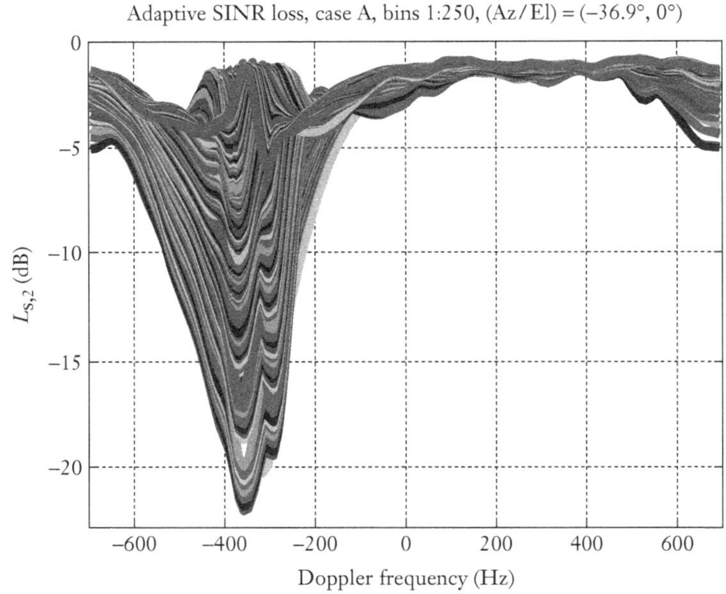

Figure 11–9 Adaptive SINR loss, $L_{s,2}$, for postdoppler STAP method, case A at array boresight, where the different lines correspond to different range bins. (After [32], © IEEE.)

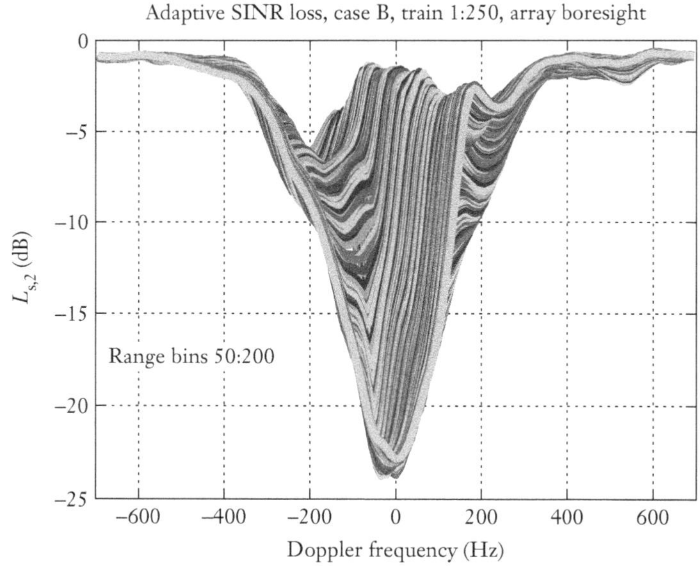

Figure 11–10 Adaptive SINR loss, $L_{s,2}$, for postdoppler STAP method, case B at array boresight. (After [32], © IEEE.)

11.4.1 Localized Processing

The appearance of only modest nonstationarity over a limited training interval centered on the range cell of interest is the fundamental premise of localized processing. RD-STAP is a cornerstone of the localized processing strategy and discussed in Section 11.2.2.3.1.

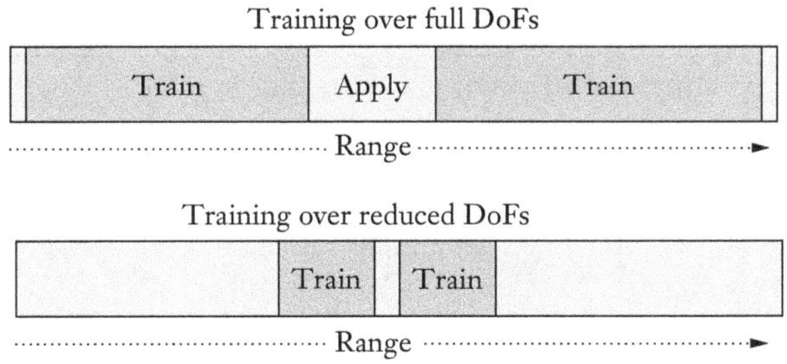

Figure 11–11 Depiction of reduced training region extent afforded through localized processing.

Table 11–2 Summary of Compensation Approaches for Clutter Nonstationarity

Method	Rationale
Localized processing [10, 11]	Local training selection minimizes nonstationarity
Time-varying weights [10, 12]	Approximate time-varying nature of unknown, optimal weight vector by allowing linear variation over range
Localized time-varying weights [13]	Impose linear variation in weight vector by restricting training range interval
Doppler warping [14, 30]	Apply deterministic, complex taper to align clutter doppler for specified cone angle, thereby enhancing STAP training set
Higher order doppler warping (HODW) [14]	Deterministically align clutter doppler over various cone angles by transforming to beam space, accomplishing the doppler alignment at multiple angles, and then inverting the transform
Angle–doppler compensation (ADC) [15]	Deterministically align maximum angle–doppler response ("spectral centers") over range to enhance STAP training set through a two-dimensional warping function
Registration of "direction–doppler (DD)" curves [16]	Use curve fitting methods to warp the power spectral density of a given range sum to a reference
Adaptive doppler compensation [31]	Similar to doppler warping, but uses least-squares estimator to determine range-varying doppler to obviate imprecise auxiliary information, thereafter aligning data to reference
Adaptive Angle–doppler Compensation (A^2DC) [17, 18]	Adaptively estimates features of dominant subspace from data, then warps the range-varying response to a reference

The RMB rule requires training data generally in excess of twice the processor's DoFs to achieve an acceptable loss relative to the optimal. Independent and identically distributed training data matched to the CUT's—cell under test's—null hypothesis is a critical assumption of the RMB rule. When the data are strongly nonstationary, commonplace in the bistatic case, the corresponding adaptive losses quickly grow. Figures 11–9 and 11–10 show the magnitude of $L_{s,2}$ for multiple ranges over doppler for a given angle of interest.

Through the use of RD-STAP, the processor projects the data into a lower dimensional subspace with fewer adaptive DoFs, and hence a requirement for substantially reduced training data. Figure 11–11 compares two typical training methods: the top figure corresponds to full-dimension

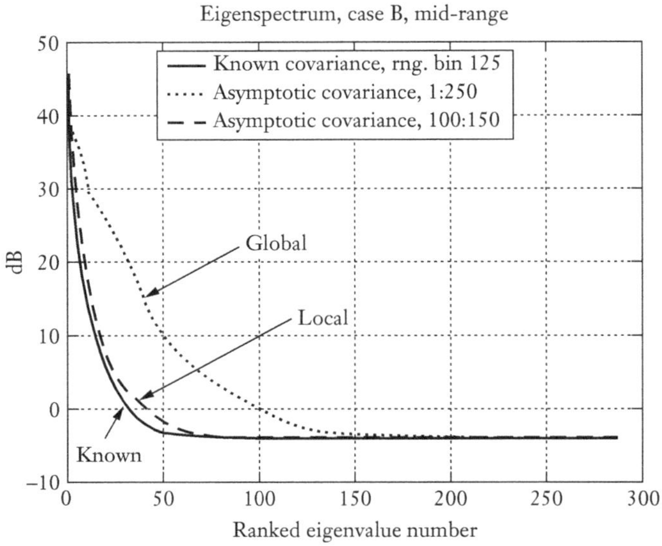

Figure 11–12 Comparison of eigenspectra at ranges with mild clutter nonstationarity, confirming potential value of localized processing for this particular region.

STAP, wherein training data taken from each side of the application region cover a broad extent; in contrast, the RD-STAP method employs training data chosen in proximity. For example, for the radar parameters of Table 11–1, twice the space–time DoFs is 576 signal vectors, whereas for the EFA example of the previous section 108 vectors represent twice the DoFs. Different STAP methods may provide even greater DoF reduction.

Issues with the localized processing approach include the following:

- The availability of adequate DoFs to cancel clutter. The bistatic case generally requires more DoFs than its monostatic counterparts, a requirement in conflict with RD-STAP.
- The availability of the right type of DoFs. For example, postdoppler beam-space processing is a popular monostatic STAP technique [35]. Some bistatic geometries require that the spatial DoFs, corresponding to adjacent beams, vary over range.
- Depending on the geometry, clutter characteristics may still vary over the local window defining the training and application region.
- Local training may emphasize certain nonideal effects—such as target-like signals in the training window—leading to performance degradation [36].

Figures 11–12 and 11–13 address this latter point. Both compare the true eigenspectrum for case B at a particular range of interest. Range bin 125, a mid-range point, is shown in Figure 11–12, whereas Figure 11–13 shows range bin 25, a near range where clutter nonstationarity is more severe. The following is also shown in each figure: the eigenspectrum for the asymptotic covariance matrix whose computation uses all available ranges (bins 1–250), thus representing global training; the eigenspectrum for the asymptotic covariance matrix using a limited number of known covariances taken at ranges about the bin of interest to characterize the performance of local training (bin numbers are given in the figure's legend). It is determined from Figure 11–12 that the local processing scheme closely approximates the eigenstructure of the CUT, suggesting much better performance potential than global training. This occurs since range bin 125 is in a region where the range-varying clutter

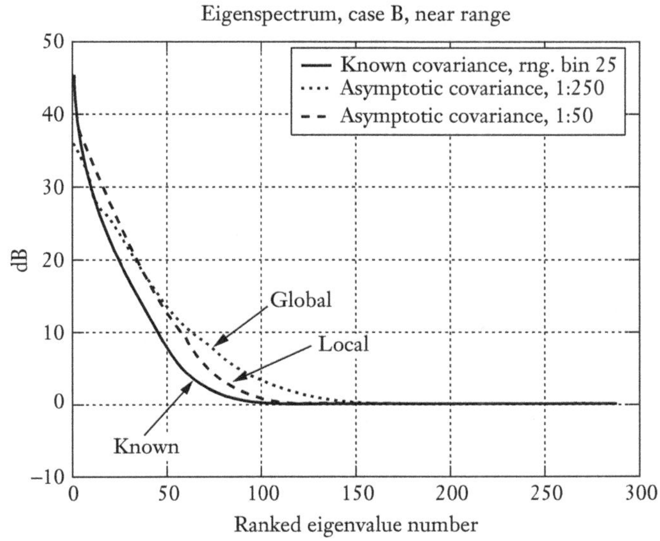

Figure 11–13 Comparison of eigenspectra at ranges exhibiting strong nonstationarity, thus indicating anticipated performance degradation of localized processing strategies.

properties are mild. In contrast, however, Figure 11–13 indicates the flaw in local processing in general: local training leads to an eigenspectrum deviating just as greatly from the CUT as the global training method. This latter situation is a result of the increased clutter nonstationarity at the nearer ranges for the particular geometry corresponding to case B.

The next section shows performance examples for locally trained STAP;

11.4.2 Time-Varying Weights

The time-varying weight (TVW) approach, also called derivative-based updating (DBU), is described in [37] and [10, 12]. Enhancing the performance of STAP applied to a rotating antenna array served as the original intent of the concept [37], with extensions to other nonstationary, adaptive filtering problems conceived thereafter.

The TVW method involves taking a Taylor series expansion of the weight vector and retaining constant and linear terms. Essentially, a doubling of the processor's DoFs occurs, but these DoFs vary over range. This allows calculation of a single weight vector of the form $\hat{\mathbf{w}}_k = \hat{\mathbf{w}}_0 + k\Delta_w$, where $\hat{\mathbf{w}}_0$ is the constant term and Δ_w is the weight vector derivative about the origin, thereby enabling a trivial modification of the weight vector by the current index k on a range cell basis. Application of the time-varying weight method to bistatic STAP was considered in [10, 12]. Divergence of the method when applied over too great a training interval was observed in [10]; this divergence suggested that variation in the weight vector was something other than linear. To avoid nonlinear weight variation, localized processing and time-varying weights were coupled in [13], yielding much improved, stable performance.

A description of TVW application follows. The Taylor series of the weight vector for range r is

$$\mathbf{w}(\mathbf{r}) = \mathbf{w}\big|_{r=0} + r\,\dot{\mathbf{w}}\big|_{r=0} + \frac{r^2}{2}\,\ddot{\mathbf{w}}\bigg|_{r=0} + \cdots \tag{11.54}$$

Truncating equation (11.54) to retain quiescent and linearly varying terms yields

$$\mathbf{w}_{\mathrm{k}} = \mathbf{w}_0 + k\Delta_{\mathrm{w}} \tag{11.55}$$

where Δ_{w} is shorthand for the time derivative. The filter output is then

$$y_k = \underbrace{\begin{bmatrix} \mathbf{w}_0^{\mathrm{H}} & \Delta_{\mathrm{w}}^{\mathrm{H}} \end{bmatrix}}_{\mathbf{z}^{\mathrm{H}}} \begin{bmatrix} \mathbf{x}_{\mathrm{k}} \\ k\mathbf{x}_{\mathrm{k}} \end{bmatrix} \tag{11.56}$$

where $\mathbf{z}$ is recognized as the new weight vector. The column vector is further denoted in equation (11.56) as the extended data vector, $\mathbf{x}_{\mathrm{k/ext}}$. The solution for $\mathbf{z}$ yielding maximal output SINR is well known, as described previously, and given by

$$\mathbf{z} = \frac{\mathbf{R}_{\mathrm{ext}}^{-1}\mathbf{s}_{\mathrm{s\text{-}t/ext}}}{\sqrt{\mathbf{s}_{\mathrm{s\text{-}t/ext}}^{H}\mathbf{R}_{\mathrm{ext}}^{-1}\mathbf{s}_{\mathrm{s\text{-}t/ext}}}} \tag{11.57}$$

where $\mathbf{R}_{\mathrm{ext}}$ is the covariance matrix of the extended data vector and $\mathbf{s}_{\mathrm{s\text{-}t/ext}}$ is the extended space–time steering vector. The denominator term is a normalization intended to place residual interference-plus-noise nominally at 0 dB. Naturally, $\mathbf{R}_{\mathrm{ext}}$ is unknown and calculated as

$$\hat{\mathbf{R}}_{\mathrm{ext}} = \frac{1}{K}\sum_{m=1}^{K}\mathbf{x}_{\mathrm{m/ext}}\mathbf{x}_{\mathrm{m/ext}}^{H} \tag{11.58}$$

Moreover, the extended space–time steering vector is $\mathbf{s}_{\mathrm{s\text{-}t/ext}} = [\mathbf{s}_{\mathrm{s\text{-}t}}^{T}(\tilde{\omega}_{\mathrm{B}}, \tilde{\omega}_{\mathrm{s}}) \;\; \mathbf{0}_{\mathrm{NM}}^{T}]^{T}$. Finally, it is common to include a normalization term, γ, prior to k in the extended data vector formulation to ensure convergence in the noise-only case [38].

The TVW method is a weight calculation strategy. It is compatible with RD-STAP and RR-STAP methods.

11.4.3 Data Warping Methods

Deterministic data warping methods include doppler warping [30], higher order doppler warping (HODW) [14], and angle–doppler compensation (ADC) [15]. Deterministic weightings "warp" the angle–doppler spectra, seeking to align the angle–doppler response at certain points over the range extent of interest. The objective of deterministic weightings is to make the training data appear stationary over specific regions of the power spectral density (PSD); using a complex modulation as the weighting function, characteristics of a given range are brought into alignment with neighboring cells.

Deterministic weightings require *a priori* knowledge of the platform velocity vectors and receiver antenna pointing direction to calculate the range-varying characteristics of the bistatic clutter angle–doppler profiles. Given this knowledge, the processor compensates the space–time snapshot as

$$\tilde{\mathbf{x}}_{\mathrm{k}} = \mathbf{h}_{\mathrm{k}} \odot \mathbf{x}_{\mathrm{k}} \in C^{NM\times 1} \tag{11.59}$$

where $\mathbf{h}_{\mathrm{k}} \in C^{NM\times 1}$ is the range-dependent warping vector aligning certain characteristics of the given range sum to an aptly chosen reference point and $\odot$ is the Hadamard (element-wise) product. For example, if $f_{\mathrm{B/ref}}$ is the dominant doppler term corresponding to a particular look angle for snapshot $\mathbf{x}_{\mathrm{ref}}$, and $f_{\mathrm{B}/k}$ is the dominant doppler corresponding to $\mathbf{x}_{\mathrm{k}}$ for the same look

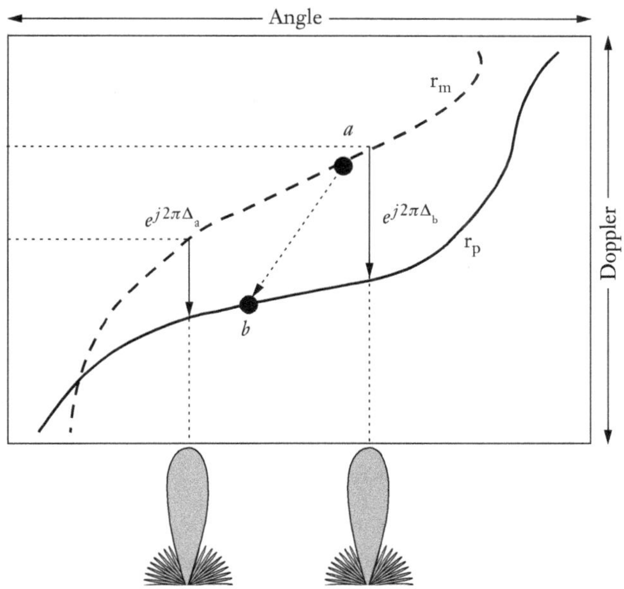

Figure 11–14 Description of data warping methods.

angle, then using

$$\mathbf{h}_k = (\mathbf{v}_t(f_{B/ref}) \otimes \mathbf{1}_M)./(\mathbf{v}_t(f_{B/k}) \otimes \mathbf{1}_M);$$
$$\mathbf{v}_t(f_{B/m}) = [1 \quad \exp(j2\pi f_{B/m}T) \quad \exp(j2\pi f_{B/m}2T) \quad \cdots \quad \exp(j2\pi f_{B/m}(N-1)T)]^T \tag{11.60}$$

in equation (11.59) aligns the dominant doppler of $\tilde{\mathbf{x}}_k$ to $f_{B/ref}$. In this case, ./ denotes element-wise division and T is the pulse repetition interval (PRI). Further note that the precise covariance matrix of $\tilde{\mathbf{x}}_k$ is then $\tilde{\mathbf{R}}_k = E[\tilde{\mathbf{x}}_k\tilde{\mathbf{x}}_k^H] = \mathbf{h}_k\mathbf{h}_k^H \odot \mathbf{R}_k$ and $\mathbf{v}_t \in C^{N\times 1}$ is the doppler steering vector.

In the doppler warping approach [30], the processor aligns the clutter angle–doppler for a *single* cone angle of interest. The previous example highlighted by equation (11.60) corresponds to doppler warping. Higher order doppler warping (HODW) enables doppler compensation at several distinct cone angles [14]. While mathematically similar in terms of the form of the data transformation, philosophically the angle–doppler compensation (ADC) method of [15] differs from either doppler warping or HODW. The approach behind ADC involves aligning "spectral centers," the angle–doppler region of the PSD exhibiting the strongest power; the warping matrix $\mathbf{h}_k$ thus modulates the data not only in doppler, but also in angle. Bistatic STAP served as the principal motivation for HODW and ADC.

Figure 11–14 describes the doppler warping, HODW, and ADC methods. In doppler warping, the modulation aligns the angle–doppler response at a particular angle and range to an arbitrary reference. Suppose range r_p in the figure is the reference. Then modulation $\mathbf{h}_k$ aligns the clutter doppler so that at a single angle, the responses for other ranges align (e.g., the modulation of $2\pi\Delta_b$ aligns the doppler for the rightmost beam—shown in the illustration—at range r_m along the vertical to the arbitrarily chosen reference doppler at range r_p to bring together the angle–doppler contours). HODW works in a similar fashion, but applies to multiple beam positions. The first step in HODW involves transforming the space–time data to beam space. The processor then applies the doppler warping strategy within each beam, corresponding to a distinct angle, to align

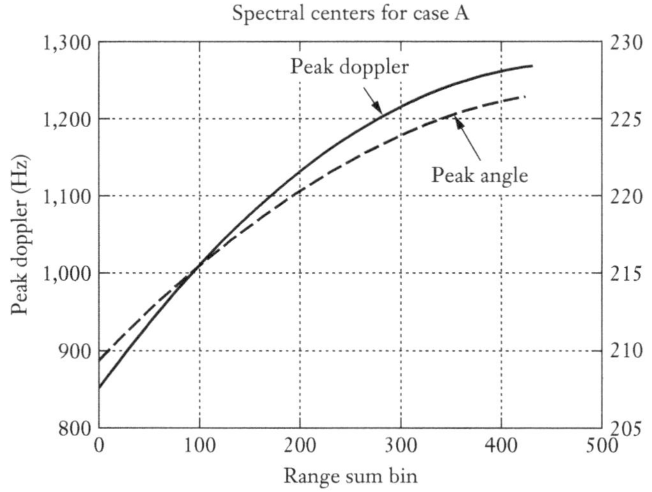

Figure 11–15 Peak doppler (solid) and angle (dashed) for case A. (After [32], © IEEE.)

sections of the clutter ridge. In Figure 11–14, a distinct modulation is applied at each of the two beam positions shown to bring those section of r_m into alignment with r_p. After the modulation step, the data are inverse transformed back to the space–time domain. Both doppler warping and HODW apply one-dimensional, temporal modulations aligning the clutter ridge in the vertical direction in Figure 11–14.

Figure 11–14 also notionally shows a spectral center as a dark circle at each range corresponding to the peak clutter response marked as points *a* and *b*. The spectral centers are evident in the simulated data shown in Figure 11–6. The ADC method aligns the range-varying spectral centers to a reference using a two-dimensional, space–time modulation. The modulation is along the diagonal line shown in the figure.

Figures 11–15 and 11–16 show the peak angle–doppler response—the so-called spectral center—for cases A and B given in Figures 11–7 and 11–8, respectively. These plots presume a homogenous terrain, and so knowledge of the platform and antenna state vectors and antenna steering direction is sufficient to calculate these curves. The angle measurement is in the receive true North coordinate system (pointing down the page in Figures 11–7 and 11–8). In the stationary case, both the doppler and spatial responses appear as straight, horizontal lines over range. However, for these bistatic cases, there is substantial variation in the peak angle–doppler response with range. This variation indicates nonstationary operating conditions and a range-varying clutter covariance matrix.

Either Figure 11–15 or Figure 11–16 can be used to describe the operation of ADC. The corresponding doppler warping or HODW implementation follows in turn. Suppose case B is chosen—with peak angle–doppler response given in Figure 11–16—and arbitrarily make range sum bin 100 the reference. The corresponding doppler frequency at this reference is near 0 Hz for a true north angle of approximately 315°. The true North angle is then converted to cone angle in the receive array coordinate system, calling this peak cone angle $\phi_{c/\text{peak}}$ (100). The goal of ADC is to apply a two-dimensional modulation to align the peak response at the other ranges to the reference as a means of partially homogenizing the data. The differences between the peak angles and doppler frequencies at each range relative to the reference thus nud to be calculated. Each pulse observation or channel sample at a particular range is then modulated by an amount

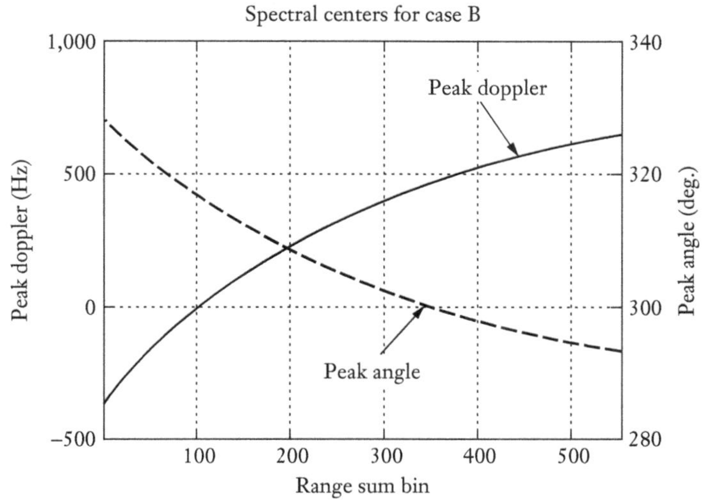

Figure 11–16 Peak doppler (solid) and angle (dashed) for case B. (After [32], © IEEE.)

equal to the difference in spatial or doppler frequency. This modulation is tantamount to aligning points *a* and *b* in Figure 11–14. If, for example, the range sum 200 is aligned to the reference bin, a modulation of approximately −225 Hz in doppler and roughly −7° in angle must be applied. The ADC approach presumes that a plot such as Figure 11–16 is available based on prior knowledge. The deterministic space–time modulation then takes the form

$$\mathbf{h}_k = (\mathbf{v}_t(f_{B/peak}(100)) \otimes \mathbf{v}_s(\tilde{\omega}_{s/peak}(100))./(\mathbf{v}_t(f_{B/peak}(k)) \otimes \mathbf{v}_s(\tilde{\omega}_{s/peak}(k)) \qquad (11.61)$$

where range sum bin 100 is the reference and $\mathbf{v}_t$ and $\mathbf{v}_s$ are temporal and spatial steering vectors, respectively.

Figure 11–17 indicates the flow diagram for doppler warping and HODW. In this case, each approach only uses a temporal modulation. The point *p* is a scatterer on the Earth's surface at a specified receive cone angle. For each point along the intersection of the receive cone angle and the Earth's surface, the processor tabulates the corresponding doppler frequency and determines the corresponding bistatic range sum bin. The processing steps generate a temporal modulation table consisting of the specified receive cone, bistatic range sum, and difference between the range bin's doppler frequency and the reference point, $\Delta f_B(k)$. $\Delta f_B(k)$ is used in a temporal-only modulation scheme to bring into alignment the clutter doppler at the specified receive cone angle for the various ranges of interest. HODW works in the same fashion, except the processor first transforms the data to beam space prior to modulating the data in each beam by the appropriate temporal modulation; the warped data are then inverse transformed back to the space–time domain. The ADC flow involves calculating the peak angle–doppler response, rather than the clutter doppler frequency for a specified cone angle.

Concerns with the deterministic warping approaches include errors in prior knowledge and range fold-over, since fold-over terms often exhibit significantly different doppler properties than characteristic of the unambiguous interval. The former concern affects calculation of $\mathbf{h}_k$, whereas the latter potentially leads to a requirement for multiple compensation points for a fixed angle, a contingency not handled by $\mathbf{h}_k$.

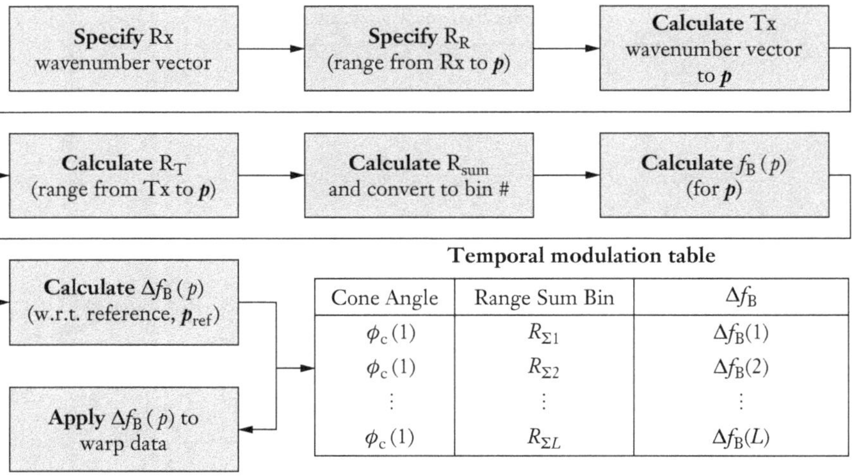

Figure 11–17 Flow diagram characterizing doppler warping and HODW methods.

11.4.4 Adaptive Data Warping Methods

Adaptive data warping methods use the measured data to estimate certain unknown parameters of the compensation approach. Advantages include the following:

- No explicit need to exchange information on transmit and receive platform and sensor state vector information to the receive processor. The methods of Section 11.4.3 employ this information to deterministically calculate the required warping modulations.
- Applicability to bistatic radar configured with a noncooperative illuminator.
- Mitigation of auxiliary data latency adversely affecting real-time processing of radar data.

Since the data are nonstationary, adaptive data warping methods must efficiently estimate unknown parameters, ideally using only a single range cell.

In [31], the authors use a least-squares method to estimate the parameters of a model describing the clutter doppler variation over range for forward-looking monostatic radar; the processor then uses the model to derive a one-dimensional, range-varying modulation to adjust for clutter nonstationarity. This approach is essentially an adaptive doppler warping algorithm, thus applying to the bistatic case. The adaptive angle–doppler compensation (A^2DC) method, introduced in [17] with further discussion in [18, 32], estimates the peak angle–doppler response on a single range cell basis, compares the estimate to surrounding ranges to avoid any bias due to outliers, then applies a two-dimensional modulation much the same as in the ADC method.

The basic steps in applying A^2DC are as follows [17]:

- *Step 1.* Adaptively estimate peak angle–doppler response for a *single* bistatic snapshot.
- *Step 2.* Compensate the peak angle–doppler response of a given bistatic snapshot to a suitably chosen reference point.
- *Step 3.* Apply STAP to mitigate ground clutter.

Figure 11–18 summarizes the process. In general, the implementation is conceptually simpler than the deterministic warping methods whose steps are given in Figure 11–17. Note that the figure only shows the steps through compensation. The following describes each of the salient steps.

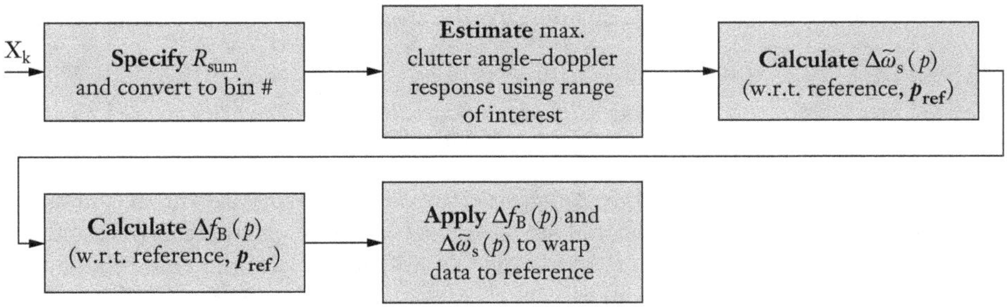

Figure 11–18 Adaptive angle–doppler compensation flow diagram.

Estimating the peak angle–doppler response of the clutter—the spectral center—using a single range sample is the goal of the estimation step. Clutter nonstationarity is the principal challenge: if the data are range varying, any estimator using fast-time samples will necessarily exhibit some degree of bias. Thus, an estimator using only information from the current cell is highly desirable. The A²DC method uses an estimator of the following form for the peak angle–doppler response,

$$(\hat{\tilde{\omega}}_{s/k}, \hat{\tilde{\omega}}_{B/k}) = g(\hat{\mathbf{R}}_{k}^{(s)}, \mathbf{v}_t, \mathbf{v}_s) \tag{11.62}$$

where $(\hat{\tilde{\omega}}_{s/k}, \hat{\tilde{\omega}}_{B/k})$ is the estimated angle–doppler peak, $g(\cdot)$ is the estimator, $\hat{\mathbf{R}}_{k}^{(s)}$ is a smoothed, single snapshot covariance estimate, and $\mathbf{v}_t$ and $\mathbf{v}_s$ are temporal and spatial steering vectors, respectively. Assuming uniform spatial and temporal sampling, the processor estimates the covariance by selecting a training set from subvectors of a single space–time snapshot [39, 40]. This approach reduces space–time aperture to avoid clutter nonstationarity and requires data symmetry in space and time. The loss of space–time aperture affects the estimation step, not the subsequent STAP filtering procedure. A least-squares fit of a polynomial to the estimates over range tends to reduce error as well. Data symmetry suggests the use of a ULA with manageable array errors; the ULA design is not overly restrictive and the most likely chosen array architecture for a variety of reasons, including simplicity and broad azimuthal coverage. It is shown in [40] that a reasonable level of array error residual only has a mild impact on the quality of $\hat{\mathbf{R}}_{k}^{(s)}$.

Next, organize the mth snapshot as the $M \times N$ matrix,

$$\bar{\mathbf{X}}_m = \begin{bmatrix} [\mathbf{x}_m]_1 & [\mathbf{x}_m]_{M+1} & [\mathbf{x}_m]_{2M+1} & & [\mathbf{x}_m]_{(N-1)M+1} \\ [\mathbf{x}_m]_2 & [\mathbf{x}_m]_{M+2} & [\mathbf{x}_m]_{2M+2} & & [\mathbf{x}_m]_{(N-1)M+2} \\ [\mathbf{x}_m]_3 & [\mathbf{x}_m]_{M+3} & [\mathbf{x}_m]_{2M+3} & & [\mathbf{x}_m]_{(N-1)M+3} \\ \vdots & \vdots & & & \\ & & & & \\ [\mathbf{x}_m]_M & [x_m]_{2M} & [\mathbf{x}_m]_{3M} & & [\mathbf{x}_m]_{NM} \end{bmatrix} \tag{11.63}$$

where the nth column is the spatial snapshot for the nth pulse. The box in the upper left corner of the matrix identifies the elements of a subvector used in calculating the smoothed covariance estimate $\hat{\mathbf{R}}_{m}^{(s)}$. The dimensions of the box determine the spatial and temporal resolution. Ideally, the size of the box is chosen so that $\hat{\mathbf{R}}_{m}^{(s)}$ is full rank.

A variety of choices exist for $g(\cdot)$. Consider, for the moment, an approximation to the MVDR spectrum,

$$\hat{M}_j(\tilde{\omega}_s, \tilde{\omega}_B) = \frac{1}{\bar{\mathbf{v}}^H_{\text{s-t}}(\tilde{\omega}_s, \tilde{\omega}_B)(\hat{\mathbf{R}}^{(s)}_j)^{-1}\bar{\mathbf{v}}_{\text{s-t}}(\tilde{\omega}_s, \tilde{\omega}_B)} \tag{11.64}$$

where $\bar{\mathbf{v}}_{\text{s-t}}$ is the reduced aperture space–time steering vector involving the Kronecker product of reduced-dimension temporal and spatial steering vectors. The calculation of equation (11.64) exhibits superresolution properties and has units of power [34]. It identifies those spatial and doppler frequencies occupied by the clutter signal. A two-dimensional search over all plausible $\tilde{\omega}_s$ and $\tilde{\omega}_B$ is necessary to find the peak: $[\hat{\tilde{\omega}}_{s/j} \quad \hat{\tilde{\omega}}_{B/j}] = \arg\max_{\tilde{\omega}_s, \tilde{\omega}_B} \hat{M}_j(\tilde{\omega}_s, \tilde{\omega}_B)$. An alternative approach is

$$\left[\hat{\tilde{\omega}}_{s/j} \quad \hat{\tilde{\omega}}_{B/j}\right] = \arg\max_{\tilde{\omega}_s, \tilde{\omega}_B} \left(\frac{1}{\bar{\mathbf{v}}^H_{\text{s-t}}(\tilde{\omega}_s, \tilde{\omega}_B)\left(\lambda^{(s)}_{\max/j}\hat{\mathbf{q}}^{(s)}_{\max/j}\left(\hat{\mathbf{q}}^{(s)}_{\max/j}\right)^H + \sigma^2_{DL}\mathbf{I}_{N_{ts}N_{ss}}\right)^{-1}\bar{\mathbf{v}}_{\text{s-t}}(\tilde{\omega}_s, \tilde{\omega}_B)} \right) \tag{11.65}$$

where $\lambda^{(s)}_{\max/j}$ is the maximum eigenvalue of $\hat{\mathbf{R}}^{(s)}_j$ with eigenvector $\hat{\mathbf{q}}^{(s)}_{\max/j}$, and N_{ts} and N_{ss} represent the size of the smoothed temporal and spatial apertures, respectively. Advantages of equation (11.65) include the following: the calculation only involves the peak part of the spectrum; a fast singular value decomposition (SVD) technique reduces computational burden and extracts only the dominant subspace with high probability; and after the fast SVD, the matrix inverse is trivial and expressible in analytic form via the matrix inversion lemma [34]. Equation (11.65) provides insight into several other strategies to improve the implementation. Another less desirable possibility for $g(\cdot)$ is an approximation to the power spectral density, $\hat{S}_j(\tilde{\omega}_s, \tilde{\omega}_B) = \bar{\mathbf{v}}^H_{\text{s-t}}(\tilde{\omega}_s, \tilde{\omega}_B)\hat{\mathbf{R}}^{(s)}_j\bar{\mathbf{v}}_{\text{s-t}}(\tilde{\omega}_s, \tilde{\omega}_B)$; this estimate is diffraction limited (i.e., the Rayleigh resolution criterion defines the achievable resolution), an issue the reduced space–time aperture corresponding to $\hat{\mathbf{R}}^{(s)}_j$ further exacerbates.

A presumption of the A^{2}DC method is the dominance of the subspace corresponding to the maximum eigenvalue by *clutter-plus-noise only*. In general this is a reasonable assumption, since the clutter RCS will typically exceed the target RCS by a substantial margin. Target contamination of the dominant subspace is a concern as system resolution increases (despite geometric dilution resulting from the bistatic geometry, which tends to increase the clutter cell size appreciably as the bistatic angle increases [2]). When the two-dimensional warping procedure adapts to a target response, the target folds into the resulting clutter null, thereby adversely affecting detection performance. Clutter discretes also hold the potential to adversely affect performance when leading to substantial misalignment of the angle–doppler profile, as might occur for a very strong side lobe clutter discrete. As the peak clutter angle–doppler response generally follows a smooth trajectory over range, any dramatic variation is attributable to strong, corruptive target or discrete energy. This undesirable circumstance is avoided by fitting a polynomial curve to the range profile of peak angle–doppler estimates.

After calculating peak spatial and doppler frequencies, the processor uniquely compensates the space–time data at each range to homogenize the data set. The compensation can take the same form as equation (11.61) for the ADC approach, except the peak spatial and doppler frequencies are estimates. Once compensated, the processor then applies one of the various STAP algorithms to the data set.

11.4.5 Influence of Geometry on Algorithm Performance

The nonstationary nature of the bistatic clutter environment degrades traditional STAP performance by corrupting the requisite covariance matrix estimate. The localized processing

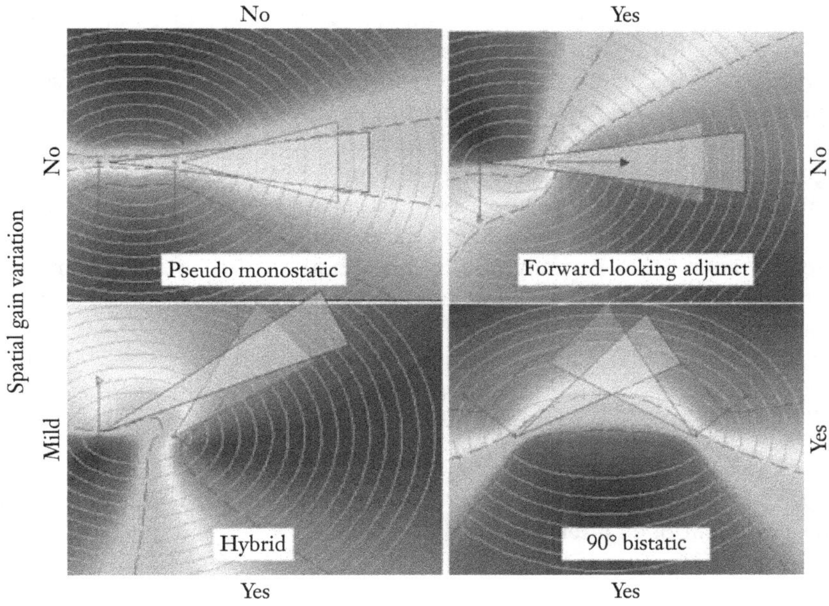

Figure 11–19 Taxonomy of bistatic geometries affecting receiver doppler and spatial gain. (After [18], © 2005 IEEE.)

method attempts to manage nonstationarity by minimizing the extent of the training window. The time-varying weighting approach allows the weight vector to vary linearly over the application region. And, the data warping methods attempt to align sections of the clutter ridge to partially homogenize the data. The effectiveness of the various approaches depends on the validity of the underlying algorithm assumptions given the nature of the clutter environment, as well as the accuracy and availability of prior knowledge.

The influence of bistatic geometry on the performance of the data warping methods is considered in [18]. Figure 11–19 depicts four scenarios with varying doppler response over range; the figure superimposes notional transmit and receive main-beam regions, the color shading characterizes the clutter doppler, and the arrows indicate directions for transmit and receive platforms. Ellipses in the figure indicate bistatic range contours superimposed on the Earth's surface. The leftmost arrow indicates the transmitter. As the figure suggests, the degree of spatial and doppler variation over bistatic range varies among the scenarios; this is seen by evaluating the change in doppler (color shading) for a chosen receive angle. A description of each scenario from the perspective of algorithm selection/performance follows:

- *Pseudomonostatic case*. The transmit and receive patterns largely overlap and the variation in doppler for a particular angle is relatively static. Traditional STAP methods employing localized processing prove adequate in this case.
- *Forward-looking adjunct*. The forward-looking adjunct configuration depicts the receiver forward of the transmitter and with orthogonal velocity vector. Otherwise, transmit and receive patterns largely overlap. Variation in doppler frequency for a specific receive angle is evident in the figure. This configuration suggests the separation of

spectral centers strictly in the doppler dimension. Hence, a one-dimensional doppler modulation provided by the doppler warping method is sufficient.

- *Hybrid configuration*. The lower left corner of the figure shows the hybrid case, where the clutter doppler varies over range for a particular receive angle and the spatial gain varies as a function of receive cone angle. The transmit and receive gains are not entirely misaligned, and so the spatial gain variation is mild. In this case, ADC and A^{2}DC should perform best, since the STAP requires information on both null placement and depth (determined by the amplitude information in the training set); the compensation mechanism of these two approaches partially homogenizes both factors over the training interval. Since the spatial variation is taken as mild, HODW and doppler warping should also prove adequate.
- *Ninety-degree bistatic case*. The lower right corner corresponds to the 90° bistatic case, where spatial gain and doppler variation over range are maximal. Since the design of both ADC and A^{2}DC compensates for both variations, these data warping methods are preferred.

It is important to note that Figure 11–19 tacitly assumes a certain degree of antenna directivity leading to the aforementioned interpretations.

11.4.6 Other Applicable Methods

A method involving alignment of "direction–doppler (DD)" (or, equivalently, angle–doppler) curves, in the context of bistatic STAP, is discussed in [16]. The essence of this method involves determining a mapping to align the DD curves of a given range to a reference point. Linear transformations are suggested as deformation mappings taking the DD characteristics of a given range cell into closer alignment with a reference DD curve. The authors of [16], recognizing that requisite parameters will not be known with certainty, extend their discussion to include estimated quantities.

In a similar fashion, the authors of [31] note that flight parameters and other information necessary when calculating deterministic $\mathbf{h}_k$ are not precisely known. For this reason, a method of estimating key parameters needed in the calculation of $\mathbf{h}_k$ is contemplated in [31] for forward-looking, monostatic radar. Specifically, presume monostatic doppler frequency is a function of quantities known with some uncertainty—viz., platform radial velocity in the direction of the main-beam scatterer, v_r, platform altitude, h_p, and waveform center wavelength, λ_0—so that

$$f_{d/k} \approx a_1 + a_2 r_k^{-2} + a_3 r_k^{-4};$$
$$a_1 = \frac{2v_r}{\lambda};\ a_2 = -\frac{v_r h_p^2}{\lambda};\ a_3 = -\frac{v_r h_p^4}{4\lambda} \tag{11.66}$$

as given in [31], where $f_{d/k}$ is the monostatic doppler and r_k is the range corresponding to the kth bin. The objective then centers on estimating the a_m using doppler measurements taken from the data to create the requisite warping function. This is accomplished in [31] using the least-squares estimator (LSE),

$$\hat{\mathbf{a}} = \left(\bar{\mathbf{Q}}^T \bar{\mathbf{Q}}\right)^{-1} \bar{\mathbf{Q}}^T \mathbf{f}_{\text{measured}} \tag{11.67}$$

where the elements of $\hat{\mathbf{a}}$ are the estimates of a_1, a_2, and a_3, respectively, $[\bar{Q}]_{k,n} = r_k^{-2(n-1)}$, with $k = 1, 2, \ldots, L$, and $n = 1, 2, 3$, and $f_{\text{measured}} = [\bar{f}_{d/1}\ \ \bar{f}_{d/2}\ \ \cdots\ \ \bar{f}_{d/L}]^T$ is the vector of estimated

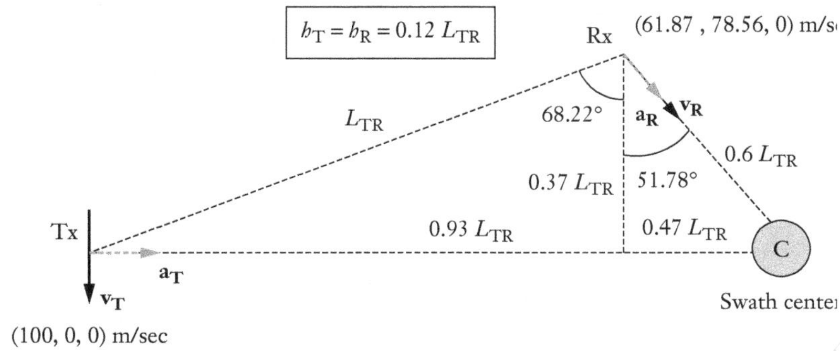

Figure 11–20 Alternate bistatic test geometry. (After [18], © 2005 IEEE.)

peak doppler frequencies for the angle of interest, taken over L range bins and based on analysis of the measured data. The processor then calculates the estimated doppler, $\hat{f}_{d/k}$, from equation (11.66) using $\hat{\mathbf{a}}$ and applies the result to construct $\mathbf{h}_k$. This method is extensible to the bistatic configuration.

11.5 BISTATIC STAP PERFORMANCE CHARACTERISTICS

In this section, we consider several numerical examples demonstrating bistatic STAP performance potential.

11.5.1 Alternate Bistatic Test Scenario

Before considering performance results corresponding to the test geometries for cases A and B of Figures 11–7 and 11–8, the geometry in Figure 11–20 taken from [18] is briefly considered. The radar parameters are the same as those given in Table 11–1. This case is used to compare the compensation mechanisms for several of the data warping methods. Findings suggest the potential for superior performance from the A^{2}DC method relative to other data warping techniques (for this specific example); thus, having established performance expectations, the capability offered by A^{2}DC relative to time-varying weights and uncompensated STAP implementations for cases A and B in Sections 11.5.2 and 11.5.3 is considered in detail.

Figure 11–21 shows the clairvoyant SINR loss, $L_{s,1}$, for receive angles of –10°, 0°, and 10°. The shape of the loss function—viz., varying loss over doppler and range for the fixed receive angle—indicates clutter nonstationarity. Moreover, the varying location of the greatest SINR loss with angle highlights the spatially varying gain due to the overlap of transmit and receive antenna gain patterns. Figure 11–21 suggests a requirement for a filter response exhibiting a range-varying doppler null as a function of angle. When left uncompensated, a covariance estimate using the data whose characteristics are shown in Figure 11–21 yields a filter response mismatched to any particular range of interest.

In contrast to Figure 11–21, Figure 11–22 shows the clairvoyant SINR loss after (a) doppler warping, (b) HODW, (c) ADC, and (d) A^{2}DC. (The corresponding warping transformations are applied to the known covariance matrix to arrive at the clairvoyant SINR loss.) This figure shows the compensation mechanism for each approach. The ideal result is a vertical line with constant loss over all range, suggesting a stationary response. Observe that all methods localize the loss in doppler. Doppler warping and HODW only provide one-dimensional, temporal modulations,

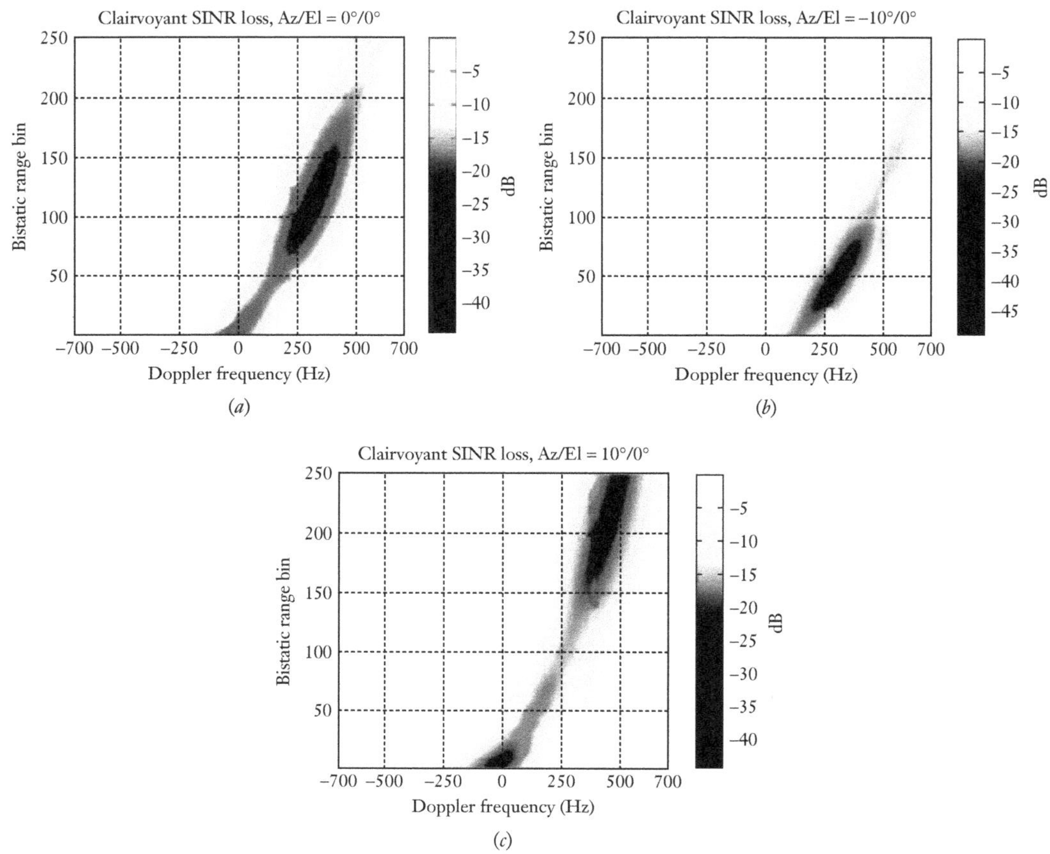

Figure 11–21 SINR loss for varying azimuth scan cases: (a) 0° azimuth, (b) −10° azimuth, and (c) 10° azimuth. (After [18], ©2005 IEEE.)

and thus do not correct for the spatially varying gain. The corresponding amplitude taper on the clutter response leads to incorrect null depth for any given range. In contrast, ADC and A^{2}DC approach the ideal response. The doppler warping, HODW, and ADC implementations all employ precisely known platform state vector and antenna pointing information; the A^{2}DC method does not require these auxiliary data, as described in Section 11.4.4, operating on the radar data alone to arrive at a response very similar to that of the ADC method.

Finally, Figures 11–23 and 11–24 show the complete SINR loss for the postdoppler EFA implementation (see Figure 11–2 for a description of EFA). In Figure 11–23, the processor employs 250 data vectors taken over the full range extent (hence the notation "1:250" in the figure legend) to train the STAP and filter the clutter signal at range bin 100 near the swath center. SINR loss results—$L_{s,1} \cdot L_{s,2}$—are given for the uncompensated STAP, as well as for doppler warping, HODW, and A^{2}DC methods. Also shown is the clairvoyant loss for the EFA method and the full space–time implementation (denoted as joint domain optimal, JDO, in the figure legend). Observe that all three data warping methods yield good performance for this scenario, whereas the uncompensated STAP yields unacceptable performance. This is not surprising: each warping method is expected to

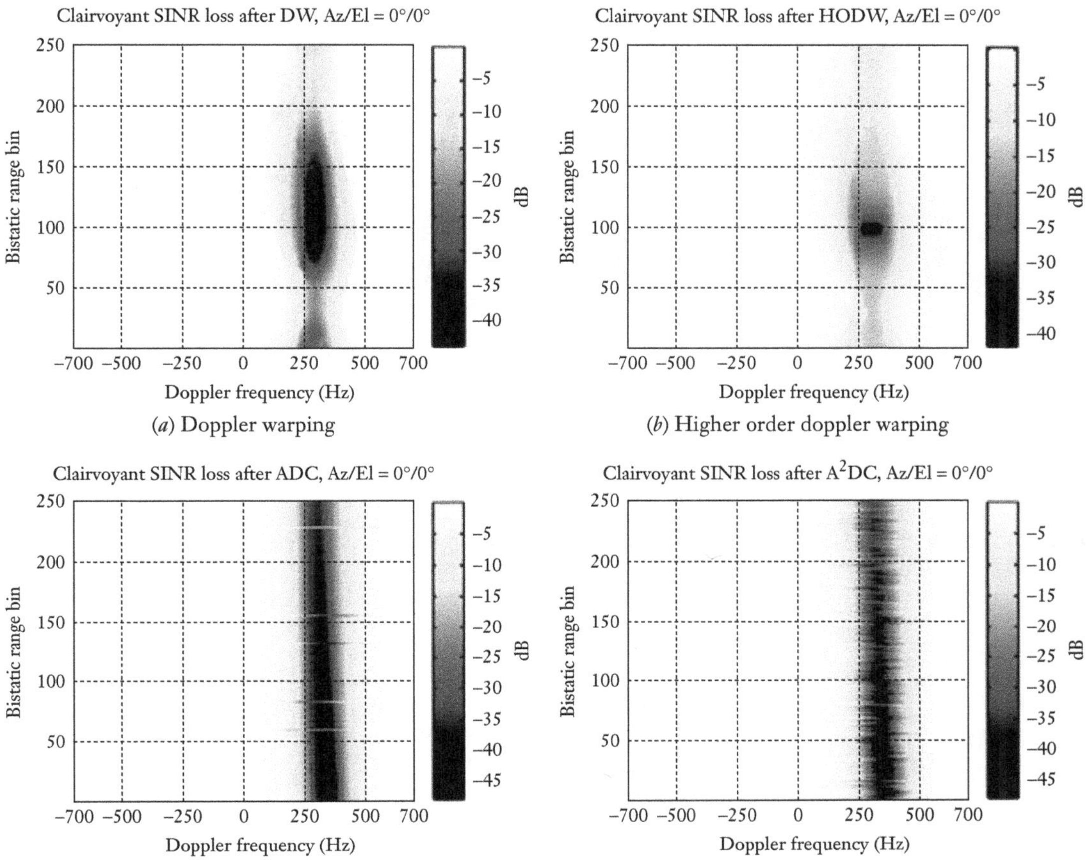

(*a*) Doppler warping (*b*) Higher order doppler warping

(*c*) Angle–doppler compensation (*d*) Adaptive-angle–Doppler compensation

Figure 11–22 Clairvoyant SINR loss after post–data warping algorithms. (After [18], © 2005 IEEE.)

generate the clutter null in the correct position—based on an examination of Figure 11–22—with perhaps only slightly reduced null depth for the doppler warping and HODW methods, whereas the uncompensated, traditional STAP approach results in a significantly mismatched filter response. Figure 11–24 shows performance for a reduced training set shifted from swath center; this selection may occur due to practical constraints, such as culturally varying clutter type or limitations in computational throughput. A²DC provides substantially better performance in this instance since it most effectively compensates the clutter nonstationarity, including the spatially varying amplitude due to overlap of transmit and receive antenna patterns.

11.5.2 Scenario A—Bistatic STAP Performance

The previous section compared the performance of several bistatic STAP methods based on data warping. Of these methods, effective performance potential for the A²DC approach was noted; for this reason, the current and subsequent sections consider A²DC in further detail. Specifically, numerical results in this section correspond to case A shown in Figure 11–7. The performance of the peak angle–doppler estimator is examined, improvements of the A²DC bistatic STAP approach

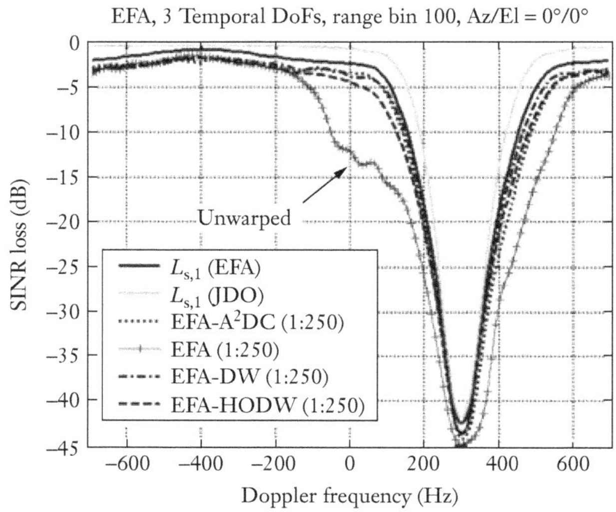

Figure 11–23 Adaptive SINR loss comparing data warping schemes with training over all ranges (bins 1:250). (After [18], © 2005 IEEE.)

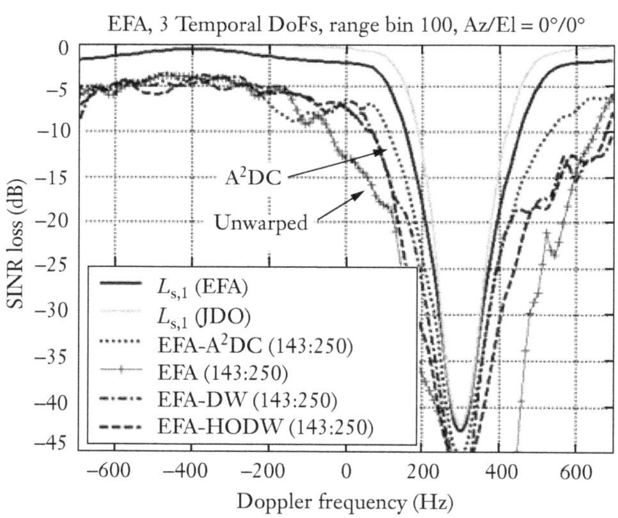

Figure 11–24 Adaptive SINR loss comparing data warping methods with reduced training set (bins 143:250). (After [18], © 2005 IEEE.)

relative to the uncompensated method are demonstrated, and performance of A^2DC and time-varying weight methods are compared.

Figure 11–25 shows the estimated and actual peak doppler responses, along with a least-squares fit using the noisy estimates. Figure 11–26 shows similar measures for the peak cone angle response. We find from these two figures that it is possible to achieve good estimates of the range-varying

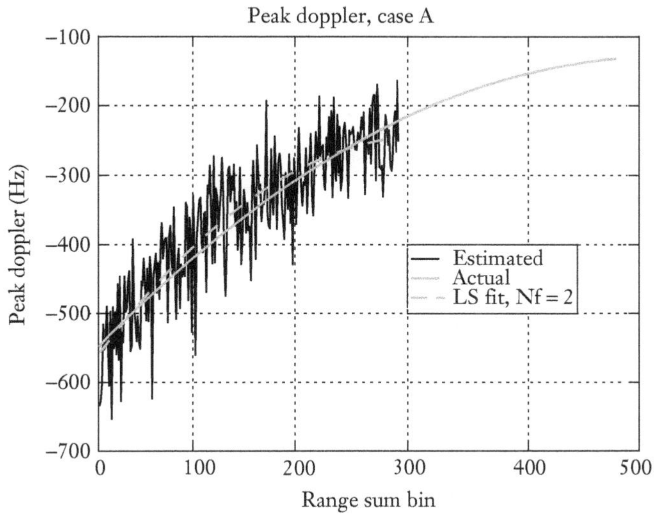

Figure 11–25 Estimated and actual peak doppler for case A. (After [32], © IEEE.)

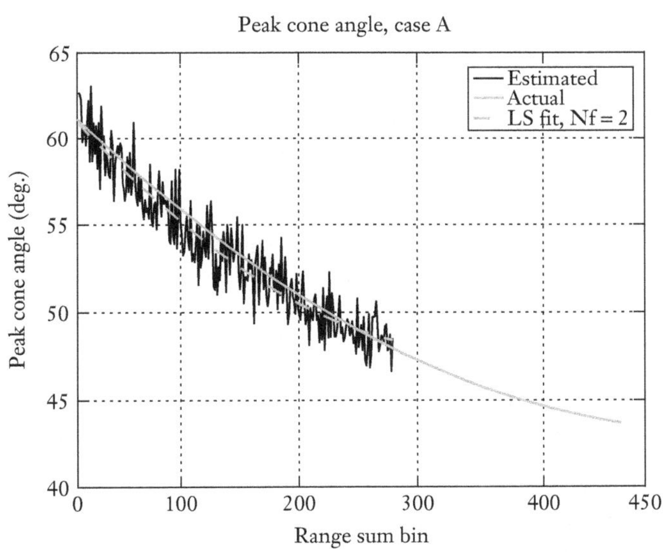

Figure 11–26 Estimated and actual peak cone angle for case A. (After [32], © IEEE.)

spectral centers—the peak angle and doppler response—using only data from a single range cell. The estimates are then used to modulate the space–time data on a range cell basis to compensate for clutter nonstationarity in the initial steps of the A^2DC process.

Next the EFA STAP method is applied—described by Figure 11–2 and applied previously to the uncompensated data—to the A^2DC compensated data. Figure 11–27 compares the average adaptive

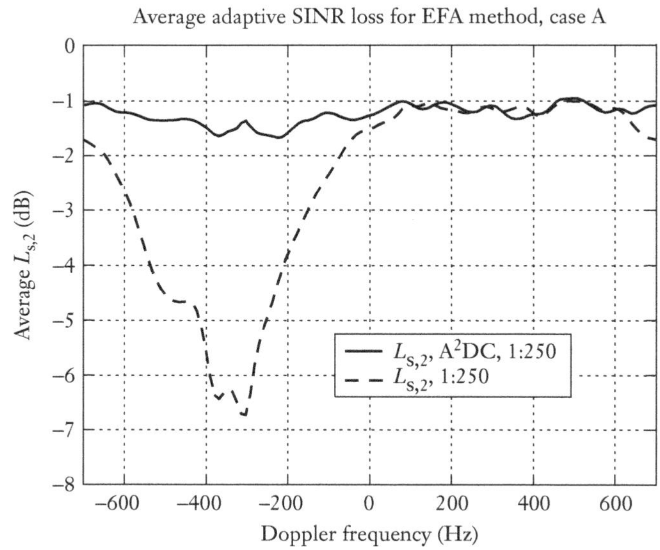

Figure 11–27 Average adaptive SINR loss, $L_{s,2}$, for postdoppler STAP method with and without A^{2}DC for case A. (After [32], © IEEE.)

losses, $L_{s,2}$, over the full range extent comprised of 250 range bins for the cases with and without A^{2}DC, as noted in the figure legend. Training data came from the full range extent of 250 space–time vectors. The average losses for the A^{2}DC case are consistent with the RMB losses present in a homogeneous environment, whereas applying EFA to the uncompensated data results in substantial loss of nearly 6 dB greater over a significant extent of the available doppler space. In other words, A^{2}DC restores performance to a level comparable to the homogeneous clutter condition. Otherwise, losses are in alignment in the exo-clutter (clutter-free) region for both methods.

Figure 11–28 shows the MVDR spectrum for range bins 1, 100, and 200 computed by applying the A^{2}DC compensation to the known covariance matrix for each bin. The asymptotic spectrum computed by averaging the A^{2}DC compensated covariance matrices are shown over all ranges. The similarities among the various spectra indicate homogenization of the data set. Interestingly, the rank-1 compensation of equation (11.59) applied to each range bin tends to align major sections of the clutter ridge over the ranges of interest.

Finally, Figure 11–29 shows the adaptive SINR losses, $L_{s,2}$, for the A^{2}DC method (as shown in Figure 11–27), time-varying weight (TVW) method combined with the EFA STAP approach, and the combination of A^{2}DC and TVW. The TVW method doubles the adaptive DoFs to enable linear variation over range. The performance averaged over range bins 1–250 (1:250) and range bins 50–250 (50:250) is considered while training over all available 250 range bins. The TVW method alone shows additional loss due to a required range variation other than linear for the nearest ranges. The combination of TVW and A^{2}DC exhibits losses consistent with the homogeneous case; however, since the adaptive DoFs are double the A^{2}DC-only case, the overall loss level is greater. This example shows the following: A^{2}DC provides good performance; TVW performance suffers due to nonlinear range variation; the combination of TVW and A^{2}DC is less useful due to the additional loss resulting from increased adaptive DoFs.

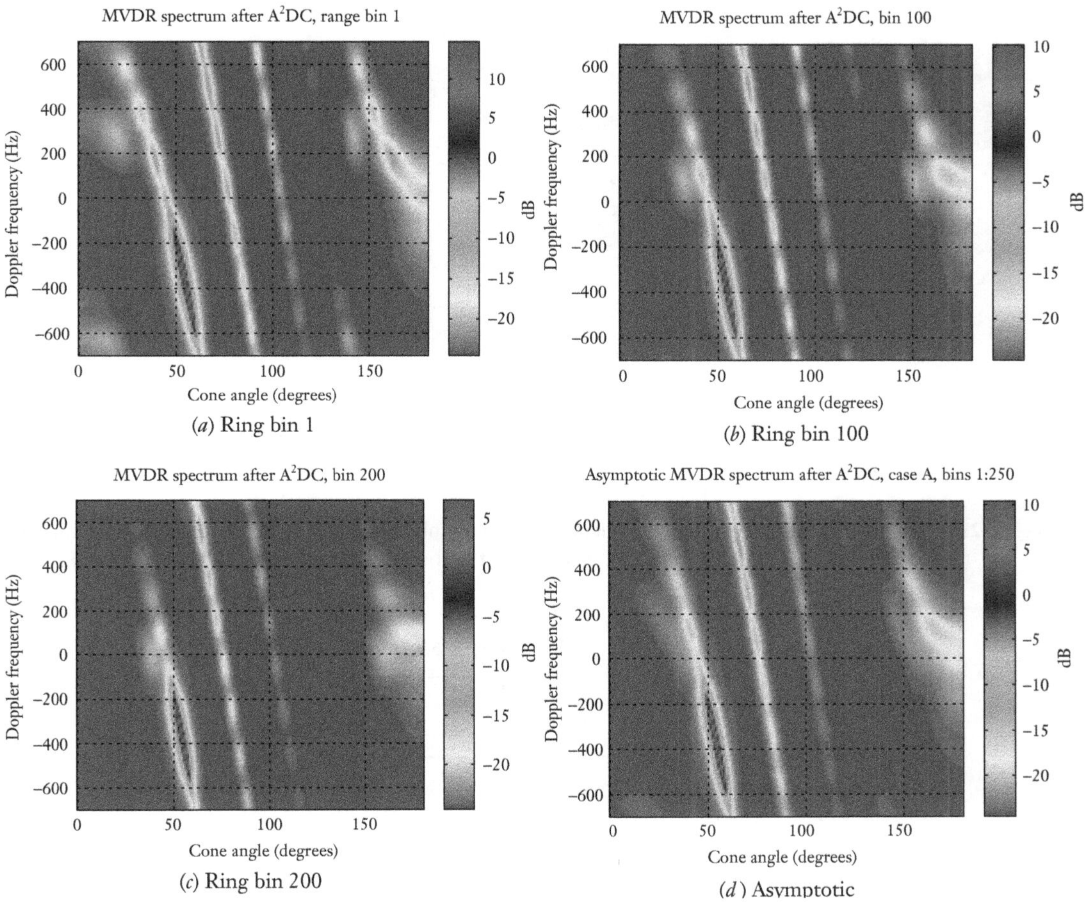

Figure 11–28 MVDR spectra after A²DC at several ranges over the swath of interest for case A. The asymptotic spectrum is also shown and indicates homogenization of the space–time data set. (After [32], © IEEE.)

11.5.3 Scenario B—Bistatic STAP Performance

Case B is considered as shown in Figure 11–8 and previously evaluated for the uncompensated case in Section 11.3.1. Otherwise, the figures and analysis approach is the same as that applied to case A in Section 11.5.2.

Figures 11–30 and 11–31 show actual, estimated, and least-squares fits for peak doppler and angle for case B. The estimator achieves very good single snapshot estimates of the spectral centers at each range, thereby forming the basis for effective A²DC performance.

Average adaptive SINR losses, $L_{s,2}$, for uncompensated and A²DC compensated STAP implementations using EFA are shown in Figure 11–32. We show performance when averaging over range bins 1–250 (1:250) and range bins 50–250 (50:250). As in the prior case, the A²DC method compensates the clutter nonstationarity such that losses approach those expected in a homogeneous environment and predicted by the RMB rule. In contrast, the uncompensated STAP losses are substantial and largely degrade target detection capability in the vicinity of the clutter spectrum.

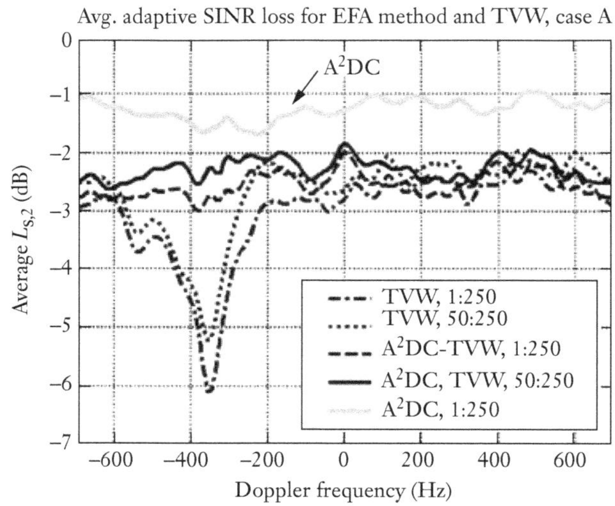

Figure 11–29 Comparison of average adaptive SINR loss for postdoppler time-varying weight implementation, with and without A^2DC, case A. Losses are in alignment in the exo-clutter region. (After [32], © IEEE.)

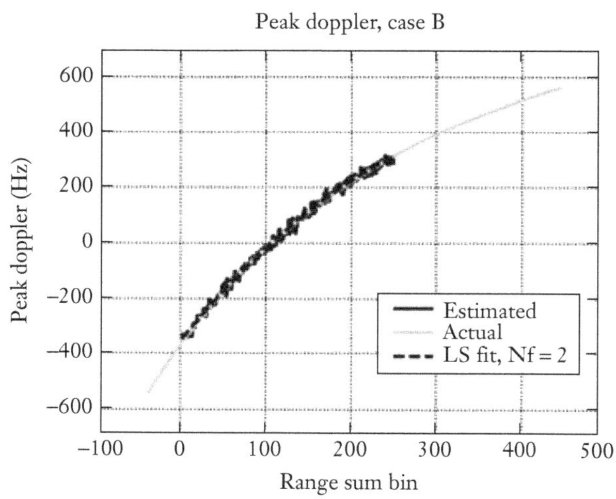

Figure 11–30 Estimated and actual peak doppler for case B. (After [32], © IEEE.)

The MVDR spectra for range bins 1, 50, and 100, as well as for the asymptotic case based on averaging over the available 250 range bins, are shown in Figure 11–33. In contrast with the uncompensated spectra in Figure 11–6, A^2DC is highly effective in homogenizing the data set. This effectiveness translates to the reduced adaptive SINR loss shown in Figure 11–32. The alignment of the clutter ridges in Figure 11–33 relative to the characteristics shown in Figure 11–6 is noteworthy.

Figure 11–34 compares the performance of A^2DC, TVW, and the combination of the two approaches applied with the EFA STAP method. In this case, A^2DC again provides the best performance potential. The deviation of the required weight vector from the linear evolution

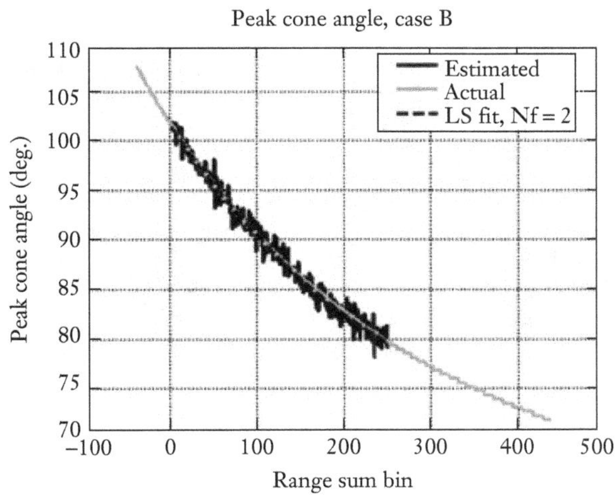

Figure 11–31 Estimated and actual peak cone angle for case B. (After [32], © IEEE.)

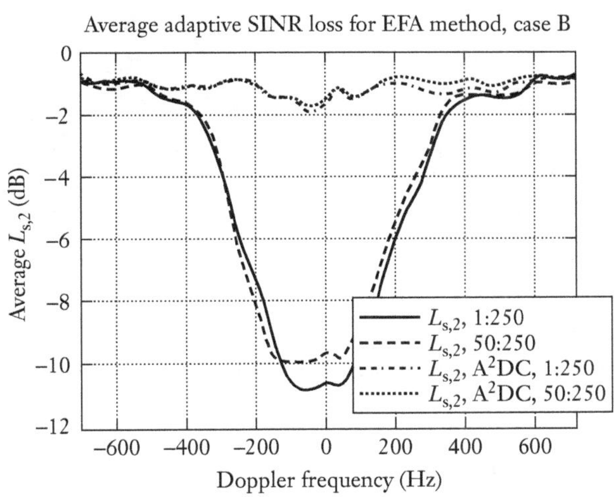

Figure 11–32 Average adaptive SINR loss, $L_{s,2}$, for postdoppler STAP method with and without A^{2}DC for case B. (After [32], © IEEE.)

presumed by the TVW implementation results in greater loss, mitigated by first applying A^{2}DC to the data set; despite this additional compensation, greater losses resulting from the combination of A^{2}DC and TVW are attributable to a doubling of the adaptive DoFs and the correspondingly increased RMB loss.

11.6 SUMMARY

Suppressing ground clutter returns at the output of the detector is an important component of aerospace bistatic MTI radar. STAP is a highly effective approach to clutter suppression in

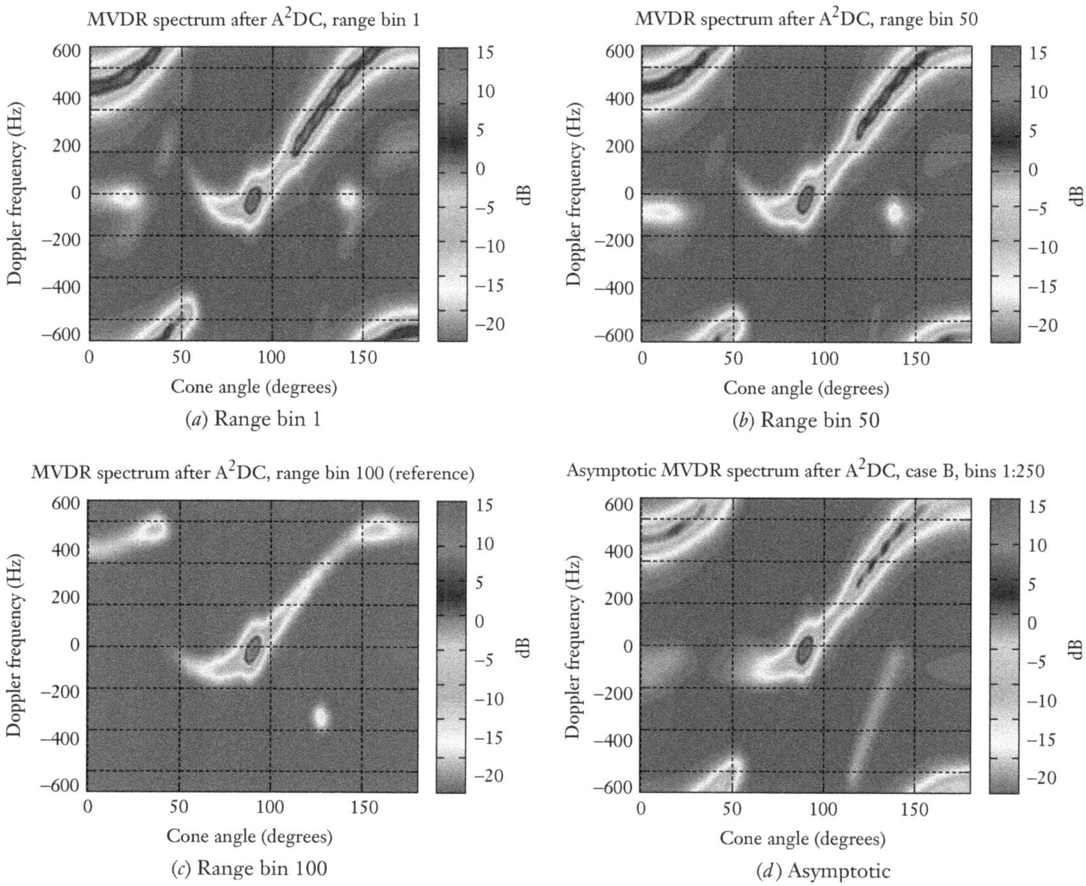

Figure 11–33 MVDR spectra after A^{2}DC at several ranges over the swath of interest for case B. The asymptotic spectrum is also shown, which indicates homogenization of the space–time data set. (After [32], © IEEE.)

monostatic radar. Extending monostatic STAP methods to the bistatic radar configuration is thus a useful pursuit.

Bistatic STAP is more complicated than its monostatic counterpart due to sensor-induced clutter nonstationarity. Specifically, the sensor geometry generally leads to range-varying angle–doppler profiles, or clutter nonstationarity (i.e., the true, but unknown clutter covariance matrix is also range varying). STAP relies on estimates of the clutter covariance matrix. Any clutter nonstationarity leads to covariance estimation error, and consequently, an inappropriate adaptive filter response. As a result of an incorrect filter response, clutter residue or signal cancellation increases and detection performance degrades. The SINR loss between the homogeneous and nonstationary cases is a key metric describing this performance loss.

The purpose of bistatic STAP is to minimize the SINR loss between the nonstationary and homogeneous cases. With this aim, three general categories of bistatic STAP were presented: (1) localized processing methods; (2) time-varying weightings; (3) data warping techniques. The localized processing approach attempts to minimize the influence of clutter nonstationarity by minimizing the training window to those ranges about the cell under test. It

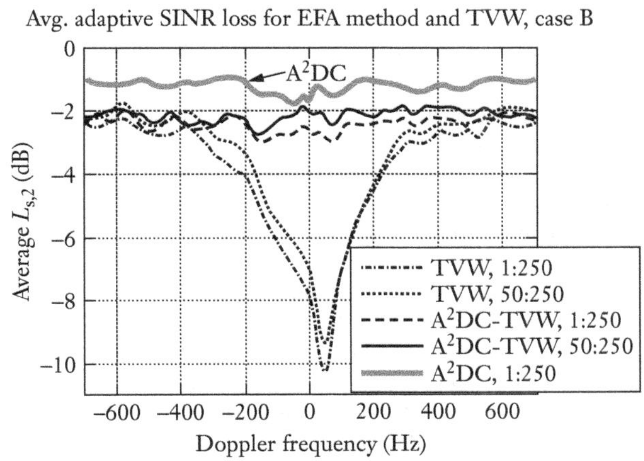

Figure 11–34 Comparison of average adaptive SINR loss for postdoppler time-varying weight implementation, with and without A²DC, case B. (After [32], © IEEE.)

is thus hoped in this case the clutter appears locally stationary. The utility of localized processing is scenario dependent and prone to underperform for certain useful bistatic geometries. Data warping methods modulate the data in space and/or time to align sections of the clutter ridge to a reference. The warping compensates for range variation. Two general data warping classes were presented in this chapter: methods requiring knowledge of the transmit and receive platform state vectors and antenna pointing directions, and methods deriving requisite details of angle–doppler variation from the data. Doppler warping, HODW, and ADC are examples of the former approach, whilst A²DC represents the latter class. The time-varying weight method allows for linear variation in weight vector elements over range in an attempt at explicitly accounting for clutter nonstationarity. Both the data warping methods and time-varying weight approach are applicable to the bistatic radar configuration.

Numerical results were included at the end of the chapter to provide a sense of the performance potential improvement of data warping and time-varying weight methods. Results indicate that effective means are available to compensate for clutter nonstationarity for STAP application to bistatic radar. Future enhancements are expected for this nascent technology.

References

1. L. E. Brennan and I. S. Reed, "Theory of Adaptive Radar," *IEEE Transactions on Aerospace and Electronic Systems*, Vol. 9, No. 2, March 1973, pp. 237–252.
2. N. J. Willis, *Bistatic Radar*, Technology Service Corporation, Silver Spring, MD, 1995.
3. L. Cantafio, *Space-Based Radar Handbook*, Artech House, Inc., Norwood, MA, 1989.
4. M. I. Skolnik (ed.) *Radar Handbook*, 2nd ed., McGraw-Hill Co., 1990.
5. R. L. Fante, "Ground and Airborne Target Detection With Bistatic Adaptive Space-Based Radar," in *Proceedings of the IEEE 1999 National Radar Conference*, Boston, MA, April 20–22, 1999, pp. 7–11.
6. G. L. Guttrich, W. E. Sievers, and N. M. Tomljanovich, "Wide Area Surveillance Concepts Based on Geosynchronous Illumination and Bistatic Unmanned Airborne Vehicles or Satellite Reception," in *Proceedings of the 1997 IEEE National Radar Conference*, Syracuse, NY, May 13–15, 1997, pp. 126–131.
7. R. Klemm, *Principles of Space-Time Adaptive Processing*, IEE Radar, Sonar, Navigation and Avionics 12, IEE Press, 2002.
8. R. Klemm, "Comparison Between Monostatic and Bistatic Antenna Configurations for STAP," *IEEE Transactions on Aerospace and Electronic Systems*, Vol. 36, No. 2, April 2000, pp. 596–608.

9. P. G. Tomlinson, "Modeling and Analysis of Monostatic/Bistatic Space-Time Adaptive Processing for Airborne and Space-Based Radar," in *Proceedings of the IEEE 1999 National Radar Conference*, Boston, MA, April 1999, pp. 102–107.
10. W. L. Melvin, M. J. Callahan, and M. C. Wicks, "Adaptive Clutter Cancellation in Bistatic Radar," in *Proceedings of the 34th Asilomar Conference*, Pacific Grove, CA, October 2000, pp. 1125–1130.
11. B. Himed, J. H. Michels, and Y. Zhang, "Bistatic STAP Performance Analysis in Radar Applications," in *Proceedings of the 2001 IEEE Radar Conference*, Atlanta, GA, May 2001, pp. 198–203.
12. S. M. Kogon and M. A. Zatman, "Bistatic STAP for Airborne Radar Systems," in *Proceedings of the IEEE SAM 2000*, Lexington, MA, March 2000.
13. W. L. Melvin, M. J. Callahan, and M. C. Wicks, "Bistatic STAP: Application to Airborne Radar," in *Proceedings of the 2002 IEEE Radar Conference*, Long Beach, CA, April 22–25, 2002.
14. F. Pearson and G. Borsari, "Simulation and Analysis of Adaptive Interference Suppression for Bistatic Surveillance Radars," in *Proceedings of the 2001 ASAP Symposium*, Lexington, MA, March 13, 2001.
15. B. Himed, Y. Zhang, and A. Hajjari, "STAP With Angle–Doppler Compensation for Bistatic Airborne Radars," in *Proceedings of the 2002 IEEE Radar Conference*, Long Beach, CA, April 22–25, 2002.
16. F. D. Lapierre, J. G. Verly, and M. Van Droogenbroeck, "New Solutions to the Problem of Range Dependence in Bistatic STAP Radars," in *Proceedings of the 2003 IEEE Radar Conference*, Huntsville, AL, May 5–8, 2003, pp. 452–459.
17. W. L. Melvin, B. Himed, and M. E. Davis, "Doubly-Adaptive Bistatic Clutter Filtering," in *Proceedings of the 2003 IEEE Radar Conference*, Huntsville, AL, May 5–8, 2003, pp. 171–178.
18. W. L. Melvin, M. J. Callahan, and M. E. Davis, "Comparison of Bistatic Clutter Mitigation Algorithms for Varying Geometries," in *Proceedings of the 2005 IEEE Int'l Radar Conference*, Alexandria, VA, May 9–12, 2005, pp. 98–103.
19. J. R. Guerci, *Space-Time Adaptive Processing for Radar*, Artech House, Inc., Norwood, MA, 2003.
20. W. L. Melvin, "A STAP Overview," *IEEE Aerospace and Electronics Systems Magazine—Special Tutorials Issue*, Vol. 19, No. 1, January 2004, pp. 19–35.
21. W. G. Carrara, R. S. Goodman, and R. M. Majewski, *Spotlight Synthetic Aperture Radar: Signal Processing Algorithms*, Artech House, Inc., 1995.
22. J. H. G. Ender, "Space-Time Processing for Multichannel Synthetic Aperture Radar," *IEE Electronics & Communication Engineering Journal*, February 1999, pp. 29–37.
23. D. H. Johnson and D. E. Dudgeon, *Array Signal Processing: Concepts and Techniques*, Prentice-Hall, Englewood Cliffs, NJ, 1993.
24. I. S. Reed, J. D. Mallett, and L. E. Brennan, "Rapid Convergence Rate in Adaptive Arrays," *IEEE Transactions on Aerospace and Electronic Systems*, Vol. 10, No. 6, November 1974, pp. 853–863.
25. W. L. Melvin, "Space-Time Adaptive Radar Performance in Heterogeneous Clutter," *IEEE Transactions on Aerospace and Electronic Systems*, Vol. 36, No. 2, April 2000, pp. 621–633.
26. J. Ward, "Space-Time Adaptive Processing for Airborne Radar," Technical Report, Lincoln Laboratory, December 1994, ESC-TR-94-109.
27. R. C. DiPietro, "Extended Factored Space-Time Processing for Airborne Radar," in *Proceedings of the 26th Asilomar Conference*, Pacific Grove, CA, October 1992, pp. 425–430.
28. W. F. Gabriel, "Using Spectral Estimation Techniques in Adaptive Processing Antenna Systems," *IEEE Transactions AP*, Vol. 34, No. 3, March 1986, pp. 291–300.
29. J. V. DiFranco and W. L. Rubin, *Radar Detection*, Artech House, Inc., Norwood, MA, 1980.
30. G. Borsari, "Mitigating Effects on STAP Processing Caused by an Inclined Array," in *Proceedings of the 1998 IEEE Radar Conference*, Dallas, TX, May 1998, pp. 135–140.
31. O. Kreyenkamp and R. Klemm, "Doppler Compensation in Forward-Looking STAP Radar," *IEE Proceedings—Radar, Sonar and Navigation*, Vol. 148, No. 5, October 2001, pp. 252–258.
32. W. L. Melvin and M. E. Davis, "Adaptive Cancellation Method for Geometry-Induced Non-Stationary Bistatic Clutter Environments," to appear in *IEEE Transactions on Aerospace and Electronic Systems*, 2007.
33. J. R. Guerci, "Theory and Application of Covariance Matrix Tapers for Robust Adaptive Beamforming," *IEEE Transactions on Signal Processing*, Vol. 47, No. 4, April 1999, pp. 977–985.
34. S. Haykin, *Adaptive Filter Theory*, 3rd ed., Prentice-Hall, Upper Saddle River, NJ, 1996.

35. H. Wang and L. Cai, "On Adaptive Spatial-Temporal Processing for Airborne Surveillance Radar Systems," *IEEE Transactions on Aerospace and Electronic Systems*, Vol. 30, No. 3, July 1994, pp. 660–670.
36. W. L. Melvin and J. R. Guerci, "Adaptive Detection in Dense Target Environments," in *Proceedings of the 2001 IEEE Radar Conference*, Atlanta, GA, May 1–3, 2001, pp. 187–192.
37. S. D. Hayward, "Adaptive Beamforming for Rapidly Moving Arrays," in *Proceedings of the CIE International Conference on Radar (IEEE Press)*, Beijing, China, October 8–10, 1996, pp. 480–483.
38. M. Zatman, "Circular Array STAP," *IEEE Transactions on Aerospace and Electronic Systems*, Vol. 36, No. 2, April 2000, pp. 510–517.
39. S. U. Pillai, Y. L. Kim, and J. R. Guerci, "Generalized Forward/Backward Subaperture Smoothing Techniques for Sample Starved STAP," *IEEE Transactions on Signal Processing*, Vol. 48, No. 12, December 2000, pp. 3569–3574.
40. M. Zatman and D. Marshall, "Forward-Backward Averaging in the Presence of Array Manifold Errors," *IEEE Transactions Antennas & Propagation*, Vol. 46, No. 11, November 1998, pp. 1700–1704.

Contributors

David K. Barton is a radar system consultant with a home and office in Hanover, New Hampshire. He was born in Greenwich, CT, in 1927, and attended Harvard College from 1944 to 1949, with a two-year interruption for service in the U.S. Army at White Sands Proving Grounds. He returned to White Sands as a civilian engineer after receiving his AB in physics in 1949. In 1953 Mr. Barton transferred to the Signal Corps Laboratories at Fort Monmouth, NJ, to initiate development of the first monopulse instrumentation radar, the AN/FPS-16. In 1955 he joined the staff at RCA, where the radar development was carried out, and participated in the test and evaluation of the system. For this work RCA awarded him, in 1958, the first David W. Sarnoff Medal for Outstanding Achievement in Engineering. Mr. Barton also scaled up that radar to create, in 1960, the AN/FPS-49 radars for the Ballistic Missile Early Warning System, only recently replaced after 40 years of operation in Alaska and Great Britain.

From 1963 to 1984, he was a consulting scientist with Raytheon Company in Wayland and Bedford, MA, where he originated the radar concepts for the AN/TPS-19 radar-controlled landing system for the U.S. Air Force, and contributed to a number of other radar and guided missile studies and projects. He joined ANRO Engineering in 1984, engaging in radar and missile system studies until his retirement in 2004.

Mr. Barton was elected a Fellow of the IEEE in 1972 and to the National Academy of Engineering in 1997. In 2002 he received the IEEE Dennis J. Picard Medal for Radar Technologies and Applications. He has served on the scientific advisory boards of the U.S. Air Force, the Defense Intelligence Agency, and the Army Research Laboratories.

He is the author of several texts on radar, the most recent of which is Radar System Analysis and Modeling (Artech House, 2005). Since 1975 he has been the editor of the Artech House Radar Library.

Hugh Griffiths received the MA degree in physics from Oxford University, and PhD and DSc(Eng) degrees from the University of London. From 1982 to 2006 he was with University College London. Since 2006, he has been Principal of the Defence College of Management and Technology, Cranfield University, which is part of the UK's Defence Academy. His research interests include radar and sonar systems, signal processing (particularly synthetic aperture radar and bistatic and multistatic radar), as well as antennas and antenna measurements. He has published over 300 papers and technical articles.

Dr. Griffiths has received numerous awards, and he serves on the IEEE AESS Board of Governors, on the IEEE AESS Radar Systems Panel, and as Editor-in-Chief of *IET Proceedings on Radar, Sonar and Navigation*. He served as Chairman of the IEE International Radar Conference RADAR 2002 in Edinburgh, UK and he has advisory roles for the UK Ministry of Defence.

He is a Fellow of the IET and the IEEE, and in 1997 he was elected to Fellowship of the Royal Academy of Engineering.

Edwin Lyon III has over 50 years' experience in design, development, research, and operation of electronics systems, including aircraft flight simulators, communications systems, radar systems, and electronic warfare suites. His specialties are radar and radar countermeasures. He has expertise in ionospheric phenomenology, wave propagation, optics, simulation and modeling of optical through VLF wave band systems, scattering and reflection, and hardware and software development. His experience in radars capable of detecting dim targets in a heavily cluttered environment extends from HF radars against cruise missiles and small boats in turbulent clutter to microwave and millimeter-wave GMTI radars for airborne surveillance of moving targets in jungle and oceanic environments.

He has worked for ITT and predecessor companies, SRI International, SAIC, and is currently working as a program manager for Science and Technology Associates, Inc., in Arlington, VA. With ITT and its predecessor companies, he developed the EARTHLING and CHECKROTE HF skywave radars, as well as many other radar, communications, and electronic warfare systems for various U.S. Government clients such as the armed services, ARPA (now DARPA), and various intelligence agencies. He has fielded radar and other sensor systems in overseas locations and aboard ships and aircraft, including installing an HF radar and many other ionospheric measurement sensors aboard the carrier USNS Core for a 99-day measurement campaign in the Indian Ocean, in 1962. As a senior engineer with SRI International, he was Technical Test Director for the test program of the USAF AN/FPS-118 over-the-horizon backscatter radar system prototype in Maine, and later supported the DoD Counterdrug Technology Development Program Office in conducting HF radar tests against dim targets and devising radar upgrades based on these tests. This work involved the Navy's ROTHR, another HF skywave radar system that is CONUS-based. This work was continued in his employment with SAIC.

In his current employment at Science and Technology Associates, Inc., Mr. Lyon supports DARPA and other agencies in novel radar (and other) detection/tracking/mitigating techniques for protecting U.S. forces from an assorted weapons spectrum.

Mr. Lyon holds the BS (1952) and MS (1954) degrees in physics, conferred by John Carroll University, Cleveland, OH. He is married to his wife of 53 years, Millie, and they reside on the slopes of Catoctin Mountain in Maryland.

William Melvin is Director of the Sensors and Electromagnetic Applications Laboratory at the Georgia Tech Research Institute and an Adjunct Professor in Georgia Tech's Electrical and Computer Engineering Department. His research expertise includes digital signal processing with application to RF sensors, including adaptive signal processing for aerospace radar detection of airborne and ground moving targets, radar applications of detection and estimation theory, electronic protection, and synthetic aperture radar. He has authored over 120 publications in his areas of research interest and holds three patents on adaptive radar technology. Dr. Melvin received the PhD in Electrical Engineering from Lehigh University in 1994, as well as the MSEE and BSEE degrees from the same institution in 1992 and 1989, respectively.

Brian D. Rigling received the BS degree in physics-computer science from the University of Dayton in 1998 and received the MS and PhD degrees in electrical engineering from The Ohio State University in 2000 and 2003, respectively. From 2000 to 2004, he was a radar systems engineer for Northrop Grumman Electronic Systems in Baltimore, MD. Since July 2004, he has been with the Department of Electrical Engineering, Wright State University, and is currently an assistant professor there. His research interests are in sensor signal processing and system engineering, array processing, autofocus, parametric modeling, and estimation.

Richard Simpson was born in 1945 and raised in New Hampshire, where he attended public schools. By 1959 he had developed an interest in radio wave propagation and had obtained his first amateur radio license. After receiving an ROTC commission in the U.S. Army Reserve and completing undergraduate studies in electrical engineering at MIT, he moved to Stanford University for graduate school. Working under bistatic radar pioneers V.R. Eshleman and G.L. Tyler, he received his PhD in 1973. He has continued at Stanford (now as a Senior Research Associate) participating in the NASA Viking, Voyager, Magellan, Mars Observer, and Mars Global Surveyor and the ESA Mars Express and Venus Express missions. He is also the Radio Science Advisor to the NASA Planetary Data System. Dr. Simpson is a member of the AAAS, AGU, AAS (and its Division for Planetary Sciences), IEEE, and URSI (Commission J). He is author or coauthor of more than two dozen professional publications. In his free time he serves as Volunteer Coordinator for the Sierra Club's backcountry ski hut system near Lake Tahoe, and is Secretary of the Winter Recreation Committee, which advises the California Department of Parks and Recreation. He, his wife, and their dog, Cedar, live in Palo Alto, CA.

John D. Sahr was born in Seattle, and grew up on a small farm in Eastern Washington State. He studied at the California Institute of Technology, graduating in June 1984 with a BS in Electrical Engineering. After one year of graduate studies at the University of California (Los Angeles), Dr. Sahr went to Cornell University. There he began an experimental program of VHF radar remote sensing of the high-latitude ionosphere with Professor Donald Farley, gaining a PhD in August 1990 (minoring in applied mathematics). Dr. Sahr spent a subsequent year as a postdoctoral fellow at Cornell, working with incoherent scatter radar data from the Jicamarca Radar Observatory.

In autumn 1991, Dr. Sahr joined the Electrical Engineering Faculty at the University of Washington, where he continues his studies of space plasma physics, scattering theory, and radar signal processing. He and his students have developed a novel passive radar for auroral physics studies. Dr. Sahr reviews manuscripts for several international journals in geophysics, radio science, and electrical engineering, and has served as an associate editor for *Geophysics Research Letters*.

Dr. Sahr has received several teaching awards as well as the Faculty Achievement Award from the UW College of Engineering. In 1996 he received the Henry Booker Prize from the International Union of Radio Science (URSI), and in 2003 he hosted the Multination Conference on Passive and Covert Radar at the University of Washington. Dr. Sahr presently serves as the Associate Dean for Undergraduate Academic Affairs at the University of Washington.

Merrill Skolnik was Superintendent of the Radar division at the US Naval Research Laboratory form 1965 to 1996. He is the author of *Introduction to Radar Systems* and the editor of *Radar Handbook*.

Melvin M. Weiner is a member of the Technical Staff (retired), The MITRE Corporation, Bedford, MA. The author of numerous professional publications including three book chapters and three books in electromagnetics, he holds five patents and is a member of the Institute of Electrical and Electronics Engineers, The American Physical Society, and the Optical Society of America. He received the SB (1956) and SM (1956) degrees in electrical engineering (cooperative honors program) from the Massachusetts Institute of Technology, Cambridge.

Nicholas J. Willis spent his career in the U.S. military–industrial complex, beginning with an NROTC scholarship to Stanford University and graduating with a degree in mathematics in 1956. He spent five years in the U.S. Navy serving on a destroyer and testing naval radars and missiles at White Sands Missile Range. He then joined industry to develop, evaluate, and test radars and electronic warfare with Philco-Ford, SRI International and Systems Control, Inc. in Silicon Valley.

In 1976 he rejoined the U.S. Government for a five-year tour at the Defense Advanced Research Projects Agency to develop radars including Pave Mover, the forerunner of JSTARS. He returned to industry for his final 17 years of designing and evaluating radars at Technology Service Corp. and Westinghouse Electric Corp., where he designed one of the F-22 radar modes. He also conducted radar and electronic warfare short courses.

Mr. Willis has published a book on bistatic radar and authored two book chapters, one on bistatic radar and one on electronic warfare, along with two award-winning radar papers in a defense journal. He is now retired and consults for government and industry.

Index

A

B

C

D

E

F

G

H

N

O

P

Q

R

S

www.ingramcontent.com/pod-product-compliance
Lightning Source LLC
LaVergne TN
LVHW062212200926
848484LV00006B/1377
9781891121487